Synergetics

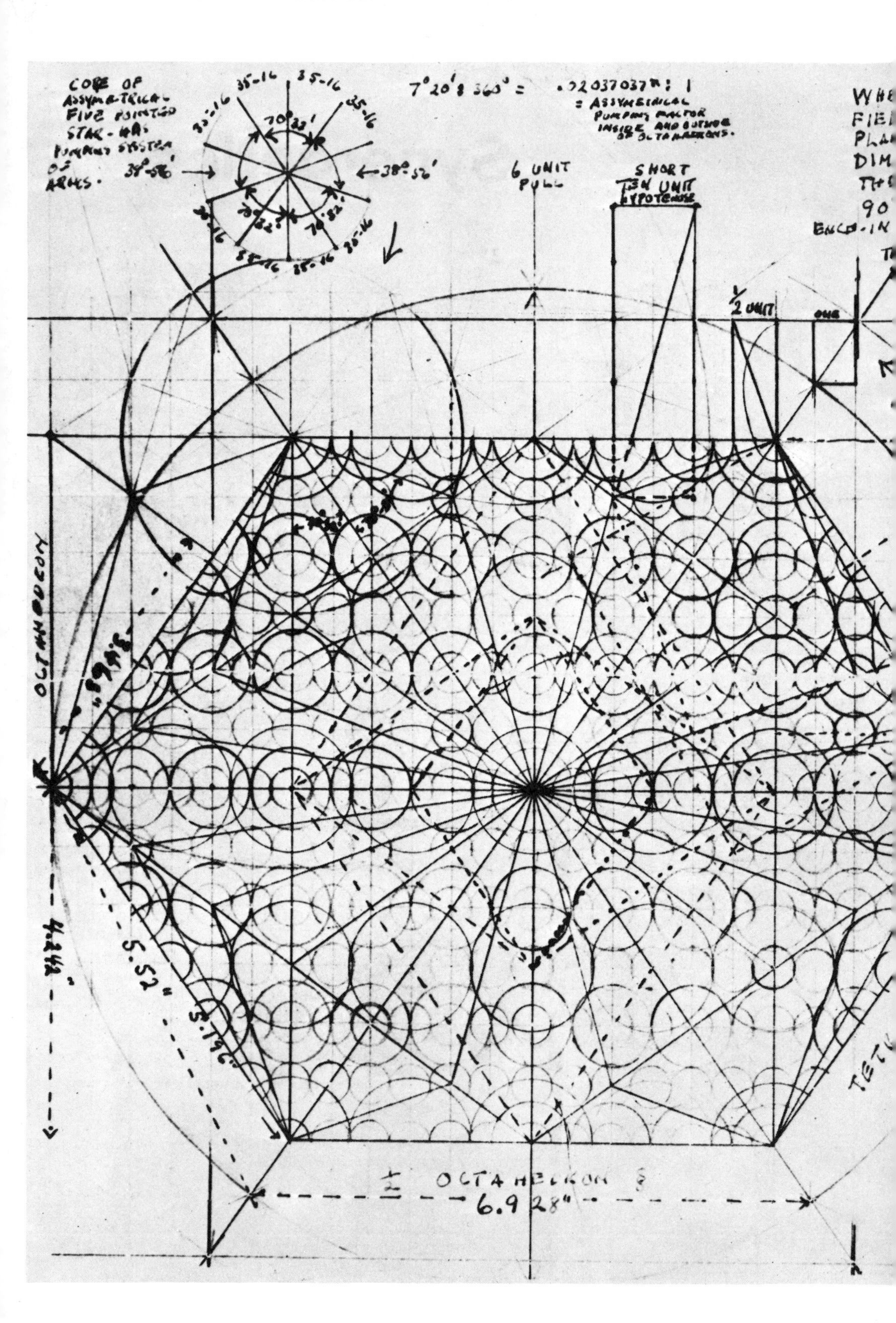
CORE OF ASSYMETRICAL FIVE POINTED STAR-HAS PUMPING SYSTEM OF ARMS.
38°-52'
35-16
70°32'
7°20' 8 360° = .02037037 : 1
= ASSYMETRICAL PUMPING FACTOR INSIDE AND OUTSIDE OF OCTAHEDRONS.
6 UNIT PULL
SHORT TEN UNIT HYPOTENUSE
1/2 UNIT
ONE
OCTAHEDRON
3.46"
4.242"
5.52"
5.196"
1/2 OCTAHEDRON
6.928"

DYMAXION KINETIC VECTOR STRUCTURE

R Buckminster Fuller
Dec 15th 1947.
6 Burns St.
Forest Hills.
N.Y.C.

GRAVITATION (DVR)
DOUBLED IN
REAT CIRCLE §)
S OF ITS COORDINATES
S. AXIS IS ROTATED
ITS VOLUMETRIC INFLU-
SED EIGHT FOLD
UCTURE ALLOWS IN
ANNER FOR
SIONAL ROTATION
O APPLIED LINE OF

SKIP THESE ANGLES
POT PROVIDES OCTAVE EXPANSION OR CONTRACTION

NOTE THAT THE HARMONIC INVERSION INCREASES ULTIMATELY TO REACH NUCLEUS OF ADJACENT SYSTEM BUT PROVIDES NO GRAVITATIONAL PATH TO CENTER OF PRIMARY NUCLEUS.

NOTE THAT SAME 1:1.41421 QUADRANGLES MAY BE FORMED ON ANY OF THESE HYPOTENUSES CONSIDERED AS LEGS BY JOINING CENTERS OF SPHERES THUS PROVIDING MEANS OF ROTATION

FOUR UNIT HYPOT

LONG TEN UNIT HYPOTENUSE

12 UNIT PUSH

THE LONG HORIZONTAL 24 BALL AXIS OF SPHERES IN CLOSE OR TANGENT ALIGNMENT IS DIAMETER OF DYMAXION POTENTIAL VECTOR STRUCTURE. WHOSE RADIUS WAS UNITY (HERE REPRESENTED BY 12-6-OR 4) FORMULA -DVF = VECTOR RADIUS OF DYMAXION (BOTH POTENTIAL AND KINETIC AND $\left(\frac{DVB}{12}\right)^2 + \left(\frac{DVB}{12}\times LENGTH\right)^2 = \left(\frac{WAVE}{LENGTH}\right)^2$

NOTE THAT WHILE THE HYPOTENUSE GROUPINGS ARE .5-1-2-4-10-20
THAT THE ALIGNMENT GROUPINGS ARE ALL IN 8-6-12 ⊙

NOTE CONSTANT RECURRENCE OF THE 10 AND 20 FACTOR IN ATOMIC STRUCTURE TO STRUCTURE $\frac{\text{MASS OF PROTON}}{10}$ = 20×92
SPEED OF ELECTRON $\frac{1}{10}$ C

A:B = B:2A = 2A:2B = 2B:4A = 3A:3B = 3B:6A

CONSTANT PROPORTION

SECTION THROUGH EQUATOR OF ANY OF SIX IDENTICAL AXIAL ROTATIONS OF DYMAXION SHOWING METHOD OF OCCURENCE OF SPHERES ON CENTERS OF GRAVITY AND AT JUNCTURES OF A. B. C. + D PARTICLES – IN A TIGHT TANGENTIAL ASSEMBLAGE OF SPHERES IN ALL DIRECTIONS.

SHORT TWENTY UNIT HYPOT. 4A WIDE 5B HIGH

< 1" >

6.928"
5.52"
5.196"
4.242"
3.463"
1.225 WAVE LENGTH 1.225"
3.675"

WAVE LENGTH DERIVED FROM UNIT MASS OF ...

NOTE THAT THERE ONLY TWO COMPLETE SET OF THREE CONCENTRIC RINGS. BEING THE CENTER ... DYMAXION ...

Books by R. BUCKMINSTER FULLER

Nine Chains to the Moon

Untitled Epic Poem on the History of Industrialization

Ideas and Integrities

No More Secondhand God

Education Automation

Utopia or Oblivion

Operating Manual for Spaceship Earth

Intuition

Earth, Inc.

Synergetics

Synergetics 2

And It Came to Pass—Not to Stay

Critical Path

SYNERGETICS

Explorations in the Geometry of Thinking

R. BUCKMINSTER FULLER

in collaboration with E. J. APPLEWHITE

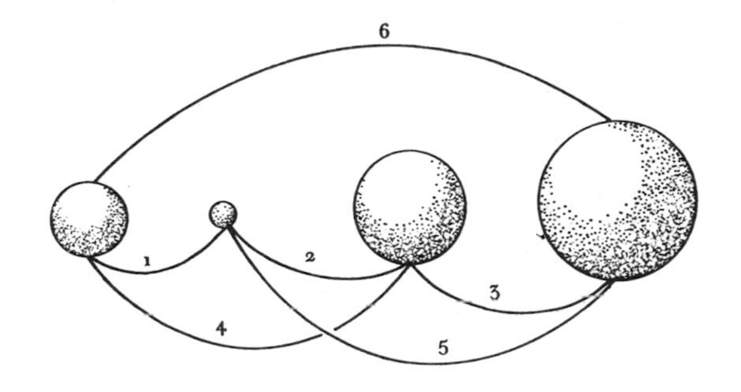

Preface and Contribution by ARTHUR L. LOEB
Harvard University

COLLIER BOOKS

MACMILLAN PUBLISHING COMPANY

NEW YORK

COLLIER MACMILLAN PUBLISHERS

LONDON

A Note to Readers: Illustrations revised from the originals may be found as follows: p. 163, figure 445.13A; p. 165, figure 450.10; p. 166, figure 450.11A; p. 167, figure 450.11B; p. 172, figure 453.01; p. 174, figure 454.06; p. 178, figure 455.20; p. 186, figure 457.40; p. 189, figure 458.12; p. 378, figure 705.01; p. 379, figure 705.02; p. 514, table 924.20; p. 534, figure 950.12; p. 569, figure 966.05; p. 776, table 1232.21. Within the unnumbered "Drawings Section," new illustrations may be found on pages 789–796, 798–808, and 811–812.

Library of Congress Cataloging in Publication Data

Fuller, R. Buckminster (Richard Buckminster), 1895–
Synergetics: explorations in the geometry of thinking.
Bibliography: p.
1. System theory. 2. Thought and thinking. 3. Mathematics—Philosophy. I. Applewhite, E. J. II. Title.
Q295.F84 1982 191 81-11798
ISBN 0-02-065320-4 AACR2

Macmillan Publishing Company
866 Third Avenue, New York, N.Y. 10022
Collier Macmillan Canada, Inc.

First Paperback Edition 1982

10 9 8 7

Synergetics is also published in hardcover edition by Macmillan Publishing Company.

Macmillan books are available at special discounts for bulk purchases for sales promotions, premiums, fund-raising, or educational use. For details, contact:

Special Sales Director
Macmillan Publishing Company
866 Third Avenue
New York, N.Y. 10022

Printed in the United States of America

Acknowledgment is gratefully made to John Entenza, the Graham Foundation, Hugh Kenner, Brooke Maxwell, Shoji Sadao, and Edwin Schlossberg for their assistance in the production of this book.

A Note on Collaboration

The discoveries, concepts, vocabulary, phraseology—every word—and entire writing style of this book originate exclusively in the mind of R. Buckminster Fuller. My role has been strictly editorial: identifying, sorting out, and organizing the presentation of five decades of Fuller's thinking, continually confronting the author with himself.

E. J. A.

THIS WORK IS DEDICATED TO

H. S. M. COXETER

PROFESSOR OF MATHEMATICS

UNIVERSITY OF TORONTO

•

To me no experience of childhood so reinforced self-
confidence in one's own exploratory faculties
as did geometry. Its inspiring effectiveness in
winnowing out and evaluating a plurality
of previously unknowns from a few given
knowns, and its elegance of proof
lead to the further discovery and comprehension of a
grand strategy for all
problem solving.
By virtue of his extraordinary life's work in mathematics,
Dr. Coxeter is the *geometer of our bestirring*
twentieth century, the spontaneously acclaimed
terrestrial curator of the historical
inventory of the science of
pattern analysis.
I dedicate this work with particular esteem for him
and in thanks to all the geometers of all time
whose importance to humanity
he epitomizes.

Contents

(A section of the author's drawings follows page 787)

Preface
by Arthur L. Loeb

THE APPEARANCE OF this sizable book is symptomatic of a considerable revival of interest in geometrics, a science of configurations. Configurations observed in the sky constitute the laboratory of our oldest science, astronomy. Patterns and regularities were discerned, and speculations regarding the influence of celestial configurations on terrestrial existence gave rise to scientific as well as mystical systems of natural philosophy. The dividing line between these two is at times surprisingly diffuse, and varies throughout history.

Platonic and Archimedean solids and such plane figures as the Pentagram were powerful tools of Applied Magic. The Age of Reason banished such configurations to the realm of superstition: their power was denied. Orthogonality prevailed, being rational and very earthbound. Interest in geometrics declined. Buckminster Fuller's search for a natural and truly rational coordinate system eventually led to the tensegrity concept and the construction of geodesic domes. Polyhedra and pentagrams, being proven useful after all, have been rescued from the limbo of superstition. Now the danger exists that geometrics will become respectable once more, and it behooves us to take a good look at the very unorthodox peregrinations of Fuller's mind before stepping into the inviting straitjacket. One of the most intriguing aspects of the present book is that there are so few ex post facto rationalizations; Fuller allows us to share his methods, his meanderings, the early influences.

Like his great aunt Margaret, Fuller is a transcendentalist: he discerns patterns and accepts their significance on faith. His is not the burden of proof: the pattern is assumed significant unless proven otherwise. If Fuller had been burdened by the necessity of proof, he would have been too hamstrung to continue looking for significant patterns. His own biographical notes in *Synergetics* show us a mind that accepts information in a highly unorthodox fashion and refuses to swallow the predigested. In rejecting the predigested, Fuller has had to discover the world all by himself. It is not surprising, in fact rather reassuring, that the obvious should emerge alongside the novel, the obscure together with the useful. Posterity will have to draw the line between the mystical and the scientific, a line that will certainly have to be redrawn from time to time.

Fuller expresses himself metaphorically: his poems sometimes convey his meaning more lucidly than his prose. Gertrude Stein's language really becomes unintelligible only when analyzed; the sentence "Entropy is not random; it is always one negative tetrahedron" (*Synergetics,* Sec. 345) is worthy of a place in American literature next to Miss Stein's. And if happiness be a warm puppy, why should not entropy be a negative tetrahedron? I have learned never to reject one of Fuller's outrageous statements without careful consideration, and even hesitated to call "Sum of angels around each vertex" in the rough manuscript a misprint. The truth has usually turned out far stranger than Buckminster Fuller!

Fuller is a Janus: grass-rooted in the past, he creates beacons for the very young. Ivy-trained, he rejected academic discipline and was highly honored by his alma mater. In each of these aspects, there is a strong parallel with an older contemporary, also a latter-day transcendentalist New Englander: composer Charles Ives. Their playful intuitive experiments nowadays inspire the most serious, abstract, and learned of the avant-garde. While there is no evidence of a mutual influence between the two, there appears to be considerable overlap among their apostles. The Compleat Geometrist might think, not of a circle of admirers, but of confocal ellipses of apostles around these two focal men, a dangerous simile if one is to conclude that the inner ellipse would be the most eccentric one!

The appearance of *Synergetics* seems to mark a watershed; Fuller has this Januslike quality, looking far into the future with an almost old-fashioned intuitive approach. The danger of respectability, alluded to previously, lies in the loss of this innocence, in the guile that kills creativity.

Fuller's hope for the future lies in doing more with less. Again and again he discovers that there is no such thing as continuity and the infinitesimal: with sufficient resolution we find that we look at a very large, but finite number of very small, but finite multiples, put together with very great, but attainable ingenuity. There always appears to be a *structure*. Computer technology has already moved in the direction of doing more with less. The cost of a modern computer would scarcely be affected if it were made of the most precious metals, for material-wise there is not very much to a computer. What counts is the knowledge of how to put things together to perform usefully. Knowledge is the tool of today and tomorrow and the hope of the day thereafter: education will be the greatest tool-making industry!

The first prerequisite for continued education is a receptiveness to one's environment. Calluses worn through a faulty environment dull the learning senses. We must educate ourselves to do more with less in creating a suitable environment. Vicious circle? Nonsense: Fuller tells us that no curve can overlap with itself! This is an upward spiral into which Fuller propels us. There is no alternative.

Moral of the Work

Dare to be naive.

Please do not refrain from reading this book because you have become suspicious that a comprehensive inventory of discovery precludes further discovery.

It is one of our most exciting discoveries that local discovery leads to a complex of further discoveries. Corollary to this we find that we no sooner get a problem solved than we are overwhelmed with a multiplicity of additional problems in a most beautiful payoff of heretofore unknown, previously unrecognized, and as-yet unsolved problems.

A complex of further discoverabilities is inherent in eternally regenerative Universe and its omni-interaccommodative complex of unique and eternal generalized principles. It is inherently potential in the integrity of eternal regeneration and the inherent complexity of unity that god is the unknowable totality of generalized principles which are only surprisingly unveiled, thereby synergetically inaugurating entirely new, heretofore unpredicted—because unpredictable—ages.

Each age is characterized by its own astronomical myriads of new, special-case experiences and problems to be stored in freshly born optimum capacity human brains—which storages in turn may disclose to human minds the presence of heretofore undiscovered, unsuspectedly existent eternal generalized principles.

—R B F

Author's Note on the Rationale for Repetition in This Work

IT IS THE writer's experience that new degrees of comprehension are always and only consequent to ever-renewed review of the spontaneously rearranged inventory of significant factors. This awareness of the processes leading to new degrees of comprehension spontaneously motivates the writer to describe over and over again what—to the careless listener or reader—might seem to be tiresome repetition, but to the successful explorer is known to be essential mustering of operational strategies from which alone new thrusts of comprehension can be successfully accomplished.

To the careless reader seeking only entertainment the repetition will bring about swift disconnect. Those experienced with the writer and motivated by personal experience with mental discoveries—co-experiencing comprehensive breakthroughs with the writer—are not dismayed by the seeming necessity to start all over again inventorying the now seemingly most lucidly relevant.

Universe factors intuitively integrating to attain new perspective and effectively demonstrated logic of new degrees of comprehension—*that's* the point. I have not forgotten that I have talked about these things before. It is part of the personal discipline, no matter how formidable the re-inventorying may seem, to commit myself to that task when inspired by intuitive glimpses of important new relationships—inspired overpoweringly because of the realized human potential of progressive escape from ignorance.

Introduction: The Wellspring of Reality

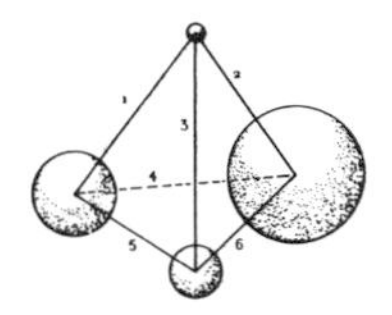

WE ARE IN an age that assumes the narrowing trends of specialization to be logical, natural, and desirable. Consequently, society expects all earnestly responsible communication to be crisply brief. Advancing science has now discovered that all the known cases of biological extinction have been caused by overspecialization, whose concentration of only selected genes sacrifices general adaptability. Thus the specialist's brief for pinpointing brevity is dubious. In the meantime, humanity has been deprived of comprehensive understanding. Specialization has bred feelings of isolation, futility, and confusion in individuals. It has also resulted in the individual's leaving responsibility for thinking and social action to others. Specialization breeds biases that ultimately aggregate as international and ideological discord, which, in turn, leads to war.

We are not seeking a license to ramble wordily. We are intent only upon being adequately concise. General systems science discloses the existence of minimum sets of variable factors that uniquely govern each and every system. Lack of knowledge concerning all the factors and the failure to include them in our integral imposes false conclusions. Let us not make the error of inadequacy in examining our most comprehensive inventory of experience and thoughts regarding the evoluting affairs of all humanity.

There is an inherently minimum set of essential concepts and current information, cognizance of which could lead to our operating our planet Earth to the lasting satisfaction and health of all humanity. With this objective, we set out on our review of the spectrum of significant experiences and seek therein for the greatest meanings as well as for the family of generalized principles governing the realization of their optimum significance to humanity aboard our Sun-circling planet Earth.

We must start with scientific fundamentals, and that means with the data of experiments and not with assumed axioms predicated only upon the misleading nature of that which only superficially seems to be obvious. It is the consensus of great scientists that science is the attempt to set in order the facts of experience. Holding within their definition, we define Universe as the aggregate of all humanity's consciously apprehended and communicated, nonsimultaneous, and only partially overlapping experiences. An aggregate of finites is finite. Universe is a finite but nonsimultaneously conceptual scenario.

The human brain is a physical mechanism for storing, retrieving, and re-storing again, each special-case experience. The experience is often a packaged concept. Such packages consist of complexedly interrelated and not as-yet differentially analyzed phenomena which, as initially unit cognitions, are potentially re-experienceable. A rose, for instance, grows, has thorns, blossoms, and fragrance, but often is stored in the brain only under the single word—*rose*.

As Korzybski, the founder of general semantics, pointed out, the consequence of its single-tagging is that the *rose* becomes reflexively considered by man only as a red, white, or pink device for paying tribute to a beautiful girl, a thoughtful hostess, or last night's deceased acquaintance. The tagging of the complex biological process under the single title *rose* tends to detour human curiosity from further differentiation of its integral organic operations as well as from consideration of its interecological functionings aboard our planet. We don't know what a rose is, nor what may be its essential and unique cosmic function. Thus for long have we inadvertently deferred potential discovery of the essential roles in Universe that are performed complementarily by many, if not most, of the phenomena we experience. But, goaded by youth, we older ones are now taking second looks at almost everything. And that promises many ultimately favorable surprises. The oldsters do have vast experience banks not available to the youth. Their memory banks, integrated and reviewed, may readily disclose generalized principles of eminent importance.

The word *generalization* in literature usually means covering too much territory too thinly to be persuasive, let alone convincing. In science, however, a generalization means a principle that has been found to hold true in *every* special case.

The principle of leverage is a scientific generalization. It makes no difference of what material either the fulcrum or the lever consists—wood, steel, or reinforced concrete. Nor do the special-case sizes of the lever and fulcrum, or of the load pried at one end, or the work applied at the lever's other end in any way alter either the principle or the mathematical regularity of the ratios of physical work advantage that are provided at progressive fulcrum-to-load increments of distance outward from the fulcrum in the opposite direction along the lever's arm at which the operating effort is applied.

Mind is the weightless and uniquely human faculty that surveys the ever larger inventory of special-case experiences stored in the brain bank and, seeking to identify their intercomplementary significance, from time to time discovers one of the rare scientifically generalizable principles running consistently through all the relevant experience set. The thoughts that discover these principles are weightless and tentative and may also be eternal. They suggest eternity but do not prove it, even though there have been no experiences thus far that imply exceptions to their persistence. It seems also to follow that the more experiences we have, the more chances there are that the mind may discover, on the one hand, additional generalized principles or, on the other hand, exceptions that disqualify one or another of the already catalogued principles that, having heretofore held ''true'' without contradiction for a long time, had been tentatively conceded to be demonstrating *eternal* persistence of behavior. Mind's relentless reviewing of the comprehensive brain bank's storage of all our special-case experiences tends both to progressive enlargement and definitive refinement of the catalogue of generalized principles that interaccommodatively govern all transactions of Universe.

It follows that the more specialized society becomes, the less attention does it pay to the discoveries of the mind, which are intuitively beamed toward the brain, there to be received only if the switches are ''on.'' Specialization tends to shut off the wide-band tuning searches and thus to preclude further discovery of the all-powerful generalized principles. Again we see how society's perverse fixation on specialization leads to its extinction. We are so specialized that one man discovers empirically how to release the energy of the atom, while another, unbeknownst to him, is ordered by his political factotum to make an atomic bomb by use of the secretly and anonymously published data. That gives much expedient employment, which solves the politician's momentary problem, but requires that the politicians keep on preparing for further warring with other political states to keep their respective peoples employed. It is also mistakenly assumed that employment is the only means by which humans can earn the right to live, for politicians have yet to discover how much wealth is available for distribution. All this is rationalized on the now scientifically discredited premise that there can never be enough life support for all. Thus humanity's specialization leads only toward warring and such devastating tools, both, visible and invisible, as ultimately to destroy all Earthians.

Only a comprehensive switch from the narrowing specialization and toward an ever more inclusive and refining comprehension by all humanity—regarding all the factors governing omnicontinuing life aboard our spaceship Earth—can bring about reorientation from the self-extinction-bound human trending, and do so within the critical time remaining before we have passed the point of chemical process irretrievability.

Quite clearly, our task is predominantly metaphysical, for it is how to get all of humanity to educate itself swiftly enough to generate spontaneous social behaviors that will avoid extinction.

Living upon the threshold between yesterday and tomorrow, which threshold we reflexively assumed in some long ago yesterday to constitute an eternal *now,* we are aware of the daily-occurring, vast multiplication of experience-generated information by which we potentially may improve our understanding of our yesterdays' experiences and therefrom derive our most farsighted preparedness for successive tomorrows.

Anticipating, cooperating with, and employing the forces of nature can be accomplished only by the mind. The wisdom manifest in the omni-interorderliness of the family of generalized principles operative in Universe can be employed only by the highest integrity of engagement of the mind's metaphysical intuiting and formulating capabilities.

We are able to assert that this rationally coordinating system bridge has been established between science and the humanities because we have made adequate experimental testing of it in a computerized world-resource-use-exploration system, which by virtue of the proper inclusion of all the parameters—as guaranteed by the synergetic start with Universe and the progressive differentiation out of all the parts—has demonstrated a number of alternate ways in which it is eminently feasible not only to provide full life support for all humans but also to permit all humans' individual enjoyment of all the Earth without anyone profiting at the expense of another and without any individuals interfering with others.

While it takes but meager search to discover that many well-known concepts are false, it takes considerable search and even more careful examination of one's own personal experiences and inadvertently spontaneous reflexing to discover that there are many popularly and even professionally unknown, yet nonetheless *fundamental,* concepts to hold true in all cases and that already have been discovered by other as yet obscure individuals. That is to say that many scientific generalizations have been discovered but have not come to the attention of what we call the educated world at large, thereafter to be incorporated tardily within the formal education processes, and even more tardily, in the ongoing political-economic affairs of everyday life. Knowledge of the existence and comprehensive significance of these as yet popularly unrecognized natural laws often is requisite to the solution of many of the as yet unsolved problems now confronting society. Lack of knowledge of the solution's existence often leaves humanity confounded when it need not be.

Intellectually advantaged with no more than the child's facile, lucid eagerness to understand constructively and usefully the major transformational events of our own times, it probably is synergetically advantageous to review swiftly the most comprehensive inventory of the most powerful human environment transforming events of our totally known and reasonably extended history. This is especially useful in winnowing out and understanding the most significant of the metaphysical revolutions now recognized as swiftly tending to reconstitute history. By such a comprehensively schematic review, we might identify also the unprecedented and possibly heretofore overlooked pivotal revolutionary events not only of today but also of those trending to be central to tomorrow's most cataclysmic changes.

It is synergetically reasonable to assume that relativistic evaluation of any of the separate drives of art, science, education, economics, and ideology, and their complexedly interacting trends within our own times, may be had only through the most comprehensive historical sweep of which we are capable.

There could be produced a synergetic understanding of humanity's cosmic functioning, which, until now, had been both undiscovered and unpredictable due to our deliberate and exclusive preoccupation only with the separate statistics of separate events. As a typical consequence of the latter, we observe our society's persistent increase of educational and employment specialization despite the already mentioned, well-documented scientific disclosure that the extinctions of biological species are always occasioned by overspecialization. Specialization's preoccupation with parts deliberately forfeits the opportunity to apprehend and comprehend what is provided exclusively by synergy.

Today's news consists of aggregates of fragments. Anyone who has taken part in any event that has subsequently appeared in the news is aware of the gross disparity between the actual and the reported events. The insistence by reporters upon having advance ''releases'' of what, for instance, convocation speakers are supposedly going to say but in fact have not yet said, automatically discredits the value of the largely prefabricated news. We also learn frequently of prefabricated and prevaricated events of a complex nature purportedly undertaken for purposes either of suppressing or rigging the news, which in turn perverts humanity's tactical information resources. All history becomes suspect. Probably our most polluted resource is the tactical information to which humanity spontaneously reflexes.

Furthermore, today's hyperspecialization in socioeconomic functioning has come to preclude important popular philosophic considerations of the synergetic significance of, for instance, such historically important events as the discovery within the general region of experimental inquiry known as virology that the as-yet popularly assumed validity of the concepts of *animate* and *inanimate* phenomena have been experimentally invalidated. Atoms and crystal complexes of atoms were held to be obviously inanimate; the protoplasmic cells of biological phenomena were held to be obviously animate. It was deemed to be common sense that warm-blooded, moist, and soft-skinned humans were clearly not to be confused with hard, cold granite or steel objects. A clear-cut threshold between animate and inanimate was therefore assumed to exist as a fundamental dichotomy of all physical phenomena. This seemingly placed life exclusively within the bounds of the physical.

The supposed location of the threshold between animate and inanimate was methodically narrowed down by experimental science until it was confined specifically within the domain of virology. Virologists have been too busy, for instance, with their DNA-RNA genetic code isolatings, to find time to see the synergetic significance to society of the fact that they have found that no physical threshold does in fact exist between animate and inanimate. The possibility of its existence vanished because the supposedly unique physical qualities of both animate and inanimate have persisted right across yesterday's supposed threshold in both directions to permeate one another's—previously perceived to be exclusive—domains. Subsequently, what was animate has become foggier and foggier, and what is inanimate clearer and clearer. All organisms consist physically and in entirety of inherently inanimate atoms. The inanimate alone is not only omnipresent but is alone experimentally demonstrable. Belated news of the elimination of this threshold must be interpreted to mean that whatever life may *be,* it has *not* been isolated and thereby identified as residual in the biological cell, as had been supposed by the false assumption that there was a separate physical phenomenon called animate within which life existed. No *life* per se has been isolated. The threshold between animate and inanimate has vanished. Those chemists who are preoccupied in synthesizing the particular atomically structured molecules identified as the prime constituents of humanly employed organisms will, even if they are chemically successful, be as remote from creating life as are automobile manufacturers from creating the human drivers of their automobiles. Only the physical connections and development complexes of distinctly "nonlife" atoms into molecules, into cells, into animals, has been and will be discovered. The genetic coding of the design controls of organic systems offers no more explanation of life than did the specifications of the designs of the telephone system's apparatus and operation explain the nature of the *life* that communicates weightlessly to *life* over the only physically ponderable telephone system. Whatever else life may be, we know it is weightless. At the moment of death, no weight is lost. All the chemicals, including the chemist's life ingredients, are present, but life has vanished. The physical is inherently entropic, giving off energy in ever more disorderly ways. The metaphysical is antientropic, methodically marshalling energy. Life is antientropic. It is spontaneously inquisitive. It sorts out and endeavors to understand.

The overconcentration on details of hyperspecialization has also been responsible for the lack of recognition by science of its inherently mandatory responsibility to reorient all our educational curricula because of the synergetically disclosed, but popularly uncomprehended, significance of the 1956 Nobel Prize-winning discovery in physics of the experimental invalidation of the concept of "parity" by which science previously had misassumed that positive-negative complementations consisted exclusively of mirror-imaged behaviors of physical phenomena.

Science's self-assumed responsibility has been self-limited to disclosure to society only of the separate, supposedly physical (because separately weighable) atomic component isolations data. Synergetic integrity would require the scientists to announce that in reality what had been identified heretofore as physical is entirely metaphysical—because synergetically weightless. Metaphysical has been science's designation for all weightless phenomena such as thought. But science has made no experimental finding of any phenomena that can be described as a solid, or as continuous, or as a straight surface plane, or as a straight line, or as infinite anything. We are now synergetically forced to conclude that all phenomena are metaphysical; wherefore, as many have long suspected—like it or not—life is but a dream.

Science has found no *up* or *down* directions of Universe, yet scientists are personally so ill-coordinated that they all still personally and sensorially see "solids" going up or down—as, for instance, they see the Sun "going down." Sensorially disconnected from their theoretically evolved information, scientists discern no need on their part to suggest any educational reforms to correct the misconceiving that science has tolerated for half a millennium.

Society depends upon its scientists for just such educational reform guidance. Where else might society turn for advice? Unguided by science, society is allowed to go right on filling its childrens' brain banks with large inventories of competence-devastating misinformation. In order to emerge from its massive ignorance, society will probably have to rely exclusively upon its individuals' own minds to survey the pertinent experimental data—as do all great scientist-artists. This, in effect, is what the intuition of world-around youth is beginning to do. Mind can see that reality is evoluting into weightless metaphysics. The wellspring of reality is the family of weightless generalized principles.

It is essential to release humanity from the false fixations of yesterday, which seem now to bind it to a rationale of action leading only to extinction.

The youth of humanity all around our planet are intuitively revolting from all sovereignties and political ideologies. The youth of Earth are moving intuitively toward an utterly classless, raceless, omnicooperative, omniworld humanity. Children freed of the ignorantly founded educational traditions and exposed only to their spontaneously summoned, computer-stored and -distributed outflow of reliable-opinion-purged, experimentally verified data, shall indeed lead society to its happy egress from all misinformedly conceived, fearfully and legally imposed, and physically enforced customs of yesterday. They can lead all humanity into omnisuccessful survival as well as entrance into an utterly new era of human experience in an as-yet and ever-will-be fundamentally mysterious Universe.

And whence will come the wealth with which we may undertake to lead world man into his new and validly hopeful life? From the wealth of the minds of world man—whence comes all wealth. Only mind can discover how to do so much with so little as forever to be able to sustain and physically satisfy all humanity.

100.00 Synergy

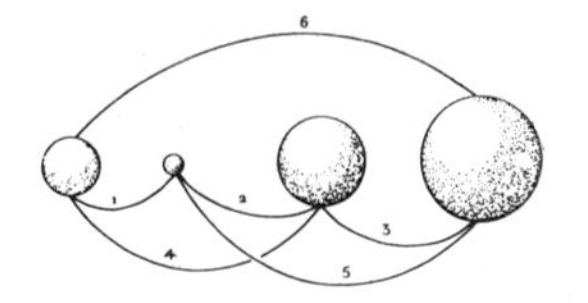

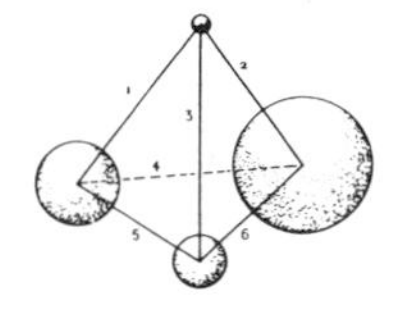

101.00 Definition: Synergy

101.01 Synergy means behavior of whole systems unpredicted by the behavior of their parts taken separately.

102.00 Synergy means behavior of integral, aggregate, whole systems unpredicted by behaviors of any of their components or subassemblies of their components taken separately from the whole.

103.00 A stone by itself does not predict its mass interattraction for and by another stone. There is nothing in the separate behavior or in the dimensional or chemical characteristics of any one single metallic or nonmetallic massive entity which by itself suggests that it will not only attract but also be attracted by another neighboring massive entity. The behavior of these two together is unpredicted by either one by itself. There is nothing that a single massive sphere will or can ever do by itself that says it will both exert and yield attractively with a neighboring massive sphere and that it yields progressively: every time the distance between the two is halved, the attraction will be fourfolded. This unpredicted, only mutual behavior is synergy. Synergy is the only word in any language having this meaning.

104.00 The phenomenon synergy is one of the family of generalized principles that only co-operates amongst the myriad of special-case experiences. Mind alone discerns the complex behavioral relationships to be cooperative between, and not consisting in any one of, the myriad of brain-identified special-case experiences.

105.00 The words synergy (*syn*-ergy) and energy (*en*-ergy) are companions. Energy studies are familiar. Energy relates to differentiating out subfunctions of nature, studying objects isolated out of the whole complex of Universe—for instance, studying soil minerals without consideration of hydraulics or of plant genetics. But synergy represents the integrated behaviors instead of all the differentiated behaviors of nature's galaxy systems and galaxy of galaxies.

106.00 Chemists discovered that they had to recognize synergy because they found that every time they tried to isolate one element out of a complex or to separate atoms out, or molecules out, of compounds, the isolated parts and their separate behaviors never explained the associated behaviors at all. It always failed to do so. They had to deal with the wholes in order to be able to discover the group proclivities as well as integral characteristics of parts. The chemists found the Universe already in complex association and working very well. Every time they tried to take it apart or separate it out, the separate parts were physically divested of their associative potentials, so the chemists had to recognize that there were associated behaviors of wholes unpredicted by parts; they found there was an old word for it—synergy.

107.00 Because synergy alone explains the eternally regenerative integrity of Universe, because synergy is the only word having its unique meaning, and because decades of querying university audiences around the world have disclosed only a small percentage familiar with the word *synergy,* we may conclude that society does not understand nature.

108.00 *Four Triangles Out of Two*

108.01 Two triangles can and frequently do associate with one another, and in so doing they afford us with a synergetic demonstration of two prime events cooperating in Universe. Triangles cannot be structured in planes. They are always positive or negative helixes. You may say that we had no right to break the triangles open in order to add them together, but the triangles were in fact never closed because no line can ever come completely back into itself. Experiment shows that two lines cannot be constructed through the same point at the same time (see Sec. 517, ''Interference''). One line will be superimposed on the other. Therefore, the triangle is a spiral—a very flat spiral, but open at the recycling point.

108.02 By conventional arithmetic, one triangle plus one triangle equals two triangles. But in association as left helix and right helix, they form a six-edged tetrahedron of *four* triangular faces. This illustrates an interference of two events impinging at both ends of their actions to give us something very fundamental: a tetrahedron, a system, a division of Universe into inside and outside. We get the two other triangles from the rest of the Universe because we are not out of this world. This is the complementation of the Universe that shows up time and again in the way structures are made and in the way crystals grow. As separate actions, the two actions and resultants were very unstable, but when associated as positive and negative helixes, they complement one another as a stable structure. (See Sec. 933.03.)

108.03 Our two triangles now add up as *one plus one equals four*. The

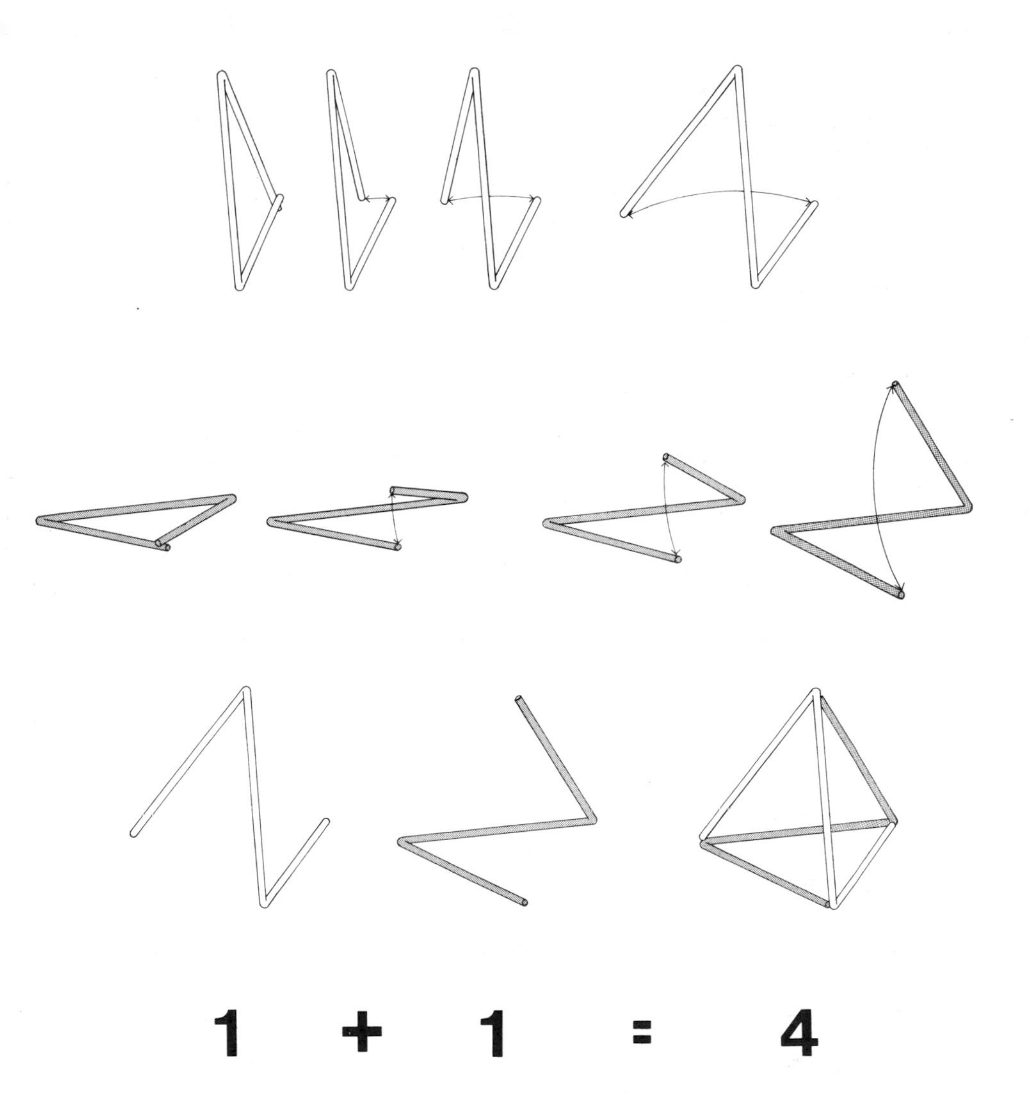

Fig. 108.01 *Triangle and Tetrahedron: Synergy* (*1+1=4*): Two triangles may be combined in such a manner as to create the tetrahedron, a figure volumetrically embraced by four triangles. Therefore one plus one seemingly equals four.

two events make the tetrahedron the four-triangular-sided polyhedron. This is not a trick; this is the way atoms themselves behave. This is a demonstration of synergy. Just as the chemists found when they separated atoms out, or molecules out, of compounds, that the separate parts never explained the associated behaviors; there seemed to be "lost" energies. The lost energies were the lost synergetic interstabilizations.

109.00 Chrome-Nickel-Steel

109.01 Synergy alone explains metals increasing their strengths. All alloys are synergetic. Chrome-nickel-steel has an extraordinary total behavior. In fact, it is the high cohesive strength and structural stability of chrome-nickel-steel at enormous temperatures that has made possible the jet engine. The principle of the jet was invented by the squid and the jellyfish long ago. What made possible man's use of the jet principle was his ability to concentrate enough energy and to release it suddenly enough to give him tremendous thrust. The kinds of heat that accompany the amount of energies necessary for a jet to fly would have melted all the engines of yesterday. Not until you had chrome-nickel-steel was it possible to make a successful jet engine, stable at the heats involved. The jet engine has changed the whole relationship of man to the Earth. And it is a change in the behavior of the whole of man and in the behavior of whole economics, brought about by synergy.

109.02 In chrome-nickel-steel, the primary constituents are iron, chromium, and nickel. There are minor constituents of carbon, manganese, and others. It is a very popular way of thinking to say that a chain is no stronger than its weakest link. That seems to be very logical to us. Therefore, we feel that we can predict things in terms of certain minor constituents of wholes. That is the way much of our thinking goes. If I were to say that a chain is as strong as the sum of the strengths of its links, you would say that is silly. If I were to say that a chain is stronger than the sum of the strengths of all of its links, you might say that that is preposterous. Yet that is exactly what happens with chrome-nickel-steel. If our regular logic held true, then the iron as the weakest part ought to adulterate the whole: since it is the weakest link, the whole thing will break apart when the weakest link breaks down. So we put down the tensile strength of the commercially available iron—the highest that we can possibly accredit is about 60,000 pounds per square inch (p.s.i.); of the chromium it is about 70,000 p.s.i.; of the nickel it is about 80,000 p.s.i. The tensile strengths of the carbon and the other minor constituents come to another 50,000 p.s.i. Adding up all the strengths of all the links we get 260,000 p.s.i. But in fact the tensile strength of chrome-nickel-steel runs to about 350,000 p.s.i. just as a casting. Here we have the behavior of the whole completely unpredicted by the behavior of the parts.

109.03 The augmented coherence of the chrome-nickel-steel alloy is accounted for only by the whole complex of omnidirectional, intermass-attractions of the crowded-together atoms. The alloy chrome-nickel-steel provides unprecedented structural stability at super-high temperatures, making possible the jet engine—one of the reasons why the relative size of our planet Earth, as comprehended by humans, has shrunk so swiftly. The performance of the alloy demonstrates that the strength of a chain is greater than the sum of the strengths of its separate links. Chrome-nickel-steel's weakest part does not adulterate the whole, allowing it to be "dissolved" as does candy when the sugar dissolves. Chains in metal do not occur as open-ended lines. In the atoms, the ends of the chains come around and fasten the ends together, endlessly, in circular actions. Because atomic circular chains are dynamic, if one link breaks, the other mends itself.

109.04 When we break one link of a circular chain continuity, it is still

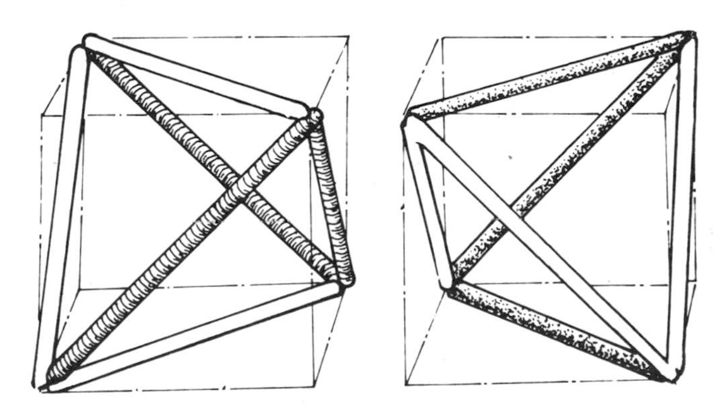

Fig. 110A

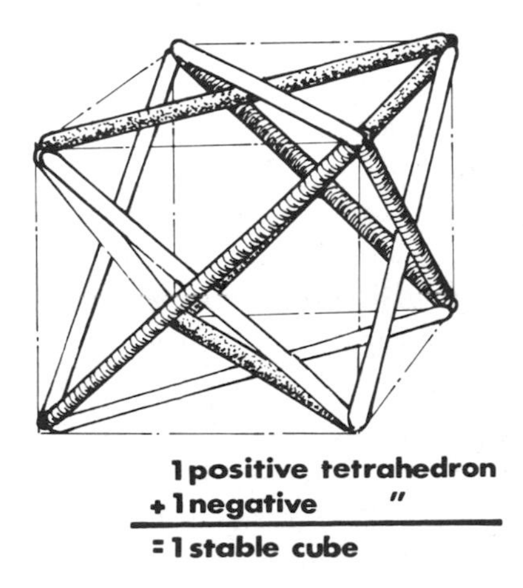

1 positive tetrahedron
+ 1 negative "
= 1 stable cube

Fig. 110B

one piece of chain. And because atomic circular chains are dynamic, while one link is breaking, the other is mending itself. Our metal chains, like chrome-nickel-steel alloys, are also interweaving spherically in a number of directions. We find the associated behaviors of various atoms complementing each other, so that we are not just talking about *one thing* and another *one thing,* but about a structural arrangement of the atoms in tetrahedral configurations.

110.00 We take one tetrahedron and associate it with another tetrahedron. Each of the two tetrahedra has four faces, four vertexes, and six edges. We interlock the two tetrahedra, as illustrated, so that they have a common center of gravity and their two sets of four vertexes each provide eight vertexes for the corners of a cube. They are interpositioned so that the vertexes are evenly spaced from each other in a symmetrical arrangement as a structurally stable cube.

111.00 Each of those vertexes was an energy star. Instead of two separate tetrahedra of four stars and four stars we now have eight stars symmetrically equidistant from the same center. All the stars are nearer to each other. There are eight stars in the heavens instead of four. Not only that, but each star now has three stars nearer to it than the old stars used to be. The stars therefore interattract one another gravitationally in terms of the second power of their relative proximity—in accordance with Newton's law of gravity. As the masses are getting closer to each other, synergy is increasing their power of interattraction very rapidly.

112.00 The distance between the stars is now in terms of the leg instead of the hypotenuse. The second power of the hypotenuse is equal to the sums of the second powers of the legs, so we suddenly discover how very much more of an attraction there is between each star to make each one more cohesive in the second power augmentation. There was no such augmentation predicted by the first power addition. Thus, it is no surprise to discover that the close interassociation of the energy stars gives us a fourfolding of the tensile strength of our strongest component of the alloy chrome-nickel-steel of 350,000 p.s.i. in relation to nickel's 80,000 p.s.i. Gravity explains why these metals, when in proper association, develop such extraordinary coherence, for we are not really dealing in a mystery—outside of the fact that we are dealing in the mystery of how there happened to be gravity and how there happened to be Universe. How there happened to be Universe is certainly a great mystery—there is no question about that—but we are not dealing with any miracle here outside of the fact that Universe is a miracle.

113.00 When we take two triangles and add one to the other to make the tetrahedron, we find that one plus one equals four. This is not just a geometrical trick; it is really the same principle that chemistry is using inasmuch as the tetrahedra represent the way that atoms cohere. Thus we discover synergy to be operative in a very important way in chemistry and in all the composition of the Universe. Universe as a whole is behaving in a way that is completely unpredicted by the behavior of any of its parts. Synergy reveals a grand strategy of dealing with the whole instead of the tactics of our conventional educational system, which starts with parts and elements, adding them together locally without really understanding the whole.

114.00 It is a corollary of synergy (see Sec. 140) that once you start dealing with the known behavior of the whole and the known behavior of some of the parts, you will quite possibly be able to discover the unknown parts. This strategy has been used—in rare breakthroughs—very successfully by man. An example of this occurred when the Greeks developed the law of the triangle: the sum of the angles is always 180 degrees, and there are six parts (three edges and three vertexes—forming three angles); thus the known behavior of the whole and the known behavior of two of the parts may give you a clue to the behavior of the other part.

115.00 Newton's concept of gravity also gave him the behavior of the whole. Other astronomers said that if he were right, they should be able to explain the way the solar system is working. But when they took the masses of the known planets and tried to explain the solar system, it didn't work out. They said you need two more planets, but we don't have them. There are either two planets we cannot see or Newton is wrong. If he was right, someday an astronomer with a powerful telescope would be able to see sufficient distance to pick up two more planets of such and such a size. In due course, they were found. The known behavior of the whole explained in terms of gravity and the known behavior of some of the parts permitted the prediction of the behavior of some of the other—at that time unknown—parts.

116.00 Physicists had predicated their grand strategies upon the experience of trying to make something like a perpetual motion machine. They found that all local machines always had friction, therefore energies were always going out of the system. They call that entropy: local systems were always losing energy to the rest of the Universe. When the physicists began to look at their total experience instead of at just one of their experiences, they found that while the energy may escape from one system, it does not go out of the Universe. It could only disassociate in one place by associating in another place. They found that this was experimentally true, and finally, by the mid-19th century, they dared to develop what they called the Law of Conservation of Energy, which said that no energy could be created and no energy could be lost. Energy is finite. Physical Universe is finite. Physical Universe is just as finite as the triangle of 180 degrees.

117.00 Dealing with a finite whole in terms of our total experience has taught us that there are different kinds of frequencies and different rates of reoccurrence of events. Some events reoccur very rapidly. Some are large events, and some small events. In a finite Universe of energy, there is only so much energy to expend. If we expend it all in two big booms, they are going to be quite far apart:

boom boom

Given the same finite amount of time, we could alternatively have a great many very small booms fairly close together:

boom boom boom boom

In others words, we can take the same amount of copper and make a propeller with just two blades, with three smaller blades, or with four much smaller blades. That is, we can with the same amount of copper invest the whole in higher frequency and get smaller wavelength. This is the quantum in wave mechanics; it is a most powerful tool that men have used to explore the nucleus of the atom, always assuming that 100 percent of the behaviors must be accounted for. We are always dealing with 100 percent finite. Experiment after experiment has shown that if there was something like .000172 left over that you could not account for, you cannot just dismiss it as an error in accounting. There must be some little energy rascal in there that weighs .000172. They finally gave it a name, the ''whatson.'' And then eventually they set about some way to trap it in order to observe it. It is dealing with the whole that makes it possible to discover the parts. That is the whole strategy of nuclear physics.

120.00 Mass Interattraction

120.01 Synergy is disclosed by the interattraction for one another of two or more separate objects. But any two masses will demonstrate that halving the distance between them will fourfold their attraction for each other. (Which is the way Newton might have said it, but did not.) He discovered the mathematical gain in attraction, but he stated it ''inversely,'' which is awkward and nonspontaneously illuminating. The inverseness led him to speak in terms of progressive diminution of the attraction: as the distance away was multiplied by two, the attraction diminished by four; ergo, he could speak of it as ''squared.'' The attraction of one mass for the other increases as the second power of the rate of increase of their proximity to one another: halve the distance and the interaction is fourfolded.

121.00 Our senses are easily deceived because mass interattraction is not explained and cannot be predicted by any characteristic of any one massive body considered alone. Local observation of mass attraction is also obscured by the overwhelming presence of Earth's gravity. For instance, two 12-inch-in-diameter spheres of so dense a material as ivory do not appear to attract each other until they are only about a paper-thin distance apart. The thickness of a paper match superimposed on a 12-inch globe represents the point at which a rocket precesses into orbit, going from its 180-degree tendency to fall into its 90-degree orbital independence as an astronomical entity. This is the critical-behavior point at which it becomes an independent entity in Universe, a satellite. Small Earth satellites orbit at an altitude of only about 100 miles, which is only about 1/80th of the diameter of the Earth. This critical proximity event of transition from 180-degree to 90-degree independence is called precession. Mass attraction is also involved in precession, another member of the family of generalized principles. But scientists still have not the slightest idea why mass attraction occurs; they only know that it does. They do not know why. This requires admission of an utter a priori mystery within which the masses demonstrate their utterly mysterious attraction for one another. It appears that no single part of the Universe can predict the behavior of the whole. As we attain greater experience and opportunity to observe the synergetic effects of Universe, there is always a greater discernment of generalized principles. The discovery of a plurality of generalized principles permits the discovery of the synergetic effect of their complex interactions.

130.00 Precession and Entropy

130.01 Critical proximity occurs where there is angular transition from "falling back in" at 180-degree to 90-degree orbiting—which is precession. (Gravity may be described as "falling back in" at 180 degrees.) The quantity of energy that ceased to "fall in" is the system's entropy. Critical proximity is when it *starts* either "falling in" or going into orbit, which is the point where either entropy or antientropy begins.

131.00 An aggregate of "falling ins" is a body. What we call an object or an entity is always an aggregate of interattracted entities; it is never a solid. And the critical proximity transition from being an aggregate entity to being a plurality of separate entities is precession, which is a "peeling off" into orbit rather than falling back in to the original entity aggregate. This explains entropy intimately. It also explains intimately the apparent energy losses in chemical transformations, associations, disassociations and high-order element disintegration into a plurality of lower-order elements—and nothing is lost. Entity has become invisible. The switch is precessional.

132.00 The unprotected far side of the Moon has more craters of the "fallen-in" asteroids. Ergo, the far side weighs more than the near side, which is shielded by the Earth. The additional far-side weight of the Moon acts centrifugally to keep the weighted side always away from the Earth around which it orbits. Ergo, there is always one side, the same side, facing us. The Moon is always oriented toward us, like a ship that has its masts pointed inwardly toward us and its weighted keel away from us. This explains why the first photographs showed a greater number of craters on the far side of the Moon. The Earth acts as a shield. On Earth, the craters are not so concentrated because the Earth gets its cosmic fallout quite evenly. Earth's weight and massive pull are progressively increased to offset the Moon's far-side weight increase and tendency otherwise to forsake Earth.

133.00 "Solids" are simply the fraternities of the "fallen-into-one-anothers."

140.00 Corollary of Synergy: Principle of the Whole System

141.00 There is a corollary of synergy known as the Principle of the Whole System, which states that the known behaviors of the whole plus the known behaviors of some of the parts may make possible discovery of the presence of other parts and their behaviors, kinetics, structures, and relative dimensionalities.

142.00 The known sum of the angles of a triangle plus the known characteristics of three of its six parts (two sides and an included angle or two angles and an included side) make possible evaluating the others. Euler's topology provides for the synergetic evaluation of any visual system of experiences, metaphysical or physical, and Willard Gibbs' phase rule provides synergetic evaluation of any tactile system.

143.00 The systematic accounting of the behavior of whole aggregates may disclose discretely predictable angle-and-frequency magnitudes required of some unknown components in respect to certain known component behaviors of the total and known synergetic aggregate. Thus the definitive identifications permitted by the Principle of the Whole System may implement conscious synergetic definition strategies with incisive prediction effectiveness.

150.00 Synergy-of-Synergies

150.01 There are progressive degrees of synergy, called synergy-of-synergies, which are complexes of behavior aggregates holistically unpredicted by the separate behaviors of any of their subcomplex components. Any subcomplex aggregate is only a component aggregation of an even greater event aggregation whose comprehensive behaviors are never predicted by the component aggregates alone. There is a synergetic progression in Universe—a hierarchy of total complex behaviors entirely unpredicted by their successive subcomplexes' behaviors. It is manifest that Universe is the maximum synergy-of-synergies, being utterly unpredicted by any of its parts.

151.00 It is readily understandable why humans, born utterly helpless, utterly ignorant, have been prone to cope in an elementary way with successive experiences or "parts." They are so overwhelmed by the synergetic mystery of the whole as to have eschewed educational strategies commencing with Universe and the identification of the separate experiences within the cosmic totality.

152.00 Synergetics is the exploratory strategy of starting with the whole and the known behavior of some of its parts and the progressive discovery of the integral unknowns and their progressive comprehension of the hierarchy of generalized principles.

153.00 Universe apparently is omnisynergetic. No single part of experience will ever be able to explain the behavior of the whole. The more experience one has, the more opportunity there is to discover the synergetic effects, such as to be able to discern a generalized principle, for instance. Then discovery of a plurality of generalized principles permits the discovery of the synergetic effects of their complex interactions. The synergetic metaphysical effect produced by the interaction of the known family of generalized principles is probably what is spoken of as wisdom.

160.00 Generalized Design Science Exploration

161.00 Science has been cogently defined by others as the attempt to set in order the facts of experience. When science discovers order subjectively, it is pure science. When the order discovered by science is objectively employed, it is called applied science. The facts of experience are always special cases. The order sought for and sometimes found by science is always eternally generalized; that is, it holds true in every special case. The scientific generalizations are always mathematically statable as equations with one term on one side of the equation and a plurality of at least two terms on the other side of the equation.

162.00 There are eternal generalizations that embrace a plurality of generalizations. The most comprehensive generalization would be that which has $U = MP$, standing for an eternally regenerative Universe of M times P, where M stands for the metaphysical and P stands for the physical. We could then have a subgeneralization where the physical $P = E^r \cdot E^m$, where E^r stands for energy as radiation and E^m stands for energy as matter. There are thus orders of generalization in which the lower orders are progressively embraced by the higher orders. There are several hundred first-order generalizations already discovered and equatingly formalized by scientist-artists. There are very few of the higher order generalizations. Because generalizations must hold true without exception, these generalizations must be inherently eternal. Though special-case experiences exemplify employment of eternal principles, those special cases are all inherently terminal; that is, in temporary employment of the principles.

163.00 No generalized principles have ever been discovered that contradict other generalized principles. All the generalized principles are interaccommodative. Some of them are synchronously interaccommodative; that is, some of them accommodate the other by synchronized nonsimultaneity. Many of them are interaccommodative simultaneously. Some interact at mathematically exponential rates of interaugmentation. Because the physical is time, the relative endurances of all special-case physical experiences are proportional to the synchronous periodicity of associability of the complex principles involved. Metaphysical generalizations are timeless, i.e., eternal. Because the metaphysical is abstract, weightless, sizeless, and eternal, metaphysical experiences have no endurance limits and are eternally compatible with all other metaphysical experiences. What is a *metaphysical experience?* It is comprehending the relationships of eternal principles. The means of communication is physical. That which is communicated, i.e., understood, is metaphysical. The symbols with which mathematics is communicatingly described are physical. A mathematical principle is metaphysical and independent of whether X,Y or A,B are symbolically employed.

164.00 The discovery by human mind, i.e., intellect, of eternally generalized principles that are only intellectually comprehendable and only intuitively apprehended—and only intellectually comprehended principles being further discovered to be interaccommodative—altogether discloses what can only be complexedly defined as a *design,* design being a complex of interaccommodation and of orderly interaccommodation whose omni-integrity of interaccommodation order can only be itself described as intellectually immaculate. Human mind (intellect) has experimentally demonstrated at least limited access to the eternal design intellectually governing eternally regenerative Universe.

165.00 *Generalized design-science exploration* is concerned with discovery and use by human mind of complex aggregates of generalized principles in specific-longevity, special-case innovations designed to induce humanity's consciously competent participation in local evolutionary transformation events invoking the conscious comprehension by ever-increasing proportions of humanity of the cosmically unique functioning of humans in the generalized design scheme of Universe. This conscious comprehension must in turn realize ever-improving implementations of the unique human functioning as well as an ever-increasingly effective concern for the relevant ecological intercomplementation involved in local Universe support of humanity's functioning as subjective discoverer of local order and thereafter as objective design-science inventor of local Universe solutions of otherwise unsolvable problems, design-science solutions of which will provide special-case, local-Universe supports of eternally regenerative generalized Universe.

166.00 The prime eternal laws governing design science as thus far accrued to that of the cosmic law of generalized design-science exploration are realizability and relative magnitude of reproducibility, which might be called the law of regenerative design: the relative physical time magnitude of reproducibility is proportional to the order of magnitude of cosmic function generalizability. Because the higher the order of synergetic function generalization, the more embracing and simple its statement; only the highest orders can embracingly satisfy the plurality of low-order interaccommodation conditions.

167.00 There are several corollaries to the prime law of regenerative design durability and amplitude of reproducibility. *Corollary A:* The simpler, the more enduringly reproducible. *Corollary B:* The special-case realizations of a given design complex correlate as the more symmetrical, the more reproducible. *Corollary C:* There being limit cases of optimum symmetry and simplicity, there are simplicities of conceptual realization. The most enduringly reproducible design entities of Universe are those occurring at the min-max limits of simplicity and symmetry.

168.00 *Corollary D:* There being unique minimum-maximum system limits governing the transformation of conceptual entities in Universe, which differentiate the conceptually unique entities of Universe into those conceptions occurring exclusively outside the system considered and all of the Universe inside of the conceptual entity, together with the structural pattern integrity system separating the inside from the outside, there being a limited minimum set of structural and operating principles eternally producing and reproducing recognizable pattern integrity. And there are likewise a minimum set of principles that interact to transform already orderly patterns into other structured patterns, and there being minimum constituent patterns that involve the complex intertransformings and structural formings of symmetrical orders and various magnitudes of asymmetrical deviations tolerated by the principles complexedly involved. There are scientifically discoverable nuclear aggregates of primary design integrity as well as complex symmetrical reassociabilities of the nuclear primary integrities and deliberately employable relationships of nuclear simplexes which designedly impose asymmetrical-symmetrical pulsative periodicities.

169.00 Corollary E: The more symmetrical and simple and nuclear, the more frequently employable; ergo, the more frequently occurring in eternally regenerative Universe's transformative problem solutions.

170.00 Corollary F: The smaller and simpler, more symmetrical, frequently occurring in Universe and the larger and more complex, less frequently originally occurring and periodically reoccurring: for example, the hydrogen minimum limit simplex constituting not only nine-tenths of physical Universe but most frequently and most omnipresent in Universe; with asymmetrical battleships (fortunately) least frequently and compatibly recurrent throughout the as yet known cosmos, being found only on one minor planet in one typical galaxy of one hundred billion stars amongst an already-discovered billion galaxies, there having been only a few score of such man-made battleships recurrent in the split-second history of humans on infinitesimally minor Earth.

171.00 All the fundamental nuclear simplexes of the 92 inherently self-regenerative physical Universe elements are a priori to human mind formulation and invention and are only discoverable by mind's intuitive initiatives. Many myriads of complex associability of chemical compounding of the nuclear simplexes can be experimentally discovered, or, after comprehending the order of the principles involved, deliberately invented by human mind. The chemical compounds are temporary and have limited associabilities. Human minds can then invent, by deliberate design, momentarily appropriate complex associative events—as, for instance, hydraulics, crystallines, and plasmics, in turn involving mechanics of a complex nature and longevity. Omniautomated self parts replacing sensingly fedback industrial complexes can be comprehensively designed by human mind, the mass reproducibility and service longevity of which will always be fundamental to the design laws, both primary and corollary.

172.00 Biological designs a priori to human alteration contriving are directly reproducible in frequency design magnitude. Blades of grass are reproduced on planet Earth in vast quantities due to the universal adequacy of Sun and other star photosynthetic impoundment. Daisies, peanuts, glowworms, etc. are reproduced in direct complement to their design complexity, which involves biological and eternal environmental interplay of chemical element simplexes and compounds under a complex of energy, heat, and pressure conditions critical to the complex of chemical associating and disassociating involved. Humans have thus far evolved the industrial complex designing which is only of kindergarten magnitude compared to the complexity of the biological success of our planet Earth. In its complexities of design integrity, the Universe is technology.

173.00 The technology evolved by man is thus far amateurish compared to the elegance of nonhumanly contrived regeneration. Man does not spontaneously recognize technology other than his own, so he speaks of the rest as something he ignorantly calls nature. Much of man's technology is of meager endurance, being comprised at the outset of destructive invention such as that of weaponry, or for something in support of the quick-profit, man-invented game of selfishly manipulative game-playing and rule-inventing for the playing of his only-ignorantly-preoccupying value systems.

174.00 The greatest and most enduring discoveries and inventions of humans on our planet are those of the scientist-artists, the name joined, or artist, or scientist. The name of artist or scientist, though often self-professed, can only be accredited to an individual by others who in retrospect discover the enduring quality of the symmetries with which the individual converted his conceptioning to the advantage of others, and realizations of increasing interadvantage in respect to survival—the gradual discovery of the function in Universe which humanity has been designed to fulfill.

200.00 Synergetics

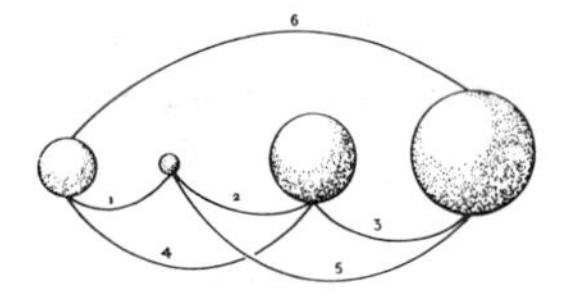

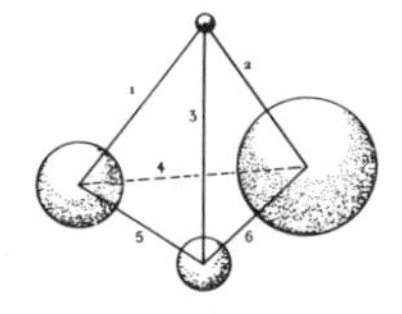

200.001 Definition: Synergetics

200.01 Synergetics promulgates a system of mensuration employing 60-degree vectorial coordination comprehensive to both physics and chemistry, and to both arithmetic and geometry, in rational whole numbers.

200.02 Synergetics originates in the assumption that dimension must be physical; that conceptuality is metaphysical and independent of size; and that a triangle is a triangle independent of size.

200.03 Since physical Universe is entirely energetic, all dimension must be energetic. Synergetics is energetic geometry since it identifies energy with number. Energetic geometry employs 60-degree coordination because that is nature's way to closest-pack spheres.

200.04 Synergetics provides geometrical conceptuality in respect to energy quanta. In synergetics, the energy as mass is constant, and nonlimit frequency is variable.

200.05 Vectors and tensors constitute all elementary definition.

201.00 Experientially Founded Mathematics

201.01 The mathematics involved in synergetics consists of topology combined with vectorial geometry. Synergetics derives from experientially invoked mathematics. Experientially invoked mathematics shows how we may measure and coordinate omnirationally, energetically, arithmetically, geometrically, chemically, volumetrically, crystallographically, vectorially, topologically, and energy-quantum-wise in terms of the tetrahedron.

201.02 Since the measurement of light's relative swiftness, which is far from instantaneous, the classical concepts of instant Universe and the mathematicians' instant lines have become both inadequate and invalid for inclusion in synergetics.

201.03 Synergetics makes possible a rational, whole-number, low-integer quantation of all the important geometries of experience because the tetrahedron, the octahedron, the rhombic dodecahedron, the cube, and the vector equilibrium embrace and comprise all the lattices of all the atoms.

202.00 Angular Topology

202.01 Synergetics is a triangular and tetrahedral system. It uses 60-degree coordination instead of 90-degree coordination. It permits conceptual modeling of the fourth and fifth arithmetic powers; that is, fourth- and fifth-dimensional aggregations of points or spheres in an entirely rational coordinate system that is congruent with all the experientially harvested data of astrophysics and molecular physics; that is, both macro- and micro-cosmic phenomena. It coordinates within one mensurational system the complete gears-interlocking of quantum wave mechanics and vectorial geometry.

202.02 Synergetics topology integrates laws of angle and volume regularities with Euler's point, area, and line abundance laws.

203.00 Scope

203.01 Synergetics explains much that has not been previously illuminated. It is not contradictory to any of the experimentally based knowledge of the classically disciplined sciences. It does not contradict the calculus or any other mathematical tool for special-case applications, although it often finds them inadequate or irrelevant.

203.02 Experientially founded synergetics clearly identifies the conceptual limitations and coordinate functionings of all the classical tools of mathematics, and it shows how their partial functioning often frustrates comprehension of experience.

203.03 Synergetics follows the cosmic logic of the structural mathematics strategies of nature, which employ the paired sets of the six angular degrees of freedom, frequencies, and vectorially economical actions and their multialternative, equieconomical action options.

203.04 Rather than refuting the bases of presently known Euclidean and non-Euclidean and hyperbolic and elliptic geometry, synergetics identifies the alternate freedoms of prime axiomatic assumption from which the present mathematical bases were selected. It embraces all the known mathematics. All of the axiomatic alternatives are logical. Thus, original assumptions eliminate the necessity for subsequent assignment of physical qualities to nonconceptual mathematical devices. Classical mathematics has, of necessity, assigned progressively discovered attributes of physical Universe to irrational relationships with the ghostly, a priori Greek geometry. The quest for a mathematics expressing nature's own design has been an elusive one. Synergetics has developed as the search for the omnirational, comprehensive, coordinate system employed by nature, i.e., Universe, throughout all its complementary and interaccommodatively transforming transactions.

203.05 As Werner Heisenberg says "if nature leads us to mathematical forms of great simplicity and beauty . . . to forms that no one has previously encountered, we cannot help thinking that they are 'true,' and that they reveal a genuine feature of nature." *

203.06 Synergetics altogether forsakes axioms as "self-evident," premicroscope, superficial beliefs. It predicates all its relationship explorations on the most accurately and comprehensively statable observations regarding direct experiences. The new set of data employed by synergetics seemingly results in sublimely facile expression of hitherto complex relationships. It makes nuclear physics a conceptual facility comprehensible to any physically normal child.

203.07 Synergetics discloses the excruciating awkwardness characterizing present-day mathematical treatment of the interrelationships of the independent scientific disciplines as originally occasioned by their mutual and separate lacks of awareness of the existence of a comprehensive, rational, coordinating system inherent in nature.

203.08 Synergetics makes possible the return to omniconceptual modeling of all physical intertransformations and energy-value transactions, as exclusively expressed heretofore—especially throughout the last century—only as algebraic, nonconceptual transactions. The conceptual modeling of synergetics does not contradict but complements the exclusively abstract algebraic expression of physical Universe relationships that commenced approximately a century ago with the electromagnetic-wave discoveries of Hertz and Maxwell. Their electrical-apparatus experiments made possible empirical verification or discard of their algebraic treatment of the measured data without their being able to see or conceptually comprehend the fundamental energy behaviors. The permitted discrete algebraic statement and treatment of invisible phenomena resulted in science's comfortable yielding to complete abstract mathematical processing of energy phenomena. The abandonment of conceptual models removed from the literary men any conceptual patterns with which they might explain the evolution of scientific events to the nonmathematically languaged public. Ergo, the lack of modelability produced the seemingly unbridgeable social chasm between the humanities and the sciences.

* *Physics and Beyond,* Harper & Row, New York, 1970, p. 68.

203.09 A study of the microbiological structures, the radiolaria, will always show that they are based on either the tetrahedron, the octahedron, or the icosahedron. The picture was drawn by English scientists almost a century ago as they looked through microscopes at these micro-sea structures. The development of synergetics did not commence with the study of these structures of nature, seeking to understand their logic. The picture of the radiolaria has been available for 100 years, but I did not happen to see it until I had produced the geodesic structures that derive from the discovery of their fundamental mathematical principles. In other words, I did not copy nature's

Fig. 203.09 *Examples of Geodesic Design in Nature.*

structural patterns. I did not make arbitrary arrangements for superficial reasons. I began to explore structure and develop it in pure mathematical principle, out of which the patterns emerged in pure principle and developed themselves in pure principle. I then realized those developed structural principles as physical forms and, in due course, applied them to practical tasks. The reappearance of tensegrity structures in scientists' findings at various levels of inquiry confirms the mathematical coordinating system employed by nature. They are pure coincidence—but excitingly valid coincidence.

203.10 Synergetics represents the coming into congruence of a mathematical system integrating with the most incisive physics findings and generalized laws. At no time am I being scientifically perverse. I am astonished by a philosophic awareness of the highest scientific order, which accommodates the most mystical and mysterious of all human experience. What we are experiencing is vastly more mystically profound by virtue of our adherence to experimentally harvested data than has ever been induced in human comprehension and imagination by benevolently implored beliefs in imagined phenomena dogmatically generated by any of the formalized religions. We are conscious of aspects of the mysterious integrity Universe which logically explains that which we experience and the integrity of the Universe to far more comprehendable degree than that occurring in the make-believe, nonscientifically founded communications of humanity.

204.00 Paradox of the Computer

204.01 Scientific entry into the present realm of nuclear competence was accomplished with the awkward irrational tools of the centimeter-gram-second (*CGS*) * measurement and the Cartesian *XYZ* 90-degree coordinate system. But the awkwardness had to be corrected by Planck's constant to produce reliable, usable information. The development and adoption of the great computers has now relieved humans of the onerous computational tasks entailed in the corrective processing by the irrational constants necessitated by the ineptness of the arithmetical rigidity of arbitrarily exclusive, three-dimensional interpretation of Euclidean geometric mensuration. These irrational constants interlink the many separately evolved quantation techniques of the separately initiated explorations of the many separate facets of universal experience—for instance, biology, crystallography, or physics are called separate, "specialized" scientific inquiries by academic departments and surrounded by NO TRESPASSING signs and electrically charged barbed wire. Because these tasks are being carried by the computers, and men are getting along all right on

* Or, more properly, the centimeter-gram-temperature-second, CG_tS measurement.

their blind-flown scientific pilgrimages, realization of the significance of the sensorially conceptual facility of dealing with nature that is opened up by synergetics has been slow.

204.02 It is a paradox that the computer, in its very ability to process nonconceptual formulae and awkwardly irrational constants, has momentarily permitted the extended use of obsolescent mathematical tools while simultaneously frustrating man's instinctive drive to comprehend his direct experiences. The computer has given man physical hardware that has altered his environmental circumstances without his understanding how he arrived there. This has brought about a general disenchantment with technology. Enchantment can only be sustained in those who have it, or regained by those who have lost it, through conceptual inspiration. Nothing could be more exciting than the dawning awareness of the discovery of the presence of another of the eloquently significant eternal reliabilities of Universe.

205.00 Vector Equilibrium

205.01 The geometrical model of energy configurations in synergetics is developed from a symmetrical cluster of spheres, in which each sphere is a model of a field of energy all of whose forces tend to coordinate themselves, shuntingly or pulsatively, and only momentarily in positive or negative asymmetrical patterns relative to, but never congruent with, the eternality of the vector equilibrium. The vectors connecting the centers of the adjacent spheres are identical in length and angular relationship. The forces of the field of energy represented by each sphere interoscillate through the symmetry of equilibrium to various asymmetries, never pausing at equilibrium. The vector equilibrium itself is only a referential pattern of conceptual relationships at which nature never pauses. This closest packing of spheres in 60-degree angular relationships demonstrates a finite system in universal geometry. Synergetics is comprehensive because it describes instantaneously both the internal and external limit relationships of the sphere or spheres of energetic fields; that is, singularly concentric, or plurally expansive, or propagative and reproductive in all directions, in either spherical or plane geometrical terms and in simple arithmetic.

205.02 When energy-as-heat is progressively extracted from systems by cryogenics, the geometries visibly approach equilibrium; that is to say, removing energy-as-heat reduces the asymmetrical pulsativeness in respect to equilibrium. As the asymmetric kinetics of energy-as-heat are removed, and absolute zero is neared, the whole field of vectors approaches identical length and identical angular interaction; that is to say, they approach the model of closest-packed spherical energy fields. The lines interconnecting the adjacent spheres' centers constitute a vectorial matrix in which all the lines and angles are identical, which is spoken of by the mathematical physicists as the isotropic vector matrix, i.e. where all the energy vectors are identical, i.e., in equilibrium: the cosmically absolute zero.

205.03 Metaphysically, the isotropic vector matrix is conceptually permitted. The difference between the physical and the metaphysical is the omnipulsative asymmetry of all physical oscillation in respect to the equilibrium. Metaphysical is equilibrious and physical is disequilibrious.

205.04 The metaphysically permitted frame of reference for all the asymmetrical physical experience of humanity is characterized by the 60-degree coordination with which synergetics explores nature's behaviors—metaphysical or physical.

205.05 The phenomenon of time entering into energy is just a metaphysical concept. It explains our slowness and our limitations. Temporality is time, and the relative asymmetries of oscillation are realizable only in time—in the time required for pulsative frequency cycling. Synergetics correlates the verities of time and eternity. The awareness of life is always a complex of cognition and recognition lags. Lags are wave frequency aberrations.

206.00 The vectorial coordinate system deriving from closest packing of spheres permits fourth- and fifth-power models of modular-volume symmetrical aggregations around single points in an omnidirectional, symmetrical, all-space-filling radial growth. (See illustration 966.05.) The unit of modular volumetric measurement is the tetrahedron, whose 60-degree angles and six equilength edges disclose omnipersistent, one-to-one correspondence of *radial wave modular growth* with *circumferential modular frequency growth* of the totally involved vectorial geometry. This means that angular and linear accelerations are identical. This is a rational convenience prohibited by 90-degree coordination, whose most economical circumferential geometries are in most cases inherently irrational.

207.00 The angular and linear accelerations of synergetics' isotropic, vectorially triangulated, omnidirectional matrix initiations are rational and uniformly modulated; whereas in the *XYZ* 90-degree coordinate analysis and plotting of the computational findings of the calculus, only the linear is analyzable and the angular resultants are usually irrationally expressed.

208.00 The frequency and magnitude of event occurrences of any system are comprehensively and discretely controllable by *valving,* that is, by angle and frequency modulation. Angle and frequency modulation exclusively define all experiences, which events altogether constitute Universe. (See Sec. 305.05.)

209.00 It is a hypothesis of synergetics that forces in both macrocosmic and microcosmic structures interact in the same way, moving toward the most economic equilibrium patternings. By embracing all the energetic phenomena of total physical experience, synergetics provides for a single coherent system of geometric principles and secures a metaphysical and evolutionary advantage for all experiential accounting and prospecting.

210.00 Synergetics provides vectorial modeling of heretofore only instrumentally apprehended phenomena—for instance, those discovered in nuclear physics. Since it discloses nature's own most economical coordinate system, it provides conceptual models for humanity to accommodate the scientists' energy experiment discoveries.

211.00 Synergetics both equates and accommodates Heisenberg's indeterminism of mensuration inherent in the omniasymmetry of wavilinear physical pulsations in respect to the only metaphysical (ergo, physically unattainable) waveless exactitude of absolute equilibrium. It is only from the vantage of eternal exactitude that metaphysical mind intuitively discovers, comprehends, and equates the kinetic integrities of physical Universe's pulsative asymmetries.

212.00 The whole theory of structure is both altered and enormously expanded and implemented by the introduction of a mathematically coordinate, comprehensively operative, discontinuous-compression, continuous-tension system as inherent to synergetics and its omnirationality of vectorial energy accounting.

213.00 The solving of problems in synergetics starts with the known behaviors of the whole system plus the known behavior of some of the system's parts, which makes possible the discovery of other heretofore unknown parts of the system and their respective behaviors. For instance, in geometry, the known sum of a triangle's angles—180 degrees—plus the known behavior of any two sides and their included angle, or vice versa, permits the discovery and measurement of the values of the other three parts.

214.00 In its search for a coordinate system of nature, synergetics has continually reexamined and reconsidered the experimentally based successive discoveries of what seemed to be a hierarchy of generalized principles possibly governing all of the physical Universe's intertransforming transactions. Thus it aims at a total epistemological reorientation and a unique philosophical reconceptioning regarding the regenerative constellar logic of Universe, making possible the formulating of more comprehensive and symmetrical statements regarding dawningly apparent natural laws.

215.00 A Geometry of Vectors

215.01 Assuming an energy Universe of curved paths generated by angular accelerations of varying intertensions, rates, and radii, resulting in orbits of high-frequency continuities, and separating time out of the compound dynamic system, there remain only the relative attractions and repulsions expressed in relative vectorial terms in respect to the radius of any one interattracted couple of the set of all the radii expressed.

215.02 In such a timeless and equilibrious instant, the remainder of the system may be discovered as a vector construction of force interrelationships between centers. A geometry composed of a system of interrelated vectors may be discovered that represents the complete family of potential forces, proclivities, and proportional morphosis by octave introversion and extroversion.

216.00 Significance of Isotropic Vector Matrix

216.01 Even with foreknowledge of the exact elegant congruences of the isotropic vector matrix (Sec. 420) with nature's eternally transforming transaction needs, we can still understand the ease with which humanity's optical-illusion-producing, minuscule stature in relation to his spherical planet magnitude made it logical for him to institute experience analysis as he did, with lines, planes, squares, perpendiculars, and cubes; and present knowledge of the significance of the isotropic vector matrix also explains lucidly why scientists' faithful measuring and calculation discovered the family of irrational mathematical constants correlating their findings as seemingly expressible only in the terms of cubism's centimeter-gram-second, *XYZ* rectilinear coordination of seemingly obvious "three-dimensional" reality.

216.02 Humanity's escape from the irrational awkwardness of the axiomatic hypothesis trap of eternal askewness which snags him, involves all young humanity's discovery of the isotropic vector matrix synergetics' elegant rational simplicity and its omniaccommodation of all experimentally founded research. Popular understanding and spontaneous employment of synergetics' isotropic vector matrix coordination involves young, popular, experience-induced, spontaneous abandonment of exclusively rectilinear *XYZ* coordination, but without loss of the *XYZ*'s uneconomically askew identity within the system—all occurring "naturally" through youth's spontaneous espousal of the most exquisitely economical comprehension of the most exquisitely economical freedoms of opportunity of individual realizations always regeneratively inspired by the inherent a priori *otherness* considerations (see Sec. 411).

216.03 Comprehension of conceptual mathematics and the return to modelability in general are among the most critical factors governing humanity's epochal transition from bumblebee-like self's honey-seeking preoccupation into the realistic prospect of a spontaneously coordinate planetary society. Insect and avian bumbling in general inadvertently cross-fertilizes all the vegetation's terrestrial impoundments of the star-radiated energy which alone regenerates all biological life around Earth planet. The vegetational impoundments would be dehydrated were they not osmotically watercooled by their root-connected hydraulic circuitry of Earth waters' atomization for return into the sky-distributed, fresh-water-regenerating biological support system; and the rooting frustrates integral procreation of the vegetation which is regeneratively cross-fertilized entirely by the insect and avian, entirely unconscious, pollen-delivering inadvertencies. It is highly probable that universal comprehension of synergetics is strategically critical to humanity's exodus from the womb of originally permitted absolute helplessness and ignorance at birth and entry into the realization that planetary society can spontaneously coordinate in universally successful life support, that is, achieve freedom from fundamental fear and political bias inherent in the ignorant assumption of life-support inadequacy.

217.00 *Prospects for Synergetics*

217.01 Synergetics recognizes the history of progressively larger and more incisive conceptionings, which have eliminated previously uncomprehended behaviors of local Universe. It recognizes that the elegant conceptionings of one period that greatly widened the horizons of human understanding reached their limits of informative capability to be progressively obsoleted by ever greater conceptioning accruing to the ever-mounting harvest of cosmic experience.

217.02 The rate of change and number of special-case self-retransformings of physical evolution tend ever to accelerate, differentiate, and multiply; while the rate of change and numbers of self-remodifyings of generalized law conceptionings of metaphysical evolution tend ever to decelerate, simplify, consolidate, and ultimately unify. (See Sec. 323.)

217.03 In the inherently endless scenario model of Einstein's Universe, truth is ever approaching a catalogue of alternate transformative options of ever more inclusive and refining degrees, wherefore the metaphysical might continually improve the scenario by conceptual discoveries of new generalized principles. (See Secs. 529.07 and 1005.50.)

217.04 Synergetics augments the prospect of humanity becoming progressively exploratory. There is clearly disclosed the desirability of commencing scientific exploration with synergy-of-synergies Universe: metaphysical and physical. While synergetics seems to open new ranges of cosmic comprehension, we assume that the time will come when the inventory of experiences that have catalyzed both its conceptioning and inception will have become overwhelmed by vaster experientially based knowledge and may well become progressively useful but, in its turn, obsolete. Because the generalized principles cannot be principles unless they are eternal, and because human experience is inherently limited, there can be no finality of human comprehension.

220.00 Synergetics Principles

220.01 Principles

220.011 The synergetics principles described in this work are experimentally demonstrable.

220.02 Principles are entirely and only intellectually discernible. The fundamental generalized mathematical principles govern subjective comprehension and objective realization by man of his conscious participation in evolutionary events of the Universe.

220.03 Pure principles are usable. They are reducible from theory to practice.

220.04 A generalized principle holds true in *every* case. If there is one single exception, then it is no longer a generalized principle. No one generalization ever contradicts another generalization in any respect. They are all interaccommodating.

220.05 The physical Universe is a self-regenerative process. Its regenerative interrelationships and intertransformings are governed by a complex code of weightless, generalized principles. The principles are metaphysical. The complex code of eternal metaphysical principles is omni-interaccommodative; that is, it has no intercontradiction. To be classifiable as "generalized," principles cannot terminate or go on vacation. If indeed they *are* generalized, they are eternal, timeless.

220.10 Reality and Eternality

220.11 What the mathematicians have been calling abstraction is reality. When they are inadequate in their abstraction, then they are irrelevant to reality. The mathematicians feel that they can do anything they want with their abstraction because they don't relate it to reality. And, of course, they *can* really do anything they want with their abstractions, even though, like masturbation, it is irrelevant to the propagation of life.

220.12 The only reality is the abstraction of principles, the eternal generalized principles. Most people talk of reality as just the afterimage effects—the realization lags that register superficially and are asymmetric and off-center and thereby induce the awareness called life. The principles themselves have different lag rates and different interferences. When we get to reality, it's absolutely eternal.

220.13 The inherent inaccuracy is what people call the reality. Man's way of apprehending is always slow: ergo, the superficial and erroneous impressions of solids and things that can be explained only in principle.

221.00 Principle of Unity

221.01 Synergetics constitutes the original disclosure of a hierarchy of rational quantation and topological interrelationships of all experiential phenomena which is omnirationally accounted when we assume the volume of the tetrahedron and its six vectors to constitute both metaphysical and physical unity. (See chart at 223.64.) (See Sec. 620.12.)

222.00 Omnidirectional Closest Packing of Spheres

222.01 Definition: The omnidirectional concentric closest packing of equal radius spheres about a nuclear sphere forms a matrix of vector equilibria of progressively higher frequencies. The number of vertexes or spheres in any given shell or layer is edge frequency (F) to the second power times ten plus two.

222.02 Equation:

$10F^2 + 2 =$ the number of vertexes or spheres in any layer,

Where,

$F =$ edge frequency, i.e., the number of outer-layer edge modules.

222.03 The frequency can be considered as the number of layers (concentric shells or radius) or the number of edge modules of the vector equilibrium. The number of layers and the number of edge modules is the same. The frequency, that is the number of edge modules, is the number of spaces between the spheres, and not the number of spheres, in the outer layer edge.

222.10 Equation for Cumulative Number of Spheres: The equation for the total number of vertexes, or sphere centers, in all symmetrically concentric vector equilibria shells is:

$$10(F_1^2 + F_2^2 + F_3^2 + \cdots + F_n^2) + 2F_n + 1$$

222.20 Characteristics of Closest Packing of Spheres: The closest packing of spheres begins with two spheres tangent to each other, rather than

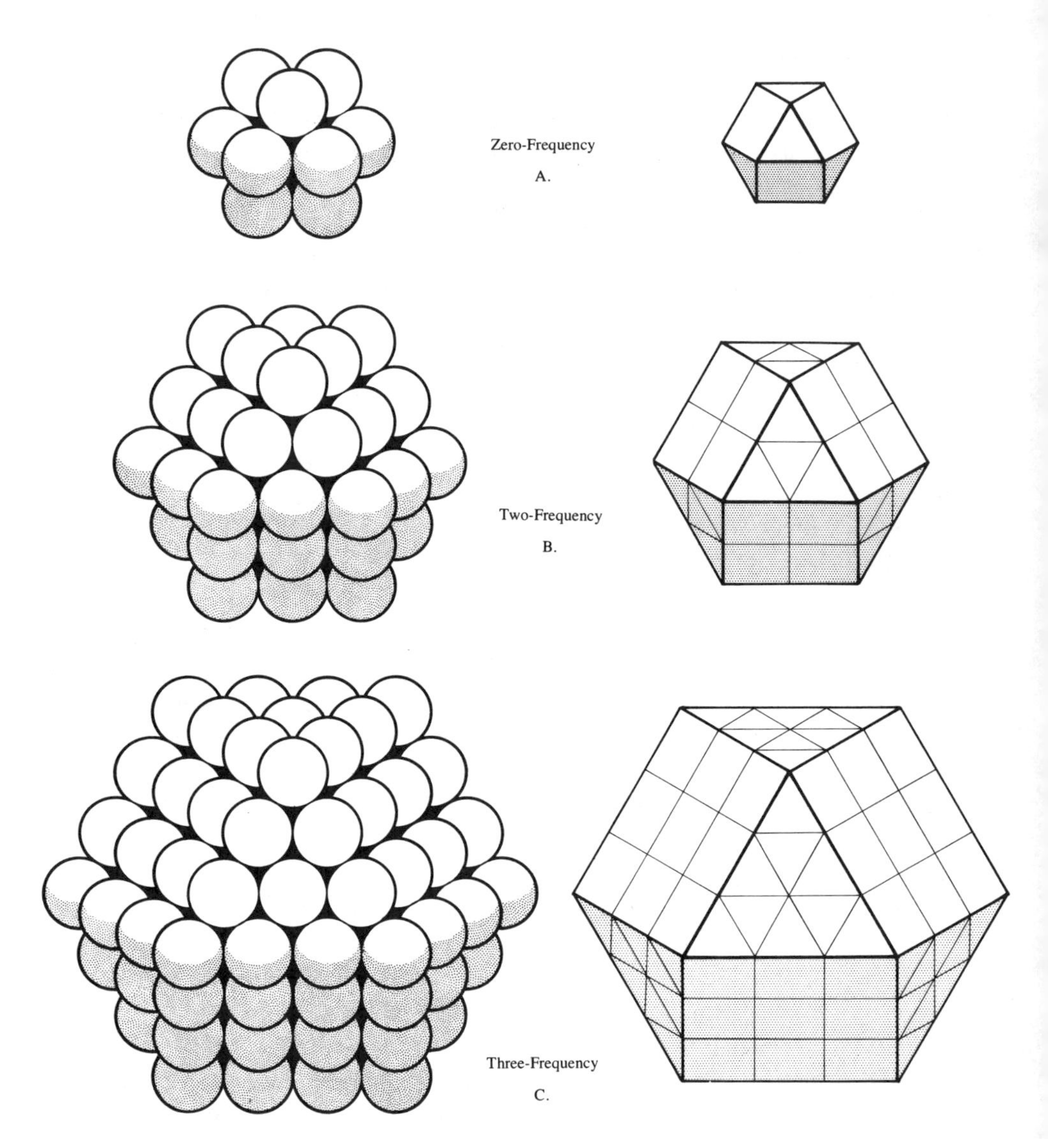

Fig. 222.01 *Equation for Omnidirectional Closest Packing of Spheres:* Omnidirectional concentric closest packings of equal spheres about a nuclear sphere form series of vector equilibria of progressively higher frequencies. The number of spheres or vertexes on any symmetrically concentric shell or layer is given by the equation $10F^2 + 2$, where $F =$ Frequency. The frequency can be considered as the number of layers (concentric shells or radius) or the number of edge modules on the vector equilibrium. A one-frequency sphere packing system has 12 spheres on the outer layer (A) and a one-frequency vector equilibrium has 12 vertexes. If another layer of spheres are packed around the one-frequency system, exactly 42 additional spheres are required to make this a two-frequency system (B). If still another layer of spheres is added to the two-frequency system, exactly 92 additional spheres are required to make the three-frequency system (C). A four-frequency system will have 162 spheres on its outer layer. A five-frequency system will have 252 spheres on its outer layer, etc.

omnidirectionally. A third sphere may become closest packed by becoming tangent to both of the first two, while causing each of the first two also to be tangent to the two others: this is inherently a triangle.

222.21 A fourth sphere may become closest packed by becoming tangent to all three of the first three, while causing each of the others to be tangent to all three others of the four-sphere group: this is inherently a tetrahedron.

222.22 Further closest packing of spheres is accomplished by the omniequiangular, intertriangulating, and omnitangential aggregating of identical-radius spheres. In omnidirectional closest-packing arrays, each single sphere finds itself surrounded by, and tangent to, at most, 12 other spheres. Any center sphere and the surrounding 12 spheres altogether describe four planar hexagons, symmetrically surrounding the center sphere.

222.23 *Excess of Two in Each Layer:* The first layer consists of 12 spheres tangentially surrounding a nuclear sphere; the second omnisurrounding tangential layer consists of 42 spheres; the third 92, and the order of successively enclosing layers will be 162 spheres, 252 spheres, and so forth. Each layer has an excess of two diametrically positioned spheres which describe the successive poles of the 25 alternative neutral axes of spin of the nuclear group. (See illustrations 450.11a and 450.11b.)

222.24 *Three Layers Unique to Each Nucleus:* In closest packing of spheres, the third layer of 92 spheres contains eight new potential nuclei which do not, however, become active nuclei until each has three more layers surrounding it—three layers being unique to each nucleus.

222.25 *Isotropic Vector Matrix:* The closest packing of spheres characterizes all crystalline assemblages of atoms. All the crystals coincide with the set of all the polyhedra permitted by the complex configurations of the isotropic vector matrix (see Sec. 420), a multidimensional matrix in which the vertexes are everywhere the same and equidistant from one another. Each vertex can be the center of an identical-diameter sphere whose diameter is equal to the uniform vector's length. Each sphere will be tangent to the spheres surrounding it. The points of tangency are always at the mid-vectors.

222.26 The polyhedral shape of these nuclear assemblages of closest-packed spheres—reliably interdefined by the isotropic vector matrix's vertexes—is always that of the vector equilibrium, having always six square openings (''faces'') and eight triangular openings (''faces'').

222.30 **Volume of Vector Equilibrium:** If the geometric volume of one of the uniform tetrahedra, as delineated internally by the lines of the isotropic vector matrix system, is taken as volumetric unity, then the volume of the vector equilibrium will be 20.

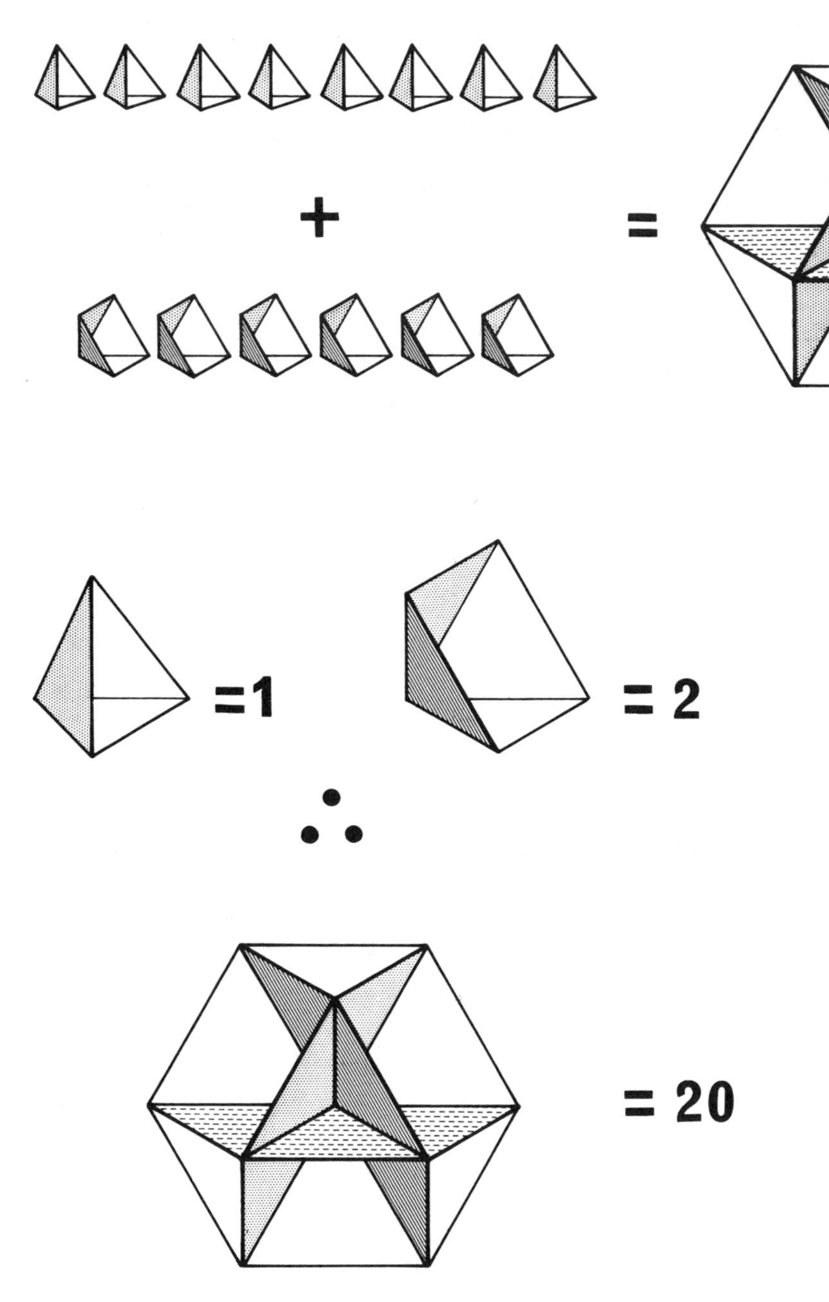

Fig. 222.30 *Volume of Vector Equilibrium:* The volume of the vector equilibrium consists of eight tetrahedra and six half-octahedra. Therefore, the volume of the vector equilibrium is exactly twenty.

222.31 The volume of any series of vector equilibria of progressively higher frequencies is always *frequency to the third power times 20.*

222.32 *Equation for Volume of Vector Equilibrium:*

Volume of vector equilibrium $= 20F^3$,

Where

$$F = \text{frequency}.$$

222.40 **Mathematical Evolution of Formula for Omnidirectional Closest Packing of Spheres:** If we take an inventory of the number of balls in successive vector equilibria layers in omnidirectional closest packing of spheres, we find that there are 12 balls in the first layer, 42 balls in the second layer, and 92 balls in the third. If we add a fourth layer, we will need 162 balls, and a fifth layer will require 252 balls. The number of balls in each layer always comes out with the number two as a suffix. We know that this system is a decimal system of notation. Therefore, we are counting in what the mathematician calls congruence in modulo ten—a modulus of ten units—and there is a constant excess of two.

222.41 In algebraic work, if you use a constant suffix—where you always have, say, 33 and 53—you could treat them as 30 and 50 and come out with the same algebraic conditions. Therefore, if all these terminate with the number two, we can drop off the two and not affect the algebraic relationships. If we drop off the number two in the last column, they will all be zeros. So in the case of omnidirectional closest packing of spheres, the sequence will read; 10, 40, 90, 160, 250, 360, and so forth. Since each one of these is a multiple of 10, we may divide each of them by 10, and then we have 1, 4, 9, 16, 25, and 36, which we recognize as a progression of second powering—two to the second power, three to the second power, and so forth.

222.42 In describing the number of balls in any one layer, we can use the term *frequency* of modular subdivisions of the radii or chords as defined by the number of layers around the nuclear ball. In the vector equilibrium, the number of modular subdivisions of the radii is exactly the same as the number of modular subdivisions of the chords (the "edge units"), so we can say that *frequency to the second power times ten plus two* is the number of balls in any given layer.

222.43 This simple formula governing the rate at which balls are agglomerated around other balls or shells in closest packing is an elegant manifest of the reliably incisive transactions, formings, and transformings of Universe. I made that discovery in the late 1930s and published it in 1944. The molecular biologists have confirmed and developed my formula by virture of which we can predict the number of nodes in the external protein shells of all the viruses, within which shells are housed the DNA-RNA-programmed design controls of all the biological species and of all the individuals within those species. Although the polio virus is quite different from the common cold virus, and both are different from other viruses, all of them employ frequency to the second power times ten plus two in producing those most powerful structural enclosures of all the biological regeneration of life. It is the structural power of these geodesic-sphere shells that makes so lethal those viruses unfriendly to man. They are almost indestructible.

222.50 **Classes of Closest Packing:** There are three classes of closest packing of unit-radius spheres:

222.51 *SYSTEMATIC Symmetrical Omnidirectional Closest Packing:* Twelve spheres closest pack omnitangentially around one central nuclear sphere. Further symmetrical enclosure by closest-packed sphere layers agglomerate in successive vector equilibria. The nucleus is inherent.

222.52 *ASYMMETRICAL Closest Packed Conglomerates:* Closest-packed conglomerates may be linear, planar, or "crocodile." Closest packed spheres without nuclear organization tend to arrange themselves as the octet truss or the isotropic vector matrix. The nuclei are incidental.

222.53 *VOLUMETRIC Symmetrical Closest Packing:* These are non-nuclear symmetrical embracements by an outer layer. The outer layer may be any frequency, but it may not be expanded or contracted by the addition inwardly or outwardly of complete closest-packed layers. Each single-layer frequency embracement must be individually constituted. Volumetric symmetrical closest packing aggregates in most economical forms as an icosahedron geodesic network. The nucleus is excluded.

223.00 Principle of Prime Number Inherency and Constant Relative Abundance of the Topology of Symmetrical Structural Systems

223.01 **Definition:** The number of vertexes of every omnitriangulated structural system is rationally and differentially accountable, first, by selecting and separating out the always *additive two* polar vertexes that must accommodate the neutral axis of spin inherent in all individual structural systems to permit and account for their independent motional freedom relationship from the rest of Universe. The number of nonpolar vertexes is called the base number. Second, we identify the always *multiplicative* duality factor of *two* characterizing the always coexistent insideness-outsideness of systems and their inherently positively and negatively congruent disparity of convexity and concavity. Third, we find the multiplicative duality factor of two to be multiplied by one of the first four prime numbers, 1, 2, 3, or 5 (multiplied by 1 if the structural system is tetrahedral, by 2 if it is octahedral, by 3 if it is the triangularly structured cube, or by 5 if it is the icosahedron or the triangularly stabilized vector equilibrium), or factored by a variety of multiples comprised of combinations of only those first four prime numbers, whether the polyhedra are in the Platonic, Archimedean, or any other progression of symmetrical structural systems. When the vector edges of the symmetrical systems are modularly subdivided, all of the foregoing products are found to be multiplied again to the second power by the frequency of uniform modular subdivisions of the vector edges of the symmetrical structural system. In respect to the original base number of nonpolar vertexes, there will always be twice as many openings ("faces") and three times as many vector edges of the symmetrical structural system, always remembering that the two polar vertexes were first extracted from the inventory of topological characteristics before multiplying the remaining number of vertexes in the manner described and in relation to which the number of nonpolar vertexes and the relative abundance of the other topological characteristics are accurately derived and operationally described.

223.02 Axis of Spin: Any two vertexes may be selected as the axis of spin, whether or not the axis described by them is immediately conceivable as the logical axis of spinnability, i.e., the axis need not be statically symmetrical. (You can take hold of a boy by his two hands and, holding one above the other and leaning backward spin him centrifugally around you. Although his two hands do not represent the symmetrical static axis of the boy's body, their dynamic positions defined the axis of your mutual spinning.)

223.03 Equation of Prime Number Inherency of All Symmetrical Structural Omnitriangulated Systems:

$$X = 2NF^2 + 2$$

Where:

X = number of vertexes (crossings) or spheres in the outer layer or shell of any symmetrical system;

N = one of the first four prime numbers: 1, 2, 3, or 5; and

F = edge frequency, i.e., the number of outer layer edge modules.

223.04 Equation of Constant Relative Abundance of Topological Aspects of All Symmetrical Structural Systems: Multiplication of one of the first four prime numbers or their powers or multiples by the constant of relative topological characteristics abundance:

$1 + 2 = 3$

1 Nonpolar vertex

2 Faces

3 Edges

In addition to the product of such multiplication of the constant relative abundance equation by one of the first four prime numbers—1, 2, 3, or 5—or their powers or multiples, there will always be two additional vertexes assigned as the poles of the axial spinnability of the system.

223.10 **Constant Relative Abundance:** Topological systems that are structurally stabilized by omnitriangulation reveal a constant relative abundance of certain fundamental characteristics deriving from the additive twoness and the multiplicative twoness of all finite systems.

223.11 The *additive* twoness derives from the polar vertexes of the neutral axis of spin of all systems. This twoness is the beginning and essence of consciousness, with which human awareness begins: consciousness of the other, the other experience, the other being, the child's mother. To describe that of which we are aware, we employ comparison to previous experience. That which we are aware of is hotter, or bigger, or sharper than the other experience or experiences. The a priori otherness of comparative awareness inherently requires time. Early humanity's concept of the minimum increment of time was the *second,* because time and awareness *begin* with the second experience, the prime *other.* If there is only one think, one think is naught. Life and Universe that goes with it begins with two spheres: you and me . . . and you are always prior to me. I have just become by my awareness of you.

223.12 The *multiplicative* twoness is inherent in the disparity of the congruent convexity and concavity of the system. The multiplicative twoness is because both you and I have insideness and outsideness, and they are not the same: one is convex and one is concave.

223.13 Conceptual systems having inherent insideness and outsideness are defined at minimum by four event foci and are, ergo, tetrahedral; at maximum symmetrical complexity, they are superficially ''spherical''—that is, they are a spherelike array of event foci too minute for casual resolution into the plurality of individual event foci of which, in experiential fact, they must consist, each being approximately equidistant from one approximately identifiable event focus at the spherical array's center. Since all the ''surface'' event foci may be triangularly interconnected with one another by chords that are shorter than arcs, all spherical experience arrays are, in fact, polyhedra. And all spheres are polyhedra. Spherical polyhedra may at minimum consist of the four vertexes of the regular tetrahedron.

223.14 We discover that the additive twoness of the two polar (and a priori awareness) spheres at most economical minimum definition of event foci are two congruent tetrahedra, and that the insideness and outsideness of complementary tetrahedra altogether represent the two invisible complementary twoness that balances the visible twoness of the polar pair. This insideouting tetrahedron is the minimum compound curve—ergo, minimum sphere. (See Sec. 624.)

223.15 When the additive twoness and the multiplicative twoness are extracted from any symmetrical and omnitriangulated system, the number of vertexes will always be a rational product of one or more of the first four prime numbers, 1, 2, 3, or 5, or their powers or multiples.

223.16 The number of openings (or "faces") will be twice that of the vertexes, minus two.

223.17 The number of vector edges will be three times the number of vertexes, minus two.

223.18 When we reduce the topological inventory of basic vertexes, areas, and edges of all omnitriangulated structural systems in Universe—whether symmetrical or asymmetrical—by taking away the two poles and dividing the remaining inventory by two, we discover a constant relative abundance of *two faces* and *three lines* for *every one vertex*. This is to say that there is a constant topological abundance characterizing all systems in Universe in which for every *nonpolar* vertex there are always *two faces* and *three (vectorial) edges*.

223.19 In an omnitriangulated structural system:

(a) the number of vertexes ("crossings" or "points") is always evenly divisible by two;

(b) the number of faces ("areas" or "openings") is always evenly divisible by four; and

(c) the number of edges ("lines," "vectors," or "trajectories") is always evenly divisible by six.

223.20 **Primary Systems:** Only four primary systems or contours can be developed by closest packing of spheres in omnisymmetrical concentric layers. The exterior contours of these points are in the chart:

	After subtracting the two Polar Vertexes: the Additive Two	And dividing by the Duality Factor Two	Outer Layer of Two Frequency	Outer Layer of Three Frequency
(a) Tetrahedron (four sides): $2+\left[(2\times\mathbf{1})\times F^2\right]=4$ vertexes (crossings)	2	1	10	20
(b) Octahedron (eight sides): $2+\left[(2\times\mathbf{2})\times F^2\right]=6$ vertexes (crossings)	4	2	18	38
(c) Cube (six Sides): $2+\left[(2\times\mathbf{3})\times F^2\right]=8$ Vertexes (crossings)	6	3	26	56
(d) Vector Equilibrium (fourteen sides): $2+\left[(2\times\mathbf{5})\times F^2\right]=12$ vertexes (crossings)	10	5	42	92

223.21 **Primary Systems: Equations:** The formulas for the number of spheres in the outer layer of closest packed spheres in primary systems is as follows:

(a) *Tetrahedron:*
$X=2F^2+2$

(b) *Octahedron:*
$X=4F^2+2$

(c) *Cube:*
$X=6F^2+2$

(d) *Vector Equilibrium* (*Icosahedron*):
$X=10F^2+2$

Where:

X = the number of spheres in the outer layer or shell of the primary system;

F = edge frequency, i.e., the number of outer-layer edge modules.

223.30 **Symmetrical Analysis of Topological Hierarchies:** Symmetrical means having no local asymmetries. Omnisymmetrical permits local asymmetries.

223.31 The following omnitriangulated systems are symmetrical:
Tetrahedron
Octahedron
Icosahedron

223.32 The following omnitriangulated systems are omnisymmetrical:
Cube
Diagonal Rhombic Dodecahedron
Rhombic Dodecahedron
Dodecahedron
Tetraxidecahedron
Triacontahedron
Enenicontrahedron

223.33 The vector equilibrium is locally mixed symmetrical and asymmetrical.

223.40 **Powering:** Second powering in the topology of synergetics is identifiable only with the vertexes of the system and not with something called the "surface area." Surfaces imply experimentally nondemonstrable continuums. There are no topologically indicated or implied *surfaces* or *solids*. The vertexes are the external points of the system. The higher the frequency of the system, the denser the number of external points. We discover then that second powering does not refer to "squaring" or to surface amplification. Second powering refers to the number of the system's external vertexes in which equating the second power and the radial or circumferential modular subdivisions of the system multiplied by the prime number *one* if a tetrahedral system; by the prime number *two* if an octahedral system; by the prime number *three* if a triangulated cubical system; and by the prime number *five* if an icosahedral system; each, mutliplied by two, and added to by two, will accurately predict the number of superficial points of the system.

223.41 This principle eliminates our dilemma of having to think of *second* and *third* powers of systems as referring exclusively to continuum *surfaces* or *solids* of the systems, neither of which states have been evidenced by experimental science. The frequencies of systems modify their prime rational integer characteristics. The second power and third power point aggregations identify the energy quanta of systems and their radiational growth or their gravitational contraction. They eliminate the dilemma in which physics failed to identify simultaneously the wave and the particle. The dilemma grew from the misconceived necessity to identify omnidirectional wave growth exclusively with the rate of a nonexperimentally existent spherical surface continuum growth, the second power of radiational growth being in fact the exterior quanta and not the spherical surface being considered as a continuum.

223.50 **Prime Number Inherency:** All structurally stabilized polyhedra are characterized by a constant relative abundance of Euler's topological aspects in which there will always be twice as many areas and three times as many lines as the number of points in the system, minus two (which is assigned to the polar axis of spin of the system).

223.51 The number of the topological aspects of the Eulerian system will always be an *even* number, and when the frequency of the edge modulation of the system is reduced to its second root and the number of vertexes is divided by two, the remainder will be found to consist exclusively of a prime number or a number that is a product exclusively of two or more intermultiplied prime numbers, which identify the prime inherency characteristics of that system in the synergetic topological hierarchy of cosmically simplest systems.

223.52 All other known regular symmetrical polyhedra (other than the tetrahedron and the octahedron) are described quantitatively by compounding rational fraction elements of the tetrahedron and the octahedron. These elements are known as the *A* and *B* Quanta Modules (see Sec. 920 through 940). They each have a volume of one-twenty-fourth of a tetrahedron.

223.60 **Analysis of Topological Hierarchies: Omnitriangulation:** The areas and lines produced by omnitriangularly and circumferentially interconnecting the points of the system will always follow the rule of constant relative abundance of points, faces, and lines.

223.61 Only triangles are structures, as will be shown in Sec. 610. Systems have insideness and outsideness—ergo, structural systems must have omnitriangulated isolation of the outsideness from the insideness. Flexibly jointed cubes collapse because they are not structures. To structure a cubical form, the cube's six square faces must be diagonally divided at minimum into 12 triangles by *one* of the two inscribable tetrahedra, or at maximum into 24 triangles by *both* the inherently inscribable positive-negative tetrahedra of the cube's six faces.

223.62 Lacking triangulation, there is no structural integrity. Therefore, all the polyhedra must become omnitriangulated to be considered in the Table. Without triangulation, they have no validity of consideration. (See Sec. 608, "Necklace.")

223.64 **Table: Synergetics Hierarchy of Topological Characteristics of Omnitriangulated Polyhedral Systems** (See pp. 46–47.)

223.65 The systems as described in *Columns 1 through 5* are in the prime state of conceptuality independent of size: metaphysical. Size is physical and is manifest by frequency of "points-defined" modular subdivisions of lengths, areas, and volumes. Size is manifest in the three variables of relative length, area, and volume; these are all expressible in terms of frequency. Frequency is operationally realized by modular subdivision of the system.

223.66 *Column 1* provides a statement of the true rational volume of the figure when the *A* and *B* Quanta Modules are taken as unity.

Column 2 provides a statement of the true rational volume of the figure when the tetrahedron is taken as unity.

Columns 1 and 2 describe the rationality by complementation of two selected pairs of polyhedra considered together. These are (a) the vector-edged icosahedron and the vector-edged cube; and (b) the vector-edged rhombic dodecahedron and the vector-edged dodecahedron.

Column 3 provides the ratio of area-to-volume for selected polyhedra.

Column 4 denotes self-packing, allspace-filling polyhedra.

Column 5 identifies complementary allspace-filling polyhedra. These are: (a) the *A* and *B* Quanta Modules in combination with each other; (b) the tetrahedron and octahedron in combination with each other; and (c) the octahedron and vector equilibrium in combination with each other.

Column 6 presents the topological analysis in terms of Euler.

Columns 7 through 15 present the topological analysis in terms of synergetics, that is, with the polar vertexes extracted from the system and with the remainder divided by two.

Column 7 accounts the extraction of the polar vertexes. All systems have axes of spin. The axes have two poles. Synergetics extracts two vertexes from all Euler topological formulas to function as the poles of the axis of spin. Synergetics speaks of these two polar vertexes as the *additive two.* It also permits polar coupling with other rotative systems. Therefore a motion system can have associability.

Column 9 recapitulates *Columns 7 and 8* in terms of the equation of constant relative abundance.

Column 10 accounts synergetics multiplicative two.

Column 11. The synergetics constants of all systems of Universe are the additive two and the multiplicative two—the Holy Ghost; the Heavenly Twins; a pair of twins.

Columns 12 and 15 identify which of the first four prime numbers are applicable to the system considered.

Column 13 recapitulates *Columns 11 and 12*.

223.70 *Planck's Constant*

223.71 Planck's constant: symbol $= h$. $h = 6.6$—multiplied by 10^{-27} grams per square centimeters per each second of time. The constant h is the invariable number found empirically by Planck by which the experimentally

Table 223.64 SYNERGETICS HIERARCHY OF TOPOLOGICAL CHARACTERISTI

				1	2	3	4	5	6
Locally Symmetrical Omnitriangulated	Locally Mixed Sym-Asym. Omnitriangulated	Locally Asymmetrical Omnitriangulated		Rational Quanta Module Volumes	Rational Regular Tetrahedral Volume or Complementary Rational Tetravolumes	Ratio: Area to Volume	Allspace Filling Systems: All-space Filler	Allspace Filling Systems: Complementary All-space Filler	Basic Topolog Characteristi Euler's Formu for Polyhedra V = Verte F = Faces E = Edges V + F = E + 2 (Euler)
			A + Quanta Module	+ 1	+ 1/24			●	
			A - Quanta Module	- 1	- 1/24			●	
			B + Quanta Module	+1	+ 1/24			●	
			B - Quanta Module	- 1	- 1/24			●	
			Mite Positive Minimum Tetrahedron	+ 3	+ 1/8		●		
			Mite Negative Minimum Tetrahedron	- 3	- 1/8		●		
			Syte Obtuse Tetrahedron	6	1/4		●		
			Syte Acute Tetrahedron	6	1/4		●		
			The Coupler Asymmetrical Octahedron	24	1		●		
			Vector - Edged Tetrahedron	24	1	1 : 1		●	4 + 4 = 6 +
			Vector - Edged Octahedron	96	4	1 : 2		● ●	6 + 8 = 12
			Vector - Diagonaled Cube	72	3		●		8 + 12 = 18
			Vector Equilibrium	480	20			●	12 + 20 = 30
			Vector - Edged Icosahedron	444.24] 648	18.510] 27	1 : 3.702			12 + 20 = 30
			Vector - Edged Cube	203.76	8.490		●		14 + 24 = 36
			Vector - Diagonaled Rhombic Dodecahedron	144	6		●		14 + 24 = 36
			Vector - Edged Rhombic Dodecahedron	623.664] 2,184.096	25.986] 91.004		●		14 + 24 = 36
			Vector - Edged Dodecahedron	1,506.432	65.018				32 + 60 = 90
			Vector - Edged Tetraxidecahedron	2,304	96		●		32 + 60 = 90
			Vector - Edged Triacontahedron						32 + 60 = 90
			Vector - Edged Enenicontahedron						92 + 180 = 270

3/20

Where one gram = 1 cubic cm. water at 4 C.

$6.6 = \frac{20}{3} = \frac{\text{Volume of Vector Equilibrium}}{\text{Volume of cube}}$

Planck's Constant = h

$h = 6.6 \times 10^{-27}$ gram cm^2/sec.

Gravitational constant

$G = 6.6 \times 10^{-8}$ gram/sec^2

[See section 223.70 for detailed exposition]

OMNITRIANGULATED POLYHEDRAL SYSTEMS (See Secs. 223.65 and 223.66 for details.)

7	8	9	10	11	12	13	14	15
ar Vertexes racted for utral Axis nergetics aration of ditive Two Permit Mo- n Freedom n Rest of verse)	Additive Operator [Euler Restated]	Remaining Characteristics after Topological Extraction of two Vertexes	Remainder divided by two after extraction of two vertexes (differentiating out the Synergetics multiplicative two to account and accommodate the concave - convex, inside - outside of all systems)	Constant relative abundance of topological characteristics after Synergetics extraction of both the additive two & the multiplicative two from Euler's formula	First four prime number characteristics of the system (unique to Synergetics topological hierarchy)	Topological abundance characteristics after extraction of the Synergetics constants (the additive two and the multiplicative two) divided by the prime number characteristic of the system	With the two faces and the three edges characterizing every nonpolar vertex in all symmetrical omnitriangulated systems, the only variables appear as the orderly hierarchy of the first four prime multipliers Q.E.D.	When you have found all the constants and the only variables of the formulas, this will require identification of the Synergetics constants (the additive two and the multiplicative two) and the unique prime number
2	+	2 + 4 = 6	$\frac{2 + 4 = 6}{2}$	1 + 2 = 3	Tetrahedron 1	$\frac{1 + 2 = 3}{1}$	1 + 2 = 3	1
2	+	4 + 8 = 12	$\frac{4 + 8 = 12}{2}$	2 + 4 = 6	Octahedron 2	$\frac{2 + 4 = 6}{2}$	1 + 2 = 3	2
2	+	6 + 12 = 18	$\frac{6 + 12 = 18}{2}$	3 + 6 = 9	Cube 3	$\frac{3 + 6 = 9}{3}$	1 + 2 = 3	3
2	+	10 + 20 = 30	$\frac{10 + 20 = 30}{2}$	5 + 10 = 15	Vector Equilibrium 5	$\frac{5 + 10 = 15}{5}$	1 + 2 = 3	5
2	+	10 + 20 = 30	$\frac{10 + 20 = 30}{2}$	5 + 10 = 15	Icosahedron 5	$\frac{5 + 10 = 15}{5}$	1 + 2 = 3	5
2	+	12 + 24 = 36	$\frac{12 + 24 = 36}{2}$	6 + 12 = 18	Vector Edged Cube 3 x 2	$\frac{6 + 12 = 18}{3 \times 2}$	1 + 2 = 3	2, 3
2	+	12 + 24 = 36	$\frac{12 + 24 = 36}{2}$	6 + 12 = 18	Vector – Diagonaled Rhombic Dodecahedron 3 x 2	$\frac{6 + 12 = 18}{3 \times 2}$	1 + 2 = 3	2, 3
2	+	12 + 24 = 36	$\frac{12 + 24 = 36}{2}$	6 + 12 = 18	Vector – Edged Rhombic Dodecahedron 3 x 2	$\frac{6 + 12 = 18}{3 \times 2}$	1 + 2 = 3	2, 3
2	+	30 + 60 = 90	$\frac{30 + 60 = 90}{2}$	15 + 30 = 45	Vector – Edged Dodecahedron 3 x 5	$\frac{15 + 30 = 45}{3 \times 5}$	1 + 2 = 3	3, 5
2	+	30 + 60 = 90	$\frac{30 + 60 = 90}{2}$	15 + 30 = 45	Tetraxidecahedron 3 x 5	$\frac{15 + 30 = 45}{3 \times 5}$	1 + 2 = 3	3, 5
2	+	30 + 60 = 90	$\frac{30 + 60 = 90}{2}$	15 + 30 = 45	Triacontahedron 3 x 5	$\frac{15 + 30 = 45}{3 \times 5}$	1 + 2 = 3	3, 5
2	+	90 + 180 = 270	$\frac{90 + 180 = 270}{2}$	45 + 90 = 135	Enenicontahedron $3^2 \times 5$	$\frac{45 + 90 = 135}{3^2 \times 5}$	1 + 2 = 3	3, 5

discovered, uniformly energized, *minimum increment of all radiation,* the *photon,* must be multiplied to equate the photon's energy value as rated by humans' energy-rating technique, with the effort expended in lifting weights vertically against gravity given distances in given times. Thus automotive horsepower or electromagnetic kilowatts per hour performance of stationary prime movers, engines, and mobile motors are rated.

223.72 Max Planck's photons of light are separately packaged at the radiation source and travel in a group-coordinated flight formation spherical surface pattern which is ever expanding outwardly as they gradually separate from one another. Every photon always travels radially away from the common origin. This group-developed pattern produces a sum-totally expanding spherical wave-surface determined by the plurality of outwardly traveling photons, although any single photon travels linearly outwardly in only one radial direction. This total energy effort is exactly expressed in terms of the exponential second-power, or areal "squaring," rate of surface growth of the overall spherical wave; i.e., as the second power of the energy effort expended in lifting one gram in each second of time a distance of one "vertical" centimeter radially outward away from the origin center.

223.73 *Whereas:* All the volumes of all the equi-edged regular polyhedra are irrational numbers when expressed in the terms of the volume of a cube = 1;

Whereas: The volume of the cube and the volumes of the other regular polyhedra, taken singly or in simple groups, are entirely rational;

Whereas: Planck's constant was evaluated in terms of the cube as volumetric unity and upon the second-power rate of surface expansion of a cube per each second of time;

Whereas: Exploring experimentally, synergetics finds the tetrahedron, whose volume is one-third that of the cube, to be the prime structural system of Universe: *prime structure* because stabilized exclusively by triangles that are experimentally demonstrable as being the only self-stabilizing polygons; and *prime system* because accomplishing the subdivision of all Universe into an interior microcosm and an external macrocosm; and doing so structurally with only the minimum four vertexes topologically defining insideness and outsideness;

Whereas: Structuring stability is accomplished by triangularly balanced energy investments;

Whereas: Cubes are unstable;

Whereas: The radial arrangement of unit tetrahedral volumes around an absolute radiation center (the vector equilibrium) constitutes a prime radiational-gravitational energy proclivity model with a containment value of 20 tetrahedra (where cube is 3 and tetrahedron 1);

Whereas: Max Planck wished to express the empirically emerged value of the photon, which constantly remanifested itself as a unit-value energy entity in the energy-measuring terms of his contemporary scientists;

Wherefore: Planck employed the *XYZ* rectilinear frame of shape, weight, volume, surface, time, distance, antigravity effort, and metric enumeration mensuration tools adopted prior to the discovery of the photon value.

223.74 Planck's constant emerged empirically, and to reconvert it to conformity with synergetics the $6.\dot{6}$ness is canceled out:

$$6.6 = \frac{20}{3} = \frac{\text{volume of vector equilibrium}}{\text{volume of cube}}$$

Therefore, to convert to synergetics accounting, we multiply Planck's $6.\dot{6} \times 3 = 20$. As seen elsewhere in synergetics' topology, the number of surface points of the identically vector-radiused and vector-chorded system's vector equilibrium—as well as of its spherical icosahedron counterpart—always multiplies at a second-power rate of the frequency (of modular subdivision of the radius vector of the system) times 10—to the product of which is added the number 2 to account for the axial rotation poles of the system, which twoness, at the relatively high megacycle frequencies of general electromagnetic wave phenomena, becomes an undetectable addition.

223.75 In synergetics' topological accounting, surface areas are always structural triangles of the systems, which systems, being vectorially structured, are inherently energy-investment systems. As synergetics' topology also shows, the number of triangular surface areas of the system increases at twice the rate of the nonpolar surface points, ergo the rate of energetic system's surface increase is accounted in terms of the number of the triangular areas of the system's surface, which rate of system surface increase is $20F^2$, where F = frequency; while the rate of volumetric increase is $20F^3$. The vector is inherent in the synergetics system since it is structured with the vector as unity. Because vectors = mass × velocity, all the factors of time, distance, and energy, as both mass and effort as well as angular direction, are inherent; and E as energy quantum of one photon = $20F^2$.

223.80 Energy Has Shape

223.81 I recognize the experimentally derived validity of the *coordinate invariant:* the result does not depend on the coordinate system used. Planck's constant is just what it says it is: an experimentally ascertained constant cosmic relationship. Planck's constant as expressed is inherently an irrational number, and the irrationality relates to the invariant quantum of energy being constantly expressed exclusively in the volume-weight terms of a special-case *shape* which, in the geometrical shape-variant field of weight-strength and surface-volume ratio limits of local structural science containment of energy, as mass or effort, by energy-as-structure, is neither maximum nor minimum. The special-case geometrical shape chosen arbitrarily by the engineering-structures-eschewing pure scientists for their energy-measurement accommodation, that of the cube, is structurally unstable; so much so as to be too unstable to be classified as a structure. Unwitting of this mensural shortcoming, Planck's constant inadvertently refers to the cube, implicit to the gram, as originally adopted to provide an integrated unit of weight-to-volume mensuration, as was the "knot" adopted by navigators as a velocity unit which integrates time-space incrementation values. The volume and weight integrate as a gram. The gram was arbitrarily assumed to be constituted by a cubic centimeter of water at a specific temperature, 4 degrees centigrade.

223.82 Relationship constants are always predicated on limits. Only limits are invariable. (This is the very essence of the calculus.) Variation is between limits. Though Planck's constant is indirectly predicated on a limit condition of physical phenomena, it is directly expressed numerically only as a prefabricated, constantly irrational number-proportionality to that limit, but it is not the inherently rational unit number of that limit condition. This is because the cube was nonstructural as well as occurring structurally between the specific limit cases of surface-to-volume ratio between whose limits of $1 \rightarrow 20$, the cube rates as 3.

223.83 Max Planck found a constant energy-value relationship emergent in all the photon-discovery experimental work of others. A great variety of exploratory work with measurements of energy behaviors in the field of radiation disclosed a hitherto unexpected, but persistent, minimum limit in relation to such energy phenomena. Planck expressed the constant, or limit condition, in the scientifically prevailing numerical terms of the physical and metaphysical equipment used to make the measuring. The measuring system included:

—the decimal system;

—the CG_tS and;

—*XYZ* coordinate analysis,

which themselves were procedurally assumed to present the comprehensively constant limit set of mensuration systems' input factors.

223.84 Let us assume hypothetically that Ponce de Leon did find the well of eternal-youth-sustaining water, and that the well had no "spring" to replenish it, and that social demand occasioned its being bailed out and poured into evaporation-proof containers; and that the scientists who bailed out that precious well of water used a cubically-shaped, fine-tolerance, machined and dimensioned one-inch-thick shelled, stainless steel bucket to do their carefully measured bailing and conserving task. They did so because they knew that cubes close-pack to fill allspace, and because water is a constant substance with a given weight per volume at a given temperature. And having ten fingers each, they decided to enumerate in the metric system without any evidence that meters are whole rational linear increments of a cosmic nature. Thus organized, the Ponce de Leon scientists soon exhausted the well, after taking out only six and two-thirds cubic bucket loads—with a little infinitely unaccountable, plus-or-minus, spillage or overestimate.

223.85 Planck's constant, *h,* denotes the minimum energy-as-radiation increment known experimentally by humans to be employed by nature, but the photon's energy value could and should be expressed in terms of a whole number as referenced directly by physical experiment to nature's limit-case transforming states.

223.86 Had, for instance, the well-of-youth-measuring scientists happened to be in a hurry and had they impatiently used a cubical container of the same size made of a thin-wall plastic such as the cubically shaped motel waste containers, they would have noticed when they stood their waterfilled plastic cube bucket on the ground beside the well that its sides bulged and that the level of the water lowered perceptibly below the container's rim; though this clearly was not caused by leaking, nor by evaporating, but because its shape was changing, and because its volume-to-container-surface ratio was changing.

223.87 Of all regular polyhedra, the sphere (i.e., the high-frequency, omnitriangulated, geodesic, spheroidal polyhedron) encloses the most volume with the least surface. Whereas the tetrahedron encloses the least volume with the most surface. The contained energy is at minimum in the tetrahedron. The structure capability is at maximum in the tetrahedron.

223.88 Planck did not deliberately start with the cube. He found empirically that the amount of the photon's energy could be expressed in terms of the CG_tS–XYZ decimal-enumeration coordinate system already employed by science as the "frame of reference"* for his photon evaluation which, all inadvertently, was characterized by awkwardness and irrationality.

223.89 Energy has shape. Energy transforms and trans-shapes in an evoluting way. Planck's contemporary scientists were not paying any attention to that. Science has been thinking shapelessly. The predicament occurred that way. It's not the size of the bucket—size is special case—they had the wrong shape. If they had had the right shape, they would have found a whole-rational-number constant. And if the whole number found was greater than

* For "frame of reference" synergetics speaks of the "multi-optioned omni-orderly scheme of behavioral reference." See Sec. 540.

unity, or a rational fraction of unity, they would simply have had to divide or multiply to find unity itself.

223.90 The multiplier 10^{-27} is required to reduce the centimeter magnitude of energy accounting to that of the tuned wavelength of the photon reception. Frequency and wave are covariably coupled; detection of one discloses the other. Since synergetics' vector equilibrium's energy converging or dispersing vector is both radially and chordally subdivided evenly by frequency—whatever that frequency may be—the frequency fractionates the unit vector energy involvement by one-to-one correspondence.

223.91 If they had taken the same amount of water at the same temperature in the form of a regular tetrahedron, they would have come out with a rational fraction of unity. They happened to be enumerating with congruence in modulo 10, which does not include any prime numbers other than 1, 2, and 5. The rational three-ness of the cube in relation to the tetrahedron is not accommodated by the decimal system; nor is the prime 7 inherent in modulo 10. Therefore, Planck's constant, while identifying a hitherto undiscovered invariant limit condition of nature, was described in the wrong frame of reference in awkward—albeit in a constantly awkward—term, which works, because it is the truth; but at the same time it befogs the otherwise lucid and rational simplicity covering this phenomenon of nature, just as does nature's whole number of utterly rational atoms exchanging rates in all her chemical combining and separating transactions accounting.

224.00 Principle of Angular Topology

224.01 Definition: When expressed in terms of cyclic unity the sum of the angles around all the vertexes of a structural system, plus 720 degrees, equals the number of vertexes of the system multiplied by 360 degrees.

224.02 All local structural systems in Universe are always accomplished by nature through the elimination of 720 degrees of angle. This is the way in which nature takes two complete 360-degree angular tucks in the illusory infinity of a plane to render systems locally and visibly finite. The difference between visually finite systems and illusory infinity is two cyclic unities.

224.03 Structural systems are local, closed, and finite. They include all geometric forms, symmetric or asymmetric, simple or complex. Structural systems can have only one inside and only one outside. Two or more structures may be concentric and triangularly interconnected to operate as one structure.

224.04 The difference between the sum of all the angles around all the vertexes of *any* system and the total number of vertexes times 360° (as angular unity) is 720°, which equals two unities. The sum of the angles of a tetrahedron always equals 720°. The tetrahedron may be identified as the 720° differential between any *definite local* geometrical system (such as Greek "solid" geometry) and *finite universe*.

224.05 **Line:** A line has two vertexes with angles around each of its vertexial ends equal to 0°. The sum of these angles is 0°. The sum of the vertexes (two) times angular unity (360°) is 720°. The remainder of 0° from 720° is 720°, or two unities, or one tetrahedron. Q.E.D.

224.06 **Triangle:** The three angles of one "face" of a planar triangle always add up to 180° as a phenomenon independent of the relative dimensional size of the triangles. One-half of definitive cyclic unity is 180°. Every triangle has two faces—its obverse and reverse. Unity is two. So we note that the angles of both faces of a triangle add up to 360°. Externally, the sum of the angles around each of the triangle's three vertexes is 120°, of which 60° is on the obverse side of each vertex; for a triangle, like a line, if it exists, is an isolatable system always having positive and negative aspects. So the sum of the vertexes around a triangle (three) times 360° equals 1080°. The remainder of 360° from 1080° leaves 720°, or one tetrahedron. Q.E.D.

224.07 **Sphere:** The Greeks defined the sphere as a surface outwardly equidistant in all directions from a point. As defined, the Greeks' sphere's surface was an absolute continuum, subdividing all the Universe outside it from all the Universe inside it; wherefore, the Universe outside could be dispensed with and the interior eternally conserved. We find local spherical systems of Universe are definite rather than infinite as presupposed by the calculus's erroneous assumption of 360-degreeness of surface plane azimuth around every point on a sphere. All spheres consist of a high-frequency constellation of event points, all of which are approximately equidistant from one central event point. All the points in the surface of a sphere may be interconnected. Most economically interconnected, they will subdivide the surface of the sphere into an omnitriangulated spherical web matrix. As the frequency of triangular subdivisions of a spherical constellation of omnitriangulated points approaches subvisibility, the *difference* between the sums of the angles around all the vertex points and the numbers of vertexes, multiplied by 360 degrees, remains constantly 720 degrees, which is the sum of the angles of two times unity (of 360 degrees), which equals one tetrahedron. Q.E.D.

224.08 **Tetrahedron:** The sum of the angles of a tetrahedron, regular or irregular, is always 720°, just as the sum of the angles of a planar triangle is always 180°. Thus, we may state two propositions as follows:

224.081 The sum of the surface angles of any polyhedron equals the number of vertexes multiplied by 360° minus one *tetrahedron;* and

224.082 The sum of the angles of any polyhedron (including a sphere) is always evenly divisible by one *tetrahedron.*

224.10 **Descartes:** Descartes is the first of record to have discovered that the sum of the angles of a polyhedron is always 720° less than the number of vertexes times 360°. Descartes did not equate the 720° with the tetrahedron or with the one unit of energy quantum that it vectorially constitutes. He did not recognize the constant, whole difference between the visibly definite system and the invisibly finite Universe, which is always exactly one finite invisible tetrahedron outwardly and one finite invisible tetrahedron inwardly.

224.11 **The Calculus:** The calculus assumes that a sphere is infinitesimally congruent with a sphere to which it is tangent. The calculus and spherical trigonometry alike assume that the sum of the angles around any point on any sphere's surface is always 360 degrees. Because spheres are not continuous surfaces but are polyhedra defined by the vectorially interconnecting chords of an astronomical number of event foci (points) approximately equidistant from one approximate point, these spherically appearing polyhedra—whose chords emerge from lesser radius midpoints to maximum radius convergences at each of the spherically appearing polyhedra's vertexes, ergo, to convex external joining—must follow the law of polyhedra by which the sum of all the angles around the vertexes of the polyhedra is always 720 degrees less than 360 degrees times the number of vertexes. The demonstration thus far made discloses that the sum of the angles around all the vertexes of a sphere will always be 720 degrees—or one tetrahedron—less than the sum of the vertexes times 360 degrees—ergo, one basic assumption of the calculus and spherical trignometry is invalid.

224.12 **Cyclic Unity:** We may also say that: where unity (1) equals 360°, 180° equals one-half unity ($^1/_2$), and that 720° equals two times unity (2); therefore, we may identify a triangle as one-half unity and a tetrahedron as cyclic unity of two. As the sum of a polyhedron's angles, 720° is unique to the tetrahedron; 720° is the angular name of the tetrahedron. 720° is two cyclic unities. The tetrahedron is the geometrical manifest of "unity is plural and, at minimum, is two." The tetrahedron is *twoness* because it is congruently both a concave tetrahedron and a convex tetrahedron.

224.13 Where cyclic unity is taken as 360 degrees of central angle, the difference between infinity and finity is always exactly *two,* or 720 degrees, or two times 360 degrees, or two times unity. Cyclic unity embraces both wave and frequency since it represents angles as well as cycles. This is topologically manifest in that the number of vertexes in any structural system multiplied by 360 degrees, minus two times 360 degrees, equals the sum of the angles around all the vertexes of the system.

224.20 Equation of Angular Topology:

$S + 720° = 360° X^n$

Where:

S = the sum of all the angles around all the vertexes (crossings)

X^n = the total number of vertexes (crossings)

	No. of Vertexes	*Sum of Angles around each Vertex*	*Sum of angles multiplied by No. of Vertexes. De-Finite*	*No. of Vertexes multiplied by 360°. Finite*	*Finite minus De-Finite*
Line	2	0° × 1 = 0°	0° ×2 = 0°	360° ×2 = 720°	720° −0 = 720°
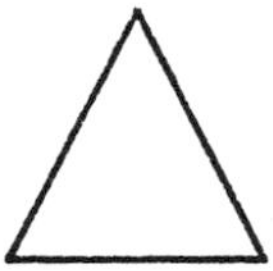 **Triangle**	3	60° × 2 = 120°	120° 3 = 360°	360° 3 = 1080°	1080° 360 = 720°
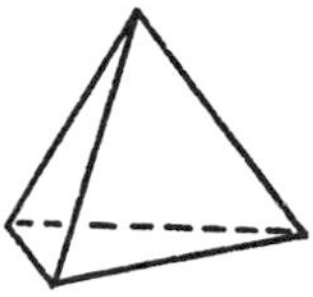 **Tetrahedron**	4	60° × 3 = 180°	180° 4 = 720°	360° 4 = 1440°	1440° 720 = 720°
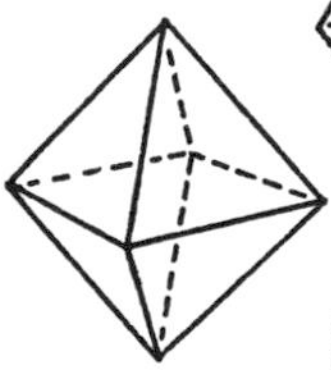 **Octa**	6	60° × 4 = 240°	240° 6 = 1440°	360° 6 = 2160°	2160° 1440 = 720°
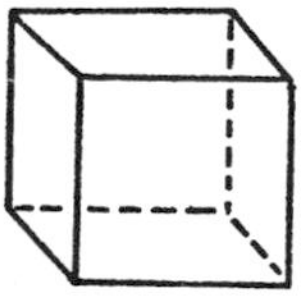 **Cube**	8	90° × 3 = 270°	270° 8 = 2160°	360° 8 = 2880°	2880° 2160 = 720°
 **Icosahedron**	12	60° × 5 = 300°	300° 12 = 3600°	360° 12 = 4320°	4320° 3600 = 720°
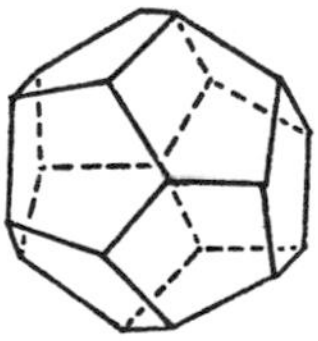 **Dodecahedron**	20	108° × 3 = 324°	324° 20 = 6480°	360° 20 = 7200°	7200° 6480 = 720°
 **Vector Equilibrium**	12	90° × 2 = 180° 60° × 2 = 120° 300°	300° 12 = 3600°	360° 12 = 4320°	4320° 3600 = 720°

Table 224.20 *Angular Topology Independent of Size.*

224.30 **Polarity:** Absolutely straight lines or an absolutely flat plane would theoretically continue outward to infinity. The difference between infinity and finity is governed by the taking out of angular sinuses, like pieces of pie, out of surface areas around a point in an absolute plane. This is the way lampshades and skirts are made. Joining the sinused fan-ends together makes a cone; if two cones are made and their open (ergo, infinitely trending) edges are brought together, a finite system results. It has two *polar* points and an equator. These are inherent and primary characteristics of all systems.

224.40 **Multivalent Applications:** Multiple-bonded bivalent and trivalent tetrahedral and octahedral systems follow the law of angular topology. Single-bonded monovalent tetrahedral and octahedral arrangements do not constitute a system; they are half systems, and in their case the equation would be:

$$S + 360° = 360° X^n$$

224.50 **Corollary: Principle of Finite Universe Conservation:** By our systematic accounting of angularly definable convex-concave local systems, we discover that the sum of the angles around each of every local system's interrelated vertexes is always two cyclic unities less than universal nondefined finite totality. We call this discovery the principle of finite Universe conservation. Therefore, mathematically speaking, all defined conceptioning always equals finite Universe minus two. The indefinable quality of finite Universe inscrutability is exactly accountable as two.

224.60 **Tetrahedral Mensuration:** The sum of the angles around all vertexes of any polyhedral system is evenly divisible by the sum of the angles of a tetrahedron. The volumes of all systems may be expressed in tetrahedra.

224.70 *Equation of Tetrahedral Mensuration:*

$$\frac{\textit{Sum of face angles}}{720°} = n \text{ tetrahedra}$$

Where:

$$720° = \text{one tetrahedron}$$

225.00 Principle of Design Covariables

225.01 **Definition:** The principle of design covariables states that angle and frequency modulation, either subjective or objective in respect to man's consciousness, discretely defines all events or experiences which altogether constitute Universe.

225.02 There are only two possible covariables operative in all design in the Universe. They are modifications of angle and frequency.

Tetrahedron	720°	$\frac{720°}{720°}$ = 1 tetrahedron
Octahedron	240° x 6 = 1440°	$\frac{1440°}{720°}$ = 2 tetrahedra
Prism	240° x 6 = 1440°	$\frac{1440°}{720°}$ = 2 tetrahedra
Cube	270° x 8 = 2160°	$\frac{2160°}{720°}$ = 3 tetrahedra
Icosahedron	300° x 12 = 3600°	$\frac{3600°}{720°}$ = 5 tetrahedra
Rhombic Dodecahedron	109°28′ x 24 = 2628° 70°32′ x 24 = 1692° 2628° x 1692° = 4320°	$\frac{4320°}{720°}$ = 6 tetrahedra
Dodecahedron	324° x 20 = 6480°	$\frac{6480°}{720°}$ = 9 tetrahedra
Triacontahedron	180° x 60 = 10,800°	$\frac{10,800°}{720°}$ = 15 tetrahedra
Two Frequency Regular Geodesic	180° x 80 = 14,400°	$\frac{14,400°}{720°}$ = 20 tetrahedra = 5×2^2
Three Frequency Alternate Geodesic	20° x 9 = 180° 180° x 180 = 32,400°	$\frac{32,400°}{720°}$ = 45 tetrahedra = 5×3^2
Four Frequency Triacon Geodesic	180° x 240 = 43,200°	$\frac{43,200°}{720°}$ = 60 tetrahedra = 15×2^2

Table 224.70A *Tetrahedral Mensuration Applied to Well-Known Polyhedra.* We discover that the sum of the angles around all vertexes of all solids is evenly divisible by the sum of the angles of a tetrahedron. The volumes of all solids may be expressed in tetrahedra.

225.03 Local structure is a set of frequency associable (spontaneously tunable), recollectible experience relationships, having a regenerative constellar patterning as the precessional resultants of concentrically shunted, periodic self-interferences, or coincidences of its systematic plurality of definitive vectorial frequency wavelength and angle interrelationships.

226.00 Principle of Functions

226.01 **Definition:** The principle of functions states that a function can always and only coexist with another function as demonstrated experimentally

	Number of Vertexes Multiplied by 360°	*Number of Triangles Multiplied by 180° Equals Sum of Angles around All Vertexes*	*Difference*
 **Regular Geodesic Two-Frequency Icosahedron**	42 × 360° = 15,120°	80 × 180° = 14,400°	15,120° − 14,400° = 720° = 1 tetrahedron
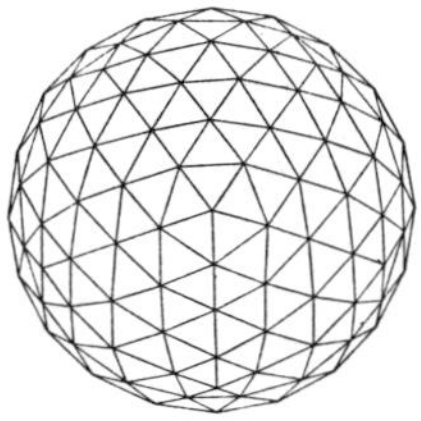 **Regular Geodesic Four-Frequency Icosahedron**	162 × 360° = 58,320°	320 × 180° = 57,600°	58,320° − 57,600° = 720° = 1 tetrahedron
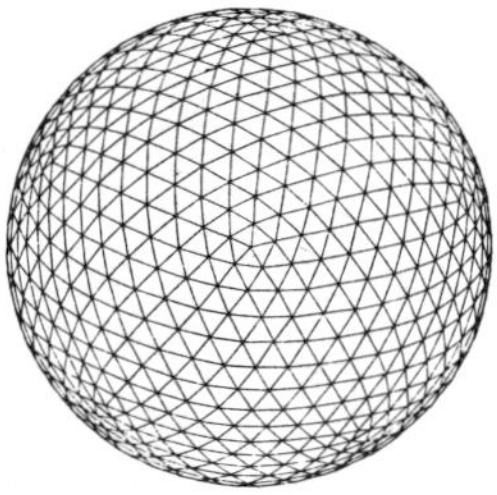 **Regular Geodesic Nine-Frequency Icosahedron**	812 × 360° = 292,320°	1620 × 180° = 291,600°	292,320° − 291,600° = 720° = 1 tetrahedron

Table 224.70B *Tetrahedral Mensuration Applied to Spheres.*

in all systems as the outside-inside, convex-concave, clockwise-counterclockwise, tension-compression couples.

226.02 Functions occur only as inherently cooperative and accommodatively varying subaspects of synergetically transforming wholes.

226.10 **Corollary: Principle of Complementarity:** A corollary of the principle of functions is the principle of complementarity, which states that two descriptions or sets of concepts, though mutually exclusive, are nevertheless both necessary for an exhaustive description of the situation.

226.11 Every fundamental behavior patterning in Universe always and only coexists with a complementary but non-mirror-imaged patterning.

UNDERLYING ORDER IN RANDOMNESS

No. of Events	Conceptuality of number of most economical relationships between events or minimum number of inter-connections of all events	No. of Relationships $\frac{n^2 - n}{2}$	Closest packed, symmetrical and most economical conceptual arrangement of number relationships.	Sum of Adjacent Relationships $(n-1)^2$	Conceptuality in closest packed Symmetry Note: This occurs as ▱ "diamonds" and not as □ "squares".	Sum of Experiences or of Events Is Always Tetrahedronal
1		0				
2	AB	1		0 + 1 = 1		
3	AB, BC, AC	3		1 + 3 = 4		
4	AB, BC, CD, AC, BD, AD	6		3 + 6 = 9		
5		10		6 + 10 = 16		
6		15		10 + 15 = 25		
7		21		15 + 21 = 36		
7	Same number of events could be in random array but minimum total of relationships are same in number.	21				

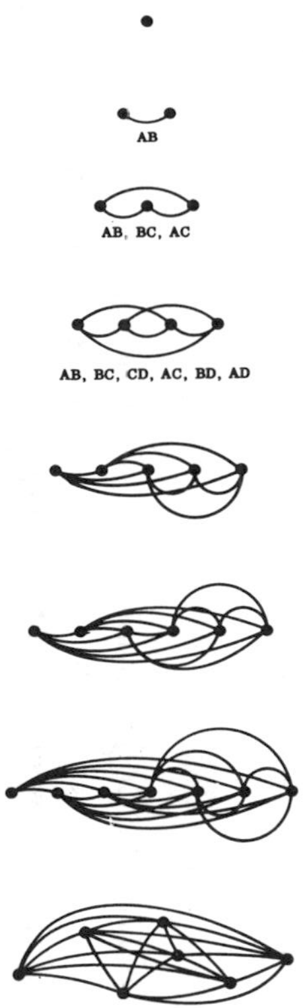

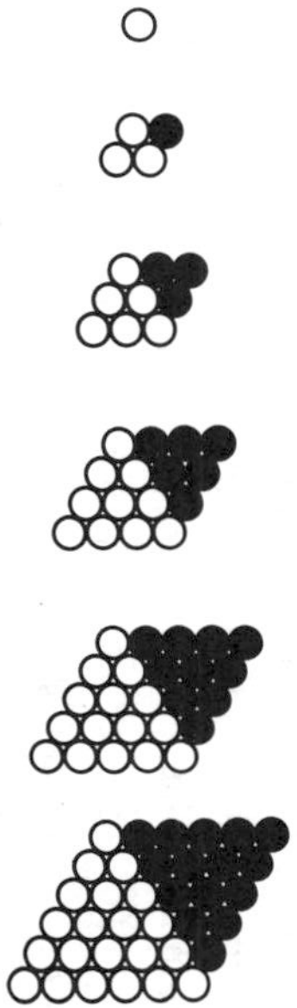

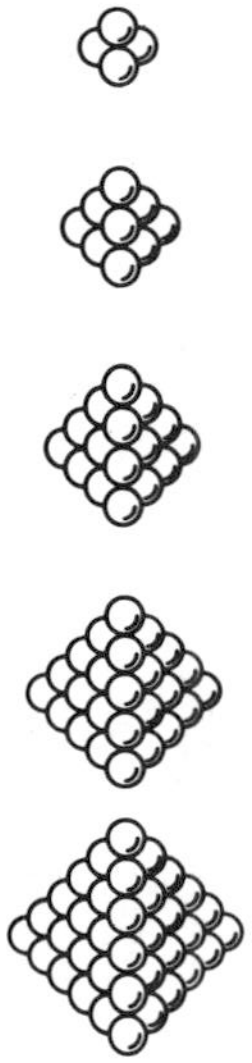

Table 227.01 *Underlying Order in Randomness.*

227.00 Principle of Order Underlying Randomness

227.01 **Definition:** The number of relationships between events is always

$$\frac{N^2 - N}{2},$$

Where:

N = the number of events of consideration

227.02 The relationships between four or more events are always greater in number than the number of events. The equation expresses the conceptuality of the number of the most economical relationships between events or the minimum number of interconnections of all events.

227.03 The number of telephone lines necessary to interequip various numbers of individuals so that any two individuals will always have their unique private telephone line is always $(N^2 - N)/2$, where N is the number of telephones. This is to say that all the special interrelationships of all experiences define comprehension, which is the number of connections necessary to an understanding of "what everything is all about." When we understand, we have all the fundamental connections between the star events of our consideration. When we add up all the accumulated relationships between all the successive experiences in our lives, they will always combine cumulatively to comprise a tetrahedron, simple or compound.

228.00 Scenario Principle

228.01 **Definition:** The scenario principle discloses that the Universe of total man experience may not be simultaneously recollected and reconsidered, but may be progressively subdivided into a plurality of locally tunable event foci or "points," of which a minimum of four positive and four negative are required as a "considerable set," that is, as the first finite subdivision of finite Universe.

228.10 **Considerable Set:** All experience is reduced to nonsimultaneously "considerable sets"; irrelevant to consideration are all those experiences that are either too large and therefore too infrequent, or too minuscule and therefore too frequent, to be tunably considerable as pertaining to the residual constellation of approximately congruent recollections of experiences.

228.11 A "considerable set" inherently subdivides all the rest of irrelevant experiences of Universe into macrocosmic and microcosmic sets immediately outside or immediately inside the considered set of experience foci.

229.00 Principle of Synergetic Advantage

229.01 **Definition:** The principle of synergetic advantage states that macro→micro does not equal micro→macro. Synergetic advantage is only to be effected by macro→micro procedure. Synergetic advantage procedures are irreversible. Micro→macro procedures are inherently frustrated.

229.02 The notion that commencing the exploration of the unknown with unity as one (such as Darwin's single cell) will provide simple and reliable arithmetical compounding (such as Darwin's theory of evolution: going from simple→complex; amoeba→monkey→man) is an illusion that as yet pervades and debilitates elementary education.

229.03 Synergy discloses that the information to be derived from micro→macro educational strategy fails completely to predict the experimentally demonstrable gravitational or mass-attraction integrities of entropically irreversible, universal scenario reality.

229.04 Human experience discloses the eminent feasibility of inbreeding biological species by mating like types, such as two fast-running horses. This concentrates the fast-running genes in the offspring while diminishing the number of general adaptability genes within the integral organism. This requires the complementary external care of the inbred specialist through invention or employment of extracorporeal environmental facilities—biological or nonbiological. It is easy to breed out metaphysical intellection characteristics, leaving a residual concentration of purely physical proclivities and evoluting by further inbreeding from human to monkey. (Witness the millions of dollars society pays for a "prizefight" in which two organisms are each trying to destroy the other's thinking mechanism. This and other trends disclose that a large segment of humanity is evoluting toward producing the next millennia's special breed of monkeys.) There is no experimental evidence of the ability to breed in the weightless, metaphysically oriented mind and its access to conceptionings of eternal generalized principles.

229.05 All known living species could be inbreedingly isolated from humans by environmental complementation of certain genetic proclivities and lethal exclusion of others, but there is no experimental evidence of any ability to compound purely physical proclivity genes to inaugurate metaphysical behaviors humanity's complex metaphysical-physical congruence with the inventory of complex behavioral characteristics of Universe.

229.06 Universe is the aggregate of eternal generalized principles whose nonunitarily conceptual scenario is unfoldingly manifest in a variety of special cases in local time-space transformative evolutionary events. Humans are each a special-case unfoldment integrity of the complex aggregate of abstract weightless omni-interaccommodative maximally synergetic non-sensorial Universe of eternal timeless principles. Humanity being a macro→micro Universe, unfolding eventuation is physically irreversible yet eternally integrated with Universe. Humanity cannot shrink and return into the womb and revert to as yet unfertilized ova.

229.10 **Corollary: Principle of Irreversibility:** The principle of irreversibility states that the evolutionary process is irreversible locally in physical "time-space"—that is, in frequency and angle definitioning, because the antientropic metaphysical world is not a mirror-imaged reversal of the entropic physical world's disorderly expansiveness.

230.00 Tetrahedral Number

230.01 **Definition:** The number of balls in the longest row of any triangular unit-radius ball cluster will always be the same as the number of rows of balls in the triangle, each row always having one more than the preceding row, and the number of balls in the complete triangular cluster will always be

$$\frac{(R+1)^2-(R+1)}{2}$$

Where:

R = the number of rows of balls, or the number of balls in the longest row

230.02 We can stack successively rowed triangular groups of balls on top of one another with one ball on the top, three below that, and six below that, as cannon balls or oranges are stacked. Such stacks are always inherently tetrahedral. We can say that the sum of all the interrelationships of all our successive experiences from birth to now—for each individual, as well as for the history of all humanity—is always a tetrahedral number.

231.00 Principle of Universal Integrity

231.01 **Definition:** The principle of universal integrity states that the wide-arc tensive or *im*plosive forces of Universe always inherently encompass the short-arc vectorial, *ex*plosive, disintegrative forces of Universe.

231.02 The gravitational constant will always be greater than the radiational constant—minutely, but always so. (For further exposition of this principle, see Secs. 251.05, 529.03, 541 and 1052.)

232.00 Principle of Conservation of Symmetry

232.01 Definition: Whereas the tetrahedron has four symmetrically interarrayed poles in which the polar opposites are four vertexes vs. four faces; and whereas the polar axes of all other symmetrical structural systems consist of vertex vs. vertex, or mid-edge vs. mid-edge, or face vs. face; it is seen that only in the case of vertex vs. face—the four poles of the tetrahedron—do the four vertexial "points" have polar face vacancies or "space" into which the wavilinear coil spring legs of the tetrahedron will permit those four vertexes to travel. The tetrahedron is the only omnisymmetrical structural system that can be turned inside out. (See Secs. 624.05, 905.18 and 905.19.)

232.02 Take the rubber glove that is green outside and red inside. Stripped off, it becomes red. The *left-handedness* is annihilated: inside-outing. You do not lose the convex-concave; all you lose is the *leftness* or the *rightness*. Whether it is a tree or a glove, each limb or finger is a tetrahedron.

232.03 Synergetics shows that the tetrahedron can be extrapolated into life in all its experience phases, thus permitting humanity's entry into a new era of cosmic awareness.

240.00 Synergetics Corollaries

240.01 Universe is finite.

240.02 Local systems are de-finite.

240.03 Unity is complex and at minimum two.

240.04 The tetrahedron is the lowest common rational denominator of Universe. The four unique quanta numbers of each and every fundamental "particle" are the four unique and minimum "stars" of every tetrahedron.

240.05 A "point" is a tetrahedron of negligible altitude and base dimensions.

240.06 A "line" (or trajectory) is a tetrahedron of negligible base dimension and significant altitude.

240.07 A "plane" (or opening) is a tetrahedron of negligible altitude and significant base dimensionality.

240.08 There are no solids or particles—no-things.

240.09 A point is an as yet undifferentiated focal star embracing a complex of local events.

240.10 There are no indivisible points.

240.11 Unities may be treated as complex star points.

240.12 For every point in Universe, there are six uniquely and exclusively operating vectors.

240.13 Vectors are size.

No vectors = No size.

No size = No vectors.

240.14 The size of a vector is its overall wavilinear length.

240.15 There are six vectors or none.

240.16 Every event has size.

240.17 Every event is six-vectored.

240.18 Six unique vectors constitute a tetrahedral event.

240.19 Each vector is reversible, having its negative alternate.

240.20 All "lines," trajectories, are the most economic vectorial interrelationships of nonsimultaneous but approximately concurrent local-event foci.

240.21 Potentially straight-line relationships require instantaneity or actions in no-time; therefore, straight lines are nondemonstrable.

240.22 The overall longitudinal length of wavilinear vectorial lines is determined by the number of waves contained.

240.23 The number of waves longitudinally accomplished in a given time constitutes frequency.

240.24 Physics has never made an experimentally demonstrable discovery of a straight line. Physics has found only waves and frequencies, i.e., angle and frequency modulation.

240.25 There are no straight lines, physical or metaphysical. There are only geodesic, i.e., most economical, interrelationships (vectors).

240.26 All "lines," trajectories, are complexedly curved.

240.27 Vectorial lines, or "trajectories," are always the most economical event interrelationships, ergo, geodesic.

240.28 Every "point" (event embryo) may articulate any one of its four event vector sets, each consisting of six positive and six negative vectors, but only one set may be operative at any one time; its alternate sets are momentarily only potential.

240.29 Potential lines are only inscrutably nonstraight; all physically realized relationships are geodesic and wavilinear.

240.30 Two energy event trajectories, or "lines," cannot go through the same point at the same time.

240.31 All geodesic lines, "trajectories," weave four-dimensionally amongst one another without ever touching one another.

240.32 It takes a minimum of six interweaving trajectories to isolate insideness from outsideness, ergo, to divide all Universe systematically into two parts—macrocosm and microcosm.

240.33 A six-trajectory isolation of insideness and outsideness has four interweaving vertexes or prime convergences of the trajectories, and four areal subdivisions of its isolation system, and constitutes a tetrahedron.

240.34 Tetrahedrons occur conceptually, independent of realized events and relative size.

240.35 Whereas none of the geodesic lines, "trajectories," of Universe touch one another, the lines, "trajectories," approach one another, passing successively through regions of most critical proximity, and diverge from one another, passing successively through regions of most innocuous remoteness.

240.36 All lines, "trajectories," ultimately return to close proximity with themselves.

240.37 Where all the local vectors are approximately equal, we have a potentially local isotropic vector equilibrium, but the operative vector complex has the inherent qualities of both proximity and remoteness in respect to any locally initiated action, ergo, a complex of relative velocities of realization lags. (See Sec. 425.01.)

240.38 Universe is a nonsimultaneously potential vector equilibrium.

240.39 All local events of Universe may be calculatively anticipated by inaugurating calculation with a local vector equilibrium frame and identifying the disturbance initiating point, directions, and energies of relative asymmetrical pulsings of the introduced action. (See Sec. 962.30.)

240.40 In the isotropic vector matrix derived from the closest packing of spheres, every vector leads from one nuclear center to another, and therefore each vector represents the operational effect of a merging of two force centers upon each other. Each vector is composed of two halves, each half belonging respectively to any two adjacent nuclear centers; each half of the vector is the unique radius of one of the tangent spheres that is perpendicular to the point of tangency. The half-vector radii of the isotropic vector matrix are always perpendicular to the points of tangency; therefore they operate as one continuous vector. Unity, as represented by the internuclear vector modulus, is of necessity always of the value of *two;* that is, unity is inherently two, for it represents union of a minimum of two energy centers.

240.41 Synergetics' six positive and six negative omnisymmetrical, potential realization, least effort interpatterning, evolutionary schemata reference frames are reinitiated and regenerated in respect to specific local energy event developments and interrelationships of Universe. (See Sec. 537.14.)

240.42 Arithmetical one-dimensionality is identified geometrically with linear (trajectory) pointal frequency.

240.43 Arithmetical two-dimensionality is identified geometrically with areal (openings) growth rate.

240.44 In a radiational (eccentric) or gravitational (concentric) wave system:

Arithmetical three-dimensionality is identified with volumetric space growth rates;

Arithmetical four-dimensionality is unidentifiable geometrically;

Synergetical *second-powering* is identified with the point population of the progressively embracing, closest-packed arrays at any given radius stated in terms of frequency of modular subdivisions of the circumferential array's radially-read concentricity layering;

Synergetical *third-powering* is identified with the cumulative total point population of all the successive wave layer embracements of the system;

Synergetical *fourth-powering* is identified with the interpointal domain volumes; and

Fifth- and *sixth-powerings* are identified as products of multiplication by frequency doublings and treblings, and are geometrically identifiable.

240.45 Synergetical six-dimensionality is identified geometrically with vectorial system modular frequency relationship.

240.46 Synergetical size dimensionality is identified geometrically with relative frequency modulation.

240.47 Dimension may be universally and infinitely altered without altering the constant vectorial integrity of the system.

240.48 There is no dimension without time.

240.49 Doubling or halving dimension increases or decreases, respectively, the magnitude of volume or force by expansive or contractive increments of eight, that is, by octave values.

240.50 Identically dimensioned nuclear systems and layer growths occur alike, relative to each and every absolutely compacted sphere of the isotropic vector matrix conglomerate, wherefore the integrity of the individual energy center is mathematically demonstrated to be universal both potentially and kinetically. (See Sec. 421.10.)

240.51 Frequency is multicyclic fractionation of unity.

240.52 A minimum of two cycles is essential to frequency fractionation.

240.53 Angle is subcyclic—that is, fractionation of one cycle.

240.54 Angular relationships and magnitudes are subcyclic; ergo, subfrequency; ergo, independent of size.

240.55 Shape is exclusively angular.

240.56 Shape is independent of size.

240.57 Abstraction means pattern relationship independent of size. Shape being independent of size is abstractable.

240.58 Abstractions may be stated in pure principle of relationship.

240.59 Abstractions are conceptually shapable!

240.60 Different shapes—ergo, different abstractions—are nonsimultaneous; but all shapes are de-finite components of integral though nonsimultaneous—ergo, shapeless—Universe.

240.61 There are no impervious surface continuums.

240.62 In a structural system, there is only one insideness and only one outsideness. (See Sec. 602.02.)

240.63 At any instant of time, any two of the evenly coupled vertexes of a system function as poles of the axis of inherent rotatability.

240.64 In a structural system:

(a) the number of vertexes (crossings) is always evenly divisible by two;

(b) the number of faces (openings) is always evenly divisible by four; and

(c) the number of edges (trajectories) is always evenly divisible by six. (See Sec. 604.01.)

240.65 The six edges of the tetrahedron consist of two sets of three vectors, each corresponding to the three-vector teams of the proton and neutron, respectively. Each of these three-vector teams is identified by nuclear physics as

one-half quantum, or

one-half Planck's constant, or

one-half spin,

with always and only co-occurring proton and neutron's combined two sets of three-vector teams constituting one quantum of energy, which in turn is vectorially identifiable as the minimum structural system of Universe.

240.66 All structural phenomena are accounted in terms of tetrahedron, octahedron, vector equilibrium, and icosahedron.

250.00 Discoveries of Synergetics

250.01 Discovery

250.02 Discoveries are uniquely regenerative to the explorer and are most powerful on those rare occasions when a generalized principle is discovered. When mind discovers a generalized principle permeating whole fields of special-case experiences, the discovered relationship is awesomely and elatingly beautiful to the discoverer personally, not only because to the best of his knowledge it has been heretofore unknown, but also because of the intuitively sensed potential of its effect upon knowledge and the consequently improved advantages accruing to humanity's survival and growth struggle in Universe. The stimulation is not that of the discoverer of a diamond, which is a physical entity that may be monopolized or exploited only to the owner's advantage. It is the realization that the newly discovered principle will provide spontaneous, common-sense logic engendering universal cooperation where, in many areas, only confusion and controversy had hitherto prevailed.

250.10 Academic Grading Variables in Respect to Science Versus Humanities

250.101 Whether it was my thick eyeglasses and lack of other personable favors, or some other psychological factors, I often found myself to be the number-one antifavorite amongst my schoolteachers and pupils. When there were disturbances in the classroom, without looking up from his or her desk, the teacher would say, "One mark," or "Two marks," or "Three marks for Fuller." Each mark was a fifteen-minute penalty period to be served after the school had been let out for the others. It was a sport amongst some of my classmates to arrange, through projectiles or other inventions, to have noises occur in my vicinity.

250.11 Where the teacher's opinion of me was unfavorable, and that, in the humanities, was—in the end—all that governed the marking of papers, I often found myself receiving lower grades for reasons irrelevant to the knowledge content of my work—such as my handwriting. In science, and particularly in my mathematics, the answers were either right or wrong. Probably to prove to myself that I might not be as low-average as was indicated by the gradings I got in the humanities, I excelled in my scientific classes and consistently attained the top grades because all my answers were correct. Maybe this made me like mathematics. But my mathematics teachers in various years would say, "You seem to understand math so well, I'll show you some more if you stay in later in the afternoon." I entered Harvard with all As in mathematics, biology, and the sciences, having learned in school advanced mathematics, which at that time was usually taught only at the college level. Since math was so easy, and finding it optional rather than compulsory at Harvard, I took no more of its courses. I was not interested in getting grades but in learning in areas that I didn't know anything about. For instance, in my freshman year, I took not only the compulsory English A, but Government, Musical Composition, Art Appreciation, German Literature, and Chemistry. However, I kept thinking all the time in mathematics and made progressive discoveries, ever enlarging my mathematical vistas. My elementary schoolwork in advanced mathematics as well as in physics and biology, along with my sense of security in relating those fields, gave me great confidence that I was penetrating the unfamiliar while always employing the full gamut of rigorous formulation and treatment appropriate to testing the validity of intuitively glimpsed and tentatively assumed enlargement of the horizon of experientially demonstrable knowledge.

250.12 My spontaneous exploration of mathematics continued after I left Harvard. From 1915 to 1938—that is, for more than twenty years after my days in college—I assumed that what I had been discovering through the post-college years, and was continuing to discover by myself, was well known to mathematicians and other scientists, and was only the well-known advanced knowledge to which I would have been exposed had I stayed at Harvard and majored in those subjects. Why I did not continue at Harvard is irrelevant to academics. A subsequent special course at the U. S. Naval Academy, Annapolis, and two years of private tutelage by some of America's leading engineers of half a century ago completed my formally acknowledged "education."

250.20 *My Independent Mathematical Explorations*

250.21 In the twentieth year after college, I met Homer Lesourd, my old physics teacher, who most greatly inspired his students at my school, Milton Academy, and who for half a century taught mathematics at Harvard. We discovered to our mutual surprise that I had apparently progressed far afield from any of the known physio-mathematical concepts with which he was familiar or of which he had any knowledge. Further inquiry by both of us found no contradiction of our first conclusion. That was a third of a century ago. Thereafter, from time to time but with increasing frequency, I found myself able to elucidate my continuing explorations and discoveries to other scientists, some of whom were of great distinction. I would always ask them if they were familiar with any mathematical phenomena akin to the kind of disclosures I was making, or if work was being done by others that might lead to similar disclosures. None of them was aware of any other such disclosures or exploratory work. I always asked them whether they thought my disclosures warranted my further pursuit of what was becoming an ever-increasingly larger body of elegantly integrated and coordinate field of omnirationally quantified vectorial geometry and topology. While they could not identify my discoveries with any of the scientific fields with which they were familiar, they found no error in my disclosures and thought that the overall rational quantation and their logical order of unfoldment warranted my further pursuing the search.

250.30 Remoteness of Synergetics Vocabulary

250.301 When one makes discoveries that, to the best of one's knowledge and wide inquiry, seem to be utterly new, problems arise regarding the appropriate nomenclature and description of what is being discovered as well as problems of invention relating to symbolic economy and lucidity. As a consequence, I found myself inventing an increasingly larger descriptive vocabulary, which evolved as the simplest, least ambiguous method of recounting the paraphernalia and strategies of the live scenario of all my relevant experiences.

250.31 For many years, my vocabulary was utterly foreign to the semantics of all the other sciences. I drew heavily on the dictionary for good and unambiguous terms to identify the multiplying nuances of my discoveries. In the meanwhile, the whole field of science was evolving rapidly in the new fields of quantum mechanics, electronics, and nuclear exploration, inducing a gradual evolution in scientific language. In recent years, I find my experiential mathematics vocabulary in a merging traffic pattern with the language trends of the other sciences, particularly physics. Often, however, the particular new words chosen by others would identify phenomena other than that which I identify with the same words. As the others were unaware of my offbeat work, I had to determine for myself which of the phenomena involved had most logical claim to the names involved. I always conceded to the other scientists, of course (unbeknownst to them); when they seemed to have prior or more valid claims, I would then invent or select appropriate but unused names for the phenomena I had discovered. But I held to my own claim when I found it to be eminently warranted or when the phenomena of other claimants were ill described by that term. For example, quantum mechanics came many years after I did to employ the term *spin*. The physicists assured me that their use of the word did not involve any phenomena that truly spun. *Spin* was only a convenient word for accounting certain unique energy behaviors and investments. My use of the term was to describe a direct observation of an experimentally demonstrable, inherent spinnability and unique magnitudes of rotation of an actually spinning phenomenon whose next fractional rotations were induced by the always co-occurring, generalized, a priori, environmental conditions within which the spinnable phenomenon occurred. This was a case in which I assumed that I held a better claim to the scientific term *spin*. In recent years, spin is beginning to be recognized by the physicists themselves as also inadvertently identifying a conceptually spinnable phenomenon—in fact, the same fundamental phenomenon I had identified much earlier when I first chose to use the word *spin* to describe that which was experimentally disclosed as being inherently spinnable. There appears to be an increasing convergence of scientific explorations in general, and of epistemology and semantics in particular, with my own evolutionary development.

250.32 Because physics has found no continuums, no experimental solids, no things, no real matter, I had decided half a century ago to identify mathematical behaviors of energy phenomena only as *events*. If there are no things, there are no nouns of material substance. The old semantics permitted common-sense acceptance of such a sentence as, "A man pounds the table," wherein a noun verbs a noun or a subject verbs a predicate. I found it necessary to change this form to a complex of events identified as *me,* which must be identified as a verb. The complex verb *me* observed another complex of events identified again ignorantly as a "table." I disciplined myself to communicate exclusively with *verbs*. There are no *wheres* and *whats;* only angle and frequency events described as *whens*.

250.40 The Climate of Invention

250.401 In the competitive world of money-making, discoveries are looked upon as exploitable and monopolizable claims to be operated as private properties of big business. As a consequence, the world has come to think of both discoveries and patents as monopolized property. This popular viewpoint developed during the last century, when both corporations and government supported by courts have required individuals working for them to assign to them the patent rights on any discoveries or inventions made while in their employ. Employees were to assign these rights during, and for two years after termination of, their employment, whether or not the invention had been developed at home or at work. The drafting of expert patent claims is an ever more specialized and complex art, involving expensive legal services usually beyond the reach of private individuals. When nations were remote from one another, internal country patents were effective protection. With today's omniproximities of the world's countries, only world-around patents costing hundreds of thousands of dollars are now effective, with the result that patent properties are available only to rich corporations.

250.41 So now the major portions of extant inventions belong to corporations and governments. However, invention and discovery are inherently individual functions of the minds of individual humans. Corporations are legal fabrications; they cannot invent and discover. Patents were originally conceived of as grants to inventors to help them recover the expenses of the long development of their discoveries; and they gave the inventor only a very short time to recover the expense. Because I am concerned with finding new technical ways of doing more with less, by which increasing numbers of humanity can emerge from abject poverty into states of physical advantage in respect to their environment, I have taken out many patent claims—first, to hold the credit of initiative for the inspiration received by humanity's needs and the theory of their best solution being that of the design revolution and not political revolution, and second, to try to recover the expense of development. But most importantly, I have taken the patents to avoid being stopped by others—in particular, corporations and governments—from doing what I felt needed doing.

250.50 Coincidental Nature of Discoveries

250.501 What often seems to the individual to be an invention, and seems also to be an invention to everyone he knows, time and again turns out to have been previously discovered when patent applications are filed and the search for prior patents begins. Sometimes dozens, sometimes hundreds, of patents will be found to have been issued, or applied for, covering the same idea. This simultaneity of inventing manifests a forward-rolling wave of logical exploration of which the trends are generated by the omni-integrating discoveries and the subsequent inventions of new ways to employ the discoveries at an accelerating rate, which is continually changing the metaphysical environment of exploratory and inventive stimulation.

250.51 I have learned by experience that those who think only in competitive ways assume that I will be discouraged to find that others have already discovered, invented, and patented that which I had thought to be my own unique discovery or invention. They do not understand how pleased I am to learn that the task I had thought needed doing, and of which I had no knowledge of others doing, was happily already being well attended to, for my spontaneous commitment is to the advantage of all humanity. News of such work of others frees me to operate in other seemingly unattended but needed directions of effort. And I have learned how to find out more about what is or is not being attended to. This is evolution.

250.52 When I witness the inertias and fears of humans caused by technical breakthroughs in the realms of abstract scientific discovery. I realize that their criteria of apprehension are all uninformed. I see the same patterns of my experience obtaining amongst the millions of scientists around the world silently at work in the realm of scientific abstract discovery, often operating remote from one another. Many are bound to come out with simultaneous discoveries, each one of which is liable to make the others a little more comprehensible and usable. Those who have paid-servant complexes worry about losing their jobs if their competitors' similar discoveries become known to their employers. But the work of pure science exploration is much less understood by the economically competitive-minded than is that of inventors. The great awards economic competitors give to the scientists make big news, but no great scientist ever did what he did in hope of earning rewards. The greats have ever been inspired by the a priori integrities of Universe and by the need of all humanity to move from the absolute ignorance of birth into a little greater understanding of the cosmic integrities. They esteem the esteem of those whom they esteem for similar commitment, but they don't work for it.

250.53 I recall now that when I first started making mathematical discoveries, years ago, my acquaintances would often say, "Didn't you know that Democritus made that discovery and said just what you are saying 2,000 years ago?" I replied that I was lucky that I didn't know that because I thought Democritus so competent that I would have given up all my own efforts to understand the phenomena involved through my own faculties and investment of time. Rather than feeling dismayed, I was elated to discover that, operating on my own, I was able to come out with the same conclusions of so great a mind as that of Democritus. Such events increased my confidence in the resourcefulness and integrity of human thought purely pursued and based on personal experiences.

250.60 Proofs

250.61 I know that many of the discoveries of synergetics in the book of their accounting, which follows, may prove in time to be well-known to others. But some of them may not be known to others and thus may be added to the ever-increasing insights of the human mind. Any one individual has inherently limited knowledge of what total Universe frontiering consists of at any one moment. My list embraces what I know to be my *own* discoveries of which I have no knowledge of others having made similar discoveries earlier than my own. I claim nothing. Proofs of some of my theoretical discoveries have been made by myself and will be made by myself. Proofs may have been made by others and will be made by others. Proofs are satisfying. But many mathematical theorems provide great living advantages for humanity over long periods of time before their final mathematical proofs are discovered. The whys and wherefores of what is rated as mathematical proof have been evolved by mathematicians; they are formal and esoteric conventions between specialists.

251.00 Discoveries of Synergetics: Inventory

251.01 The ability to identify all experience in terms of only angle and frequency.

251.02 The addition of angle and frequency to Euler's inventory of crossings, areas, and lines as absolute characteristics of all pattern cognizance.

251.03 The omnirational accommodation of both linear and angular acceleration in the same mathematical coordination system.

251.04 The discovery that the pattern of operative effectiveness of the gravitational constant will always be greater than that of the radiational constant—the excess effectiveness being exquisitely minute, but always operative, wherefore the disintegrative forces of Universe are effectively canceled out and embraced by the integrative forces.

251.05 The gravitational is comprehensively embracing and circumferentially contractive—ergo, advantaged over the centrally radiational by a 6 : 1 energy advantage; i.e., a circumference chord-to-radius vectorial advantage of contraction versus expansion, certified by the finite closure of the circumference, ergo, a cumulative series versus the independent, disassociating disintegration of the radii and their separating and dividing of energy effectiveness. (This is an inverse corollary of the age-old instinct to divide and conquer.) (See Secs. 529.03, 541 and 1052.)

251.06 The gravitational-radiational constant $10F^2 + 2$.

251.07 The definition of gravity as a spherically circumferential force whose effectiveness has a constant advantage ratio of 12 to 1 over the radial inward mass-attraction.

251.10 The introduction of angular topology as the description of a structural system in terms of the sum of its surface angles.

251.11 The definition of structure as the pattern of self-stabilization of a complex of events with a minimum of six functions as three edges and three vertexes, speaking both vectorially and topologically.

251.12 The introduction of angular topology as comprised entirely of central-angle and surface-angle phenomena, with the surface angles accounting for concavity and convexity, and the thereby-derived maximum structural advantage of omni-self-triangulating systems.

251.13 As a result of the surface-angle concave-convex take-outs to provide self-closing finiteness of insideness and outsideness, central angles are generated, and they then function in respect to unique systems and differentiate between compoundings of systems.

251.14 One of the differences between atoms and chemical compounds is in the number of central-angle systems.

251.20 The discovery of the mathematically regular, three-way, great-circle, spherical-coordinate cartographic grid of an infinite frequency series of progressive modular subdivisions, with the spherical radii that are perpendicular to the enclosing spherical field remaining vertical to the corresponding planar surface points of cartographic projection; and the commensurate identification of this same great-circle triangulation capability with the icosahedron and vector equilibrium, as well as with the octahedron and the tetrahedron. (See Secs. 527.24 and 1009.98.)

251.21 The development of the spherical triangular grid bases from the spherical tetrahedron, spherical cube, spherical octahedron, and the spherical vector equilibrium and its alternate, the icosahedron, and the discovery that there are no other prime spherical triangular grids. All other spherical grids are derivatives of these.

251.22 The spherical triangular grids are always identified uniquely only with the first four prime numbers 1, 2, 3 and 5: with the tetrahedron always identifying with the prime number 1; the octahedron with 2, the face-triangulated cube with 3; and the vector equilibrium and icosahedron with the prime number 5; with the other Platonic, Archimedean, and other symmetrical polyhedra all being complex compoundings and developments of these first four prime numbers, with the numbers compounded disclosing the compounding of the original four base polyhedra.

251.23 The number of the external crossings of the three-way spherical grids always equals the prime number times the frequency of modular subdivision to the second power times two, plus the two extra crossings always assigned to the polar axis functioning to accommodate the independent spinnability of all systems.

251.24 The mathematical regularity identifies the second power of the linear dimensions of the system with the number of nonpolar crossings of the comprehensive three-way great circle gridding, in contradistinction to the previous mathematical identification of second powering exclusively with surface areas.

251.25 The synergetic discovery of the identification of the surface points of the system with second powering accommodates quantum mechanics' discrete energy packaging of photons and elucidates Einstein's equation, $E=Mc^2$, where the omnidirectional velocity of radiation to the second power—c^2—identifies the rate of the rational order growth of the discrete energy quantation. This also explains synergetics' discovery of the external point growth rate of systems. It also elucidates and identifies the second-power factoring of Newton's gravitational law. It also develops one-to-one congruence of all linear and angular accelerations, which are factorable rationally as the second power of wave frequency.

251.26 The definition of a system as the first subdivision of finite but nonunitary and nonsimultaneous conceptuality of the Universe into all the Universe outside the system, and all the Universe inside the system, with the remainder of the Universe constituting the system itself, which alone, for the conceptual moment, is conceptual.

251.27 The definition of Universe as a scenario of nonsimultaneous and only partially overlapping events, all the physical components of which are ever-transforming, and all the generalized metaphysical discoveries of which ever clarify more economically as eternally changeless.

251.28 The vector model for the magic numbers, which identifies the structural logic of the atomic isotopes in a symmetrical synergetic hierarchy.

251.30 The rational identification of number with the hierarchy of all the geometries.

251.31 The *A* and *B* Quanta Modules.

251.32 The volumetric hierarchy of Platonic and other symmetrical geometricals based on the tetrahedron and the *A* and *B* Quanta Modules as unity of coordinate mensuration.

251.33 The identification of the nucleus with the vector equilibrium.

251.34 Omnirationality: the identification of *triangling* and *tetrahedroning* with second- and third-powering factors.

251.35 Omni-60-degree coordination versus 90-degree coordination.

251.36 The identification of waves with vectors as waviform vectors; the deliberately nonstraight line.

251.37 The comprehensive, closed-system foldability of the great circles and their identification with wave phenomena.

251.38 The accommodation of odd or even numbers in the shell-generating frequencies of the vector equilibrium.

251.39 The hierarchy of the symmetrically expanding and contracting pulsations of the interpolyhedral transformations, and their respective circumferentially and radially covarying states. (Also described as the symmetrical contraction, "jitterbugging," and pumping models.)

251.40 The provision for the mathematical treatment of the domains of interferences as the domains of vertexes (crossings).

251.41 Mathematical proof of the four-color map theorem.

251.42 The introduction of the tensegrity structural system of discontinuous compression and continuous tension.

251.43 The identification of tensegrity with pneumatics and hydraulics.

251.44 The discovery of the number of primes factorial that form the positives and negatives of all the complex phenomena integratively generated by all possible permutations of all the 92 regenerative chemical elements.

251.45 The disclosure of the rational fourth-, fifth-, and sixth-powering modelability of nature's coordinate transformings as referenced to the 60° equiangular, isotropic vector equilibrium.

251.46 The discovery that once a closed system is recognized as exclusively valid, the list of variables and the degrees of freedom are closed and limited to six positive and six negative alternatives of action for each local transformation event in Universe.

251.47 The discovery of the formula for the rational-whole-number expression of the tetrahedral volume of both the spherical and interstitial spaces of the first- and third-power concentric shell-growth rates of nuclear closest-packed vector equilibria.

251.50 The integration of geometry and philosophy in a single conceptual system providing a common language and accounting for both the physical and metaphysical.

300.00 Universe

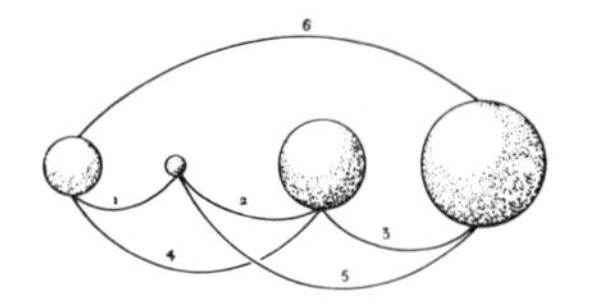

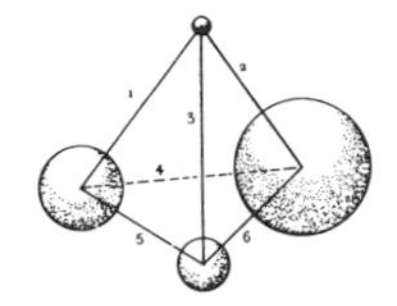

301.00 Definition: Universe

301.10 Universe is the aggregate of all humanity's consciously apprehended and communicated nonsimultaneous and only partially overlapping experiences.

302.00 *Aggregate* means sum-totally but nonunitarily conceptual as of any one moment. *Consciousness* means an awareness of otherness. *Apprehension* means information furnished by those wave frequencies tunable within man's limited sensorial spectrum. *Communicated* means informing self or others. *Nonsimultaneous* means not occurring at the same time. *Overlapping* is used because every event has duration, and their initiatings and terminatings are most often of different duration.* Neither the set of all *experiences* nor the set of all the words used to describe them are instantly reviewable nor are they of the same length. Experiences are either involuntary (subjective) or voluntary (objective), and all experiences, both physical and metaphysical, are finite because each begins and ends.

303.00 Universe is the comprehensive, historically synchronous, integral-aggregate system embracing all the separate integral-aggregate systems of all men's consciously apprehended and communicated (to self or others) nonsimultaneous, nonidentical, but always complementary and only partially overlapping, macro-micro, always-and-everywhere, omnitransforming, physical and metaphysical, weighable and unweighable event sequences. Universe is a dynamically synchronous scenario that is unitarily nonconceptual as of any one moment, yet as an aggregate of finites is sum-totally finite.

* The complex of event sequences is most often characterized by overlappings. A man is born, grows up, has children and grandchildren. His life overlaps that of his grandfather and father and that of his children and grandchildren. But his grandfather's life did not overlap his children's nor his grandchildren's lives. Hence, *partially* overlapping.

305.00 Synergetic Advantage: Macro→Micro

305.01 Universe is the starting point for any study of synergetic phenomena. The Principle of Synergetic Advantage (see Sec. 229) requires that we return to the Universe as our starting point in all problem consideration. We assiduously avoid all the imposed disciplines of progressive specialization. We depend entirely upon our innate facilities, the most important of which is our intuition, and test our progressive intuitions with experiments.

305.02 Universe is, inferentially, the biggest system. By starting with Universe, we automatically avoid leaving out any strategically critical variables. In the Universe, everything is always in motion and everything is always moving in the directions of least resistance. When we are dealing always in terms of a finite Universe or totality of behavior, we are able to work from the generalized whole to the particular or special-case manifestation of the generalized accounting. This is the basis of the grand philosophic accounting of quantum mechanics.

305.03 I have found during my whole life that I have had a problem dealing with society because I am dealing with that which is not obvious to the rest of society. I have found that Universe is actually operating in an entirely different way from the way society thinks it is. Society is living in a sort of "fault." The kind of fault I am thinking of is like an Earth fault, like a great cliff, a great discontinuity in Earth's surface. There are great discontinuities in the public's image of what the Universe *is*.

305.04 We find no record of man having defined the Universe—scientifically and comprehensively—to include both the metaphysical and the physical. The scientist was able to define physical Universe by virtue of the experimentally verified discovery that energy can neither be created nor lost; therefore that energy is conserved; therefore it is finite. Thus, man has been able to define successfully physical Universe—but not, as yet, the metaphysical Universe.

305.05 Our definition of Universe includes both the objective and the subjective, i.e., all voluntary experiences (experiments) as well as all involuntary experiences (happenings). The total of experiences is integrally synergetic. Universe is the comprehensive, a priori synergetic integral. Universe continually operates in comprehensive, coordinate patternings that are transcendental to the sensorially minuscule apprehension and mental-comprehension and prediction capabilities of mankind, consciously and inherently preoccupied as he is only with special local and nonsimultaneous pattern considerations. Angle and frequency modulations, either subjective or objective in respect to man's consciousness, discretely define all events or experiences which altogether constitute Universe. (See Secs. 208 and 503.)

305.06 To each of us, Universe must be all that isn't me, plus me.

306.00 Universe and Self

306.01 People say to me, "I think you have left something out of your definition of Universe." That statement becomes part of my experience. But never will anyone disprove my working hypothesis because it will take experimental proof to satisfy me, and the experiment will always be part of the experience of my definition, ergo included. This gives me great power because my definition of Universe includes not only the physical but also the metaphysical experiences of Universe, which the physicists thought they had to exclude from their more limited definition of the finite physical portion of Universe. The metaphysical embraces all the weightless experiences of thought, including all the mathematics and the organization of data regarding all the physical experiments, science itself being metaphysical.

306.02 My Universe is that portion of the intercommunicated aggregate of all conscious and operationally described experiences of all history's beings, including my own, which is now totally recallable only in fragments as progressively and spontaneously tunable within my own angular orientation and zonal discernment limits of the multidirectional and multimagnitude, sensorial-frequency-spectrum inventory of the frequently accumulating, integrating, and accommodatingly rearranging memory album of all discernibly unique patternings whatsoever. While in many ways similar, each of humanity's individual's Universes must always seem to differ in some total experience inventory aspects.

306.03 The scenario events of Universe are the regenerative interactions of all otherness and me.

307.01 Universe is the ultimate collective concept embracing all intelligible, inherently separate evolutionary events, which apparently occur always and only through differentiating considerations that progressively isolate the components of whole and inclusive sets, supersets, and subsets of generalized conceptioning in retrospectively abstracted principles of relationships. (See Sec. 1056.15.)

307.02 As defined, Universe is inherently unitarily inconceivable.

307.03 There may be no absolute division of energetic Universe into isolated or noncommunicable parts. There is no absolutely enclosed surface, and there is no absolutely enclosed volume. Universe means "toward oneness" and implies a minimum of twoness.

307.04 Because of the fundamental nonsimultaneity of universal structuring, a single, simultaneous, static model of Universe is inherently both nonexistent and conceptually impossible as well as unnecessary. Ergo, Universe does not have a shape. Do not waste your time, as man has been doing for ages, trying to think of a unit shape "outside of which there must be something," or "within which, at center, there must be a smaller something."

308.00 Universe is finite because it is the sum total of finitely furnished experiences. The comprehensive set of all experiences synergetically constituting Universe discloses an astronomically numbered variety of subset event-frequency rates and their respective rates of conceptual tunability comprehension. It takes entirely different lengths of time to remember to "look up" different names or facts of past events. Universe, like the dictionary, though integral, is ipso facto nonsimultaneously recollectable; therefore, as with the set of all the words in the dictionary, it is nonsimultaneously reviewable, ergo, is synergetically incomprehensible, as of any one moment, yet is progressively revealing.

309.00 Universe is the minimum of intertransformings necessary for total self-regeneration.

310.01 Our modern concept of Universe is as a comprehensive system of energy processes. Universe is a nonsimultaneously potential vector equilibrium. The integrity of Universe is implicit in the external finiteness of the entirely embracing circumferential set of the integrated vectors of the vector equilibrium that always enclose the otherwise divisive, disintegrative, entirely embraced internal radial set of omnidirectional vectors. Universe is tensional integrity.

310.02 The star tetrahedron's entropy may be the basis of irreversible radiation, whereas the syntropic vector equilibrium's reversibility—inwardly, outwardly—is the basis for the gravitationally maintained integrity of Universe. The omnidirectional, omniwave, propagating pulsivity of Universe realizations is eternally potential and implicit in the vector equilibrium.

310.03 Universe and its experiences cannot be considered as being physical, for they balance out as weightless. Every positively weighted "particle" has its negatively weighted complementary, but non-mirror-imaged, counterpart behavior. The integrated weights of physical Universe add up to zero. The weightless experience is metaphysical—physical phenomena having been identified by the physicists as being always uniquely weighable, that is, ponderable, that is, detected by the mass-attracted levering of an indicator needle.

311.00 Humans as Local Universe Technology

311.01 Of all the subcosmic, integrally interpatterning complexes that we know of in our Universe, there is no organic complex that in any way compares with that of the human being. We have only one counterpart of total complexity, and that is Universe itself.* That such a complex miniature Universe is found to be present on this planet, and that it is "born" absolutely ignorant, is part of the manifold of design integrities.

311.02 Universe is technology—the most comprehensively complex technology. Human organisms are Universe's most complex local technologies.

311.03 Universe is the aggregate of eternal generalized principles whose nonunitarily conceptual scenario is unfoldingly manifest in a variety of special-case, local, time-space transformative, evolutionary events. Humans are each a special-case unfoldment-integrity of the multi-alternatived complex aggregate of abstract, weightless, omni-interaccommodative, maximally synergetic, non-sensorial, eternal, timeless principles of Universe. Humanity being a macro→micro Universe-unfolding eventuation is physically irreversible yet eternally integrated with Universe. Humanity cannot shrink and return into the womb and revert to as yet unfertilized ova. Humanity can only evolve toward cosmic totality, which in turn can only be evolvingly regenerated through new-born humanity.

320.00 Scenario Universe

321.01 When people say of Universe, "I wonder what is outside its outside?" they are trying to conjure a unitary conception and are asking for a single picture of an infinitely transforming, nonsimultaneous scenario. Therefore, their question is not only unanswerable but unrealistic, and indicates that they have not listened seriously to Einstein and are only disclosing their ignorance of its significance when they boastfully tell you that the speed of light is 186,000 miles per second.

321.02 You cannot get out of Universe. Universe is not a system. Universe is not a shape. Universe is a scenario. You are always in Universe. You can only get out of systems.

* Apparently, man matches the Universe in displaying the same relative abundance of the 92 self-regenerative chemical elements.

Fig. 321.01 *Universe as "A Minimum of Two Pictures":* Evolution as a transformation of nonsimultaneous events: the behavior of "Universe" can only be shown with a minimum of two pictures. *Unity* is plural and at minimum two. (*Drawings courtesy Mallory Pearce*)

322.01 Universe can only be thought of competently in terms of a great, unending, but finite scenario whose as yet unfilled film strip is constantly self-regenerative. All experiences are terminated, ergo finite. An aggregate of finites is finite. Our Universe is finite but nonsimultaneously conceptual: a moving-picture scenario of nonsimultaneous and only partially overlapping events. One single picture—one frame—does not tell the story. The single-frame picture of a caterpillar does not foretell or imply the transformation of that creature, first, into the chrysalis stage and, much later, into the butterfly phase of its life. Nor does one picture of a butterfly tell the viewer that the butterfly can fly.

322.02 In scenarios, you have to have a pretty long sequence run in order to gain any clue at all as to what is going on. You cannot learn what it is all about from a single picture. You cannot understand life without much experience.

323.00 In the endless but finite and never exactly repeating (Heisenberged) ''film-strip'' scenario of evolutionary Universe, after the film strip has been projected, it goes through a dissolved phase and re-forms again to receive the ever-latest self-intertransforming patterning just before being again projected. The rate of change and the numbers of special-case self-retransformings of *physical evolution* tend ever to accelerate, differentiate, and multiply; while the rate of change and the numbers of self-remodifyings of generalized law conceptionings of *metaphysical evolution* tend ever to decelerate, simplify, consolidate, and ultimately unify.

324.00 Finite structures are mostly nonconceptual in any momentary sense, though certain local structures in Universe are momentarily conceptual; for instance, the continually transforming, momentarily residual aggregate of men's experiences packaged together in the words *Planet Earth*.

325.00 The speeds of all the known different phases of measured radiation are apparently identical despite vast differences in wavelength and frequency. Einstein's adoption of electromagnetic radiation expansion—omnidirectionally in vacuo—as *normal speed* suggests a top speed of omnidirectional entropic disorder increase accommodation at which radiant speed reaches highest velocity. This highest velocity is reached when the last of the eternally regenerative Universe cyclic frequencies of multibillions of years have been accommodated, all of which complex of nonsimultaneous transforming, multivarietied frequency synchronizations is complementarily balanced to equate as zero by the sum totality of locally converging, orderly, and synchronously concentrating energy phases of scenario Universe's eternally pulsative, and only sum-totally synchronous, disintegrative, divergent, omnidirectionally exporting, and only sum-totally synchronous, integrative, convergent, and discretely directional individual importings.

330.00 Universe as Minimum Perpetual Motion Machine

331.00 The physical Universe is a machine. In fact, Universe is the minimum and *only* perpetual motion machine. Universe is the minimum as well as the maximum closed system of omni-interacting, precessionally transforming, complementary transactions of synergetic regeneration.

332.00 Universe is finite. Local systems are de-finite. Local perpetual motion machines are impossible since Universe is the minimum regenerative set of perpetually intercomplementary, transformative functioning.

333.00 There is a minimum set of patterns, which is a consequence of one set of patterns reacting with another set of patterns. In order to have a monkey wrench, you also have to make one or buy one at a store, you have to have other things, and these procurements in turn have antecedent event requirements. Each event of Universe leads back to all the great complex of events, and we get then to a minimum set of complementary events whereby the system regenerates itself, and we thus come to Universe. This tends to be a clearly defined inventory of relative abundance of the various chemical element patterns in Universe which needs a large amount of the pattern hydrogen while apparently not as much of the pattern uranium.

334.00 The significance of Einstein's electromagnetic radiation's *top speed* unfettered in vacuo is that there is a cosmic limit accommodation point of complete regeneration by which our Universe is the only and minimum perpetually self-regenerative system. It is a self-regenerative Universe of fantastic complexities and design of great integrity in which the sum total of running through the total film takes hundreds of billions of years before it accomplishes its remotest re-*wow*.

340.00 Expanding Universe

341.00 In the most comprehensive picture of Universe, we find physical Universe consisting entirely of energy. As the Second Law of Thermodynamics shows, every local physical system continually loses energy to surrounding systems in physical Universe. This loss of energy is called entropy.

Because all the local systems of Universe are in constant motion and transformation, the energies are given off nonsimultaneously in multidirections and with increasing diffusion. The scientists call this the law of the increase of the random element. The increasing random element brings about physical Universe enlargement. These nonsimultaneous enlargements bring about *expanding physical Universe*. The expansion is verified by the astronomers' discovery of the red shifts in remote galaxies.

342.00 Entropy is the measurement of disorder within a closed system. Entropy measures the lack of information about a structure in a system. In Eddington's proof of irreversibility, he dumps a box of wooden matches on the table. Each one splinters the others a little; therefore, there are little hairs and fibers sticking out. They could never be put back in the box the same way without pressing, that is, without investing more energy. This is how the law of the increase of the random element operates. The cycle keeps on time and again, from dust to atoms to proton to neutron. This is what nature is doing in high- and low-pressure pulsations. . . . And then, after maximum dispersal, comes reassociation, because Universe is regenerative. One hundred million years later they will all be back in the box again. After the last *Wow!*

343.00 While energy leaves one local system after another, it does so only by joining other local systems. The energy is always 100 percent accountable. The energies are given off in an ever-increasing diffusion as all the different and nonsimultaneous transformations and reorientations occur. The energies given off alter the environment irreversibly. The biologicals take on and give off more energy than the nonbiologicals.

344.00 Universe expands through progressively differentiating out and multiplying discrete considerations.

345.00 All the differences between de-finite conceptual systems and finite, yet nonconceptual total Universe seem to provide a fundamental means of identifying the physical phenomenon entropy. Entropy no longer means inherent escape of energy from any local system, or decrease of local order, or increase to disorder. Entropy now means the invisible extraction from any local definitive system of the negative conceptual entity; i.e., one negative tetrahedron deposited into Universe balance of energy conservation, permitting the local extraction of any visible, orderly conceptual system. Entropy is not random: it is always one negative tetrahedron. It can account finitely for any discrete rate of energy loss. (See Secs. 620.12 and 625.03.)

350.00 Negative Universe

351.00 Negative Universe is the complementary but invisible Universe.

352.00 Those subsequently isolated elements beyond the 92 prime chemical elements constitute superatomics. They are the non-self-regenerative chemical elements of negative Universe.

353.00 The star tetrahedron may explain the whole negative phase of energetic Universe.

360.00 Universe: System: Conceptuality: Structure

361.00 Universe itself is simultaneously unthinkable. You cannot think about the Universe sum-totally except as a scenario. Therefore, for further examination and comprehension, you need a thinkable set, or first subdivision of Universe, into systems.

362.00 Our original definition of Universe is a finite but nonsimultaneously occurring aggregate of all human experiences, which is, therefore, a nonconceptual total Universe. It is logical to proceed from this definition to discover the patterning characteristics of the first conceptual subdivision of Universe into a structural system. After we subdivide Universe into systems, we will make further reductions into basic event experiences and to quantum units. We will then come to the realization that all structuring can be identified in terms of tetrahedra and of topology. (See Sec. 603.01.)

363.00 The Principle of Synergetic Advantage requires that we proceed from macro to micro. Definition requires conceptuality. Conceptuality requires the generalization of patterns gleaned from special-case experiences. Systems are geometrically definable. Self-stabilizing systems are structures. Conceptuality defines the basic event experiences and quantum unit measurement, which altogether constitute structure.

400.00 System

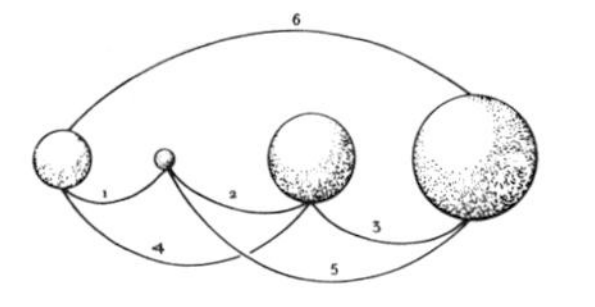

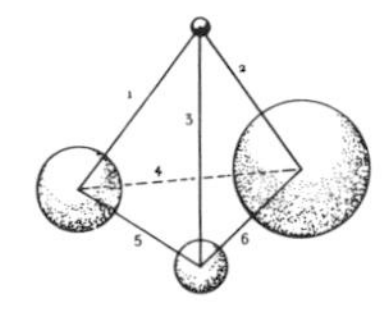

400.01 Definition: System

400.011 A system is the first subdivision of Universe. It divides all the Universe into six parts: first, all the universal events occurring geometrically outside the system; second, all the universal events occurring geometrically inside the system; third, all the universal events occurring nonsimultaneously, remotely, and unrelatedly prior to the system events; fourth, the Universe events occurring nonsimultaneously, remotely, and unrelatedly subsequent to the system events; fifth, all the geometrically arrayed set of events constituting the system itself; and sixth, all the Universe events occurring synchronously and or coincidentally to and with the systematic set of events uniquely considered.

400.02 A system is the first subdivision of Universe into a conceivable entity separating all that is nonsimultaneously and geometrically outside the system, ergo irrelevant, from all that is nonsimultaneously and geometrically inside and irrelevant to the system; it is the remainder of Universe that conceptually constitutes the system's set of conceptually tunable and geometrical interrelatability of events.

400.03 Conceptual tuning means occurring within the optical "rainbow" range of human's sensing within the electromagnetic spectrum and wherein the geometrical relationships are imaginatively conceivable by humans independently of size and are identifiable systematically by their agreement with the angular configurations and topological characteristics of polyhedra or polyhedral complexes.

400.04 All systems are polyhedra. Systems having insideness and outsideness must return upon themselves in a plurality of directions and are therefore interiorally concave and exteriorally convex. Because concaveness reflectively concentrates radiation impinging upon it and convexity diffuses radiation impinging upon it, concavity and convexity are fundamentally different, and therefore every system has an always and only coexisting inward and outward functionally differentialed complementarity. Any one system has only one insideness and only one outsideness.

400.05 In addition to possessing inherent insideness and outsideness, a system is inherently concave and convex, complex, and finite. A system may be either symmetrical or asymmetrical. A system may consist of a plurality of subsystems. Oneness, twoness, and threeness cannot constitute a system, as they inherently lack insideness and outsideness. Twoness constitutes wavilinear relatedness. Threeness constitutes planar relatedness, which is inherently triangular. Three triangular planes alone cannot differentiate, distinguish, or constitute a system. At minimum, it takes four triangular planes having inherent fourness of vertexes to constitute differential withinness and withoutness. Fourness of geometrically contiguous and synchronous event foci and their coincidentally defined four triangular planes, along with their six common edges provided by the six wavilinear vectors connecting the four event foci, altogether inherently differentiate, distinguish, initially institute, and constitute prime or minimum withinness and withoutness.

400.06 Thought is systemic. Cerebration and intellection are initiated by differential discernment of relevance from nonrelevance in respect to an intuitively focused-upon complex of events which also intuitively suggests inherent and potentially significant system interrelatedness.

400.07 Human thoughts are always conceptually and definitively confined to system considerability and comprehension. The whole Universe may not be conceptually considered by thought because thinkability is limited to contiguous and contemporary integrity of conformation of consideration, and Universe consists of a vast inventory of nonsynchronous, noncontiguous, noncontemporary, noncoexisting, irreversibly transforming, dissimilar events.

400.08 *Unit* means system integrity. *Organic* means regenerative system integrity. As minimum or prime systems consist of four event foci and their always and only coexisting fourness of triangularly defined planar facets, along with their sixness of a wavilinearly defined minimum set of unique componentation relatedness, unity is inherently plural. Unity is plural. A system is a local phenomenon in the Universe. Each of the conceivable or imaginable awareness or thinkability entities or phenomena inducing or producing onenesses or twonesses are subvisible and potentially further subdivisible, or as yet unresolved, ergo unrecognized systems. Functions always and only cooccur as subsystem relativistics, characteristics, inherencies, and proclivities. Functions occur only as parts of systems. Universe is constituted of a complex plurality of nonsimultaneous and only partially overlappingly occurring systems, not one system.

400.09 All the interrelationships of system foci are conceptually representable by vectors (see Sec. 521). A system is a closed configuration of vectors. It is a pattern of forces constituting a geometrical integrity that returns upon itself in a plurality of directions. Polyhedral systems display a plurality of polygonal perimeters, all of which eventually return upon themselves. Systems have an electable plurality of view-induced polarities. The polygons of polyhedra peregrinate systematically and sometimes wavilinearly around three or more noncongruent axes.

400.10 Absolutely straight lines or absolutely flat planes would, theoretically, continue onwardly or spread areally outward to infinity. The difference between infinity and finity is governed by the taking out of angular sinuses, like pieces of pie cut out of surface areas around a point in an otherwise absolute and infinitely extendable plane, and joining together the open gap's radial edges. This is the way lampshades and skirts are made. Joining the sinused fan-edges together makes a cone. If two cones are made and their respective open circle edges are brought together, a finite or closed system results. It has two poles and two polar domains. The two poles and their polar cone surface domains, as well as the defined insideness and outsideness, are inherent and primary characteristics of all systems.

400.11 All systems are continually importing as well as exporting energy. Physics has found only myriad pattern integrities of comprehensively nonsimultaneous and only partially overlapping evolution; of disintegrative "heres" and reintegrative "theres," which are omnilocal vari-intertransformabilities of limited duration identities of an apparently eternal, physical Universe regenerating mathematically treatable energy quanta.

400.20 **Comprehensibility of Systems:** All systems are subject to comprehension, and their mathematical integrity of topological characteristics and trigonometric interfunctioning can be coped with by systematic logic.

400.21 A system is the antithesis of a nonsystem. A nonsystem lacks omnidirectional definition. Nonsystems such as theoretical planes or straight lines cannot be found experimentally. We are scientifically bound to experientially discovered and experimentally demonstrable systems thinking.

400.22 General systems theory treats with phenomena that are holistically comprehensible. The objects of our experience are finite systems. Their superficial outlines close back upon themselves multidirectionally as a systematic continuity of relevantly contiguous events.

400.23 Maximum system complexity consists of a dissimilarly quantified inventory of unique and nonintersubstitutable components. That is, Euler's irreducible-system aspects of *vertexes, areas,* and *edges* exhibit the respective dissimilar quantities *4, 4,* and *6* in the minimum prime system, the tetrahedron. This demonstrates the inherent synergy of all systems, since their minimum overall inventory of inherent characteristics is unpredicted and unpredictable by any of the parts taken separately. Systems are unpredicted by oneness, twoness, or threeness. This explains how it happens that general systems theory is a new branch of science. (See Sec. 537.30.)

400.24 General systems theory is another example of evolution by inadvertence. It developed fortuitously to accommodate the unprecedented and vastly complex undertakings of the late twentieth century, such as the 10 million separate and only partially overlapping "critical path" tasks that had to be accomplished and tested to foolproof reliability en route to countdown to eventual blastoff, Moon landing, and safe return to Earth, which found all conventional mathematical theory wanting. It required the development of the computer and star-focused instruments and computer programming arts together with operational research, which guess-improvises the inventory of parameter of variables that must be progressively programmed into the system in order further to reduce the magnitude of tolerated errors consequent to trial "bird" (rocket vehicle) "flight" (trajectory) control as the vehicles are progressively zeroed-in to progressive target rendezvous with celestial entities. Neither differential and integral calculus, nor "probability" statistics, nor any branch of specialized hard science has accredited synergy as an a priori assumption. General systems theory, which recognizes synergy as inherent, was discovered and named by the biologically inspired Ludwig von Bertalanffy.

400.25 Every system, as a subdivision of the total experience of Universe, must accommodate traffic of inbound and outbound events and inward-outward relationships with other systems' aspects of Universe. Effective thinking is systematic because intellectual comprehension occurs only when the interpatternings of experience events' star foci interrelationships return upon themselves. Then the case history becomes "closed." A system is a patterning of enclosure consisting of a conceptual aggregate of recalled experience items, or events, having inherent insideness, outsideness, and omniaroundness.

400.26 Systems are aggregates of four or more critically contiguous relevant events having neither solidity nor surface or linear continuity. Events are systemic.

400.30 Tiger's Skin: Typical of all finitely conceptual objects, or systems, the tiger's skin can be locally pierced and thence slotted open. Thereafter, by elongating the slot and initiating new subslots therefrom in various directions, the skin gradually can be peeled open and removed all in one piece. Adequate opening of the slots into angular sinuses will permit the skin to lie out progressively flat. Thus, the original lunar gash from the first puncture develops into many subgashes leading from the original gash into any remaining

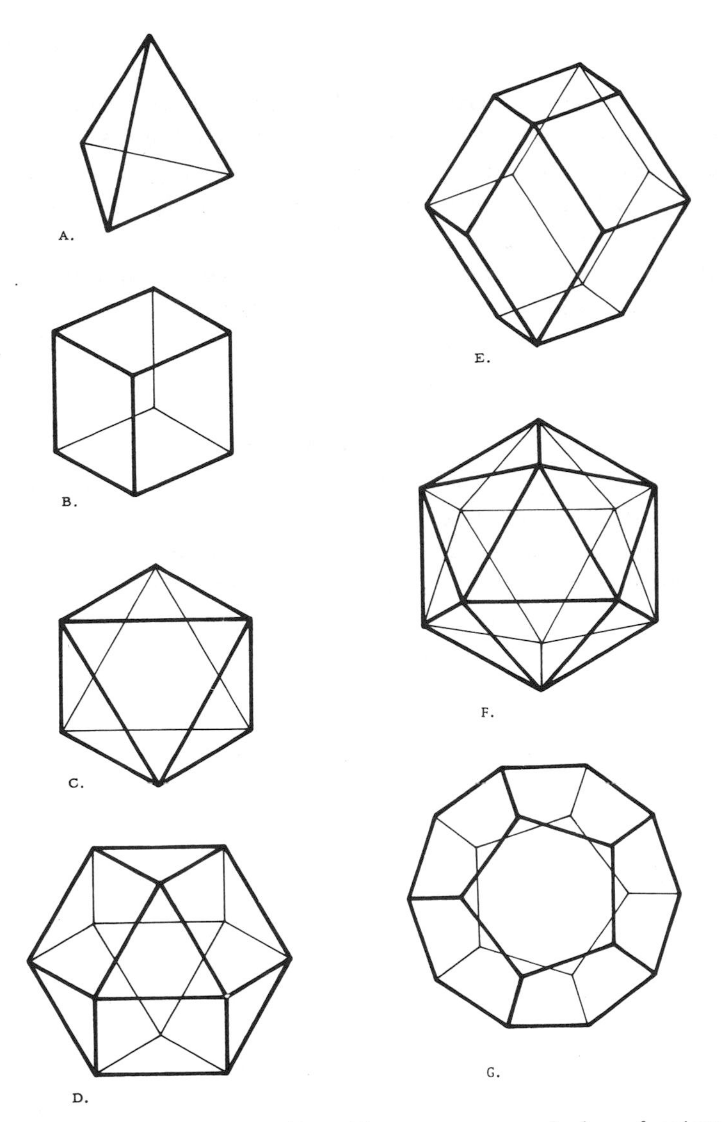

Fig. 400.30 *Topological relationships of faces, vertexes, and edges of various polyhedra:*

A. Tetrahedron: 4 faces, 4 vertexes, 6 edges.
B. Cube: 6 faces, 8 vertexes, 12 edges.
C. Octahedron: 8 faces, 6 vertexes, 12 edges.
D. Vector Equilibrium (cuboctahedron): 14 faces, 12 vertexes, 24 edges.
E. Rhombic dodecahedron: 12 faces, 14 vertexes, 24 edges.
F. Icosahedron: 20 faces, 12 vertexes, 30 edges.
G. Pentagonal dodecahedron: 12 faces, 20 vertexes, 30 edges.

Euler's topological formula is: $f + (v - 2) = e$, or $f + v = e + 2$. In any system, two vertexes may be considered polarized. These vertexes are then subtracted to balance the equation. This suggests the inherent twoness of Universe.

domical areas of the skin. The slitting of a paper cone from its circular edge to its apex allows the paper to be laid out as a flat ''fan'' intruded by an angular sinus. A sinus is the part of an angle that is *not* the angle's diverging sides. *Sinus* means in Latin a ''withoutness''—an opening out—a definitively introduced ''nothingness.''

400.31 The surface contour of any object or system—be it the skin of a complex creature such as a crocodile, or the skin of a simple prune, or a sugar-cube wrapping, or a dodecahedron, or any formal angular polyhedra—can thus be ''skinned'' and laid out flat.

400.40 **Finiteness of Systems:** Definition: Single systems occurring initially and minimally as four synchronously related event foci—ergo, inherently as tetrahedra, regular or irregular—are omnitriangulated and may be either symmetrical or asymmetrical. In single symmetrical systems, all the vertexes are equidistant radially from their common volumetric centers, and the centers of area of all their triangular facets are also equidistant from the system's common volumetric center.

400.41 The minimum single symmetrical system is the regular tetrahedron, which contains the least volume with the most surface as compared to all other symmetrical single systems. There are only three single symmetrical systems: the regular tetrahedron, with a ''unit'' volume-to-skin ratio of 1 to 1; the regular octahedron, with a volume-to-surface ratio of 2 to 1; and the regular icosahedron, with a volume-to-surface ratio of 3.7 to 1. Single asymmetrical systems contain less volume per surface area of containment than do symmetrical or regular tetrahedra. The more asymmetrical, the less the volume-to-surface ratio. Since the structural strength is expressed by the vector edges, the more asymmetrical, the greater is the containment strength per unit of volumetric content.

400.42 Since the minimum system consists of two types of tetrahedra, one symmetrical (or regular) and the other asymmetrical (or irregular); and since also the asymmetrical have greater enveloping strength per units of contained event phenomena, we will differentiate the two minimum-system types by speaking of the simplest, or minimum, single symmetrical system as the *mini-symmetric* system; and we will refer to the minimum asymmetric system as the *mini-asymmetric* system. And since the mini-symmetric system is the regular tetrahedron, which cannot be compounded face-to-face with other unit-edged symmetric tetrahedra to fill allspace, but, in order to fill allspace, must be compounded with the tetrahedron's complementary symmetrical system, the octahedron, which is not a minimum system and has twice the volume-to-surface ratio of the tetrahedron of equal edge vector dimension; and since, on the other hand, two special-case minimum asymmetric tetrahedra, the *A* Quanta Modules and the *B* Quanta Modules (see Sec. 920), have equal volume and may be face-compounded with one another to fill allspace, and are uniquely the highest common volumetric multiple of allspace-filling; and since the single asymmetrical tetrahedron formed by compounding two symmetrical tetrahedral *A* Modules and one asymmetrical tetrahedral *B* Module will compound with multiples of itself to fill all positive space, and may be turned inside out to form its noncongruent negative complement (which may also be compounded with multiples of itself to fill all negative space), this three-module, minimum asymmetric (irregular) tetrahedral system, which accommodates both positive or negative space and whose volume is exactly $^1/_8$ that of the regular tetrahedron; and exactly $^1/_{32}$ the volume of the regular octahedron; and exactly $^1/_{160}$ the volume of the regular vector equilibrium of zero frequency; and exactly $^1/_{1280}$ the volume of the vector equilibrium of the initial of all frequencies, the integer 2, which is to say that, expressed in the omnirational terms of the highest common multiple allspace-filling geometry's *A* or *B* Modules, the minimum realizable nuclear equilibrium of closest-packing symmetry of unit radius spheres packed around one sphere—which is the vector equilibrium (see Sec. 413)—consists of 1,280 *A* or *B* Modules, and $1{,}280 = 2^8 \times 5$.

400.43 Since the two-*A*-Module, one-*B*-Module minimum asymmetric system tetrahedron constitutes the generalized nuclear geometrical limit of rational differentiation, it is most suitably to be identified as the prime minimum rational structural system: also known as the MITE (see "Modelability," Sec. 950). The MITE is the mathematically demonstrable microlimit of rational fractionation of both physically energetic structuring and metaphysical structuring as a single, universal, geometrically discrete system-constant of quantation. The MITE consists of two *A* Modules and one *B* Module, which are mathematically demonstrable as the minimum cosmic volume constant, but not the geometrical shape constant. The shape differentiability renders the volume-to-surface ratio of the *B* Modules more envelopingly powerful than the volume-to-surface ratio of the *A* Modules; ergo, the most powerful local-energy-impounding, omnirationally quantatable, microcosmic structural system.

400.44 The MITE may be turned inside out by having each of its two *A* Modules and one *B* Module turn themselves inside out and recombine to fill all negative space. It is also to be observed that one all-negative-space-filling and one all-positive-space-filling MITE may be face-associated structurally to produce yet another single minimum system asymmetric tetrahedral, all-*positive-and-negative* space filler whose modular volumetric unity value of six corresponds with the sixness of vectorial edges of the minimum system's tetrahedral four foci event relationships.

400.45 It is characteristic of a single prime system that the aggregate of angles convergent around its vertexes must be concave or convex with respect to the position from which they are viewed—concave if viewed from the inside, convex when viewed from outside.

400.46 There are in all systems the *additive twoness* of the poles and the *multiplicative twoness* of the coexistent concavity and convexity of the system's insideness and outsideness.

400.47 Planet Earth is a system. You are a system. The "surface," or minimally enclosing envelopmental relationship, of any system such as the Earth is finite.

400.50 **Other Characteristics of Systems:** *Prime Rational Integer Characteristics:* Electromagnetic frequencies of systems are sometimes complex but always exist in complementation of gravitational forces to constitute the prime rational integer characteristics of physical systems.

400.51 Systems may be symmetrical or asymmetrical.

400.52 Systems are domains of volumes. Systems can have nuclei, and prime volumes cannot.

400.53 **Interconnection of Systems:** If two adjacent systems become joined by one vertex, they still constitute two systems, but universally interjointed. If two adjacent systems are interconnected by two vertexes, they remain two systems, interlocked by a hinge. If two adjacent systems become adjoined by three vertexes, they become one complex system because they have acquired unit insideness and outsideness.

400.54 If two adjacent systems are interpositioned with their respective centers of volume congruent and all their respective vertexes equidistant from their common center of volumes, they become one system. If their respective vertexes are at different distances radially from their common center of volumes, they become one complex system. If the complex system's respective interpositioned systems are all symmetric, then they become one complex symmetric system.

400.60 **Motion of Systems:** Systems can spin. There is at least one axis of rotation of any system.

400.61 Systems can orbit. Systems can contract and expand. They can torque; they can turn inside out; and they can interprecess their parts.

400.62 Systems are, in effect, spherical gears. Their internal-external pulsating and rotating "teeth" consist in reality of both circumferential and radial waves of various frequencies of subdivision of spherical unity. They often fail to mesh with other local systems. Some of them mesh only in special aspects. The universally frequent nonmeshing of geometrical sizes and rates of wavelengths and frequencies produces an omnicondition in which the new system's center, as each is created, must continually occupy an omnidirectionally greater domain.

401.00 Twelve Vectors of Restraint, Six Positive and Six Negative, Define Minimum System (*See drawings section.*)

401.01 At the top of Illus. 401.01 (see also Illus. 401.00 in drawings section), we see something like a ping-pong ball attached to a string. The ping-pong ball represents *me,* and the string is gravity (or mass attraction), with its other end tethered to a *point* that represents all the rest of Universe that isn't me. Because of mass attraction, the one thing that I cannot do is escape absolutely from Universe. I may try to exert terrific acceleration and shoot out through a hole in the Galaxies, beyond the Pleiades, where the stars seemingly are so scarce that as I look back from fantastically far out, all the stars of Universe seem to be collected in approximately one bright spot. The single string of the model, long though it may be, represents the combined mass attraction exerted upon me by all the stars of Universe.

401.02 Tetherball: There is an old game called tetherball, played by tennis players lacking a tennis court. There is a tennis ball fastened powerfully to a strong, slender cord 19 feet long suspended from the top of a pole 22 feet in height above the ground level. There is a circular marker on the pole at the 11′ 4″ height. The server bats the ball in a clockwise circumferential direction around the pole, attempting to wind its cord completely around the pole above the 11′ 4″ mark. The opponent can intercept and attempt to wind the ball counterclockwise. Obviously, a tethered ball on a long string is free to describe any omnigeometric forms of circles, spheres, or giraffes, but it cannot get away from the Universe. This is called *one restraint:* the fundamental "otherness" essential to initial "awareness" of the observer. (Nothing to observe: no awareness: only nothingness.) Otherness always imposes a minimum of one restraint, weak though it may be, on all awareness, which is the beginning of "Life."

401.03 But the imagined experience of cosmically long journeys teaches me that the possibility of finding such a hole in the celestial myriadicity to attain such a unified paralactic bunching of all island nebulae is a futile search. Therefore, I resign myself to acknowledgment of at least two a priori restraints that inescapably affect my relative cosmic freedom. Hoping to save myself vast cosmic time, and accepting my present position in Universe, I try to process the known data on the mass dimensions of all the known stars and

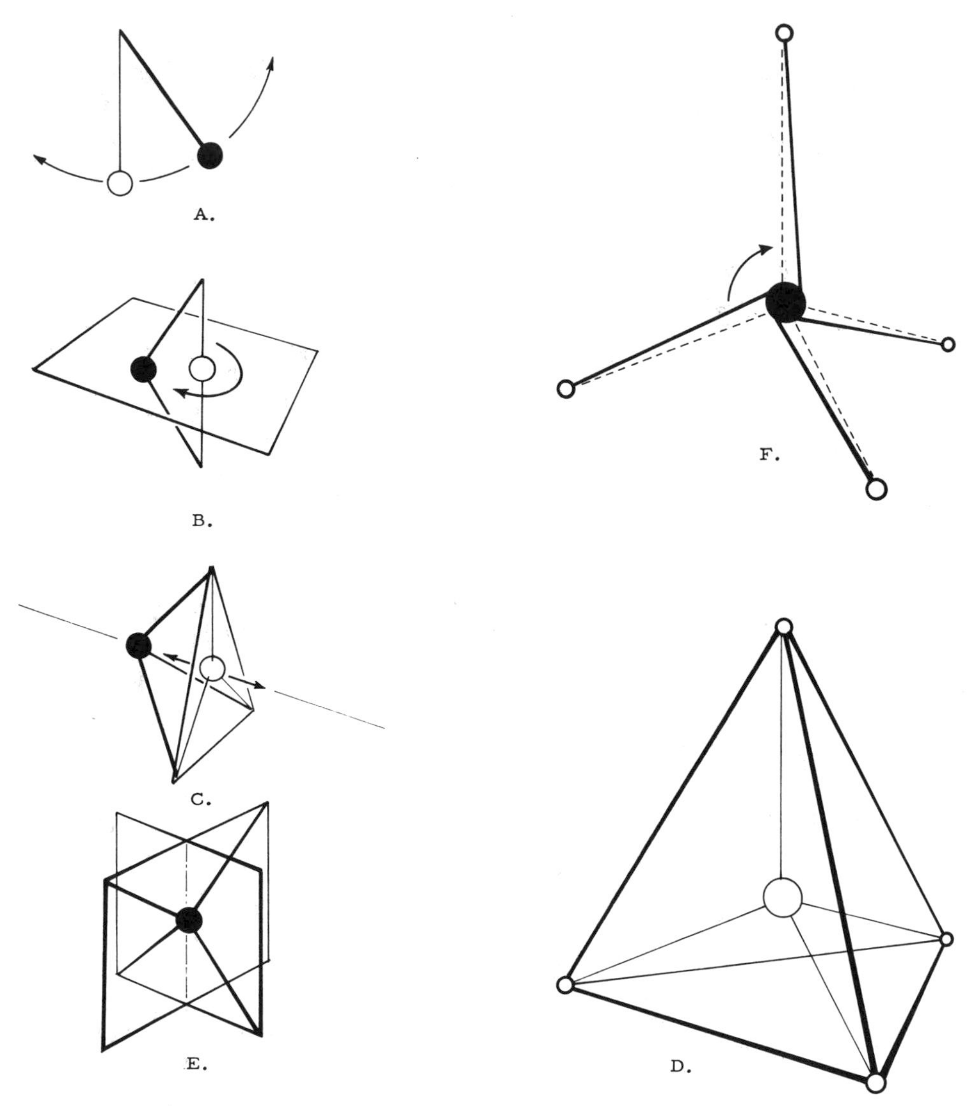

Fig. 401.01 *Four Vectors of Restraint Define Minimum System:*

Investigation of the requirements for a minimum system.

A. One vector of restraint allows ball to define complete sphere—a three-dimensional system.
B. Two vectors: a plane—a two-dimensional system.
C. Three vectors: a line—a one-dimensional system.
D. Four vectors: a point—no displacement.
E. Note the possibility of turbining with the position otherwise fixed by the four vectors of restraint.
F. The four vectors define the tetrahedron: the first identifiable ''system''—a primary or minimum subdivision of Universe. The ball lies at the center of gravity within the tetrahedron.

try to divide them all into two opposite hemispherical teams—those to my right and those to my left. Then assuming all the right-hand hemisphere group's mass attraction to be accumulatively resolved into one mass-attractive restraint tied to my right arm's wrist, and all the other cosmic hemisphere's equal tensions tied to my left arm's wrist, I find myself used like a middleman in a ropeless tug of war, liable to have my arms pulled out of my armpits. So I tie both the tension lines around my waist. Now I am in the same dynamic situation as a ping-pong ball suspended in the middle of a one-string fiddle. Because all strings, no matter how tautly strung, can still vibrate, I can still move. But I find that with two restraints I can move about in circles, cloverleafs, or figures-of-eight, but always and only in a plane that is perpendicular to the string of which I am in the middle.

401.04 Now I conclude that the various motions of the stars make it illogical to assume any persistence of the two hemispherical star sphere groupings. The star accelerations produce the inertial advantage of awayness to which my mass-attraction tethers were attached. I therefore conclude that it is more probable that such dynamic inertia will persist in three groups. Now I have three restraints, and the ping-pong ball "me" acts as if it were in the middle of a drumhead, or as tensilely suspended at the center of area of a triangle by three strings fastened at the triangle's corners. The ping-pong ball "me" can still move, but only in a line perpendicular to the plane of the drumhead or web triangle. I am constrained by three converging lines as I oscillate to and fro between the opposite apexes of two dynamically described, base-to-base, positive or negative tetrahedra formed by the resonating drumhead's terminal oscillations.

401.05 With only one restraint, the ball was moving omnidirectionally or multidimensionally. With two restraints, it was moving in a plane; with three restraints, it moves only in a line. I now conclude that it is more probable that I can concentrate all the restraints operating upon me from all the stars because of the multidirectional pull of all the stars actually pulling me. I conclude that there is much redundancy but that four restraints is closer to a matter of reality than three restraints. When we attach a fourth restraint perpendicular to the center of the drumhead and pull it only in the "fro" direction, the ping-pong ball "me" seems at last to be immobilized. With four restraints the ping-pong ball "me" can no longer move either toward or away from any other parts of the Universe. But the ball can twist locally, that is, it can rotate in place around an axis, and that axis itself can incline at many angles, as does the gyroscope top, without alteration of its volumetric center position in respect to the four vertexial star groups. Because the vectors are coming together in nonequilateral quadrangles, i.e., in trapezoids, the restraints are not intertriangulated, and we have learned experimentally that

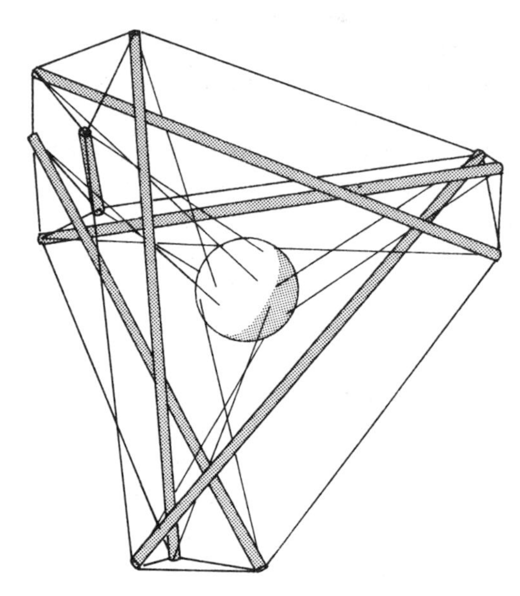

Fig. 401.05 The six compression members are the acceleration vectors trying to escape from Universe at either end by action and reaction, while the ends of each would-be escapee are restrained by three tensors; while the ball at the center is restrained from local torque and twist by three triangulated tensors from each of the four corners tangentially affixed.

only triangles are stable. (See Sec. 610, Triangulation.) Therefore, it is possible for the ball to "turbine," rotate, and precess locally in place without altering the geometrical position at volume center of the celestial tetrahedron from whose four corners the four vectors of restraint were imposed. The six edges of the celestial spherical tetrahedron represent the three mass-attraction restraints imposed on each of the tetrahedron's four corner mass centers as each being in normal acceleration is precessionally restrained from exiting from Universe. Each of the four corners' group massiveness is restrained by all three of the other tetrahedral corner mass centers. Any one of the mass-moment acceleration tendencies to part company with the others is overpowered three-to-one by the three others. Thus the cohering integrity of Universe is manifest to us by consideration of the celestial advantage points from which our four central restraints were mounted. Though the ping-pong "me" ball can be twisted and torqued in place, it cannot be moved from its tetrahedral center position. To prevent local in-position twist and torque, each of the four corner tensional restraints will have to be multiplyingly replaced by three restraints, all springing from three external points at each of the four tetrahedral corners; and each of the three tensions from any one of the four corners must cross the others triangularly and be attached tangentially to the ball at the center. These 12 now completely restrain any motion of the central ball in relation to the other four.

401.06 The purpose of our investigation was to find the requirements of a minimum system. Our experimental model demonstrates that it takes four vectors to define a point with the ping-pong ball at the center of gravity and center of volume of the regular tetrahedron. It takes 12 such vectors to both position and locally immobilize. It takes six external push vectors and six external pull vectors to define the minimum nuclear structural system: a primary subdivision of Universe. To summarize, the celestial tetrahedron has six positive and six negative internal vectors and six positive and six negative external vectors.

401.07 Four external ''star'' foci effecting complete immobilization of the ''me'' ball are the same four event foci that we learned earlier (Sec. 400.05) always constitute the minimum number of events necessary to define the insideness and outsideness of a system.

402.00 *Tetrahedron as System*

402.01 The tetrahedron as a real system consists of one concave tetrahedron and a second convex tetrahedron, plus a third tetrahedron for all the Universe outside the system-as-tetrahedron, and a fourth tetrahedron complementarily accounting for all Universe inside the system-as-tetrahedron. All the angles are the same on the inside as on the outside.

402.02 A tetrahedron is a triangularly faceted polyhedron of four faces. It is unique as a system, for it is the minimum possible system.

403.00 *Stable and Unstable Systems*

403.01 There are stable systems and unstable systems. (For a discussion of stable and unstable structures, see Sec. 608, Stability: Necklace.)

403.02 Conceptuality: Unstable systems are conceptual as momentary positional relationships of unstructured-component event aggregates; for example, amongst the stars comprising the Big Dipper—in Ursa Major—the second and third stars in the dipper's handle are, respectively, 100- and 200-light-years away from Earth and, though seemingly to us in the same plane, are not all so; and they are both moving in opposite directions and so in due course they will no longer seem to be in the same constellation. In the same way, four airplanes flying in different directions may be within visible range of one another, but are far too remote for mass inter-attraction to become critical and pull them into one another. Stable systems are conceptual as structured, which means componently omni-intertriangulated critical-proximity, interrelevant, coordinate, constellar event aggregates.

403.03 **Generalized Principles:** If the only momentary and optically illusory system consideration proves to be unstable, it does not manifest generalized principle. If systems are stable, they are inherent in and accommodate all generalized principles.

410.00 Closest Packing of Spheres

410.01 Nature's Coordination

410.011 About 1917, I decided that nature did not have separate, independently operating departments of physics, chemistry, biology, mathematics, ethics, etc. Nature did not call a department heads' meeting when I threw a green apple into the pond, with the department heads having to make a decision about how to handle this biological encounter with chemistry's water and the unauthorized use of the physics department's waves. I decided that it didn't require a Ph.D. to discern that nature probably had only one department and only one coordinate, omnirational, mensuration system.

410.02 I determined then and there to seek out the comprehensive coordinate system employed by nature. The omnirational associating and disassociating of chemistry—always joining in whole low-order numbers, as for instance H_2O and never $H_\pi O$—persuaded me that if I could discover nature's comprehensive coordination, it would prove to be omnirational despite academic geometry's fortuitous development and employment of transcendental irrational numbers and other "pure," nonexperimentally demonstrable, incommensurable integer relationships.

410.03 I was dissatisfied with abstract, weightless, unstable, ageless, temperatureless, straight-line-defined squares and cubes as models for calculating our omnicurvilinear experience. I was an early rebel against blockheads and squares. Reviewing the history of chemical science, I became intuitively aware that the clue to vectorial, volumetric, geometrical coordination with physical reality and all the fundamental energetic experiences of reality, such as temperature, time, and force, might be found in Avogadro's experimental proof of his earlier hypothesis, which stated that *all gases under identical conditions of heat and pressure will always disclose the same number of mol-*

ecules per given volume. Here was disclosed a ''Grand Central Station'' accommodating all comers; despite ''fundamental'' or elementarily unique differences of identity, all accommodated on a common volume (space)-to-number basis. One molecule of any element: One space. A cosmic democracy.

410.04 I felt intuitively that inasmuch as the variety of gases experimented with often consisted of only one unique chemical element, such as hydrogen or oxygen, and that inasmuch as these gases also could be liquefied, and also inasmuch as most of the elements are susceptible to some heat- or pressure-produced transformation between their liquid, crystalline, and vapor, or incandescent, states, it might also be hypothetically reasonable to further generalize Avogadro's hypothesis by assuming that, *under identical energy conditions,* all elements may disclose the same number of ''somethings'' per given volume. Such a generalized concept is not limited to pressure and heat: we wanted to be much more inclusive, so we said we assume that *all the conditions of energy are identical;* this includes not only the *pressure* and *heat* conditions of thermodynamics, which developed before electromagnetics became an applied realization, but also the conditions of electromagnetics as constant.

410.05 I went on from there to reason that vectors, being the product of physical energy constituents, are ''real,'' having *velocity* multiplied by *mass* operating in a specific direction; velocity being a product of *time* and *size;* and mass being a volume-weight relationship. On impact, *mass* at *velocity* transforms into *heat* and *work.* These energy factors can be translated not only into work, but into heat or into time as well. Furthermore, electromagnetic scientists had found that all their EMF (electromotive force) problems could be graphed vectorially; the fact that ''graphable'' or ''modelable'' vectors can interact modelably in real Universe space seemed to promise that the equations of nature's omnicoordinate transactions, expressed in omni-space-intruding vectorial models, might produce real models of reality of nature's Grand Central Station of omnicoordination.

410.06 So I then went on to say that, *if all the energy conditions were everywhere the same, then all the vectors would be the same length and all of them would interact at the same angle.* I then explored experimentally to discover whether this ''isotropic vector matrix,'' as so employed in matrix calculus, played with empty sets of symbols on flat sheets of paper, could be *realized* in actual modeling. Employing equilength toothpicks and semi-dried peas, as I had been encouraged to do in kindergarten at the age of four (before receiving powerful eyeglasses and when I was unfamiliar with the right-angled structuring of buildings as were the children with normal vision), I fumbled tactilely with the toothpicks and peas until I could feel a stable structure, and thus assembled an omnitriangulated complex and so surprised the teachers that their exclamations made me remember the event in detail. I thus rediscovered the octet truss whose vertexes, or convergent foci, were all sixty-degree-angle interconnections, ergo omniequilateral, omniequiangled, and omni-intertriangulated; ergo, omnistructured. Being omnidirectionally equally interspaced from one another, this omni-intertriangulation produced the isotropic matrix of foci for omni-closest-packed sphere centers. This opened the way to a combinatorial geometry of closest-packed spheres and equilength vectors.

410.07 Over and over again, we are confronted by nature obviously formulating her structures with beautiful spherical agglomerations. The piling of oranges, coconuts, and cannonballs in tetrahedral or half-octahedral pyramids has been used for centuries and possibly for ages. Almost a half-century ago, F. W. Aston, the British scientist first identified for physics the most economical uniradius spherical interagglomerations as the "closest packing of spheres," which had fresh interest for the physicist and crystallographer because of the then recently discovered microscopic realization that nature frequently employed omni-intertriangulated systems, which hold mathematical clues to the principles of symmetrical coordination governing natural structure, the dynamic vectorial geometry of the atomic nucleus, as well as of the atoms themselves.

410.10 Omnitriangulation of Sixty Degrees

410.11 The closest packing of unit radius spheres always associates in omnitriangulations of 60 degrees, whether in planar or omnidirectional arrays. Six unit radius spheres pack most tightly around one sphere on a billiard table. Twelve unit radius spheres compact around one sphere in omnidirectional closest packing.

410.12 If we take three billiard balls on a flat table, we find that they compact beautifully into a triangle. If we arrange four of them on a billiard table in a square, they tend to be restless and roll around each other. If compacted into a condition of stability, the four form a 60-degree-angled diamond shape made of two stable triangles.

411.00 Four Spheres as Minimum System

411.01 The South Seas islander piling his coconuts, the fruit dealer selling oranges, and the cannoneer stacking equiradius cannonballs, or the much earlier round-rock-slinging soldiers were probably the first to learn about the closest packing of spheres. The stacking of balls in symmetrical rows and layers leads inevitably to a stable pyramidal aggregate.

411.02 Closest packing of spheres does not begin with a nucleus. Closest packing begins with two balls coming together.

411.03 One ball cannot zoom around alone in Universe. Without otherness, there is no consciousness and no direction. If there were only one entity—say it is a sphere called "*me*"—there would be no Universe: no otherness: no awareness: no consciousness: no direction. Once another entity—let's say a *sphere*—is sighted, there is awareness and direction. There is no way to tell how far away the other sphere may be or what its size may be. Size sense comes only with a plurality of comparative experiences.

411.04 As a single sphere, now aware of an otherness sphere somewhere out there, "*me*" has to rotate about without restraint and can observe its rotation in relation to the otherness, but could misassume the otherness sphere to be zooming around it, as there are no third, or more, othernesses by which to judge.

411.05 When moved unknowingly toward another by mass attraction, the "other" ball and the "*me*" ball, either or both of which could have been rotating as they approach one another, each misassumes the other to be growing bigger, until finally they touch each other. Now they can roll around upon one another, and they might be cotraveling together, but there is not as yet anywhere to go because there is no otherness than their joint selves relative to which to travel. They can't go through each other and they can't get away from each other. They are only free to rotate upon each other, and because of friction they must do so cooperatively.

411.06 When a third ball looms into sight, providing a sense of direction for the tangently rolling-upon-each-other first couple, the third one and the first two are mass-attracted toward each other and finally make contact. The newcomer third sphere may roll around on one of the first couple until it rolls into the valley between the first two. The third ball then gets locked into the valley between the first two by double mass attraction, and now becomes tangent to both of the first two.

411.07 The three balls, each one tangent to both of the others, now form an equiangular triangular group with a small opening at their center. The friction of each of their double contacts with the other two gives them a geartooth interaction effect. With two gears, one can turn clockwise and the other can turn counterclockwise. Even numbers of gears reciprocate; odd numbers of gears block one another. Thus our three balls can no longer roll circumferentially around each other; they can only rotate cooperatively on the three axes formed between each of their two tangent contact points. The friction be-

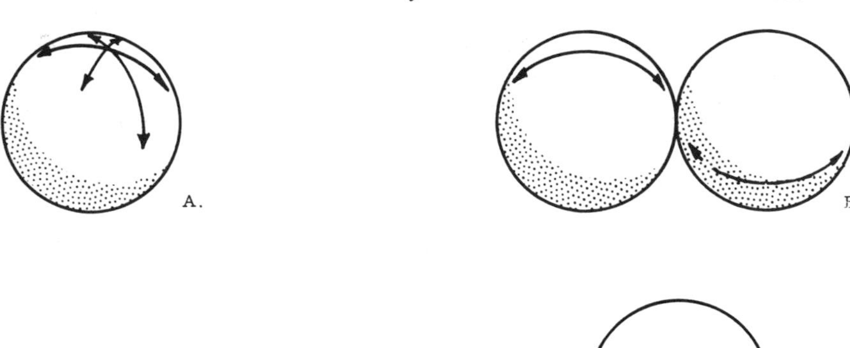

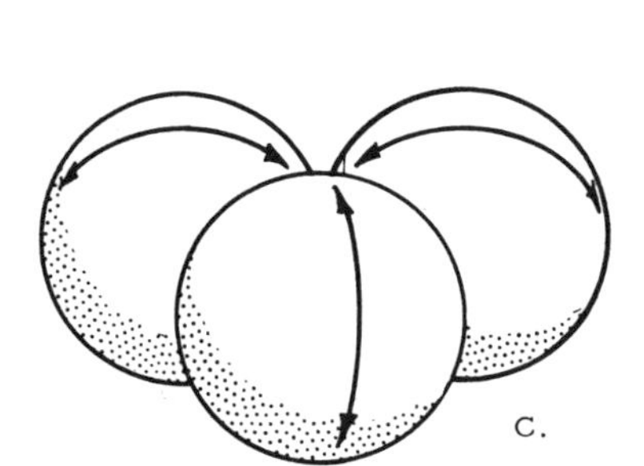

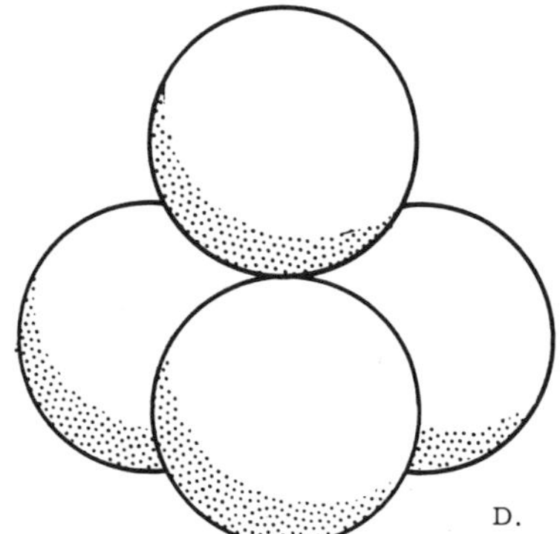

Fig. 411.05 *Four Spheres Lock as Tetrahedron:*

A. A single sphere is free to rotate in any direction.

B. Two tangent spheres although free to rotate in any direction must do so cooperatively.

C. Three spheres can rotate cooperatively only about respective axes which are parallel to the edges of the equilateral triangle defined by joining the sphere centers, i.e. each sphere rotates toward the center of the triangle.

D. Four spheres lock together. No rotation is possible, making the minimum stable closest-packed-sphere system: the tetrahedron.

tween their surface contacts forces all three to rotate unidirectionally about their respective contact axes, which are parallel to the edges of the equilateral triangle defined by the three sphere centers; i.e., the three spheres can now only co-rotate *over and into* the hole at the center of the triangle, and *out and away* upwardly again from the bottom of the center hole. Thus the three balls can involute or evolute axially, like a rubber doughnut in respect to the hole at their triangular center, but they cannot rotate circumferentially.

411.08 Finally, a fourth ball appears in Universe and is mass-attracted to tangency with one of the three previously triangulated spheres, then rolls on one of the joined balls until it falls into the valley formed by the hole at the top center of the triangulated three; but being of equal diameter with the other three balls, it cannot fall through that hole whose radius is less than those of the associated spheres. Thus nested in the central valley, the fourth sphere now touches each of the other three and vice versa. The four closest-packed spheres make a closest-packing array. They are mass-attractively locked together as a tetrahedron.* No further interrotation is possible. As a tetrahedron, they form the minimum *stable structural system* and provide the nuclear matrix for further mass attraction and closest-packed growth of additional spheres falling into their four-surface triangular nests. This produces increasing numbers of closest-packed nests. Thus do atoms agglomerate in closest packing in tetrahedrally conformed arrays, often truncated asymmetrically at corners and along edges to obscure the tetrahedral origin of the collection.

411.10 **Unpredicted Degrees of Freedom:** Reviewing the history of self-discoveries of restraints progressively and mutually interimposed upon one another by the arrival of and association with successive othernesses, self may discover through progressive retrospections and appreciate the significance of theretofore unrecognized and unrealized degrees of cosmic freedom successively and inadvertently deducted from the original total inventory of unexpended self-potentials with which the individual is initially, always and only, endowed. Only with the progressive retrospection inventoryings induced only by otherness-developed experience of awareness does the loss of another degree of freedom become consciously subtracted from the previous experience inventory of now consciously multiplying rememberable events.

411.11 With the discovery of principles through progressive deprivations dawns new awareness of the elective employability of the principles initially separated out from their special-case experiencing to self-control more and more of the pattern of events.

411.12 Only through relationships with otherness can self learn of principles; only by discovery of the relationships existing between self and othernesses does inspiration to employ principles objectively occur. There is nothing in self per se, or in otherness per se, that predicts the interrelatedness behaviors and their successively unique characteristics. Only from realization of the significance of otherness can it be learned further that only by earnest commitment to others does self become inadvertently behaviorally advantaged to effect even greater commitment to others, while on the other hand all self-seeking induces only ever greater self-loss.

* There is an alternate sequence, which is perhaps more likely, in which the balls would first join as two pairs—like dumbbells "docking" in space. Then our old friend precession would cause them to form a momentary square, only to have one pair revolve until its axis is at 90 degrees to the other couple's axis, whereafter they dock to interlock as a tetrahedron.

411.20 **Discovery as a Function of Loss:** It is a basic principle that you only discover what you had had by virtue of losing it. Due to our subconscious organic coordination, you don't know what you are losing until you lose it. Naught can be so advantageous as thoughtfully considered loss and resolve to employ the principles thereby discovered for others. You don't know how much you have to give until you start trying to give. The more you try to give effectively to advantage others, the more you will possess to give, and vice versa.

411.21 Retrospective awareness of losses can bring preoccupation with self, blinding self to recognition of the synergetic gains that, by virtue of the second-power law, have brought group advantage gains in which the individual has attained fourth-power continuance potential often way offsetting individual freedom losses, particularly in view of the group's discovery that as a group it can enjoy all the original freedoms individually lost but never realized by the individual to exist—ergo, unemployable consciously by the individual, who was more of a victim of the unknown freedoms than an enjoyer.

411.22 Only as group structuring occurs do the discovered cosmic freedoms become consciously employable—employable effectively only for all and not for self. It is when this retrospective discovering is made by the grouped-in individual and he tries to employ the freedoms exclusively for self or exclusive subdivisions of the group, that his attempts become inherently unfulfillable and scheduled for ultimate failure.

411.23 Self-seeking brings a potential loss that engenders first caution, then fear: fear of change; change being inexorable, fear increases and freezes. Self-seeking always eventuates in self-destruction through inability to adapt.

411.30 **Intergeared Mobility Freedoms:** Only with the arrival of the second otherness do individuals become aware of the loss of mutual anywhere-around-one-another rollability, and then discover that they have also lost the ability to go in all directions, for they cannot go through each other. It is inferred that they haven't lost mutually accomplishable omnidirectional mobility. (Here is an example of one of those comprehendings from an apprehending.) With the acquisition of the second otherness, self discovers what it has lost which self didn't know it had, until the loss brought retrospective awareness of the lost freedom: an inter-anywhere-roll-aroundness with the first other.

411.31 With the mutually interattracted threeness, each having two contacts with their two otherness partners, they learn, as a fourth otherness nests into their triangular opening, that they have now lost a frictionally intergeared mutually evoluting-involuting rotational freedom (torus). Now blocked by the frictionally intergeared fourth otherness, the mutual omnidirectionality of the structural system so produced by that structural system can be discovered only by the self-observation of the realization of another structural system's cumulative repetition repeating the evolutionary accumulation of its own fourfolding; observing that the other structural system can move omnidirectionally, their observed rotations and magnitude changes can be explained only by the omnidirectional freedom only mutually experienceable by the whole individual structural system as it had been originally and only subconsciously experienced by the individual self. Naught has been lost. Much has been mutually gained. Each can take off from and return to the others.

411.32 The variations of the features of second-structural-system otherness can be explained only by the self-structural system assumption of increasing distance of travel of the otherness to and away, or by the other system's experiencing a freedom theretofore unself-realized: that of individual expansion and contraction. For any one of the four members of a structural-system team can expand and contract coordinatedly at individual rates, mutual rates, or interpaired rates, provided one does not become so small as to "fall through" the triangular opening at its nest bottom. (If it fell through, one otherness would start rotating hingewise around the axis of tangent contact with the next-largest, without touching the fourth and next-smallest-to-self.)

411.33 Thus the self-structural system discloses by observation of otherness's system changing features that its own system had been enjoying, as with freedoms of which it was previously unconscious.

411.34 Our inventory of intergeared mobility freedoms is fourfold. It is four-dimensional:

(1) omnidirectionality of united movement;
(2) roll-aroundness (orbiting);
(3) polarized evoluting-and-involuting, and polarized spin; and
(4) inward-outward expandability singly, doubly, three- or four-partite.

411.35 The inward-outward expandability is the basis of convergence-divergence and radiation-gravitation pulsation—which seems furthest from man's awareness. This is what science has discovered: a world of waves in which waves are interpenetrated by waves in frequency modulation. There is a systemic interrelationship of basic fourness always accompanied by a sixness of alternatives of freedoms.

411.36 When a sphere gets so small that it can roll through a hole between other close packed spheres, the omnidirectionality of any one individual would not be impeded under the following circumstances:

(a) The individual mass-attracted by any threeness being drawn through the hole of any other threeness.
(b) Where a fourth otherness could be attracted by a momentary critical-proximity threeness.

(c) We learn there is individuality and magnitude change; then we learn that, due to the energy losses and gains of systems occasioned by the continual variations of omnidirectional proximities and omnivariability of expansion-contraction system accumulating rates, there is a degree of freedom phenomena *rate* as well as a *terminal* condition.

411.37 *Rate* occurs only when there is *terminal.* Rate is a modulation between terminals. With termination, a system's integrity is brought about by the individually covarying magnitudes and the omnidirectional experience pulls on the system.

411.38 The degree of freedom that is lost is discoverable only retrospectively by the very fact of the loss. It is an inverse synergetic behavior wherein no feature of the self part predicted the successive behaviors of the whole and where the individual part freedoms were only mutually disclosed by their subsequently realized loss.

412.00 Closest Packing of Rods

412.01 Just as six balls may be closest packed around a nuclear ball in a plane, six rods or wires may be closest packed around a nuclear rod or wire in a cluster. When the seven wires are thus compacted in a parallel bunch, they may be twisted to form a cable of hexagonal cross section, with the nuclear wire surrounded by the other six. The hexagonal pattern of cross section persists as complete additional layers are symmetrically added to the cluster. These progressive symmetrical surroundments constitute circumferentially finite integrities in universal geometry.

412.02 Surface Tension Capability: We know by conclusive experiments and measuring that the progressive subdivision of a given metal fiber into a plurality of approximately parallel fibers provides tensile behavior capabilities of the smaller fibers at increased magnitudes up to hundreds- and thousandsfold that of the unit solid metal section. This is because of the increased surface-to-mass ratios and because all high tensile capability is provided by the work hardening of the surfaces. This is because the surface atoms are pressed into closer proximity to one another by the drawing tool through which the rod and wire are processed.

413.00 Omnidirectional Closest Packing

413.01 In omnidirectional closest packing of equiradius spheres around a nuclear sphere, 12 spheres will always symmetrically and intertangentially surround one sphere with each sphere tangent to its immediate neighbors. We

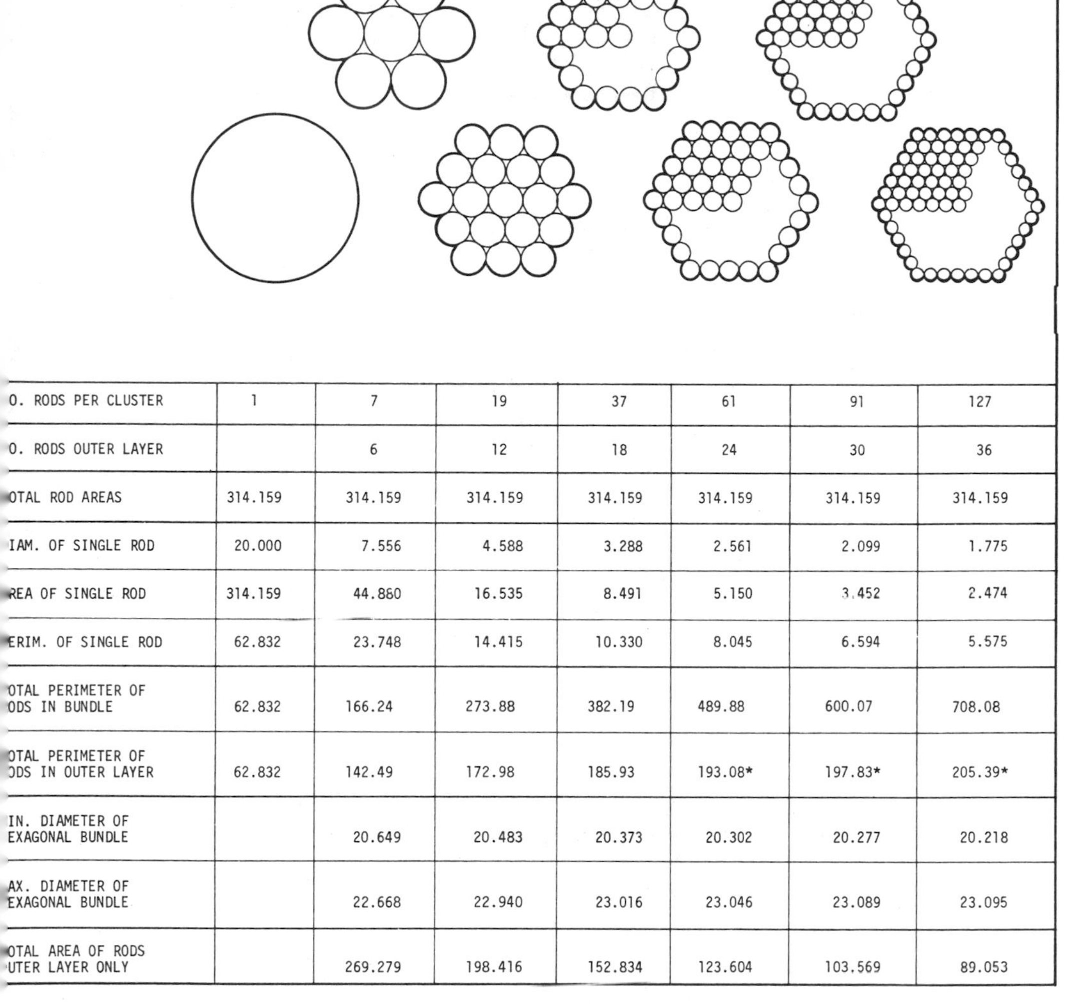

O. RODS PER CLUSTER	1	7	19	37	61	91	127
O. RODS OUTER LAYER		6	12	18	24	30	36
OTAL ROD AREAS	314.159	314.159	314.159	314.159	314.159	314.159	314.159
IAM. OF SINGLE ROD	20.000	7.556	4.588	3.288	2.561	2.099	1.775
REA OF SINGLE ROD	314.159	44.860	16.535	8.491	5.150	3.452	2.474
ERIM. OF SINGLE ROD	62.832	23.748	14.415	10.330	8.045	6.594	5.575
OTAL PERIMETER OF ODS IN BUNDLE	62.832	166.24	273.88	382.19	489.88	600.07	708.08
OTAL PERIMETER OF ODS IN OUTER LAYER	62.832	142.49	172.98	185.93	193.08*	197.83*	205.39*
IN. DIAMETER OF EXAGONAL BUNDLE		20.649	20.483	20.373	20.302	20.277	20.218
AX. DIAMETER OF EXAGONAL BUNDLE		22.668	22.940	23.016	23.046	23.089	23.095
OTAL AREA OF RODS UTER LAYER ONLY		269.279	198.416	152.834	123.604	103.569	89.053

*NOTE THAT PERIMETERS OF OUTER LAYER RODS ALONE EXCEEDS THREE TIMES PERIMETER OF LARGE ROD.

Fig. 412.01 *Closest Packing of Rods.*

may then close-pack another symmetrical layer of identical spheres surrounding the original 13. The spheres of this outer layer are also tangent to all of their immediate neighbors. This second layer totals 42 spheres. If we apply a third layer of equiradius spheres, we find that they, too, compact symmetrically and tangentially. The number of spheres in the third layer is 92.

413.02 Equiradius spheres closest packed around a nuclear sphere do not

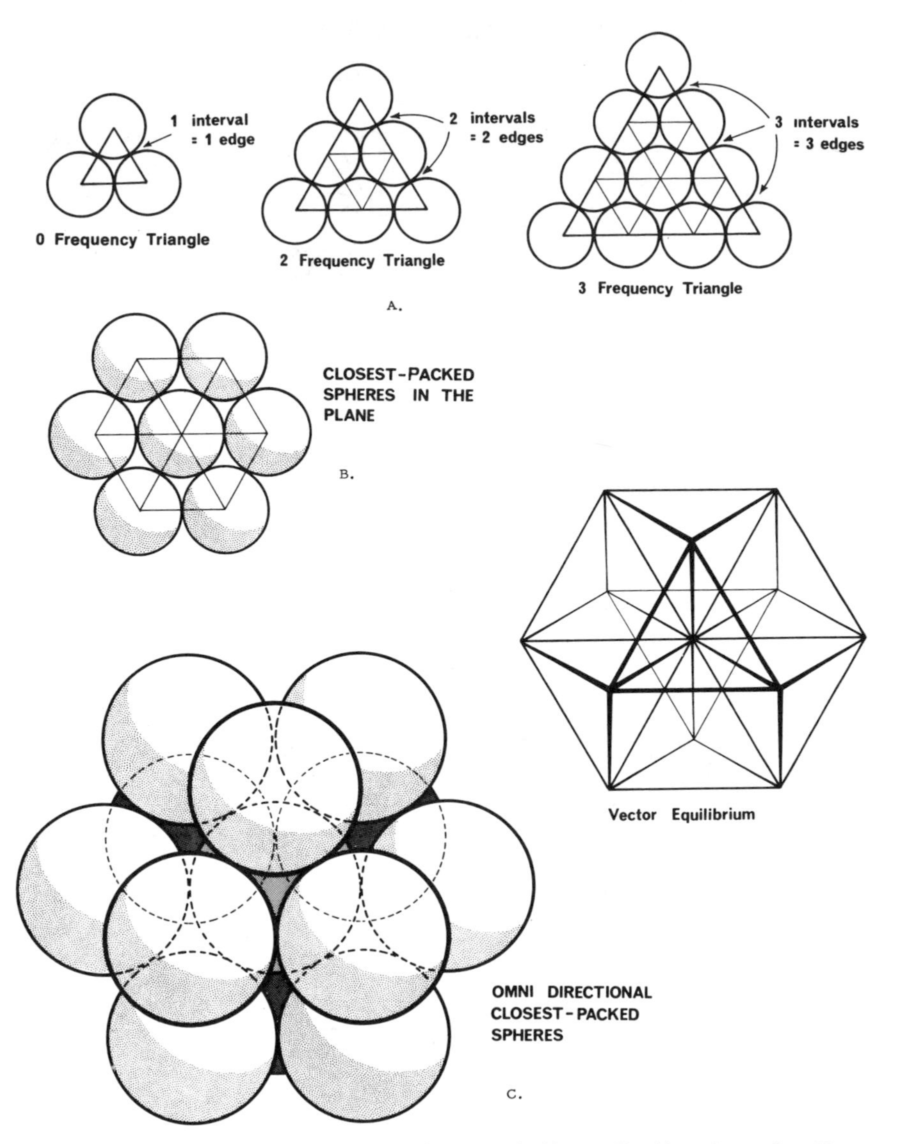

Fig. 413.01 *Vector Equilibrium: Omnidirectional Closest Packing Around a Nucleus:* Triangles can be subdivided into greater and greater numbers of similar units. The number of modular subdivisions along any edge can be referred to as the *frequency* of a given triangle. In triangular grids each vertex may be expanded to become a circle or sphere showing the inherent relationship between closest packed spheres and triangulation. The frequency of triangular arrays of spheres in the plane is determined by counting the number of intervals (A) rather than the number of spheres on a given edge. In the case of concentric packings or spheres around a nucleus the frequency of a given system can either be the edge subdivision or the number of concentric shells or layers. Concentric packings in the plane give rise to hexagonal arrays (B) and omnidirectional closest packing of equal spheres around a nucleus (C) gives rise to the vector equilibrium (D).

form a supersphere, as might be expected. They form a symmetrical polyhedron of 14 faces: the vector equilibrium.

413.03 If we add on more layers of equiradius spheres to the symmetrical polyhedron of 14 faces close-packed around one sphere, we find that they always compact symmetrically and tangentially, and that this process of enclosure may seemingly be repeated indefinitely. Each layer, however, is in itself a finite or complete and symmetrical embracement of spheres. Each of these embracing layers of spheres constitutes a finite system. Each layer always takes the 14-face conformation and consists of eight triangular and six square faces. Together with the layers they enclose and the original sphere center, or *nucleus,* these symmetrically encompassing layers constitute a concentric finite system.

413.04 As additional layers are added, it is found that a symmetrical pattern of concentric systems repeats itself. That is, the system of three layers around one sphere, with 92 spheres in the outer layer, begins all over again and repeats itself indefinitely with successively enclosing layers in such a way that the successive layers outside of the 92-sphere layer begin to penetrate the adjacent new nuclear systems. We find then that only the concentric system of spheres within and including the layer of 92 are *unique* and individual systems. We will pursue this concept of a finite system in universal geometry still further (see Sec. 418, et seq.) in order to relate it to the significance of the 92 self-regenerative chemical elements.

414.00 Nucleus

414.01 In closest packing of equiradius spheres, a nucleus by definition must be tangentially and symmetrically surrounded. This means that there must be a ball in every possible tangential and optically direct angular relationship to the nucleus. This does not happen with the first layer of 12 balls or with the second layer of 42 balls. Not until the third layer of 92 balls is added are all the tangential spaces filled and all the optically direct angles of nuclear visibility intercepted. We then realize a nucleus.*

414.02 It will also be discovered that the third layer of 92 spheres contains eight new potential nuclei; however, these do not become realized nuclei until each has two more layers enveloping it—one layer with the nucleus in it and two layers enclosing it. *Three layers are unique to each nucleus.* This tells us that the nuclear group with 92 spheres in its outer, or third, layer is the limit of unique, closest-packed symmetrical assemblages of unit wavelength and frequency. These are nuclear symmetry systems.

* This does not apply to an asymmetrical or single-axis system, e.g., hydrogen, where a nucleus may be encircled by action within a single plane and where the surround is generated by a single orbit.

414.03 It is characteristic of a nucleus that it has at least two surrounding layers in which there is no nucleus showing, i.e., no potential. In the third layer, however, eight potential nuclei show up, but they do not have their own three unique layers to realize them. So the new nuclei are not yet realized, they are only potential.

414.04 The nucleus ball is always two balls, one concave and one convex. The two balls have a common center. Hydrogen's one convex proton contains its own concave nucleus.

415.00 Concentric Shell Growth Rates

415.01 Minimal Most Primitive Concentric Shell Growth Rates of Equiradius, Closest-Packed, Symmetrical Nucleated Structures: Out of all possible symmetrical polyhedra produceable by closest-packed spheres agglomerating, only the vector equilibrium accommodates a one-to-one arithmetical progression growth of *frequency number* and *shell number* developed by closest-packed, equiradius spheres around one nuclear sphere. Only the vector equilibrium—"equanimity"—accommodates the symmetrical growth or contraction of a nucleus-containing aggregate of closest-packed, equiradius spheres characterized by either even or odd numbers of concentric shells.

415.02 Odd or Even Shell Growth: The hierarchy of progressive shell embracements of symmetrically closest-packed spheres of the vector equilibrium is generated by a smooth arithmetic progression of both even and odd frequencies. That is, *each* successively embracing layer of closest-packed spheres is in exact frequency and shell number atunement. Furthermore, additional embracing layers are accomplished with the least number of spheres per exact arithmetic progression of higher frequencies.

415.03 Even-Number Shell Growth: The tetrahedron, octahedron, cube, and rhombic dodecahedron are nuclear agglomerations generated only by even-numbered frequencies:

Nuclear tetrahedron:	$F=4$	(34 around one)
	$F=8$	(130 around one)
Nuclear octahedron:	$F=2$	(18 around one)
	$F=4$	(66 around one)
Nuclear cube:	$F=4$	(210 around one)
	$F=6$	(514 around one)
Nuclear rhombic dodecahedron:	$F=4$	(74 around one)
	$F=8$	(386 around one)

Chart 415.03 Rate of Occurrence of Symmetrically Nucleated Polyhedra of Closest Packing

Shell	Vector Equilib. Outer Shell:	$10F^2 + 2$ Vector Equilib. Cumulative: All Shells	$4F^2 + 2$ Octahedron $4(F+2)^2 + 2$	$2F^2 + 2$ Tetrahedron $2(F+4)^2 + 2$	$12F^2 + 2$ Rhombic Dodecahedron Octa = $^1/_4$ Tet × 8	$6F^2 + 2$ Cube Vector Equilib. + $^1/_8$ Octa × 8	Icosahedron & Dodecahedron are Inherently Non-Nuclear at All Frequencies
0	zero = 2	zero = 2	zero = 2	zero = 2	zero = 2	zero = 2	
1	12	12					
2	42	54	18				
3	92	146					
			Outer shell 66	34	74	210	
4	162	308	Cumulative 84		92	364	
5	252	560					
						Outer shell 514	
6	362	922				Cumulative 1098	
7	492	1414					
				Outer shell 130	386		
8	642	2056		Cumulative 164	470		
9	812	2868					
10	1002	3870					

415.10 **Yin-Yang As Two (Note to Chart 415.03):** Even at zero frequency of the vector equilibrium, there is a fundamental twoness that is not just that of opposite polarity, but the twoness of the concave and the convex, i.e., of the inwardness and outwardness, i.e., of the microcosm and of the macrocosm. We find that the nucleus is really two layers because its inwardness turns around at its own center and becomes outwardness. So we have the congruence of the inbound layer and the outbound layer of the center ball.

$10F^2 + 2$
$F = 0$
$10 \times 0 = 0$
$0 + 2 = 2$ (at zero frequency)

Because people thought of the nucleus only as oneness, they for long missed the significant twoness of spherical unity as manifest in the atomic weights in the Periodic Table of the Elements.

415.11 When they finally learned that the inventory of data required the isolation of the neutron, they were isolating the concave. When they isolated the proton, they isolated the convex.

415.12 As is shown in the comparative table of closest-packed, equiradius nucleated polyhedra, the vector equilibrium not only provides an orderly shell for each frequency, which is not provided by any other polyhedra, but also gives the nuclear sphere the first, or earliest possible, polyhedral symmetrical enclosure, and it does so with the least number—12 spheres; whereas the octahedron closest packed requires 18 spheres; the tetrahedron, 34; the rhombic dodecahedron, 92; the cube, 364; and the other two symmetric Platonic solids, the icosahedron and the dodecahedron, are inherently, ergo forever, devoid of equiradius nuclear spheres, having insufficient radius space within the triangulated inner void to accommodate an additional equiradius sphere. This inherent disassociation from nucleated systems suggests both electron and neutron behavior identification relationships for the icosahedron's and the dodecahedron's requisite noncontiguous symmetrical positioning outwardly from the symmetrically nucleated aggregates. The nucleation of the octahedron, tetrahedron, rhombic dodecahedron, and cube very probably plays an important part in the atomic structuring as well as in the chemical compounding and in crystallography. They interplay to produce the isotopal Magic Number high point abundance occurrences. (See Sec. 995.)

415.13 The formula for the nucleated rhombic dodecahedron is the formula for the octahedron with frequency plus four (because it expands outwardly in four-wavelength leaps) plus eight times the closest-packed central angles of a tetrahedron. The progression of layers at frequency plus four is made only when we have one ball in the middle of a five-ball edge triangle, which always occurs again four frequencies later.

415.14 The number of balls in a single-layer, closest-packed, equiradius triangular assemblage is always

$$\frac{N^2 - N}{2} + 2.$$

415.15 To arrive at the cumulative number of spheres in the rhombic dodecahedron, you have to solve the formula for the octahedron at progressive frequencies *plus four,* plus the solutions for the balls in the eight triangles.

415.16 The first cube with 14 balls has no nucleus. The first cube with a nucleus occurs by the addition of 87-ball corners to the eight triangular facets of a four-frequency vector equilibrium.

415.20 **Organics:** It could be that organic chemistries do not require nuclei.

415.21 The first closest-packed, omnitriangulated, ergo structurally stabilized, but non-nuclear, equiradius-sphered, cubical agglomeration has 14 spheres. This may be Carbon 14, which is the initially closest-packed, omnisymmetrical, polyhedral fourteenness, providing further closest-packability surface nests suitable for structurally mounting hydrogen atoms to produce all organic matter.

415.22 The cube is the prime minimum omnisymmetrical allspace filler. But the cube is nonstructural until its six square faces are triangularly diagonaled. When thus triangularly diagonaled, it consists of one tetrahedron with four one-eighth octahedra, of three isosceles and one equilateral-faced tetrahedron, outwardly applied to the nuclear equilateral tetrahedron's four triangular faces. Thus structurally constituted, the superficially faced cube is prone to closest-packing self-associability. In order to serve as the carbon ring (with its six-sidedness), the cube of 14 spheres (with its six faces) could be joined with six other cubes by single atoms nestable in its six square face centers, which singleness of sphericity linkage potential is providable by Hydrogen 1.

415.23 In the atoms, we are always dealing in equiradius spheres. Chemical compounds may, and often do, consist of atomic spheres with a variety of radial dimensions. Since each chemical element's atoms are characterized by unique frequencies, and unique frequencies impose unique radial symmetrics, this variety of radial dimensionality constitutes one prime difference between nuclear physics and chemistry.

415.30 **Eight New Nuclei at Fifth Frequency of Vector Equilibrium:** Frequency five embraces nine nuclei: the original central nucleus plus eight new nuclei occurring at the centers of volume of the eight tetrahedra symmetrically surrounding the nucleus, with each of the nine enclosed with a minimum of two layers of spheres.

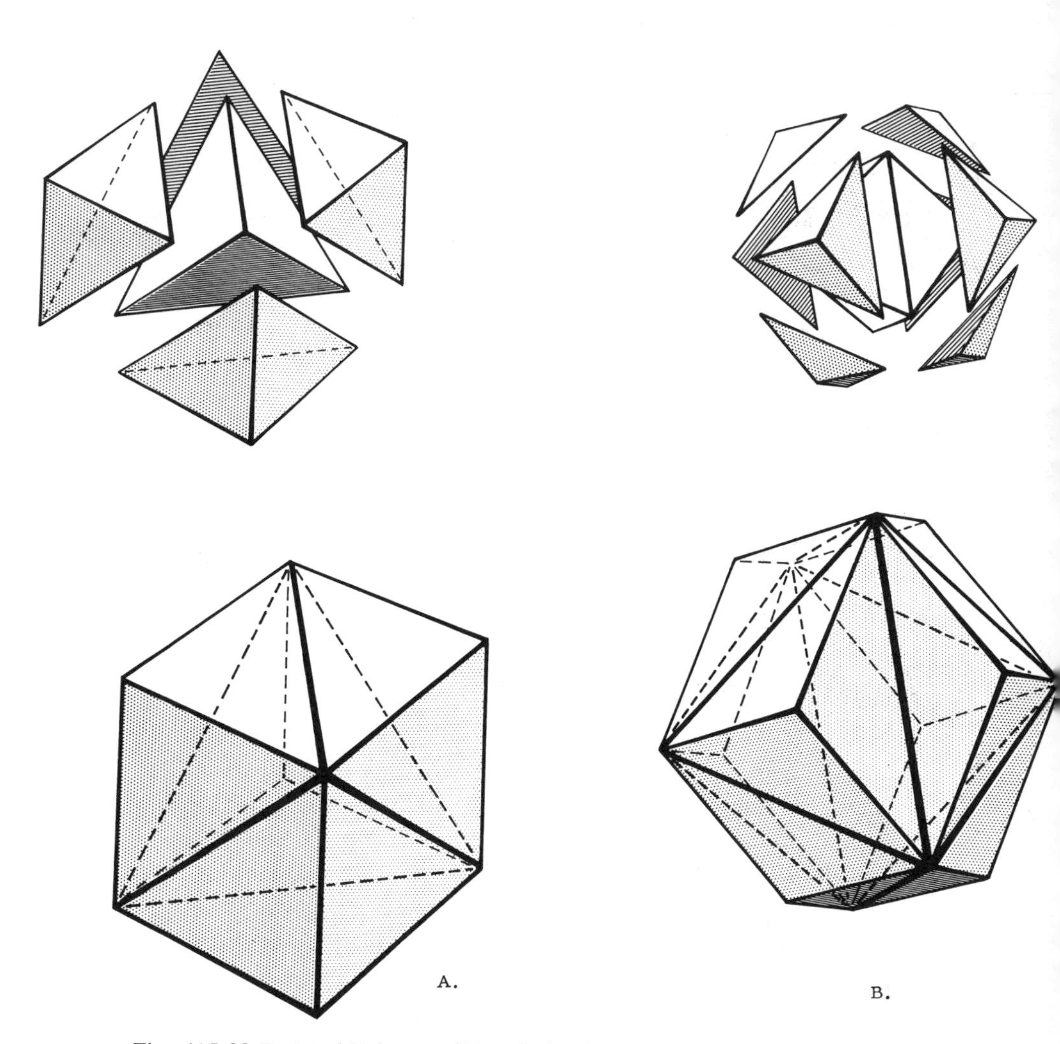

Fig. 415.22 *Rational Volumes of Tetrahedroning:*

A. The cube may be formed by placing four $^1/_8$-octahedra with their equilateral faces on the faces of a tetrahedron. Since tetrahedron volume equals one, and $^1/_8$-octahedron equals $^1/_2$, the volume of the cube will be: $1 + 4(^1/_2) = 3$.

B. The rhombic dodecahedron may be formed by placing eight $^1/_4$-tetrahedra with their equilateral faces on the faces of an octahedron. Since the octahedron volume equals four and $^1/_4$-tetrahedron equals $^1/_4$, the volume of the rhombic dodecahedron will be: $4+8(^1/_4)=6$.

415.31 The vector equilibrium at $f^0 = 12$; at $f^2 = 42$; $f^3 = 92$; $f^4 = 162$ spheres in the outer shell; and at $f^5 = 252$ we get eight new nuclei. Therefore, their eightness of "begetness" relates to the eight triangles of the vector equilibrium.

415.32 Six nucleated octahedra with two layer omni-enclosure of their nuclei does not occur until $f^6 = 362$ in the outer shell of the vector equilibrium. At this stage we have six new nuclei, with 14 nuclei surrounding the 15th, or original, nucleus.

415.40 **Begetted Eightness:** The "begetted" *eightness* as the system-limit number of nuclear uniqueness of self-regenerative symmetrical growth may well account for the fundamental octave of unique interpermutative integer effects identified as plus one, plus two, plus three, plus four, as the interpermutated effects of the integers one, two, three, and four, respectively; and as minus four, minus three, minus two, minus one, characterizing the integers five, six, seven, and eight, respectively. The integer nine always has a neutral, or zero, intermutative effect on the other integers. This permutative, synergetic or interamplifying or dimensioning effect of integers upon integers, together with the octave interinsulative accommodation produced by the zero effect of the nineness, is discussed experientially in our section on *Indigs* in Chapter 12, Numerology.

415.41 The regenerative initial *eightness* of first-occurring potential nuclei at the frequency-four layer and its frequency-five confirmation of those eight as constituting true nuclei, suggest identity with the third and fourth periods of the Periodic Table of Chemical Elements, which occur as

1st period = 2 elements
2nd period = 8 elements
3rd period = 8 elements

415.42 Starting with the center of the nucleus: plus one, plus two, plus three, plus four, outwardly into the last layer of nuclear uniqueness, whereafter the next pulsation becomes the minus fourness of the outer layer (fifth action); the sixth event is the minus threeness of canceling out the third layer; the seventh event is the minus twoness canceling out the second layer; the eighth event is the minus oneness returning to the center of the nucleus—all of which may be identified with the frequency pulsations of nuclear systems.

415.43 The *None* or *Nineness/Noneness* permits wave frequency propagation cessation. The *Nineness/Zeroness* becomes a shutoff valve. The *Zero/Nineness* provides the number logic to account for the differential between potential and kinetic energy. The *Nineness/Zeroness* becomes the number identity of vector equilibrium, that is, energy differentiation at zero. (See Secs. 1230 et seq. and the Scheherazade Number.)

415.44 The eightness being nucleic may also relate to the relative abundance of isotopal magic numbers, which read 2, 8, 20, 50, 82, 126. . . .

415.45 The inherent zero-disconnectedness accounts for the finite energy packaging and discontinuity of Universe. The vector equilibria are the empty set tetrahedra of Universe, i.e., the tetrahedron, being the minimum structural system of Universe independent of size, its four facet planes are at maximum remoteness from their opposite vertexes and may have volume content of the third power of the linear frequency. Whereas in the vector equilibrium all four planes of the tetrahedra pass through the same opposite vertex—which is the nuclear vertex—and have no volume, frequency being zero: F^0.

415.50 **Vector-Equilibrium Closest-Packing Configurations:** The vector equilibrium has four unique sets of axes of symmetry:

(1) The three intersymmetrical axes perpendicular to, i.e., normal to, i.e., joining, the hemispherically opposite six square faces;
(2) The four axes normal to its eight triangular faces;
(3) The six axes normal to its 12 vertexes; and
(4) the 12 axes normal to its 24 edges.

The tetrahedron, vector equilibrium, and octahedron, with all their planes parallel to those of the tetrahedron, and therefore derived from the tetrahedron, as the first and simplest closest-packed, ergo omnitriangulated, symmetrical structural system, accept further omnidirectional closest packing of spheres. Because only *eight* of its 20 planar facets are ever parallel to the four planes of the icosahedron, the icosahedron refuses angularly to accommodate anywhere about its surface further omnidirectional closest packing of spheres, as does the tetrahedron.

415.51 Consequently, the (no-nucleus-accommodating) icosahedron formed of equiradius, triangularly closest-packed spheres occurs only as a one-sphere-thick shell of any frequency only. While the icosahedron cannot accommodate omnidirectionally closest-packed multishell growth, it can be extended from any one of its triangular faces by closest-packed sphere agglomerations. Two icosahedra can be face-bonded.

415.52 The icosahedron has three unique sets of axes of symmetry:

(1) The 15 intersymmetric axes perpendicular to and joining the hemispherically opposite mid-edges of the icosahedron's 30 identical, symmetrically interpatterned edges;
(2) The 10 intersymmetric axes perpendicular to the triangular face centers of the hemispherically opposite 20 triangular faces of the icosahedron; and

(3) The six intersymmetric axes perpendicularly interconnecting the hemispheric opposites of the icosahedron's 12 vertexes, or vertexial corner spheres of triangular closest packing.

415.53 While the 15-axes set and the 6-axes set of the icosahedron are always angularly askew from the vector equilibrium's, *four* out of its 10 axes of symmetry are parallel to the set of four axes of symmetry of the vector equilibrium. Therefore, the icosahedron may be face-extended to produce chain patterns conforming to the tetrahedron, octahedron, vector equilibrium, and rhombic dodecahedron in omnidirectional, closest-packing coordination—but only as chains; for instance, as open linear models of the octahedron's edges, etc.

415.55 **Nucleus and Nestable Configurations in Tetrahedra:** In any number of successive planar layers of tetrahedrally organized sphere packings, every third triangular layer has a sphere at its centroid (nucleus). The dark ball rests in the valley between three balls, where it naturally falls most compactly and comfortably. The next layer is three balls to the edge, which means two-frequency. There are six balls in the third layer, and there very clearly is a nest right in the middle. There are ten balls in the fourth layer: but we cannot nest a ball in the middle because it is already occupied by a dark centroid ball. Suddenly the pattern changes, and it is no longer nestable.

415.56 At first, we have a dark ball at the top; then a second layer of three balls with a nest but no nucleus. The third layer with six balls has a nest but no nucleus. The fourth layer with ten balls has a dark centroid ball at the nucleus but no nestable position in the middle. The fifth layer (five balls to the edge; four frequency) has 15 balls with a nest again, but no nucleus. This 35-sphere tetrahedron with five spheres on each edge is the lowest frequency tetrahedron system that has a central sphere or nucleus. (See Fig. A, illustration 415.55.)

415.57 The three-frequency tetrahedron is the highest frequency single-layer, closest-packed sphere shell without a nuclear sphere. This three-frequency, 20-sphere, empty, or nonsphere nucleated, tetrahedron may be enclosed by an additional shell of 100 balls; and a next layer of 244 balls totaling 364, and so on. (See Fig. B, illustration 415.55.)

415.58 **Basic Nestable Configurations:** There are three basic nestable possibilities shown in Fig. C. They are (1) the regular tetrahedron of four spheres; (2) the one-eighth octahedron of seven spheres; and (3) the quarter tetrahedron, with a 16th sphere nesting on a planar layer of 15 spheres. Note that this "nesting" is only possible on triangular arrays that have no sphere at

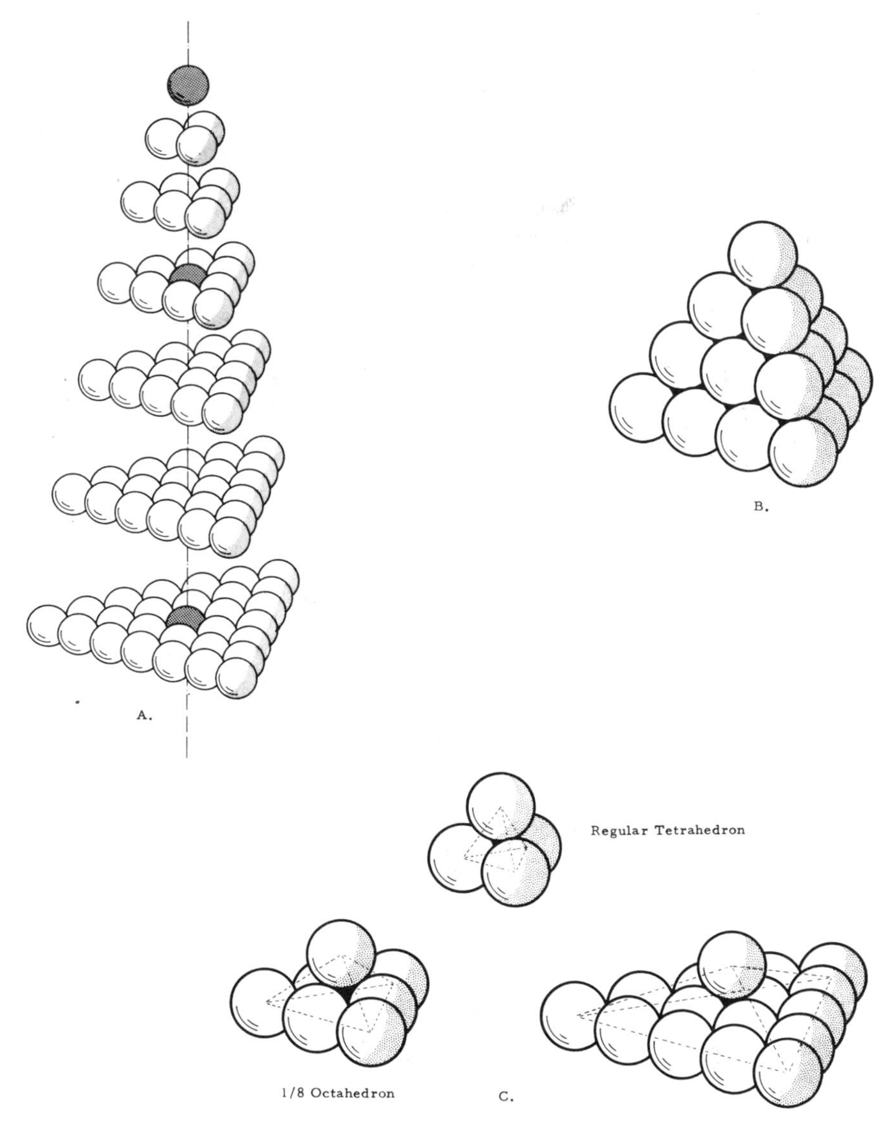

Fig. 415.55 *Tetrahedral Closest Packing of Spheres: Nucleus and Nestable Configurations:*

A. In any number of successive planar layers of tetrahedrally organized sphere packings, every third triangular layer has a sphere at its centroid (a nucleus). The 36-sphere tetrahedron with five spheres on an edge (four-frequency tetrahedron) is the lowest frequency tetrahedron system which has a central sphere or nucleus.

B. The three-frequency tetrahedron is the highest frequency without a nucleus sphere.

C. Basic "nestable" possibilities show how the regular tetrahedron, the $^1/_4$-tetrahedron and the $^1/_8$-octahedron may be defined with sets of closest packed spheres. Note that this "nesting" is only possible on triangular arrays which have no sphere at their respective centroids.

their respective centroids. This series is a prime hierarchy. One sphere on three is the first possibility with a central nest available. One sphere on six is the next possibility with an empty central nest available. One sphere on 10 is impossible as a ball is already occupying the geometrical center. The next possibility is one on 15 with a central empty nest available.

415.59 Note that the 20-ball empty set (see Fig. B, illustration 415.55) consists of five sets of four-ball simplest tetrahedra and can be assembled from five separate tetrahedra. The illustration shows four four-ball tetrahedra at the vertexes colored "white." The fifth four-ball tetrahedron is dark colored and occupies the central octahedral space in an inverted position. In this arrangement, the four dark balls of the inverted central tetrahedron appear as center balls in each of the four 10-ball tetrahedral faces.

416.00 Tetrahedral Precession of Closest-Packed Spheres

416.01 You will find, if you take two separate parallel sets of two tangent equiradius spheres and rotate the tangential axis of one pair one-quarter of a full circle, and then address this pair to the other pair in such a manner as to bring their respective intertangency valleys together, that the four now form a tetrahedron. (See Fig. B, illustration 416.01.)

416.02 If you next take two triangles, each made of three balls in closest packing, and twist one of the triangles 60 degrees around its center hole axis, the two triangular groups now may be nested into one another with the three spheres of one nesting in the three intersphere tangency valleys of the other. We now have six spheres in symmetrical closest packing, and they form the six vertexes of the octahedron. This twisting of one set to register it close-packedly with the other, is the first instance of two pairs internested to form the tetrahedron, and in the next case of the two triangles twisted to internestability as an octahedron, is called *interprecessing* of one set by its complementary set.

416.03 Two pairs of two-layer, seven-ball triangular sets of closest-packed spheres precess in a 60-degree twist to associate as the cube. (See Fig. A, illustration 416.01.) This 14-sphere cube is the minimum cube that may be stably produced by closest-packed spheres. While eight spheres temporarily may be tangentially glued into a cubical array with six square hole facades, they are not triangulated; ergo, are unstructured; ergo, as a cube are utterly unstable and will collapse; ergo, no eight-ball cube can be included in a structural hierarchy.

416.04 The two-frequency (three spheres to an edge), two-layer tetrahedron may also be formed into a cube through 90-degree interprecessional effect. (See Fig. A.)

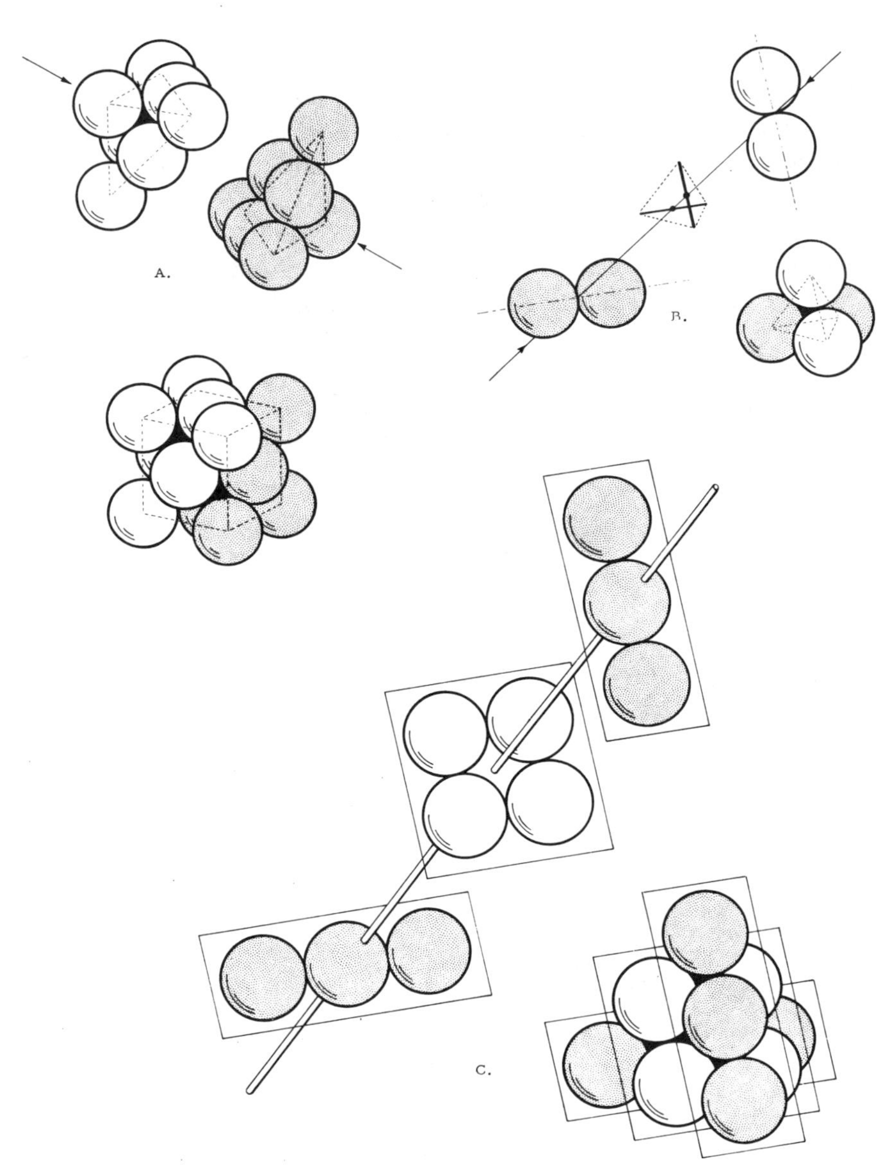

Fig. 416.01 *Tetrahedral Precession of Closest Packed Spheres:*

A. Two pairs of seven-ball, triangular sets of closest packed spheres precess in 60 degree twist to associate as the cube. This 14-sphere cube is the minimum structural cube which may be produced by closest-packed spheres. Eight spheres will not close-pack as a cube and are utterly unstable.

B. When two sets of two tangent balls are self-interprecessed into closest packing, a half-circle inter-rotation effect occurs. The resulting figure is the tetrahedron.

C. The two-frequency (three-sphere-to-an-edge) square-centered tetrahedron may also be formed through one-quarter-circle precessional action.

417.00 Precession of Two Sets of 60 Closest-Packed Spheres

417.01 Two identical sets of 60 spheres in closest packing precess in 90-degree action to form a seven-frequency, eight-ball-to-the-edge tetrahedron with a total of 120 spheres; exactly 100 spheres are on the outer shell, exactly 20 spheres are in the inner shell, and there is no sphere at the nucleus. This is the largest possible double-shelled tetrahedral aggregation of closest-packed spheres having no nuclear sphere. As long as it has the 20-sphere tetrahedron of the inner shell, it will never acquire a nucleus at any frequency.

417.02 The 120 spheres of this non-nuclear tetrahedron correspond to the 120 basic triangles that describe unity on a sphere. They correspond to the 120 identical right-spherical triangles that result from symmetrical subdividing of the 20 identical, equilateral, equiangular triangles of either the spherical or planar-faceted icosahedron accomplished by the most economical connectors from the icosahedron's 12 vertexes to the mid-edges of the opposite edges of their respective triangles, which connectors are inherently perpendicular to the edges and pass through one another at the equitriangles' center and divide each of the equilaterals into six similar right triangles. These 120 triangles constitute the highest common multiple of system surface division by a single module unit area, as these 30°, 60°, 90° triangles are not further divisible into identical parts.

417.03 When we first look at the two unprecessed 60-ball halves of the 120-sphere tetrahedron, our eyes tend to be deceived. We tend to look at them "three-dimensionally," i.e., in the terms of exclusively rectilinear and perpendicular symmetry of potential associability and closure upon one another. Thus we do not immediately see how we could bring two oblong quadrangular facets together with their long axes crossing one another at right angles.

417.04 Our sense of exclusively perpendicular approach to one another precludes our recognition that in 60-degree (versus 90-degree) coordination, these two sets precess in 60-degree angular convergence and not in parallel-edged congruence. This 60-degree convergence and divergence of mass-attracted associabilities is characteristic of the four-dimensional system.

418.00 Analogy of Closest Packing, Periodic Table, and Atomic Structure

418.01 The number of closest-packed spheres in any complete layer around any nuclear group of layers always terminates with the digit 2. First layer, 12; second, 42; third, 92 . . . 162, 252, 362, and so on. The digit 2 is

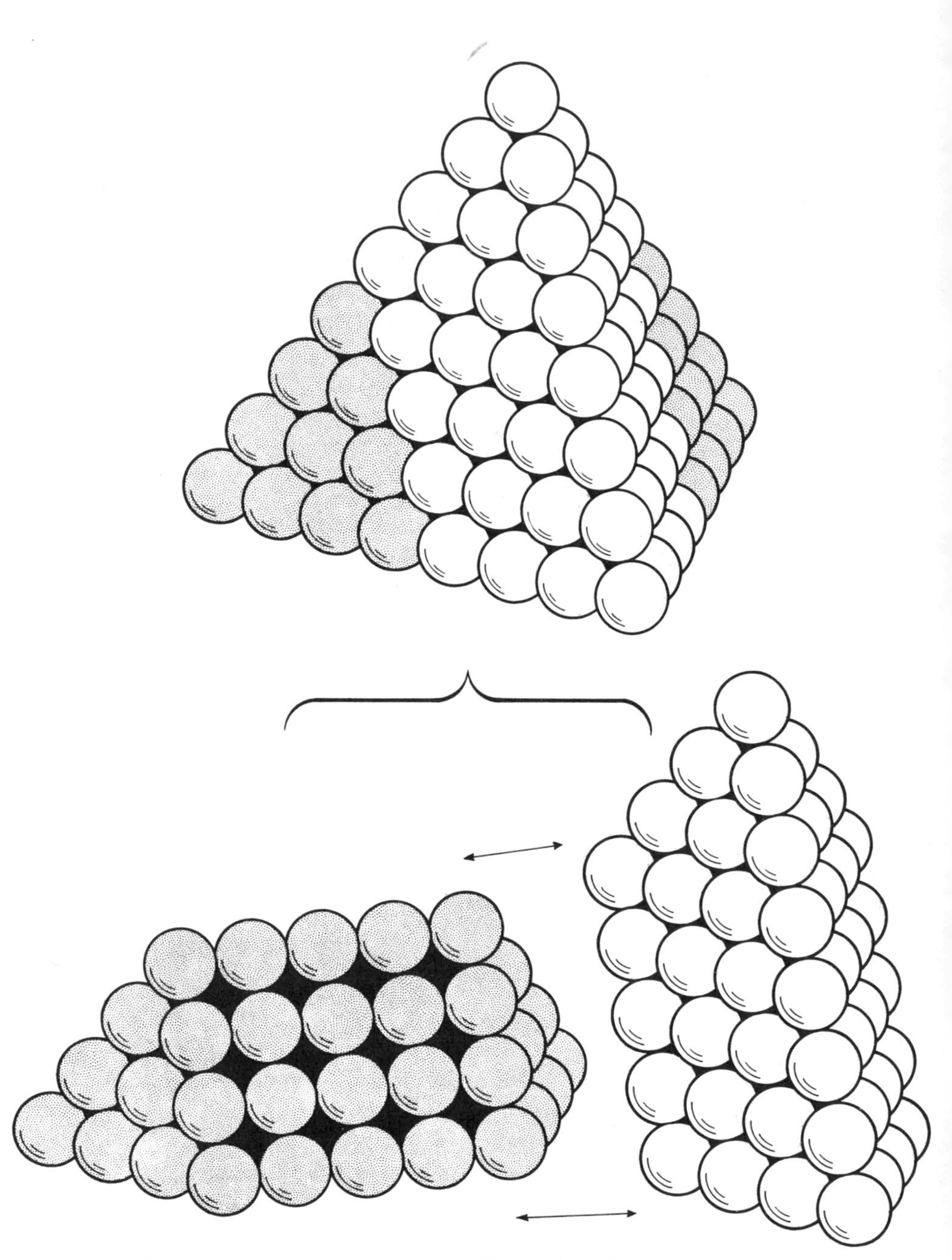

Fig. 417.01 *Precession of Two Sets of 60 Closest-Packed Spheres as Seven-Frequency Tetrahedron:* Two identical sets of 60 spheres in closest packing precess in 90-degree action to form a seven-frequency, eight-ball-edged tetrahedron with a total of 120 spheres, of which exactly 100 spheres are on the surface of the tetrahedron and 20 are inside but have no geometrical space accommodation for an equiradius nuclear sphere. The 120-sphere, nonnucleated tetrahedron is the largest possible double-shelled tetrahedral aggregation of closest-packed spheres having no nuclear sphere.

always preceded by a number that corresponds to the second power of the number of layers surrounding the nucleus. The third layer's number of 92 is comprised of the 3 multiplied by itself (i.e., 3 to the second power), which is 9, with the digit 2 as a suffix.

418.02 This third layer is the outermost of the symmetrically unique, nuclear-system patterns and may be identified with the 92 unique, self-regenerative, chemical-element systems, and with the 92nd such element—uranium.

418.03 The closest-sphere-packing system's first three layers of 12, 42, and 92 add to 146, which is the number of neutrons in uranium—which has the highest nucleon population of all the self-regenerative chemical elements; these 146 neutrons, plus the 92 unengaged mass-attracting protons of the outer layer, give the predominant uranium of 238 nucleons, from whose outer layer the excess two of each layer (which functions as a neutral axis of spin) can be disengaged without distorting the structural integrity of the symmetrical aggregate, which leaves the chain-reacting Uranium 236.

418.04 All the first 92 chemical elements are the finitely comprehensive set of purely abstract physical principles governing all the fundamental cases of dynamically symmetrical, vectorial geometries and their systematically self-knotting, i.e., precessionally self-interfered, regenerative, inwardly shunting events.

418.05 The chemical elements are each unique pattern integrities formed by their self-knotting, inwardly precessing, periodically synchronized self-interferences. Unique pattern evolvement constitutes elementality. What is unique about each of the 92 self-regenerative chemical elements is their nonrepetitive pattern evolvement, which terminates with the third layer of 92.

418.06 Independent of their isotopal variations of neutron content, the 92 self-regenerative chemical elements belong to the basic inventory of cosmic absolutes. The family of prime elements consists of 92 unique sets of from one to 92 electron-proton counts inclusive, and no others.

419.00 Superatomics

419.01 Those subsequently isolated chemical elements beyond the 92 prime self-regenerative chemical elements constitute super-atomics. They are the non-self-regenerative chemical elements of negative Universe.

419.02 Negative Universe is the complementary but invisible Universe. To demonstrate negative Universe, we take one rubber glove with an external green surface and an internal red surface. On the green surface a series of 92 numbers is patterned; and on the red surface a continuance of 93, 94, through to 184, with number 184 at the inside end of the pinky—each of the inner surface numbers being the inner pole of the outer pole point number positionings. The positions of the numbers on the inside correspond to the positions of the numbers on the outside. The numbering starts with the position of the five fingernails, then their successive first joints, and then their successive second joints from the tips: 5, 10, 15, and 20 numbers accommodated by the digits. The other 62 members are arranged in four rows of 12 each around the back and front of the palm of the hand. There is a final row of 14 at the terminal edge of the glove opening—this makes a total of 92. Now we can see why the 92 numbers on the outside were discoverable in a random manner requiring very little physical effort. It was just a matter of which part of your gloved hand you happened to be looking at. But if we become curious about what may be on the inside of the glove we discover that the glove is powerfully resilient. It takes a great deal of power to turn it up, to roll back the open edge—and it takes increasing amounts of power to cope with the increasing thickness of the rubber that rolls up as the glove opens. The elements from 93 on are revealed progressively by the numbers.

419.03 The discovery of the first 92 self-regenerative chemical elements was not by the numbers starting with one, but in a completely random sequence. In the super-atomics, beyond Uranium, number 92, the split-second-lived chemical elements have been discovered in a succession that corresponds to their atomic number—for example, the 94th discovery had the atomic weight of 94; the 100th discovery was atomic weight 100, etc.

419.04 This orderly revelation is in fundamental contrast to the discoveries of the 92 self-regenerative elements and their naturally self-regeneratively occurring isotopes. The discovery of the post-uranium elements has involved the employment of successively greater magnitudes of energy concentration and focusing. As each of the super-atomic trans-uranium elements was isolatingly discovered, it disintegrated within split seconds. The orderliness of the succession of the discovery of super-atomics corresponds to the rate of increase of the magnitudes of energy necessary to bring them into split-second identifiability before they revert to their inside—ergo, invisible to outside—position.

419.05 Every layer of a finite system has both an interior, concave, associability potential and an exterior, convex, associability potential. Hence the *outer* layer of a vector-equilibrium-patterned atom system always has an additional full number ''unemployed associability'' count. In the example cited above (Sec. 418.03), an additional 92 was added to the 146 as the sum of the number of spheres in the first three shells. The total is 238, the number of nucleons in uranium, whose atomic weight is 238. Four of the nucleons on the surface of one of the square faces of the vector equilibrium's closest-packed aggregation of nucleons may be separated out without impairing the structural-stability integrity of the balance of the aggregate. This leaves a residue of 236 nucleons, which is the fissionable state of uranium—which must go on chain-reacting due to its asymmetry.

420.00 Isotropic Vector Matrix

420.01 When the centers of equiradius spheres in closest packing are joined by most economical lines, i.e., by geodesic vectorial lines, an isotropic vector matrix is disclosed—"isotropic" meaning "everywhere the same," "isotropic vector" meaning "everywhere the same energy conditions." This matrix constitutes an array of equilateral triangles that corresponds with the comprehensive coordination of nature's most economical, most comfortable, structural interrelationships employing 60-degree association and disassociation. Remove the spheres and leave the vectors, and you have the octahedron-tetrahedron complex, the octet truss, the isotropic vector matrix. (See Secs. 650 and 825.28.)

420.02 The isotropic vector matrix is four-dimensional and 60-degree-coordinated. It provides an omnirational accounting system that, if arbitrarily accounted on a three-dimensional, 90-degree basis, becomes inherently irrational. The isotropic vector matrix demonstrates the ability of the symmetrically and asymmetrically terminaled, high-frequency energy vectors to accommodate the structuring of any shape. (See Sec. 923.)

420.03 Our extension of the Avogadro hypothesis (Sec. 410) generalizes that all energy conditions are the same. Inasmuch as vectors describe energy conditions, this would mean a volumetric aggregation of vectors in a structural complex in which all of the interacting vectors would have to be of the same length and all of their intersecting angles would have to be the same. This state of omnisameness of vectors stipulates the "isotropic," meaning everywhere the same. This prescribes an everywhere state of equilibrium.

420.04 Equilibrium

420.041 Nature is said to abhor an equilibrium as much as she abhors a perfect vacuum or a perfect anything. Heisenberg's indeterminism and quasi-precision mechanics' recognition of inherent inaccuracy of observation or articulation seems to suggest that the asymmetric deviations and aberrations relative to equilibrium are inherent in the imperfection of a *limited* life of humans with a tightly limited range of perceptible differentiation of details of

1250 A.D. 1270 1290 1310 1330 1350 1370 1390 1410 1430 1450 1470 1490 1510 1530 1550 1570 1590 1610

EARTH ORBIT IN MAN MADE ENVIRONMENT CONTROL: PRODUCT OF SUCCESSFUL APPLICATION OF HIGH PERFORMANCE PER UNIT OF INVESTED RESOURCES

PROFILE OF THE INDUSTRIAL REVOLUTION AS EXPOSED BY THE CHRONOLOGICAL RATE OF ACQUISITION OF THE BASIC INVENTORY OF COSMIC ABSOLUTES—THE 92 ELEMENTS

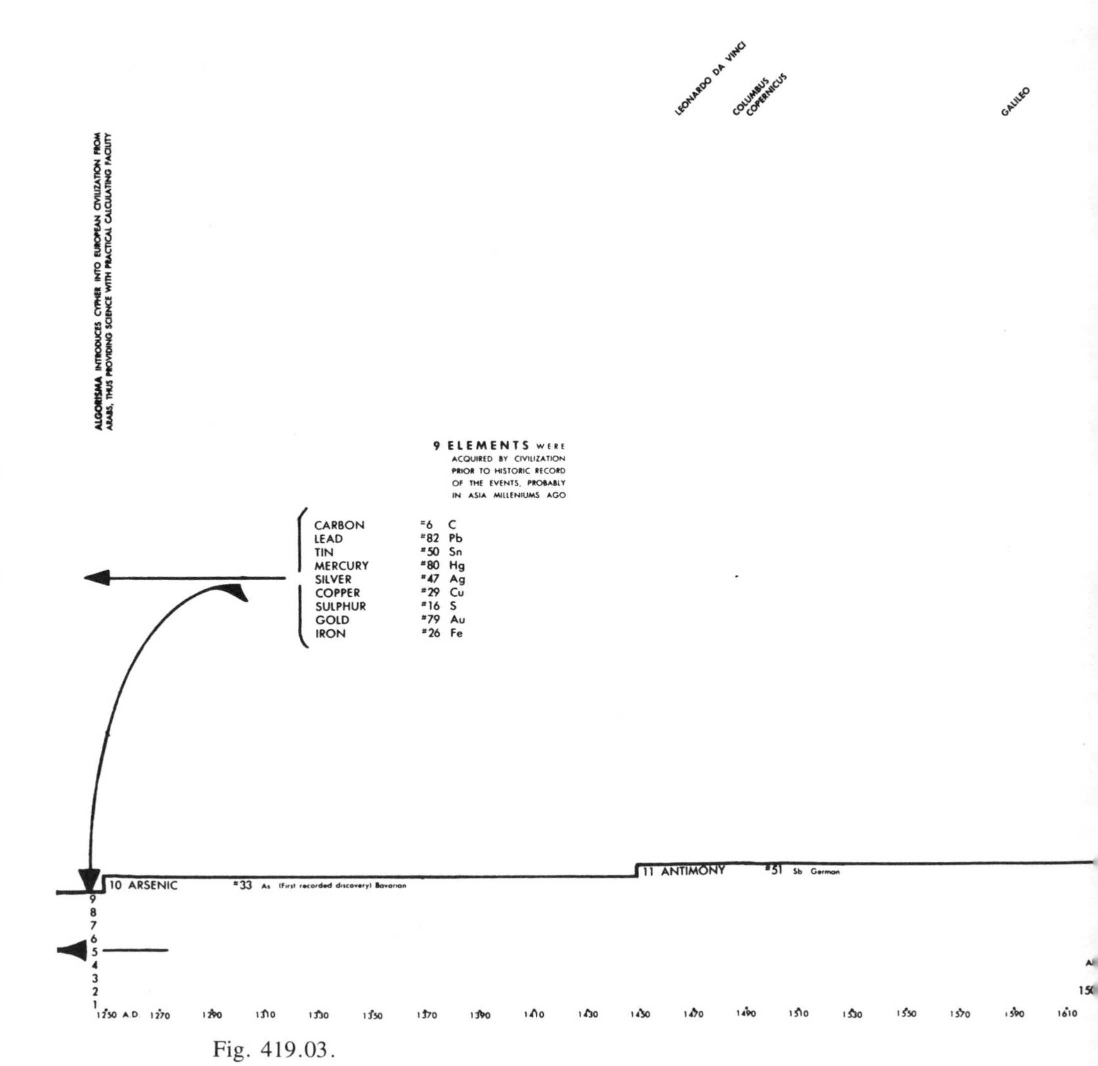

Fig. 419.03.

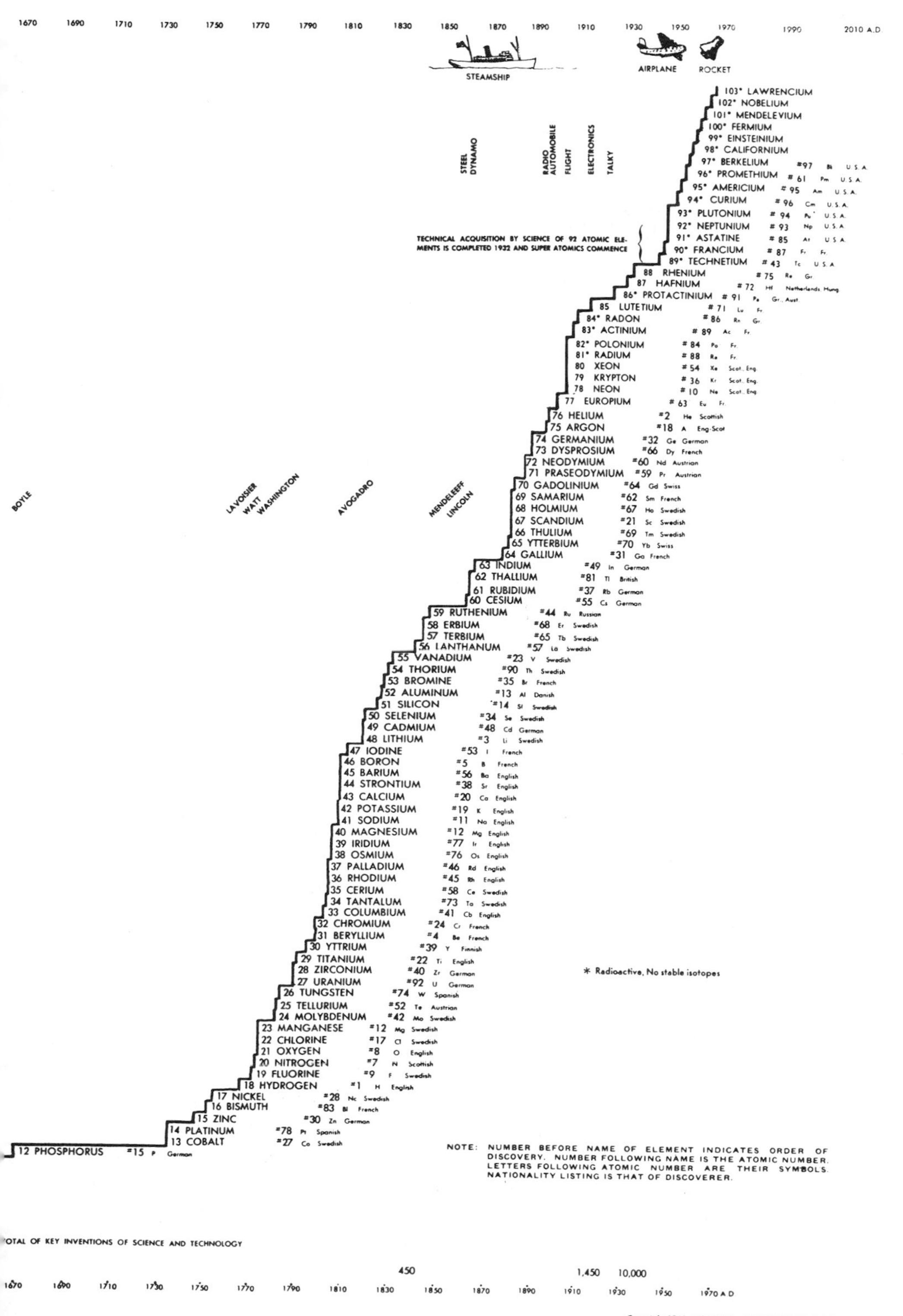

1670 1690 1710 1730 1750 1770 1790 1810 1830 1850 1870 1890 1910 1930 1950 1970 1990 2010 A.D.
STEAMSHIP
AIRPLANE
ROCKET
STEEL
DYNAMO
RADIO
AUTOMOBILE
FLIGHT
ELECTRONICS
TALKY
TECHNICAL ACQUISITION BY SCIENCE OF 92 ATOMIC ELEMENTS IS COMPLETED 1932 AND SUPER ATOMICS COMMENCE
BOYLE
LAVOISIER
WATT
WASHINGTON
AVOGADRO
MENDELEEFF
LINCOLN
103* LAWRENCIUM
102* NOBELIUM
101* MENDELEVIUM
100* FERMIUM
99* EINSTEINIUM
98* CALIFORNIUM
97* BERKELIUM #97 Bk U.S.A.
96* PROMETHIUM #61 Pm U.S.A.
95* AMERICIUM #95 Am U.S.A.
94* CURIUM #96 Cm U.S.A.
93* PLUTONIUM #94 Pu U.S.A.
92* NEPTUNIUM #93 Np U.S.A.
91* ASTATINE #85 At U.S.A.
90* FRANCIUM #87 Fr Fr.
89* TECHNETIUM #43 Tc U.S.A.
88 RHENIUM #75 Re Gr.
87 HAFNIUM #72 Hf Netherlands, Hung.
86* PROTACTINIUM #91 Pa Gr., Aust.
85 LUTETIUM #71 Lu Fr.
84* RADON #86 Rn Gr.
83* ACTINIUM #89 Ac Fr.
82* POLONIUM #84 Po Fr.
81* RADIUM #88 Ra Fr.
80 XEON #54 Xe Scot., Eng.
79 KRYPTON #36 Kr Scot., Eng.
78 NEON #10 Ne Scot., Eng.
77 EUROPIUM #63 Eu Fr.
76 HELIUM #2 He Scottish
75 ARGON #18 A Eng.-Scot.
74 GERMANIUM #32 Ge German
73 DYSPROSIUM #66 Dy French
72 NEODYMIUM #60 Nd Austrian
71 PRASEODYMIUM #59 Pr Austrian
70 GADOLINIUM #64 Gd Swiss
69 SAMARIUM #62 Sm French
68 HOLMIUM #67 Ho Swedish
67 SCANDIUM #21 Sc Swedish
66 THULIUM #69 Tm Swedish
65 YTTERBIUM #70 Yb Swiss
64 GALLIUM #31 Ga French
63 INDIUM #49 In German
62 THALLIUM #81 Tl British
61 RUBIDIUM #37 Rb German
60 CESIUM #55 Cs German
59 RUTHENIUM #44 Ru Russian
58 ERBIUM #68 Er Swedish
57 TERBIUM #65 Tb Swedish
56 LANTHANUM #57 La Swedish
55 VANADIUM #23 V Swedish
54 THORIUM #90 Th Swedish
53 BROMINE #35 Br French
52 ALUMINUM #13 Al Danish
51 SILICON #14 Si Swedish
50 SELENIUM #34 Se Swedish
49 CADMIUM #48 Cd German
48 LITHIUM #3 Li Swedish
47 IODINE #53 I French
46 BORON #5 B French
45 BARIUM #56 Ba English
44 STRONTIUM #38 Sr English
43 CALCIUM #20 Ca English
42 POTASSIUM #19 K English
41 SODIUM #11 Na English
40 MAGNESIUM #12 Mg English
39 IRIDIUM #77 Ir English
38 OSMIUM #76 Os English
37 PALLADIUM #46 Pd English
36 RHODIUM #45 Rh English
35 CERIUM #58 Ce Swedish
34 TANTALUM #73 Ta Swedish
33 COLUMBIUM #41 Cb English
32 CHROMIUM #24 Cr French
31 BERYLLIUM #4 Be French
30 YTTRIUM #39 Y Finnish
29 TITANIUM #22 Ti English
28 ZIRCONIUM #40 Zr German
27 URANIUM #92 U German
26 TUNGSTEN #74 W Spanish
25 TELLURIUM #52 Te Austrian
24 MOLYBDENUM #42 Mo Swedish
23 MANGANESE #12 Mg Swedish
22 CHLORINE #17 Cl Swedish
21 OXYGEN #8 O English
20 NITROGEN #7 N Scottish
19 FLUORINE #9 F Swedish
18 HYDROGEN #1 H English
17 NICKEL #28 Ni Swedish
16 BISMUTH #83 Bi French
15 ZINC #30 Zn German
14 PLATINUM #78 Pt Spanish
13 COBALT #27 Co Swedish
12 PHOSPHORUS #15 P German
* Radioactive, No stable isotopes
NOTE: NUMBER BEFORE NAME OF ELEMENT INDICATES ORDER OF DISCOVERY. NUMBER FOLLOWING NAME IS THE ATOMIC NUMBER. LETTERS FOLLOWING ATOMIC NUMBER ARE THEIR SYMBOLS. NATIONALITY LISTING IS THAT OF DISCOVERER.
TOTAL OF KEY INVENTIONS OF SCIENCE AND TECHNOLOGY
450
1,450
10,000
1670 1690 1710 1730 1750 1770 1790 1810 1830 1850 1870 1890 1910 1930 1950 1970 A.D.

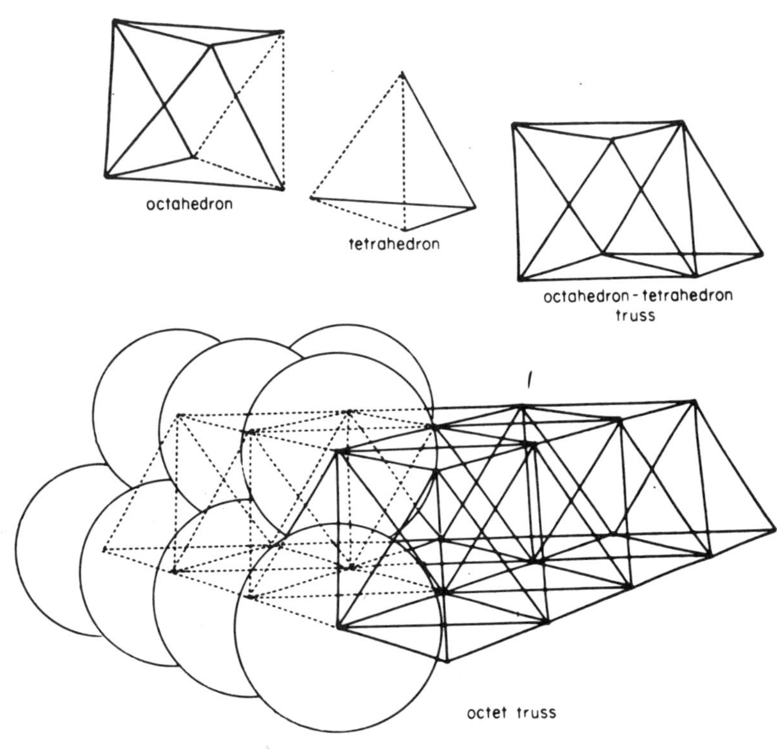

Fig. 420.01 *Octet Truss.*

its experience. Nature demonstrates her abhorrence of equilibrium when an airplane in flight slows to a speed that reduces the airfoil ''lift'' and brings the airplane's horizontal flight forces into equilibrium with Earth gravity's vertically Earthward pull. The plane is said then to stall, at which moment the plane's indeterminate direction makes it unmanageable because the rudder and elevator surfaces lack enough passing air to provide steerability, and the plane goes swiftly through equilibrium and into an Earthward-spinning plunge. Despite the untenability of equilibrium, it seemed to me that we could approach or employ it referentially as we employed a crooked line—the deliberately nonstraight (see Sec. 522) line that approaches but never reaches the perfect or exact. A comprehensive energy system could employ the positive and negative pulsations and intertransformative tendencies of equilibrium. The vector equilibrium became the logical model of such omnidimensional, omniexperience-accommodation studies. Because we have learned that scientists have experimental evidence only of waves and wavilinearity and no evidence of straight lines, it became evident that the radial and circumferential

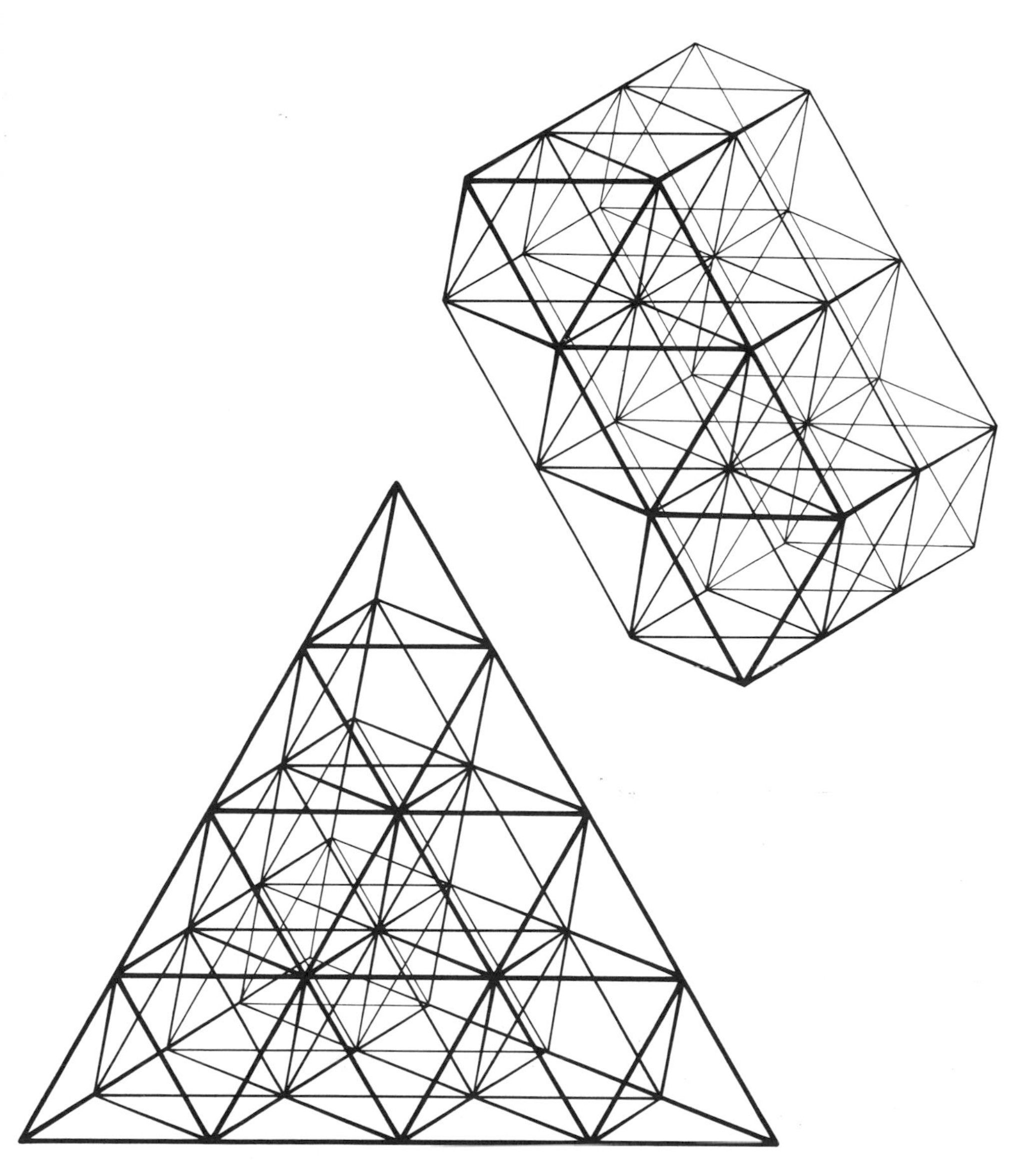

Fig. 420.02 When the centers of equiradius spheres in closest packing are joined with lines, an isotropic vector matrix is formed. This constitutes an array of equilateral triangles which is seen as the comprehensive coordination frame of reference of nature's most economical, most comfortable structural interrelationships employing 60-degree association and disassociation. This provides an omnirational accounting system which, if arbitrarily accounted on a 90-degree basis, becomes inherently irrational. The isotropic vector matrix demonstrates the capability of accommodating all symmetrically and asymmetrically terminaled, high-frequency energy vectors of any structural shaping.

vectors of the vector equilibrium must be wavilinear, which meant that as coil springs when compressioned will lessen in length and when tensed will be increased in length—ergo, the explosive disintegrative radial forces of Universe would compress and lessen in outward disintegrative length and would be well inside the closed-back-on-itself, hexagonally tensed, embracing vectors, indicating a higher effectiveness of tensile integrity of Universe over any locally disintegrative forces. The comprehensive vector-equilibrium system would also have to recognize all the topological interpatterning characteristics and components; also, as a quasi-equilibrious system, all of its structural component vectors would have to be approximately the same length; therefore, all the interangulation would have to be in aberration increments relative to 60 degrees as the equilibrious norm.

420.05 The closest-packing-of-spheres model coincides with the observed real world's atomic packing of like atoms with their own counterparts.

420.06 We find that the space compartmentation formed by the vectors connecting the sphere centers always consists only of tetrahedra and octahedra. The spheres in closest packing coincide with the Eulerian vertexes; the vectors between the sphere centers are the Eulerian edges; and the triangles so formed are the "faces."

420.07 All of the polygons formed by the interacting vectors of the isotropic vector matrix consist entirely of equilateral triangles and squares. The squares occur as equatorial cross sections of the octahedra. The triangles occur as the external facets of both the tetrahedra and the octahedra.

420.08 All the polygons are reducible to triangles and are not further reducible. All polyhedra are reducible to triangulation, i.e., to trusses and are not further reducible. Infinite polyhedra are infinitely faceted by basic trusses.

421.00 Function of Nucleus in Isotropic Vector Matrix

421.01 Because the spacing of absolutely compacted spheres is tangential and hexagonal in great-circle cross section around any one sphere, the contact points are always spaced equidistant from the centers of the spheres and from their immediately neighboring points, respectively; wherefore the dimensions of a system of lines joining each and all adjacent spherical centers are identical to the universal radii of the identical spheres and, therefore, to each other. Such a universal system of identically dimensioned lines, growing outwardly from any one nuclear vertex, constitutes a universal vector system in dynamic equilibrium, for all the force lines are of equal magnitude.

421.02 In the isotropic vector matrix, every vector leads from one nuclear center to another, and therefore represents the operational effect of a merging of any two or more force centers upon each other. Each vector is composed of two halves, each half belonging respectively to any two adjacent nuclear centers. Each half of the interconnecting vectors represents the radius of one of the two spheres tangent to one another at the vector midpoints.

421.03 Unity as represented by the internuclear vector modulus is of necessity always of the value of two, for it represents union of a minimum of two energy centers. (See Sec. 240.40.)

421.04 Each nuclear ball can have a neutral function among the aggregates. It is a nuclear ball whether it is in a planar array or in an omnidirectional array. It has a unique function in each of the adjacent systems that it bonds.

421.05 The nucleus can accommodate wave passage without disrupting the fundamental resonance of the octaves. The tetrahedron is the minimum, ergo prime, non-nucleated structural system of Universe. The vector equilibrium is the minimum, ergo prime, nucleated structural system of Universe.

421.10 Corollary: Identically dimensioned nuclear systems and layer growths occur alike, relative to each and every absolutely compacted sphere of the isotropic vector matrix conglomerate, wherefore the integrity of the individual energy center is mathematically demonstrated to be universal both potentially and kinetically (Sec. 240.50).

422.00 Octet Truss

422.01 In an isotropic vector matrix, there are only two clear-space polyhedra described internally by the configuration of interacting vectors: these are the regular tetrahedron and the regular octahedron operating as complementary space fillers. The single octahedron-tetrahedron deep truss system is known in synergetics as the *octet truss*.

422.02 The octet truss, or the isotropic vector matrix, is generated by the asymmetrical closest-packed sphere conglomerations. The nuclei are incidental.

422.03 When four tetrahedra of a given size are symmetrically intercombined by single bonding, each tetrahedron will have one of its four vertexes uncombined, and three combined with the six mutually combined vertexes symmetrically embracing to define an octahedron; while the four noncombined vertexes of the tetrahedra will define a tetrahedron twice the edge length of the four tetrahedra of given size; wherefore the resulting central space of the double-size tetrahedron is an octahedron. Together, these polyhedra comprise a common octahedron-tetrahedron system.

422.04 The tetrahedronated octahedron and all other regular symmetrical polyhedra known are described repetitiously by compounding two types of rational fraction asymmetric elements of the tetrahedron and octahedron. These elements are known in synergetics as the *A* and *B* Quanta Modules. (See Sec. 920.)

422.10 Force Distribution: In the three-way grid octet truss system, concentrated energy loads applied to any one point are distributed radially outward in nine directions and are immediately diffused into the finite hexagonally arranged six vectors entirely enclosing the six-way-distributed force. Each of the hexagon's six vertexes distribute the loads 18 ways to the next outwardly encircling vectors, which progressively diffusing system ultimately distributes the original concentrated energy force equally to all parts of the system as with a pneumatic tire. Thus the system joins together synergetically to distribute and inhibit the forces.

422.20 Geometry of Structure: Considered solely as geometry of structure, the final identification of the octet truss by the chemists and physicists as closest packing also identifies the octet truss and vector equilibria structuring as amongst the prime cosmic principles permeating and facilitating all physical experience.

423.00 60-Degree Coordination

423.01 In the octet truss system, all the vectors are of identical length and all the angles around any convergence are the same. The patterns repeat themselves consistently. At every internal convergence, there are always 12 vectors coming together, and they are always convergent at 60 degrees with respect to the next adjacent ones.

423.02 There are angles other than 60-degrees generated in the system, as for instance the square equatorial mid-section of the octahedron. These angles of other than 60-degrees occur between nonadjacently converging vectorial connectors of the system. The prime structural relationship is with the 60-degree angle.

423.03 Fundamental 60-degree coordination operates either circumferentially or radially. This characteristic is lacking in 90-degree coordination, where the hypotenuse of the 90-degree angles will not be congruent and logically integratable with the radials.

423.04 When we begin to integrate our arithmetical identities, as for instance n^2 or n^3, with a 60-degree coordination system, we find important coincidence with the topological inventories of systems, particularly with the isotropic vector matrix which makes possible fourth- and fifth-power modeling.

423.10 **Hexagon as Average of Angular Stabilizations:** The irrational radian and *pi* (π) are not used by nature because angular accelerations are in finite package impellments * which are chordal (not arcs) and produce hexagons because the average of all angular stabilizations from all triangular interactions average at 60 degrees—ergo, radii and 60-degree chords are equal and identical; ergo, six 60-degree chords equal one frequency cycle; ergo, one quantum. Closest packed circles or spheres do not occupy all area or space, but six-triangled, nucleated hexagons do constitute the shortest route cyclic enclosure of closest-packed nucleation and do uniformly occupy all planar area or volumetric space.

424.00 *Transformation by Complementary Symmetry*

424.01 The octet truss complex is a precessionally nonredundant, isotropic vector-tensor evolutionary relationship whose energy transformation accountings are comprehensively rational—radially and circumferentially—to all chemical, biological, electromagnetic, thermodynamic, gravitational, and radiational behaviors of nature. It accommodates all transformations by systematic complementary symmetries of concentric, contractile, involutional, turbo-geared, rational, turbulence-accommodating, inside-outing, positive-to-negative-to-equilibrium, pulsative coordinate displacements.

424.02 Thus we see both the rational energy quantum of physics and the topological tetrahedron of the isotropic vector matrix rationally accounting all physical and metaphysical systems and their transformative transactions. (See Sec. 620.12.)

424.03 This indefinitely extending vector system in dynamic equilibrium provides a rational frame of reference in universal dimension for measurement of any energy conversion or any degree of developed energy factor disequilibrium or its predictable reaction developments—of impoundment or release—ergo, for atomic characteristics.

425.00 *Potentiality of Vector Equilibrium*

425.01 Where all the frequency modulations of the local vectors are approximately equal, we have a potentially local vector equilibrium, but the operative vector frequency complexity has the inherent qualities of accommodating both proximity and remoteness in respect to any locally initiated actions, ergo, a complex of relative frequencies and velocities of realization lags are accommodated (*Corollary* at Sec. 240.37).

* For a related concept see Secs. 1009.50, Accleration, and 1009.60, Hammerthrower.

426.00 Spherics

426.01 An isotropic vector matrix can be only omnisymmetrically, radiantly, and "broadcastingly" generated, that is, propagated and radiantly regenerated, from only one vector equilibrium origin, although it may be tuned in, or frequency received, at any point in Universe and thus regenerate local congruence with any of its radiantly broadcast vector structurings.

426.02 An isotropic vector matrix can be only radiantly generated at a "selectable" (tunable) propagation frequency and vector-size (length) modular spacing and broadcast omnidirectionally or focally beamed outward from any vector-center-fixed *origin* such that one of its symmetrically regenerated vector-convergent fixes will be congruent with any other identical wavelength and frequency attuned and radiantly reachable vector-center fixes in Universe.

426.03 In time-vectorable Universe, the maximal range of radiant-regenerative reachability in time is determined by the omnidirectional velocity of all radiation: c^2, i.e., $(186{,}000)^2$.*

426.10 Definition of a Spheric: A "spheric" is any one of the rhombic dodecahedra symmetrically recurrent throughout an isotropic-vector-matrix geometry wherein the centers of area of each of the rhombic dodecahedra's 12 diamond facets are exactly and symmetrically tangent at 12 omnisymmetrically interarrayed points lying on the surface of any one complete sphere, entirely contained within the spheric-identifying rhombic dodecahedra, with each of any such rhombic dodecahedra's tangentially contained spheres symmetrically radiant around *every other,* i.e., every omnidirectionally alternate vertex of every isotropic vector matrix, with the 12 points of spherical tangency of each of the rhombic dodecahedra exactly congruent also with the 12 vertexes of the vector equilibrium most immediately surrounding the vertex center of the sphere, each of whose 12 vector equilibrium radii are the special set of isotropic vector matrix vectors leading outwardly from the sphere's center vertex to the 12 most immediately surrounding vertexes.

426.11 These 12 vertexes, which are omni-equidistant from every other vertex of the isotropic vector matrix, also occur at the diamond-face centers of the "spheric" rhombic dodecahedra and are also the points of tangency of 12 uniradius spheres immediately and omni-intertangentially surrounding (i.e.,

* Within a week after this paragraph was drafted *The New York Times* of 22 November 1972 reported that the National Bureau of Standards laboratories at Boulder, Colorado, had determined the speed of light as "186,282.3960 miles per second with an estimated error margin no greater than 3.6 feet a second . . . Multiplying wavelength by frequency gives the speed of light."

closest-packing) the sphere first defined by the first rhombic dodecahedron. Each rhombic dodecahedron symmetrically surrounds every radiantly alternate vertex of the isotropic vector matrix with the other radiantly symmetrical unsurrounded set of vertexes always and only occurring at the diamond-face centers of the rhombic dodecahedra.

426.12 One radiantly alternate set of vertexes of the isotropic vector matrix always occurs at the spheric centers of omni-closest-packed, uniradius spheres; whereas the other radiantly alternate set of vertexes of the isotropic vector matrix always occurs at the spheric intertangency points of omni-closest-packed, uniradius spheres.

426.20 **Allspace Filling:** The rhombic dodecahedra symmetrically fill allspace in symmetric consort with the isotropic vector matrix. Each rhombic dodecahedron defines exactly the unique and omnisimilar domain of every radiantly alternate vertex of the isotropic vector matrix as well as the unique and omnisimilar domains of each and every interior-exterior vertex of any aggregate of closest-packed, uniradius spheres whose respective centers will always be congruent with every radiantly alternate vertex of the isotropic vector matrix, with the corresponding set of alternate vertexes always occurring at all the intertangency points of the closest-packed spheres.

426.21 The rhombic dodecahedron contains the most volume with the least surface of all the allspace-filling geometrical forms, ergo, rhombic dodecahedra are the most economical allspace subdividers of Universe. The rhombic dodecahedra fill and symmetrically subdivide allspace most economically, while simultaneously, symmetrically, and exactly defining the respective domains of each sphere as well as the spaces between the spheres, the respective shares of the inter-closest-packed-sphere -interstitial space. The rhombic dodecahedra are called "spherics," for their respective volumes are always the unique closest-packed, uniradius spheres' volumetric domains of reference within the electively generatable and selectively "sizable" or tunable of all isotropic vector matrixes of all metaphysical "considering" as regeneratively reoriginated by any thinker anywhere at any time; as well as of all the electively generatable and selectively tunable (sizable) isotropic vector matrixes of physical electromagnetics, which are also reoriginatable physically by anyone anywhere in Universe.

426.22 The rhombic dodecahedron's 12 diamond faces are the 12 unique planes always occurring perpendicularly to the midpoints of all vector radii of all the closest-packed spheres whenever and wherever they may be metaphysically or physically regenerated, i.e., perpendicular to the midpoints of all vectors of all isotropic vector matrixing.

426.30 **Spherics and Modularity:** None of the rhombic dodecahedra's edges are congruent with the vectors of the isotropic vector matrix, and only six of the rhombic dodecahedra's 14 vertexes are congruent with the symmetrically co-reoccurring vertexes of the isotropic vector matrix. The other eight vertexes of the rhombic dodecahedra are congruent with the centers of volume of the eight edge-interconnected tetrahedra omnisymmetrically and radiantly arrayed around every vertex of the isotropic vector matrix, with all the edges of all the tetrahedra always congruent with all the vectors of the isotropic vector matrix, and all the vertexes of all the tetrahedra always congruent with the vertexes of the isotropic vector matrix, all of which vertexes are always most economically interconnected by three edges of the tetrahedra.

426.31 A spheric is any one of the rhombic dodecahedra, the center of each of whose 12 diamond facets is exactly tangent to the surface of each sphere formed equidistantly around each vertex of the isotropic vector matrix.

426.32 A spheric has 144 *A* and *B* modules, and there are 24 *A* Quanta Modules (see Sec. 920 and 940) in the tetrahedron, which equals $^1/_6$th of a spheric. Each of the tetrahedron's 24 modules contains $^1/_{144}$th of a sphere, plus $^1/_{144}$th of the nonsphere space unique to the individual domain of the specific sphere of which it is a $^1/_{144}$th part, and whose spheric center is congruent with the most acute-angle vertex of each and all of the *A* and *B* Quanta Modules. The four corners of the tetrahedron are centers of four embryonic (potential) spheres.

426.40 **Radiant Valvability of Isotropic-Vector-Matrix-Defined Wavelength:** We can resonate the vector equilibrium in many ways. An isotropic vector matrix may be both radiantly generated and regenerated from any vector-centered fixed origin in Universe such that one of its vertexes will be congruent with any other radiantly reachable center fix in Universe; i.e., it can communicate with any other noninterfered-with point in Universe. The combined reachability range is determined by the omnidirectional velocity of all radiation, c^2 within the availably investable time.

426.41 The rhombic dodecahedron's 144 modules may be reoriented within it to be either radiantly disposed from the contained sphere's center of volume or circumferentially arrayed to serve as the interconnective pattern of six $^1/_6$th-spheres, with six of the dodecahedron's 14 vertexes congruent with the centers of the six individual $^1/_6$th spheres that it interconnects. The six $^1/_6$th spheres are completed when 12 additional rhombic dodecahedra are close-packed around it.

426.42 The fact that the rhombic dodecahedron can have its 144 modules oriented as either introvert-extrovert or as three-way circumferential provides its valvability between broadcasting-transceiving and noninterference relaying. The first radio tuning crystal must have been a rhombic dodecahedron.

426.43 Multiplying wavelength by frequency equals the speed of light. We have two experimentally demonstrable radiational variables. We have to do whatever we do against time. Whatever *we* may be, each *we* has only so much commonly experienceable time in scenario Universe within which to articulate thus and so. Therefore, the vector equilibrium's radiant or gravitational "realizations" are always inherently geared to or tuned in with the fundamental time-sizing of $\approx$186,000 mps, which unique time-size-length increments of available time can be divided into any desirable frequency. One second is a desirable, commonly experienceable increment to use, and within each unit of it we can reach $\approx$ 186,000 miles in any non-frequency-interfered-with direction.

426.44 Wavelength times frequency is the speed of all radiation. If the frequency of the vector equilibrium is four, its vector radius, or basic wavelength = 186,000/4 miles reachable within one second = 46,500 reach-miles. Electromagnetically speaking, the unarticulated vector equilibrium's one-second vector length is always 186,282.396 miles.

426.45 We multiply our frequency by the number of times we divide the vector of the vector equilibrium, and that gives c^2; our reachable points in Universe will multiply at a rate of $F^2 \times 10 + 2$.

426.46 All the relative volumetric intervaluations of all the symmetric polyhedra and of all uniradius, closest-packed spheres are inherently regenerated in omnirational respect to isotropic vector matrixes, whether the matrixes are inadvertently—i.e., subjectively—activated by the size-selective, metaphysical-consideration initiatives, whether they are objectively and physically articulated in consciously tuned electromagnetic transmission, or whether they are selectively tuned to receive on that isotropic-vector-matrix-defined "wavelength."

426.47 Humans may be quite unconscious of their unavoidable employment of isotropic vector matrix fields of thought or of physical articulations; and they may oversimplify or be only subconsciously attuned to employ their many cosmically intertunable faculties and especially their conceptual and reasoning faculties. However, their physical brains, constituted of quadrillions times quadrillions of atoms, are always and only most economically interassociative, interactive, and intertransforming only in respect to the closest-packed isotropic vector matrix fields which altogether subconsciously accommodate the conceptual geometry picturing and memory storing of each individual's evolutionary accumulation of special-case experience happenings, which human inventories are accumulatingly stored isotropic-vector-matrix-wise in the brain and are conceptually retrievable by brain and are both subconsciously and consciously reconsidered reflexively or by reflex-shunning mind.

427.00 Nuclear Computer Design

427.01 Though I have found an omnidirectional vector equilibrium matrix and the complex of momentarily positively and negatively asymmetrical intertransformabilities pulsating through the equilibrious state, I knew that nature would never allow temporal humans to omniarrest cosmic kinetics at the timeless, i.e., eternal equilibrium zero. But experimenting in cryogenics, taking energy-as-heat out of the insulatingly isolated liquefied gaseous element system approaching absolute zero, we learn that as the temperature gets lower and lower, an increasingly orderly and an increasingly symmetrical, microgeometrical patterning occurs—the Platonic solids appear to become more symmetrically uniform. Contrariwise, when energy-as-heat is progressively reintroduced, the kinetics increase and the complex of conceptual behavior becomes progressively asymmetric. At lowest cryogenic temperatures the omnigrametric interpatterning approaches isotropic vector matrix equilibrium.

427.02 The progressive energy-starving experimental strategy reveals that nature always transforms through, and relative centrally to, the omni-isotropic-vector-matrix equilibrium, while kinetically emphasizing the mildly off-center asymmetric aspects. Nature grows her crystals positively or negatively askew—she twists and spirals around the local, three-way great-circle grid systems in the alternate positive-negative geodesic complementations. Such kinetic considerations of closest packing are significant.

427.03 The isotropic vector matrix equilibrium multiplies omnidirectionally with increasing frequency of concentric, vector-equilibrium-conformed, closest-packed uniradius sphere shells, conceptually disclosing the cosmically prime unique sequence of developed interrelationships and behaviors immediately surrounding a prime nucleus. While the physicist processes his nuclear problems with nonconceptual mathematics, the conceptual isotropic vector matrix equilibria model provides a means of comprehending all the electromagnetic and nonelectromagnetic energy valving and angular shunting controls of the solid state transistors.

427.04 With one layer of spheres around the nuclear sphere we will get one set of angular interrelationships of the surrounding spheres with the nucleus and with one another. With two layers of spheres around the nuclear sphere a different angular relationship between the nuclear sphere and its intersurrounding spheres occurs (see Sec. 415). At the third layer of enclosure some of the angular interrelationship patternings begin to repeat themselves. Thus we are able to inventory what we are going to call *a nuclear set of unique interrelationship patterns*.

427.05 The isotropic vector matrix multiplies concentrically. But because vectors are discrete, the isotropic vector matrix's lines do not go to infinity. Their length must always represent sum-totally the total energy of eternally regenerative physical Universe. No matter how high the internal frequency of finite Universe, the overall vector equilibrium is of unit magnitude. This magnitude corresponds to that of the speed of radiation uninterfered with in vacuo. We find that the different frequencies in their phases of symmetry identify precisely with what we now call the Magic Numbers identifying the successively reoccurring five peaks in relative abundance of atomic isotopes. (See Sec. 995.)

427.06 I am confident that I have discovered and developed the conceptual insights governing the complete family of variables involved in realization by humanity of usable access to the ultimate computer . . . ultimate meaning here: the most comprehensive, incisive and swiftest possible information-storing, retrieving, and variably processing facility with the least possible physical involvement and the least possible investment of human initiative and cosmic energization.

427.07 Science evolved the name "solid state" physics when, immediately after World War II, the partial conductors and partial resistors—later termed "transistors"—were discovered. The phenomena were called "solid state" because, without human devising of the electronic circuitry, certain small metallic substances accidentally disclosed electromagnetic pattern-holding, shunting, route-switching, and frequency-valving regularities, assumedly produced by the invisible-to-humans atomic complexes constituting those substances. Further experiment disclosed unique electromagnetic circuitry characteristics of various substances without any conceptual model of the "subvisible apparatus." Ergo, the whole development of the use of these invisible behaviors was conducted as an intelligently resourceful trial-and-error strategy in exploiting invisible and uncharted-by-humans natural behavior within the commonsensically "solid" substances. The addition of the word "state" to the word "solid" implied "regularities" in an otherwise assumedly random conglomerate. What I have discovered goes incisively and conceptually deeper than the blindfolded assumptions and strategies of solid state physics—whose transistors' solid state regularities seemingly defied discrete conceptuality and scientific generalization and kinetic omnigramming.

427.10 Invisible Circuitry of Nature

427.11 We have here the disclosure of a new phase of geometry employing the invisible circuitry of nature. The computer based on such a design could be no bigger than the subvisibly dimensioned domain of a pinhead's glitter, with closures and pulsations which interconnect at the vector equilibrium stage and disconnect at the icosahedron stage in Milky-Way-like remoteness from one another of individual energy stars.

427.12 As we get into cryogenics—taking energy-as-heat out of the system—the geometries become more regular and less asymmetric, thus fortifying the assumptions of synergetics because the geometrically ''twinkling'' asymmetries of kinetics progressively subside and approach, but do not quite attain, absolute cessation at the isotropic vector equilibrium state.

427.13 The atomically furnished isotropic vector matrix can be described as an omnidirectional matrix of ''lights,'' as the four-dimensional counterpart of the two-dimensional light-bulb-matrix of the Broadway-and-Forty-second-Street, New York City billboards with their fields of powerful little light bulbs at each vertex which are controlled remotely off-and-on in intensity as well as in color. Our four-dimensional, isotropic vector matrix will display all the atom ''stars'' concentrically matrixed around each isotropic vector equilibrium's nuclear vertex. By ''lighting'' the atoms of which they consist, humans' innermost guts could be illustrated and illuminated. Automatically turning on all the right lights at the right time, atomically constituted, center-of-being light, ''you,'' with all its organically arranged ''body'' of lights omnisurrounding ''you,'' could move through space in a multidimensional way just by synchronously activating the same number of lights in the same you-surrounding pattern, with all the four-dimensional optical effect (as with two-dimensional, planar movies), by successively activating each of the lights from one isotropic vector vertex to the next, with small, local ''movement'' variations of ''you'' accomplished by special local matrix sequence programmings.

427.14 We could progressively and discretely activate each of the atoms of such a four-dimensional isotropic vector matrix to become ''lights,'' and could move a multidimensional control ''form'' through the isotropic multidimensional circuitry activating field. The control form could be a ''sphere,'' a ''vector equilibrium,'' or any other system including complex you-and-me, et al. This multidimensional scanning group of points can be programmed multidimensionally on a computer in such a manner that a concentric spherical cluster of four-dimensional ''light'' points can be progressively ''turned on'' to comprise a ''substance'' which seemingly moves from here to there.

427.15 This indeed may be what Universe is doing! Employing a scanner of each of our atoms, this is one way humans could have been radio-transmitted and put aboard Earth from any place in Universe. The naked human eye cannot differentiate visually the separate dots of a matrix when their frequency of uniform-moduled spaced occurrence is greater than one hundred to the linear inch, or ten thousand to the square inch, or one million to the cubic inch. Let us radiantly activate isotropically and modularly grouped local atoms of a human's physical organism in such a manner that only one million per cubic inch out of all the multibillions of actual atoms per cubic inch of which humans consist, are radiationally, ergo visibly, activated. The human, thus omni-internally illumined by the local one-in-one-million atomic ''street lamps,'' could be realistically scanned by discrete ''depth-sounding'' devices and programmed to move ''visibly'' through an omnidimensional, high-frequency, isotropic light matrix field ''mass.''

427.16 Employing as broadcastable channels the 25 great circles of the vector equilibrium all of which pass through all the ''*K*'' (kissing) points of intertangency of all uniform radius, closest-packed spheres of all isotropic vector matrixes; and employing as local holding patterns the 31 great circles of the icosahedron; and employing as a resonance field all the intertransforming spheres and between-sphere spaces; and employing the myriadly selectable, noninterfering frequencies of such propagatable intertransformation resonance; it is evidenced that the isotropic vector matrixes of various atomic elements may be programmed to receive, store, retrieve, and uniquely constellate to provide computer functioning of unprecedented capacity magnitude within approximately invisible atomic domains. The control mechanism for the operational programming of such microcosmic ''computers'' will be visible and dextrous and will be keyed by the Mite orientations of the prime-number-one-volumed ''Couplers.'' *

427.17 The ultra micro computer (UMC) employs step-up, step-down, transforming visible controls between the invisible circuitry of the atomic computer complex pinhead-size programmer and the popular outdoor, high-in-the-sky, ''billboard'' size, human readability.

430.00 Vector Equilibrium

430.01 Definition

430.011 The geometric form most compactly developed from the closest packing of spheres around one nuclear sphere is not that of a composite

* For an exposition of the behavior of Mites and Couplers see Sec. 953 and 954.

sphere, but is always a polyhedron of 14 faces composed of six squares and eight triangles, with 12 vertexes extending in tangential radius from the original 12 spheres surrounding the nucleus sphere. (See illustration 413.01.)

430.02 It is called the vector equilibrium because the radials and the circumferentials are all of the same dimension and the tendencies to both explode and implode are symmetrical. That the explosive and implosive forces are equal is shown by the four-dimensional hexagonal cross sections whose radial and circumferential vectors balance. The eight triangular faces reveal four opposite pairs of single-bonded tetrahedra in a positive and negative tetrahedral system array with a common central vertex and with coinciding radial edges. The four hexagonal planes that cross each other at the center of the vector-equilibrium system are parallel to the four faces of each of its eight tetrahedra. Six square faces occur where the six half-octahedra converge around the common vector-equilibrium nuclear vertex.

430.03 In terms of vectorial dynamics, the outward radial thrust of the vector equilibrium is exactly balanced by the circumferentially restraining chordal forces: hence the figure is an equilibrium of vectors. All the edges of the figure are of equal length, and this length is always the same as the distance of any of its vertexes from the center of the figure. The lines of force radiating from its center are restrainingly contained by those binding inward arrayed in finite closure circumferentially around its periphery—barrel-hooping. The vector equilibrium is an omnidirectional equilibrium of forces in which the magnitude of its explosive potentials is exactly matched by the strength of its external cohering bonds. If its forces are reversed, the magnitude of its contractive shrinkage is exactly matched by its external compressive archwork's refusal to shrink.

430.04 The vector equilibrium is a truncated cube made by bisecting the edges and truncating the eight corners of the cube to make the four axes of the four planes of the vector equilibrium. The vector equilibrium has been called the "cuboctahedron" or "cubo-octahedron" by crystallographers and geometers of the non-experimentally-informed and non-energy-concerned past. As such, it was one of the original 13 Archimedean "solids."

430.05 The vector equilibrium is the common denominator of the tetrahedron, octahedron, and cube. It is the decimal unit within the octave system. Double its radius for octave expansion.

430.06 The vector equilibrium is a system. It is not a structure. Nor is it a *prime volume,* because it has a nucleus. It is the *prime nucleated system.* The eight tetrahedra and the six half-octahedra into which the vector equilibrium may be vectorially subdivided are the volumes that are relevantly involved.

431.00 Volume

431.01 The vector equilibrium consists of six one-half octahedra, each with a volume of two ($6 \times 2 = 12$), and eight tetrahedra each with a volume of one, so $8 + 12 = 20$, which is its exact volume. (See illustration 222.30.)

431.02 The volume of a series of vector equilibria of progressively higher frequencies is always frequency to the third power times 20, or $20F^3$, where $F =$ frequency. When the vector equilibrium's frequency is one (or radiationally inactive), its volume is $20 \times 1^3 = 20$.

431.03 But *frequency,* as a word key to a functional concept, never relates to the word *one* because frequency obviously involves some plurality of events. As a one-frequency, ergo sub-frequency, system, the vector equilibrium is really subsize, or a size-independent, conceptual integrity. Therefore, frequency begins with two—where all the radials would have two increments. When the edge module of a cube is one, its volume is one; when the edge module of a cube is two, its volume is eight. But when the edge module of a vector equilibrium is one, its volume is 20. A nuclear system is subsize, subfrequency. Equilibrious unity is 20; its minimum frequency state is $160 = 2^5 \times 5$. This is one of the properties of 60-degree coordination.

431.04 Looking at a two-frequency vector equilibrium (with all the radials and edge units divided into two) and considering it as the domain of a point, we find that it has a volume of 480 *A* and *B* Modules. The formula of the third power of the frequency tells us the exact number of quanta in these symmetrical systems, in terms of quantum accounting and in terms of the *A* and *B* Modules (see Chapter 9, Modelability).

432.00 Powering

432.01 The vector equilibrium makes it possible to make conceptual models of fourth-, fifth-, and sixth-dimensional omniexperience accounting by using tetrahedroning. If we have a volume of 20 around a point, then two to the fourth power (16) plus two to the second power (4) equals 20. We can then accommodate these powerings around a single point.

432.02 Using frequency to the third power with a no-frequency nucleus, the vector equilibrium models all of the first four primes. For instance, the number 48 (in 480) is 16×3. Three is a prime number, and 16 is two to the fourth power: that is 48, and then times 10. Ten embraces the prime numbers five times the number two; so instead of having 16 times 2, we can call it 32, which is two to the fifth power. The whole 480-moduled vector equilibrium consists of the prime number one times two to the fifth power, times three, times five ($1 \times 2^5 \times 3 \times 5$). These are the first four prime numbers.

432.03 Using frequency to the third power with a two-frequency nucleus, we have $2^3 \times 2^5 = 2^8$. If the frequency is two, we have two to the eighth power in the model times three times five ($2^8 \times 3 \times 5$).

432.04 In a three-frequency system, we would have three to the third power times three, which makes three to the fourth power, which we would rewrite as $2^5 \times 3^4 \times 5$. We get two kinds of four-dimensionality in here. There is a prime dimensionality of three to the fourth power (3^4). And there is another kind of four-dimensionality if the frequency is four, which would be written $2^5 \times 3 \times 5$. But since it is frequency to the third power, and since four is two times two (2×2) or two to the second power (2^2), we would add two to make two to the seventh power (2^7), resulting in $2^7 \times 3 \times 5$. If the frequency is five, it would then be two to the fifth power (2^5) times three, because frequency is to the third power times five, which makes five to the fourth power. Quite obviously, multidimensionality beyond three dimensions is experienceably, i.e., conceptually, modelable in synergetics accounting.

433.00 Outside Layer of Vector Equilibrium

433.01 The unique and constantly remote but-always-and-only co-occurring geometrical ''starry'' surroundment ''outsideness'' of the nucleated vector equilibrium is always an icosahedron, but always occurring only as a single layer of vertexes of the same frequency as that of the nuclear vector equilibrium's outermost vertexial layer.

433.02 There may be multilayer vector equilibria—two-frequency, three-frequency, four-frequency, or whatever frequency. The circumferential vector frequency will always be identical to that of its radial vector frequency contraction of the vector equilibrium's outer layer of unit radius spheres by local surface rotation of that outer layer's six square arrays of non-closest-single-layer packing of tangent spheres inter-rearranging into closest triangular packing as in the vector equilibrium's eight triangular facets, thus transforming the total outer layer into the icosahedron of equal outer edge length to that of the vector equilibrium, but of lesser interior radius than the vector equilibrium of the same outer edge length, and therefore of lesser interior volume than that of the vector equilibrium, ergo unable to accommodate the same number of interiorally-closest-packed, nuclear-sphere-centered unit radius spheres as that of the vector equilibrium. The icosahedron's multifrequenced outer layer surface arrays of unit radius, closest-planar-packed spheres cannot accommodate either concentric layers of unit radius closest-packed spheres nor—even at zero frequency—can the icosahedron's 12-ball, omni-intertangentially triangulated outer shell accommodate one nuclear sphere of the same radius as that of its shell spheres. Icosahedral outer shell arrays of identical frequency to that of the vector equilibria of the same frequency, can therefore only occur as single-layer, symmetrical, enclosure arrays whose individual spheres cannot be tangent to one another but must be remotely equipositioned from one another, thus to form an omni-intertriangulated, icosahedrally conformed starry array, remotely and omnisurroundingly occupying the vector equilibrium's sky at an omnistar orbit-permitting equidistance remoteness around the vector equilibrium whose outer shell number of spheres exactly corresponds to the number of the icosahedron's "stars." This geometrical dynamically interpositioning integrity of relationship strongly suggests the plurality of unique electron shell behaviors of all the chemical elements' atoms, and the identical number relationships of the atoms' outer layer protons and its electrons; and the correspondence of the vector equilibrium's number of concentric closest-packed, nucleus-enclosing layers with the number of quantum-jump-spaced electron orbit shells; and finally the relative volume relationship of equi-edged vector equilibria and icosahedra, which is, respectively, as 20 is to 18.51, which suggests the relative masses of the proton and the electron, which is as 1 : $^{1}/_{1836}$.

440.00 Vector Equilibrium as Zero Model

440.01 Equilibrium between positive and negative is zero. The vector equilibrium is the true zero reference of the energetic mathematics. Zero pulsation in the vector equilibrium is the nearest approach we will ever know to eternity and god: the zerophase of conceptual integrity inherent in the positive and negative asymmetries that propagate the differentials of consciousness.

440.02 The vector equilibrium is of the greatest importance to all of us because all the nuclear tendencies to implosion and explosion are reversible and are always in exact balance. The radials and the circumferentials are in balance. But the important thing is that the radials, which would tend to explode since they are outwardly pushing, are always frustrated by the tensile finiteness of the circumferential vectors, which close together in an orderly manner to cohere the disorderly asundering. When the radial vectors are tensilely contractive and separately implosive, they are always prevented from doing so by the finitely closing pushers or compressors of the circumferential set of vectors. The integrity of Universe is implicit in the external finiteness of the circumferential set and its surfaçe-layer, close-packing, radius-contracting proclivity which always encloses the otherwise divisive internal radial set of omnidirectional vectors.

440.03 All the internal, or nuclear, affairs of the atom occur internally to the vector equilibrium. All the external, or chemical, compoundings or associations occur externally to the vector equilibrium. All the phenomena external to—and more complex than—the five-frequency vector equilibria relate to chemical compounds. Anything internal to—or less complex than—the five-frequency vector equilibrium relates principally to single atoms. Single atoms maintain omnisymmetries; whereas chemical compounds may associate as polarized and asymmetrical chain systems.

440.04 The vector equilibrium is the anywhere, anywhen, eternally regenerative, event inceptioning and evolutionary accommodation and will never be seen by man in any physical experience. Yet it is the frame of evolvement. It is not in rotation. It is sizeless and timeless. We have its mathematics, which deals discretely with the chordal lengths. The radial vectors and circumferential vectors are the same size.

440.05 The vector equilibrium is a condition in which nature never allows herself to tarry. The vector equilibrium itself is never found exactly symmetrical in nature's crystallography. Ever pulsive and impulsive, nature never pauses her cycling at equilibrium: she refuses to get caught irrecoverably at the zero phase of energy. She always closes her transformative cycles at the maximum positive or negative asymmetry stages. See the delicate crystal asymmetry in nature. We have vector equilibriums mildly distorted to asymmetry limits as nature pulsates positively and negatively in respect to equilibrium. Everything that we know as reality has to be either a positive or a negative aspect of the omnipulsative physical Universe. Therefore, there will always be positive and negative sets that are ever interchangeably intertransformative with uniquely differentiable characteristics.

440.06 The vector equilibrium is at once the concentric push-pull interchange, vectorial phase or zone, of neutral resonance which occurs between outwardly pushing wave propagation and inwardly pulling gravitational coherence.

440.07 All the fundamental forms of the crystals are involved in the vector equilibrium. It is a starting-point—not anything in its own right—if it is a vector equilibrium.

440.08 As the circumferentially united and finite great-circle chord vectors of the vector equilibrium cohere the radial vectors, so also does the metaphysical cohere the physical.

441.00 Vector Equilibrium as Zero Tetrahedron

441.01 **Emptiness at the Center:** All four planes of all eight tetrahedra, i.e., 32 planes in all, are congruent in the four visible planes passing through their common vector equilibrium center. Yet you see only four planes. Both the positive and the negative phase of the tetrahedra are in congruence in the center. They are able to do this because they are synchronously discontinuous. Their common center provides the locale of an absolutely empty event.

441.02 Vector equilibrium accommodates all the intertransformings of any one tetrahedron by polar pumping, or turning itself inside out. Each vector equilibrium has four directions in which it could turn inside out. It uses all four of them through the vector equilibrium's common center and generates eight tetrahedra. The vector equilibrium is a tetrahedron exploding itself, turning itself inside out in four possible directions. So we get eight: inside and outside in four directions. The vector equilibrium is all eight of the potentials.

441.03 **Terminal Condition:** The formula for the number of balls in any one of the concentric layers of the vector equilibrium is always $10F^2+2$. The center ball of a vector equilibrium is the zero layer. The layer frequency is zero just as in the first layer the frequency is one. So zero times 10 is zero; to the second power is zero; plus two is two. So the center ball has a value of two. The significance is that it has its concavity and its convexity. It has both insideness and outsideness. Its center is as far as you can go inward. You turn yourself inside out and come out in the outside direction. Its inbound shell and its outbound shell are equally valid, and though you see them as congruent and as one, they are two. This central sphere center is a cosmic terminal condition.

441.04 Let us consider a tetrahedron, which also always has an externality and an internality. At its internal center is its terminal turn-around-and-come-outward-again condition. This is exactly why in physics there is a cosmic limit point at which systems turn themselves inside out. They get to the outside and they turn themselves inside out and come the other way. This is why radiation does not go off into a higher velocity. Radiation gets to a maximum velocity unrestrained in vacuo and then turns itself inward again—it becomes gravity. Then gravity comes to its maximum concentration and turns itself around and goes outward—becomes radiation again.

441.05 This Boltzmann's import-export-import-export; entropy-syntropy-entropy-syntropy, cosmically complementary, human-heartlike, eternally pulsative, evolutionary regeneration system, also locally manifests itself in the terrestrial biosphere as the ever alternatively, omni-interpulsing, barometric *highs* and *lows* of the weather.

441.10 **Coordinate Symmetry:** In coordinate symmetry, as the faces of the tetrahedron move in toward the opposite vertex, the volume gets less at an exponential velocity of the third power, its surfaces diminish at a second-power rate of change, and its lines shorten at a covariation rate of the first power. When all four of the tetrahedral faces come to congruence with the same common nucleus of the vector equilibrium, all three of these different rates of size change come synchronously to common zero size. The constant tetrahedral fourness of vertexes and faces, sixness of edges, insideness and outsideness, convexity and concavity—these integrated constants of conceptuality never change.

441.20 **Turbining:** In looking at a tetrahedron, we see that there are around any one vertex three faces and three edges in beautiful synchronization; we say that it all looks simple and logical. We find, however, that the inventory of three faces around each vertex comes out of a total inventory of four that are always available in the tetrahedron. On the other hand, the inventory of three edges around each vertex comes out of a total inventory of six that are available. So the sixness and the fourness are from very different total quantity inventories. Somehow, around any one vertex of the same system nature has arranged to synchronize them in a neat three-to-three balance while using them all in a total symmetry despite their being supplied from their differing inventories.

441.21 Consider the case of the cheese tetrahedron (see Sec. 623.20), where we push one of the faces toward the opposite vertex. We can move that face in until it is congruent with the opposite vertex. There is now no volume, but we have agreed that the condition of symmetry is a constant of the abstractly conceptual system, the tetrahedron: the sixness and the fourness are still there, but they are empty. With one face congruent with the opposite vertex, we have all four planes of the tetrahedron going through the same exact point at the same time, or theoretically as close as we can ever get to exactly. We also have six edges of the tetrahedron going through the same point at the same time. We have agreed that this is a condition that can never happen in reality, but in the vector equilibrium, where there is no size, we have the only possible time when this would seem to occur.

441.22 So we have the total inventory of four faces and six edges going through the same theoretical point at the same moment. We have said that this is a vector equilibrium and in a zero condition and it is nonreality. Nature would not permit it. But a moment later, those six edges *turbine* around that point one way or another—and we have seen plenty of models of the lines turbining around—but we will have to say that there had to be a moment when this plane went from being a positive tetrahedron to being a negative tetrahedron, and it had theoretically to pass through that point.

441.23 Very clearly, vector equilibrium is a zero-size tetrahedron. We have already had tetrahedron as an indestructible phenomenon independent of size. And then we have it getting into its own true zero vector equilibrium. It is a condition that nature apparently does not permit in our life, but what we call physical reality is always a positive and negative pulsating aberration of the whole—a multifrequency-accommodating, vector equilibrium aberratability whole.

442.00 Polarization of Vector Equilibrium

442.01 In closest packing of spheres, there are always between the spheres alternate spaces that are not being used, and so triangular space and space-available groups can be alternated from their original positions 60 degrees to alternate nestable positions. We find that you can take two halves of the vector equilibrium and rotate one of the halves 60 degrees. Instead of having the vector-equilibrium condition of alternate faces in symmetrical array around it, you will get a polarized system around the equatorial zone of which you will get a square and a square side by side and a triangle and a triangle side by side. By rotating the system 60 degrees, you will get a top polar triangle and a bottom polar triangle in the same orientation. If you rotate the vector equilibrium to the next 60-degree nestable position, suddenly it is omnisymmetrical again.

442.02 It is in this polarized condition that a section through the vector equilibrium makes the famous chemical hex that the chemists have used for years. The chemists recognized it as a polarized system, but they did not recognize it as the vector equilibrium because chemists had not had any internal atomic experience like that. Apparently, then, all the chemical compounding in the organic chemistry relates to polarized systems.

443.00 Vector Equilibrium as Equanimity Model

443.01 In order to reduce the concept of vector equilibrium to a single-name identity, we employ the word *equanimity* as identifying the eternal metaphysical conceptuality model that eternally tolerates and accommodates all the physically regenerative, intertransforming transactions of eternal, inexorable, and irreversible evolution's complex complementations, which are unitarily unthinkable, though finite.

443.02 The equanimity model permits metaphysically conceptual thinkability and permits man to employ the package-word *Universe*. Equanimity, the epistemological model, is the omni-intertransformative, angle- and frequency-modulatable, differential accommodator and identifies the direction toward the absolute, completely exquisite limit of zero-error, zero-time omnicomprehension toward which our oscillatory, pulsating reduction of tolerated cerebrally reflexed aberrations trends.

443.03 Humanity's physical brains' inherent subjective-to-objective time lag reflexing induces the relatively aberrated observation and asymmetrical articulation tolerated by ever more inclusively and incisively demanding mind's consciousness of the absolute exactitude of the eternally referential centrality at zero of the *equanimity model*. Thus mind induces human consciousness of evolutionary participation to seek cosmic zero. Cosmic zero is conceptually but sizelessly complex, though full-size-range accommodating.

443.04 In the equanimity model, the physical and the metaphysical share the same design. The whole of physical Universe experience is a consequence of our not seeing instantly, which introduces time. As a result of the gamut of relative recall time-lags, the physical is always the imperfect experience, but tantalizingly always ratio-equated with the innate eternal sense of perfection.

445.00 Frequency of Vector Equilibrium

445.01 As the most compact spherical agglomeration of unit radius spheres around a unit radius nuclear sphere, the vector equilibrium is indefinitely expandable, either by additional unit radius spheres, colonied, layer-embracement multiplication, or by uniform time-size increase. If expanded by unit radius, sphere-colonized, omni-embracing, concentric layer multiplication, additional new locally operative nuclei are progressively born with every four successive concentric generations of symmetrical, omni-embracing layer multiplication. We use here the concept of multiplication only by division of the conceptual sizeless whole in a greater number of coordinate parts.

445.02 The eight-triangled, six-square, planar-faceted space volume of the vector equilibrium is always frequency to the third power times 20—with frequency being of omnisymmetric, radial-circumferential, modular subdivisioning of the whole. The geometrical conceptioning of volume as used here is in contradistinction to the number of vari-frequenced, vector-equilibrium-forming, closest-packed spheres and their nonoccupied concave octahedra and concave vector equilibria-shaped intersphere interstices: these being the all and only shapes of inter-closest-packed unit radius sphere agglomeration interstices. (See illustration 1032.31.)

445.03 When the frequency of the vector equilibrium is *one* (or radiationally inactive), its initial volume is always $20 \times F^3 = 20$; wherefore, with $F = 1$, $1^3 = 1$, $1 \times 20 = 20$. When the frequency is, for instance, eight, the $8^3 = 512$; $512 \times 20 = 10{,}240$.

445.04 The relative size of the vector equilibrium begins with the initial zerosize integrity of conceptuality and its omnidimensional modular subdivisibility for accounting any frequency of geometrical configuration transformative accounting. Vector equilibria, as with the tetrahedra and other polyhedra, are conceptually valid as vector equilibria or tetrahedra, independent of size. Size is where relativity becomes generated. The eternality of synergetics is the experienceable conceptuality whose imaginability is independent of individual human life's successive special-case experiences of time and size relativity.

445.05 Considering vector equilibrium as initial unity, 20 in respect to tetrahedral unity of one, it constitutes the total volumetric domain unique to any universal focus or point.

445.06 We find that the vector equilibrium has a volume of 480 in terms of the *A* and *B* Quanta Modules.

445.07 When the frequency of the vector equilibrium is 50 (that is, with 50 edge intervals and with radii at 50 intervals), the volume is frequency to the third power times unity of 480 *A* and *B* Quanta Modules: or $50^3 \times 480 = 375{,}000$ *A* and *B* Quanta Modules. This will give us the exact number of quanta at any symmetrical stage growth. Here we witness experientially the quantum propagation of radiant wave after radiant wave identifiable with given wavelengths and frequencies of embracements.

445.10 Frequency inherently involves a plurality of events, which means that frequency begins with two or more event experiences. Where frequency is one, it means frequency is none = N-one = frequency-is-not-one, because frequency *is* two-or-more. (None is Latin for number nine, which is also numerologically a zero.) A frequency of one in the vector equilibrium is like the vector equilibrium itself, a *zero state,* i.e., energetically inactive. Frequency of experience inherently involves *intervals* between experience. Intervals that are nonexperiences are not *nothingness:* they are number integrities like zero state. This zero-state integrity is brought home to human cognition by numerological arithmetic disclosure (see Chapter 12, Numerology) that every nine experiences in arithmetical system integrity (integer) accounting always has a zero operational effect in all intersystem multiplying or dividing calculations, and that nature's number coordination coincides with its geometrical operational evolution, having only octave periodicity in which four positive entities accrue followed by the accrual of four negative entities and an interval zero state nine, which positive-negative accrual and intervalling accommodates rationally and elegantly both the wavilinear and discontinuity characteristics of all-experimentally acquired information to date regarding the electromagnetics and chemistry of both physical and metaphysical Universe. (See Sec. 1012 and Fig. 1012.14B.)

445.11 At eternal "outset," the vector equilibrium's frequency is none—non, which is inactive, which is different in meaning from nonexistent. Zero is the inside-out phase of conceptual integrity; it is the eternal complementation of system. Quite the contrary to "nonexistent," it means only "eternally existent" in contrast to "temporarily existent." Experience is all temporary. Between experiences is the forever eternal metaphysical, which cannot be converted into existent. Zerophase, i.e., the absolute integrity, is a metaphysical potential in pure principle but is inherently inactive. The inactivity of zerophase can be converted into activity only by pure principle of energetic geometrical propagation of successive positive-negative-positive-negative aberrational pulsations which intertransform locally initiated Universe through vector-equilibrium complex frequency accommodations in pure principle. The propagative pulsations are unopposed by the inherent but eternal, limitless, unoccupied outwardness of absolute metaphysical integrity. The unlimited metaphysical conceptual equilibrium integrity permits the limited special-case realizations. The limited cannot accommodate the unlimited. The unlimited metaphysical can and does accommodate the limited and principles-dependent physical; but the physical, which is always experienceable and special-case, cannot accommodate the metaphysical independence and unlimited capability.

445.12 The first layer of nuclear sphere embracement $= 10F^2 + 2 = 12$. Twelve balls of the first layer. The center ball has a value of two for the outwardness-inwardness, concave-convex terminal condition. But the center ball's frequency is zero.

445.13 The number of vertexes in the vector equilibrium is always the same as the number of spheres in omnidirectional closest packing: but frequency is identifiable only as the interval between the sphere centers. Two spheres have only one interval, ergo, $F = 1$. Frequency to the second power times ten plus two—$10F^2 + 2 = 12$.

445.14 The vector equilibrium is the common denominator of the tetrahedron, octahedron, and cube. It is the modular domain of the nine-zero-punctuated octave system. Double the vector equilibrium's modules radius for octaval volume accommodation.

445.15 When we compare the two-frequency-edge moduled cube's volume as quantized exclusively with cubes, to the vector equilibrium's volume as quantized exclusively with tetrahedra, we find that the volume of the two-frequency cube equals eight—which is two to the third power, expressed as 2^3. Whereas the volume of the two-frequency vector equilibrium equals 160, which is the integer two raised to the fifth power, $2^5 = 32$, and then multiplied by five, $5 \times 32 = 160$, expressed as $5 \cdot 2^5$. (See illustration 966.05.)

445.16 In 60-degree vector equilibrium accounting, when the edge module reads two, we have an energy potential (20) converted to an energy realized value of $20F^3$, F being 2, ergo, $2^3 = 8$, ergo, $20 \times 8 = 160$, which is also expressible most economically in prime numbers as two to the fifth power times five.

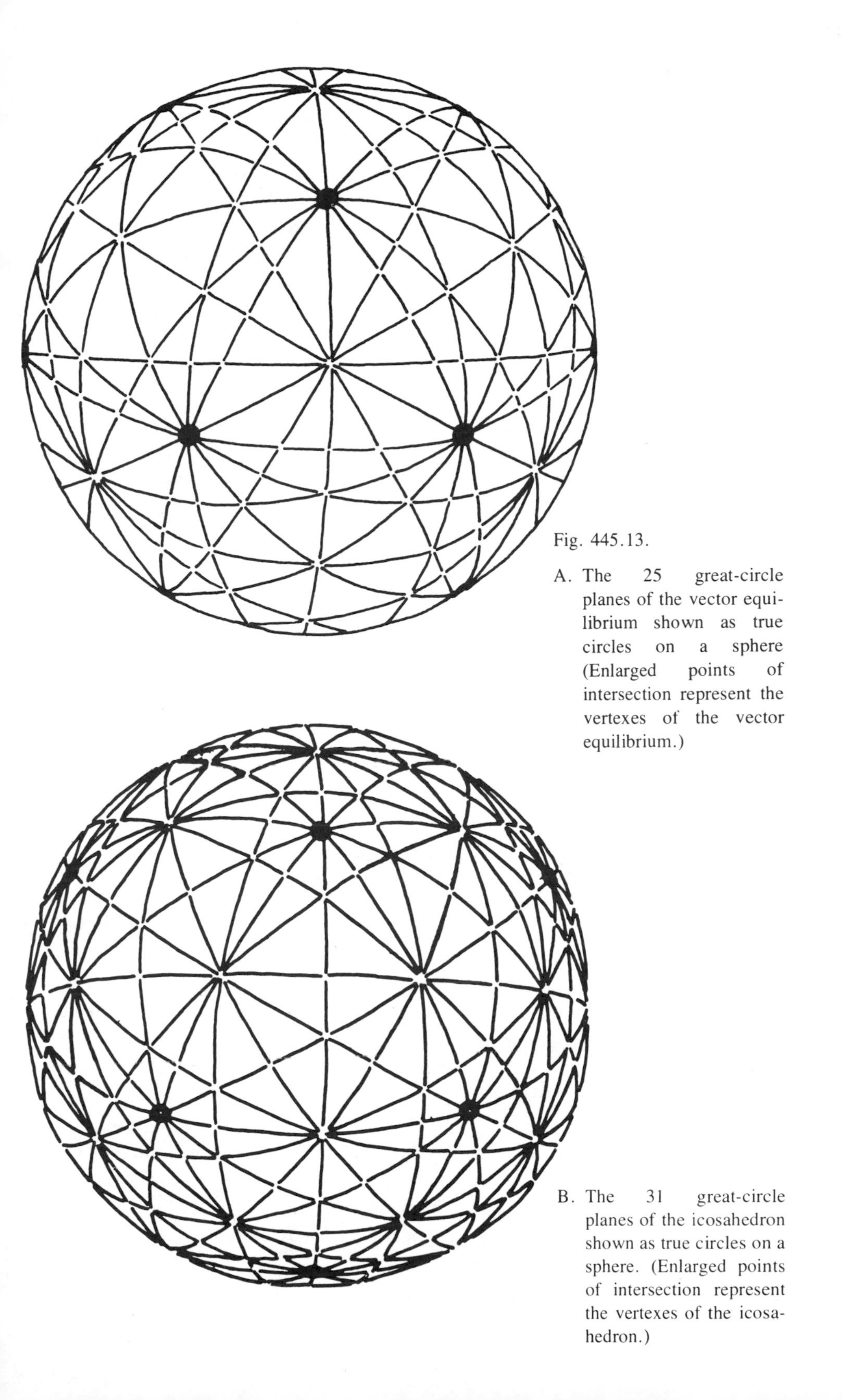

Fig. 445.13.

A. The 25 great-circle planes of the vector equilibrium shown as true circles on a sphere (Enlarged points of intersection represent the vertexes of the vector equilibrium.)

B. The 31 great-circle planes of the icosahedron shown as true circles on a sphere. (Enlarged points of intersection represent the vertexes of the icosahedron.)

445.17 We thus understand the misassumption of mid-19th century science that the *fourth* and *fifth* dimensions were inherently nonmodelable. The misassumption was occasioned by science's identification of dimensional uniqueness only with the rectilinearity of the cube, instead of with the vector equilibrium's omni-sixty-degreeness and the isotropic vector matrix's most economically mass-interattracted closest-self-packing unit radius sphere nucleation.

445.18 Generalized omni-intertransformable modelability now faithfully permits popular human comprehension of all experimentally derived scientific knowledge regarding physical phenomena heretofore translated only into exclusively abstract mathematical schemes of notationally formalized and formulated treatment, study, discovery; objective physical application now becomes modelable with energy-vectored tetrahedroning. This means that omniconceptuality of the geometrical intertransformations of eternally self-regenerative Universe now returns to science. This also means that the omnirational quantation we are discovering here means that children can conceptualize nuclear geometry even in their kindergarten years. This means in turn that nuclear physics will become lucidly explorable by humanity in its elementary spontaneously conceptioning and reasoning years.

450.00 Great Circles of the Vector Equilibrium and Icosahedron

450.10 Great Circles of the Vector Equilibrium

450.11 **Four Sets of Axes of Spin:** The omni-equi-edged and radiused vector equilibrium is omnisymmetrical, having 12 vertexes, six square faces, eight triangular faces, and 24 edges for a total of 50 symmetrically positioned topological features. These four sets of unique topological aspects of the vector equilibrium provide four different sets of symmetrically positioned polar axes of spin to generate the 25 great circles of the vector equilibrium. The 25 great circles of the vector equilibrium are the equators of spin of the 25 axes of the 50 unique symmetrically positioned topological aspects of the vector equilibrium.

450.12 Six of the faces of the vector equilibrium are square, and they are only corner-joined and symmetrically arrayed around the vector equilibrium in respect to one another. We can pair the six opposite square faces so that there are three pairs, and we can interconnect their opposite centers of area to pro-

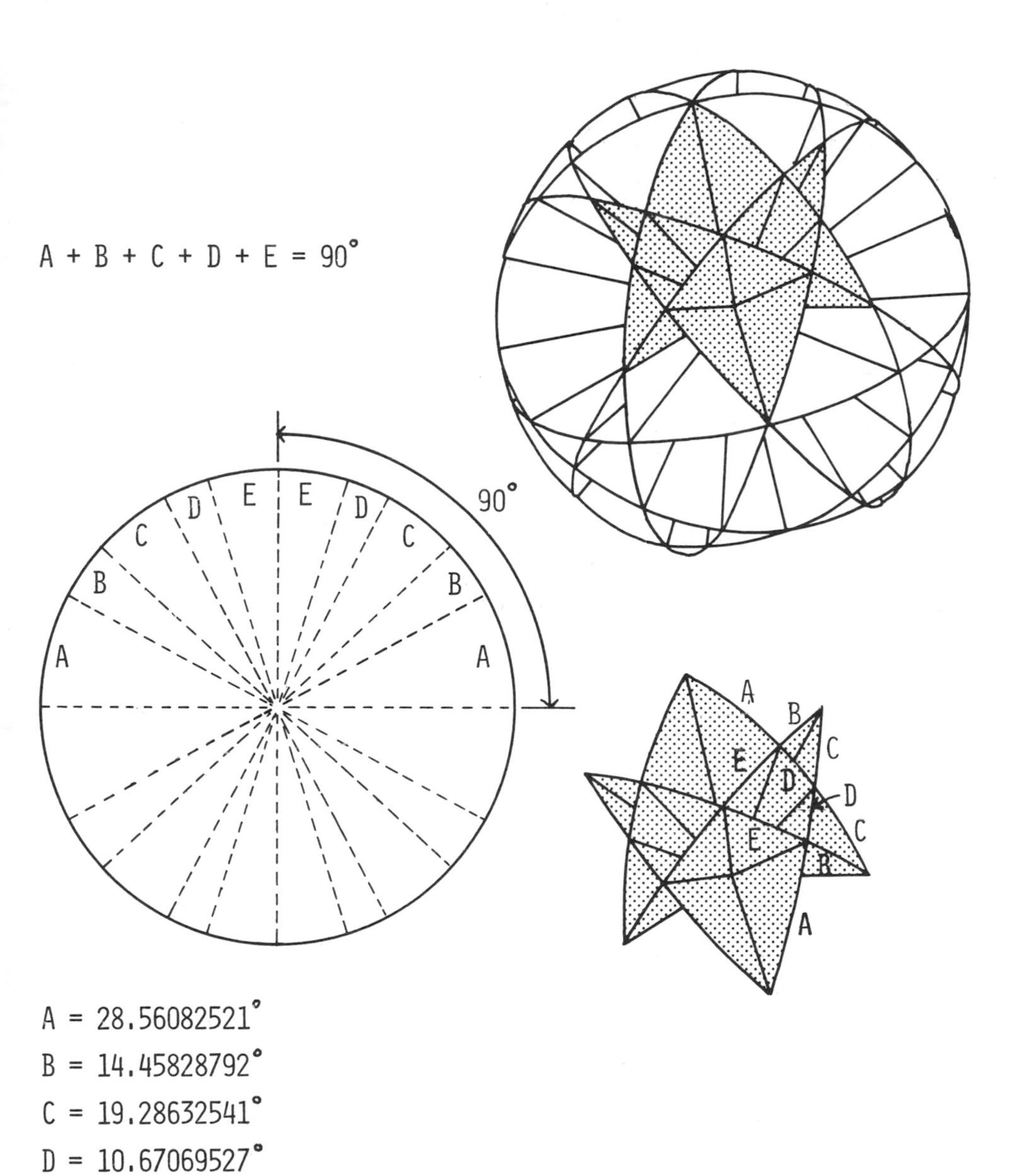

Fig. 450.10 *The 12 Great Circles of the Vector Equilibrium Constructed from 12 Folded Units (Shown as Shaded).*

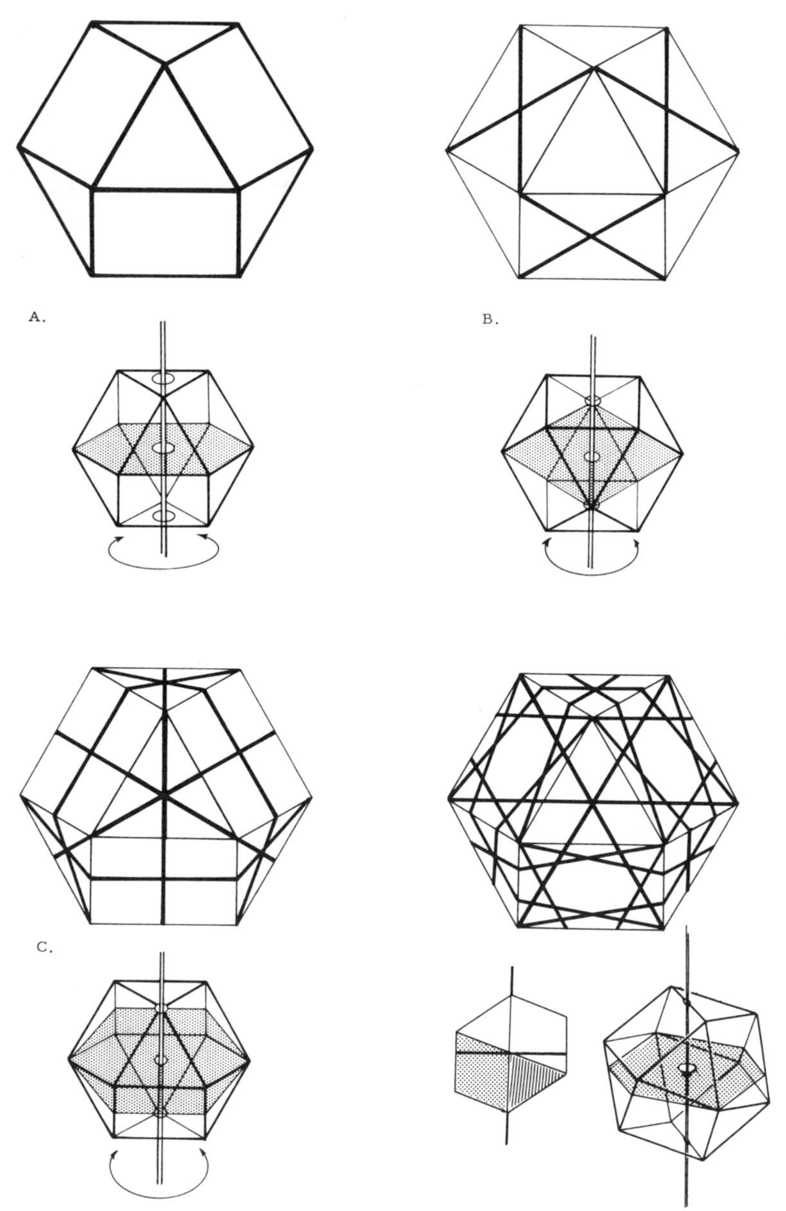

Fig. 450.11A *Axes of Rotation of Vector Equilibrium:*

A. Rotation of vector equilibrium on axes through centers of opposite triangular faces defines four equatorial great-circle planes.

B. Rotation of the vector equilibrium on axes through centers of opposite square faces defines three equatorial great-circle planes.

C. Rotation of vector equilibrium on axes through opposite vertexes defines six equatorial great-circle planes.

D. Rotation of the vector equilibrium on axes through centers of opposite edges defines twelve equatorial great-circle planes.

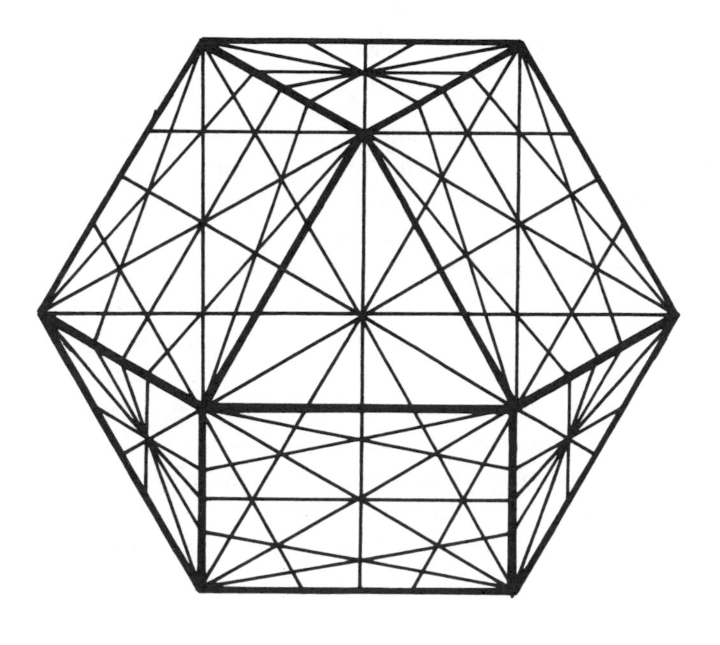

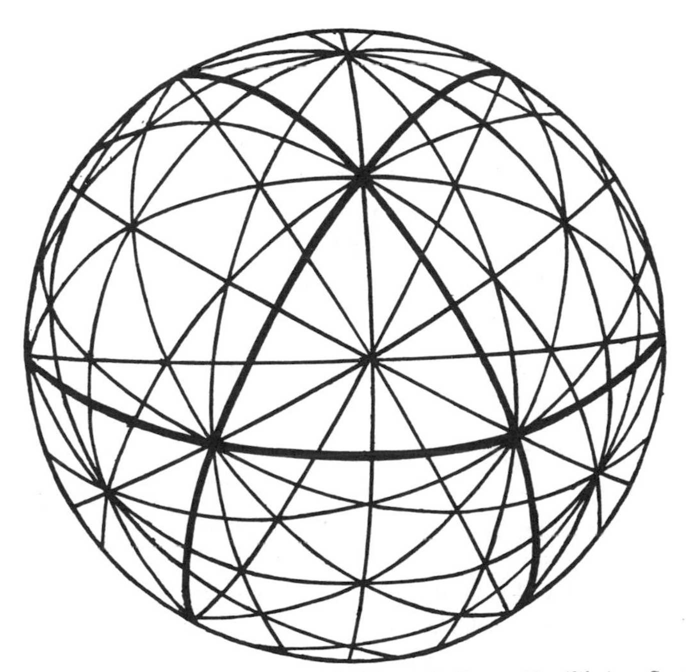

Fig. 450.11B *Projection of 25 Great-Circle Planes in Vector Equilibrium System:* The complete vector equilibrium system of 25 great-circle planes, shown as both a plane faced-figure and as the complete sphere (3 + 4 + 6 + 12 = 25). The heavy lines show the edges of the original 14-faced vector equilibrium.

vide three axes, corresponding to the *XYZ* coordinates of Cartesian geometry. We can spin the vector equilibrium on each of these three intersymmetrically positioned axes of square symmetry to produce three equators of spin. These axes generate the set of *three intersymmetrical great-circle equators* of the vector equilibrium. Together the three great circles subdivide the vector equilibrium into eight octants.

450.13 There are also eight symmetrically arrayed triangular faces of the vector equilibrium. We can pair the symmetrically opposite triangular faces so that there are four pairs, and we can interconnect their opposite centers of area to provide four intersymmetrically positioned axes. We can spin the vector equilibrium on each of these four axes of symmetry to produce four intersymmetrical equators of spin. These axes generate the set *four intersymmetrical great-circle equators* of the vector equilibrium.

450.14 When the 12 intersymmetrically positioned vertexes of the vector equilibrium are polarly interconnected, the lines of most economical interconnection provide six symmetrically interpositioned axes of spin. These six axes generate the set of *six intersymmetrical great-circle equators* of the vector equilibrium.

450.15 We may also most economically interconnect the 24 polarly opposed midpoints of the 24 intersymmetrically arrayed edges of the vector equilibrium to provide 12 sets of intersymmetrically positioned axes of spin. These axes generate the set of *twelve intersymmetrical great-circle equators* of the vector equilibrium.

450.16 As described, we now have sum-totally *three* square-face-centered axes, plus *four* triangular-face-centered axes, plus *six* vertex-centered axes, plus *12* edge-centered axes ($3+4+6+12=25$). There are a total of 25 complexedly intersymmetrical great circles of the vector equilibrium.

451.00 Vector Equilibrium: Axes of Symmetry and Points of Tangency in Closest Packing of Spheres

451.01 It is a characteristic of all the 25 great circles that each one of them goes through two or more of the vector equilibrium's 12 vertexes. Four of the great circles go through six vertexes; three of them go through four vertexes; and 18 of them go through two vertexes.

451.02 We find that all the sets of the great circles that can be generated by all the axes of symmetry of the vector equilibrium go through the 12 vertexes, which coincidentally constitute the only points of tangency of closest-packed, uniform-radius spheres. In omnidirectional closest packing, we always have 12 balls around one. The volumetric centers of the 12 uniform-radius balls closest packed around one nuclear ball are congruent with the 12 vertexes of the vector equilibrium of twice the radius of the closest-packed spheres.

451.03 The network of vectorial lines most economically interconnecting the volumetric centers of 12 spheres closest packed around one nuclear sphere of the same radius describes not only the 24 external chords and 12 radii of the vector equilibrium but further outward extensions of the system by closest packing of additional uniform-radius spheres omnisurrounding the 12 spheres already closest packed around one sphere and most economically interconnecting each sphere with its 12 closest-packed tangential neighbors, altogether providing an isotropic vector matrix, i.e., an omnidirectional complex of vectorial lines all of the same length and all interconnected at identically angled convergences. Such an isotropic vector matrix is comprised internally entirely of triangular-faced, congruent, equiedged, equiangled *octahedra* and *tetrahedra*. This isotropic matrix constitues the omnidirectional grid.

451.04 The basic gridding employed by nature is the most economical agglomeration of the atoms of any one element. We find nature time and again using this closest packing for most economical energy coordinations.

452.00 Vector Equilibrium: Great-Circle Railroad Tracks of Energy

452.01 The 12 points of tangency of unit-radius spheres in closest packing, such as is employed by any given chemical element, are important because energies traveling over the surface of spheres must follow the most economical spherical surface routes, which are inherently great circle routes, and in order to travel over a series of spheres, they could pass from one sphere to another only at the 12 points of tangency of any one sphere with its closest-packed neighboring uniform-radius sphere.

452.02 The vector equilibrium's 25 great circles, all of which pass through the 12 vertexes, represent the only "most economical lines" of energy travel from one sphere to another. The 25 great circles constitute all the possible "most economical railroad tracks" of energy travel from one atom to another of the same chemical elements. Energy can and does travel from sphere to sphere of closest-packed sphere agglomerations only by following the 25 surface great circles of the vector equilibrium, always accomplishing the most economical travel distances through the only 12 points of closest-packed tangency.

452.03 If we stretch an initially flat rubber sheet around a sphere, the outer spherical surface is stretched further than the inside spherical surface of the same rubber sheet simply because circumference increases with radial increase, and the more tensed side of the sheet has its atoms pulled into closer radial proximity to one another. Electromagnetic energy follows the most highly tensioned, ergo the most atomically dense, metallic element regions, wherefore it always follows great-circle patterns on the convex surface of metallic spheres. Large copper-shelled spheres called Van De Graaff electrostatic generators are employed as electrical charge accumulators. As much as two million volts may be accumulated on one sphere's surface, ultimately to be discharged in a lightninglike leap-across to a near neighbor copper sphere. While a small fraction of this voltage might electrocute humans, people may walk around inside such high-voltage-charged spheres with impunity because the electric energy will never follow the concave surface paths but only the outer convex great-circle paths for, by kinetic inherency, they will always follow the great-circle paths of greatest radius.

452.04 You could be the little man in Universe who always goes from sphere to sphere through the points of intersphere tangencies. If you lived inside the concave surface of one sphere, you could go through the point of tangency into the next sphere, and you could go right through Universe that way always inside spheres. Or you could be the little man who lives on the outside of the spheres, always living convexly, and when you came to the point of tangency with the next sphere, you could go on to that next sphere convexly, and you could go right through Universe that way. Concave is one way of looking at Universe, and convex is another. Both are equally valid and cosmically extensive. This is typical of how we should not be fooled when we look at spheres—or by just looking at the little local triangle on the surface of our big sphere and missing the big triangle * always polarly complementing it and defined by the same three edges but consisting of all the unit spherical surface area on the outer side of the small triangle's three edges. These concave-convex, inside-out, and surface-area complementations are beginning to give us new clues to conceptual comprehending.

452.05 As was theoretically indicated in the foregoing energy-path discoveries, we confirm *experimentally* that electric charges never travel on the concave side of a sphere: they always stay on the convex surface. In the phenomenon of electroplating, the convex surfaces are readily treated while it is almost impossible to plate the concave side except by use of a close matrix of local spots. The convex side goes into higher tension, which means that it is stretched thinner and tauter and is not only less travel-resistant, but is more readily conductive because its atoms are closer to one another. This means that electromagnetic energy automatically follows around the outside of convex surfaces. It is experimentally disclosed and confirmed that energy always seeks the most economical, ergo shortest, routes of travel. And we have seen

* See Sec. 810, "One Spherical Triangle Considered as Four."

that the shortest intersphere or interatom routes consist exclusively of the 25 great-circle geodesic-surface routes, which transit the 12 vertexes of the vector equilibrium, and which thus transit all the possible points of tangency of closest-packed spheres.

452.06 There always exists some gap between the closest-packed spheres due to the nuclear kinetics and absolute discontinuity of all particulate matter. When the 12 tangency gaps are widened beyond voltage jumpability, the eternally regenerative conservation of cosmic energy by pure generalized principles will reroute the energies on spherically closed great-circle "holding patterns" of the 25 great circles, which are those produced by the central-angle foldings of the four unique great-circle sets altogether comprising the vector equilibrium's 25 great circles.

452.07 High energy charges in energy networks refuse to take the longest of the two great-circle arc routes existing between any two spherical points. Energy always tends to "short-circuit," that is, to complete the circuit between any two spherical surface points by the shortest great-circle arc route. This means that energy automatically triangulates via the diagonal of a square or via the triangulating diagonals of any other polygons to which force is applied. Triangular systems represent the shortest, most economical energy networks. The triangle constitutes the self-stabilizing pattern of complex kinetic energy interference occasioned angular shuntings and three-fold or more circle interaction averaging of least-resistant directional resultants, which always trend toward equiangular configurations, whether occurring as free radiant energy events or as local self-structurings.

453.00 Vector Equilibrium: Basic Equilibrium LCD Triangle

453.01 The system of 25 great circles of the vector equilibrium defines its own lowest common multiple spherical triangle, whose surface is exactly $^1/_{48}$th of the entire sphere's surface. Within each of these $^1/_{48}$th-sphere triangles and their boundary arcs are contained and repeated each time all of the unique interpatterning relationships of the 25 great circles. Twenty-four of the 48 triangles' patternings are "positive" and 24 are "negative," i.e., mirror-images of one another, which condition is more accurately defined as "inside out" of one another. This inside-outing of the big triangles and each of their contained triangles is experimentally demonstrable by opening any triangle at any one of its vertexes and holding one of its edges while sweeping the other two in a 360-degree circling around the fixed edge to rejoin the triangle with its previous outsideness now inside of it. This is the basic equilibrium LCD triangle; for a discussion of the basic disequilibrium LCD triangle, see Sec. 905.

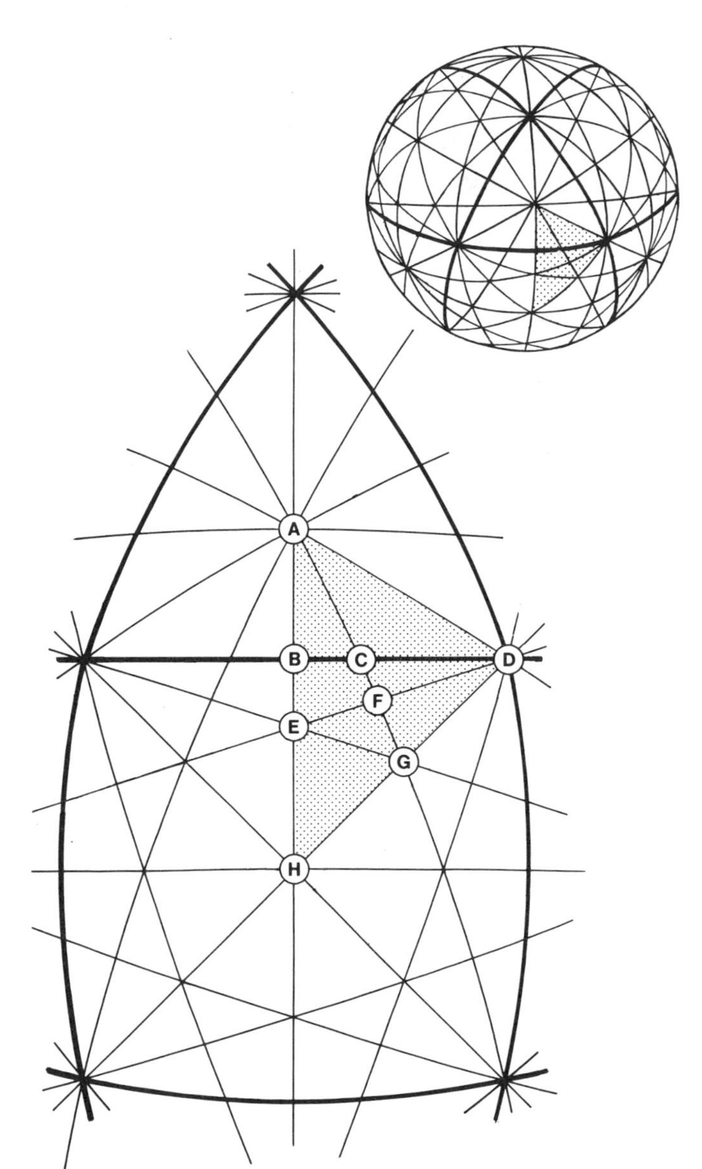

CENTRAL ANGLES		
19.47122063°	AB	19° 28' 16.394"
35.26438968	AD	35 15 51.803
22.20765430	AC	22 12 27.555
10.89339465	BC	10 53 36.221
19.10660535	CD	19 06 23.779
10.02498786	BE	10 01 29.956
6.35317091	CF	6 21 11.415
14.45828792	EF	14 27 29.837
17.02386618	FD	17 01 25.918
19.28632541	EG	19 17 10.771
10.67069527	FG	10 40 14.503
25.23940182	EH	25 14 21.847
26.56505118	HG	26 33 54.184
18.43494882	GD	18 26 5.816
31.48215410	DE	31 28 55.755
30.	BD	30 00 00
45.	DH	45 00 00
54.73561031	AH	54 44 8.197

FACE ANGLES		
30.°	BAC	30° 00' 00.000"
30.	CAD	30 00 00.000
90.	ABC	90 00 00.000
61.87449430°	ACB	61 52 28.179
118.1255057	ACD	118 7 31.821
35.26438968	ADC	35 15 51.803
90.	EBC	90 00 00.000
118.1255057	BCF	118 7 31.821
73.22134512	BEF	73 13 16.842
80.40593179	CFE	80 24 21.354
61.87449430	FCD	61 52 28.179
19.47122063	CDF	19 28 16.394
99.59406821	CFD	99 35 38.646
73.22134512	HEG	73 13 16.842
65.90515745	EGH	65 54 18.567
45.	EHG	45 00 00.000
99.59406821	EFG	99 35 38.646
33.55730977	FEG	33 33 26.315
48.18968511	FGE	48 11 22.866
80.40593179	GFD	80 24 21.354
35.26438969	FDG	35 15 51.803
65.90515745	FGD	65 54 18.567

Fig. 453.01 *Great Circles of Vector Equilibrium Define Lowest Common Multiple Triangle: 1/48th of a Sphere:* The shaded triangle is 1/48th of the entire sphere and is the lowest common denominator (in 24 rights and 24 lefts) of the total spherical surface. The 48 LCD triangles defined by the 25 great circles of the vector equilibrium are grouped together in whole increments to define exactly the spherical surface areas, edges, and vertexes of the spherical tetrahedron, spherical cube, spherical octahedron, and spherical rhombic dodecahedron. The heavy lines are the edges of the four great circles of the vector equilibrium. Included here is the spherical trigonometry data for this lowest-common-denominator triangle of the 25-great-circle hierarchy of the vector equilibrium.

453.02 **Inside-Outing of Triangle:** The inside-outing transformation of a triangle is usually misidentified as "left vs. right," or "positive and negative," or as "existence vs. annihilation" in physics (see drawings section).

453.03 The inside-outing is four-dimensional and often complex. It functions as complex intro-extroverting.

454.00 Vector Equilibrium: Spherical Polyhedra Described by Great Circles

454.01 The 25 great circles of the spherical vector equilibrium provide all the spherical edges for five spherical polyhedra: the tetrahedron, octahedron, cube, rhombic dodecahedron, and vector equilibrium, whose corresponding planar-faceted polyhedra are all volumetrically rational, even multiples of the tetrahedron. For instance, if the tetrahedron's volume is taken as unity, the octahedron's volume is four, the cube's volume is three, the rhombic dodecahedron's is six, and the vector equilibrium's is 20 (see drawings section).

454.02 This is the hierarchy of rational energy quanta values in synergetics, which the author discovered in his youth when he first sought for an omnirational coordinate system of Universe in equilibrium against which to measure the relative degrees of orderly asymmetries consequent to the cosmic myriad of pulsatively propagated energetic transactions and transformations of eternally conserving evolutionary events. Though almost all the involved geometries were long well known, they had always been quantized in terms of the cube as volumetric unity and its edges as linear unity; when employed in evaluating the other polyhedra, this method produced such a disarray of irrational fraction values as to imply that the other polyhedra were only side-show geometric freaks or, at best, "interesting aesthetic objets d'art." That second-powering exists today in academic brains only as "squaring" and third-powering only as cubing is manifest in any scientific blackboard discourse, as the scientists always speak of the x^2 they have just used as "x squared" and likewise always account x^3 as "x cubed" (see drawings section).

454.03 The spherical tetrahedron is composed of four spherical triangles, each consisting of 12 basic, least-common-denominator spherical triangles of vector equilibrium.

454.04 The spherical octahedron is composed of eight spherical triangles, each consisting of six basic-vector-equilibrium, least-common-denominator triangles of the 25 great-circle, spherical-grid triangles.

454.05 The spherical cube is composed of six spherical squares with corners of 120° each, each consisting of eight basic-vector-equilibrium, least-common-denominator triangles of the 25 great-circle spherical-grid triangles.

454.06 The spherical rhombic dodecahedron is composed of 12 spherical

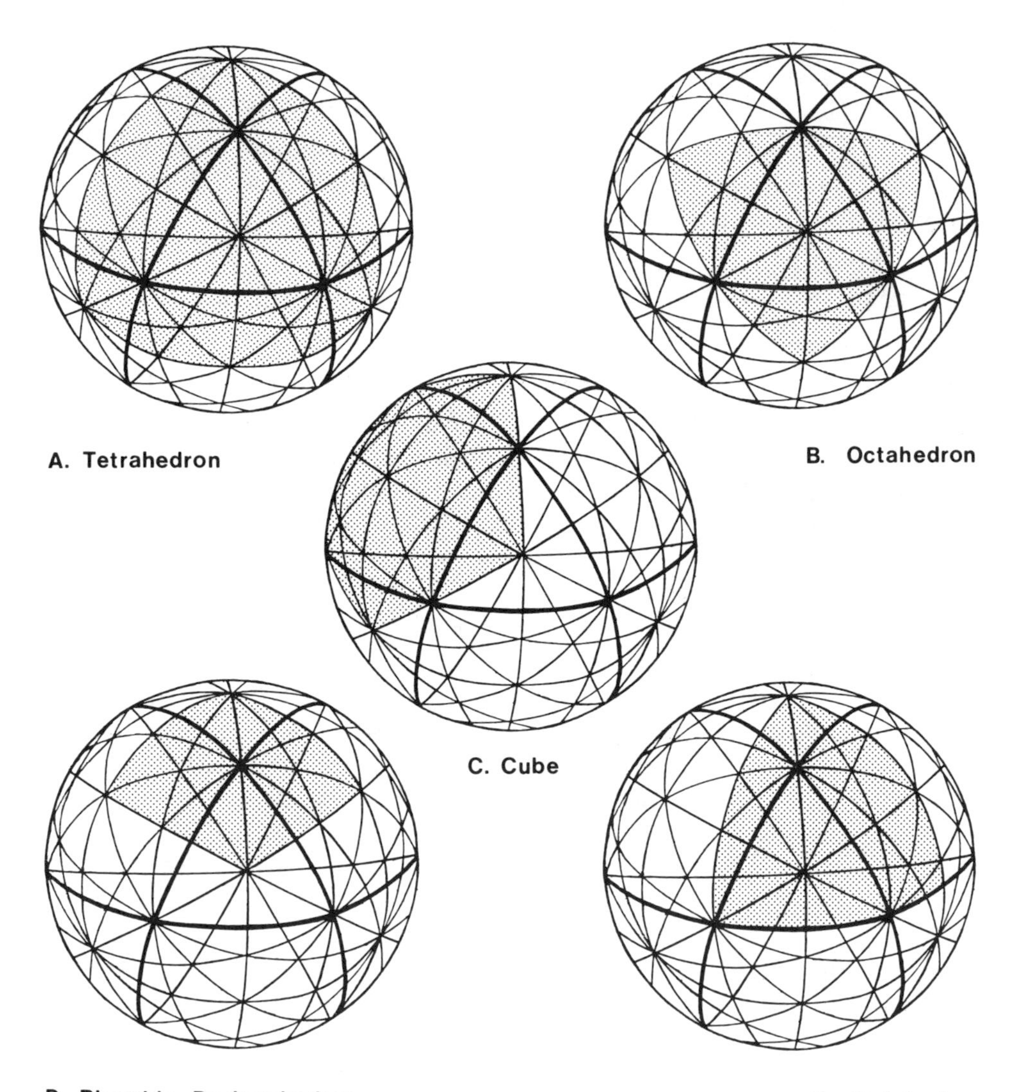

Fig. 454.06 *Definition of Spherical Polyhedra in 25-Great-Circle Vector Equilibrium System:* The 25 great circles of the spherical vector equilibrium provide all the spherical edges for four spherical polyhedra in addition to the vector equilibrium whose edges are shown here as heavy lines. The shading indicates a typical face of each as follows:

A. The edges of one of the spherical tetrahedron's four spherical triangles consists of 12 VE basic LCD triangles.
B. The edges of one of the spherical octahedron's eight spherical triangles consists of six VE basic LCD triangles.
C. The edges of one of the spherical cube's six spherical squares consists of eight VE basic LCD triangles.
D. The edges of one of the spherical rhombic dodecahedron's 12 spherical rhombic faces consists of four VE basic LCD triangles.
E. The edges of one of the spherical octahedron's eight spherical triangles consists of a total area equal to six VE basic LCD triangles.

diamond-rhombic faces, each composed of four basic-vector-equilibrium, least-common-denominator triangles of the 25 great-circle, spherical-grid triangles.

455.00 Great-Circle Foldabilities of Vector Equilibrium

455.01 **Foldability of Vector Equilibrium Four Great-Circle Bow Ties:** All of the set of four great circles uniquely and discretely describing the vector equilibrium can be folded out of four whole (non-incised), uniform-radius, circular discs of paper, each folded radially in 60-degree central angle increments, with two diametric folds, mid-circle, hinge-bent together and locked in radial congruence so that their six 60-degree arc edges form two equiangled spherical triangles, with one common radius-pairing fastened together at its external apex, that look like a *bow tie*. The pattern corresponds to the external arc trigonometry, with every third edgefold being brought into congruence to form great-circle-triangled openings at their top with their pointed lower ends all converging ice-cream-cone-like at the center of the whole uncut and only radially folded great circles. When the four bow ties produced by the folded circles are assembled together by radii congruence and locking of each of their four outer bow-tie corners to the outer bow-tie corners of one another, they will reestablish the original four great-circle edge lines of the vector equilibrium and will accurately define both its surface arcs and its central angles as well as locating the vector-equilibrium axes of symmetry of its three subsets of great-circle-arc-generating to produce, all told, 25 great circles of symmetry. When assembled with their counterpart foldings of a total number corresponding to the great-circle set involved, they will produce a whole sphere in which all of the original great circles are apparently restored to their completely continuing-around-the-sphere integrity.

455.02 The sum of the areas of the four great-circle discs elegantly equals the surface area of the sphere they define. The area of one circle is πr^2. The area of the surface of a sphere is $4\pi r^2$. The area of the combined four folded great-circle planes is also $4\pi r^2$ and all four great-circle planes go through the exact center of the sphere and, between them, contain no volume at all. The sphere contains the most volume with the least surface enclosure of any geometrical form. This is a cosmic limit at maximum. Here we witness the same surface with no volume at all, which qualifies the vector equilibrium as the most economic nuclear "nothingness" whose coordinate conceptuality rationally accommodates all radiational and gravitational interperturbational transformation accounting. In the four great-circle planes we witness the same surface area as that of the sphere, but containing no volume at all. This too, is cosmic limit at zero minimumness.

455.03 It is to be noted that the four great-circle planes of the vector equilibrium passing exactly through its and one another's exact centers are parallel to the four planes of the eight tetrahedra, which they accommodate in the eight triangular bow-tie concavities of the vector equilibrium. The four planes of the tetrahedra have closed on one another to produce a tetrahedron of no volume and no size at all congruent with the sizeless center of the sphere defined by the vector equilibrium and its four hexagonally intersected planes. As four points are the minimum necessary to define the insideness and outsideness unique to all systems, four triangular facets are the minimum required to define and isolate a system from the rest of Universe.

455.04 Four is also the minimum number of great circles that may be folded into local bow ties and fastened corner-to-corner to make the whole sphere again and reestablish all the great circles without having any surfaces double or be congruent with others or without cutting into any of the circles.

455.05 These four great-circle sets of the vector equilibrium demonstrate all the shortest, most economical railroad "routes" between all the points in Universe, traveling either convexly or concavely. The physical-energy travel patterns can either follow the great-circle routes from sphere to sphere or go around in local holding patterns of figure eights on one sphere. Either is permitted and accommodated. The four great circles each go through six interspherical tangency points.

455.10 **Foldability of Vector Equilibrium Six Great-Circle Bow Ties:** The foldable bow ties of the six great circles of the vector equilibrium define a combination of the positive and negative spherical tetrahedrons within the spherical cube as well as of the rhombic dodecahedron.

455.11 In the vector equilibrium's six great-circle bow ties, all the internal, i.e., central angles of 70° 32′ and 54° 44′, are those of the surface angles of the vector equilibrium's four great-circle bow ties, and vice versa. This phenomenon of turning the inside central angles outwardly and the outside surface angles inwardly, with various fractionations and additions, characterizes the progressive transformations of the vector equilibrium from one great-circle foldable group into another, into its successive stages of the spherical cube and octahedron with all of their central and surface angles being both 90 degrees even.

455.20 **Foldability of 12 Great Circles into Vector Equilibrium:** We can take a disc of paper, which is inherently of 360 degrees, and having calculated with spherical trigonometry all the surface and central angles of both the associated and separate groups of 3–4–6–12 great circles of the vector equilibrium's 25 great circles, we can lay out the spherical arcs which

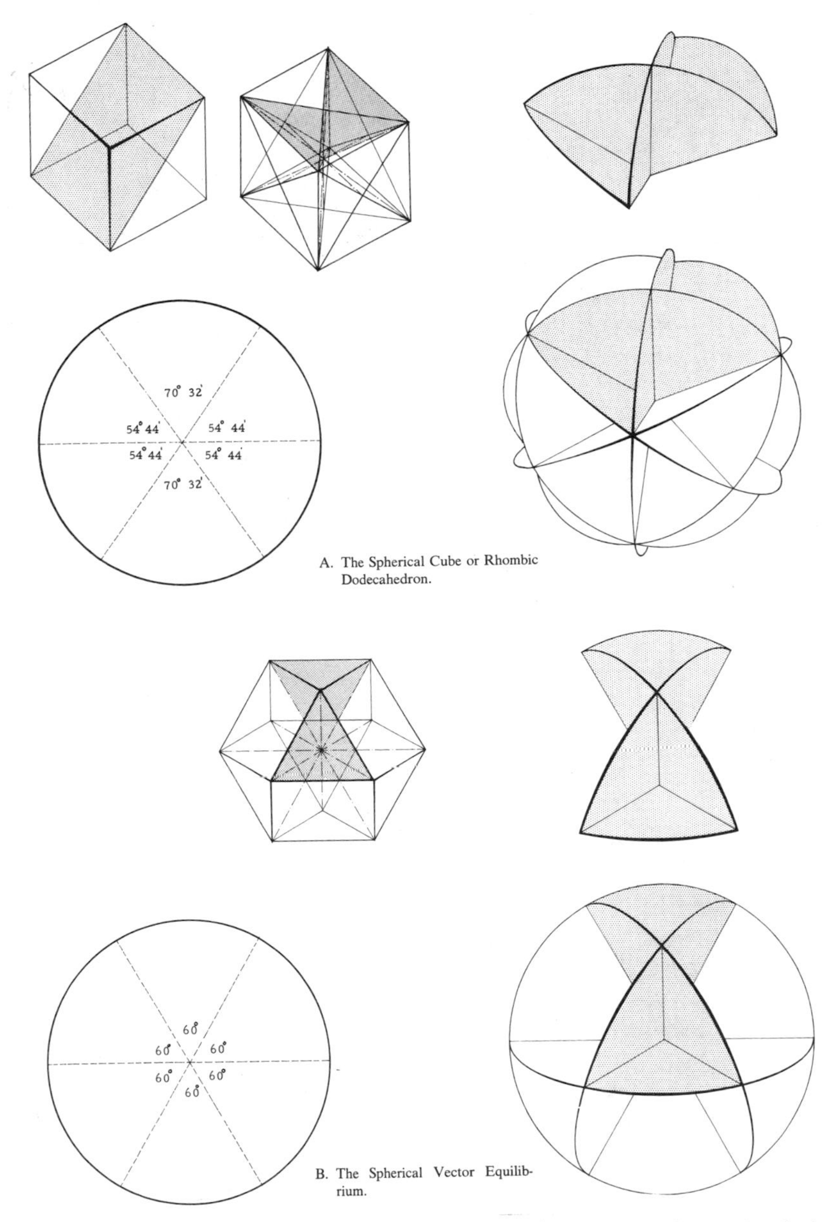

A. The Spherical Cube or Rhombic Dodecahedron.

B. The Spherical Vector Equilibrium.

Fig. 455.11 *Folding of Great Circles into Spherical Cube or Rhombic Dodecahedron and Vector Equilibrium: Bow-Tie Units:*

A. This six-great-circle construction defines the positive-negative spherical tetrahedrons within the cube. This also reveals a spherical rhombic dodecahedron. The circles are folded into "bow-tie" units as shown. The shaded rectangle in the upper left indicates the typical plane represented by the six great circles.

B. The vector equilibrium is formed by four great circles folded into "bow-ties." The sum of the areas of the four great circles equals the surface area of the sphere. ($4r^2$).

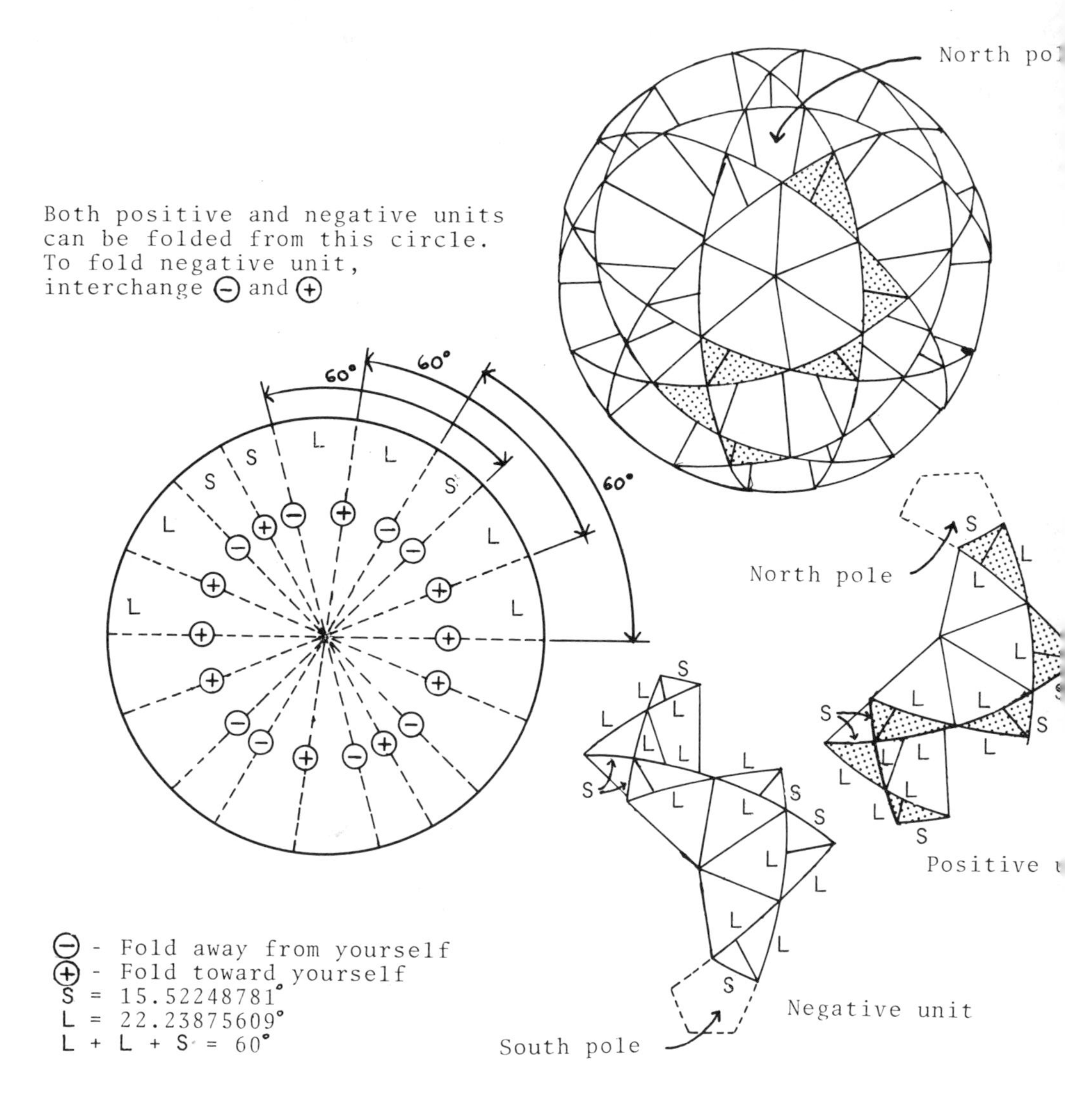

Fig. 455.20 *The 10 great circles of the Icosahedron Constructed from 10 folded units (5 positive units + 5 negative units).*

always subtend the central angles. The 25 great circles interfere with and in effect "bounce off" or penetrate one another in an omnitriangulated, nonredundant spherical triangle grid. Knowing the central angles, we can lay them out and describe foldable triangles in such a way that they make a plurality of tetrahedra that permit and accommodate fastening together edge-to-edge with no edge duplication or overlap. When each set, 3–4–6–12, of the vector equilibrium is completed, its components may be associated with one another to produce complete spheres with their respective great-circle, 360-degree integrity reestablished by their arc increment association.

455.21 The 25 folded great-circle sections join together to reestablish the 25 great circles. In doing so, they provide a plurality of 360-degree local and long-distance travel routes. Because each folded great circle starts off with a 360-degree disc, it maintains that 360-degree integrity when folded into the bow-tie complexes. It is characteristic of electromagnetic wave phenomena that a wave must return upon itself, completing a 360-degree circuit. The great-circle discs folded or flat provide unitary-wave-cycle circumferential circuits. Therefore, folded or not, they act like waves coming back upon themselves in a perfect wave control. We find their precessional cyclic self-interferences producing angular resultants that shunt themselves into little local 360-degree, bow-tie "holding patterns." The entire behavior is characteristic of generalized wave phenomena.

455.22 In the case of the 12 great circles of the vector equilibrium, various complex transformative, anticipatory accommodations are manifest, such as that of the 12 sets of two half-size pentagons appearing in the last, most complex great-circle set of the vector equilibrium, which anticipates the formation of 12 whole pentagons in the six great-circle set of the 31 great circles of the icosahedron into which the vector equilibrium first transforms contractively.

456.00 Transformation of Vector Equilibrium into Icosahedron

456.01 While its vertical radii are uniformly contracted from the vector equilibrium's vertexial radii, the icosahedron's surface is simultaneously and symmetrically askewed from the vector equilibrium's surface symmetry. The vector equilibrium's eight triangles do not transform, but its six square faces transform into 12 additional triangles identical to the vector equilibrium's original eight, with five triangles cornered together at the same original 12 vertexes of the vector equilibrium.

456.02 The icosahedron's five-triangled vertexes have odd-number-imposed, inherent interangle bisectioning, that is, extensions of the 30 great-circle edges of any of the icosahedron's 20 triangles automatically bisecting the apex angle of the adjacently intruded triangle into which it has passed. Thus extension of all the icosahedron's 20 triangles' 30 edges automatically bisects all of its original 60 vertexial-centered, equiangled 36-degree corners, with all the angle bisectors inherently impinging perpendicularly upon the opposite mid-edges of the icosahedron's 20 equilateral, equiangled 72-degree-cornered triangles. The bisecting great-circle extensions from each of all three of the original 20 triangles' apexes cross inherently (as proven elsewhere in Euclidian geometry) at the areal center of those 20 original icosahedral triangles. Those perpendicular bisectors subdivide each of the original 20 equiangled triangles into six right-angled triangles, which multiplies the total surface subdivisioning into 120 "similar" right-angled triangles, 60 of which are *positive* and 60 of which are *negative,* whose corners in the spherical great-circle patterning are 90°, 60°, and 36°, respectively, and their chordally composed corresponding planar polyhedral triangles are 90, 60, and 30 degrees, respectively. There is exactly 6 degrees of "spherical excess," as it is formally known, between the 120 spherical vs. 120 planar triangles.

456.03 This positive-negative subdivision of the whole system puts half the system into negative phase and the other half into positive phase, which discloses an exclusively external "surface" positive-negative relationship quite apart from that of the two surface polar hemispheres. This new aspect of complementarity is similar to the systematic omnicoexistence of the concave and convex non-mirror-imaged complementarity whose concavity and convexity make the 60 positive and 60 negative surface triangle subdivisions of spherical unity inherently noninterchangeable with one another when turned inside out, whereas they are interchangeable with one another by inside-outing when in their planar-faceted polyhedral state.

456.04 We thus find the split-phase positive-and-negativeness of odd-number-of-vertexial-angle systems to be inherently askewed and inside-outingly dichotomized omnisymmetries. This surface phase of dichotomization results in superficial, disorderly interpatterning complementation. This superficially disarrayed complementation is disclosed when the 15 great circles produced by extension of all 30 edges of the icosahedron's 20 triangles are folded radially in conformity to the central interangling of the 120 triangles' spherical arc edges.

456.05 The 15 great circles of the icosahedron interact to produce 15 "chains" of three varieties of four corner-to-corner, sausage-linked, right triangles, with four triangles in each chain. These 15 chains of 60 great-circle triangles are each interconnectible corner-to-corner to produce a total spherical surface subdivided into 120 similar spherical triangles. An experiment with 15 unique coloring differentiations of the 15 chains of three sequential varieties of four triangles each, will exactly complete the finite sphere and the 15 great-circle integrities of total spherical surface patterning, while utterly frustrating any systematically orderly surface patterning. The 15 chains' 60 triangles' inadvertent formation of an additional 60 similar spherical triangles occurring between them, which exactly subdivides the entire spherical surface into 120 symmetrically interpatterned triangles—despite the local surface disorder of interlinkage of the three differently colored sets of four triangles composing the 15 chains—dramatically manifests the half-positive, half-negative, always and only coexisting, universal non-mirror-imaged complementarity inherently permeating all systems, dynamic or static, despite superficial disorder, whether or not visibly discernible initially.

456.10 **Icosahedron as Contraction of Vector Equilibrium:** The icosahedron represents the 12-way, omniradially symmetrical, transformative, rotational contraction of the vector equilibrium. This can be seen very appropriately when we join the 12 spheres tangent to one another around a central nuclear sphere in closest packing: this gives the correspondence to the vector equilibrium with six square faces and eight triangular faces, all with 60-degree internal angles. If we had rubber bands between the points of tangency of those 12 spheres and then removed the center sphere, we would find the 12 tangent spheres contracting immediately and symmetrically into the icosahedral conformation.

456.11 The icosahedron is the vector equilibrium contracted in radius so that the vector equilibrium's six square faces become 12 ridge-pole diamonds. The ridge-pole lengths are the same as those of the 12 radii and the 24 outside edges. With each of the former six square faces of the vector equilibrium now turned into two equiangle triangles for a total of 12, and with such new additional equiangled and equiedged triangles added to the vector equilibrium's original eight, we now have 20 triangles and no other surface facets than the 20 triangles. Whereas the vector equilibrium had 24 edges, we now have added six more to the total polyhedral system as it transforms from the vector equilibrium into the icosahedron; the six additional ridge poles of the diamonds make a total of 30 edges of the icosahedron. This addition of six vector edge lengths is equivalent to one great circle and also to one quantum. (See Sec. 423.10.)

456.12 We picture the location of the vector equilibrium's triangular faces in relation to the icosahedron's triangular faces. The vector equilibrium could contract rotatively, in either positive or negative manner, with the equator going either clockwise or counterclockwise. Each contraction provides a different superposition of the vector equilibrium's triangular faces on the icosahedron's triangular faces. But the centers of area of the triangular faces remain coincidental and congruent. They retain their common centers of area as they rotate.

456.13 We find that the 25 great circles of the icosahedron each pass through the 12 vertexes corresponding to the 25 great circles of the vector equilibrium, which also went through the 12 vertexes, as the number of vertexes after the rotational contraction remains the same.

456.20 **Single-Layer Contraction:** The icosahedron, in order to contract, must be a single-layer affair. You could not have two adjacent layers of vector equilibria and then have them collapse to become the icosahedron. But take any single layer of a vector equilibrium with nothing inside it to push it outward, and it will collapse into becoming the icosahedron. If there are two layers, one inside the other, they will not roll on each other when the radius contracts. The gears block each other. So you can only have this contraction in a single layer of the vector equilibrium, and it has to be an outside layer remote from other layers.

456.21 The icosahedron has only the outer shell layer, but it may have as high a frequency as nature may require. The nuclear center is vacant.

456.22 The single-shell behavior of the icosahedron and its volume ratio of 18.63 arouses suspicions about its relation to the electron. We appear to have the electron kind of shells operating in the nucleus-free icosahedron and are therefore not frustrated from contracting in that condition.

457.00 Great Circles of Icosahedron

457.01 **Three Sets of Axes of Spin:** The icosahedron has three unique symmetric sets of axes of spin. It provides 20 triangular faces, 12 vertexes, and 30 edges. These three symmetrically interpatterned topological aspects—faces, vertexes, and mid-edges—provide three sets of axes of symmetric spin to generate the spherical icosahedron projection's grid of 31 great circles.

457.02 The icosahedron has the highest number of identical and symmetric exterior triangular facets of all the symmetrical polyhedra defined by great circles.

457.10 When we interconnect the centers of area of the 20 triangular faces of the icosahedron with the centers of area of their diametrically opposite faces, we are provided with 10 axes of spin. We can spin the icosahedron on any one of these 10 axes to produce 10 equators of spin. These axes generate the set of *10 great-circle* equators of the icosahedron. We may also interconnect the midpoints of the 30 edges of the icosahedron in 15 sets of diametrically opposite pairs. These axes generate the *15 great-circle* equators of the icosahedron. These two sets of 10 and 15 great circles correspond to the 25 great circles of the vector equilibrium.

457.20 **Six Great Circles of Icosahedron:** When we interconnect the 12 vertexes of the icosahedron in pairs of diametric opposites, we are provided with six axes of spin. These axes generate the *six great-circle* equators of the icosahedron. The six great circles of the icosahedron go from mid-edge to mid-edge of the icosahedron's triangular faces, and they do not go through any of its vertexes.

457.21 The icosahedron's set of six great circles is unique among all the seven axes of symmetry (see Sec. 1040), which include both the 25 great circles of the vector equilibrium and the 31 great circles of the icosahedron. It is the only set that goes through none of the 12 vertexes of either the vector equilibrium or the icosahedron. In assiduously and most geometrically avoiding even remote contact with any of the vertexes, they represent a new behavior of great circles.

457.22 The 12 vertexes in their "in-phase" state in the vector equilibria or in their "out-of-phase" state in the icosahedra constitute all the 12 points of possible tangency of any one sphere of a closest-packed aggregate with another sphere, and therefore these 12 points are the only ones by which energy might pass to cross over into the next spheres of closest packing, thus to travel their distance from here to there. The six great circles of the icosahedron are the only ones not to go through the potential intertangency points of the closest-packed unit radius spheres, ergo energy shunted on to the six icosahedron great circles becomes locked into local holding patterns, which is not dissimilar to the electron charge behaviors.

457.30 **Axes of Symmetry of Icosahedron:** We have now described altogether the 10 great circles generated by the 10 axes of symmetry occurring between the centers of area of the triangular faces; plus 15 axes from the midpoints of the edges; plus six axes from the vertexes. $10+15+6=31$. There is a total of 31 great circles of the icosahedron.

457.40 **Spherical Polyhedra in Icosahedral System:** The 31 great circles of the spherical icosahedron provide spherical edges for three other polyhedra in addition to the icosahedron: the rhombic triacontrahedron, the octahedron, and the pentagonal dodecahedron. The edges of the spherical icosahedron are shown in heavy lines in the illustration.

457.41 The spherical rhombic triacontrahedron is composed of 30 spherical rhombic diamond faces.

457.42 The spherical octahedron is composed of eight spherical triangles.

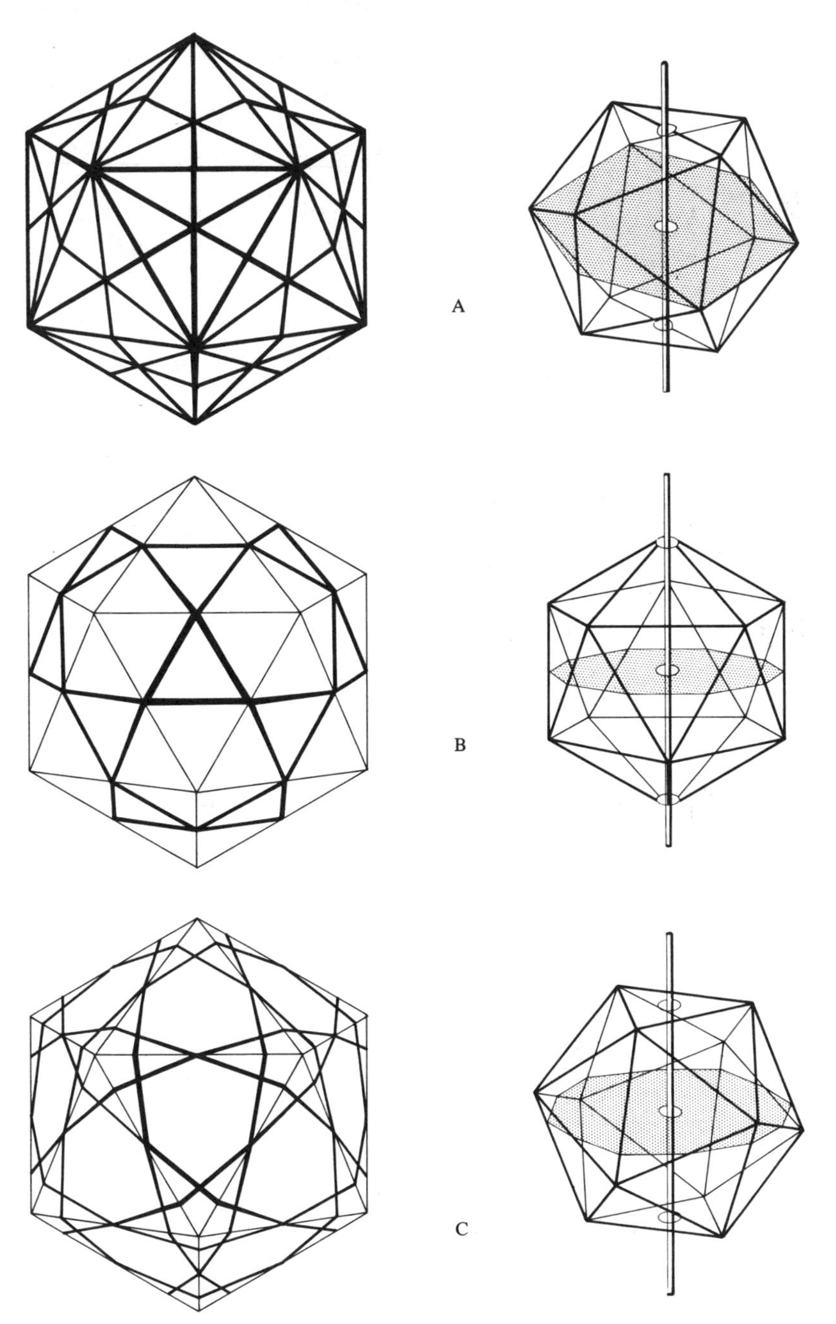

Fig. 457.30A *Axes of Rotation of Icosahedron:*

A. The rotation of the icosahedron on axes through midpoints of opposite edges define 15 great-circle planes.
B. The rotation of the icosahedron on axes through opposite vertexes define six equatorial great-circle planes, none of which pass through any vertexes.
C. The rotation of the icosahedron on axes through the centers of opposite faces define ten equatorial great-circle planes, which do not pass through any vertexes.

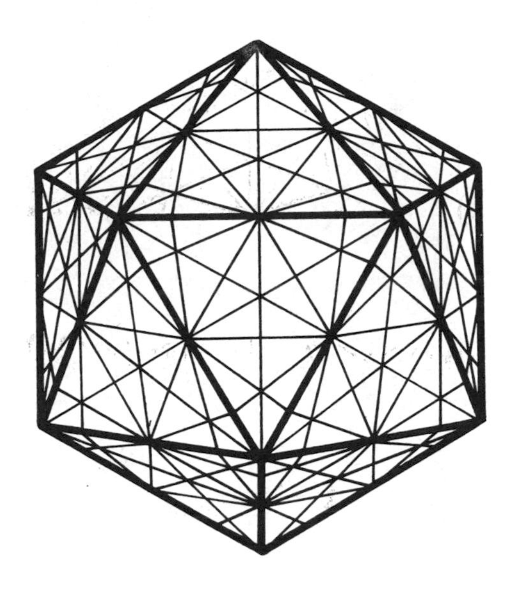

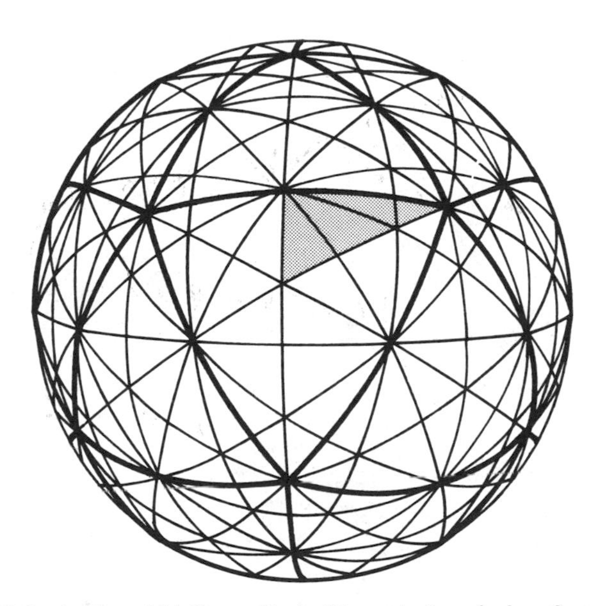

Fig. 457.30B *Projection of 31 Great-Circle Planes in Icosahedron System:* The complete icosahedron system of 31 great-circle planes shown with the planar icosahedron as well as true circles on a sphere (6 + 10 + 15 = 31). The heavy lines show the edges of the original 20-faced icosahedron.

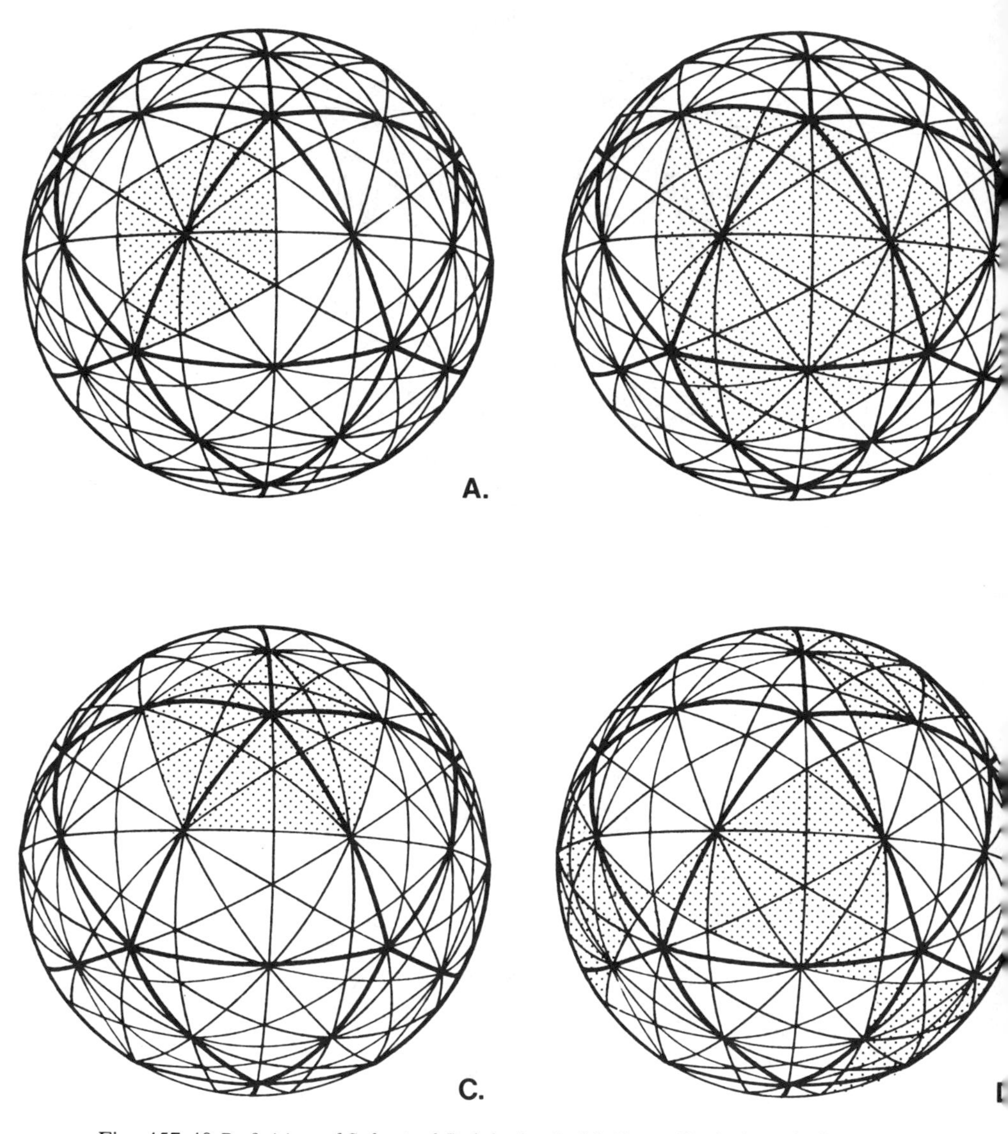

Fig. 457.40 *Definition of Spherical Polyhedra in 31-Great-Circle Icosahedron System:* The 31 great circles of the spherical icosahedron provide spherical edges for three other polyhedra in addition to the icosahedron itself, whose edges are shown as heavy lines. The shading indicates a typical face, as follows:

A. The rhombic triacontrahedron with 30 spherical rhombic faces, each consisting of four basic, least-common-denominator triangles.
B. The octahedron with 15 basic, least-common-denominator spherical triangles.
C. The pentagonal dodecahedron with ten basic, least-common-denominator spherical triangles.
D. Skewed spherical vector equilibrium.

457.43 The spherical pentagonal dodecahedron is composed of 12 spherical pentagons.

458.00 *Icosahedron: Great Circle Railroad Tracks of Energy*

458.01 Whereas each of the 25 great circles of the vector equilibrium and the icosahedron goes through the 12 vertexes at least twice; and whereas the 12 vertexes are the only points of intertangency of symmetric, unit-radius spheres, one with the other, in closest packing of spheres; and inasmuch as we find that energy charges always follow the convex surfaces of systems; and inasmuch as the great circles represent the most economical, the shortest distance between points on spheres; and inasmuch as we find that energy always takes the most economical route; therefore, it is perfectly clear that energy charges passing through an aggregate of closest-packed spheres, from one to another, could and would employ only the 25 great circles as the great-circle railroad tracks between the points of tangency of the spheres, ergo, between points in Universe. We can say, then, that the 25 great circles of the vector equilibrium represent all the possible railroad tracks of shortest energy travel through closest-packed spheres or atoms.

458.02 When the nucleus of the vector equilibrium is collapsed, or contracted, permitting the 12 vertexes to take the icosahedral conformation, the 12 points of contact of the system go out of register so that the 12 vertexes that accommodate the 25 great circles of the icosahedron no longer constitute the shortest routes of travel of the energy.

458.03 The icosahedron could not occur with a nucleus. The icosahedron, in fact, can only occur as a single shell of 12 vertexes remote from the vector equilibrium's multi-unlimited-frequency, concentric-layer growth. Though it has the 25 great circles, the icosahedron no longer represents the travel of energy from any sphere to any tangent sphere, but it provides the most economical route between a chain of tangent icosahedra and a face-bonded icosahedral structuring of a "giant octahedron's" three great circles, as well as for energies locked up on its surface to continue to make orbits of their own in local travel around that single sphere's surface.

458.04 This unique behavior may relate to the fact that the volume of the icosahedron in respect to the vector equilibrium with the rational value of 20 is 18.51 and to the fact that the mass of the electron is approximately one over 18.51 in respect to the mass of the neutron. The icosahedron's shunting of energy into local spherical orbiting, disconnecting it from the closest-packed railroad tracks of energy travel from sphere to sphere, tends to identify the icosahedron very uniquely with the electron's unique behavior in respect to nuclei as operating in remote orbit shells.

458.05 The energy charge of the electron is easy to discharge from the surfaces of systems. Our 25 great circles could lock up a whole lot of energy to be discharged. The spark could jump over at this point. We recall the name *electron* coming from the Greeks rubbing of amber, which then discharged sparks. If we assume that the vertexes are points of discharge, then we see how the six great circles of the icosahedron—which never get near its own vertexes—may represent the way the residual charge will always remain bold on the surface of the icosahedron.

458.06 Maybe the 31 great circles of the icosahedron lock up the energy charges of the electron, while the six great circles release the sparks.

458.10 **Icosahedron as Local Shunting Circuit:** The icosahedron makes it possible to have individuality in Universe. The vector equilibrium never pauses at equilibrium, but our consciousness is caught in the icosahedron when mind closes the switch.

458.11 The icosahedron's function in Universe may be to throw the switch of cosmic energy into a local shunting circuit. In the icosahedron energy gets itself locked up even more by the six great circles—which may explain why electrons are borrowable and independent of the proton-neutron group.

458.12 The vector-equilibrium railroad tracks are trans-Universe, but the icosahedron is a locally operative system.

459.00 Great Circle Foldabilities of Icosahedron

459.01 The great circles of the icosahedron can be folded out of circular discs of paper by three different methods: (a) 15 multi-bow ties of four tetrahedra each; (b) six pentagonal bow ties; and (c) 10 multi-bow ties. Each method defines certain of the surface arcs and central angles of the icosahedron's great circle system, but all three methods taken together do not define all of the surface arcs and central angles of the icosahedron's three sets of axis of spin (see drawings section).

459.02 The 15 great circles of the icosahedron can be folded into multi-bow ties of four tetrahedra each. Four times 15 equals 60, which is half the number of triangles on the sphere. Sixty additional triangles inadvertently appear, revealing the 120 identical spherical triangles which are the maximum number of like units which may be used to subdivide the sphere.

459.03 The six great circles of the icosahedron can be folded from central angles of 36 degrees each to form six pentagonal bow ties. (See illustration 458.12.)

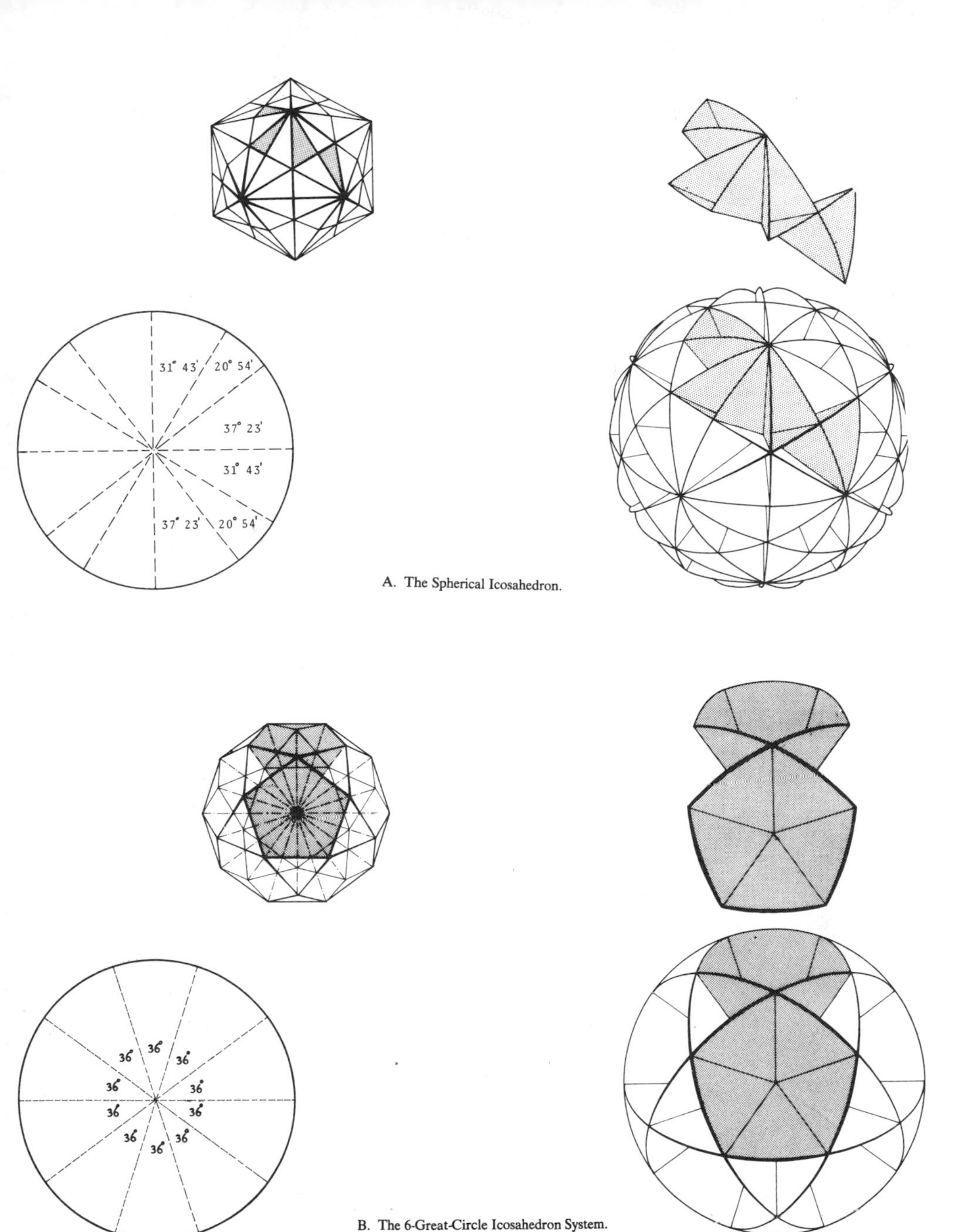

Fig. 458.12 *Folding of Great Circles into the Icosahedron System:*

A. The 15 great circles of the icosahedron folded into "multi-bow-ties" consisting of four tetrahedrons each. Four times 15 equals 60, which is ½ the number of triangles on the sphere. Sixty additional triangles inadvertently appear, revealing the 120 identical (although right- and left-handed) spherical triangles, which are the maximum number of like units that may be used to subdivide the sphere.

B. The six great-circle icosahedron system created from six pentagonal "bow-ties."

460.00 *Jitterbug: Symmetrical Contraction of Vector Equilibrium*

460.01 Definition

460.011 The "jitterbug" is the finitely closed, external vector structuring of a vector-equilibrium model constructed with 24 struts, each representing the push-pull, action-and-reaction, local compression vectors, all of them cohered tensionally to one another's ends by flexible joints that carry only tension across themselves, so that the whole system of only-locally-effective compression vectors is comprehensively cohered by omniembracing continuous four closed hexagonal cycles' tension.

460.02 When the vector-equilibrium "jitterbug" assembly of eight triangles and six squares is opened, it may be hand-held in the omnisymmetry conformation of the vector equilibrium "idealized nothingness of absolute middleness." If one of the vector equilibrium's triangles is held by both hands in the following manner—with that triangle horizontal and parallel to and above a tabletop; with one of its apexes pointed away from the holder and the balance of the jitterbug system dangling symmetrically; with the opposite and lowest triangle, opposite to the one held, just parallel to and contacting the tabletop, with one of its apexes pointed toward the individual who is hand-holding the jitterbug—and then the top triangle is deliberately lowered toward the triangle resting on the table without allowing either the triangle on the table or the triangle in the operator's hands to rotate (keeping hands clear of the rest of the system), the whole vector equilibrium array will be seen to be both rotating equatorially, parallel to the table but not rotating its polar-axis triangles, the top one of which the operating individual is hand-lowering, while carefully avoiding any horizontal rotation of, the top triangle in respect to which its opposite triangle, resting frictionally on the table, is also neither rotating horizontally nor moving in any direction at all.

460.03 While the equatorial rotating results from the top triangle's rotationless lowering, it will also be seen that the whole vector-equilibrium array is contracting symmetrically, that is, all of its 12 symmetrically radiated vertexes move synchronously and symmetrically toward the common volumetric center of the spherically chorded vector equilibrium. As it contracts comprehensively and always symmetrically, it goes through a series of geometrical-transformation stages. It becomes first an icosahedron and then an octahedron, with all of its vertexes approaching one another symmetrically and without twisting its axis.

460.04 At the octahedron stage of omnisymmetrical contraction, all the vectors (strut edges) are doubled together in tight parallel, with the vector equilibrium's 24 struts now producing two 12-strut-edged octahedra congruent with one another. If the top triangle of the composite octahedron (which is the triangle hand-held from the start, which had never been rotated, but only lowered with each of its three vertexes approaching exactly perpendicularly toward the table) is now rotated 60 degrees and lowered further, the whole structural system will transform swiftly into a tetrahedron with its original 24 edges now quadrupled together in the six-edge pattern of the tetrahedron, with four tetrahedra now congruent with one another. Organic chemists would describe it as a quadrivalent tetrahedral structure.

460.05 Finally, the model of the tetrahedron turns itself inside out and oscillates between inside and outside phases. It does this as three of its four triangular faces hinge open around its base triangle like a flower bud's petals opening and hinging beyond the horizontal plane closing the tetrahedron bud below the base triangle.

460.06 As the tetrahedron is opened again to the horizontal four-triangle condition, the central top triangle may again be lifted, and the whole contractive sequence of events from vector equilibrium to tetrahedron is reversed; the system expands after attaining the octahedral stage. When lifting of the top-held, nonhorizontally rotated triangle has resulted in the whole system expanding to the vector equilibrium, the equatorial rotational momentum will be seen to carry the rotation beyond dead-center, and the system starts to contract itself again. If the operating individual accommodates this momentum trend and again lowers the top triangle without rotating it horizontally, the rotation will reverse its original direction and the system will contract through its previous stages but with a new mix of doubled-up struts. As the lowering and raising of the top triangle is continuously in synchronization with the rotating-contracting-expanding, the rotation changes at the vector equilibrium's "zero"—this occasions the name jitterbug. The vector equilibrium has four axial pairs of its eight triangular faces, and at each pair, there are different mixes of the same struts.

460.07 The jitterbug employs only the external vectors of the vector equilibrium and not its 12 internal radii. They were removed as a consequence of observing the structural stability of 12 spheres closest packed around a nuclear sphere. When the nuclear sphere is removed or mildly contracted, the 12 balls rearrange themselves (always retaining their symmetry) in the form of the icosahedron. Removal of the radial vectors permitted contraction of the model—and its own omnisymmetrical pulsation when the lowering and raising patterns are swiftly repeated. It will be seen that the squares accommodate the jitterbug contractions by transforming first into two equiangular triangles and then disappearing altogether. The triangles do not change through the transformation in size or angularity. The original eight triangles of the vector equilibrium are those of the octahedron stage, and they double together to form the four faces of the tetrahedron.

460.08 In the jitterbug, we have a sizeless, nuclear, omnidirectionally pulsing model. The vector-equilibrium jitterbug is a conceptual system independent of size, ergo cosmically generalizable. (See Secs. 515.10 and 515.11.)

461.00 Recapitulation: Polyhedral Progression in Jitterbug

461.01 If the vector equilibrium is constructed with circumferential vectors only and joined with flexible connectors, it will contract symmetrically, due to the instability of the square faces. This contraction is identical to the contraction of the concentric sphere packing when the nuclear sphere is removed. The squares behave as any four balls will do in a plane. They would like to rest and become a diamond, to get into two triangles. They took up more room as a square, and closer packing calls for a diamond. The 12 vertexes of the vector equilibrium simply rotate and compact a little. The center ball was keeping them from closer packing, so there is a little more compactibility when the center ball goes out.

461.02 Icosahedron: The icosahedron occurs when the square faces are no longer squares but have become diamonds. The diagonal of the square is considerably longer than its edges. But as we rotate the ridge pole, the diamonds become the same length as the edge of the square (or, the same length as the edge of the tetrahedron or the edge of the octahedron). It becomes the octahedron when all 30 edges are the same length. There are no more squares. We have a condition of omnitriangulation.

461.03 We discover that an icosahedron is the first degree of contraction of the vector equilibrium. We never catch the vector equilibrium in its true existence in reality: it is always going one way or the other. When we go to the icosahedron, we get to great realities. In the icosahedron, we get to a very prominent fiveness: around every vertex you can always count five.

461.04 The icosahedron contracts to a radius less than the radii of the vector equilibrium from which it derived. There is a sphere that is tangent to the other 12 spheres at the center of an icosahedron, but that sphere is inherently smaller. Its radius is less than the spheres in tangency which generate the 12 vertexes of the vector equilibrium or icosahedron. Since it is no longer

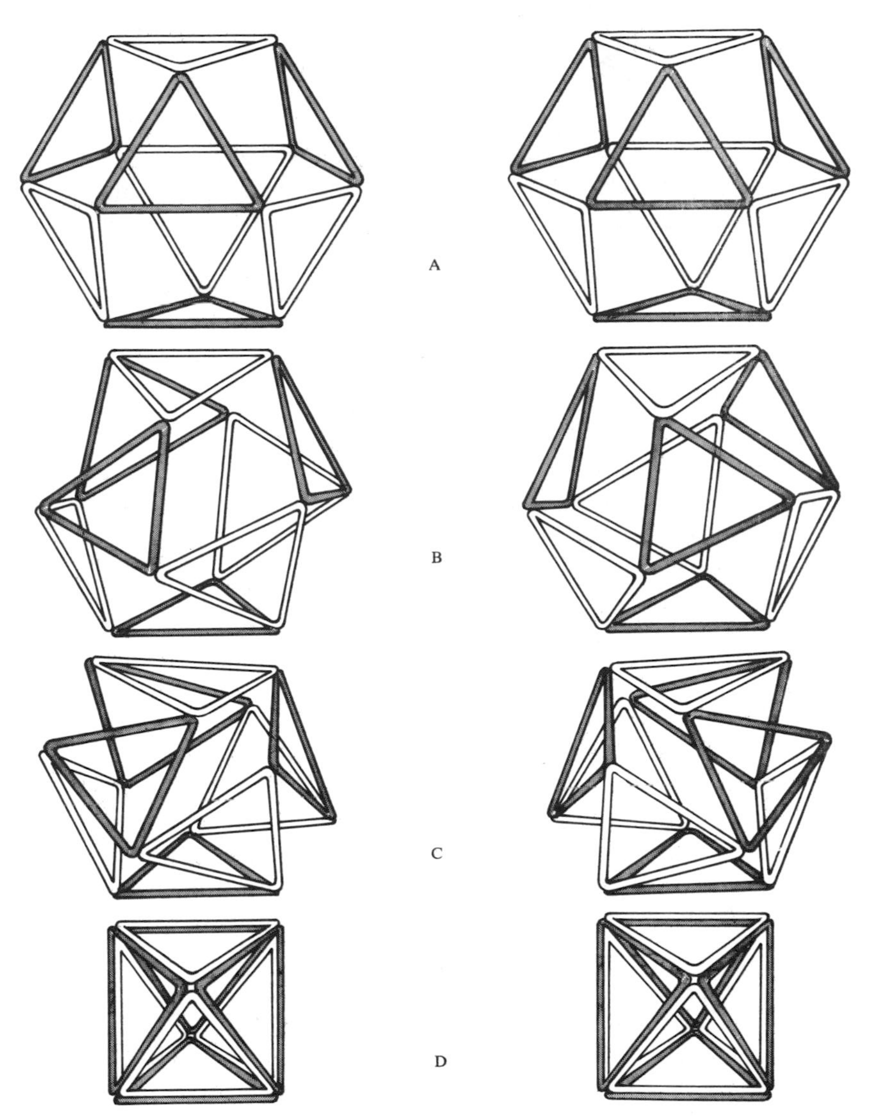

Fig. 460.08 *Symmetrical Contraction of Vector Equilibrium: Jitterbug System:* If the vector equilibrium is constructed with circumferential vectors only and joined with flexible connectors, it will contract symmetrically due to the instability of the square faces. This contraction is identical to the contraction of the concentric sphere packing when its nuclear sphere is removed. This system of transformation has been referred to as the "jitterbug." Its various phases are shown in both left- and right-hand contraction:

A. Vector equilibrium phase: the beginning of the transformation.
B. Icosahedron phase: When the short diagonal dimension of the quadrilateral face is equal to the vector equilibrium edge length, 20 equilateral triangular faces are formed.
C. Further contraction toward the octahedron phase.
D. Octahedron phase: Note the doubling of the edges.

the same-size sphere, it is not in the same frequency or in the same energetic dimensioning. The two structures are so intimate, but they do not have the same amount of energy. For instance, in relation to the tetrahedron as unity, the volume of the icosahedron is 18.51 in respect to the vector equilibrium's volume of 20. The ratio is tantalizing because the mass of the electron in respect to the mass of the neutron is one over 18.51. That there should be such an important kind of seemingly irrational number provides a strong contrast to all the other rational data of the tetrahedron as unity, the octahedron as four, the vector equilibrium as 20, and the rhombic dodecahedron as six: beautiful whole rational numbers.

461.05 The icosahedron goes out of rational tunability due to its radius being too little to permit it having the same-size nuclear sphere, therefore putting it in a different frequency system. So when we get into atoms, we are dealing in each atom having its unique frequencies.

461.06 In the symmetrical jitterbug contraction, the top triangle does not rotate. Its vertex always points toward the mid-edge of the opposite triangle directly below it. As the sequence progresses, the top triangle approaches the lower as a result of the system's contraction. The equator of the system twists and transforms, while the opposite triangles always approach each other rotationlessly. They are the polar group.

461.07 Octahedron: When the jitterbug progresses to the point where the vector edges have doubled up, we arrive at the octahedron. At this stage, the top triangle can be pumped up and down with the equatorial vectors being rotated first one way and then the other. There is a momentum of spin that throws a twist into the system—positive and negative. The right-hand octahedron and the left-hand octahedron are not the same: if we were to color the vectors to identify them, you would see that there are really two different octahedra.

461.08 Tetrahedron: As the top triangle still plunges toward the opposite triangle, the two corners, by inertia, simply fold up. It has become the tetrahedron. In the octahedron stage, the vectors were doubled up, but now they have all become fourfold, or quadrivalent. The eight tetrahedra of the original vector equilibrium are now all composited as one. They could not escape from each other. We started off with one energy action in the system, but we have gone from a volume of 20 to a volume of one.* The finite clo-

* In vectorial geometry, you have to watch for the times when things double up. The vectors represent a mass and a velocity. Sometimes they double up so they represent twice the value—or four times the value—when they become congruent.

Fig. 461.08 *Jitterbug System Collapses into Tetrahedron: Polarization:* The "jitterbug" system, after reaching the octahedron phase, may be collapsed and folded into the regular tetrahedron. Note that because the vector equilibrium has 24 edges the tetrahedra have accumulated four edges at each of their six normal edges. The "jitterbug" can also be folded into a larger but incomplete tetrahedron. Note that in this case the two sets of double edges suggest polarization.

sure of the four-great-circle, six-hexagon-vector "necklaces" were never "opened" or unfastened.

461.09 We have arrived at the tetrahedron as a straight precessional result. The quadrivalent tetrahedron is the limit case of contraction that unfolds and expands again symmetrically only to contract once more to become the other tetrahedron (like the pumping of the positive and negative octahedron). All of the jitterbug sequence was accomplished within the original domain of the vector equilibrium. The tensional integrity survives within the internal affairs domain of atoms.

462.00 Rotation of Triangle in Cube

462.01 To comprehend the complex of transformings demonstrated by the jitterbug we may identify each of the eight triangles of the vector equilibrium with the eight small cubes which comprise a two-frequency large cube's eight corners. When the jitterbug transforms into an octahedron, the jitterbug vector equilibrium's six square faces disappear leaving only the eight triangles of the vector equilibrium, each of which has moved inwardly at a symmetrical rate toward the common center of the vector equilibrium as the squares disappear and the triangles approach one another until their respective three edges each become congruent with one another, thus doubling their vector edges together in paralleled congruence. Since each of the eight triangles behaved the same way as the others we can now study how one behaved and we find that each triangle "did its thing" entirely within the domain of one of the eight cubes of the two-frequency big cube. Thus we learn that a triangle can rotate within the topological lines of a cube with the triangle's three corners being guided by the cube's edges.

462.02 Wave-propagating action is cyclically generated by a cube with a triangle rotating in it.

463.00 Diagonal of Cube as Wave-Propagation Model

463.01 There are no straight lines, only waves resembling them. In the diagram, any zigzag path from *A* to *C* equals the sum of the sides *AB* and *BC*. If the zigzag is of high frequency, it may look like a diagonal that should be shorter than *ABC*. It is not.

463.02 As the triangle rotates in the cube, it goes from being congruent with the positive tetrahedron to being congruent with the negative tetrahedron. It is an oscillating system in which, as the triangles rotate, their corners describe arcs (see Sec. 464.02) which convert the cube's 12 edges from quasi-straight lines to 12 arcs which altogether produce a dynamically described

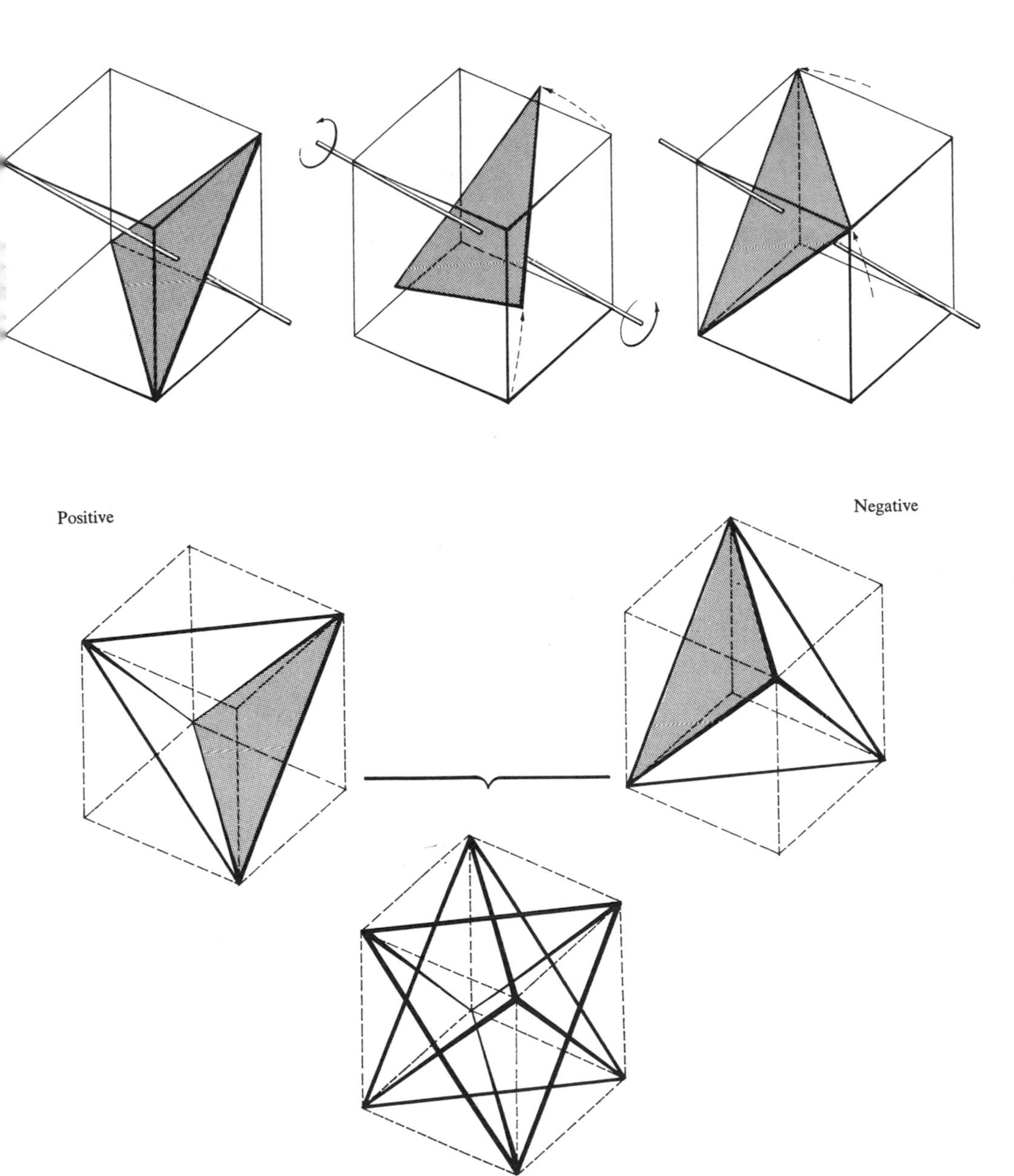

Fig. 462.00 The triangle formed by connecting diagonals of three adjacent faces of the cube is the face of the tetrahedron within the cube. If the triangle is rotated so that its vertexes move along the edges of the cube, its position changes from the positive to the negative tetrahedron. Two equal tetrahedra (positive and negative) joined at their common centers define the cube. The total available energy of a system is related to its surface area, involving the second power (square) of the radius. $E = Mc^2$: The conjunction of any two similar systems results in a synergetic relationship: the second power of individual totals of cohesiveness of the systems.

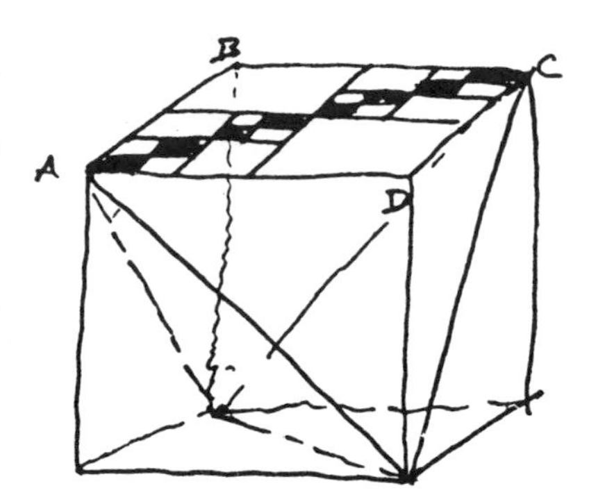

Fig. 463.01 There are no straight lines, only waves resembling them. In this diagram, any zigzag path from *A* to *C* equals the sum of the sides *AB* and *BC*. If zigzag is infinitely small, it looks like a diagonal that should be shorter than *ABC*. It is not.

sphere (a spherical cube) which makes each cube to appear to be swelling locally. But there is a pulsation arc-motion lag in it exactly like our dropping a stone in the water and getting a planar pattern for a wave (see Sec. 505.30), but in this model we get an omnidirectional wave pulsation. This is the first time man has been able to have a conceptual picture of a local electromagnetic wave disturbance.

463.03 The cube oscillates from the static condition to the dynamic, from the potential to the radiant. As it becomes a wave, the linear becomes the second-power rate of growth. The sum of the squares of the two legs = the square of the hypotenuse = the wave. The 12 edges of the cube become the six diagonals of the tetrahedron by virtue of the hypotenuse: the tetrahedron is the normal condition of the real (electromagnetic) world. (See Sec. 982.21.)

463.04 There is an extraordinary synergetic realization as a consequence of correlating (a) the arc-describing, edge-pulsing of cubes generated by the eight triangles rotating in the spheres whose arcs describe the *spherical cube* (which is a sphere whose volume is 2.714—approximately three—times that of the cube) and (b) the deliberately nonstraight line transformation model (see Sec. 522), in which the edges of the cube become the six wavilinear diagonals of the cube, which means the cube transforming into a tetrahedron. Synergetically, we have the tetrahedron of volume one and the cube of volume three—as considered separately—in no way predicting that the cube would be transformed into an electromagnetic-wave-propagating tetrahedron. This is an energy compacting of $3 \rightarrow 1$; but sum-totally this means an energetic-volumetric contraction from the spherical cube's volume of 8.142 to the tetrahedron's one, which energetic compacting serves re-expansively to power the electromagnetic-wave-propagating behavior of the wavilinear-edged tetrahedron. (See Sec. 982.30.)

463.05 We really find, learning synergetically, from the combined behaviors of the tetrahedron, the cube, and the deliberately-nonstraight-line cubical transformation into a tetrahedron, how the eight cubical corners are self-truncated to produce the vector equilibrium within the allspace-filling cubical isotropic-vector-matrix reference frame; in so doing, the local vacatings of the myriad complex of closest-packing cube truncations produce a "fallout" of all the "exterior octahedra" as a consequence of the simultaneous truncation of the eight corners of the eight cubes surrounding any one point. As we learn elsewhere (see Sec. 1032.10), the *exterior* octahedron is the contracted vector equilibrium and is one of the spaces between spheres; the octahedron thus becomes available as the potential alternate new sphere when the old spheres become spaces. The octahedra thus serve in the allspace-filling exchange of spheres and spaces (see Sec. 970.20).

464.00 Triangle in Cube as Energetic Model
(*see drawings section*)

464.01 The triangle *CDE* formed by connecting the diagonals of the three adjacent square faces surrounding one corner, *A,* of the cube defines the *base triangular face* of one of thc two tetrahedra always coexisting within, and structurally permitting the stability of, the otherwise unstable cubic form. The triangle *GHF* formed by connecting the three adjacent faces surrounding the *B* corner of the same cube diametrically, i.e. polarly, opposite the first triangulated corner, defines the triangular face *GHF* of the other of the two tetrahedra always coexisting within that and all other cubes. The plane of the green triangle *CDE* remains always parallel to the plane of the red triangle *GHF* even though it is rotated along and around the shaft *AB* (see drawings section).

464.02 If the first triangle *CDE* defined by the three diagonals surrounding the *A* corner of the cube is rotated on the axis formed by the diagonal leading from that corner of the cube inwardly to its polarly opposite and oppositely triangled *B* corner, the rotated triangle maintains its attitude at right angles to its axis, and its three vertexes move along thc three edges of the cube until the green triangle reaches and become congruent with the red base triangle of the axially opposite corner. Thereafter, if the rotation continues in the same circular direction, the same traveling triangle will continue to travel pulsatingly, back and forth, becoming alternately the base triangle of the positive and then of the negative tetrahedron. As the triangle returns from its first trip away, its corners follow three additional edges of the cube. As the vertexes of the shuttling triangle follow the six cube edges, their apexes protrude and describe spherical arcs outwardly along the cubes' edges running from cube corner to cube corner. Swift rotation of the triangle's shaft not only causes the triangle to shuttle back and forth, but also to describe six of the 12 edges of the spherical cube producing an equatorially spheroid pulsation. The two equal tetrahedra are not only oppositely oriented, but their respective volumetric centers (positive and negative) are congruent, being joined at their common centers of volume, which coincide with that of the containing cube. Because each cube in the eight-cube, two-frequency big cube has both a positive and a negative tetrahedron in it, and because each tetrahedron has four triangular faces, each cube has eight equilateral triangular edges corresponding to the 12 diagonalling hypotenuses of each cube's six faces.

464.03 Each cube has four pairs of polarly opposite corners. There are four co-occurring, synchronously operative, triangularly shuttleable systems within each cube; with all of them synchronously operative, the cube's 12 edges will be synchronously accommodating— $4 \times 6 = 24$ —edge-arcings traveling 12 positively and 12 negatively, to produce the profile of two spherical cubes, one positive and one negative.

464.04 Each vector equilibrium, when complemented by its coexistent share of one-eighth of its (concave) external octahedra, embraces eight cubes, each of which has four activable, axially shuttleable, electromagnetic-energy-generating potentials.

464.05 Eight of these triangular shuttle cubes may be completed on each of the vector equilibrium's eight triangular faces by adding one $^1/_8$th-Octa corner to each of them. Each $^1/_8$th-Octa corner consists of six *A* and six *B* modules. As one such $^1/_8$th-Octa, 6*A*-6*B* moduled, 90°-apexed, equiangle-based, isosceles tetrahedron is added to any of the vector equilibrium's eight triangular faces, which contain the potential new nucleus—which thus becomes a newborn active nucleus—when so double-layer covered by the 12 *A*'s and 12 *B*'s energy modules, which altogether produce a total of 24 energy modules whenever the rotating triangle alternates its position, which combined 24 modules correspond to the 24 energy modules of one whole regular tetrahedral event, which *is* the quantum in nuclear physics.

464.06 The vector equilibrium's jitterbugging conceptually manifests that any action (and its inherent reaction force) applied to any system always articulates a complex of vector-equilibria, macro-micro jitterbugging, involving all the vector equilibria's ever cosmically replete complementations by their always co-occurring internal and external octahedra—all of which respond to the action by intertransforming in concert from "space nothingnesses" into closest-packed spherical "somethings," and vice versa, in a complex three-way shuttle while propagating a total omniradiant wave pulsation operating in unique frequencies that in no-wise interfere with the always omni-co-occurring cosmic gamut of otherly frequenced cosmic vector-equilibria accommodations.

464.07 In contradistinction to the sphere, the tetrahedron has the most surface with the least volume of any symmetrical form. The total available energy of a system is related to its surface area, involving the second power of the radius. $E = Mc^2$. The mass congruence of any two similar systems results in a synergetic relationship with a second-powering of cohesiveness of the joined systems. This releases the fourfolded energy, which no longer has the two tetrahedra's mass-interattraction work to do, and this in turn releases the energies outward to the tetrahedra's highest-capacity surfaces. And since surface functions as the electromagnetic-energy carrier, and since the energy relayed to their surfaces alternates from the positive to the negative tetrahedron, and since the distance between their surface centers is only two A Module altitude wavelengths (each of which two A Module altitudes constitute and serve as one generalized electromagnetic wavelength with generatable frequency beginning at two), the rotation of the triangle within the cube passes through the common energy centers of the two tetrahedra and delivers its content to the other base surface, after which it pulses through center delivery of the opposite charge to the other surface, which altogether propagates potentially exportable, frequency-determinate, electromagnetic energy. The six cube-edge travelings of the triangles' vertexes accomplished with each cycle of the triangle-in-cube shuttle coincides in number and is akin to the six vector edges comprising one tetrahedral quantum; the sixness of wavilinear and sometimes reangularly redirected traveling employs also the six basic degrees of freedom articulated by each and every one *cosmic event.*

464.08 Thus we realize conceptually the ever-self-regenerative, omni-idealized, eternal integrity of the utterly metaphysical, timeless, weightless, zerophase geometric frame of transformations referencing function, which is served by the vector equilibrium in respect to which all the aberrational dimensioning of all realization of the variety of relative durations, sensorial lags, recalls, and imaginings are formulatingly referenced to differentiate out into the special-case local experiences of the eternal scenario Universe, which each of us identifies to ourselves as the "Shape of Things" and which each individual sees differently yet ever intuits to be rigorously referenced to an invisibly perfect prototype in pure principle, in respect to which only approachable but never realizable "understanding" of one of us by others occurs: "And it Came to Pass."

465.00 Rotation of Four Axes of Vector Equilibrium: Triangles, Wheels, and Cams

465.01 We can have a vector equilibrium model made out of a tubular steel frame with each of the eight triangular faces connected by four axes with a journal to slide on the shafts and with each of the rods being perpendicular to two of the eight triangular faces. This is a four-dimensional, four-axis system. Just as a regular tetrahedron has four unique faces, so there are four unique perpendiculars to them, making a four-dimensional system.

465.02 We can put a little rivet through the centers of area of the eight triangles, and we can let the brass rod run through the journals and slide on a wire. We can tie the corners of the triangles together with nylon threads. If we spin the model rapidly on one of the axes, all the triangles slide outwardly to form the vector equilibrium. If next we touch a finger or a pencil to any midface of one triangle in the spinning system, the whole system will contract symmetrically until it becomes an octahedron. But when we take the finger or pencil off again, centrifugal force will automatically open up the system to the vector-equilibrium condition again. The oscillating motion makes this an expanding and contracting system.

465.03 We see that every one of the triangles in the vector equilibrium can shuttle back and forth, so that all the edges of the cube would be arced outwardly with pairs of arcing triangle corners shuttling in opposite directions by each other. With a swiftly oscillating system and a pulsating spherical expansion-contraction going on everywhere locally, the whole system becomes an optically pulsating sphere. We find that each one of the little triangles rotates as if it were swelling locally. Each one of their vertexes brings about a further spherical condition, so that in the whole system, all the wires locally bend outwardly temporarily to accommodate the whole motion. We may now put together a large omnidirectional complex of the sets of four-axis and eight vertex-interconnected transparent plastic triangles with alternate sets of red transparent and uncolored transparent plastic triangles. We can interconnect the triangles from set to set. We then find experimentally that if one force, such as a pencil, is applied to one triangle of one open vector equilibrium, that vector equilibrium closes to become an octahedron, and vice versa, throughout the whole system. Every vector equilibrium will become an octahedron and every octahedron will become a vector equilibrium. (Which is to say that every space becomes a sphere and every sphere becomes a space.)

465.04 Since there is a force distribution lag in the system, it is exactly like dropping a stone into water and getting a planar pattern for a wave, but in this one, we get an omnidirectional wave. We can see the electromagnetic

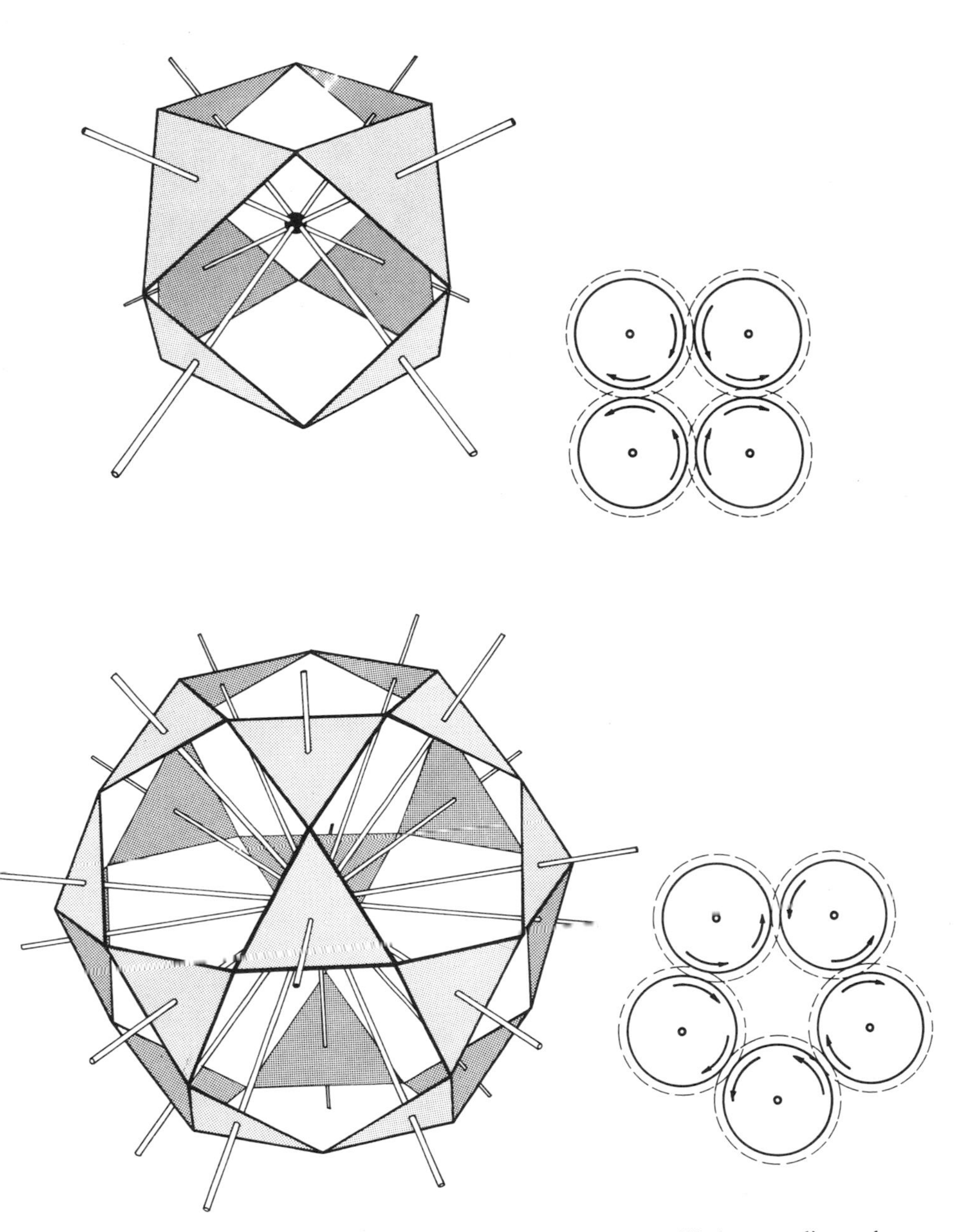

Fig. 465.00 Note that the eight triangular faces of the vector equilibrium are disposed about four-sided openings, i.e. square faces. It is possible to arrange 20 triangles in similar fashion around five-sided openings, i.e. pentagons. The shape is the icosidodecahedron. When a model is constructed with 20 spokes, i.e. ten axes, meeting at its center, which pass through the centers of each triangle, an unexpected behavior results. In the vector equilibrium model the triangles will rotate and contract towards its center, however, with the icosidodecahedron the entire structure remains fixed. It is not capable of contraction due to the fact that there is an odd number of triangles surrounding each opening. The diagrams show clearly why this is so. Any odd-numbered array of interlocked gears will not be free to rotate.

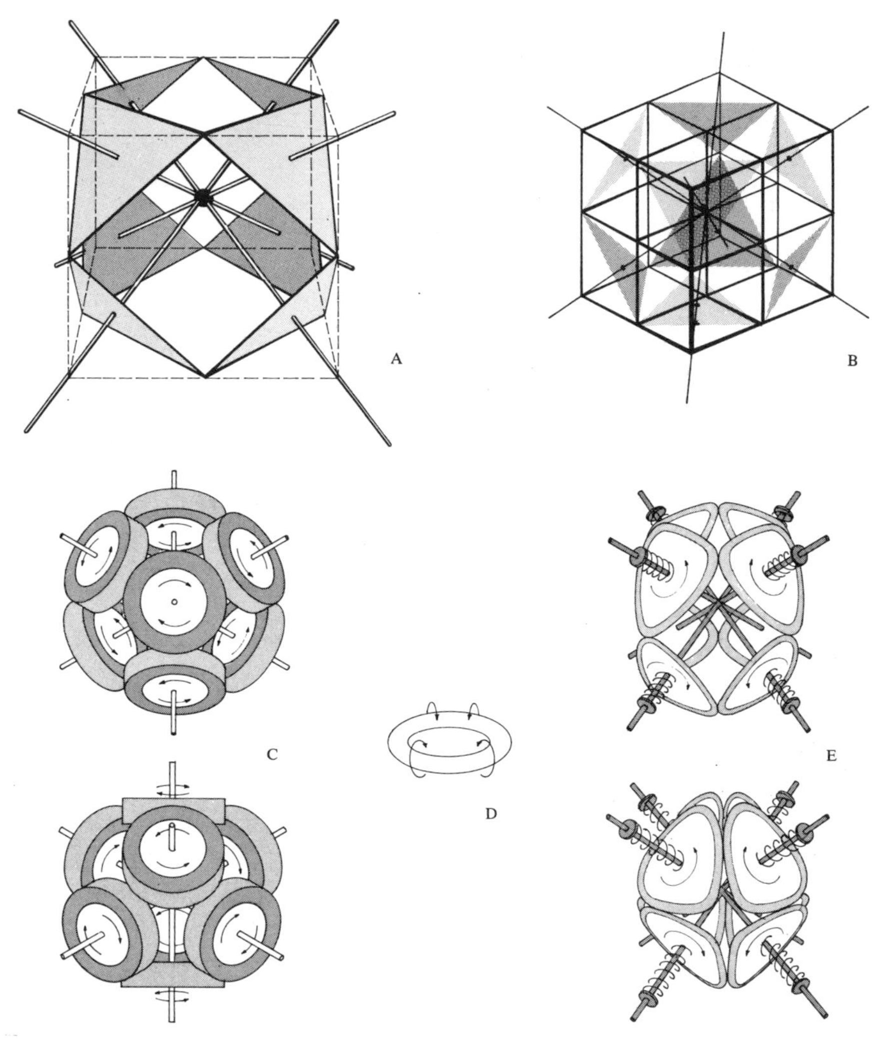

Fig. 465.01 *Four Axes of Vector Equilibrium with Rotating Wheels or Triangular Cams:*

A. The four axes of the vector equilibrium suggesting a four-dimensional system. In the contraction of the "jitterbug" from vector equilibrium to the octahedron, the triangles rotate about these axes.
B. Each triangle rotates in its own cube.
C. The four axes of the vector equilibrium shown with wheels replacing the triangular faces. The wheels are tangent to one another at the vertexes of the triangles, and when one wheel is turned, the others also rotate. If one wheel is immobilized and the system is rotated on the axes of this wheel, the opposite wheel remains stationary, demonstrating the polarity of the system.
D. Each wheel can be visualized as rotating inwardly on itself thereby causing all other wheels to rotate in a similar fashion.
E. If each wheel is conceived as a triangular cam shape, when they are rotated a continuous "pumping" or reciprocating action is introduced.

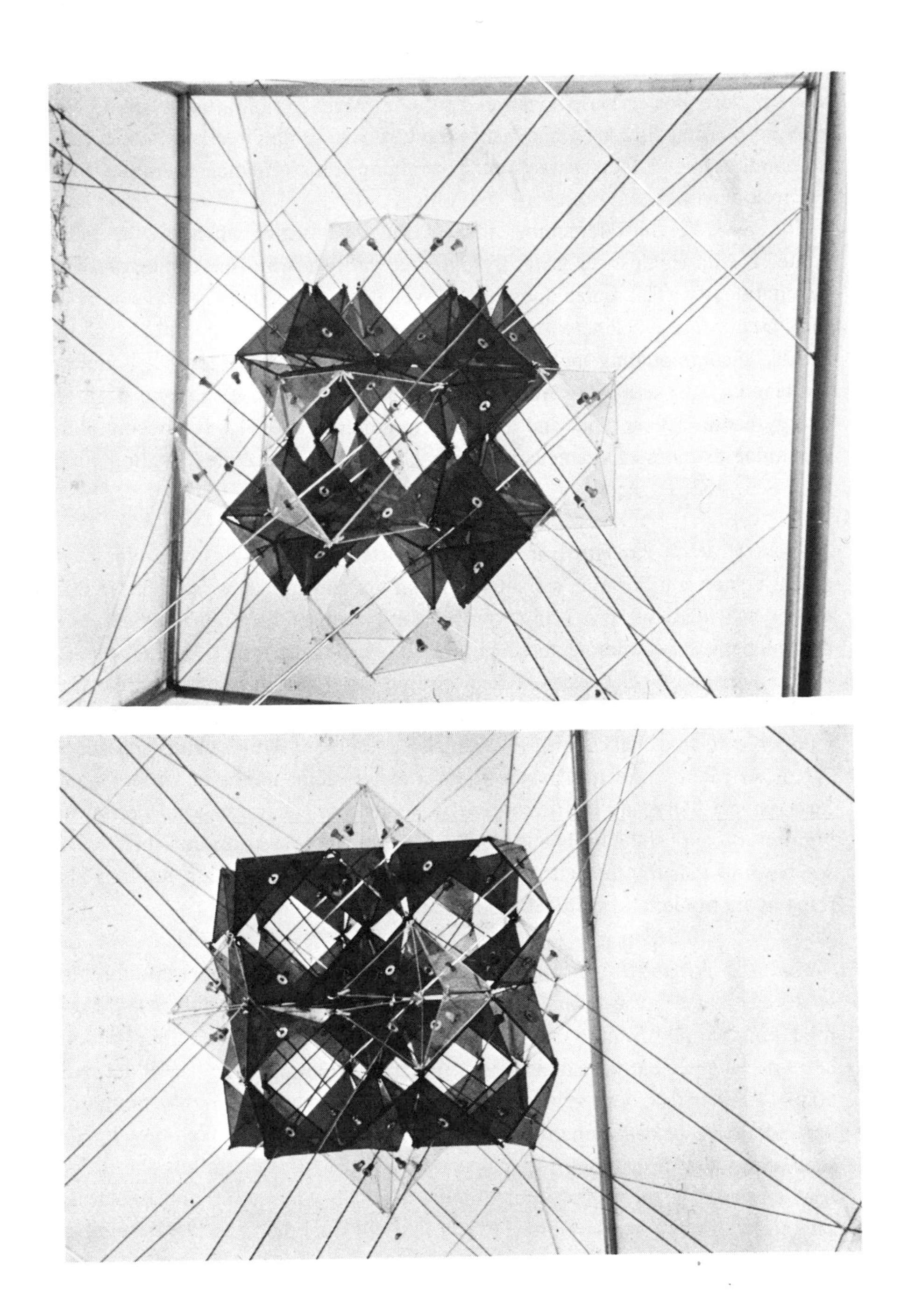

Fig. 465.03 *Rotation of Four Axes of Vector Equilibrium: Articulation of Eight Triangular Faces.*

wave pattern as clearly demonstrated by one energy action in the system. This may be the first time man has been provided with omnidirectional conceptual comprehension of the separate and combining transformation events of local electromagnetic-wave-propagation events.

465.05 We must remember that in the local water where we drop the stone, the molecules run inwardly and outwardly toward the center of Earth gravitationally. The water does not move; it accommodates a wave moving through it. A wave inherently goes outward in a pattern without any of the locally accommodating molecules or atoms migrating elsewhere. It is not simultaneous; we are using our memory and afterimage. We make a single energy action at one point and a complete omnidirectional wave occurs. This is similar to the steel-frame cube with all the many triangles rotating in it. (See Sec. 462, et seq.)

465.10 **Wheels: Rubber Tires:** If, instead of the eight triangular faces of the vector equilibrium, we substitute on the same shaft a little automobile tire on a wheel, we can bring tires in until each of the tires is frictionally touching the other tires at three points. If we have a train of gears, as one wheel goes one way, the next wheel can go the other way very comfortably. Around any hole there are four gears, and since there are four—an even number—we find that the trains reciprocate. There is no blocking anywhere. When we rotate one wheel in the light-wheel system, the other wheels rotate responsively. They are in friction with one another. Or we can hold on to the bottom of one of the wheels and turn the rest of the system around it. If we do so, we find that the top wheel polarly opposite the one we are holding also remains motionless while all the other six rotate.

465.20 **Torus:** If one of the mounted tires were just a rubber doughnut, it could be rotated inwardly like a torus; or it could be rotated outwardly like a big atomic-bomb mushroom cloud, opening in the center and coming in at the bottom. This is what we call an evoluting and involuting torus (see illustration 505.41, Pattern). These rubber tires of the eight-wheel assembly could not only rotate around on each other, but it is quite possible to make one wheel in such a way that it has little roller bearings along its rim that allow the rubber tires to rotate in the rim so that the tire could be involuting and evoluting. Therefore, if any one tire started to evolute, all the other tires would reciprocate.

465.21 If we hold only an axis in our hand, we can rotate the system around it. But as we rotate it around, all the wheels are rolling. As we saw in the pumping vector equilibrium, the opposite triangles never torque in relation to each other. The opposite wheel of the one we are holding does the same.

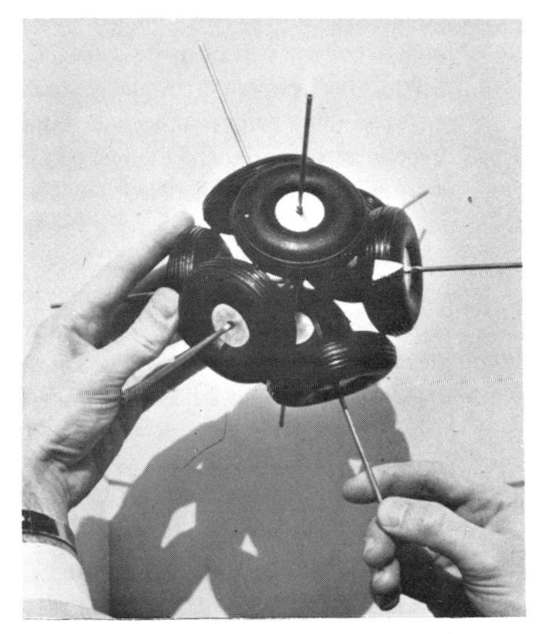

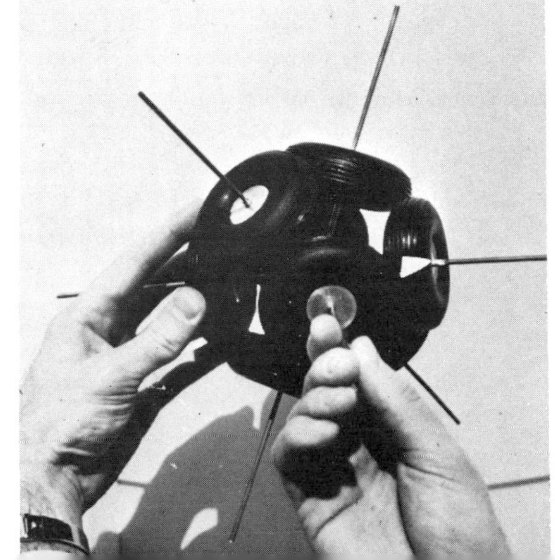

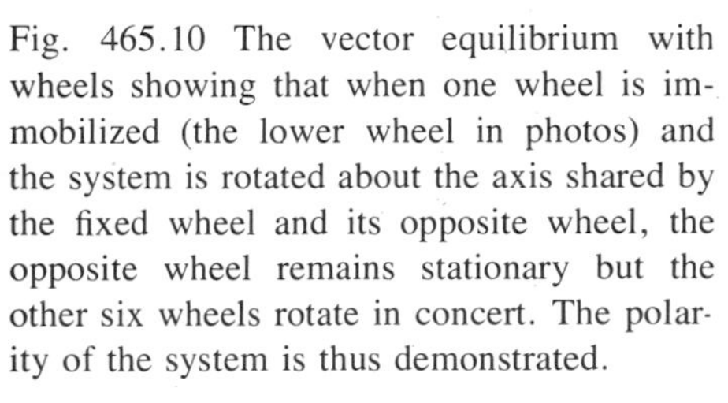

Fig. 465.10 The vector equilibrium with wheels showing that when one wheel is immobilized (the lower wheel in photos) and the system is rotated about the axis shared by the fixed wheel and its opposite wheel, the opposite wheel remains stationary but the other six wheels rotate in concert. The polarity of the system is thus demonstrated.

With the bottom wheel stationary on the ground and another wheel immobilized by one holding it, we can rotate the system so that one wheel rolls around the other. But we find that no matter how much we move it equatorially, if we immobilize one wheel in our fingers, the one opposite it becomes immobilized, too. If we not only hold a wheel immobilized while another is turning, but also squeeze and evolute it, all of the wheels will also involute and evolute.

465.22 It is quite possible to make an automobile tire and mount it in such a way that it looks triangular; that is, it will have a very small radius in its corners. I can take the same rubber and stretch it onto a triangular frame and also have the same little roller bearings so that it can involute and evolute. We will have a set of triangular tires that will pump from being the vector equilibrium into being the octahedron and back again. If we were then to immobilize one part of it, i.e., not let it involute and evolute, the rest of the system, due to rotation, would contract to become an octahedron so that it makes all the others reciprocate involuting and evoluting. We are able then to immobilize one axis, and the rest of the system except our opposite pole will both rotate and involute-evolute pulsatively.

465.30 **Four-Dimensional Mobility:** We are now discovering that in omnimotional Universe, it is possible to make two moving systems that move four-dimensionally, comfortably, the way we see four sets of wheels (eight wheels altogether) moving quite comfortably. But if we fasten one vector equilibrium to another by a pair of wheels—immobilizing one of them and having an axis immobilized—the rest of the system can keep right on rolling around it. By fastening together two parts of the Universe, we do not stop the rest of the four-dimensional motion of Universe. In all other non-four-dimensional mechanical systems we run into a ''three-dimensional'' blockage: if anything is blocked, then everything is blocked. But in a four-dimensional system, this is not at all the case. We can have two atoms join one another perfectly well and the rest of Universe can go right on in its motion. Nothing is frustrated, although the atoms themselves may do certain polarized things in relation to one another, which begins to explain a lot of the basic experiences.

470.00 Allspace-Filling Transformations of Vector Equilibrium

470.01 In the closest packing of spheres, there are only two symmetric shapes occurring in the spaces between the spheres. They are what we call the concave octahedron and the concave vector equilibrium. One is an open condition of the vector equilibrium and the other is a contracted one of the octahedron. If we take vector equilibria and compact them, we find that the triangular faces are occupying a position in closest packing of a space and that the square faces are occupying the position in closest packing of a sphere. (For a further exposition of the interchange between spheres and spaces, see illustrations at Sec. 1032, "Convex and Concave Sphere Packing Voids.")

470.02 When we compact vector equilibria with one another, we find that two of their square faces match together. Within a square face, we have a half octahedron; so bringing two square faces together produces an internal octahedron between the two of them. At the same time, a set of external octahedra occurs between the triangular faces of the adjacent vector equilibria.

480.00 Tetrahedron Discovers Itself and Universe

481.00 The initial self-and-other spherical associability (see Sec. 401 and illustration 401.01) produced first, associability; next, triangulation as structure; and then, tetrahedron as system. The inherent self-stretch-apartness reaction identified as mass attraction, and the inherent otherness of awareness, and the discovery of the self through the otherness, as a consequence of which awareness of relatedness, and curiosity about the interrelatedness of further unprecedented self-and-otherness discoveries, all initiate the tetrahedron's self-discovering that it can turn itself inside out by employing the mass-coherent integrity of any three spheres's intergeared frictionality to swallow involutingly the fourth sphere through the three's central passage and to extrude it evolutingly outward again on the other side.

482.00 Thus tetrahedron discovers that each of its four vertexes can be plunge-passaged through its innards to be extended on the opposite side of its four triangular faces. This automatically develops eight tetrahedral, self-transformation awarenesses and produces eight common nuclear-vertex tetrahedra of the vector equilibrium.

483.00 Further self-examination of the tetrahedron discovers that the geometry of its insideness proves to be a concave octahedron, with four of the octahedron's triangular faces represented by the four triangular windows at the face centers of the tetrahedron and the other triangular faces hidden from the view of outsiders, but clearly viewable from inside the tetrahedron's system as the spherical triangular areas of the interior surfaces of the tetrahedron's four corner balls; the edges of the triangles are defined as the great-

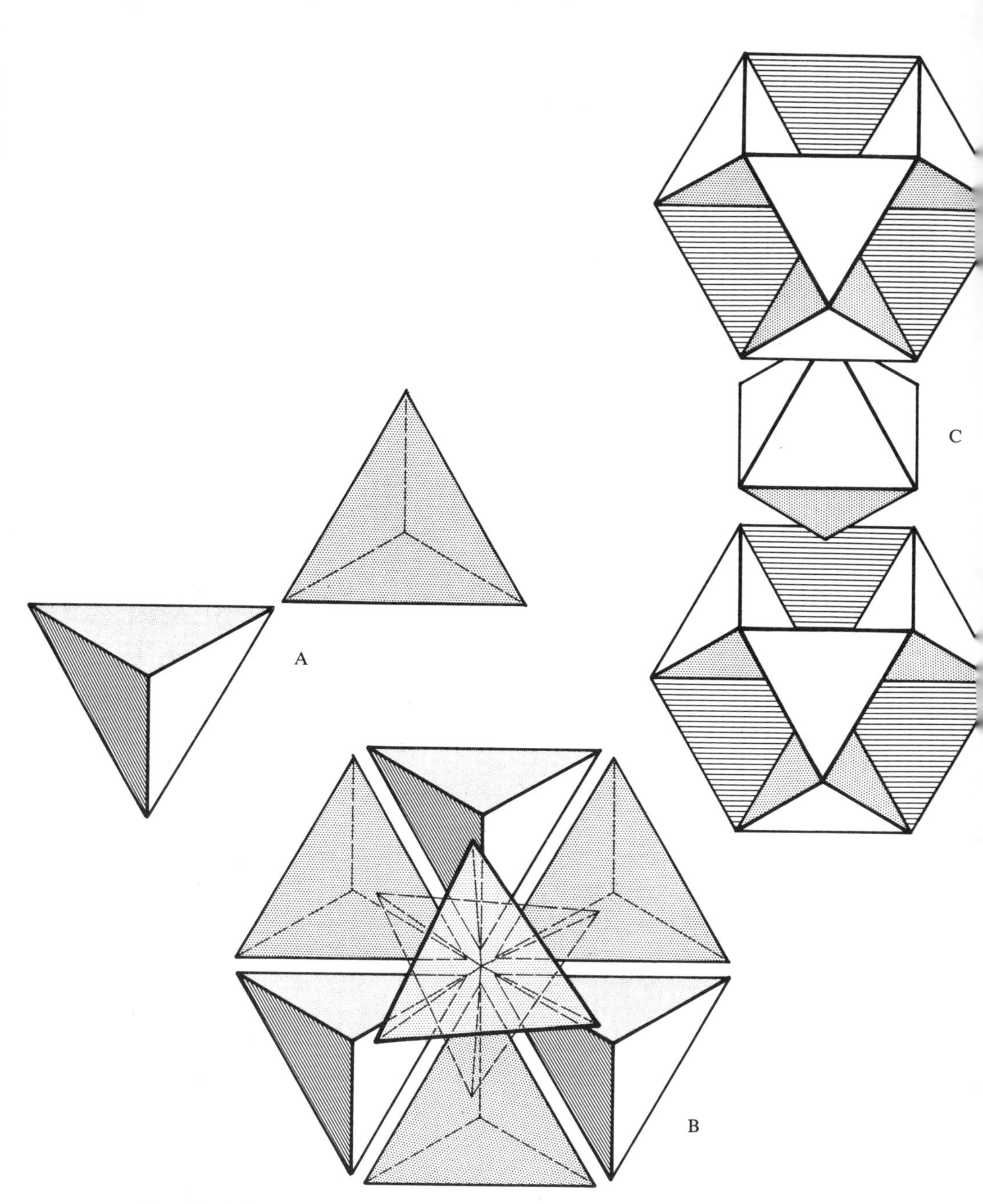

Fig. 470.02A *Role of Tetrahedra and Octahedra in Vector Equilibrium:*

A. Positive-negative tetrahedron system.

B. Vector equilibrium formed by four positive-negative tetrahedron systems with common central vertex and coinciding radial edges. Equilibrium of system results from positive-negative action of double radial vectors.

C. The relationship of space-filling tetrahedra and octahedra to the vector equilibrium defined by eight radially disposed tetrahedra.

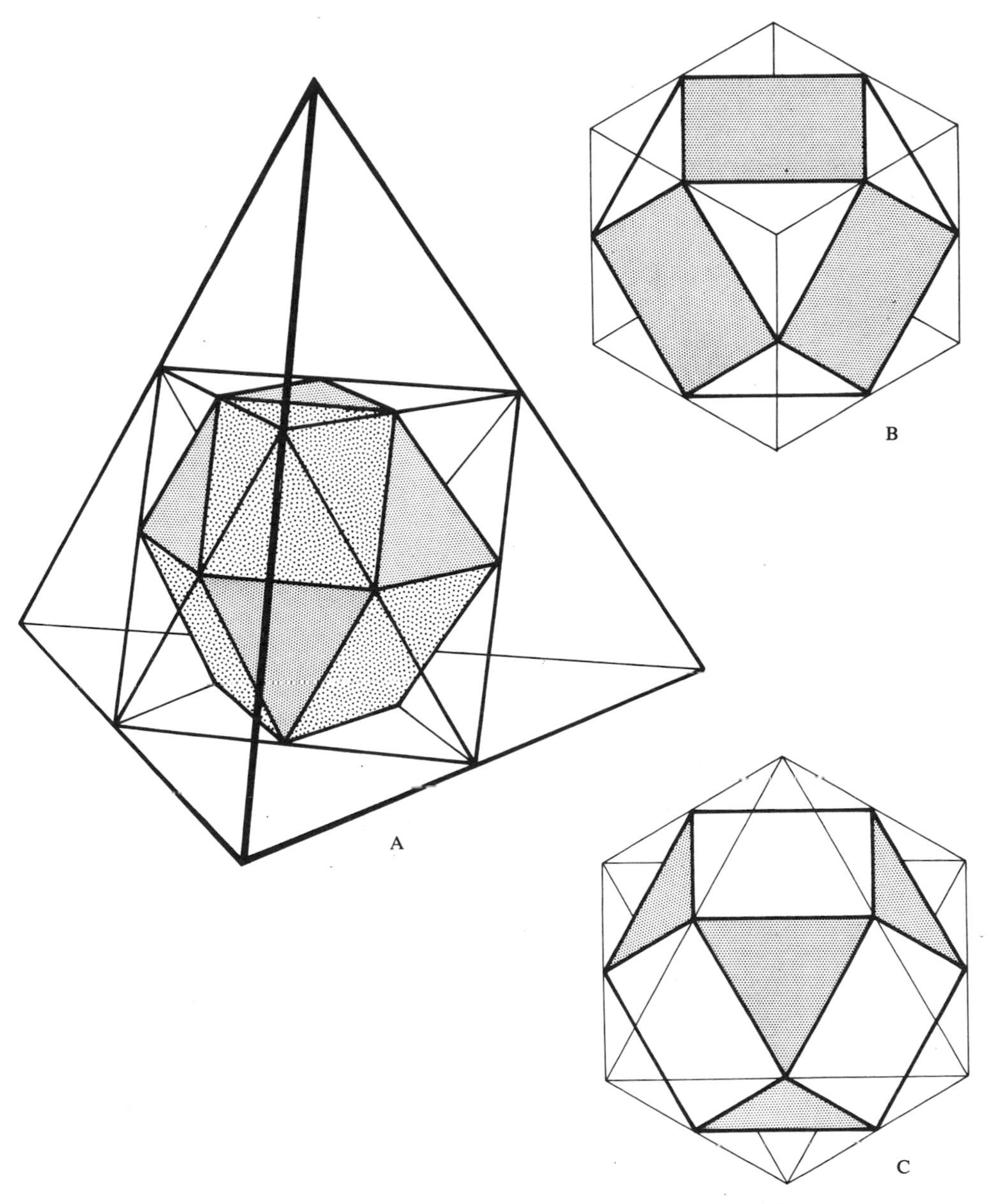

Fig. 470.02B *Relationship of Vector Equilibrium to Cube and Octahedron:*

A. Joining and interconnecting the midpoints of tetrahedron edges results in the octahedron. Joining and interconnecting the midpoints of the octahedron edges results in the vector equilibrium.
B. Relationship of vector equilibrium to cube.
C. Relationship of vector equilibrium to octahedron.

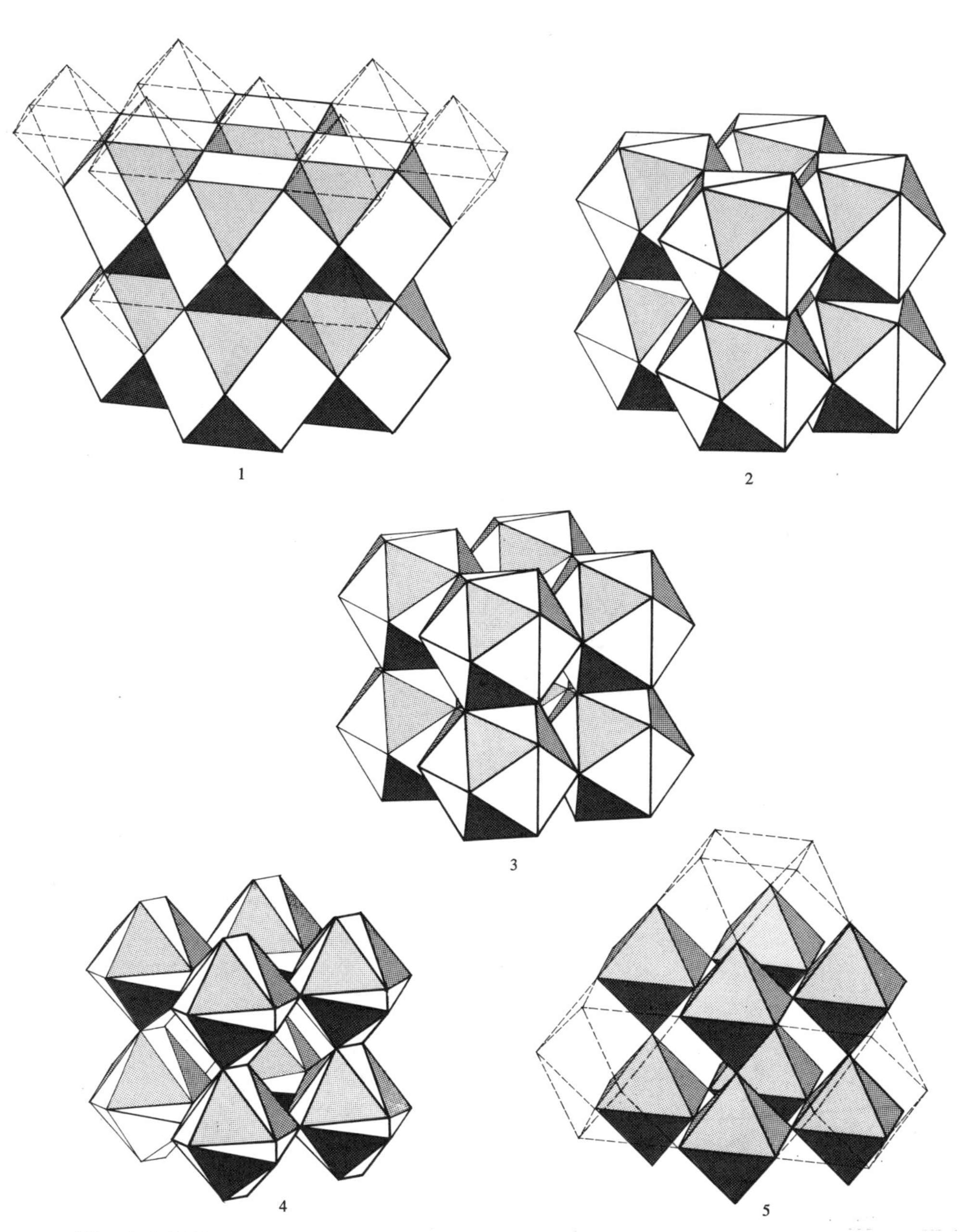

Fig. 470.02C *Transformation of Vector Equilibrium and Octahedron as Space-Filling Jitterbug:* Because the vector equilibrium and the octahedron will fill space, it is possible to envision a space-filling "jitterbug" transformation. If we combine vector equilibria on their square faces in a space-filling arrangement, the triangular faces form octahedral voids (1). As the vector equilibria contract, just as in the single "jitterbug," they transform through the icosahedron phase (3) and end at the octahedron phase (5).

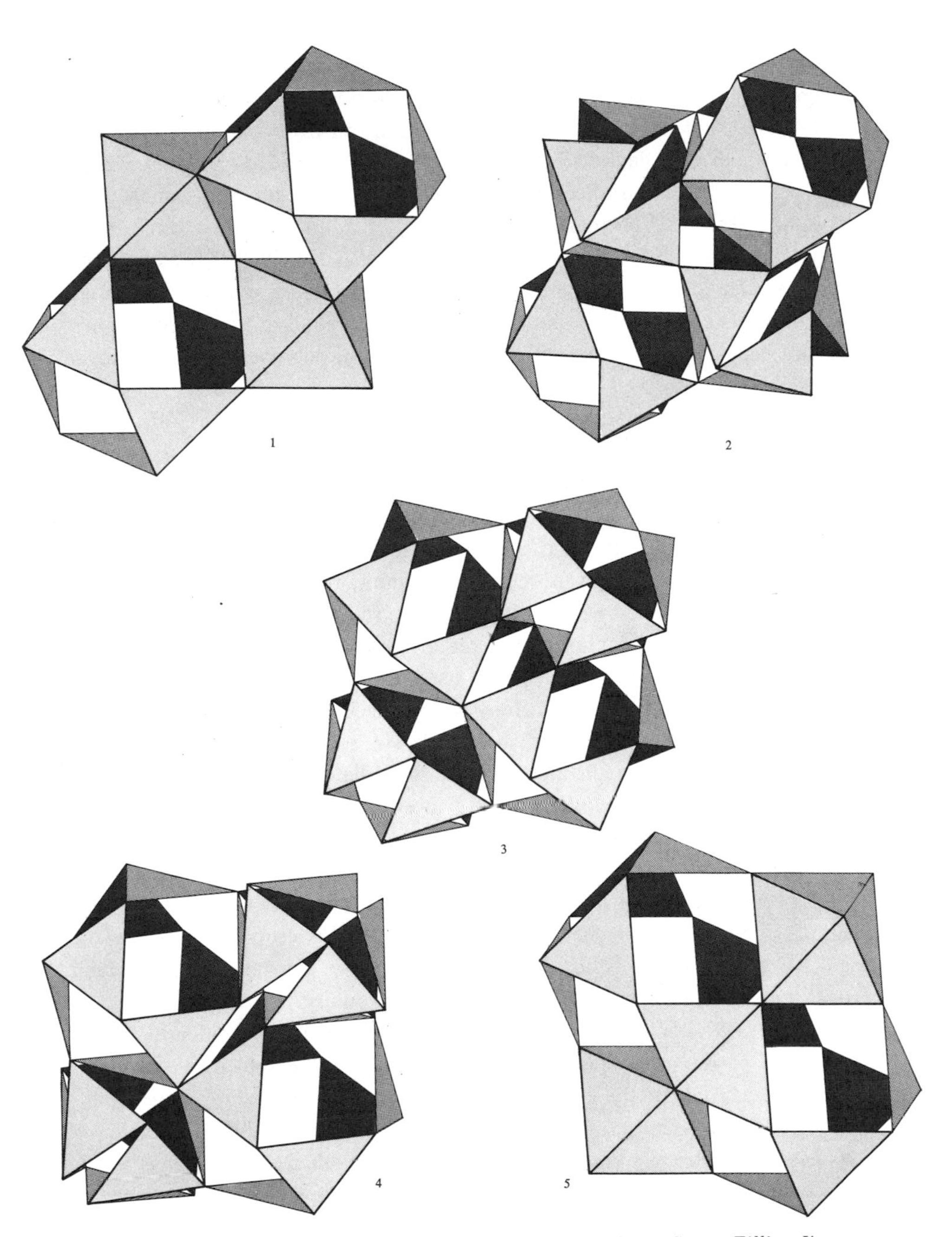

Fig. 470.02D *Reciprocity of Vector Equilibrium and Octahedra in Space-Filling Jitterbug:* In the space-filling ''jitterbug'' transformation, the vector equilibria contract to become octahedra, and, because in space filling array there are equal numbers of octahedra and vector equilibria, the original octahedra expand and ultimately become vector equilibria. There is a complete change of the two figures.

circle arcs leading most economically between each ball's three interior tangent contact points with each of the other three balls, respectively.

484.00 And now we have the octahedron self-examinatingly discovered and more sharply defined by the 12 chords of the great-circle arcs being realized as shorter distances than arcs, as lines of sight, between their six common vertexes. Thereafter, awareness of the fourfold equatorial square symmetry of four of those octahedral, equidistanted six vertexes of the octahedron, and discovery of the three *XYZ* axes crossing one another at the octahedron's center within the shortest distance centrally apart from the three sets of opposite vertexes or poles, thus establishing the one-quarter full circle as well as the one-sixth full circle angular self-fractionating as the octa-and-tetra interpump through the phases of the icosa into full extension of the vector equilibrium: "equanimity" of all potential systems and the extreme local domain of its local self-realization.

485.00 Thereafter, self-recognition of its six half-octahedra aspects of its own six polar potentials, and thence the self-discovery of its integral four great-circle symmetry, and its vector equanimity of effectively opposed disintegrative propensities by its mass-attractive, full-circle closingness at high-leverage advantage of radius of lever arm self-wrapping around itself, as being more powerfully effective than its self-disintegratively employed equipotential disintegrativeness; whereby the ever self-multiplications at the second-power arithmetical rate of its associative propensities are realized by their initiation, in contradistinction to the immediate second-root rate of diminution of the energy potential whenever it even starts to disintegrate.

486.00 Thus the self-discovery of the tetrahedral structural system and subsequent evolutionary realization of its inherent octahedral symmetry goes on further to discover its tetrahedron-octahedron complementarity of allspace-filling, and its development thereby of the universal isotropic vector matrix as a self-referring frame comprehensive of its relative aberrations of realizable exactitude which only approaches its ideal equanimity.

487.00 Whereafter the self-discovery process goes on to identify all the hierarchy of geometrical intertransformings that are the subject of this book, and proceeds inherently, by synergetic strategy of commencing with totality of Universe self-realization, to its progressive omnirational differentiation of its ever symmetrically equated potentials. And all other geometrical proofs of the Greeks and their academic successors aboard our self-realizing planet are herewith usably embraced; and all the rules of geometrical self-development proofs in terms of a priori self-realization proofs are discovered to be germane but always holistically embraced in omnirational identity. Self is not a priori evident. Thus we have avoided mathematical axioms that hold certain recognized a priori self-recognized conditions to be self-evidently irreducible by further analysis.

488.00 Instead of starting with parts—points, straight lines, and planes—and then attempting to develop these inadequately definable parts into omnidirectional experience identities, we start with the whole system in which the initial "point" turned out to be self, which inherently embraced all of its parameters wrapped tightly in that initial underdeveloped, self-focused aspect of self and went on to self-develop through successively discovered relative awarenesses whereby the proof of totality and omni-integrity is not only always inherent, but all the rules of operational procedure are always totally observed.

500.00 Conceptuality

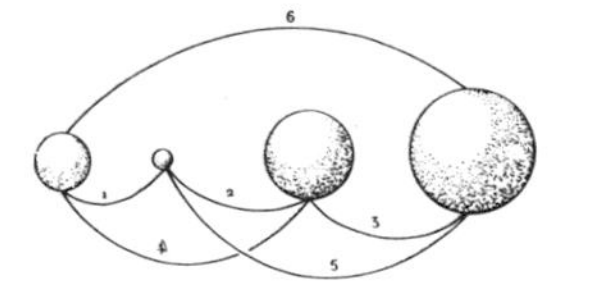

500.00 CONCEPTUALITY

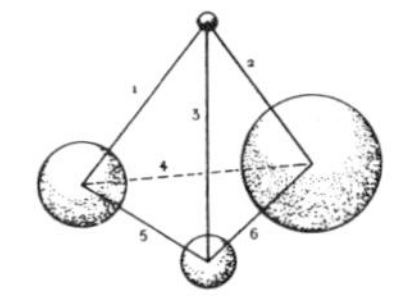

501.00 Definition: Conceptuality

501.01 The greatest of all the faculties is the ability of the imagination to formulate conceptually. Conceptuality is subjective; realization is objective. Conceptuality is metaphysical and weightless; reality is physical.

501.02 Definition requires conceptuality. All local systems are conceptual. Conceptual totality is inherently prohibited. There is systematic conceptuality within the totality, but it is always cosmically partial.

501.03 The artist was right all the time. Nature is conceptual. This is the difference between visibility and invisibility. The invisible does not mean nonconceptual. Conceptuality is independent of visibility or invisibility. You can have conceptuality, or an understanding of the principles, independent of size, which makes it possible to conceive of events as they occur at magnitudes that would be subvisible or supravisible. Conceptuality operates experimentally, independent of size. Size alone can come to zero, not conceptuality.

501.04 Conceptuality requires the generalization of patterns gleaned from special-case experiences and thus defines the basic event experiences that constitute structure.

501.05 We may think conceptually of assemblies of triangles or basic, generalized, structural arrangements that will hold true at either an atomic-nucleus size or a super-galaxy size, because all angularly defined systems are conceptually independent of the relative sizes of special-case experiences.

501.06 There is no half-profile of you. All conceptuality is systemic; it has to be finitely closed. Conceptuality has to have both frequency and angle. The angle part has to do with the circuitry design.

501.07 *Momentarily conceptual* means standing dynamically together like star groups. (See Sec. 324, Scenario Universe, and Sec. 510, Star Events.) *Aggregate* means nonunitarily conceptual as of any one moment.

501.08 We may hypothesize that information as it increases exponentially—explodes. Conceptuality implodes, becoming increasingly more simplified.

501.10 Omnidirectional Halo

501.101 Any conceptual thought is a system and is structured tetrahedrally. This is because all conceptuality is polyhedral. The sums of all the angles around all the vertexes—even crocodile, or a 10,000-frequency geodesic (which is what the Earth really is)—will always be 720 degrees less than the number of vertexes times 360 degrees.

501.11 The difference between nonconceptual, nonsimultaneous Universe and *thinkability* is always two tetrahedra: one as macro, to complete the convex localness outside the system, and one as micro, to complete the concave localness inside the system, to add up to finite but nonconceptual Universe. Thus the thinkable system takeout from Universe has a ''left-out'' outside irrelevancy tetrahedron and a ''left-in'' inside irrelevancy tetrahedron.

501.12 You have to have the starkly nonvisible to provide the complementary tetrahedron to account for the visibility, since concave and convex are not the same. That stark invisible reality of the nonconceptual macro- and micro-tetrahedra also have to have this 720-degree elegance. But the invisible outside tetrahedron was equally stark. The finite but nonconceptual inness and outness: that is the Omnidirectional Halo.

501.13 Complementarity requires that where there is conceptuality, there must be nonconceptuality. The explicable requires the inexplicable. Experience requires the nonexperienceable. The obvious requires the mystical. This is a powerful group of paired concepts generated by the complementarity of conceptuality. Ergo, we can have annihilation and yet have no energy lost; it is only locally lost.

501.14 The invisibility of negative Universe may seem a discrepancy, but only because the conceptual is such a fantastically limited part of the total, not just in the electromagnetic spectrum range, but in metaphysical, cosmic thinkability itself.

502.00 Experience

502.01 Experience is the raw material of science.

502.02 It is the nature of all our experiences that they begin and end. They are packaged. Our experiences, both physical and metaphysical, are all finite because they all begin and end. Experience is always special-case. Special cases are all biterminal, i.e., having both beginning and ending.

502.10 Many years ago, I developed a system of question-asking in which I ruled that I must always answer the questions from experience. My answers must not be based on hearsay, beliefs, axioms, or seeming self-evidence.

502.11 It has been part of my experience that there are others who, while experiencing what I was experiencing, were able to describe what we mutually were experiencing as well as, or better than, I could. Therefore, my experience taught me that I could trust the reporting of some others as reliable data to be included in my "answering" resources. For instance, I could include the experimentally derived data of scientists.

502.12 I am willing to accredit the experiences of other men when I am convinced by my experiences that they communicate to me faithfully; that is, I am able to enlarge my experience by the experience of others.

502.13 Certainly, my experience and your experience includes the fact that we dream. This doesn't have to be realized reality. There may be people who lie to you, but manipulation of the data doesn't alter your determination to rely upon experience, for it now becomes a part of your experience that some people lie, and you learn which ones are reliable as suppliers of your experience inventorying. Our experience includes the becoming. It includes the multiplication of experiences. It includes dichotomy.

502.20 All experiences are finitely furnished with differentiated cognitions, recognitions, and comprehensions. The finite furniture consists of widely ranging degrees of comprehensive constellar complexities.

502.21 Experience is inherently discontinuous and islanded. Each special experience represents a complex of generalized principles operative in local-angle and frequency-modulated realizations.

502.22 Among the irreversible succession of self-regenerative human events are experiences, intuitions, speculations, experiments, discoveries, and productions. Because experience always alters previous experience, the process is both irreversible and nonidentically repetitive.

502.23 Since experience is finite, it can be stored, studied, directed, and turned with conscious effort to human advantage. This means that evolution pivots on the conscious, selective use of cumulative human experience and not on Darwin's hypothesis of chance adaptation to survival nor on his assumption of evolution independent of individual will and design.

502.24 Consciousness is experience. Experience is complex consciousness of being, of self coexisting with all the nonself. Experience is plural and nonsimultaneous. Experience is recurrent consciousness of sequences of self reexperiencing similar events. Reexperienced consciousness is recognition.

Re-cognitions generate identifications. Re-cognition of within-self rhythms of heartbeat or other identities generates a matrix continuum of time consciousness upon which, as on blank music lines, are superimposed all the observances by self of the nonself occurrences.

502.25 Experience is inherently omnidirectional; ergo, there is not just one "other." There are always at least twelve "others." The connection between the six degrees of freedom and omnidirectionality is, of course, the vector equilibrium. Pulsation in the vector equilibrium is the nearest thing we will ever know to eternity and god: the zerophase of conceptual integrity inherent in the positive and negative asymmetries that propagate the problems of consciousness evolution. Our inherently limited perceptivity requires these definitions of the asymmetric emphasis of experience. Experience is inherently terminal, partial, and differentiable: the antithesis of eternal integrity.

502.30 Experimentally Demonstrable vs. Axiomatic

502.31 The difference between synergetics and conventional mathematics is that it is derived from experience and is always considerate of experience, whereas conventional mathematics is based upon "axioms" that were imaginatively conceived and that were inconsiderate of information progressively harvested through microscopes, telescopes, and electronic probings into the nonsensorially tunable ranges of the electromagnetic spectrum. Whereas *solids, straight lines, continuous surfaces,* and *infinity* seemed imaginatively obvious, i.e., axiomatic; physics has discovered none of the foregoing to be experimentally demonstrable. The imaginary "abstraction" was so logical, valid, and obviously nonsolid, nonsubstantial in the preinstrumentally-informed history of the musings of man that the mathematician assumed abstraction to be systemic conceptuality, i.e., metaphysical absolutely devoid of experience: He began with oversight.

502.40 The "Purely Imaginary Straight Line"

502.41 In speaking of his "purely imaginary straight line," the mathematician uses four words, all of which were invented by man to accommodate his need to communicate his experience to self or others:

(a) *Purely:* This word comes from the relativity of man's experiences in relation to impurities or "undesirable presences."

(b) *Imaginary:* "Image-inary" means man's communication of what he thinks it is that he thinks his brain is doing with the objects of his experience. His discovery of general conceptual principles characterizing all of his several experiences—as the rock having insideness and outsideness, the many pebbles having their corners knocked off and developing roundness—means that there could be pure "roundness" and thus he *imagined* a perfect sphere.

(c) *Straight:* Man's experiences with curvilinear paths suggested that the waviness could be reduced to straightness, but there was naught in his experience to validate that nonexperienced assumption. Physics finds only waves. Some are of exquisitely high frequency, but inherently discontinuous because consisting of separate event packages. They are oscillating to and from negative Universe, that is to say, in pulsation.

(d) *Line:* Line is a *leading,* the description of man's continual discovery of the angularly observable directional sequences of events. Lines are trajectories or traceries of event happenings in respect to the environmental events of the event happening.

502.50 Experiment

502.51 A voluntary experience is an *experiment.* To be experimental, we must have an observer and the observed, the articulator and the articulated. Experiences include experiments: there are experimentally demonstrable cyclic regularities, such as frequencies of the occurrence of radiation emissions of various atomic isotopes, which become the fundamental time increment references of relative size measurement of elemental phenomena.

502.60 Happening

502.61 An involuntary experience is a *happening.* To be experiential, to have a happening, we must have an observer and the observed.

503.00 Happenings

503.01 A happening is an involuntary experience. You cannot program "happen."

503.02 Happenings contradict probability. That's why they are happenings. Probability is not a reliable anticipatory tool; it is stronger than "possibility" but crude in comparison to "navigation" and "astronomy." If probability were *reliable,* there would not be a stock market or a horse race.

503.03 The vector equilibrium is the minimum operational model of happenings.

503.04 Evolution is the scenario of happenings permitted by nature's precise external laws governing angular degrees and frequencies of event freedoms.

504.00 Special Case

504.01 Experience is always special case.

504.02 The human brain apprehends and stores each sense-reported bit of information regarding each special-case experience. Only special-case experiences are recallable from the memory bank.

504.03 There is in Universe a vast order. It never forsakes. I throw a coin in the air, and it returns and hits the floor *every time*. Nature is never at a loss about what to do about anything. Nature never vacillates in her decisions. The rolling oceans cover three-quarters of Earth. Along the beaches, the surf is continually pounding on the shore. No two successive local surf poundings have ever been the same nor will they ever be the same. They typify the infinitude of individualism of every special-case event in the Universe.

504.04 Weightless, abstract human mind reviews and from time to time discovers mathematically reliable and abstractly statable interrelationships existing between and amongst, but not "in" or "of," any of the special-case experience components of the relationship. When a long-term record of testing proves the relationship to persist without exception, it is rated as a scientifically generalized principle. Whenever human mind discovers a generalized principle to exist amongst the special-case experience sets, the *discovery event* itself becomes a *new* special-case experience to be stored in the brain bank and recalled when appropriate. Amongst a plurality of brain-stored, *newly understood experiences,* mind has, from time to time, discovered greater and more significant understandings, which in their turn as discoveries, which are "experiences," constitute further *very* special-case experiences to be stored in the recallable and reconsiderable brain bank's wealth of special-case experiences.

505.00 Pattern

505.01 When we speak of pattern integrities, we refer to generalized patterns of conceptuality gleaned sensorially from a plurality of special-case pattern experiences that have been proven experimentally to be existent always, without exception, in every special case within the required class of experiences.

505.02 Special-case events may appear to be both continuous and "linear"—but only as locally and momentarily experienced. For all experimental observations of at first seemingly "continuous" and "straight-line" experiences (subjective) or of experimental experiences (objective), when projected or prolonged, are always discovered to have been short increments of larger multidirectionally peregrinating, curvilinear, wave actions of discontinuous events (stars) in Milky Way–like, stepping-stone, "linear" arrays.

505.03 All experiences are omnidirectionally oriented. Omnidirectional experiences resolve themselves scientifically into discrete angle and frequency patterns. That is life! Relationships are local to pattern. Patterns are comprehensive to relationships.

505.04 In a comprehensive view of nature, the physical world is seen as a patterning of patternings whose constituent functions are fields of force, each of which compenetrates and influences other localized fields of force.

505.05 Action and interaction of events are accompanied by relative omnidirectional displacements and accommodations of other events. In considering a total inventory of the relative abundance of different patterns, it becomes apparent that patterns are reciprocal.

505.06 The artist frequently conceives of a unique pattern in his imagination before the scientist finds it objectively in nature.

505.10 Euler: Minimum Aspects of Pattern

505.101 Euler said that we are dealing in pattern. Mathematics is pattern, and there are irreducible aspects of pattern. That is, the patterns represent events. A line is a unique kind of pattern. If I have two lines, where the two lines cross is distinctly different from where the two lines do not cross. Euler called this the vertex, the convergence. He saw this as absolute pattern uniqueness. (See Sec. 523.)

505.11 Euler showed that all optical experiences that we can pattern or form are composed exclusively of three patterning elements: lines, vertexes, and areas—or *trajectories, crossings,* and *openings,* as they are known in synergetics. These incontrovertible minimum aspects of pattern are all that is necessary to analyze and inventory all parts of all optically apprehended patterns as well as of all whole patterns. And Euler disclosed three algebraic formulae characterizing the constant relative-abundance relationships of these three fundamental topological elements in all patterns.

505.12 All happening patterns consist of experience recalls. The recallable ingredients of experiences consist inherently of paired-event quanta of six-vectored positive and negative actions, reactions, and resultants.

505.20 Pattern Integrity

505.201 A pattern has an integrity independent of the medium by virtue of which you have received the information that it exists. Each of the chemical elements is a pattern integrity. Each individual is a pattern integrity. The pattern integrity of the human individual is evolutionary and not static.

505.21 Each of the chemical elements is a uniquely complex pattern of energy event interrelatednesses which interact inter-interferingly to continually relocalize the involved quantity of energy. These self-interference patterns of atomic element components are in many ways similar to the family of knots that are tied with rope by sailors to produce various local behaviors, all of which, however, result in further contraction of the knot as the two ends of the rope immediately outside the knot are pulled away from one another by forces external to the knot—and thus all the attractive forces of Universe operating upon the atoms may result precessionally in keeping the atomic knots pulled together. (See Sec. 506.14.)

505.30 Waves

505.301 When we drop a stone into water, we see a wave emanate outwardly in a plane. We agree that it is not water but that we are seeing a wave in pure principle. It is not simultaneous: therefore to conceptualize we are using our memory and afterimage. We can never have static waves; they have nothing to do with statics. We see a wave operative in time and in pure principle. If we initiate wave-propagating energy action at one point, a complete omnidirectional wave develops.

505.31 When a stone is dropped into a tank of water, the stone does not penetrate the water molecules. The molecules are jostled; they "accommodate" the stone and in the process jostle their neighboring molecules, which, in turn, jostle their own outwardly surrounding water molecule neighbors. Thus waves of relayed jostling are propagated. Each relayed wave, although a composite of locally forwarded actions, provides a synergetic continuity scenario of those actions. The consequence is a pattern of events that has an integrity of its own, independent of the local displacement accommodations (which are innocent with respect to the overall synergetic pattern).

505.32 The same stone dropped successively in pools of water, milk, or gasoline will generate the same wave patterns. Yet the waves are essences neither of milk nor of water nor of gasoline. The waves are distinct and measurable pattern integrities in their own right, visibly growing and traveling outwardly as each locally involved molecule of the liquids develops a narrow vertical ellipse circuitry returning to where it started, unless a powerful wind operating parallel to and above the liquid blows the top molecules free as bubbles to tumble down the wave side like water on a hillside. (See Sec. 1005.14.)

505.33 Individuals regenerate their own sound and air displacement waves and ripples in the physical environment just as stones create waves and ripples in the different liquids into which they are thrown. They also propagate metaphysical wave patterns that develop local pattern displacements in the human affairs cosmos. They also propagate both conscious and unconscious electromagnetic waves. The wave is as abstract as the concept of an angle. Waves are weightless patterns.

505.34 The room we sit in is permeated by thousands of weightless waves, each of unique character. You can tune in hundreds of wide-frequency-range radios within your room, and each can bring in a different program from a different part of the world because the individual, weightless waves flow through trees and house walls. That extraordinary world of weightless, invisible waves is governed by mathematical laws, not by the opinions of men. The magnificent orderliness of that ever individually and uniquely patterning weightless wave Universe is not of man's contriving. The infinite variety of evolutionary complexities, inherent to the orderliness of complementary principles operative in Universe, is of unending synergetic uniqueness.

505.40 *Wave or Particle*

505.41 One of the things we have to make clear for society is the intellectual dilemma of the Max Planck–descended scientists: the way they do their problems, they can have either a wave or a particle but not both simultaneously. Heisenberg has the same dilemma. They make the error of thinking of a wave as a physical continuity rather than as a metaphysical, weightless pattern integrity, experimentally detectable only by virtue of the medium of the locally displaced, frequency tuned, physical phenomenon—a principle operating utterly independent of any physical medium. (See Secs. 973.30 and 1009.36.)

506.00 Knot

506.01 A knot in a spliced rope consisting successively of manila, cotton, wool, or nylon may be progressively slipped along the spliced-together

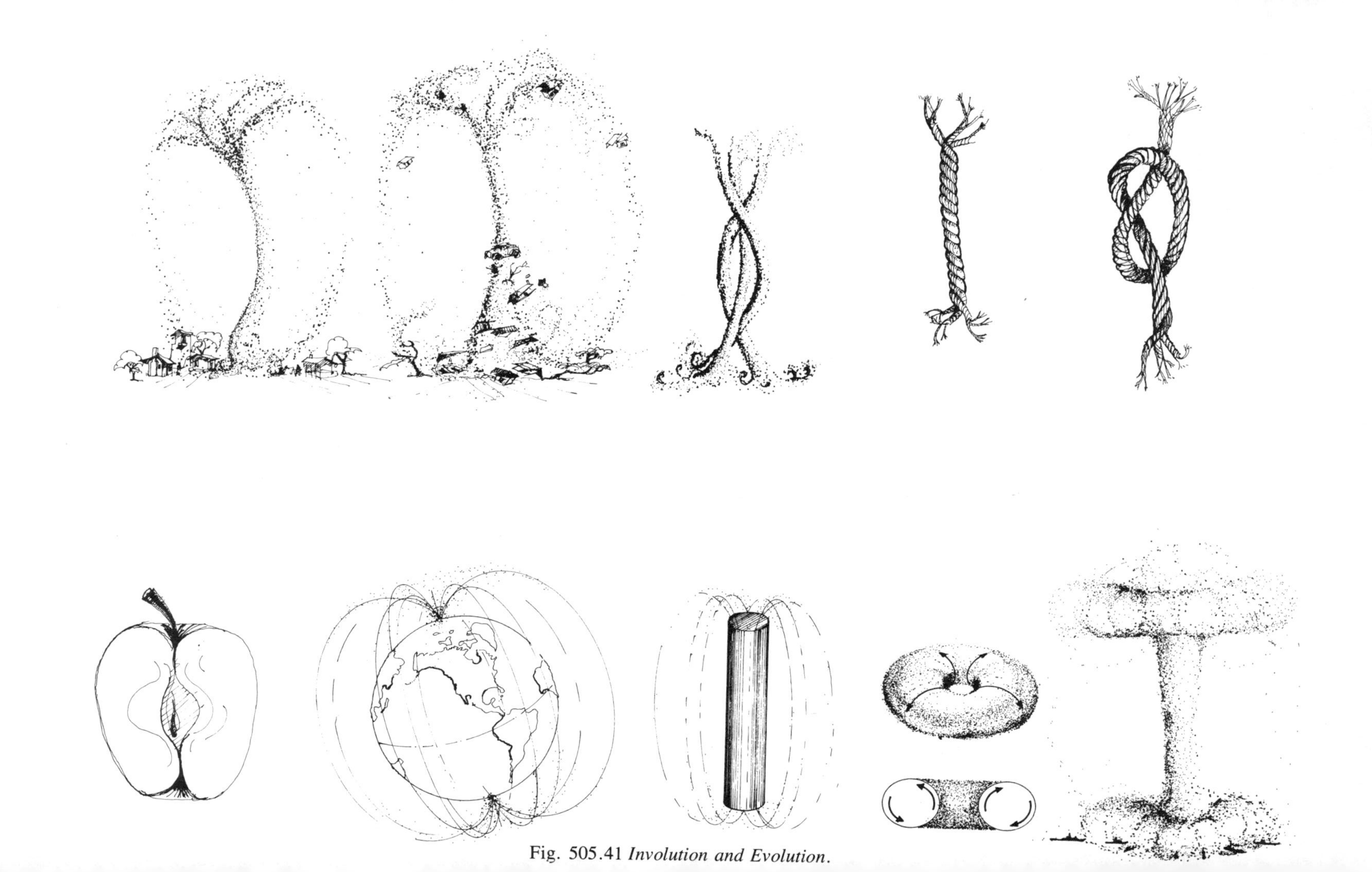

Fig. 505.41 *Involution and Evolution.*

rope with all the latter's material changes of thickness, color, and texture along its length. We agree that the "knot" is not really any of these locally traversed substances. They were just so many colors and tactile experiences whose pattern displacement reported something moving through as a locally recurring pattern configuration. The knot is not the rope; it is a weightless, mathematical, geometric, metaphysically conceptual, pattern integrity tied momentarily into the rope by the knot-conceiving, weightless mind of the human conceiver—knot-former.

506.02 What we call the rope itself turns out to be wave phenomena. The fibers themselves were humanly twisted into a spiral wave phenomenon. We are beginning to discover that there is not too much difference between the tactile superficiality of apprehension and the real frequency phenomena that we cannot see in the intervals between the waves. The actual fact is that the water wave and the manila wave are frequencies nontunable within the electromagnetic frequency range of the human organism's optical faculties, wherefore human cognition of the water waves is provided exclusively by the human brain's afterimage lag and the brain's successive recall apprehending of static picture frames of successively different pattern states as moving pictures.

506.10 I'll bet a monkey can't invent a knot. If they could, they would tie the whole jungle up in knots. What would the behaviorists say? Mind saw the knot; monkey did not. The monkeys hold hands, but they have not yet discovered that the handshake is two circles knotted through one another.

506.11 You cannot have a knot with less than two circles (two finite unities). The mind tells the brain to control the muscles in a knot-tying event scenario as follows: one hand grasps the rope end and describes the first circle. When the first circle is complete, the second hand holds the completed circle as the first hand continues to lead the rope end through the center of the first circle in an orbital plane different from that of the first circle. (If they were both in the same plane, they would generate a coil or a spiral and fail to knot.) The perimeter of the second circle should go through the center of the first circle. One has to capture the other in an interference pattern.

506.12 The rope with the knot in it is a physical memory pattern tracery of where your hands have led its end. The hand-led rope end and its pulled-through rope section form a visibly sustained trajectory of the conceptual patterning employed by mind in negotiating its visual realization by the brain-coordinated sensing of self or others. Like the contrails of jet planes, in the sky, the smoke trails of skywriting airplanes, or the extruded plastic threads of spiders, the roped knot represents a long-lasting memorandum of the abstract, weightless mind's weightless conceptioning in pure principle.

506.13 Each circle has 360 degrees; the two interference circles that comprise the minimum knot always involve 720 degrees of angular change in the hand-led pattern, just as the total angles of the four triangles of a tetrahedron add up to 720 degrees. The hands describe circles nonsimultaneously; the result is a progression. The knot is the same 720-degree angular value of a minimum structural system in Universe, as is the tetrahedron.

506.14 Pulling on the two ends of the knotted rope causes the knot to contract. This is a form of interference wave where the wave comes back on itself, and as a consequence of any tension in it, the knot gets tighter. This is one of the ways in which the energy-mass patterns begin to tighten up. It is self-tightening. This is the essence of ''matter'' as a consequence of two circles of 720 degrees tending to annihilate or lose one's self. Tetrahedron creates an insideness. Knot attempts to annihilate it. The knot is a tetrahedron or a complex of tetrahedra. Yin-Yang is a picture of a minimum tetrahedron knot interference tying. (See Sec. 505.21.)

506.15 At the end of the piece of rope, we make a metaphysical disconnect and a new set of observations is inaugurated, each consisting of finite-quanta integral ingredients such as the time quality of all finite-energy quanta.

506.20 The metabolic flow that passes through a man is not the man. He is an abstract pattern integrity that is sustained through all his physical changes and processing, a knot through which pass the swift strands of concurrent ecological cycles—recycling transformations of solar energy.

506.30 As curves—lines—cannot reenter or ''join back into themselves,'' the circling line can only wrap around or pass over or under another ''part'' of its continuity self, as the knot-making sailors says it. Because of a line's inability to reenter itself, when circles are followed around and around upon themselves, the result is a coil—which is a mildly asymmetric spiral wave accumulation that may be piled upon its micro-diameter self only as long as intellect wishes to pursue such an experiential investigation.

507.00 Parity

507.01 The rubber glove, with its red exterior and green interior, when stripped inside-outingly from off the left hand as red, now fits the right hand as green. First the left hand was conceptual and the right hand was nonconceptual—then the process of stripping off inside-outingly *created* the right hand. And then vice versa as the next strip-off occurs. Strip it off the right hand and there it is left again.

507.02 That is the way our Universe is. There are the visibles and the invisibles of the inside-outing nonsimultaneity. What we call thinkable is always outside out. What we call space is just exactly as real, but it is inside out. There is no such thing as right and left.

507.03 The always and only coexisting convex and concave demonstrates that unity is plural and at minimum two, in which only one is spontaneously accounted as obvious.

507.04 The positive (right) spiral and negative (left) spiral make one tetrahedron: $1 + 1 = 4$: therefore no parity. But there is parity in the internal complementary macrocosmic tetrahedron of the sum of the angles around all the convex vertexes of the system, and an internal complementary microcosmic tetrahedron of the sum of the angles around all the concave vertexes of the system.

507.05 When physics finds experimentally that a unique energy patterning—erroneously referred to in archaic terms as a particle—is annihilated, that annihilation is only of the inside-outing rubber-glove kind. The positive becomes the negative and the positive only *seems* to have been annihilated. We begin to realize conceptually the finite, yet nonsensorial, outness continuum integrity that can be converted into sensorial inness by the inside-outing process, but only at the expense of losing afterimage of the previous sense-experienced conceptual fixation.

507.06 The complementary of parity is disparity and not a reflective image.

508.00 Number

508.01 Numbers are experiences. You have one experience and another experience, which, when reviewed, are composited. Numbers have unique experiential *meaning*. The minimum structural systems of Universe, the tetrahedron and the thinkable set, both consist of four points and their six unique interrelatednesses. Even the development of sets derives from experience. Mathematics is generalization, a third-degree generalization that is a generalization of generalizations. But generalization itself is sequitur to experience where intuition and mind discover the synergetic interbehavior that is not implicit in any single item of the empirical data of the past.

508.02 Intuition and mind apprehend that which is comprehensively between, and not of, the parts.

508.03 The mathematician talks of "pure imaginary numbers" on the false assumption that mathematics could cerebrate a priori to experience. "Lines" are definitions of experiences—of graven traceries, or of erosively deposited tracks, or of gaseous fallout along a trajectory—and the symbols for number extractions, such as X and Y, are always and only experientially conceived devices.

508.04 All number awareness is discovered through experiences, which are all special cases. Every time you write a number—every time you say, write, or read a number—you see resolvable clusters of light differentiation. And clusters are an experience. Conscious thoughts of numbers, either subjective or objective, are always special-case.

508.10 Before topology, mathematicians erroneously thought that they had attained utter abstraction or utter nonconceptuality—ergo, "pure" nonsensoriality—by employing a series of algebraic symbols substituted for calculus symbols and substituted for again by "empty-set" symbols. They overlooked the fact that even their symbols themselves were conceptual patterns and only recognizable that way. For instance, numbers or phonetic letters consist of physical ingredients and physical-experience recalls, else they would not have become employable by the deluding, experience-immersed "purists."

508.20 $\frac{N^2-N}{2}$ is always a triangular number as, for instance, the number of balls in the rack on a pool table. A telephone connection is a circuit; a circuit is a circle; two people need one circuit and three people need three circles, which make a triangle. Four people need six circuits, and six circuits cluster most economically and symmetrically in a triangle. Five people need 10 private circuits, six people need 15, and seven people need 21, and so on: all are triangular numbers. (See Sec. 227, Order Underlying Randomness, and illustration 227.01.)

508.30 Successive stackings of the number of relationships of our experiences are a stacking of triangles. The number of balls in the longest row of any triangular cluster will always be the same number as the number of rows of balls in the triangle, each row always having one more than the preceding row. The number of balls in any triangle will always be $\frac{(R+1)^2-(R+1)}{2}$ where $R=$ the number of rows (or the number of balls in the longest row). (See Sec. 230, Tetrahedral Number.)

509.00 Considerable Set

509.01 The conceptual process is never static. Thinking does not consist of the insertion of invented images into an otherwise empty vacuum-tube

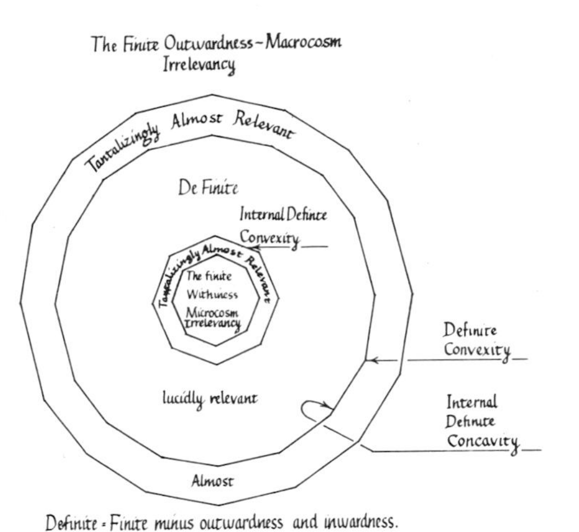
The Finite Outwardness-Macrocosm
Irrelevancy
Tantalizingly Almost Relevant
De Finite
Internal Definite
Convexity
Tantalizingly Almost Relevant
The finite
Withinness
Microcosm
Irrelevancy
Definite
Convexity
lucidly relevant
Internal
Definite
Concavity
Almost
Definite = Finite minus outwardness and inwardness.

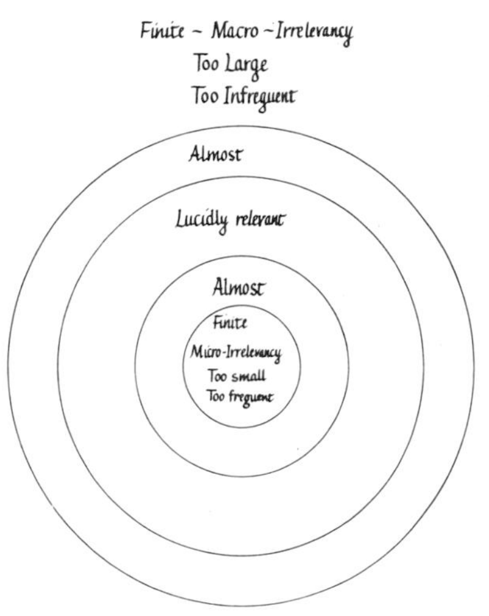
Finite ~ Macro ~ Irrelevancy
Too Large
Too Infrequent
Almost
Lucidly relevant
Almost
Finite
Micro-Irrelevancy
Too small
Too frequent
Thinking is frequency modulation-tuning out finite irrelevancies
into two main classes
Micro, Macro which leaves residual definated system as lucidly relevant

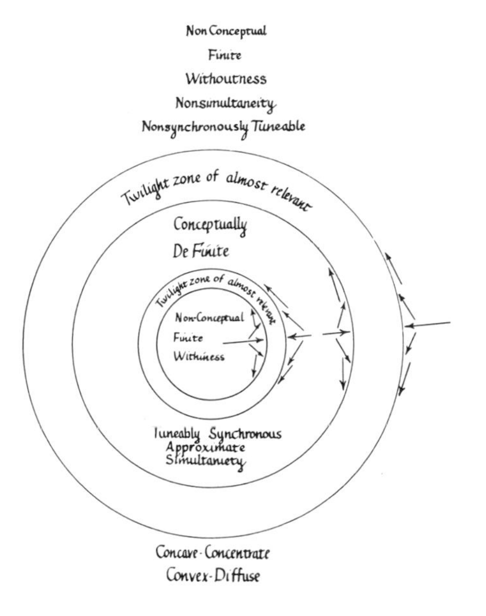
Non Conceptual
Finite
Withoutness
Nonsimultaneity
Nonsynchronously Tuneable
Twilight zone of almost relevant
Conceptually
De Finite
Twilight zone of almost relevant
Non-Conceptual
Finite
Withinness
Tuneably Synchronous
Approximate
Simultaneity
Concave-Concentrate
Convex-Diffuse

Fig. 509.01 A,B,C. *Patterns of Thought.*

chamber called brain. Thinking is the self-disciplined process of preoccupied consideration of special-case sets of feedback answers selected out of the multitude of high-frequency alternating transceiver brain traffic. This traffic consists of omniexperienced and processed answers to present or past questions, formulated either by the conscious or subconscious coordinating initiative of the individual or possibly the individual's overlapping generation of group memory.

509.02 A considerable set is a locally definitive system within Universe that returns upon its considerability in all circumferential directions and therefore has an inherent withoutness and withinness; the latter two differentiable functions inherently subdivide all Universe into the two unique extremes of macro- and micro-frequencies.

509.03 For instance, we find that all irrelevancies fall into two main categories, or *bits*. One set embraces all the events that are irrelevant because they are too large in magnitude and too delayed in rate of reoccurrence to have any effect on the set of relationships we are considering. The other set of irrelevancies embraces all the events that are too small and too frequent to be differentially resolved at the wavelength to which we are tuned, ergo, in any discernible way to alter the interrelationship values of the set of experience relationships we are considering. Having dismissed the two classes of irrelevancies, there remains the *lucidly relevant set* to be studied.

509.04 Because of the varying depths of storage of past experiences, some answers come back swiftly, some slowly. The recollectibility rates are unpredictable. Ergo, the returning-answers traffic is heterogeneous. Many answers come to questions we have forgotten that we asked ourselves. Conceptually systematic tuning of questions and feedback answers, comparatively considered in the brain, results in temporary, tunably valved exclusion of all other incoming signals. Discrete tuning admits consideration of only those recollections that are *clearly relevant* to the omnidirectional rounding out of systematic comprehension of the special-case set of events intuitively selected for momentary focal consideration. Thinking consists, then, of a self-disciplined deferment of conscious consideration of any incoming information traffic other than that which is lucidly relevant to the experience-intuited quest for comprehension of the significance of the vividly emergent pattern under immediate priority of consideration.

509.05 Neither the set of all experiences, nor the set of all words that describe them, nor the set of all the generalized conceptual principles harvested from the total of experiences is either instantly or simultaneously reviewable. "What was that man's name?" Our answering service may take five seconds, five hours, five days, or five generations to reply. Our conscious, orderly reconsideration of our variable-lag experiences discloses subconsciously coordinated regularities of feedback rates governing the recall phenomena.

509.06 What we do when we think is to dismiss momentarily all the *irrelevant* thoughts as we would part the grass to right and left in order to find a path. Thinking is high-frequency interception and very temporary diversion to a local holding pattern outside our consideration of all the irrelevant inbound feedback—just as inbound airplanes are "stacked up" in the sky near airports by the ground control when too many come in at about the same time and may interfere with each discretely safe landing operation. Landing is a slowing operation and an exact timing operation. Having isolated a finite set of experiences—spontaneously grouped for comprehensive consideration—by dismissing the irrelevancies, we may proceed to comprehend or "land" the isolated system by applying the theory of *bits,* which breaks up finite wholes into finite parts.

509.07 We may now say that what we do in thinking, after deliberately excluding the irrelevancies and thereby inadvertently isolating the considered set, is to further subdivide Universe into four parts:

(1) All of the parts of Universe that are externally irrelevant because too large and too infrequent;

(2) all the events of Universe that are internally irrelevant because too small and too frequent to be resolvable and discretely differentiated out for inclusion in our interrelationship considerations;

(3) all of the lucidly relevant remainder of Universe, which constitutes the considered and reconsidered set of experiences as viewed from outside the set; and

(4) the lucidly relevant set as viewed from inside the set.

Part 1 is the untuned, macrocosmic, long-wavelength, low-frequency, high-energy set. Part 2 is the untuned, microcosmic, short-wavelength, high-frequency, low-energy set. Parts 3 and 4 are the tuned, plus (+) and minus (−), interface sets.

509.10 The thinking process results in varying degrees of lucidity of the arrayed residue of focal-event patterns uniquely consequent to the disciplined deferment of irrelevancies. Thinking is a putting-aside, rather than a putting-in, discipline. Thinking is FM—frequency modulation—for it results in the tuning out of irrelevancies (static) as a result of definitive resolution of the exclusively tuned-in or accepted feedback messages' pattern differentiability. And as the exploring navigator picks his channel between the look-out-detected rocks, the intellect picks its way between irrelevancies of feedback messages. Static and irrelevancies are the same.

509.11 There are two inherent twilight zones of "tantalizingly almost-relevant recollections" spontaneously fed back in contiguous frequency bands: the macro-twilight and the micro-twilight. They inherently subdivide all Universe into the two unique extremes of macro- and micro-frequencies.

509.20 So I find that *you* and *I* and the *lamppost* and its *lamp* are basic subdivisions of Universe. You and I and *complex it* are either all of the Universe that is *inside,* all of the Universe that is *outside,* or all the remaining Universe, which comprises a given recognizable system or set. The residual constellation to be reconsidered constitutes a local conceptual system.

509.30 You cannot program the unknowns you are looking for because they are the relationship connections and not the things. The only thing you *can* program is the dismissal of irrelevancies.

509.31 When we say "we think," our feedback has variable lags that may take overnight or months of time, for all we know. Because we want to understand—that is, to know the interrelationships of clusters of experiences—our first great discovery is dismissing irrelevancies, the macro-micro characteristics. Add: forgotten questions; different rates of feedback; persons' names; random questionings; the challenging set you would like to understand; our friend intuition.

510.00 Star Events

510.01 A star is an exquisitely concentrated coordination of events that your optical tuning facilities are unable to resolve differentially into separately identifiable events. We may call a star a point. Playing Euler's game, "stars," or "points," are "crossings," or "fixes" as navigators would say it. As so considered, a "star-point-crossing" does not have an outsideness and an insideness. It is the point of superimposed crossing of trajectories or of their interferences. A point fix is a potential embryo consideration, a potential thought, a potential system.

510.02 Thinking is the consideration of different experiences and of inherently separate sets of events, and trying to find out what their relatedness is. Each one is a star. How many stars does it take to develop a geometry of outwardness and inwardness? What is the minimum number of stars needed to divide the Universe into outwardness and inwardness? I find it takes a minimum of four; you can't do it with three. Four very clearly has an outsideness and an insideness. This is what we call the tetrahedron, which has these four stars and six sets of interrelatedness. This comes in very interestingly in mathematics with the generalization that you don't have to worry too much about the shape, but the four stars are the minimum we can have for a thought. If I can at first discover only three stars in a thought challenge, there must be at least a fourth star lurking somewhere in the critical neighborhood. In fact, I discover that the total number of stars that could possibly be related is always subdivisible by four. The mathematics shows this up very clearly as complexes of tetrahedra. Tetrahedron becomes the minimum thinkable set, the minimum reconsiderable set, and it turns out to be the fundamental increment out of which all thoughts are constructed.

510.03 The minimum set that may form a system to divide Universe into macro- and micro-cosmos is a set of four items of consideration. Four nonsimultaneously bursting rockets in a unitarily considerable set of overlapping visibility durations.

510.04 The stars of the four rocket bursts constitute the four vertexes of a tetrahedron—the fundamental quantum of Universe's structuring. There is a tetrahedral structuring *interrelationship* between (a) *the day before yesterday,* (b) *yesterday,* (c) *today,* and (d) *tomorrow.* Though we speak of them as "the four balls in the air"—maintained there successively by a juggler using five balls to do his trick—they are not the same balls, and the four are never in the same positions; nonetheless, there are always and only six fundamental *interrelationships* between "the four balls in the air"—i.e., *ab, ac, ad, bc, bd,* and *cd,* although *a, b, c,* and *d* are nonsimultaneous events. Universe structures most frequently consist of the physical interrelationship of nonsimultaneous events.

510.05 A star is the focal point of an as yet undifferentiated concentration of events, ergo, considerable or constellar patterning: an exploratory grouping of stars or complex-idea entities that seem to man's limited tunability to stand out together. The word *consideration* comes from *con,* together, and *sidus,* the Latin "star." When we have found all the relationships between the number of items of our consideration, we have what we speak of as "understanding." When we understand, we have all the fundamental connections between the *star events* of our consideration. They then become a constellation. They *stand* clearly *together*.

510.06 When n stands for the number of stars or items of consideration, the number of connections necessary to understanding is always $\frac{n^2-n}{2}$.

510.07 Four is the minimum number of stars having an inherent arrangement of withoutness and withinness. The minimum conceptually considerable, generalized-experiences-set affording macro-micro separation of Universe is a set of four local-event foci. Between the four stars that form the vertexes of the tetrahedron, there are six edges that constitute all the possible relationships between those four stars. The four stars have an inherent sixness of interrelationship. The four-foci, six-relationship set is definable as the

Fig. 510.03 *Four Rocket Bursts.*

tetrahedron. This minimum fourness of relevant frequency—ergo, thinkable—"stars" coincides with quantum mechanics' requirement of four unique quanta numbers per each uniquely considerable "particle."

510.08 The regenerative patterns of structural events may be described as *constellar* because their component events interinterfere tensively in high-frequency, dynamic, self-regenerative patternings which only superficially seem to stand together as "static" structures. Star groupings "fly" in celestial formation, though seeming to hang motionless in the celestial theater. Any event patternings that become locally regenerative are constellar patterns. They are momentarily conceptual.

510.09 Until the present age, people thought that all of their faculties were simultaneously and instantly coordinate and operating at equal velocities. Einstein showed that neither *simultaneous* nor *instant* are valid, i.e., experimentally demonstrable. Observe that when we send up four rockets one-half second apart, their afterimages are approximately simultaneous. So we say that we see four rockets "at the same time." The illusion of simultaneity is one of the most important illusions for us to consider. Musicians may be able to comprehend nonsimultaneity better than do others. Einstein emphasized the importance of attempted spontaneous comprehension of the nonsimultaneity of all the events of Universe—a concept akin to our discovery that in our Universe, none of the lines can ever go simultaneously through the same points (See Sec. 517 et seq.). What Einstein is telling us is that there is no conceptual validity to the notion that everything in Universe is actually in simultaneous static array.

510.10 All words in the dictionary do not make one sentence; all the words cannot be simultaneously *considered,* yet each of the words is valid as a tool of communication; and some words combine in a structure of meaning. All the words are memoranda of all of humanity's attempts to communicate to self or to others their understanding of the unique evolvement of their separately viewed experiences. The dictionary is the inventory of unique aspects of the totally composited experiences known as Universe.

511.00 Energy Event

511.01 A single event is integrally complex. As angles are conceptual, independent of size, events are conceptual, independent of frequency of occurrence. An "original" or "prime" energy event is conceptual. An energy event is inherently complex. It is a nuclear component, but it is not the nucleus. Nuclei—complexedly composed of prime or original energy events —are themselves "prime" and "original," originality being inherently complex integrals. Energy transactions occur between nuclei as an extramural complex of events—as a "chemical compound."

511.02 All energy-event experimentation discloses omnioptimally economic, behavioral patterning of physical events. Every physical event in nonsimultaneous scenario Universe is characterized by three multidimensionally interlinked vectors that interact precessionally, i.e., at angles other than 180 degrees to one another, as in the multidimensional, helically zig-zagging pattern of lightning.

511.03 There are six positive and six negative degrees of fundamental transformation freedoms, which provide 12 alternate ways in which nature can behave most economically upon each and every energy-event occurrence. You have six vectors or none for every energy event.

511.04 One set of three-vector groups corresponds to the proton (with its electron and anti-neutrino), and the other set of three-vector groups corresponds to the neutron (with its positron and neutrino). Each of these three vector teams is identified by nuclear physics as

one-half Planck's constant; or
one-half spin; or
one-half quantum.

When we bring together these two sets of three vectors each, they integrate as six vectors and coincidentally also make one tetrahedron (of six vector edges). The tetrahedron is the veritably conceptualizable unit of one energy quantum.

511.10 The open-ended tripartite spiral can be considered as one energy event consisting of an action, reaction, and resultant. Two such tripartite-vectored "spirals," one negative and one positive, combine to form the tetrahedron. (See illustration 108.01.) The tripartite vector set looks like an "erected cobra" *Z*, that is, with two of its interlinked vector lines on the ground and one erected. One erected *Z* cobra erects its third vector member clockwise, and the other *Z* cobra erects its third vector in a counter-clockwise direction in respect to its base. (See Sec. 620.)

511.11 We find that the triangular *Z* cobra, is not operating in a plane because there is no such thing as a plane. Therefore, one of the legs sticks up a bit. We have a positive *Z* cobra and a negative *Z* cobra, and one cannot nest in the other. They will never be congruent with one another, but they can complement one another to become the tetrahedron. An event is a triangle. A triangle is an event. Two of them together make the tetrahedron.

511.12 Each of the three-vector, action, reaction, and resultant, minimum event Z cobras has two open ends and two internal angles. The two Z cobras have together four ends and four internal angles. We will call the open ends

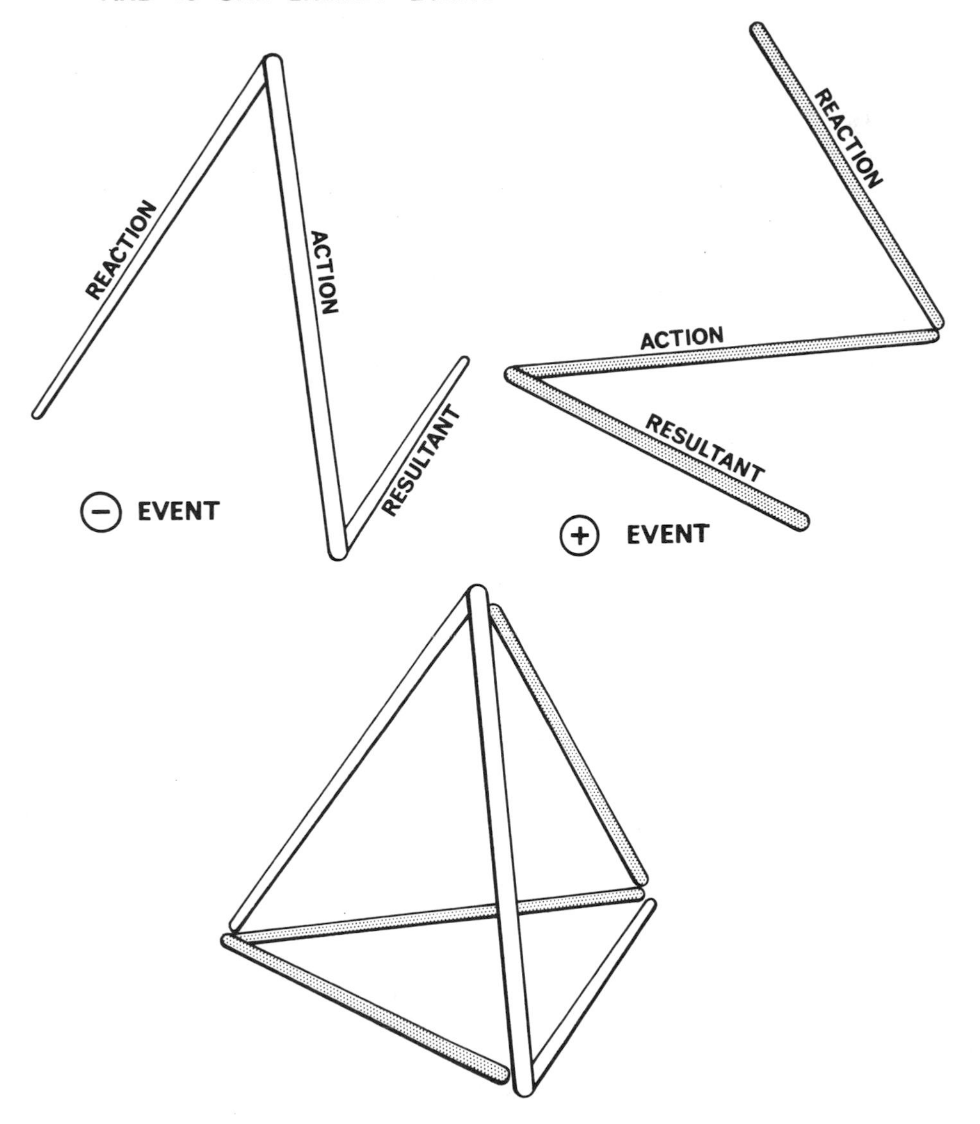

Fig. 511.10 *Two Triangular Energy Events Make Tetrahedron:* The open-ended triangular spiral can be considered one "energy event" consisting of an action, reaction and resultant. Two such events (one positive and one negative) combine to form the tetrahedron.

male and the internal angles *female*. We can marry the two Z cobra, half quantum events in an always consistent, orderly manner, by always having a male end interconnected with an internal female angle. When all four such marriage ceremonies have been consumated, we have produced one tetrahedron, i.e., one quantum, i.e., one prime minimum structural system of Universe. When the end of one energy action comes over the middle of another energy vector, there is a precessional effect, a tensional effect. One energy event gets angularly precessed, the next energy event goes by the center of another mass, and each one of them interaffects the other. It is a basketry interweaving, where each one precesses the other angularly so that they hold together very much as a cotton ball.

511.13 The energy event of an action, a reaction, and a resultant is inherently precessional.

511.20 An energy event is illustrated by a diagram of a man jumping from one boat to another. At the top of the picture, a man standing in one boat jumps. He does not glide horizontally: he jumps.* That is, he goes outwardly from the center of Earth, and that is a vector. That is an energy action in itself. He jumps. He is the action. The action was not just horizontal, it was also vertical. It was mildly vertical in that he went outwardly. As he jumps, the boat goes into reaction and shoots off the other way. A moment later, he lands, and the second boat moves in a complex that is both horizontal and vertical. There is a reaction and a result, so there really is a fourfoldedness going on. It may appear as threefolded because the man does not jump very high. We should consider it as a tetrahedron of very low altitude.

511.21 At the outset, the boats are more or less parallel to one another. As the man jumps from the stern of the boat, it turns and whirls around, so that the reaction is following the resultant. They are not going in opposite directions. The reaction and resultant run into each other. Notice that it begins to look like a triangle, but with a vertical component, so it ends up as our friend, tetrahedron.

511.22 Engineers have been proud of pointing out that the difference between engineers and lay society is that engineers know that every action has its reaction and that lay society thinks only of the actions. Before the speed of light was measured, light seemed, to all humanity, to be instantaneous. Since we now know experientially that neither light nor any other phenomenon is instantaneous, we may conclude that an action and the vectors that it creates are neither simultaneously occurring nor instantaneous. Because vectors have discrete length, whose dimension represents the energy mass multiplied by its velocity, every action vector has two terminals—a ''beginning'' and an ''end-

* In this way, we begin to discover that force diagrams in engineering result from oversimplification.

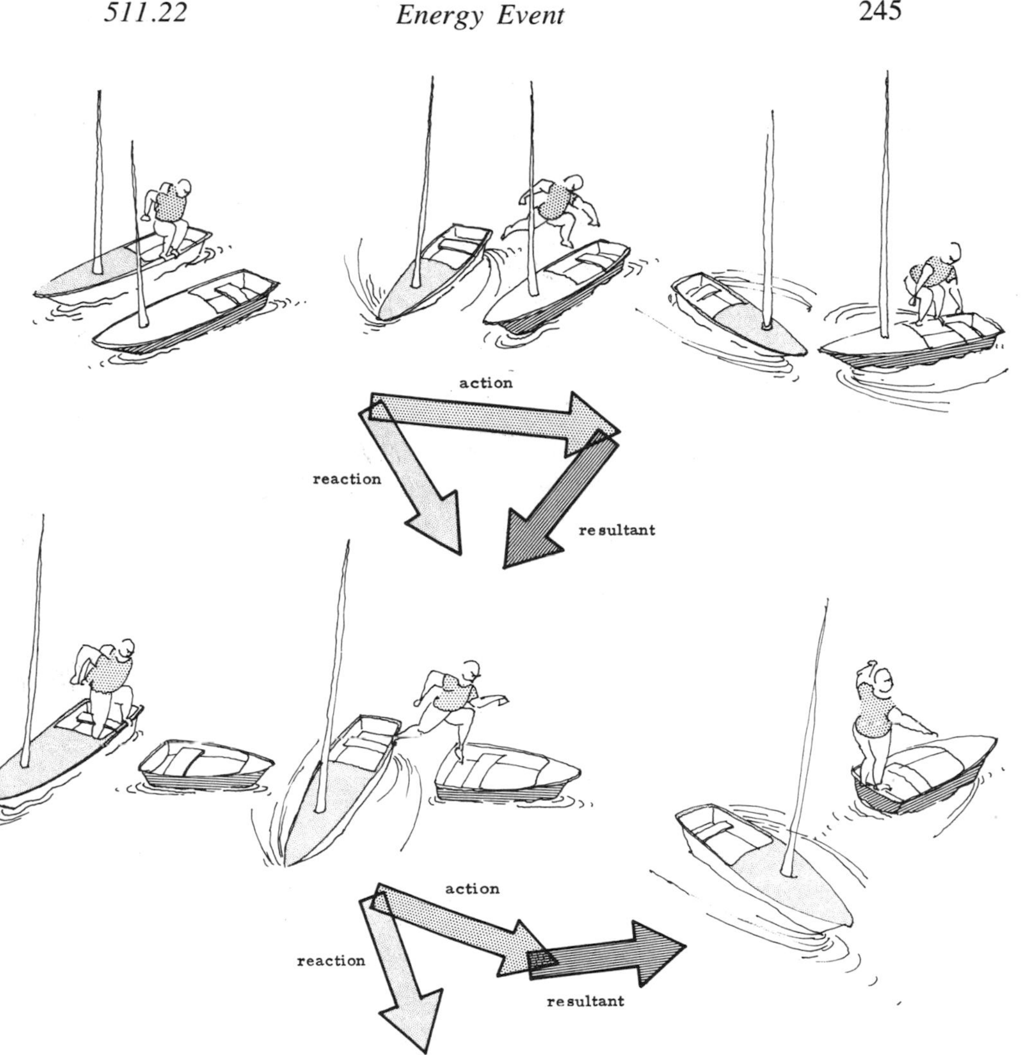

Fig. 511.20 *One Energy Event: Action, Reaction, and Resultant:* One energy event as demonstrated by the man jumping from the boat. His action always demonstrates the action, reaction, and resultant of the open-ended triangular spiral.

ing'' at the end of its noninstantaneous action. The beginnings and the endings are nonsimultaneously occurrent. Therefore, the ''ending'' terminal of an action's vector occurs later than its ''beginning.'' Therefore, every action must have a reaction vector at its ''beginning'' terminal and a resultant vector at its ''ending'' terminal. The reaction vectors and the resultant vectors are never angled at 180 degrees to the action vectors. They are always angled precessionally at other than 180 degrees.

512.00 Locality

512.01 Mechanically and chemically, a steerable rocket embraces a complex of internal and external events. Both airplanes and steerable rockets are complexes of internal and external energy-event transactions and omni-interacting, resultant "motions" in Universe transcendental to Earth motions, where the observer-articulator is extraterrestrially positioned. Since the Earth is moving as a dependent motion-complex in respect to the Sun's and other planets' motions, and since the Sun is engaged in a plurality of internal and external motions in respect to the galactic system, and since the galactic system is a complex of motions in respect to other galaxies and supergalaxies, and so on, and since the whole set of motion events are nonsimultaneous and of uniquely variant durations, and since the intereffects of the events vary vastly in respect to eons of time, it is obvious that any thinkably meaningful conceptual coordination of event interrelationships in the meager lifetime limits of humans is inherently limited to a relatively local set within Universe and within a time sense, and the relationships may be measured only in respect to the angle and frequency magnitude characteristics of any one subsystem of the totality.

513.00 Vectorial Orientation and Observation

513.01 The angles of orientation and the dimensional fixes of vectorial energy-event manifestations are always conceptually oriented and positioned in respect to the optionally selected axis of conceptual observation.

513.02 Fixes consist of both angular and dimensional observations.

513.03 To be *experiential,* we must have an observer and the observed.

513.04 To be *experimental,* we must have the *articulator,* the *articulated,* and the *observer*.

513.05 The vectorial angulation of both the experientially observed and the experimentally articulated is always referential to the axis of conceptual observation of the observer or the articulator, respectively. These always and only coexisting functions of experience and experiments embrace the fundamental parameters of operational science.

513.06 "My life" is the progressive harvestings of the information unpredictably accruing in the attempt to be both adequate and accurate. The harvest is stored in the brain bank. Life consists of alternate observing and articulating interspersed with variable-recall rates of "retrieved observations" and variable rates of their reconsideration to the degrees of "understandability."

513.07 Resonantly propagated evolution oscillatingly induces tetrahedral quanta—both metaphysical and physical—formulated vectorially between four "star-event" phases

(1) observation,
(2) consideration,
(3) understanding, and
(4) articulation,

or

(1) recall;
(2) reconsideration;
(3) understanding;
(4) articulation.

514.00 Axis of Reference

514.01 The axis of reference is the axis of conceptual observation. The axis of observer reference frequently occurs spontaneously: as the line between the nose and the navel.

514.02 The direction of a vector is an angular one in respect to an omnidirectional coordinate system having a specific central point and a specific set of external points at specific angles and distances from one another and from the central point.

514.03 Our definition of an opening is that it is framed by trajectories. Every trajectory in a system has at least two crossings, but these crossings are as *viewed,* because the lines could be at different levels from other points of observation.

515.00 Frequency

515.01 Definition

515.011 Because there are no experimentally known "continuums," we cannot concede validity to the concept of continuous "surfaces" or of continuous "solids." The dimensional characteristics we used to refer to as "areas" and "volumes," which are always the *second-* and *third-*power values of linear increments, we can now identify experimentally, arithmetically, and geometrically only as quantum units that aggregate as points, both in system-embracing areal aggregates and within systems as volume-occupant aggregates. The areal and volumetric quanta of separately islanded "points" are always accountable numerically as the second and third powers of the *frequency of modular subdivision of the system's radial or circumferential vectors.*

515.02 The frequency of any system is determined by the isotropic, omni-intertriangulated, omnidirectionally considerate, vectorially moduled, subdivision enumeration of the system's radial and geodesically chorded circumferential closure's totally relevant involvement limits taken in respect to the system's independent, event-regenerating center. Because of the required omnitriangulation and isotropicity, systems are inherently moduled only by equiangular-equilateral triangles, and their generative center is that of the vector equilibrium wherefore the radial and circumferentially chorded time-size, i.e., frequency-wavelength modules subdivisions, by which alone system frequency may be determined, are always identical.

515.10 Angles

515.101 Because angles are parts of only one cycle, they are inherently subcyclic. Because size must be predicated Einsteinianly upon local-experience time cycles, relative size is measured in cyclic units. Therefore, angles, which are less than one cycle, are inherently less than one unit of size. Angles are inherently "subsize" consideration. Because angles are subcyclic, they are "subsize." Therefore, we are permitted to think independently of size in respect to triangles, which consist of three separate angles.

515.11 We may think independently of size in respect to tetrahedra, which consist of 12 separate angles. Triangles and tetrahedra and all varieties of polyhedra are thinkable independently of size. The cyclic-module measurement of the time of experiencing or generating the length of the edge of any triangulated special-case system can represent the basic "standard" of relative size-comparisoning to other object experiences. Each cyclic "sizing" increment is one unit of frequency and each cyclic increment inherently constitutes one unit of experienced physical energy.

515.12 When man employs nature's basic designing tools, he needs only generalized angles and special-case frequencies to describe any and all omnidirectional patterning experience subjectively conceived or objectively realized.

515.13 For how many cycles of relative-experience timing shall we go in each angular direction before we change the angle of direction of any unique system-describing operation? *

515.14 Angular fractionation is absolute. Triangles can be equiangular—one-fourth of a cycle or one-fiftieth of one cycle of unity—but they cannot be equilateral. Angles are constant and independent of size. Size is always special-case experience. Angles are generalized. Only eternal constants can be generalized. We do not know the length of the edges. Edges can be any length permitted by time. The length of the edges is frequency, while the angle is subfrequency.

515.20 Energy

515.21 The physical Universe is an aggregate of frequencies. Each chemical element is uniquely identifiable in the electromagnetic spectrum by its own unique set of separately unique frequencies. None of the chemical-element sets or individual frequencies is the same as those of any of the other chemical elements' frequencies. The different frequencies of one element's set produce unique cyclic-frequency interactions whose resonances are similar to musical chords. The electromagnetic spectrum of physical Universe embraces the full spectrum range of as yet discovered and identified radiation frequencies of all the first 92 self-regenerative, as well as the only split-second enduring elements beyond the 92 self-regeneratives thus far discovered by experimental physics. The macro/micro-cosmic electromagnetic spectrum chart discloses a cosmic orchestration that ranges from those of the microcosmic to the very complex macrocosmic-embracing whole celestial Universe nebulae. The human senses are able to tune in no more than one-millionth of the total known frequency range limits of the presently known electromagnetic spectrum. Whether expressed in foot-pounds per minute or kilowatt-hours, the total physical work done by all the muscles of all humans in all the two and one-half million years of known presence of humans aboard our planet Earth, amounts to less than the energy released in one second of time by one hurricane; one hurricane's released energy equals the total energy of the combined atomic bombs thus far produced and stockpiled by the Russians and the U.S.A. In contradistinction to this minuscule energy involvement of all history's human muscle, the invisible, weightless, but cosmically magnificent minds of humans have thus far discovered, quantized, and catalogued the relative abundance of each and all of the 92 regenerative chemical elements oc-

* Now that we understand this much, we may understand how man, consisting of a vast yet always inherently orderly complex of *wave angles* and *line frequencies,* might be scanningly transmitted from *any here* to *any there* by radio.

curring on all the visible stars of known Universe. Thus emerges human awareness of the physical-energy-mastering potential of the metaphysical mind's extraordinary information-sorting and -analyzing capability.

515.30 Frequency is plural unity. Frequency is a multicyclic fractionation of unity. A minimum of two cycles is essential to frequency fractionation. Frequency means a discrete plurality of cycles within a given greater cyclic increment.

515.31 In closest packing of spheres, frequency is the number of spaces between the balls, not the number of balls. In closest packing, frequency is equal to radius.

515.32 Electromagnetic frequencies of systems are sometimes complex but always constitute the prime rational integer characteristic of physical systems. (See Secs. 223.41 and 400.50.)

515.33 Wave magnitude and frequency are experimentally interlocked as cofunctions, and both are experimentally gear-locked with energy quanta.

516.00 Frequency Modulation

516.01 There are only two possible covariables operative in all design in Universe: they are the modifications of angle and of frequency.

516.02 Frequency means a discrete plurality of cycles within a greater cyclic increment. An angle is an angle independent of the length of its sides. An angle is inherently a subdivision of a single cycle and is conceptually independent of linear, areal, and volumetric size considerations. A triangle *is* a triangle independent of size. A tetrahedron *is* a tetrahedron independent of size.

516.03 By designedly synchronized frequency of reoccurrence of their constituent event patternings, a machine gun's bullets may be projected through a given point in the rotational patterning of an airplane's propeller blades. Such purposeful synchronization of a succession of alternate occupations at a point, first by a bullet and then by a discretely angled propeller blade, and repeat, is called angle and frequency modulation; together, they avoid interferences. All physical phenomena, from the largest to the smallest, are describable as frequencies of discrete angular reoccurrence of intimately contiguous but physically discontinuous events. All physical phenomena are subject to either use or nonuse of angular- and frequency-modulating interference capabilities.

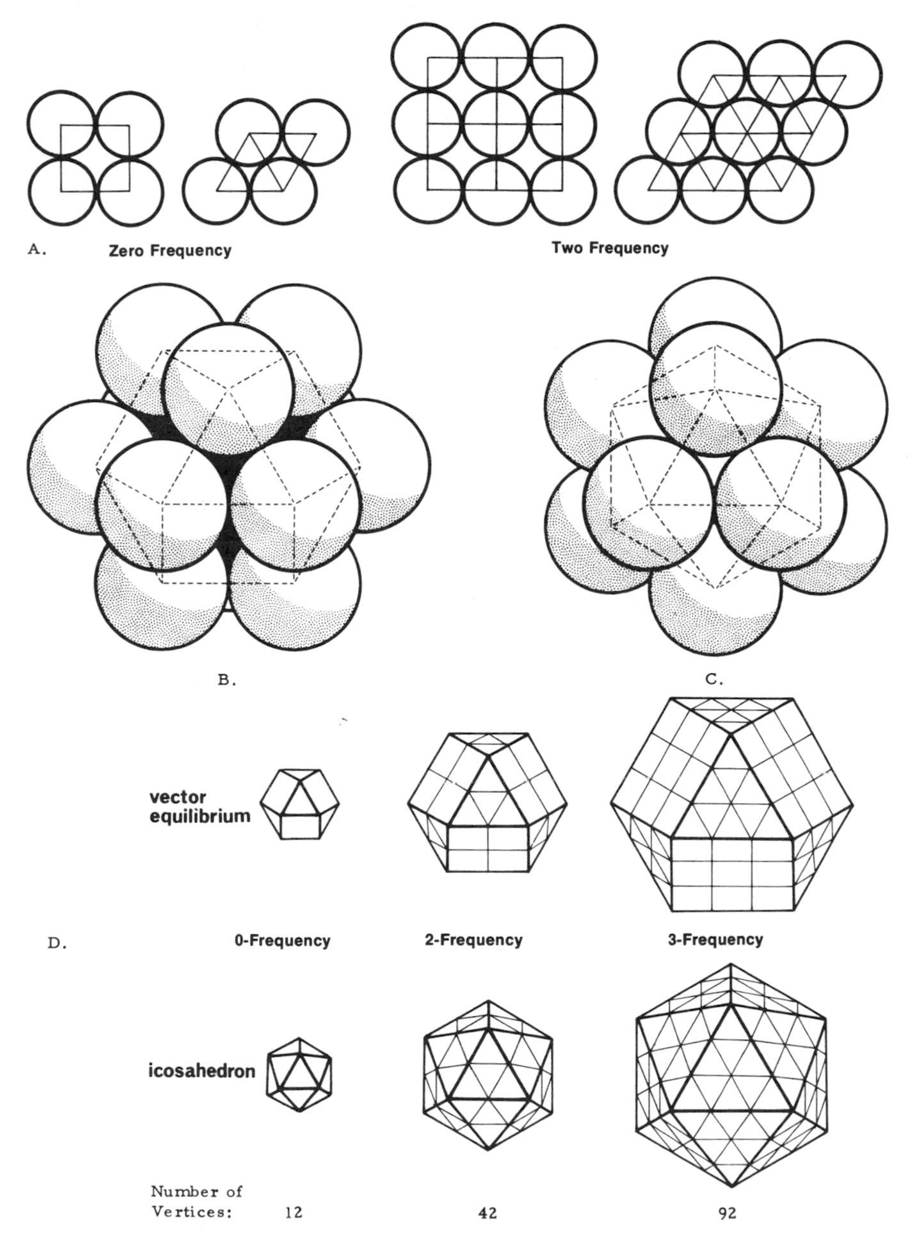

Fig. 516.03 *Frequency:* A square of any frequency is topologically equivalent to two adjacent triangles of corresponding frequency, i.e. a square has the same number of vertexes as two adjacent triangles (A). When the central sphere is removed from the vector equilibrium (B), it contracts symmetrically to a more compact arrangement (C), which is the icosahedron. The vector equilibrium has eight triangular faces and six square faces. The six square faces shift to become 12 triangular faces: $12+8=20$ triangular faces for the icosahedron. Outer shells of the vector equilibrium and icosahedron of the same frequency will always have the same number of vertexes or spheres (D). Therefore the equation $10F^2+2$ applies to both figures.

517.00 Interference

517.01 Two different energy events articulated as invisibly modulated, spiraled, vectorial lines each represent their respective masses multiplied by their velocities, and each has a unique angular direction in respect to the observer's axis. They cannot pass through the same point at the same time. When one energy event is passing through a given point and another impinges upon it, there is an *interference.*

517.02 Speaking operationally, lines are products of the energy interactions of two or more separate systems. The local environment is a system. A line is always formed by an alteration of the local environment by another system. "Lines" are the patterns of consequences of one system altering another system either by *adding to* it or *taking away* from it. The event leaves some kind of tracery—either additively, as with a vapor trail or a chalk mark, or reductively, as with a chiseled groove or a pin scratch, as a crack opened between two parts of a formerly unit body, or as a coring through an apple.

517.03 We find experimentally that two lines cannot go through the same point at the same time. One can cross over or be superimposed upon another. Both Euclidian and non-Euclidian geometries misassume that a plurality of lines can go through the same point at the same time. But we find experimentally that two or more lines cannot physically go through the same point at the same time.

517.04 When a physicist bombards a group of atoms in a cloud chamber with a neutron, he gets an interference. When the neutron runs into a nuclear component: (1) it separates the latter into smaller components; (2) they bounce acutely apart (reflection); (3) they bounce obliquely (refraction); (4) they combine, mass attractively. The unique angles in which they separate or bounce off identify both known or unknown atomic-nucleus components.

517.05 There is a unique and limited set of angle and magnitude consequences of interfering events. These resultants may always be depicted as vectors in the inward-and-outward, omnidirectional, multifrequency-ranging, circumferential-or-radial relativistic system patternings, which altogether constitute the comprehensively combined metaphysical and physical "reality" that is reported into and is processed by our brain and is reconsidered by our thoughts as referenced conceptually to various optimally selected observational axes and time-module durations.

517.06 When there is an interference of two energy events of similar magnitude, there is a coequal pattern of interference resultants, as when two knitting needles slide tangentially by one another. But when one converging body of an interfering pair is much larger than the other, the little one ''seems'' to do all the resultant moving as viewed by an observer small enough to see the small converger's motion—as, for instance, human beings see a tennis ball hit the big ball Earth and see only the tennis ball bounce away, the Earth ball being too big to be seen as a ball by the viewer and the relative bounce-off deflection of Earth's orbit from the tennis ball point of impact being too small for detection. As the magnitudes of energy vectors are products of the mass multiplied by the velocity, the velocity may be high and the mass small, or vice versa, and the vectors remain the same length or magnitude. A little body moving at sufficient velocity could have the same effect upon another body with which it interferes as could a big body moving at a slower rate. With these vectorial variables in mind, we see that there are three fundamental preconditions of the interference vectors: where one is larger than the other; one is the same; or one is smaller in energy magnitude than the other.

517.10 Six Interference Resultants

517.101 There are six fundamentally unique patterns of the resultants of interferences. The first is a tangential avoidance, like knitting needles slipping by one another. The second is modulated noninterference, as in frequency modulation. The third is reflection, which results from a relatively direct impact and a rebound at an acute angle. The fourth, which is refraction, results from a glancing impact and an obtuse angle of deflection. The fifth is a smash-up, which results in several parts of one or the other interfering bodies going away from one another in a plurality of angular directions (as in an explosion). The sixth is a going-the-same-way, ''critical-proximity,'' attraction link-up such as that established between the coordinated orbiting of Earth and Moon around the Sun.

517.11 Summary of Interference Phenomena

517.12
(a) Tangential avoidance
(b) Modulated noninterference (Frequency Modulation)
(c) Reflection
(d) Refraction
(e) Smash-up (Explosion)
(f) Critical Proximity (The Minimum Knot)

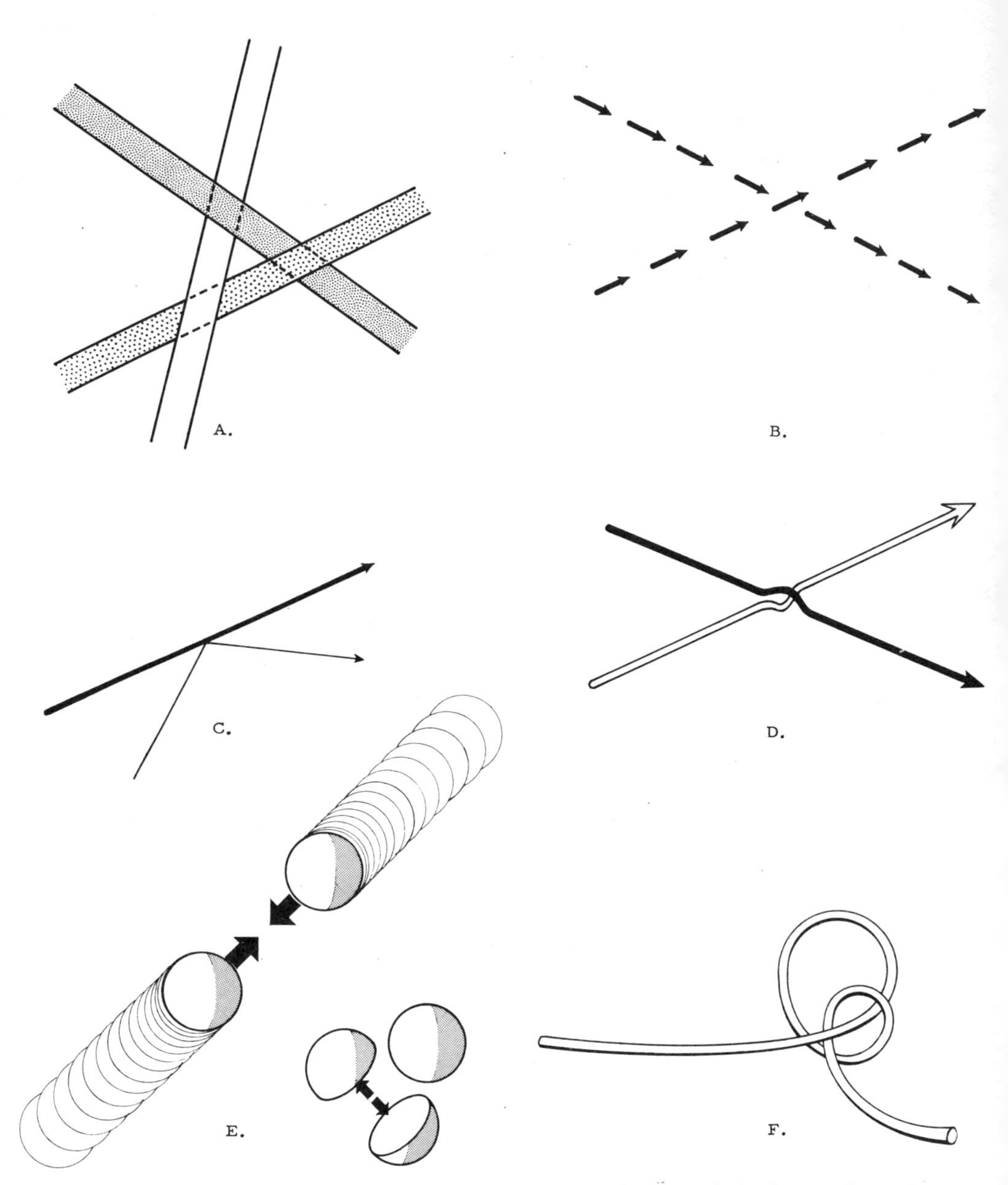

Fig. 517.10 *Interference Phenomena: Lines Cannot Go Through the Same Point at the Same Time:* Interference phenomena: No two actions can go through the same point at the same time. The consequences of this can be pictured as follows:

A. Tangential avoidance (like knitting needles).
B. Modulated noninterference.
C. Reflection.
D. Refraction.
E. Smash-up.
F. The minimum knot or critical proximity.

517.13 All three of these vectorial conditions and all six of these resultants are manifest in cloud chambers, in which the physicist can view with his naked eye the photographed resultants of angular directions and energy-magnitude lengths of the interference patternings that occur when, for instance, they bombard a group of atoms with an accelerated neutron that moves at such velocity as probably to interfere with one or another of billions times billions of atoms present in the elemental "gas" aggregation. From these cloud-chamber interference patterns, physicists are able to calculate much information regarding the interfering components. The cloud chamber makes it obvious that two lines, which are always experimentally proven to be energy vectors, cannot pass through the same point at the same time.

517.20 Tetrahedron of Interferences

517.201 A machine gun is shooting through a swiftly revolving airplane propeller. It is automatedly timed to shoot between every blade, or every second blade, or every third blade—with a sonic "wow" every time it goes between the propeller blades. We are synchronizing purposefully. Unautomated by human mind and brain's anticipatory designing, bullets would produce a random sequence of patterns as they hit the propeller blades; some would at first bounce off precessionally, while others would knock off sections of the propeller blades.

517.21 Let us assume two machine guns firing from two different positions, one of them due north of a point in space and the other due west of the same point. One is aimed south and the other is aimed east, which means they are both firing through a common point in space. They are synchronized so that their bullets will not interfere with one another. The bullets all weigh the same. If they were nonsynchronized, they would frequently meet and be precessionally deflected.

517.22 Now place three machine guns at the three corners of an equilateral triangle. From the center of area (sometimes miscalled the center of gravity) of the equilateral triangle, one of the three corners lies in a bearing of 0° (i.e., 360°) in a northerly direction; the second bears at 120°; the third at 240° from the triangle's center. We then aim all three machine guns toward the center of the triangle and elevate their aim to 35° 16′. We synchronize their firing periods to coincide. We thus introduce an interference at the center of gravity of a regular tetrahedron whose triangular base corners are occupied by the three guns. Precession will take place, with the result that all three bullets precess into a vertical trajectory as a triangular formation team through the apex of the regular tetrahedron whose base corners are identified by the three guns.

517.23 Every action has an equal and opposing reaction. So now let us assume that instead of machine guns firing in one direction only, we have three bazookas in which both action and reaction are employed in two directions. The double-ended openness of the bazooka sees the rocket missile projected in one direction while a blast of air is articulated in an opposing conical zone of directions. The cone's *inertia* provides the shove-off for the projectile by the explosion. Inertia is dynamic—as sensed in the orbital course integrity of the enormous mass of Earth going around the Sun at 60,000 m.p.h. so that the little man on board it, who is also going around the Sun at 60,000 m.p.h., and is also walking around Earth at four m.p.h. and as he steps around Earth's surface he pushes Earth in the opposite direction to his walking, but so negligibly that the little man does not conceive of his Earth as movable and so has invented the concept of completely inert, or "at rest." Our deceptive fixity of celestial position as a standing still in Universe is fortified by the absolute silence of travel in vacuo around the Sun.

517.24 Now we take two bazookas (not three!) firing in different parallel planes and not at the same level. One is aimed north-south in respect to the North Star and the Southern Cross. The other bazooka is in a parallel plane but remote; it is aiming east-west. They are fired, and at each of their two terminals, we get four precessional effects of the reactions and resultants occurring at 55 degrees in respect to their respective parallel planes. The result will be six vectors interacting to form the tetrahedron, a *tetrahedron of interferences*.

518.00 Critical Proximity

518.01 Though lines (subvisibly spiraling and quantitatively pulsative) cannot go through the same point at the same time, they can sometimes get nearer or farther from one another. They can get into what we call "critical proximity." Critical proximity is the distance between interattracted masses—when one body starts or stops "falling into" the other and instead goes into orbit around its greater neighbor, i.e., where it stops yielding at 180 degrees and starts yielding to the other at 90 degrees. (See Sec. 1009.)

518.02 Critical proximity would be, for instance, the relative interpositioning of the distances of the Moon-Earth team's Sun co-orbiting wherein there is a complex mass-attraction hookup. When at critical proximity the 180-degree mass attraction takes over and one starts falling into the other—with the attraction fourfolded every time the distance between them is halved—they establish a mass-attraction, relative-proximity "contact" bond and interoperate thereafter as a "universal joint"—or a locally autonomous motion freedoms' joint. Either body is free to carry on individual, local, angular-relationship-changing motions and transformations by itself, such as revolving and precessing. But without additional energy from elsewhere being applied to their interrelationship, they cannot escape their critical proximity to one another as they co-orbit together around the Sun—with which they are in common critical proximity.

518.03 Critical proximity occurs at the precessional moment at which there is a 90-degree angular transition of interrelationship of the two bodies from a 180-degree falling-back-in to a 90-degree orbiting direction, *or vice versa*. (See Sec. 1009.63.)

518.04 The transition of physical phenomena from being an apparent unit entity to being an apparent complex, or constellation of a plurality of entities, is that of the individual components reaching the critical proximity precessional condition and "peeling off" into individual orbits from their previous condition of falling back into one another under nonangularly differentiable entity conditions. This is the difference between an apparent "stone" and its crushed-apart "dust" parts.

518.05 Critical proximity explains mass-attraction coherence. It accounts for all the atoms either falling into one another or precessing into local orbits. This accounts for the whole Universe as we observe it, the collections of things and matter and noncontiguous space intervals. The coming-apart phase of critical proximity is radiation. The coming-together and holding-together phase is emphasized in our ken as gravity.

518.06 Critical proximity is a threshold, the absolute vector equilibrium threshold; if it persists, we call it "matter."

519.00 Point

519.01 What we really mean by a point is an unresolved definition of an activity. A point by itself does not enclose. There are no indivisible points.

519.02 Without insideness, there is no outsideness; and without either insideness or outsideness, there is only a locus fix. Ergo, "points" are inherently nondemonstrable, and the phenomena accommodated by the packaged word *point* will always prove to be a focal center of differentiating events. A locus fix constitutes conceptual genesis that may be realized in time. Any conceptual event in Universe must have insideness and outsideness. This is a fundamentally self-organizing principle.

519.03 Points are complex but only as yet nondifferentiably resolvable by superficial inspection. A star is something you cannot resolve. We call it a point, playing Euler's game of crossings. One locus fix does not have an insideness and an outsideness. It takes four to define insideness and outsideness. It is called a point only because you cannot resolve it. Two remotely crossing trajectories have no insideness or outsideness, but do produce optically observable crossings, or locus fixes, that are positionally alterable in respect to a plurality of observation points. A point's definitively unresolved event relationships inherently embrace potential definitions of a complex of local events. When concentrically and convergently resolved, the "point" proves to be the "center"—the zero moment of transition from going inwardly and going outwardly.

519.10 Physical points are energy-event aggregations. When they converge beyond the critical fall-in proximity threshold, they orbit coördinatedly, as a Universe-precessed aggregate, as loose pebbles on our Earth orbit the Sun in unison, and as chips ride around on men's shoulders. A "point" often means "locus of inflection" when we go beyond the threshold of critical proximity and the *inness* proclivity prevails, in contradistinction to the differentiable other fallen-in aggregates orbiting precessionally in only mass-attractively cohered remoteness outwardly beyond the critical-proximity threshold.

519.20 If light or any other experiential phenomenon were instantaneous, it would be less than a point.

519.21 A point on a sphere is never an infinitesimal tangency with a plane.

519.22 The domains of vertexes are spheres.

519.30 For every event-fixed locus in Universe, there are six uniquely and exclusively operative vectors. (See Sec. 537, Twelve Universal Degrees of Freedom.)

520.00 Wavilinearity: Fixes

520.01 Linear does not mean straight. Lines are energy-event traceries, mappings, trajectories. Physics has found no straight lines: only waves consisting of frequencies of directional inflections in respect to duration of experience.

520.02 Calculus treats discretely and predictively with frequency change rates and discrete directions of angles of change of the omnicurvilinear event quanta's successively occurring positionings: *fixes.*

520.03 Fixes consist of both angular and frequency (size) observations. Coincidental angle and dimension observations provide fixes.

520.10 Spiralinearity

520.101 Regenerative precession imposes wavilinearity on vectors and tensors. Wavilinearity is spiralinear.

520.11 All actions are spiral because they cannot go through themselves and because there is time. The remote aspect of a spiral is a wave because there are no planes.

520.12 As with coil springs, in tensors and vectors of equal magnitude, the spiralinearity of the vector is shorter in overall spatial extent than is the spiralinearity of the tensor. Compressed lines or rods tend to arcs of diminishing radius; tensed lines or rods tend to arcs of increasing radius.

521.00 Vectors: Trajectories

521.01 A vector manifests a unique energy event—either potential or realized—expressed discretely in terms of direction, mass, velocity, and distance. A vector is a partial generalization, being either metaphysically theoretical or physically realized, and in either sense an abstraction of a special case, as are numbers both abstract (empty sets) or special-case (filled sets).

521.02 A vector always has unique direction relative to other events. It is discrete because it has a beginning and an end. Its length represents energy magnitude, the produce of its velocity and its mass. The direction is angular in respect to the axis of reference of the observer or in respect to an omnidirectional coordinate system.

521.03 Vectors are wavilinear lines of very high frequency regeneration of events whose high frequencies and whose short wavelengths only superficially appear to be "straight." Since neither light nor any other experiential phenomena are instantaneous. They are "linear." If they were instantaneous, they would be less than a point. The terminal of an action's vector occurs "later."

521.04 Vectors are spearlike lines representing the integrated velocities, directions, and masses of the total aggregate of nonredundant forces operating complexedly within a given energy event as it transpires within a generalized environment of other experiences whose angular orientations and interdistance relationships are known.

521.05 Vectors always and only coexist with two other vectors, whether or not expressed; i.e., every event has its nonsimultaneous action, reaction, and resultant. (See Sec. 511, Energy Event.) But every event has a cosmic complementary; ergo, every vector's action, reaction, and resultant have their cosmic tripartite complementaries.

521.06 A vector has two vertexes with angles around each of its vertexial ends equal to 0 degrees. Every vector is reversible, having its negative alternate. For every point in Universe, there are six uniquely and exclusively operative vectors. (See Sec. 537, Twelve Universal Degrees of Freedom.)

521.07 Every event is six-vectored. There are six vectors or none.

521.08 Vectors are size. The size of a vector is its overall wavilinear length.

521.09 A vector is one-twelfth of relevant system potential.

521.10 Tensors

521.101 Vectors and tensors constitute all elementary dimension. A vector represents an expelling force and a tensor an impelling force.

521.20 Lines

521.201 Pure mathematics' axiomatic concepts of straight lines are completely invalid. Lines are vector *trajectories*.

521.21 The word *line* was nondefinable: infinite. It is the axis of intertangency of unity as plural and minimum two. Awareness begins with two. This is where epistemology comes in. The "line" becomes the axis of spin. Even two balls can exhibit both axial and circumferential degrees of freedom. (See Sec. 517.01, Sec. 537.22, and Sec. 240, Synergetics Corollaries, Subsec. 06, 13, 14, 15, 20, 21, 22, 24, 25, 26, 27, 29, 30, 31, 35, 36.)

521.22 A line is a directional experience. A line is specific like *in,* while *out* is anydirectional. Lines are always curvilinearly realized because of universal resonance, spinning, and orbiting.

521.23 A point is not a relationship. A line is the simplest relationship. Lines are relativity. A line is the first order of relativity: the basic sixness of minimum system and the cosmically constant sixness of relationship identifies lines as the relativity in the formula $\frac{N^2-N}{2}$.

521.30 *Omnidirectional Force Vectors*

521.30 Galileo's parallelogram of forces is inadequate to account for resultants other than in the special-case, one-plane, billiard-table situation. Force vectors must express the omnidirectional interaction of forces, with lengths proportional to their mass times the velocity, and indicating that there are unique directions in Universe.

521.31 When we vector the course of one ship on a collision course with a second ship, the resultant of forces in Galileo's diagram would have them waltzing off together some 12 miles to the north-northeast. But all sane men can see such behavior is just what ships do not display after a collision. One of the two ships colliding on the wavy surface of spherical Earth may go a few hundred feet in the direction of Galileo's resultant of forces, but not 12 miles. But the other one probably goes in toward the center of Earth—which isn't in the diagram at all.

521.32 When ships run into each other, they actually first rise outwardly from Earth's center because in acceleration both were trying to leave Earth. (If they could accelerate faster, like rockets, they would leave Earth.) In reality, there are four forces operating. Two rise outwardly against gravity, accelerating conically together before they subside, when one or both go to the bottom. In addition to the vector for each ship, there is gravity plus the resultant. We are operating omnidimensionally, and this is what the minimum set of forces is. The pattern of force lines looks very much like a music stand: three vectorial legs spread out with a fourth vertical vector. (See Secs. 621.20 and 1012.37.)

522.00 *Deliberately Nonstraight Line*

522.01 The so-called pure mathematician's straight line must be the "impossible"; it must be *instantly infinite* in two infinitely remote opposite directions. All of its parts must be absolutely, uniformly nothing and simultaneously manifest as discretely, and infinitely divisible, increments. It may not be generated progressively or drawn physically, in time, as an experimentally produced action trajectory of one system modifying another. Microscopic inspection of the impressed, graven, deposited, or left-behind trails of all physical Universe's action trajectories always discloses a complex of gross, noninfinite, nonstraight, non-equal-magnitude irregularities. Progressively closer inspections of experimentally attempted demonstrations by pure mathe-

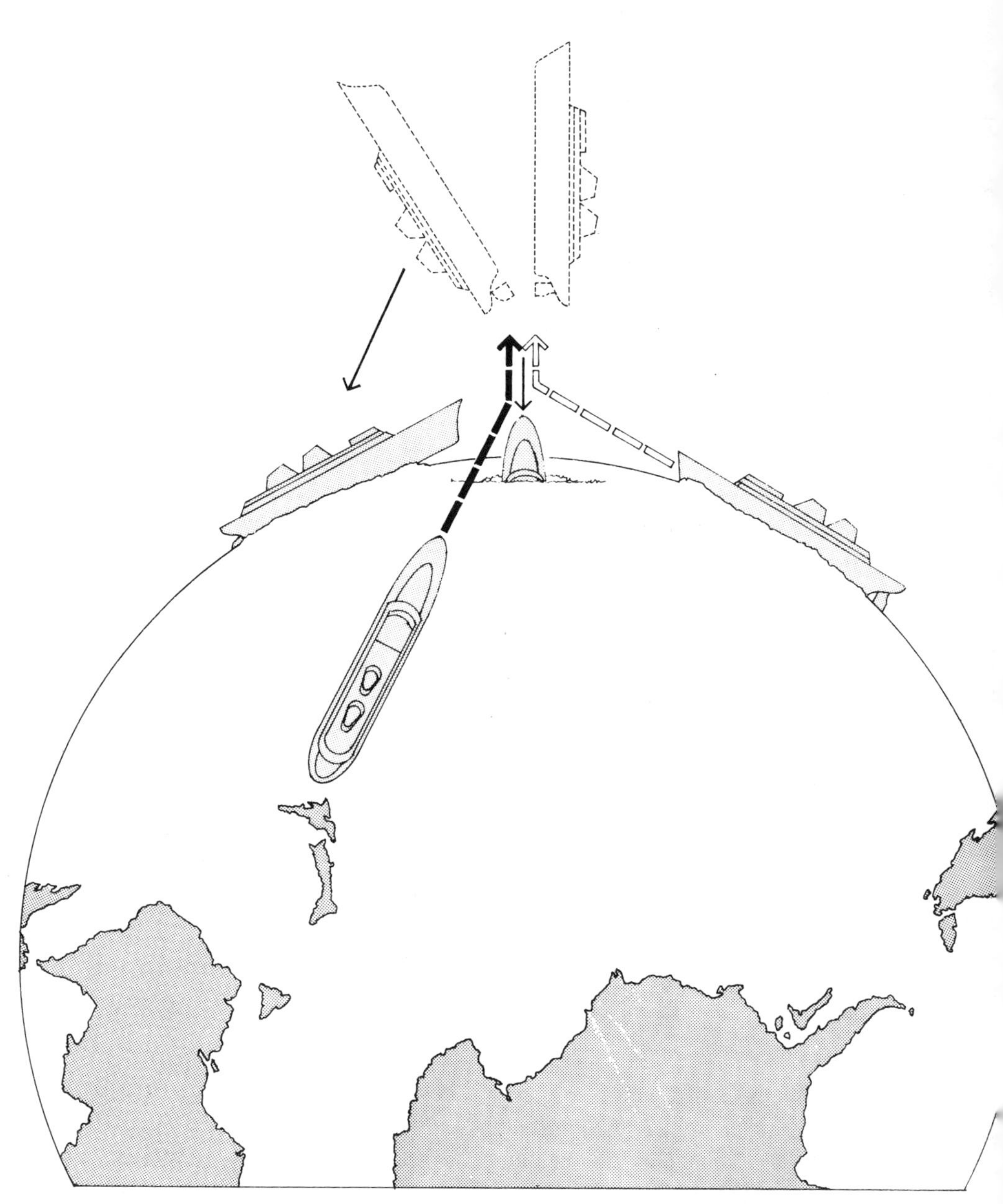

Fig. 521.30 *Omnidirectional Lines of Force:* Ships colliding on the globe after sudden acceleration reveal the inadequacy of parallelogram force diagrams for explaining the omnidirectional interaction of forces.

maticians of their allegedly "straight" lines disclose increasingly volumetric aberration and angular digressions from straightness.

522.02 "Straight lines" may be axiomatically invoked but are nonrealizable in pure imagination: *image-ination* involves reconsidered and hypothetically rearranging the "furniture" of remembered experience as retrieved from the brain bank. Straight lines are axiomatically self-contradictory and self-canceling hypothetical ventures. Physics has found only waves, no straight lines. Physics finds the whole physical Universe to be uniquely differentiated and locally defined as "waves."

522.03 The deliberately nonstraight line of synergetics employs the mathematicians' own invention for dealing with great dilemmas: the strategy of reductio ad absurdum. Having moments of great frustration, the mathematician learned to forsake looking for local logic; he learned to go in the opposite direction and deliberately to choose the most absurd. And then, by progressively eliminating the degrees of absurdity, he could work back to the not too absurd. In hunting terms, we call this *quarrying* his objective. Thus he is able at least to learn where his quarry is within a small area.

522.04 To develop methodically a very much less crooked line than that of conventional geometry, we start to produce our deliberately nonstraight line by taking a simple piece of obviously twisted rope. We will use Dacron, which is nonstretchable (nylon will stretch, and manila is very offensively stretchable). We then take the two ends of our rope and splice them into each other to form a loop. This immediately contradicts the definition of a straight line, which is that it never returns upon itself. We can take the two parts of the rope loop that are approximately parallel to one another and hold these two parts in our hands. We may call this pairing. Holding one hand on one of the pairs, we can slide the rope on the other hand, continually pairing it away from the point of first pairing. As we massage the two parts along, our hand finally gets to where the rope comes into a very sharp little loop and turns to come back on itself. We can hold it very tight at this point and put a little ribbon on the bend, the arch where it bends itself back. Sliding our hands the other way, holding and sliding, holding and sliding, massaging the rope together, we come to the other looping point and carefully put a ribbon marker in the bend of the arch. Having carefully made a rope that returns upon itself, we have now divided unity into two approximately equal halves.

522.05 Heisenberg makes it experientially clear that we cannot be absolutely exact. The act of measuring alters that which is measured. But with care we can be confident that we have two experiencially satisfactory halves of the total rope circuit existing between our two ribbon markers. Proceeding further, we can bring the ribbon-marked, half-points together, thus to divide the rope into four equal parts of unity. We can separately halve each of those quarter-lengths of the rope's closed-circuit unity to produce one-eighth unity length, while avoiding compounding of error. Each time we halve a local fraction, we halve any residual error. We can evenly subdivide our deliberately nonstraight line into as many small fractions as may be desirable.

522.06 We now ask four friends each to take hold of a half- or a quarter-point in the rope, and then ask them to walk away from each other until the rope unity is taut. We ask them to lower their four-sided geometrical figure to the floor and ask another friend to drive nails into the floor inside the four tightly stretched corners of the rope. A diamond rope pattern is thus produced with its corners marked *A, B, C,* and *D*. We are provided with plenty of proofs about equilateral parallelograms; we know that if the sides are equal in length, we can assume them to be approximately parallel because the wall we have nailed them to is an approximate plane. It may be pretty rough as the mathematicians talk about planes, but it is nonetheless a satisfactory plane for our purposes.

522.07 We next put in more nails in the floor at the ribbon-marked eighth points. *C* is the right-hand corner of the diamond, and *D* is the top of the diamond. We can call the bottom half of the diamond a *V,* and we can call the top half of the diamond a *lambda*. Putting nails at the one-eighth points means that halfway down from *A* to *B* there is a nail and halfway from *B* upward to *C* there is a nail. Halfway from *C* upward to *D* we put a nail at the eighth point. Then halfway down from *D* back to *A* again we put another nail at the eighth point. We then take the rope off *D* and place it over those one-eighth nails. The rope now changes from a *lambda* pattern into an "M" form. Because it is an equilateral parallelogram, we know that the new middle loop must be at the center of the diamond. We place a nail at this center of the diamond and mark it *O*. We next go from *C*, which is at the extreme right-hand corner of the diamond, down to take the rope off *B*. Taking the rope off the *V* (which used to be *ABC*), we convert the *V* to a *W*—with the bottom points of the *W* at the one-eighth-point nails. We then move the rope off *B* and up to the center of the diamond also. This gives us two diamonds, two little diamonds strung end-to-end together at the center of the big diamond. Their extreme ends are at *A* and *C*. Because we know that these are all equilateral parallelograms, we know that the length of the new letter *M* is the same as the length of the new letter *W*. We can now give these new one-eighth points the designations *E, F, G,* and *H*. So it now reads *AHOGC* and *AEOFC*. And we have two beautiful diamonds.

522.08 From now on, all we have to do is convert each of these diamonds in the same manner into two smaller ones. We convert the two diamonds into four. And then the four into eight. And the eight into 16. But the chain of diamonds always remains *A–C* in overall length. Both the altitudes and lengths of the diamonds are continually halving; thus what we are doing is simply increasing the frequency of the modular subdivision of the original unity of the rope. As the frequency of the wavelike subdivisions is multiplied, the deliberately nonstraight line approaches contractively toward straight behaviors. The rope remains exactly the same length, but its two parts are getting closer and closer to one another. The plane of the floor is really an illusion. As we get to a very high frequency of diamonds, we realize that instead of doing it the way we did, we could simply have twisted the original rope so that it would be a series of spirals of the same number as that of the chain of diamonds. We look at the profile of the rope and realize that all we are seeing is twice as many twists every time—at every progression. This gives us a very intimate concept of what actually happens in wave phenomena.

522.09 The old-fashioned physicist used to put one nail in the wall, fasten a rope to it, and stand back and throw a whip into the rope. The whip goes to the nail on the wall and then comes back to his hand and stops. That is the prime characteristic of waves. They always make a complete cycle. That is why, for instance, gears are always whole circles. A gear is a fundamental wave phenomenon. Electromagnetic waves always close back upon themselves. Deliberately nonstraight lines are round-trip circuits.

522.10 Our deliberately nonstraight-line model provides us whatever frequency of modular subdivision we want in unity, which is the cycle. This is what we mean by frequency of modular subdivision, whether unity is a sphere or a circle. What is going on in our rope, the way we have handled it, we make it into unity as a cycle. We see these waves going in a round-trip trajectory pattern from *A* to the extreme point *C* and back again to *A*. The overall distance traveled by any of the routes remains the same. So what we see on the floor or in the diamond chain diagram is a true model of basic wave phenomena. As we double the frequency and halve the wavelength of positive and negative waves, we swiftly arrive at a visibly far less crooked condition and approach relative straightness. We can see quite clearly that we do not have to increase the subdividing of those diamonds many times before they tend to look like a straight line as far as your eye and my eye can see. This concept agrees elegantly with fundamental wave theory as predicated on electromagnetic experimentation.

522.11 For instance, on an engineer's scale you and I can see 50 divisions of an inch. We can see $^1/_{50}$th of an inch, but $^1/_{100}$th of an inch goes gray and blurred. When we get to where an inch of the deliberately nonstraight line has more than 100 subdivisions, it looks like an absolutely straight line. When we get into the kinds of frequencies that characterize light waves, we get into very, very high numbers, and we can understand that what we call a line of

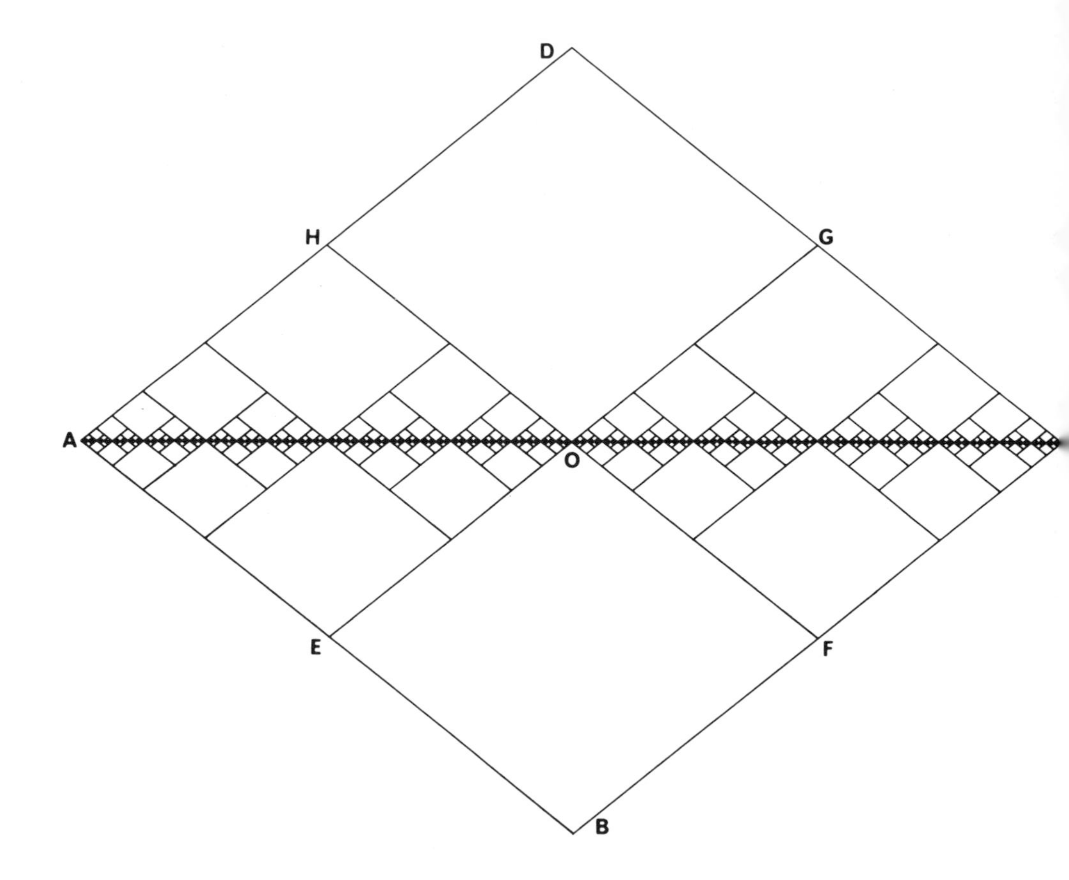

Fig. 522.09 *The Deliberately Nonstraight Line:* Quasi-''straight'' lines: *ABCDA* = unit wave; *AEOFCGOHA* = *ABCDA*. As we double the frequency and halve the wavelength of positive and negative waves, we approach relative straightness: proof that two deliberately nonstraight lines between points *A* and *C* approach relative straightness to more effective degree than attainable by an assumed straight construction.

sight has become so thin that it is invisible altogether. So we can understand that when the mathematician asked for a line of sight, which felt so good to him, he was asking for something that is really very beautifully imaginary. It was always a deliberately nonstraight line.

522.20 All experiments show that with ever closer inspections, the mathematicians' "straight" lines become obviously ever less straight. On the other hand, the quasi-straight line, which is demonstrated here as the deliberately nonstraight line, does get progressively straighter. Tending toward a greater straightness than that which is physically demonstrable, the deliberately nonstraight line thus serves all the finite geometries heretofore employed schematically by the mathematicians' alleged but unprovable straight lines, i.e., to demonstrate proof of the Euclidian and non-Euclidian geometrical propositions.

522.21 "Lines of sight" taken with transits are truer than string lines or penciled lines. Sight approaches "straight" behaviors. Lines of sight are high-frequency energy-wave interactions. Because the truest lines of sight are energy-wave quanta, they are always finite. The mathematician might say, "Oh, I mean a much straighter line than you can draw, I mean as straight and intangible as a line of sight." Then you remind the mathematician that when you have your transit's telescope focused on the "kissing point," as Earth's horizon becomes tangent to Sun's disc at daylight's end, you must remember that it takes eight minutes for the light to reach us from the Sun. Wherefore, the Sun has not been there for eight minutes, and you must admit that you are "seeing" the Sun around and beyond the horizon, which proves that your "line of sight" is curved, not straight. Due to the lag in the speed of light, Sun has not been there in a direct line of sight for eight minutes, so you are looking around the horizon through a curved "pipe" of light. This is what Einstein referred to as curved space.

522.22 To provide a more accurate identity of the only apparently straight-line phenomenon that the pure mathematician had erroneously thought of as "the shortest distance between two points," Einstein reinvoked the elliptical geometry of the mathematician Riemann and instituted the present concept of geodesic lines, which we may describe experimentally as "the most economic relationships between two event foci."

522.23 To comprehend and apprehend experimentally such "most economic relationships," all that you need do is to attempt to hit a flying object with a bullet fired by you from a gun. If you fire at the flying object where it is at the moment you fire, you will not hit it. You must fire at where you figure it is going to be at a later moment when it would most probably collide with your bullet. Gravity will start curving your bullet toward Earth as soon as it leaves your gun. The amount of curvature may be imperceptible to you, but it is easily detected by using a camera and a tracer firing charge. The air is always in motion, and your bullet will corkscrew ever so mildly between you and the flying object. This corkscrewing of the geodesic line, which is the most economical time-distance-effort relationship between the gun, the firer, and the flying object he hits, is dramatically shown in night photography of dogfights of World-War-II airplanes firing machine-gun tracer bullets at one another, with one being hit while the photographs are taken by a third plane flying in close vicinity of the dogfight.

522.30 Reduction by Bits

522.301 What the mathematicians thought was a straight line is not a straight line; it is an ultravisible, high-frequency, linearly articulated, spiral-wave event. The binary-mathematics methodology of progressive halving, or cybernetic "bitting," not only explains linear-wave phenomena but also identifies Pythagoras's halving of the string of a musical instrument to gain an exact musical octave—or his "thirding" of the musical string to produce the musical fifths of progression of flat and sharp keys.

522.31 The computer programmed to employ the cybernetic bits of binary mathematics progressively subdivides until one of its peak or valley parts gets into congruence with the size and position of the unit we seek. The identification process is accounted for in the terms of how many bits it takes to locate the answer, i.e., to "tune in."

522.32 Starting with whole Universe, we quickly reach any local system within the totality by differentiating it out temporarily from the whole for intimate consideration. We do so by the process of *reduction by bits.*

522.33 All irrelevancies fall into two main categories, or bits. Bits break up finite wholes into finite parts.

522.34 Once you state what your realistic optimum recognition of totality consists of, then you find how many bits or subdivision stages it will take to isolate any items within that totality. It is like the childhood game of Twenty Questions: You start by saying, "Is it physical or metaphysical?" Next: "Is it animate or inanimate?" (One *bit.*) "Is it big or little?" (Two *bits.*) "Is it hot or cold?" (Three *bits.*) It takes only a few bits to find out what you want. When we use bit subdivision to ferret out the components of our problems, we do exactly what the computer is designed to do. The computer's mechanism consists of simple go–no go, of yes and no circuit valves, or binary-mathematics valves. We keep "halving" the halves of Universe until we refine out the desired *bit.* In four halvings, you have eliminated 94 percent of irrelevant Universe. In seven halvings, you have removed 99.2 percent of irrelevant Universe. Operating as fast as multithousands of halvings per second, the computer seems to produce instantaneous answers.

522.35 Thus we learn that our naturally spontaneous faculties for acquiring comprehensive education make it easy to instruct the computer and thus to obtain its swift answers. Best of all, when we get the answers, we have comprehensive awareness of the relative significance, utility, and beauty of the answers in respect to our general universal evolution conceptioning.

522.36 Our method of demonstrating the nature of the special-case experiences out of which the pure mathematicians' imaginary generalized case of his pure straight line was evolved, also contains within it the complete gears-interlocking of quantum-wave mechanics and vectorial geometry, which are coordinately contained in synergetics with computer binary "bitting."

523.00 Vertexes: Crossings

523.01 Euler showed that where we have two lines—any kind of lines, crooked or not so crooked—where the lines cross is distinctly different from where the lines do not cross. The pattern of two or more lines crossing one another is also completely distinguishable from any single line by itself. We call this crossing or convergence of lines a *vertex*. This is absolute pattern uniqueness.

523.02 Crossings are superimposed lines. They do not go through each other. They are just a *fix*—what physicists call points.

523.03 In a structural system, the number of vertexes is always divisible by four and the number of triangle edges is always divisible by six. Edges and vertexes do not come out as the same number systems, but you can describe the world both ways and not be redundant.

524.00 Novent

524.01 We experience events and no-events. Ergo, we invent *novent*. Novents characterize the finite but nonsensorial remote masses' interattraction, i.e., the gravitational continuum.

524.02 Seeming "space" is the absence of energy events. The word *space* as a noun misleadingly implies properties that are altogether lacking.

524.03 All of our experiences are periodically terminated: the termination characterizes both the physical and the metaphysical aspects of our observing faculties and the observed phenomena. There are no experimentally known continuums. Physics has found no "solids." We have only *awake* or *asleep—experience* or *nonexperience*—occurrence durations and nonoccurrence intervals; either discrete and unique packages of energy or thought, on the one hand, or of nonenergy or nonthought, on the other hand. Each and all of these are as uniquely differentiable, and as separable, from one another as are the individual stars of the Milky Way.

524.04 The nonevent continuum is the novent. The novent continuum permeates the finitely populated withinness and comprises the finite novent withoutness. Novent is the finite but nonsensorial continuum. (See Sec. 905.20.)

524.10 In and Out

524.101 There are no specific directions or localities in Universe that may be opposingly designated as *up* or *down*. In their place, we must use the words *out* and *in*. We move *in* toward various individual energy-event concentrations, or we move *out* from them. But the words *in* and *out* are not mirror-image opposites. *In* is a specific direction toward any one local individual system of Universe. *Out* is not a direction; *out* is nondirectional because it is anydirectional.

524.11 You are always *in* Universe. You cannot get *out* of Universe. All the word *out* means is that you are not inside a system. You can only get *out* of systems.

524.12 *In* designates individual experience foci. Foci are *in,* because focusable, but always, as entropy shows, temporary. Relationships exist between the *ins* because they are definable. *Out* is common to all; *out* is timeless; *out* is not really packaged.

524.13 *In* is discrete; *out* is general. The *ins* are discontinuous; the *outs* are continuous. *Out* is nothingness, i.e., nonexperience. Only the nonexperience nothingness constitutes continuum.

524.14 *In* is temporal; *out* is eternal. *Ins* are knowable; *outs* are unknowable. *In* is individually, uniquely identifiable; *out,* though total, inherently integral, and finite, is nonidentifiable. *In* is individually, uniquely directional; *out* is any, all, and no direction. *Out* is all directions; even when temporarily inward toward center, it passes beyond the center to eventual outness.

524.20 Areas: Faces

524.21 It is experimentally demonstrable that an apparent "plane" is a "surface" area of some structural system. There are no experimentally demonstrable continuums. All that has been found is discontinuity, as in star constellations or atomic nuclear arrays. Areas are discontinuous by constructional definition. Areas, as system "faces," are inherently empty of actions or events, and therefore are not "surfaces."

524.30 Openings

524.31 There are no surfaces. Therefore, there are no areas. So Euler's topological aspects have to be altered to read: "Lines" = *trajectories;* "vertexes" = *crossings;* and "areas" = *openings,* i.e., where there are no trajectories or crossings.

524.32 When three or more "lines," "vectors," or *trajectories* each cross two others, we have an *opening.* Our definition of an opening is that it is surrounded, i.e., framed, by trajectories.

524.33 Every trajectory in a system will have to have at least two crossings. These are always *as viewed,* because the lines could be at different levels from other points of observation.

525.00 Solids: Matter

525.01 If subvisibly modulated spiraling wave lines cannot go through the same point at the same time, there can be no continuous, perfectly level planes. Planes are not experimentally demonstrable. Solids are not experimentally demonstrable. Physical experiment has never discovered any phenomena other than discontinuous discrete-energy events, each uniquely identifiable amongst the gamut of frequencies of cyclic discontinuity of all the physical phenomena, as comprehensively and overlappingly arrayed in the vast frequency ranges of the electromagnetic spectrum. The electromagnetic spectrum "reality" has been found experimentally to embrace all known physical phenomena: visible, subvisible, or ultravisible thus far detected as present in Universe. There are no solids. The synergetic behavior of structures satisfactorily explains as discontinuous that which we have in the past superficially misidentified as "solid."

525.02 For a microscopic example of our spontaneous and superficial misapprehending and miscomprehending environmental events, we must concede that both theoretically and experimentally we have now learned and "know" that there are no "solids," no continuous surfaces, only Milky Way-like aggregations of remotely interdistanced atomic events. Nonetheless, society keeps right on seeing, dealing, and superficially cerebrating in respect to "things" called "solids" or "matter."

525.03 Take the simple word *solid.* We have physics of the "solid state," a very late phase of physics very improperly called "solid." Even in solid state, the voids between the atoms are as voids of interstellar space. The nucleus itself is as empty as space itself. But the concept "solid" was a comfortable kind of concept, not easy to jettison.

526.00 Space

526.01 There is no universal space or static space in Universe. The word *space* is conceptually meaningless except in reference to intervals between high-frequency events momentarily "constellar" in specific local systems. There is no shape of Universe. There is only omnidirectional, nonconceptual "out" and the specifically directioned, conceptual "in." We have time relationships but not static-space relationships.

526.02 Time and space are simply functions of velocity. You can examine the time increment or the space increment separately, but they are never independent of one another.

526.03 Space is the absence of events, metaphysically. Space is the absence of energy events, physically.

526.04 The atmosphere's molecules over any place on Earth's surface are forever shifting position. The air over the Himalayas is enveloping California a week later. The stars now *overhead* are *underfoot* twelve hours later. The stars themselves are swiftly moving in respect to one another. Many of them have not been where you see them for millions of years; many burnt out long ago. The Sun's light takes eight minutes to reach us. We have relationships—but not space.

526.05 You cannot get out of Universe. You are always in Universe. (See Sec. 321, Scenario Universe. See Sec. 524, Novent.)

527.00 Dimension

527.01 There is no dimension without time.

527.02 Dimension is experiential; it is of time; ergo, must be physical; ergo, must be energetic. Vector and tensor matrixes embrace all the elementary data governing the size dimensioning of the frequency and angle interactions in respect to an axis of reference and a cyclic norm.

527.03 Dimensions may be expressed only in magnitudes of time, energy, frequency concentrations, and angular modulations. What we call "length" is a "duration" of experience and is always measured in time.

527.04 The energetic juxtaposition of compression (radiation) and tension (gravity) provides dimension—the basis of "self" awareness or "other" awareness—of awareness of life itself.

527.05 Dimension may be universally and infinitely altered without altering the symmetrical relationship of the whole system.

527.06 All dimensions are simultaneously considerable.

527.07 All dimensions are definitively and intercoordinatably manifest in the isotropic vector matrix. (See Sec. 960.)

527.10 Three Unique Dimensional Abundances

527.11 Polar points, nonpolar points, areas, and lines have uniquely different cosmic abundances. In addition to every system's two polar points, there are three uniquely coexistent topological characteristics. For every one nonpolar point there are always two areas and three lines, and there are always an even number of each:

×1		×2		×3
Nonpolar points	–	Areas	–	Lines

527.20 Nonpolar Points

527.21 Polar points are two dimensional: plus and minus, opposites.

527.22 Nonpolar points, or localities, are four-dimensional—there is the inside-out (i.e., concave and convex) dimension and three symmetrically interacting, great-circle-ways-around—producing spherical octation, with eight tetrahedra having three internal (central) angles and three external spherical surface triangles' angles each.

527.23 The spherical octahedron's three inside-out, symmetrically unique diameters and the three unique external chords produce two unique sets of three nonparallel lines each, but with one set coordinating at 60 degrees and the other set coordinating at 90 degrees.

527.24 The nonpolar points are not fixable or structurally stabilized until occurring at the crossings of a three-way-great-circled-triangular-spherical-surface grid, generated symmetrically in respect to the polar axis of the system.

527.30 Areas

527.31 The octahedron's *planar* system is four-dimensionally referenced, being parallel to the four symmetrically interacting planes of the tetrahedron, vector equilibrium, and isotropic vector matrix. *Planar* and *nonpolar-vertex four-dimensionality* accommodates and imposes the four positive, four negative, and neutral (nineness) of the operational interwave behavior of number.

527.40 Lines

527.41 *Linear,* as manifest in the tetrahedron, the simplest structural system of Universe, is six-dimensional, providing for the six degrees of universal freedom and the operational six-wave phenomenon of number.

527.50 Fiveness: Five-Dimensionality

527.51 We know that the sphere points on the outer shell of the vector equilibrium and the icosahedron (between which states the pulsative propagation of electromagnetic waves oscillates) isolate the icosahedron and the vector equilibrium, but the number of points remains the same: $10F^2+2$.

527.52 $10F^2$—which *ten,* being divisible by the concave-convex twoness, brings in the prime number *five* to the hierarchy of low-order prime numbers characterizing synergetics. The polar twoness is the additive twoness. The twoness in the ten is the basic multiplicative twoness; it is the concave-convex unity-is-twoness inherent in the nuclear sphere and in the number of outer spheres in the vector-equilibrium-icosahedron's regenerative system, which always equates as $10F^2+2$.

527.53 The fundamental fiveness is introduced with the *initial* (frequency is $^1/_2$, i.e., in equilibrium, that is, poised between $^1/_2$ positive and $^1/_2$ negative) vector equilibrium interiorly defining the nuclear sphere where the vector equilibrium's volume = 2.5 (i.e., $^5/_2$) and the two-frequency's eightfold volumetric increase is 20.

527.54 Five-dimensionality is realized by the pulsation of the positive-negative *VE*—Icosa—*VE*—as 2.5—five. (Where VE stands for vector equilibrium and Icosa stands for icosahedron. Compare the interaction of the 31 great circles of the icosahedron and the 25 great circles of the vector equilibrium. See Sec. 1042.)

527.60 *Dimensionality and Constant Relative Abundance*

527.61 The rhombic dodecahedron *six* is entirely outside, but twelvefoldedly tangential to, the initial sphere. The cube, part inside and part outside the sphere, is *three*. The octahedron, mostly outside but partly inside the nuclear sphere, is *four*. Vector equilibrium is 2.5 and is entirely inside the sphere, with its 12 external vertexes congruent with the surface of the nuclear sphere at the same 12 points of tangency inside the sphere as the 12 points of the same initial sphere at which the rhombic dodecahedron is externally tangential; and the initial vector equilibrium's central vertexes are congruent with the volumetric center of the initial, i.e., nuclear sphere.

527.62 It was our synergetics' discovery and strategy of taking the two poles out of Euler's formula that permitted disclosure of the omnirational, constant relative abundance of *V*s, *F*s, and *E*s, and the disclosure of the initial additive twoness and multiplicative twoness, whereby the unique prime-number relationships of the prime hierarchy of omnisymmetric polyhedra occurred, showing tetra = 1; octa = 2; cube = 3; *VE* or Icosa = 5.

528.00 Size

528.01 Conceptuality operates independent of size. Whether referring to the size of an object in respect to other objects or the sizes of any one object's subdivisions, *size* emerges exclusively as a cyclic-frequency concept, uniquely differentiating out each special-case experience.

528.02 Size is a measure of three kinds of energetic experience: a measure of relative magnitude of separate linear, areal, and volumetric rates of change, and each one has a differently rated change velocity. Size and size alone can come to zero.

528.03 Size and time are synonymous. Frequency and size are the same phenomenon.

528.04 Size is the concept of one experience's relationship to another experience defined in terms of cyclic repetition of any one experimentally demonstrable, self-terminating, or single-cycle experience. A triangle or a tetrahedron or a sphere is a triangle or a tetrahedron or a sphere conceptually independent of size. An angle is an angle independent of the length of its edges. All of Plato's solids may have the same length edges because their differences are entirely angular. An angle is inherently a subdivision of a single cycle. Therefore, an angle is subsize.

528.05 Size begins with one specific cycle's completion. As the linear size of an object is doubled, surface is fourfolded and volume is eightfolded; ergo, areas increase at a velocity of the second power, and volumes increase at a velocity of the third power. Ergo, size-variation relationships are deceptive and not superficially predictable by any one experience.

528.06 Size and intensity are relative sensorial comparing functions of the special-case experience by brain, not by mind. Mind is concerned only with principles that hold true independent of size yet govern all relative size relationships.

529.00 Time

529.01 Time is experience. Time can be *expressed* only in relative magnitude ratios of relevant experiences. Time can be *defined* only in terms of the relative frequency of reoccurrence of relative angular changes of the observer's environment, the relative frequency-of-occurrence rate being referenced to any constantly recycling behavior of any chosen subsystem of Universe.* All experiential realizations are conceptually definable in degrees of angulation change and in relative frequency-of-occurrence rates in respect to the observer's optionally chosen axis of conceptuality and to his specifically identified time-recycling rate.

529.02 Distance is measured in time. Time increments are calculated in respect to a variety of cyclic regularities manifest in our environmental experiences. Experimentally demonstrable cyclic regularities, such as the frequencies of the reoccurrence of radiation emissions of various atomic isotopes, become the fundamental time-increment references of relative size measurement of elemental phenomena.

529.03 Newton said that time was a very specific phenomenon, assuming that there was a specific and finite time that permeated Universe and that everything observable in Universe was occurring at the same time. It was Einstein who discerned that time might be relative to the individual observer. A majority of academic people and the vast majority of nonscientists are still thinking in terms of the classical Newtonian scientific conceptioning of "instant Universe." While light's speed of approximately 700 million miles per

* E.g., a clock.

hour is very fast in relation to automobiles, it is very slow in relation to the "no time at all" of society's obsolete instant-Universe thinking. It was part of the classical scientist's concept of instant Universe that Universe is a system in which all parts affect one another *simultaneously* in varying degrees. Contemporary science as yet assumes that all local systems in physical Universe are instantly and simultaneously affecting one another in widely ranging degrees of influence. (And the degrees of influence are governed by relative proximity.) Whereas radiation, i.e., entropy, casts shadows and gravity, syntropy, does not; and whereas the tensional integrity of Universe and all its substructurings is continuous and omniembracing—while compression is islanded and discontinuous—it may also be that while light and radiation has a velocity, gravity is timeless and eternally instant. (See Secs. 231.01, 251.05, 541 and 1052.)

529.04 All the time phenomena of the physicists are expressed in linear data coordinates, but all cyclic actions are spirals because there are no straight lines and also because lines cannot "go through" or "return into" themselves. There can be no experientially demonstrable circles as continuous lines "returning into" themselves. Lines cannot return into themselves. Therefore matter is a cyclic self-interfering *knotting;* whereas radiation's waves are non-self-interfering *spirals*.

$$\frac{\text{matter}}{\text{radiation}} = \frac{\text{knots}}{\text{coils}}$$

Which reads: matter is to radiation as knots of rope are to coils of rope. Because there are no planes, a wave is a spiral. A spiral articulated in a direction perpendicular to our observation presents an illusory, wavilinear, planar profile.

529.05 Generalized principles are often called constants by the semantics of scientific specialization, whose viewpoint is myopic. Constancy is a time concept. Time is relative and cyclically terminal. Time is energetic, physical, and ever finitely evolving—which is the opposite of "constant."

529.06 Minimal consciousness evokes a nonsimultaneous sequence, ergo *time*. Time is not the fourth dimension and should not be so identified. Time is only a relative observation, a set of local sequences of experience afterimage formulation lags of the brain. Time is not a function of space. We can discuss time as if there were no time. It exists in weightless, metaphysical conceptuality. There is a metaphysical timeless time, just as there is a difference between physical tetrahedron and metaphysically conceptual but weightless, substanceless tetrahedron. Instantaneity would eliminate otherness, time, and self-and-other-awareness. Instantaneity and eternity are both timeless: they are the same.

529.07 The concept of being alive may be inherent only in the eternal principle of differentiability, and of a theoretical number system, and of complexes of different numbers. Seeming consciousness and life may well be inherent only in mind-conceivable theories of differentiations. To perceive of "truth" involves the concept of "nontruth," ergo, of differentiation. An intellectual integrity of Universe evokes its own theoretical evolvement of a Universe of ever-multiplying problems and pure-principles solutions and regeneration of multiplying problem-solving. (See Secs. 217.03 and 1005.50.)

529.08 The measuring act always involves time increments of our totally available time of life and may be conceived of only in respect to local events in nonsimultaneous Universe, there being no overall *largest* size to be referred to. Einstein was able to show that every individual's every-time employed yardstick of time (that is, the cyclic increment of imaginary reference) is always unique and different from every other's, a difference that amplifies greatly as we enter into astronomical observing by individual instruments, whose progressively designed reduction of tolerated error is always unique and only calculable relative to each experience.

529.09 It is a consequence of the phenomenon time and a consequence of the phenomena we call afterimage, or thinking, or reconsideration, which has inherent lags in the time rates of recallability of the various special-cases and types of experiences. So the very consequence of awareness is to impose the phenomenon time upon eternal, timeless Universe.

529.10 It is one of the strange facts of experience that when we try to think into the future, our thoughts jump backward. It may well be that nature has some fundamental metaphysical law by which opening up what we call the future also opens up the past in equal degree. The metaphysical law corresponds to the physical law of engineering that "every action has an equal and opposite reaction." (See Sec. 1031.16.)

529.11 The future is not linear. Time is wavilinear. Experience is expansive, omnidirectionally including and refining the future. It probably consists of omnidirectional wave propagations. We seem to be talking about a greater range of known cycling. It is both a subjective "now" and an objective "now"; a forward-looking now and a backward-looking now which combine synergetically as one complete "now." Because every action has both a reaction and a resultant, every now must have both a fading past and a dawning future.

529.20 No-Time-at-All

529.201 Intellect is top speed, which is instantaneous, being vastly faster than the speed of light and all radiation. Radiation's 700 million miles an hour is very slow in comparison to 700 million miles a minute and infinitely slower than 700 million miles in no-time-at-all—which is the rate at which intellect operates, being able to jump instantly to consideration of stars that are operating millions of years ago and thousands of light-years away.

529.21 The top speed of radiation is simply the minimum operational lag before making the cosmic leap to the eternal no-speed, where the instantaneity spontaneous to a child's conceptioning is normal and eternal. Not that it is ever lost. None of the differentiation of the generalized principles is lost. Many principles as yet undiscovered are nonetheless operative. Understanding is exquisitely total. Understanding includes a large increment of intuition to account for the as-yet-undiscovered but nonetheless operative generalized principles. (See Sec. 1056.03.)

529.22 Motion is not relative to *standing still.* Motion is relative to eternity, which is no-time-at-all. No-time-at-all is inherent in the generalized principles which, to be valid, must have no exceptions and be eternal, thus eternally true. The beginning of awareness, of intellect, is otherness. The whole complex of different and nonintercontradictory, all-interaccommodative, generalized principles is eternal. Complexity is eternal. The principle of mass interattraction of complex otherness is eternal and relates all the eternal complexity to our eternal system interfunctionings.

529.23 Newton's norm, as disclosed in his first phrase of his first law of motion, was "at rest." Newton's stars were "fixed." The planets and the moons of the planets, as well as comets, were in motion because hurled into motion by explosion from "fixed" stars.

529.24 Einstein's philosophy did not hold the speed of radiation unfettered in vacuo to be "very fast." It assumed this speed to be normal, and all other lesser speeds manifest in physical Universe to be occasioned by local interferences, shunting independent phenomena into local circuit repatternings.

529.30 Eternal Instantaneity

529.301 We have a new norm. The phenomenon lag is simply due to the limited mechanism of the brain; we have to wait for the afterimage to be realized.

529.31 The norm of Einstein is absolute speed instead of "at rest" . . . what we called instantaneous in our innocence of yesterday. We get to lesser and lesser lags, and we then approach eternal instantaneity—no lag at all. We have now learned, however, from our generalizations of the great complexity of the interactions of principles, that as we are disembarrassed of our local exclusively physical chemistry, our local information-sensing devices, what will be realized is an eternal and instantaneous awareness of all the potentials that ever existed. All the great metaphysical integrity of the individual, which is potential in the complex of interactions of the generalized principles, will always and only coexist eternally. I am saying that the arrival rate of intellect vs. the top-speed of radiation manifests the minimum lag short of no lag at all, i.e., "eternal."

529.32 Intuition derives from the approximate instantaneity of intellect, which is much faster than any physical phenomenon like the brain lags. Intuition is the absolute-velocity insistence of the intellect upon the laggingly reflexed brain to call its attention to significance of various special-case, brain-registered, experience relationships. Intuition is intellect coming instantly in at highest speed into dominance over lower-speed, lagging brain reflexing.

529.33 Eternity is simply highest speeds: not "at rest," because it gets there in no-time-at-all: Complete intellection + Otherness + No-time-at-all.

529.34 Differentiation of functions is inherently eternal and implicit to the plurality of generalized principles, which are everywhere nonredundant, redundancy being a temporal consequence of brain-lagged dullness of comprehension and ignorance.

530.00 Nonsimultaneity

530.01 Thought discovers that we divide Universe into an "outwardness and inwardness," so thinking is the first subdivision of Universe, because Universe, we discovered, was finite. Thinking is a nonsimultaneously recallable aggregate of inherently finite experiences and finite experience furniture—such as photons of light. One of the most important observations about our thought is the discovery that experiences are nonsimultaneous. Nonsimultaneity is a fundamental characteristic, and if experiences are nonsimultaneous, you cannot have simultaneous reconsideration.

530.02 All the words of all the vocabularies could be said to represent all the formalized attempts of men to communicate all their experiences. So we could set out to examine all the dictionaries of the world. We can pick up any one dictionary and discover that it is a nice finite package. We can open one page, but we cannot look at all the words at once. If we cannot look at all the words even on one page, we certainly cannot look at all the words of a whole dictionary at once. It does not make the dictionary infinite because we cannot look at all the words at once or think about all the words at once. The inability to think about everything at once does not mean that experience or consideration of experience is infinite. It is perplexing that one of the most persistent contemplations of human beings has been predicated on a static concept of

Universe, the kind of Universe that went out with classical Newtonian mechanics. We cannot think of Universe as a fixed, static picture, which we try to do when people ask where the outwardness of Universe ends. Humans try to get a finite unit package. We have a monological propensity for the *thing,* the *key,* the *building block* of Universe. What we discover here is that it is not possible to think about all Universe at once. It is nonsimultaneously conceptual. This in no way mitigates against its finiteness and thinkableness.

530.03 The parents tell the child he cannot have both the Sun and the Moon in the picture at the same time. The child says that you can. The child has the ability to coordinate nonsimultaneity. The parents have lost the ability to coordinate nonsimultaneity. One of our great limitations is our tendency to look only at the static picture, the one confrontation. We want one-picture answers; we want key pictures. But we are now discovering that they are not available.

530.04 We can think about all the experiences progressively and successively. And we can coordinate our thoughts about our experiences. Our very ability to think is such a propensity for the coordination of the reconsiderations of relationships. We do not have to be simultaneous to be interconnected. We can telephone across the international date line from Sunday back to Saturday.

530.05 We have had a tendency in our thinking to say that what is finite is statically conceptual as a one-unit glimpse, so we have been seemingly frustrated in trying to understand Universe, which is an omnidirectional experience, and so we feel there ought to be an outwardness of this sphere. That is a static concept. We are not dealing with such a sphere at all, because we have all these nonsimultaneous reports, and all we have is the interconnectedness of the nonsimultaneity. One of Einstein's most intellectual discoveries was this nonsimultaneity, which he apparently could have come upon by virtue of his experience in examining the thoughts of inventors and their patent claims regarding timekeeping devices, watches, and clocks.

530.06 The speed-of-light measurements plus Planck's quantum mechanics and Einstein's relativity showed that Universe is an aggregate of nonsimultaneous events. Their experiments showed that as each of the nonsimultaneous events lost their energy, they lost it to newly occurring events. Thus energy always became 100 percent accounted for.

530.07 All Universe is in continual transformation. The geology of our Spaceship Earth makes it very clear how very severe have been the great transformations of history. The movement of topsoils around the surface of the Earth is very new, geologically speaking. As Einstein interpreted the speed-of-light information and the observation of the Brownian movement of the constant motion in water, he then posited a Universe in which we knew that light takes eight minutes to get to us from Sun and two and a half years to

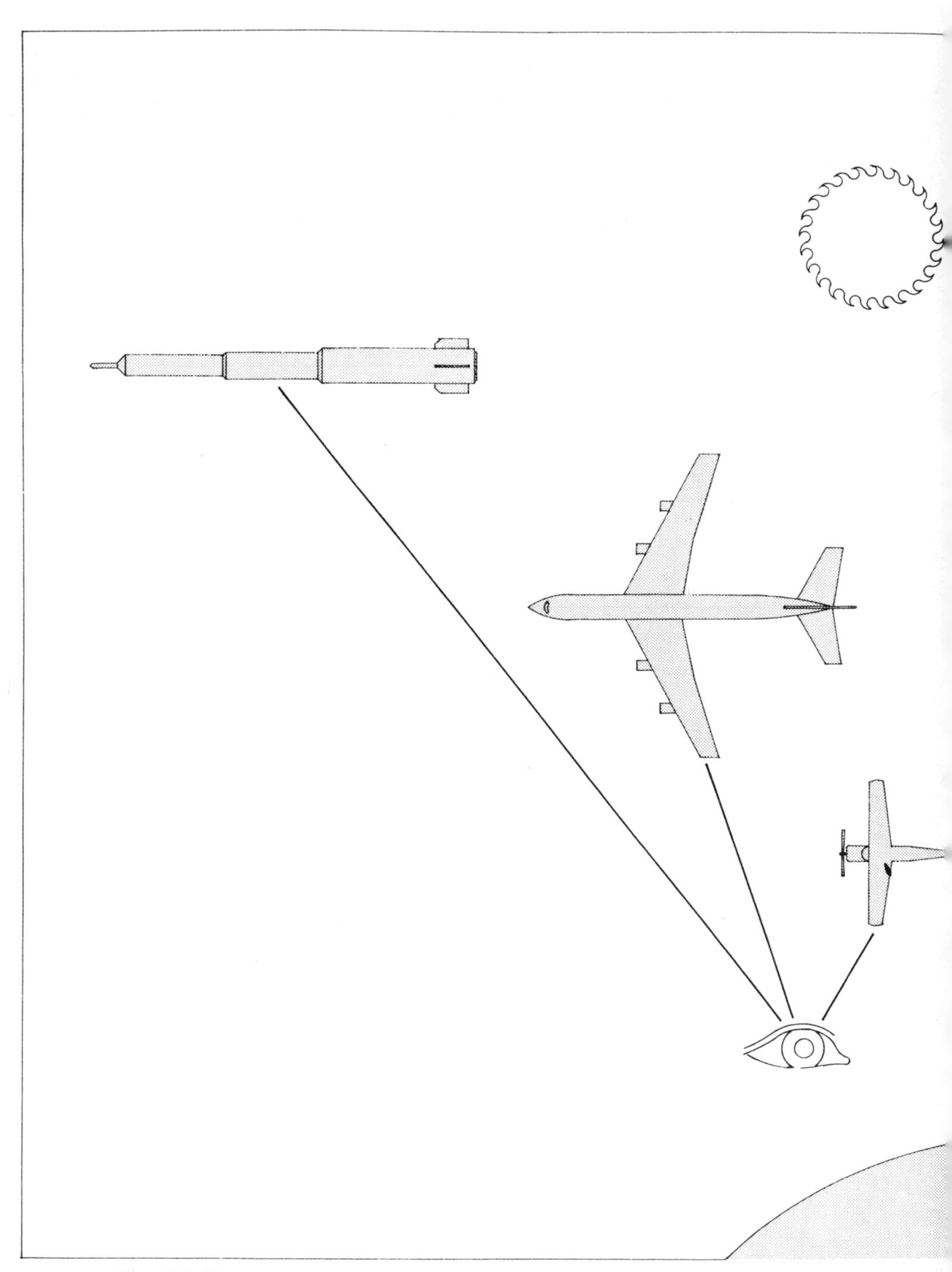

Fig. 530.07 *Simultaneous and Instant Are Nondemonstrable: Simultaneous* and *instant* cannot be experimentally demonstrated.

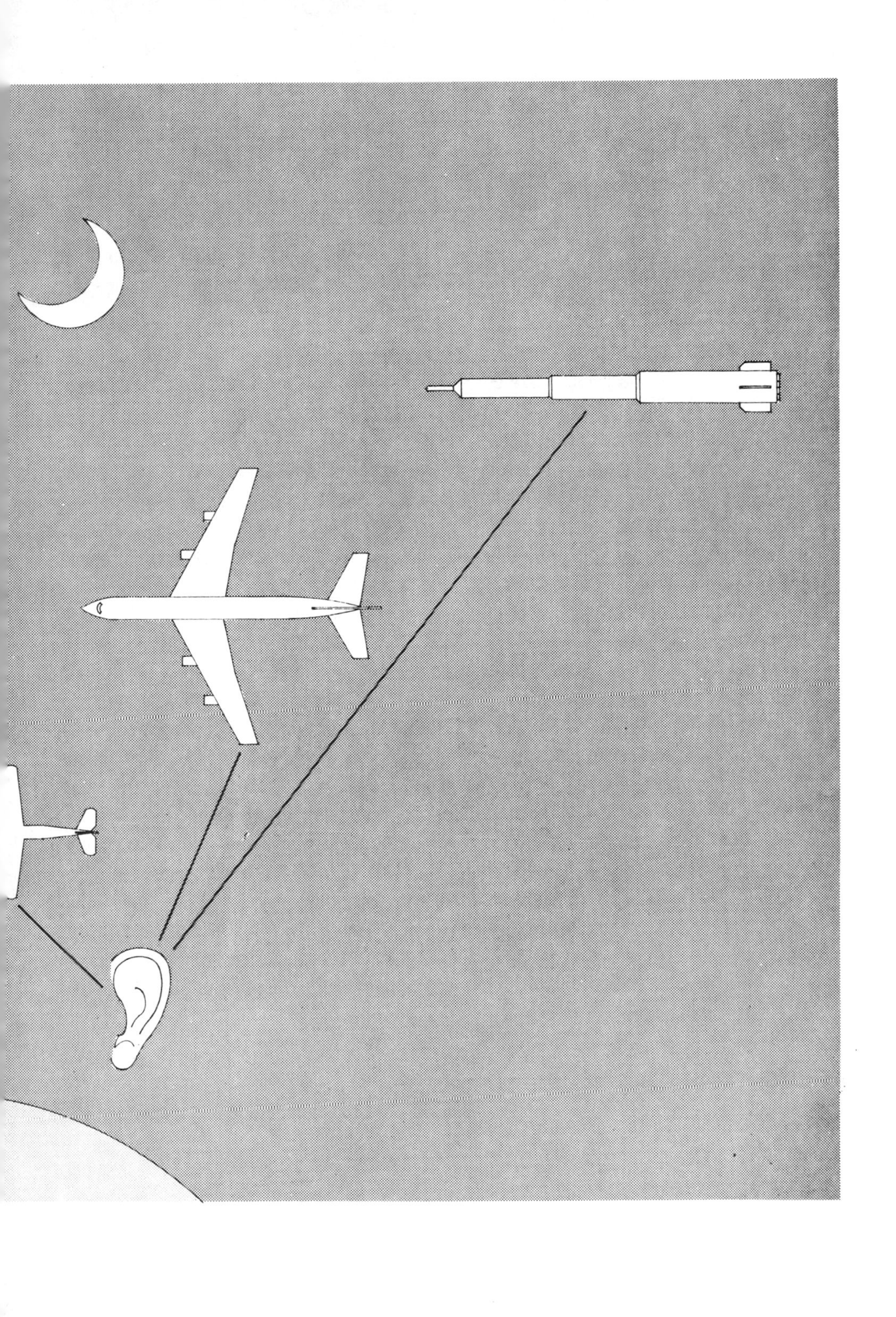

get to us from the nearest star; astronomical information shows that some of the stars we are looking at are live shows coming in from 100 years ago, others from 1,000 years ago, while the light from some of the stars we are looking at started on its way over a million years ago. With that kind of information, Einstein had to say physical Universe is quite obviously an aggregate of nonsimultaneous and only partially overlapping transformation events.

531.00 Life

531.01 Life is the eternal present in the temporal. Life is the *now* event with reaction *past* and resultant *future*. Each individual life is a special-case articulation of the infinite variety of "scenarios" to be realized within the multidegrees of freedom and vast range of frequencies of actions that are accommodated by the generalized laws governing Universe. With death, the individual probably loses nothing but gains the insight and knowledge of all others as well.

531.02 The average of all plus (+) and minus (−) weights of Universe is zero weight. The normal is weightless, eternal. What we call life is a complex of multidimensional oscillations and palpitations among various degrees of positive and negative asymmetries, whose multivariant lags in conceptioning bring about what *seems* to be temporal substance and time. The complex *woof* of a plurality of lag rates—of afterimages and recalls—produces pure, weightless, metaphysical images, produces the awareness we speak of as life.

531.03 Lags are intervals—nothing. Instantaneity would eliminate otherness, time, and self-and-other-awareness. Instantaneity and eternity are both timeless: they are the same. Eternity contains time; time does not contain eternity. The relationship is irreversible. The contained time of eternity provides eternal awareness.

532.00 Symmetry

532.01 Definition

532.02 It has been customarily said by the public journals, assumedly bespeaking public opinion, that "the scientists wrest order out of chaos." But the scientists who have made the great discoveries have been trying their best to tell the public that, as scientists, they have never found chaos to be anything other than the superficial confusion of innately a priori human ignorance at birth—an ignorance that is often burdened by the biases of others to remain gropingly unenlightened throughout its life. What the scientists have always found by physical experiment was an a priori orderliness of nature, or Universe always operating at an elegance level that made the discovering scientists' own working hypotheses seem crude by comparison. The discovered reality made the scientists' exploratory work seem relatively disorderly.

532.10 Oscillation of Symmetry and Asymmetry

532.11 We may say that nature proceeds from the obviously orderly and symmetrical to the nonobviously (but always) orderly transformation phases known as asymmetries, which, having gone through their maximum or peak positive-phase asymmetry, only *seem* (to the uninformed brain) to be disorderly; they always return transformatively thereafter through an orderly progression of decreasing asymmetry to a fleeting passing through the condition of *obvious* symmetry or equilibrium popularly recognized as "order," thereafter deviating asymmetrically to the negative phase of balancing limits of oscillation.

532.12 This transformative progression in dynamically and oscillatively produced orderliness is dealt with incisively by the calculus and is the fundamental pulsating principle governing omnidirectional electromagnetic-wave propagation.

532.13 There is no true "noise" or "static." There are only as yet undifferentiated and uncomprehended frequency and magnitude orders. Chaos and ignorance are both conditions of the brain's only-sense-harvested and stored information as yet unenlightenedly reviewed and comprehendingly processed by the order-seeking and -finding mind.

532.14 Asymmetry is the reason that Heisenberg's measurement is always indeterminate. Asymmetry is physical. Symmetry is metaphysical.

532.15 All most-economic-pattern systems, asymmetric as well as symmetric, are resolvable into symmetric components in synergetic accounting.

532.16 Our seeability is so inherently local that we rarely see anything but the asymmetries. Sociologists have trouble because they are o'erwhelmed by the high frequency of asymmetries (rather than the only synergetically discoverable principles).

532.20 Dynamic Symmetry

532.21 Within every equilateral triangle, we can inscribe a three-bladed propeller, its tips protruded into the three corners. The propeller blades are approximately pear-shaped, and each of the blades is the same shape as the others. The pear-shaped propeller blade is locally asymmetrical. We call this revolvable omnibalanced asymmetry *dynamic symmetry*.

532.22 We then have three pear-shaped blades at 120 symmetrical degrees from one another. They act as three perpendicular bisectors of an equilateral triangle, crossing each other at the triangle's center of area and dividing the total triangle into six right triangles, of which three are positive and three are negative. So there are six fundamentals of the triangle that make possible dynamic symmetry. (One part may look like a scalene, but it doesn't matter because it is always in balance.) Each corner is balanced by its positive and negative—like four streetcorners. This is called dynamic balance. Literally, all machinery is dynamically balanced in this manner.

532.23 Let me take one propeller blade by itself. I am going to split it longitudinally and get an S curve, one in which the rates are changing and no power of the curve is the same. So it is asymmetrical by itself: it is repeated six times: positive, negative, positive, negative . . . and the six blades come round in dynamic symmetry. The energy forces involved are in beautiful absolute balance. We have energetic balance.

532.30 Symmetrical and Omnisymmetrical

532.31 The difference between symmetrical and omnisymmetrical is that in symmetrical we have no local asymmetries as we do in any one of the propeller blades taken by themselves. *Symmetrical* means having no local asymmetries, whereas in contradistinction, *omnisymmetrical* and *dynamic symmetry* both permit local or momentary asymmetries, or both.

532.32 Universe is omnisymmetrical as well as dynamically symmetrical in its evolutionarily transformative regeneration of scenario Universe.

532.40 Three Basic Omnisymmetrical Systems

532.41 There are only three possible cases of fundamental omnisymmetrical, omnitriangulated, least-effort structural systems in nature: the tetrahedron, with three triangles at each vertex; the octahedron, with four triangles at each vertex; and the icosahedron, with five triangles at each vertex. (See illus. 610.20 and Secs. 724, 1010.20, 1011.30 and 1031.13.)

533.00 *Precession*

533.01 The effects of all components of Universe in motion upon any other component in motion is precession, and inasmuch as all the component patterns of Universe seem to be motion patterns, in whatever degree they affect one another, they are interaffecting one another precessionally, and they are bringing about angular resultants other than the 180 degreenesses. Precess means that two or more bodies move in an interrelationship pattern of other than 180 degrees.

533.02 Precession is the effect of any moving system upon any other moving system; the closer the proximity, the more powerful the effect. Mass attraction is inherent in precession. Mass attraction is to precession as a single note is to music. We do not pay much attention to precession because we think only of our own integral motions instead of those of Universe, though we are precessing Universe every time we take a step.

533.03 All the intergravitation effects are precessional angular modulations. Precessional effects are always angular and always something other than 180 degrees; they are very likely to be 90 degrees or 60 degrees.

533.04 Precession is regenerative, and that is why you have the wave. When the stone drops in the water, it impinges on the molecules and their atoms; everything is set in motion, and immediately there is a resultant at 90 degrees. The resultant is the wave; the 90 degreeness begets another 90 degreeness, this 90 degreeness begets another 90 degreeness, and so on until you have a series of 90 degreeness interrelationships, i.e., an omnilocally-orbiting system, as with all the electrons of all the atoms, and all the stars of the galaxies.

533.05 The elliptic orbiting of the Sun's planets as well as the Solar System's motion relative to the other star groups of the galactic nebulae are all and only accounted for by precession.

533.06 Precession is describable vectorially in terms of physically realized design expressed differentially as relative modification of angle, velocity, and mass in respect to an axis. (See Sec. 130, Precession and Entropy.)

534.00 Doppler Effect

534.01 Definition

534.011 There is the phenomenon known as the Doppler effect, of which humans took much note in the early days of the steam locomotive. The high tone of the locomotive's whistle as it approached changed to an increasingly low pitch as the locomotive went by. This is because the sound waves of perturbed air coming toward us at about 700 miles per hour from the approaching locomotive were crowded together, piled up, by the locomotive's own independent speed of about 60 miles per hour. Similarly, the waves were thinned out by the locomotive's speeding away.

534.02 The Doppler effect also may be operating in our historical-event-cognition system in such a manner that the relative frequency and wavelengths of approaching historical events are compacted and receding ones are thinned out. It could be that by traveling mentally backward in history as far as we have any information, humans could—like drawing a bowstring—impel our thoughts effectively into the future.

534.03 The Doppler effect, or wave-reception frequency modulation caused by the relative motions of the observer and the observed, are concentric wave systems that compound as fourth- and fifth-power accelerations. In the summary of synergetic corollaries (Sec. 240.44), fifth- and sixth-powering are identified as products of multiplication by frequency doublings and treblings, etc., in radiational or gravitational wave systems.

534.04 The Doppler effect is usually conceived of as an approximately "linear" experience. "You," the observer, stand beside a railway track (which is a "linear" model); a swift train approaches with whistle valve held open (at a constant-frequency pitch as heard "on board" by the engineer "blowing" the whistle). The whistle sound comes to you at the atmospheric sound-wave speed of approximately 700 linear miles per hour, but the train is speeding toward you at an additional 60 linear miles per hour. The train's motion reduces the interval between the successive wave emissions, which in effect decreases the wavelength, which gives it "higher" pitch as heard at your remote and "approached" hearing position. After the train goes by, the train runs away from each successive wave emission, thus increasing the interval between wave "crests" and therefore lengthening the wave-reception intervals, which apparently "lowers" the pitch as *you* hear it, but not as others elsewhere may hear it. This is pure observational hearing relativity. But the real picture of the Doppler effect is not linear; it is omnidirectional.

534.05 The Doppler effect may also be explained in omnidirectional, experience-patterning conceptionality, which is more informative than the familiar linear conceptioning of the railroad train and "you" at the crossing. Suppose "you" were flying in an air transport that exploded; because of the sudden change in pressure differential between your innards and your out'ards at high altitude, you personally have just been "exploded" into many separate parts, which are receding from one another at high velocity. A series of secondary explosions follows elsewhere from exploding "you" and at various locales in the center of the galaxy of exploding debris, as one item after another of the late airplane's explosive cargo is reached by progressive local-conflagration-heat concentrations. The sound waves of the successive explosions speed after your receding parts, amongst which are your two ear diaphragms, as yet "stringily" interconnected with your exploding brain cells, which "hear" the explosion's sound waves first at low pitch. But as your parts explode from one another at a decelerating rate because of air friction, etc., the waves of remote-explosion sounds "shorten" and pitches go "up." Now consider many separate, nonsimultaneous, secondary explosions of your various exploding parts, all of varying intensities of energetic content and in varying degrees of remoteness, and realize that the decelerations and accelerations of Doppler effects will render some of the explosive reverberations infra and some ultra to your tuning-range limits of hearing, so that the sum total of *heard* events provides very different total conceptioning as heard from various points in the whole galaxy of exploding events, whose separate components would tend to new grouping concentrations.

534.06 Because the humanly "heard" events are geared directly to the atmospheric waves with an average speed of 700 miles per hour, and the humanly "seen" events operate a million times faster and are geared to electromagnetic fields operating independent of and beyond the atmospheric biosphere of Earth, the visual and hearable information is macrocosmically so far out of synchronization that the stroboscopic effect, which can make the wheels of automobiles sometimes appear to be going backward in amateur moving pictures, can cause society to misinterpret the direction and speed of vital events—some may be seen as going in the opposite direction from the realities of universal evolution.

535.00 Halo Concept

535.01 The phenomenon "infinity" of the calculus is inherently finite (see Sec. 224.11). Universe is nonsimultaneous but finite, because all experiences *begin* and *end,* and being terminal, are finite; ergo, Universe as the sum of finites is finite.

535.02 Nonsimultaneous Universe is finite but conceptually undefinable; local systems are definable. We discover that Universe is finite and a local system is *definite;* every definite local system has inherent, always and only co-occurring twoness of polar axis spinnability and twoness of concave-convex complementary disparity of energy interaction behavior,* plus two invisible tetrahedra (or two unities), altogether adding together as equal finitely fourfold symmetry Universe. The difference between Universe and any local system is always two invisible tetrahedra. Every local system may be subdivided into whole tetrahedra.

535.03 Finite minus de-finite means four tetrahedra minus two tetrahedra. Finite Universe equals eight cyclic unities. Every tetrahedron equals two, having inside-outingness oscillatory transformability unavailable to any structural system other than the tetrahedron.

535.04 Halo conceptioning discloses the minute yet finitely discrete inaccuracy of the fundamental assumption upon which calculus was built; to wit, that for an infinitesimal moment a line is congruent with the circle to which it is tangent and that a plane is congruent with the sphere to which it is tangent. Calculus had assumed 360 degrees around *every* point on a sphere. The sum of a sphere's angles was said to be infinite. The halo concept and its angularly generated topology proves that there are always 720 degrees, or two times unity of 360 degrees, *less* than the calculus' assumption of 360 degrees times every point in every "spherical" system. This 720 degrees equals the sum of the angles of a tetrahedron. We can state that the number of vertexes of any system (including a "sphere," which must, geodesically, in universal-energy conservation, be a polyhedron of *n* vertexes) minus two times 360 degrees equals the sum of the angles around all the vertexes of the system. Two times 360 degrees, which was the amount subtracted, equals 720 degrees, which is the angular description of the tetrahedron. We have to take angular "tucks" in the nonconceptual finity (the calculus infinity). The "tucks" add up to 720

* Concave concentrates radiation; convex diffuses radiation.

degrees, i.e., one tetrahedron. The difference between conceptual de-finity and nonconceptual finity is one nonconceptual, finite tetrahedron.

535.05 In the general theory of variables, it has been recognized that the set of all the variables may be divided into two classes: (1) the class of all the inclusive variables within a given system, the *interior relevants,* and (2) the class of all those operative exclusive of the system, the *exterior relevants.* It has been further recognized that the variables outside the system may affect the system from outside. In varying degrees, specific levels of subclasses of these "background" or outside variables are identified as *parameters.* But the "background" concept is fallaciously inadequate; dealing with insideness and outsideness for "background" is limited to the two-dimensional or flat-projection concept, which inherently lacks insideness—ergo, cannot also have outsideness, which always and only coexists with insideness. Ergo, all two-dimensional copings with systems are inherently inadequate and prophetically vitiated.

535.06 Our omnioriented halo concept converts the *parameter* consideration to symmetrically conceptual four-dimensionality and discloses a set of parameters *inside* as well as *outside* the zone of lucidly considered system stars. And the parameters are, at minimum, fourfold:

(1) the concave twilight zone of inward relevancy;
(2) the convex twilight zone of outward relevancy;
(3) the *stark,* nonconceptual irrelevancy inward; and
(4) the stark, nonconceptual irrelevancy outward.

Parameter 1 is a visible tetrahedron. Parameter 2 is a visible tetrahedron. Parameter 3 is an invisible tetrahedron. Parameter 4 is an invisible tetrahedron.

535.07 The *considered* relevancy within the zone of lucidity consists of one tetrahedron or more. For each "considered tetrahedron," there are three complementary always and only co-occurring parametric tetrahedra. We discover that our omnihalo epistemological accounting consists entirely of rational tetrahedral quantation.

535.08 By the omnidirection, star-studded halo reasoning, the development of a conceptual tetrahedron automatically changes a negative yet invisible tetrahedron into the nonsimultaneous, *nonconceptual, finite* Universe, comprehensive to the local de-finite conceptual system.

535.09 The halo concept is that of an omnidirectional, complex, high-frequency, Doppler-effected, hypothetical zone experience in an omnidirectional, universal maelstrom of nonsimultaneous near and far explosions and their interaccelerating and refractive wave-frequency patternings and complex, precessionally-induced, local orbitings. The omni-interactions impinge on your nervous system in all manner of frequencies, some so "high" as to appear as "solid" things, some so slow as seeming to be "absolute voids."

535.10 Spherical Structures

535.11 Because spherical sensations are produced by polyhedral arrays of interferences identified as points approximately equidistant from a point at the approximate center, and because the mass-attractive or -repulsive relationships of all points with all others are most economically shown by chords and not arcs, the spherical array of points is all interconnected triangularly by the family of generalized principles being operative as Universe, which produces very-high-frequency, omnitriangulated geodesic structures, which are an aggregate of chords triangularly interconnecting all the nearestly-surrounding points whose vertexly-converging angles always add up to less than 360°.

536.00 Interference Domains of Structural Systems

536.01 As distinct from other mathematics, synergetics provides *domains of interferences* and *domains of crossings*. In the isotropic vector matrix, the domains of vertexes are spheres, and the domains of spheres are rhombic dodecahedra. These are all the symmetries around points. Where every vertex is the domain of a sphere we have closest-rhombic-dodecahedral-packing.

536.02 The coordinate system employed by nature uses 60 degrees instead of 90 degrees, and no lines go through points. There are 60-degree convergences even though the lines do not go through a point. The lines get into critical proximities, then twist-pass one another and there are domains of the convergences.

536.03 In a polyhedral system, critical-proximity-interference domains are defined by interconnecting the adjacent centers of area of all the separate superficial faces, i.e., "external areas" or "openings," surrounding the vertex, or "crossing." The surface domain of a surface vertex is a complex of its surrounding triangles: a hexagon, pentagon, or other triangulated polygon. (See Sec. 1006.20.)

536.10 Domains of Volumes

536.11 There are domains of the tetrahedron interfaced (triple-bonded) with domains of the octahedron. The domains of both are rationally sub-

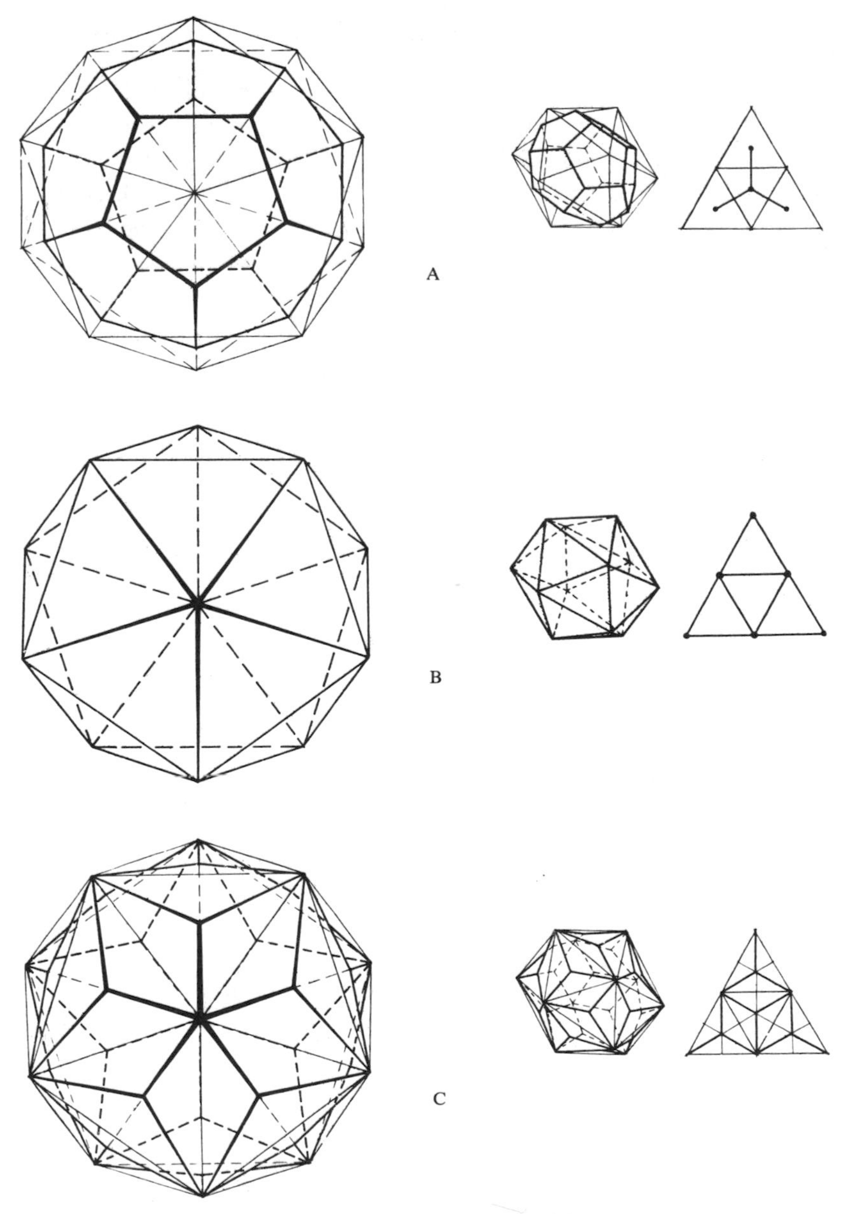

Fig. 536.03 *Domains of Vertexes, Faces, and Edges of Systems:*

A. The domain of the vertex of a system: the domain of each vertex of the icosahedron is a pentagon whose edges connect the centers of gravity of five icosahedron face triangles. The resulting figure is the pentagonal dodecahedron.

B. The domain of the face of a system: The domain of each face of the icosahedron is the triangular face itself.

C. The domain of the edge of a system: the domain of each edge of the icosahedron is a diamond formed by connecting the vertexes of two adjacent icosahedron face triangles with their centers of gravity.

divided into either *A* or *B* Modules. There is the center of volume (or gravity) of the tetrahedron and the center of volume (or gravity) of the octahedron, and the volumetric relationship around those centers of gravity is subdivisible rationally by *A* and *B* Quanta Modules* in neat integer whole numbers. I can then speak of these domains quantitatively without consideration of now obsolete (superficial) face surfaces, i.e., polyhedra. Even though the cork is not in the bottle, I can speak quantitatively about the contents of the bottle. This is because it is a domain even though the edge-surrounded opening is uncorked. So we have no trouble topologically considering tensegrity mensuration. It is all open work, but its topological domains are clearly defined in terms of the centers of the systems involved having unique, centrally angled *insideness* and surface-angle-defined *outsideness*.

536.20 Domain of an Area

536.21 Areas do not have omnidirectional domains. The domain of an area is the area itself: it is the superficial one that man has looked at all these centuries. The domain of a face is a triangle in the simplest possible statement. Thus the domain of each face of the icosahedron is the triangular face itself.

536.30 Domain of a Line

536.31 The domains of the vector edges are defined by interconnecting the two centers of area of the two surface areas divided by the line with the ends of the line. The edge dominates an area on either side of it up to the centers of area of the areas it divides. Therefore, they become diamonds, or, omnidirectionally, octahedra. The domains of lines are two tetrahedra, not one octahedron.

536.32 The domains of lines must be two triple-bonded (face-bonded) tetrahedra or one octahedron. There could be two tetrahedra base-to-base, but they would no longer be omnisymmetrical. You can get two large spheres like Earth and Moon tangent to one another and they would seem superficially to yield to their mass attractiveness dimpling inward of themselves locally to have two cones base to base. But since spheres are really geodesics, and the simplest sphere is a tetrahedron, we would have two triangles base to base—ergo, two tetrahedra face-bonded and defined by their respective central angles around their two gravity centers.

536.33 The domain of each edge of the icosahedron is a diamond formed

* See Sec. 920.

by connecting the vertexes of two adjacent icosahedron-face triangles with their centers of area.

536.40 Domain of a Point

536.41 Looking at a vector equilibrium as unity, it is all the domain of a point with a volume of 480.

536.42 The domains of points as vertexes of systems are tetrahedra, octahedra, or triangulated cubes. Or they could be the *A* and *B* Modules formed around the respective polyhedra.

536.43 The most complete description of the domain of a point is not a vector equilibrium but a rhombic dodecahedron, because it would have to be allspace filling and because it has the most omnidirectional symmetry. The nearest thing you could get to a sphere in relation to a point, and which would fill all space, is the rhombic dodecahedron.

536.44 A bubble is only a spherical bubble by itself. The minute you get two bubbles together, they develop a plane between them.

536.50 Domains of Actions

536.51 There are critical proximities tensionally and critical proximities compressionally—that is, there are attractive fields and repelling fields, as we learn from gravity and electromagnetics. There are domains or fields of actions. In gases under pressure, the individual molecules have unique atomic component behaviors that, when compressed, do not allow enough room for the accelerated speeds of their behavior; the crowded and accelerating force impinges upon the containing membrane to stretch that membrane into maximum volume commensurate with the restraints of its patterned dimensions.

537.00 Twelve Universal Degrees of Freedom

537.01 Nothing stands in a vacuum of Universe. Nothing can change locally without changing everything else. We have to look for conditions where there is permitted transformability and where there is some really great unanimity of degrees of freedom. We see that certain kinds of patterns accrue from certain numbers of restraints. You could see how planar things could happen as a consequence of two restraints and how linear things could happen as a consequence of three restraints. (See Sec. 401, Twelve Vectors of Restraint Define Minimum System.) We see, then, that we are in a Universe where there is a certain limited number of permitted freedoms. Synergetics discovers that whatever is rigidly related to anything else discloses 12 restraints. There are a minimum of 12 restraints in developing anything we might call a rigidly related set of events.

537.02 We start with Universe as a closed system of complementary patterns—i.e., regenerative, i.e., adequate to itself—that has at any one moment for any one of its subpatterns 12 degrees of freedom. There is an enormous complexity of choice. We start playing the game, the most complicated game of chess that has ever been played. We start to play the game Universe, which requires absolute integrity. You start with 12 alternate directions and multibillions of frequency options for your first move and from that move you have again the same multioptions at each of your successive moves. The number of moves that can be made is unlimited, but the moves must always be made in absolute respect for all the other moves and developments of evoluting Universe.

537.03 The game of Universe is like chess with 92 unique men, each of which has four different frequencies available, and it works on 12 degrees of freedom instead of a planar checkerboard. The vector equilibrium becomes the omnidirectional checker frame and you can change the frequencies to suit conditions. But you must observe and obey the complexity of mass attraction and the critical proximity between precessing and falling in. And there are also electromagnetic attractions and repulsions built into the game.

537.04 In order to be able to think both finitely and comprehensively, in terms of total systems, we have to start off with Universe itself. We must include all the universal degrees of freedom. Though containing the frequently irrational and uneconomic *XYZ* dimensional relationships, Universe does not employ the three-dimensional frame of reference in its ever-most-economical, omnirational, coordinate-system transactions. Nature does not use rectilinear coordination ir its continual intertransforming. Nature coordinates in 12 alternatively equieconomical degrees of freedom—six positive and six negative. For this reason, 12 is the minimum number of spokes you must have in a wire wheel in order to make a comprehensive structural integrity of that tool. You must have six positive and six negative spokes to offset all polar or equatorial diaphragming and torque. (See illustration 640.40.)

537.05 Once a closed system is recognized as exclusively valid, the list of variables and the degrees of freedom are closed and limited to six positive and six negative alternatives of action for each local transformation event in Universe.

537.10 Six Vectors for Every Point

537.11 Each of the six positive and six negative energy lines impinging on every nonpolarized point ("focal event") in Universe has a unique and symmetrical continuation beyond that point. The six positive and six negative vectors are symmetrically arrayed around the point. Consequently, all points in Universe are inherently centers of a local and unique isotropic-vector-matrix domain containing 12 vertexes as the corresponding centers of 12 closest-packed spheres around a nuclear sphere. (See Synergetics Corollaries at Secs. 240.12, 240.15, and 240.19.)

537.12 Experiments show that there are six positive and six negative degrees of fundamental transformation freedoms, which provide 12 alternate ways in which nature can behave most economically upon each and every energy-event occurrence. Ergo, there is not just one "other"; there are always at least 12 "others." (See Secs. 502.25 and 511.03.)

537.13 We find that in the 12 degrees of freedom, the freedoms are all equal and they are all realizable with equal "minimum effort."

537.14 Basic Event

537.15 A basic event consists of three vectorial lines: the action, the reaction, and the resultant. This is the fundamental tripartite component of Universe. One positive and one negative event together make one tetrahedron, or one quantum. The number of vectors (or force lines) cohering each and every subsystem of Universe is always a number subdivisible by six, i.e., consisting of one positive and one negative event on each of three vectors, which adds up to six. This holds true topologically in all abstract patterning in Universe as well as in fundamental physics. The six vectors represent the fundamental six, and only six, degrees of freedom in Universe. Each of these six, however, has a positive and a negative direction, and we can therefore speak of a total of 12 degrees of freedom. These 12 degrees of freedom can be conceptually visualized as the radial lines connecting the centers of gravity of the 12 spheres, closest packed around one sphere, to the center of gravity of that central sphere. The 12 degrees of freedom are also identified by the push-pull alternative directions of the tetrahedron's six edges.

537.20 High-Tide Aspects

537.21 Spheres in closest packing are high-tide aspects of vertexes. It is easy to be misled into thinking that there are no lines involved when you see two spheres in tangency, because the lines are hidden inside the spheres and between the points of tangency. And if you do realize that there is a force line between the two spheres' centers, you could assume that there is only one line between the two. This is where you see that unity is two, because the line breaks itself into radii of the two spheres.

537.22 In synergetics, a "line" is the axis of intertangency of unity as plural and minimum two. The line becomes the axis between two tangent balls which, without disturbing that single-axis aspect, can articulate both axial and circumferential degrees of freedom.

537.30 General Systems Applications

537.31 The 12 universal degrees of freedom govern the external and internal motions of all independent systems in Universe. In order to take synergetic strategy advantage and thereby to think comprehensively and anticipatorily, in terms of total systems, we have to start off with Universe itself as a closed finite system that misses none of the factors. We must also include all the universal degrees of freedom, and the approximately unlimited range of frequencies in the use thereof, which cover all variable interrelationships of Universe. They become the controlling factors governing general systems and, thereby govern such supercomplex systems design as that of a nation's navy or a fundamental program for comprehensively considerate and efficiently effective use of all world resources. The general systems approach starts with the differentiation of Universe, including both metaphysical and physical, and permits progressive subdivisions in cybernetical bits to bring any local pattern of any problem into its identification within the total scheme of generalized system events. Problem solving starts with Universe and thereafter subdivides by progressively discarding irrelevancies thereby to identify the "critical path" priorities and order of overlapping developments that will most economically and efficiently and expeditiously realize the problem's solution by special local problem identification and location within the totality of the problem-solving scenario.

537.32 Because of our overspecialization and our narrow electromagnetic spectrum range of our vision, we have very limited integrated comprehension of the significance of total information. For this reason, we see and comprehend very few motions among the vast inventory of unique motions and transformation developments of Universe. Universe is a nonsimultaneous complex of unique motions and transformations. Of course, we do not "see" and our eyes cannot "stop" the 186,000-miles-per-second kind of motion. We do not see the atomic motion. We do not even see the stars in motion, though they move at speeds of over a million miles per day. We do not see the tree's or child's moment-to-moment growth. We do not even see the hands of a clock in motion. We remember where the hands of a clock were when we last looked and thus we accredit that motion has occurred. In fact, experiment shows that we see and comprehend very little of the totality of motions.

537.33 Therefore, society tends to think statically and is always being surprised, often uncomfortably, sometimes fatally by the omni-inexorable motion of Universe. Lacking dynamic apprehension, it is difficult for humanity to get out of its static fixations and to see great trends evolving. Just now, man is coming into technical discovery of general systems theory. The experimental probing of the potentials of the computers awakened man to a realization of the vast complexes of variables that can be mastered by general systems theory. So far, man has dealt but meagerly and noncomprehensively with his powerful planning capability. So far, he has employed only limited systems theory in special open-edged systems—"tic-tac-toe" rectilinear grid systems and planar matrixes. The arbitrary open parameters of infinite systems can never be guaranteed to be adequate statements of all possible variables. Infinite systems engender an infinite number of variable factors. Unless one starts with Universe, one always inadvertently starts with open infinite systems. Only by starting with finite Universe and progressively dismissing finite irrelevancies can one initiate finite, locally limited, general systems theory to assured satisfaction in problem solving.

537.34 The Dymaxion airocean world map is only one of many devices that could provide man with a total information-integrating medium. We are going to have to find effective ways for all of humanity to see total Earth. Nothing could be more prominent in all the trending of all humanity today than the fact that we are soon to become world man; yet we are greatly frustrated by all our local, static organizations of an obsolete yesterday.

538.00 *Probability*

538.01 Nature's probability is not linear or planar, but the mathematical models with which it is treated today are almost exclusively linear. Real Universe probability accommodates the omnidirectional, interaccommodative transformating transactions of universal events, which humanity identifies superficially as environment. Probability articulates locally in Universe in response to the organically integral, generalized, omnidirectional *in, out, inside out, outside in,* and *around* events of the self-system as well as with the self-system's extraorganic travel and externally imposed processing around and amongst the inwardly and outwardly contiguous forces of the considered system as imposed by both its synchronously and contiguously critically near macrocosmic and microcosmic neighbors.

538.02 Real Universe's probability laws of spherically propagative whole systems' developments are intimately and finitely conditioned by the three-way great-circle spherical grids inherently embracing and defining the nonredundant structuring of all systems as formingly generated by critical proximity interferences of the system's components' behaviors and their dynamical self-triangulations into unique system-structuring symmetries whose configurations are characterized by the relative abundance patterning laws of topological crossing points, areas, and lines of any considered system as generally disclosed by the closed-system hierarchy of synergetics.

538.03 Synergetics, by relating energy and topology to the tetrahedron, and to systems, as defined by its synergetic hierarchy, replaces randomness with a rational hierarchy of omni-intertransformative phase identifications and quantized rates of relative intertransformations.

538.10 Probability Model of Three Cars on a Highway

538.11 I am tying up the social experience, often observed, in which three independently and consistently velocitied automobiles (and only three) come into close proximity on the highway—often with no other cars in sight. Mathematically speaking, three points—and only three—define both a plane and a triangle. The cars make a triangle; and because it is mathematically discovered that the total number of points, or areas, or lines of a system are always even numbers; and that this divisibility by two accommodates the polar-and-hemispherical positive-negativeness of all systems; and because the defining of one small triangle on the surface of a system always inadvertently defines a large triangle representing the remainder of the whole system's surface; and this large triangle's corners will always be more than 180 degrees each; ergo, the triangle is an "inside-out," i.e., negative, triangle; and to convert it to positive condition requires halving or otherwise fractionating each of its three corners by great circle lines running together somewhere within the great negative triangle; thus there develops a minimum of four positive triangles embracing the Earth induced by such three-car convergences.

538.12 The triangle made by the three cars is a complementarity of the three other spherical triangles on the Earth's surface. The triangle formed by the two cars going one way, and one the other way, gets smaller and smaller and then reverses itself, getting ever larger. There is always a closer proximity between two of the three. This is all governed by topological "pattern integrity."

538.13 Probability is exclusively abstract mathematics: theoretically calculated points on curves. The statisticians think almost exclusively in lines or planes; they are what I call planilinear. Willard Gibbs in evolving his phase rule was engaged in probability relating to chemistry when he inadvertently and intuitively conceived of his phase rule for explaining the number of energetic freedoms necessary to introduce into a system, complexedly constituted of crystals, liquids, and gases, in order to unlock them into a common state of liquidity. His discovered phase rule and topology are the same: they are both synergetic. Despite the synergetic work of such pioneers as Euler and Gibbs, all the different chemistries and topologies still seem to be random. But synergetics, by relating energy and topology to the tetrahedron, and to systems as defined, and by its synergetic hierarchy, replaces randomness with a rational cosmic, shape-and-structural-system hierarchy. This hierarchy discloses a constant relative abundance of the constituents; i.e., for every nonpolar point there are always two faces and three edges. But systems occur only as defined by four points. Prime structural systems are inherently tetrahedral, as is also the quantum.

538.14 A social experience of three cars: they make a triangle changing from scalene to equilateral to scalene. The triangles are where the cars don't hit. (These are simply the windows.) But you can't draw less than four triangles. The complementarity of the three triangles makes the spherical tetrahedron—which makes the three-way grid. The little spherical triangle window is visible to human observers in greatest magnitude of human observability and awareness of such three-car triangles at 15 miles distance, which is 15 minutes of spherical arc of our Earth. Such dynamically defined Earth triangulation is not a static grid, because the lines do not go through the same point at the same time; lines—which are always action trajectories—never do. All we have is patterning integrity of critical proximities. There is always a nonviolated intervening boundary condition. This is all that nature ever has.

538.15 Nature modulates probability and the degrees of freedom, i.e., frequency and angle, leading to the tensegrity sphere; which leads to the pneumatic bag; all of which are the same kind of reality as the three automobiles. All the cosmic triangling of all the variety of angles always averages out to 60 degrees. That is the probability of all closed systems, of which the Universe is the amorphous largest case. Probability is not linear or planar, but it is always following the laws of sphericity or whole systems. Probability is always dependent upon critical proximity, omnidirectional, and only dynamically defined three-way gridding pattern integrity, and with the concomitant topologically constant relative abundance of points, areas, and lines, all governed in an orderly way by low-order, prime-number, behavioral uniqueness as disclosed by synergetics.

539.00 Quantum Wave Phenomena

539.01 We say that Universe is design and that design is governed exclusively by frequency and angular modulations, wherefore the "angle" and "frequency" must be discretely equatable with quantum mechanics which deals always synergetically with the totality of Universe's finite energy.

539.02 The relative acutenesses and the relative obtusenesses of the angle and frequency modulating must relate discretely to the relative mass experienciabilities of Universe.

539.03 Quantum wave phenomena's *omni-wholeness* of required a priori accountability and persistent consideration is always systematically conceivable as a sphere and may be geodesically fractionated into great-circle-plane subsets for circular plane geometry considerability. Quantum waves always complete their cycles (circles). The circle can be divided into any number of arc increments as with the teeth of a circular gear—many little teeth or a few big teeth. In quantum wave phenomena we may have a few big, or many small, differentiated events, but they always add up to the same whole.

539.04 The rate of angular change in a big wave is very much slower than the rate of angular change in a small wave, even though they look superficially to be the same forms—as do two circles of different size appear to be the same form. The difference in the wave that is big and the wave that is small, is always in relation to the dimensioning of the observer's own integral system, and determines the discrete difference (i.e., the "relativity") of the wave angle.

539.05 What is "the most economical relationship" or "leap" between the last occurred event and the next occurring event? It is the chord (identifiable only by central angle) and the rate of the central-angle reorientation-aiming most economically toward that event, which is the angular (momentum) energy change involved in the angular and frequency modulation of all design of all pattern integrity of Universe.

539.06 Let us say that you are progressively leaping—"pacing"—around the perimeter of two circles: one small, six feet in diameter; the other large, 600 feet in diameter, leaping clockwise as seen from above. You are six feet long—"tall." On the small circle you will be turning, or angularly reorienting your direction to the right, much more obtusely in relation to your last previous direction of leap-accomplished facing and pacing.

539.07 Your rate of angular change in direction will be apprehendable in relation to the angles and overall direction of "you," as the observer, and as the criteria of the "rate of angle modulation."

539.08 Newton's first law: A body persists in a straight line except as affected by other bodies. But the 1974 era of physics' discoveries of "prime otherness" must add to Newton that: All bodies are always being affected by other bodies, and the intereffects are always precessional. The intereffects are angular-momentum aberrating. The angular momentum alterations are all determined by the angle and frequency modulating.

539.09 We may think of our leaps as describing the circular chords between the successive circular circumference points leaped-to. With our relative leap-size taken as that of our height—six feet—the chord of "our self"—either leaping around or lying down—in a small circle will represent the chord of the arc of a much larger central angle than it would constitute in respect to a large circle. The relative *angular difference* is that of the respective central angle changes as subtended by each use of self (the observer) as the chord of a circle of given size. This ground-contact-discontinuing chordal "leap" of self relates to quantum mechanics employment in experimental physics wherein no absolute continuum is manifest.

539.10 If a six-foot man lies down in a six-foot circle he becomes the diameter and the central angle is 180 degrees. If a six-foot man lies down in a 600-foot-diameter circle, he will be a chord subtending a central angle of approximately one degree—a chord whose arc altitude is so negligible that the observing self's height of six-feet will be a chord so relatively short as to lie approximately congruent with the one-degree arc of the circle. When the relative circle size in respect to the observer is of macro-differential magnitudes, such as that of the circumference of the galactic system in respect to each planet observer, then the central-angle magnitude of the subtended macrocosmic arc becomes undetectable, and the astronomer and navigator assume parallelism—parallax—to have set in, which produces a constant factor of error which must be incorporated in mathematical formulation of system descriptions. In quantum accounting and analysis of energy events and transformative transactions, this parallelism separates one quantum tetrahedron from its three surrounding tetrahedra.

540.00 *Frame of Reference*

540.01 The system generates itself whenever there is an event. The system actually regenerates itself: it is an eternal rebirth system.

540.02 The octet truss is not a priori. The octet truss is simply the most economical way of behaving relative to unity and to self. The octet truss is the evolutionary patterning, intervectoring, and intertrajectory-ing of the ever-recurrent 12 alternative options of action, all 12 of which are equally the most economical ways of self-and-otherness interbehaving—all of which interbehavings we speak of as Universe.

540.03 Starting with whole Universe as consisting always of *observer* plus the *observed,* we can subdivide the unity of Universe. In synergetics—as in quantum mechanics—we have multiplication only by division.

540.04 I do not like the word *frame*. What we are talking about is the multi-optioned omni-orderly scheme of behavioral reference; simply the most economic pattern of evolvement. Pattern of evolvement has many, many equieconomical intertransformability options. There are many transformation patterns, but tetrahedron is the absolute minimum limit case of structural system interself-stabilizing. A tetrahedron is an omnitriangulated, four-entity, six-vector interrelationship with system-defining insideness and outsideness independent of size; it is not a rigid frame and can be any size. ''Rigid'' means ''sized''—arbitrarily sized. ''Rigid'' is always special-case. Synergetics is *sizeless generalization*.

540.05 Synergetics is not a frame at all, but a pattern of most omnieconomic (ergo, spontaneous) interaccommodation of all observed self-and-otherness interexperiencing (ergo, geodesic—geodesic being the most economical interrelationships of a plurality of events).

540.06 Prime otherness demands identification of the other's—initially nebulous—entity integrity, which entity and subentities' integrities first attain cognizable self-interpatterning stabilization, ergo, discrete considerability, only at the tetrahedron stage of generalizable entity interrelationships. Resolvability and constituent enumerability, and systematic interrelationship cognition of entity regeneration presence, can be discovered only operationally. (See Secs. 411.00, 411.10, 411.20, and 411.30.) After the four-ball structural interpatterning stability occurs, and a fifth ball comes along, and, pulled by mass attraction, it rolls into a three-ball nest, and there are now two tetrahedra bonded face-to-face.

540.07 Because of discontinuity, the otherness points and subpoints may be anywhere. We start always with any point—event points being as yet noncomprehended; ergo, initially only as an apprehended otherness entity. Synergetics, as a strategy of converting apprehension to discrete comprehension, always proceeds vectorially.

540.08 The only difference between experience and nonexperience is time. The time factor is always radial, outwardly, inwardly, and chordally around; always accounted only in most economical to self-experience, energy-time relationship (i.e., geodesic) units. The vector is time-energy incrementation, embracing both velocity and relative mass, as well as the observer's angulation of observation—strictly determined in relation to the observer's head-to-toe axis and time, relative, for instance, to heartbeat and diurnal cyclic experience frequencies.

540.09 A vectorial evolvement in no way conforms to a rigid rectilinear frame of the *XYZ* coordinate analysis which arbitrarily shuns most economical directness and time realizations—by virtue of which calculus is able only awkwardly to define positions rectilinearly, moving only as the chessman's knight. Nature uses rectilinear patterns only precessionally; and precession brings about orbits and not straight lines.

541.00 Radiation and Gravity

(*See Drawings Section*)

541.01 Radiation distributes energy systems outwardly in omnidiametric directions. Radiation fractionates whole systems into multidiametrically dispatched separate packages of the whole. The packaging of spherical unity is accomplished by radii-defined, central-angle partitioning of the spherical whole into a plurality of frequency-determined, simplest, central divisioning, thus producing a plurality of three-sided cornucopias formed inherently at minimum limit of volumetric accommodation by any three immediately adjacent central angles of any sphere or of any omnitriangulated polyhedron.* The threefold central-angle vertex surroundment constitutes the inner vertex definition of a radially amplified tetrahedral pack of energy; while the three inner faces of the energy package are defined by the interior radial planes ** of the sphere of omnidiametric distribution; and the fourth, or outermost, face is the spherical triangle surface of the tetrahedron which always occurs at the radial distance outwardly traveled from the original source at the speed of radiation, symbolized as lower-case *c*.

541.02 Radiation is omni-outwardly and omnidiametrically *distributive;* its fractionally packaged radiations are angularly and pulsatively precessed by the universal otherness frequency effects, ergo, in wavilinearly-edged tetrahedral packages. Radiation is wavilinearly amplifying and radially distributive and is defined by the central-angle-partitioning into discontinuous, not-everywhere entities.

* The spherical tetrahedra, octahedra, and icosahedra are the only symmetrical omnitriangulated systems. (See Sec. 532.40.)

** There is a great-circle plane common to any two radii.

541.03 Gravity is omnipresent, omniembracing, and omnicollective: shadowless and awavilinear. Awavilinear means nonwavilinear or antiwavilinear. Gravity counteracts radiation; it is progressively and centrally focusing; and it is always apparently operative in the most economical, i.e., radially-contractive, transformation—the radii being the shortest distances between a sphere's surface and its volumetric center; ergo, employing the absolute straight-nothingness, radial line of direction, which, as such, is inherently invisible.

541.04 Radiation is pushive, ergo tends to increase in curvature. Gravity is tensive, ergo tends to decrease its overall curvature. The ultimate reduction of curvature is no curvature. Radiation tends to increase its overall curvature (as in the "bent space" of Einstein). The pushive tends to arcs of ever lesser radius (microwaves are the very essence of this); the tensive tends to arcs of ever greater radius. (See Sec. 1009.56.)

541.05 The omni-inbound gravity works collectively toward the invisibility of the central zero-size point. The outbound, tetrahedrally packaged, fractional point works toward and reaches the inherent visibility phases of radiation. Radiation is disintegrative; gravity is integrative.

541.06 Gravity's omniembracing collectiveness precessionally generates circumferential surface foldings—waves (earthquakes)—consequent to the second-power rate of surface diminution in respect to the radially-measured, first-power linear rate of system contraction. Gravity is innocent of wave. Gravity is innocent of radial; i.e., linear aberration waves; i.e. gravity is nonwavilinear. The most economical interterminal relationship is always that with the least angular aberration. Gravity is the geodesic—most economical—relationship of events.

541.07 Gravity's awavilinear, collective, integrative, economical effectiveness is always greater than that of the radiation's disintegrative, wavilinear distributiveness; ergo, gravity guarantees the integrity of eternally regenerative omni-intertransformative Universe.

541.08 Radiation is wavilinearly and radially distributive; ergo, it is central-angle partitioned. Circularly, it means a single central angle. Spherically, it means a minimum of three central angles: those of a tetrahedron formed with a circumferential limit of the surface of the speed-of-light radial reach.

541.09 Radiation is tetrahedral. A tetrahedron is a tetrahedron independent of size. There are points and no-points. They are both tetrahedral.

541.10 Gravity is circumferentially omniembracing and is never partial, but always whole. Radiation is always packaged. Gravity is the inside-outness of energy-as-matter: the integrity of Universe. It is the sum of all the no-points embracing all the points; and it compounds at the surface-embracing, second-power rate of the linear proximity gains. All the no-points (novents) are always embracing all the points. All the quanta are local-system, center-of-event activity, focal points—fractionations of the whole point: what are minimally, ergo, most economically, packaged, and expanded outwardly and omnidiametrically as three-central-angle-defined tetrahedra. (See Secs. 251.05 and 529.03.)

541.20 Solution of Four-Color Theorem

541.21 Polygonally all spherical surface systems are maximally reducible to omnitriangulation, there being no polygon of lesser edges. And each of the surface triangles of spheres is the outer surface of a tetrahedron where the other three faces are always congruent with the interior faces of the three adjacent tetrahedra. Ergo, you have a four-face system in which it is clear that any four colors could take care of all possible adjacent conditions in such a manner as never to have the same colors occurring between two surface triangles, because each of the three inner surfaces of any tetrahedron integral four-color differentiation must be congruent with the same-colored interior faces of the three and only adjacent tetrahedra; ergo, the fourth color of each surface adjacent triangle must always be the one and only remaining different color of the four-color set systems.

541.30 Photon as Tetrahedral Package (*see also drawings section*)

541.31 An ice cream cone (Fig. A).
A cornucopia (Fig. B).
A cone which, in its flattened state, has zero interior volume. In its flattened state it is two-sided (Fig. C).
The three-sided state has the volume of a tetrahedron (Fig. D).
The six-sided state (Fig. E)
and the 12-sided state (Fig. F)
have progressively greater volume with the same surface area.

541.32 The seemingly circular—but inscrutably multifaceted—state, the conic (Fig. G) has most volume with the same surface as that of its tetrahedral cornucopia state.

541.33 Circular cornucopia can be tangentially assembled around interior points to form a "spherical bouquet," a spherical array, but the tangent circle areas do not constitute the total surface of the sphere. There are concave triangle interstices (see Fig. 541.30H in drawings section).

541.34 But three-sided, triangle-mouthed cornucopia will, together, subdivide the total sphere (see Fig. 541.30I in drawings section). Therefore, as the three-sided tetrahedral packages become outwardly separated from one another, they will inherently yield to their greater volume and, being spun precessionally (Fig. K) by their sum-total cosmic otherness, will, by centrifu-

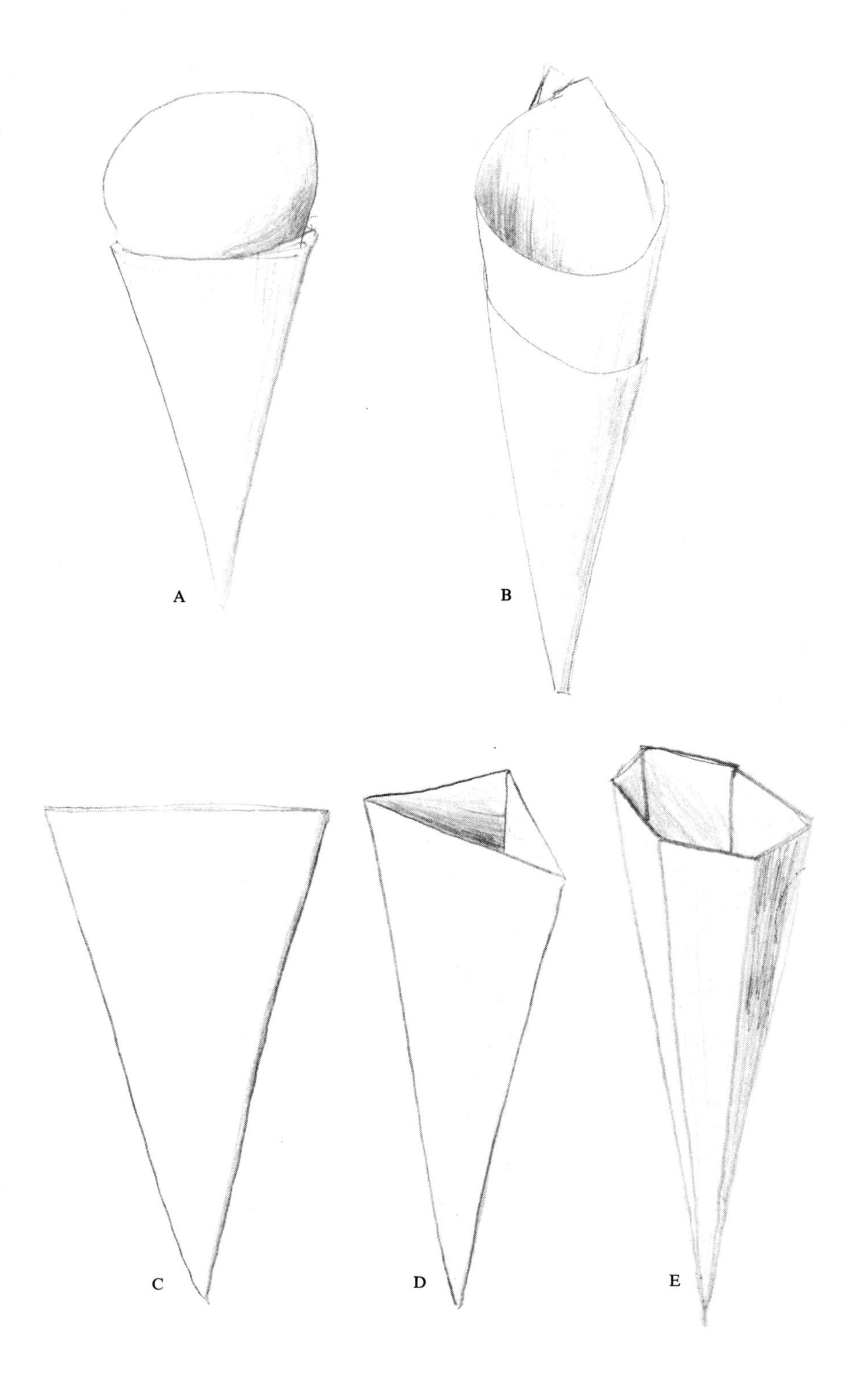
A
B
C
D
E

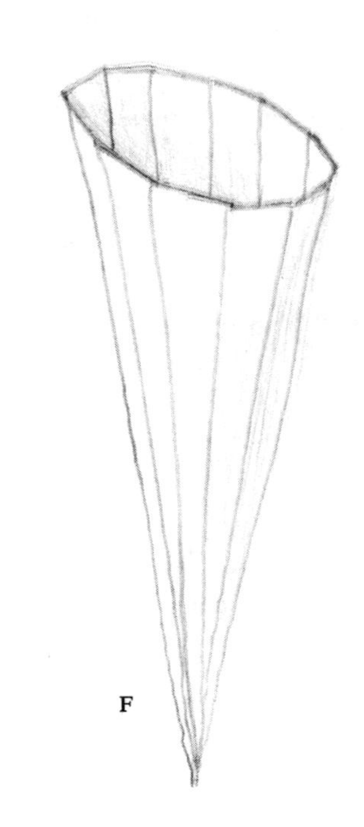

F

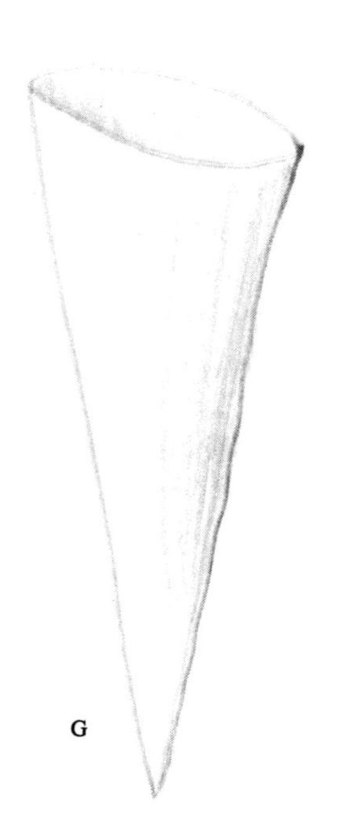

G

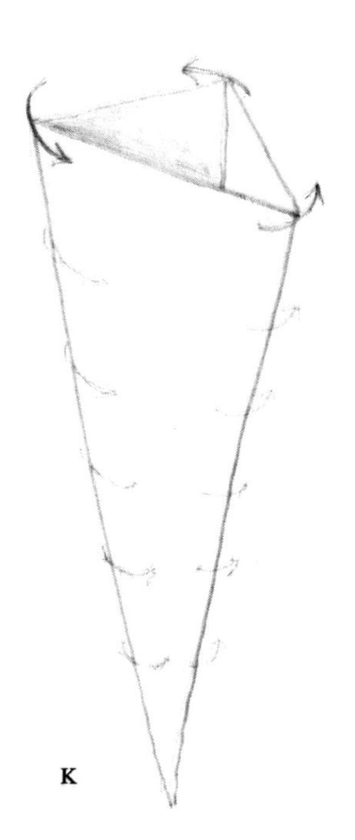

K

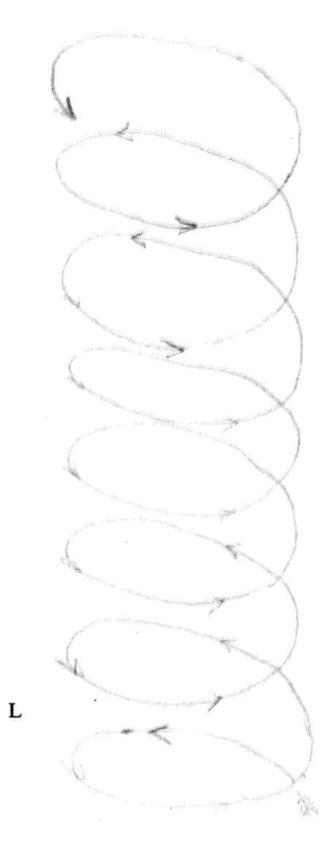

L

gal force, become *cones;* which, rotating on their long conic axis will generate a cylindrical, spiral, wave pattern: (Fig. L). By their radially outward dispatch, they rotationally describe the cornucopia, or cone of gradual beam spread—the spread rate being negligible in relation to the axial speed-of-light rate of travel (See Fig. 541.30M).

541.35 Progressively the four tetrahedron-defining vertexial components of the photon packages spiralling precession results in an equilibrium-seeking of the inward-outward-and-around proclivities of the four separate interattractivenesses which spontaneously generate the four great circles of the vector equilibrium and establish its tactical energy center as the four planes of the zero-size phase of the tetrahedron.

541.36 The total vector equilibrium spherical package becomes an export photon. Though superficially amorphous, radiation is inherently tetrahedrally and spherically packaged, and is discretely accountable as such. The tetrahedron is the quantum model. (See Secs. 620 and 1106.23.)

600.00 Structure

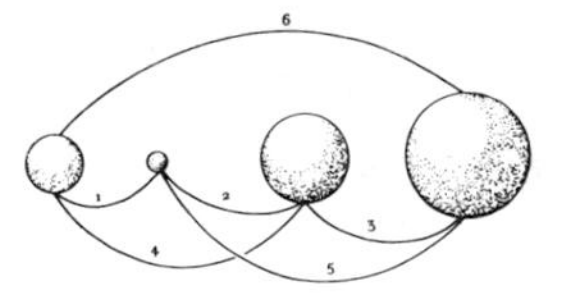

600.00 STRUCTURE

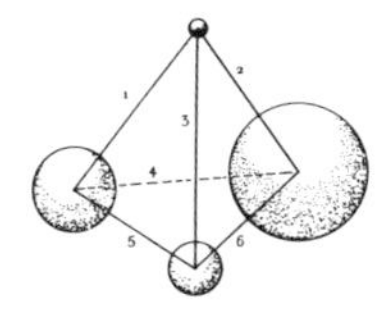

600.01 Definition: Structure

600.02 A structure is a self-stabilizing energy-event complex.

600.03 A structure is a system of dynamically stabilized self-interfering and thus self-localizing and recentering, inherently regenerative constellar association of a minimum set of four energy events.

600.04 *Stability* means angular invariability. *Inherent* means behavior principles that man discovers to be reliably operative under given conditions always and anywhere in Universe. *Regenerative* means local energy-pattern conservation. *Constellar* means an aggregation of enduring, cosmically isolated, locally co-occurring events dynamically maintaining their interpositioning: e.g., macroconstellations such as the Big Dipper, Orion, and the Southern Cross and microconstellations such as matter in general, granite, cheese, flesh, water, and atomic nuclei.

601.00 Pattern Conservation

601.01 It is a tendency for patterns either to repeat themselves locally or for their parts to separate out to join singly or severally with other patterns to form new constellations. All the forces operative in Universe result in a complex progression of most comfortable—i.e., least effort, rearrangings in which the macro-medio-micro star events stand dynamically together here and there as locally regenerative patterns. Spontaneously regenerative local constellations are cosmic, since they appear to be interoriented with angular constancy.

601.02 Structures are constellar pattern conservations. These definitions hold true all the way from whole Universe to lesser and local pattern differentiations all the way into the atom and its nuclear subassemblies. Each of the families of chemical elements, as well as their most complex agglomerations as super-star Galaxies, are alike cosmic structures. It is clear from the results of modern scientific experiments that *structures are not things*. Structures are *event constellations*.

602.01 Structural systems are cosmically localized, closed, and finite. They embrace all geometric forms—symmetric and asymmetric, simple and complex.

602.02 Structural systems can have only one insideness and only one outsideness.

602.03 Two or more structures may be concentric and/or triangularly—triple-bondedly—interconnected to operate as one structure. Single-bonded (universally jointed) or double-bonded (hinged) means that we have two flexibly interconnected structural systems.

603.01 All structuring can be topologically identified in terms of tetrahedra. (See Sec. 362.)

604.00 Structural System

604.01 In a *structural system:*

(1) the number of vertexes (crossings) is always evenly divisible by two;
(2) the number of faces (openings) is always evenly divisible by four; and
(3) the number of edges (trajectories) is always evenly divisible by six.

605.01 Inasmuch as there are always and everywhere 12 fundamental degrees of freedom (six positive and six negative), and since every energy event is characterized by a threefold vectoring—an action, a reaction, and a resultant—all structures, symmetrical or asymmetrical, regular or irregular, simple or compound, will consist of the twelvefoldedness or its various multiples.

606.01 "Mathematics is the science of structure and pattern in general." * Structure is defined as a locally regenerative pattern integrity of Universe. We cannot have a total structure of Universe. Structure is inherently only local and inherently regenerative.

606.02 Structures most frequently consist of the physical interrelationships of nonsimultaneous events.

606.03 One of the deeply impressive things about structures is that they cohere at all—particularly when we begin to know something about the atoms and realize that the components of atoms are really very remote from one another, so that we simply have galaxies of events. Man is deceiving himself when he sees anything "solid" in structures.

* From the Massachusetts Institute of Technology's 1951 official catalog of the self-definition by M.I.T. Mathematics Department.

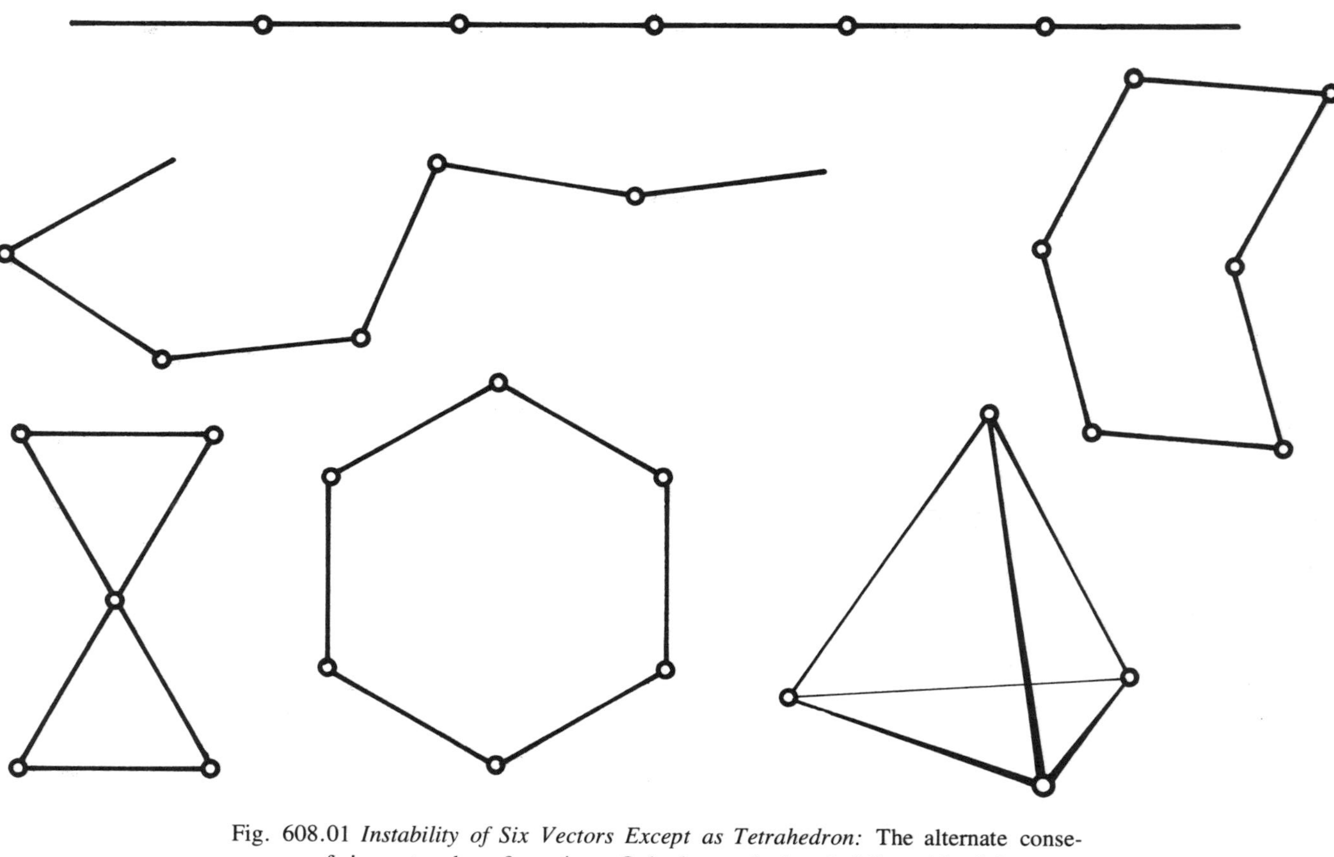

Fig. 608.01 *Instability of Six Vectors Except as Tetrahedron:* The alternate consequences of six vectored configurations. Only the tetrahedron is fully stable. It is synergetic.

608.00 Stability: Necklace

608.01 A necklace is unstable. The beads of a necklace may be superficially dissimilar, but they all have similar tubes running through them with the closed tension string leading through all the tubes. The simplest necklace would be one made only of externally undecorated tubes and of tubes all of the same length. As the overall shape of the necklace changes to any and all polygonal shapes and wavy drapings, we discover that the lengths of the beads in a necklace do not change. Only the angles between the tubes change. Therefore, *stable* refers only to angular invariability.

608.02 A six-edged polygon is unstable; it forms a drapable necklace. If we make a five-sided polygon, i.e., a pentagonal necklace, it is unstable. It, too, is a drapable necklace and is structurally unstable. Why? A necklace of three rigid tubes also has three flexible angle-accommodating tension joints. Here are six separate parts, each with its unique behavior characteristics which self-interfere to produce a stable pattern. How and why? We are familiar with the principle of lever advantage gained per length of lever arm from the fulcrum. We are familiar with the principle of the shears in which two levers share a common fulcrum, and the stronger and longer the shear arms, the more powerfully do they cut. Steel-bolt cutters have long lever arms.

608.03 In every triangle each corner angle tension connector serves as the common interfulcrum of the two push-pull, rigid lever arms comprising two of the three sides of the triangle adjacent to their respectively common angular corners; each pair of the triangle's tubular necklace sides, in respect to a given corner of the triangle, represent levers whose maximum-advantage ends are seized by the two ends of the third, rigid, push-pull, tubular side of the triangle, whose rigidity is imposed by its command of the two lever arm ends upon the otherwise flexible opposite angle. Thus we find that each of the necklace's triangular rigid tube sides stabilizes its opposite angle with minimum effort by controlling the ends of the two levers fulcrumed by that opposite tension fastening of the triangle. Thus we find the triangle to be not only the unique pattern-self-stabilizing, multienergied complex, but also accomplishing pattern stabilization at minimum effort, which behavior coincides with science's discovery of the omni-minimum-effort behavior of all physical Universe.

608.04 The six independent energy units of the triangle that interact to produce pattern stability are the only plural polygon-surrounding, energy-event complexes to produce stabilized patterns. (The necklace corners can be fastened together with three separate tension-connectors, instead of by the string running all the way through the tubes, wherefore the three rigid tubes and the three flexible tension connectors are six unique, independent, energy events.)

608.05 We may say that structure is a self-stabilizing, pattern-integrity complex. Only the triangle produces structure and structure means only triangle; and vice versa.

608.06 Since tension and compression always and only coexist (See Sec. 640) with first one at high tide and the other at low tide, and then vice versa, the necklace tubes are rigid with compression at visible high tide and tension at invisible low tide; and each of the tension-connectors has compression at invisible low tide and tension at visible high tide; ergo, each triangle has both a positive and a negative triangle congruently coexistent and each visible triangle is two triangles: one visible and one invisible.

608.07 Chain-linkage necklace structures take advantage of the triangulation of geodesic lines and permit us to encompass relatively large volumes with relatively low logistic investment. Slackened necklace geodesic spheres can be made as compactable as hairnets and self-motor-opened after being shot into orbit.

608.08 It is a synergetic characteristic of minimum structural systems (tetrahedra) that the system is not stable until the last strut is introduced. Redundancy cannot be determined by energetic observation of behaviors of single struts (beams or columns) or any chain-linkage of same, that are less than six in number, or less than tetrahedron.

608.10 **Necklace Polygons and Necklace Polyhedra:** Tetrahedral, octahedral, and icosahedral necklace structures are all stable. Necklace cubes, rhombic dodecahedra, pentadodecahedra, vector equilibria, and tetrakaidecahedra are all unstable. Only necklace-omnitriangulated, multifrequency geodesic spheres are stable structures, because they are based entirely on omnitriangulated tetra-, octa-, and icosahedral systems.

608.11 The number of vertexes of the omnitriangulated spherical tetra-, octa-, or icosahedral structures of multifrequency geodesic spheres corresponds exactly with the number of external layer spheres of closest-packed unit radius spherical agglomeration of tetrahedra, octahedra, or icosahedra:

Tetrahedra	$2F^2 + 2$
Octahedra	$4F^2 + 2$
Icosahedra	$10F^2 + 2$

Only tetrahedral, octahedral, and icosahedral structural systems are stable, i.e., complete, nonredundant, self-stabilizing. (See Sec. 223.22.)

609.00 Instability of Polyhedra from Polygons of More Than Three Sides

609.01 Any polygon with more than three sides is unstable. Only the triangle is inherently stable. Any polyhedron bounded by polygonal faces with more than three sides is unstable. Only polyhedra bounded by triangular faces are inherently stable.

610.00 Triangulation

610.01 By structure, we mean a self-stabilizing pattern. The triangle is the only self-stabilizing polygon.

610.02 By structure, we mean omnitriangulated. The triangle is the only structure. Unless it is self-regeneratively stabilized, it is not a structure.

610.03 Everything that you have ever recognized in Universe as a pattern is re-cognited as the same pattern you have seen before. Because only the triangle persists as a constant pattern, any recognized patterns are inherently recognizable only by virtue of their triangularly structured pattern integrities. Recognition is as dependent on triangulation as is original cognition. Only triangularly structured patterns are regenerative patterns. Triangular structuring is a pattern integrity itself. This is what we mean by *structure*.

610.10 Structural Functions

610.11 Triangulation is fundamental to structure, but it takes a plurality of positive and negative behaviors to make a structure. For example:

—always and only coexisting push and pull (compression and tension);

—always and only coexisting concave and convex;

—always and only coexisting angles and edges;

—always and only coexisting torque and countertorque;

—always and only coexisting insideness and outsideness;

—always and only coexisting axial rotation poles;

—always and only coexisting conceptuality and nonconceptuality;

—always and only coexisting temporal experience and eternal conceptuality.

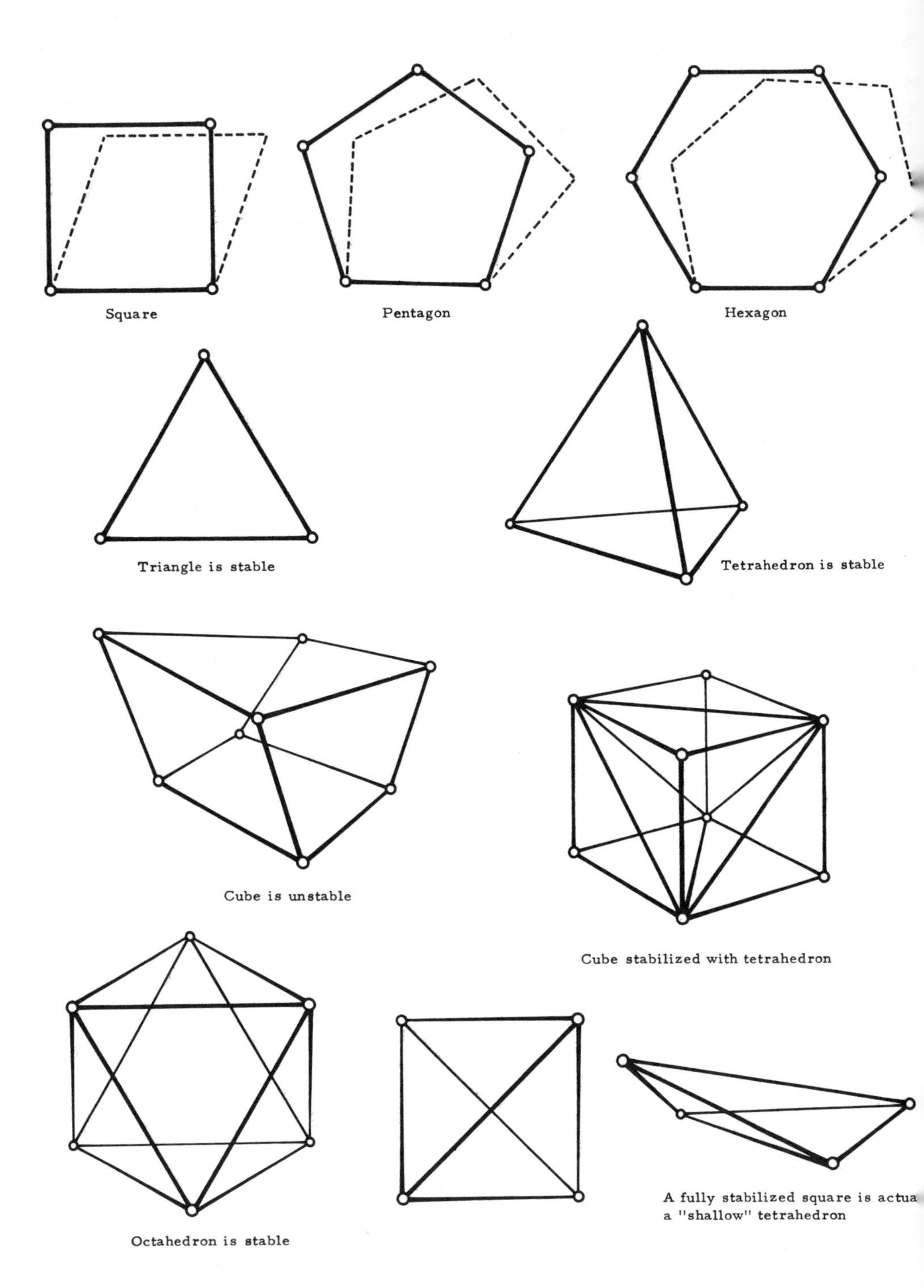

Fig. 609.01 *Instability of Polyhedra from Polygons of More than Three Sides.*

610.12 If we want to have a structure, we have to have triangles. To have a structural *system* requires a minimum of four triangles. The tetrahedron is the simplest structure.

610.13 Every triangle has two faces: obverse and reverse. Every structural system has omni-intertriangulated division of Universe into insideness and outsideness.

610.20 Omnitriangular Symmetry: Three Prime Structural Systems

610.21 There are three types of omnitriangular, symmetrical structural systems. We can have three triangles around each vertex; a tetrahedron. Or we can have four triangles around each vertex; the octahedron. Finally we can have five triangles around each vertex; the icosahedron. (See Secs. 532.40, 610.20, 724, 1010.20, 1011.30 and 1031.13.)

610.22 The tetrahedron, octahedron, and icosahedron are made up, respectively, of one, two, and five pairs of positively and negatively functioning open triangles.

610.23 We cannot have six symmetrical or equiangular triangles around each vertex because the angles add up to 360 degrees—thus forming an infinite edgeless plane. The system with six equiangular triangles "flat out" around each vertex never comes back upon itself. It can have no withinness or withoutness. It cannot be constructed with pairs of positively and negatively functioning open triangles. In order to have a system, it must return upon itself in all directions.

611.00 Structural Quanta

611.01 If the system's openings are all triangulated, it is structured with minimum effort. There are only three possible omnisymmetrical, omnitriangulated, least-effort structural systems in nature. They are the tetrahedron, octahedron, and icosahedron. When their edges are all equal in length, the volumes of these three structures are, respectively, *one,* requiring one structural quantum; *four,* requiring two structural quanta; and *18.51,* requiring five structural quanta. Six edge vectors equal one minimum-structural system: 6 edge vectors = 1 structural quantum.

611.02 Six edge vectors = one tetrahedron. One tetrahedron = one structural quantum.

1 Tetrahedron (volume 1) = 6 edge vectors = 1 structural quantum;

1 Octahedron (volume 4) = 12 edge vectors = 2 structural quanta;

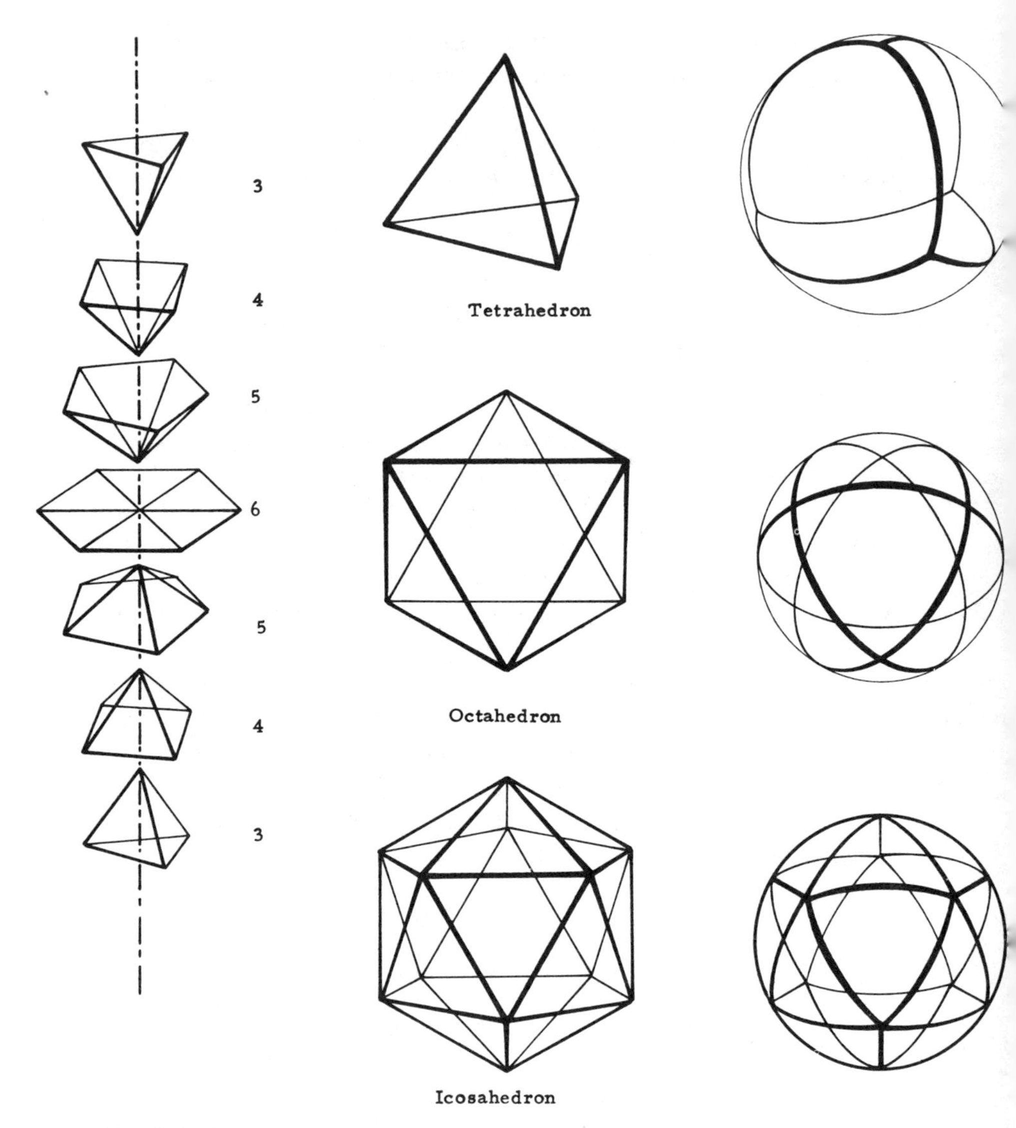

Fig. 610.20 *The Three Basic Structural Systems in Nature with Three, Four or Five Triangles at Each Vertex:*

There are only three possible cases of fundamental omnisymmetrical, omnitriangulated, least-effort structural systems in nature: the tetrahedron with three triangles at each vertex, the octahedron with four triangles at each vertex, and the icosahedron with five triangles at each vertex. If there are six equilateral triangles around a vertex we cannot define a three-dimensional structural system, only a "plane." The left column shows the minimum three triangles at a vertex forming the tetrahedron through to the six triangles at a vertex forming an "infinite plane." The center column shows the planar polyhedra. The right column shows the same polyhedra in spherical form.

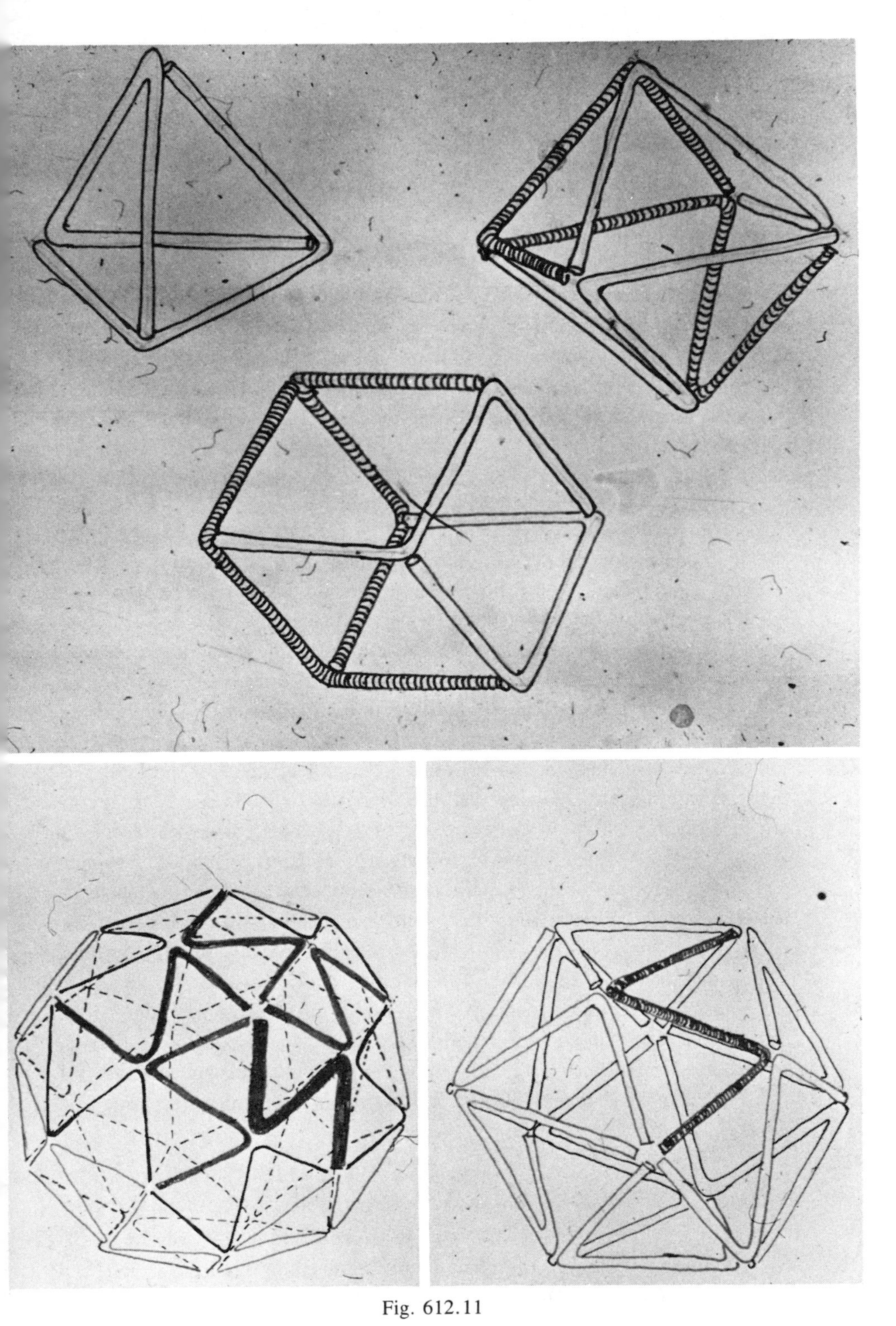

Fig. 612.11

1 Icosahedron (volume 18.51) = 30 edge vectors = 5 structural quanta.
Therefore:
with tetrahedron, 1 structural quantum provides 1 unit of volume;
with octahedron, 1 structural quantum provides 2 units of volume;
with icosahedron, 1 structural quantum provides 3.7 units of volume.

612.00 Subtriangulation: Icosahedron

612.01 Of the three fundamental structures, the tetrahedron contains the most surface and the most structural quanta per volume; it is therefore the strongest structure per unit of volume. On the other hand, the icosahedron provides the most volume with the least surface and least structural quanta per units of volume and, though least strong, it is structurally stable and gives therefore the most efficient volume per units of invested structural quanta.

612.10 Units of Environment Control: The tetrahedron gives one unit of environment control per structural quantum. The octahedron gives two units of environment control per structural quantum. The icosahedron gives 3.7 units of environment control per structural quantum.

612.11 That is the reason for the employment of the triangulated icosahedron as the most efficient fundamental volume-controlling device of nature. This is the way I developed the multifrequency-modulated icosahedron and geodesic structuring. This is probably the same reason that nature used the multifrequency-modulated icosahedron for the protein shells of the viruses to house most efficiently and safely all the DNA-RNA genetic code design control of all biological species development. I decided also to obtain high local strength on the icosahedron by *subtriangulating* its 20 basic Icosa LCD spherical triangles with locally superimposed tetrahedra, i.e., an octahedron-tetrahedron truss, which would take highly concentrated local loads or impacts with minimum effort while the surrounding rings of triangles would swiftly distribute and diminishingly inhibit the outward waves of stress from the point of concentrated loading. I had also discovered the foregoing structural mathematics of structural quanta topology and reduced it to demonstrated goedesic dome practice before the virologists discovered that the viruses were using geodesic spheres for their protein shell structuring. (See Sec. 901.)

613.00 Triangular Spiral Events Form Polyhedra

613.01 Open triangular spirals may be combined to make a variety of different figures. Note that the tetrahedron and icosahedron require both left- and right-handed (positive and negative) spirals in equal numbers, whereas the

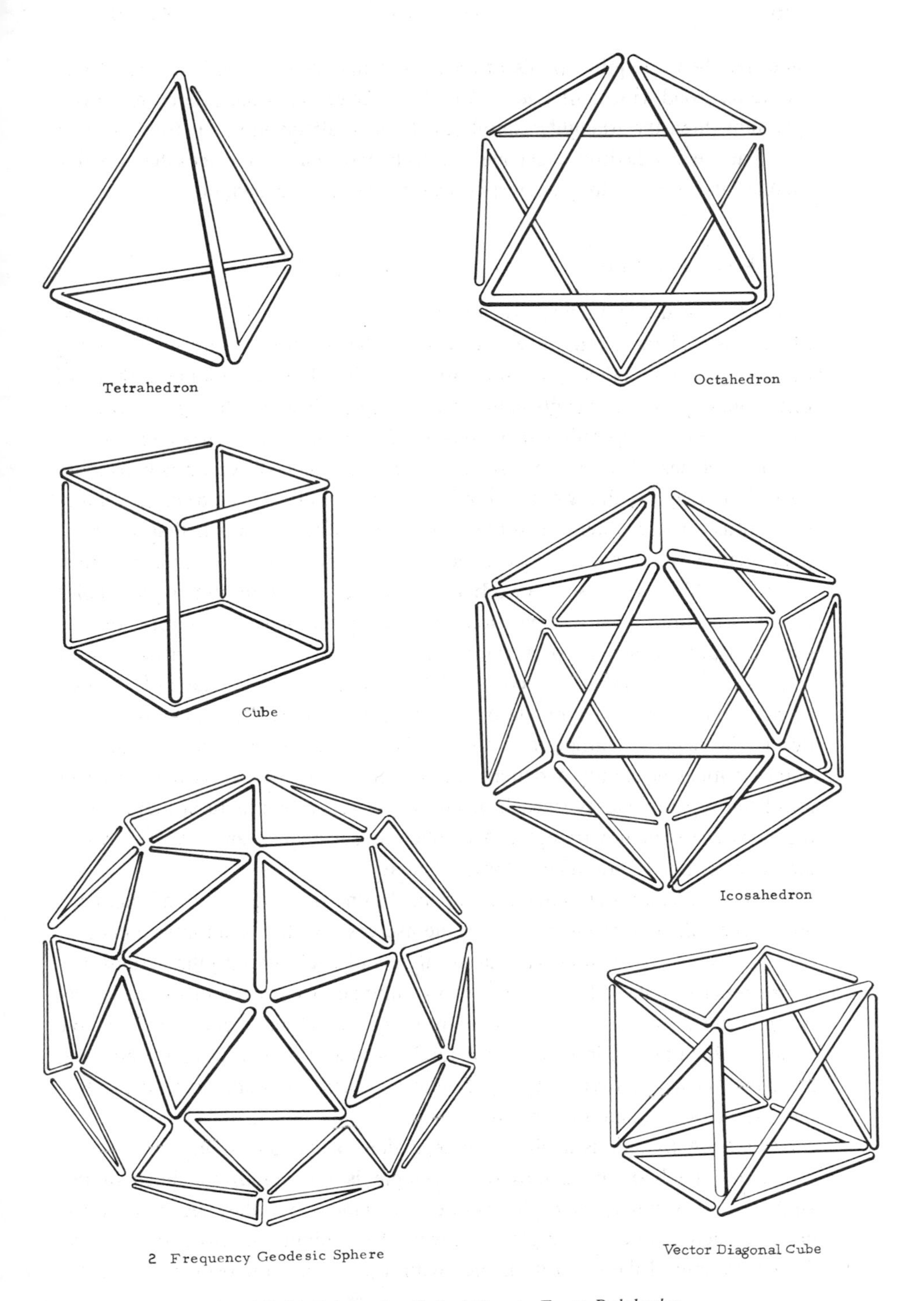

Fig. 613.01 *Triangular Spiral Events Form Polyhedra.*

other polyhedra require spirals of only one-handedness. (See Sec. 452, Great Circle Railroad Tracks of Energy.) If the tetrahedron is considered to be one quantum, then the triangular spiral equals one-half quantum. It follows from this that the octahedron and cube are each two quanta, the icosahedron five quanta, and the two-frequency spherical geodesic is 15 quanta.

614.00 Triangle

614.01 A triangle's three-vector parts constitute a basic event. Each triangle consists of three interlinked vectors. In the picture, we are going to add one triangle to the other. (See illustration 511.10.) In conventional arithmetic, one triangle plus one triangle equals two triangles. The two triangles represent two basic events operating in Universe. But experientially triangles do not occur in planes. They are always omnidimensional positive or negative helixes. You may say that we do not have any right to break the triangles' three-sided rims open in order to add them together, but the answer is that the triangles were never closed, because no line can ever come completely back "into" or "through" itself. Two lines cannot be passed through a given point at the same time. One will be superimposed on the other. Therefore, the superimposition of one end of a triangular closure upon another end produces a spiral—a very flat spiral, indeed, but openly superimposed at each of its three corners, the opening magnitude being within the critical limit of mass attraction's 180-degree "falling-in" effect. The triangle's open-ended ends are within critical proximity and mass-attractively intercohered, as are each and all of the separate atoms in each of all the six separate structural members of the necklace-structure triangle. All coherent substances are "Milky Way" clouds of critically proximate atomic "stars."

614.02 Triangles are inherently open. As one positive event and one negative event, the two triangles arrange themselves together as an interference of the two events. The actions and the resultants of each run into the actions and the resultants of the other. They always impinge at the ends of the action as two interfering events. As a tetrahedron, they are fundamental: a structural system. It is a tetrahedron. It is structural because it is omnitriangulated. It is a system because it divides Universe into an outsideness and an insideness—into a macrocosm and a microcosm.

614.03 A triangle is a triangle independent of its edge-sizing.

614.04 Each of the angles of a triangle is interstabilized. Each of the angles was originally amorphous—i.e., unstable—but they become stable because each edge of a triangle is a lever. With minimum effort, the ends of the levers control the opposite angles with a push-pull, opposite-edge vector.

A triangle is the means by which each side stabilizes the opposite angle with minimum effort.

614.05 The stable structural behavior of a whole triangle, which consists of three edges and three individually and independently unstable angles (or a total of six components), is not predicted by any one or two of its angles or edges taken by themselves. A triangle (a structure) is synergetic: it is a behavior of a whole unpredicted by the behavior of any of its six parts considered only separately.

614.06 When a bright light shines on a complex of surface scratches on metal, we find the reflection of that bright light upon the scratched metal producing a complex of concentric scratch-chorded circles. In a multiplicity of omnidirectional actions in the close proximity of the viewable depth of the surfaces, structurally stable triangles are everywhere resultant to the similarly random events. That triangles are everywhere is implicit in the fact that wherever we move or view the concentric circles, they occur, and that there is always one triangle at the center of the circle. We could add the word *approximately* everywhere to make the everywhereness coincide with the modular-frequency characteristics of any set of random multiplicity. Because the triangles are structurally stable, each one imposes its structural rigidity upon its neighboring and otherwise unstable random events. With energy operative in the system, the dominant strength of the triangles will inherently average to equilateralness.

614.07 When we work with triangles in terms of total leverage, we find that their average, most comfortable condition is equilateral. They tend to become equilateral. Randomness of lines automatically works back to a set of interactions and a set of proximities that begin to triangulate themselves. This effect also goes on in depth and into the tetrahedra or octahedra.

615.00 Positive and Negative Triangulation of Cube and Vector Equilibrium

615.01 To be referred to as a rememberable entity, an object must be membered with structural integrity, whether maple leaf or crystal complex. To have structural integrity, it must consist entirely of triangles, which are the only complex of energy events that are self-interference-regenerating systems resulting in polygonal pattern stabilization.

615.02 A vectorial-edged cube collapses. The cube's corner flexibility can be frustrated only by triangulation. Each of the four corners of the cube's six faces could be structurally stabilized with small triangular gussets, of which there would be 24, with the long edge structurals acting as powerful levers against the small triangles. The complete standard stabilization of the cube can be accomplished with a minimum of six additional members in the form of six structural struts placed diagonally, corner to corner, in each of the six square faces, with four of the cube's eight corner vertexes so interconnected. These six, end-interconnected diagonals are the six edges of a tetrahedron. The most efficiently stabilized cubical form is accomplished with the prime structural system of Universe: the tetrahedron.

615.03 Because of the structural integrity of the blackboard or paper on which they may be schematically pictured, the cubically profiled form can exist, but only as an experienceable, forms-suggesting picture, induced by lines deposited in chalk, or ink, or lead, accomplished by the sketching individual with only 12 of the compression-representing strut edge members interjoined by eight flexible vertex fastenings.

615.04 The accomplishment of experienceable, structurally stabilized cubes with a minimum of nonredundant structural components will always and only consist of one equiangled and equiedged "regular" tetrahedron on each of whose four faces are congruently superimposed asymmetrical tetrahedra, one of whose four triangular faces is equiangled and therefore congruently superimposable on each of the four faces of the regular tetrahedron; while the four asymmetrical superimposed tetrahedra's other three triangular—and outwardly exposed—faces are all similar isosceles triangles, each with two 45-degree-angle corners and one corner of 90 degrees. Wherefore, around each of the outermost exposed corners of the asymmetrical tetrahedra, we also find three 90-degree angles which account for four of the cube's eight corners; while the other four 90-degree surrounded corners of the cube consist of pairs of 45-degree corners of the four asymmetric tetrahedra that were superimposed upon the central regular tetrahedron to form the stabilized cube. More complex cubes that will stand structurally may be compounded by redundant strutting or tensioning triangles, but redundancies introduce micro-invisible, high- and low-frequency, self-disintegrative accelerations, which will always affect structural enterprises that overlook or disregard these principles.

615.05 In short, structurally stabilized (and otherwise unstable) cubes are always and only the most simply compact aggregation of one symmetrical and four asymmetrical tetrahedra. Likewise considered, a dodecahedron may not be a cognizable entity-integrity, or be rememberable or recognizable as a regenerative entity, unless it is omnistabilized by omnitriangulation of its systematic subdivision of all Universe into either and both insideness and outsideness, with a small remainder of Universe to be discretely invested into the system-entity's structural integrity. No energy action in Universe would bring about a blackboard-suggested pentagonal necklace, let alone 12 pentagons collected edge to edge to superficially outline a dodecahedron. The dodecahedron is a demonstrable entity only when its 12 pentagonal faces are subdivided into five triangles, each of which is formed by introducing into each pentagon five struts radiating unitedly from the pentagons' centers to their five corner vertexes, of which vertexes the dodecahedron has 20 in all, to whose number when structurally stabilized must be added the 12 new pentagonal center vertexes. This gives the minimally, nonredundantly structural dodecahedron 32 vertexes, 60 faces, and 90 strut lines. In the same way, a structural cube has 12 triangular vertexes, 8 faces, and 18 linear struts.

615.06 The vector equilibrium may not be referred to as a stabilized structure except when six struts are inserted as diagonal triangulators in its six square faces, wherefore the topological description of the vector equilibrium always must be 12 vertexes, 20 (triangular) faces, and 30 linear struts, which is also the topological description of the icosahedron, which is exactly what the six triangulating diagonals that have hypotenusal diagonal vectors longer than the square edge vectors bring about when their greater force shrinks them to equilength with the other 24 edge struts. This interlinkage transforms the vector equilibrium's complex symmetry of six squares and eight equiangled triangles into the simplex symmetry of the icosahedron.

615.07 Both the cube and the vector equilibrium's flexible, necklacelike, six-square-face instabilities can be nonredundantly stabilized as structural integrity systems only by one or the other of two possible diagonals of each of their six square faces, which diagonals are not the same length as the unit vector length. The alternate diagonaling brings about positive or negative symmetry of structure. (See illustration 464.01 and 464.02 in color section.) Thus we have two alternate cubes or icosahedra, using either the red diagonal or the blue diagonal. These alternate structural symmetries constitute typical positive or negative, non-mirror-imaged intercomplementation and their systematic, alternating proclivity, which inherently propagate the gamut of frequencies uniquely characterizing the radiated entropy of all the self-regenerative chemical elements of Universe, including their inside-out, invisibly negative-Universe-provokable, split-second-observable imports of transuranium, non-self-regenerative chemical elements.

616.00 Surface Strength of Structures

616.01 The highest capability in strength of structures exists in the triangulation of the system's enclosing structure, due to the greater action-reaction leverage distance that opposite sides of the system provide. This is what led men to hollow out their buildings.

616.02 The structural strength of the exterior triangles is not provided by the "solid" quality of the exterior shell, but by triangularly interstabilized lines of force operating within that shell. They perforate the shell with force lines. The minimum holes are triangular.

616.03 The piercing of the shells with triangular holes reduces the solid or continuous surface of second-power increase of the shells. This brings the rate of growth of structures into something nearer an overall first-power or linear rate of gain—for the force lines are only linear. (See also Sec. 412, Closest Packing of Rods: Surface Tension Capability, and Sec. 750, Unlimited Frequency of Geodesic Tensegrities.)

617.00 Cube

617.01 If the cubic form is stable, it has 18 structural lines. If a dodecahedron is stable, it has 32 vertexes, 60 faces, and 90 structural lines. (The primes 5 and 3 show up here to produce our icosahedral friend 15.)

617.02 Whenever we refer to a stable entity, it has to be structurally valid; therefore, it has to be triangulated. This does not throw topology out.

617.03 A nonstructurally triangulated cube exists only by self-deceptive topological accounting: someone shows you a paper or sheet-metal cube and says, "Here is a structurally stable cube without any face diagonaling." And you say, "What do you call that sheet metal or paper that is occupying the square faces without which the cube would not exist? The sheet metal or paper does diagonal the square but overdoes it redundantly."

617.04 A blackboard drawing of a 12-line cube is only an imaginary, impossible structure that could not exist in this part of Universe. It could temporarily hold its shape in gravity-low regions of space or in another imaginary Universe. Because we are realistically interested only in this Universe, we find the cube to be theoretical only. If it is real, the linear strut cube has 12 isosceles, right-angle-apexed, triangular faces.

618.00 Dimpling Effect

618.01 Definition: When a concentrated load is applied (toward the center) of any vertex of any triangulated system, it tends to cause a dimpling effect. As the frequency or complexity of successive structures increases, the dimpling becomes progressively more localized, and proportionately less force is required to bring it about.

618.02 To illustrate dimpling in various structures, we can visualize the tetrahedron, octahedron, and icosahedron made out of flexible steel rods with rubber joints. Being thin and flexible, they will bend and yield under pressure.

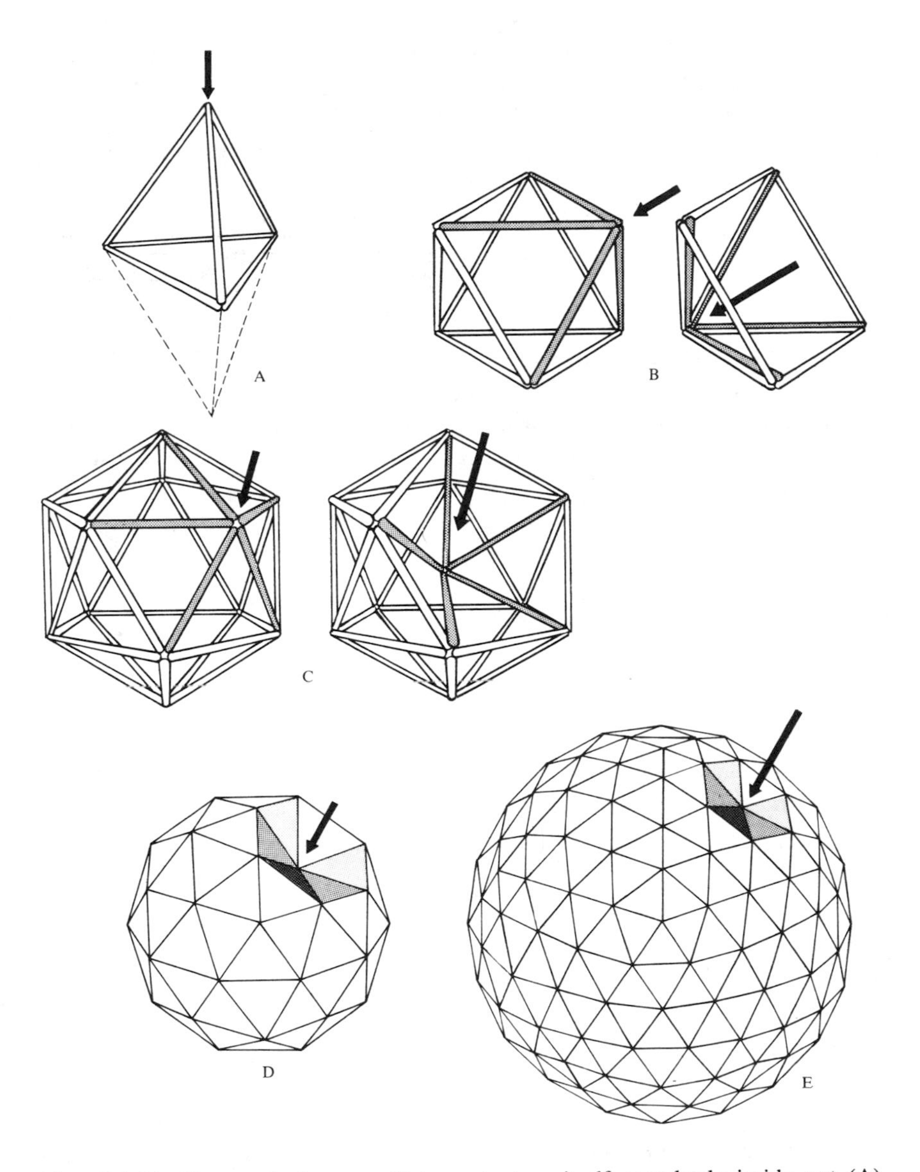

Fig. 618.01: The tetrahedron would have to turn itself completely inside out (A), and as this constitutes a complete change in the entire structure (with no localized effect in evidence) the tetrahedron clearly has the greatest resistance of any structure to externally applied concentrated load. The octahedron dimples in on itself (B), and the icosahedron (C), although dimpling locally, does reduce its volume considerably when doing so, implying that it still has good resistance to concentrated load. The geodesic spheres (D and E) exhibit "very local" dimpling as the frequency increases, suggesting much less resistance to concentrated loads but very high resistance to distributed loads.

618.10 **Tetrahedron:** Beginning with the tetrahedron as the minimum system, it clearly will require proportionately greater force to create a "dent." In order to dimple, the tetrahedron will have to turn itself completely inside out with no localized effect in evidence. Thus the dimpling forces a complete change in the entire structure. The tetrahedron has the greatest resistance of any structure to externally applied concentrated load. It is the only system that can turn itself inside out. Other systems can have very large dimples, but they are still local. Even a hemispherical dimple is still a dimple and still local.

618.20 **Octahedron:** If we apply pressure to any one of the six vertexes of the octahedron, we will find that one half will fit into the other half of the octahedron, each being the shape of a square-based Egyptian pyramid. It will nest inside itself like a football being deflated, with one half nested in the other. Although the octahedron dimples locally, it reduces its volume considerably in doing so, implying that it still has a good resistance to concentrated load.

618.30 **Icosahedron:** When we press on a vertex of the icosahedron, five legs out of the thirty yield in dimpling locally. There remains a major part of the space in the icosahedron that is not pushed in. If we go into higher and higher triangulation—into geodesics—the dimpling becomes more local; there will be a pentagon or hexagon of five or six vectors that will refuse to yield in tension and will pop inwardly in compression, and not necessarily at the point where the pressure is applied. (See Sec. 905.17.)

620.00 Tetrahedron

620.01 In the conceptual process of developing the disciplines for carrying on the process of consideration, the process of temporarily putting aside the irrelevancies and working more closely for the relationships between the components that are considered relevant, we find that a geometry of configuration emerges from our awareness of the minimum considered components. A minimum constellation emerges from our preoccupation with getting rid of the irrelevancies. The geometry appears out of pure conceptuality. We dismiss the irrelevancies in the search for understanding, and we finally come down to the minimum set that may form a system to divide Universe into macrocosm and microcosm, which is a set of four items of consideration. The minimum consideration is a four-star affair that is tetrahedral. Between the four stars that form the vertexes of the tetrahedron, which is the simplest system in Universe, there are six edges that constitute all the possible relationships between those four stars.

620.02 The tetrahedron occurs conceptually independent of events and independent of relative size.

620.03 By tetrahedron, we mean the minimum thinkable set that would subdivide Universe and have interconnectedness where it comes back upon itself. The four points have six interrelatednesses. There are two kinds of number systems involved: four being prime number two and six being prime number three. So there are two very important kinds of oscillating quantities numberwise, and they begin to generate all kinds of fundamentally useful mathematics. The basic structural unit of physical Universe quantation, tetrahedron has the fundamental prime number oneness.

620.04 Around any one vertex of the tetrahedron, there are three planes. Looking down on a tetrahedron from above, we see three faces and three edges. There are these three edges and three faces around any one vertex. That seems very symmetrical and nice. You say that is logical; how could it be anything else? But if we think about it some more, it may seem rather strange because we observe three faces and three edges from an inventory of four faces and six edges. They are not the same inventories. It is interesting that we come out with symmetry around each of the points out of a dissimilar inventory.

620.05 The tetrahedron is the first and simplest subdivision of Universe because it could not have an insideness and an outsideness unless it had four vertexes and six edges. There are four areal subdivisions and four interweaving vertexes or prime convergences in its six-trajectory isolation system. The vertexial set of four local-event foci coincides with the requirement of quantum mathematics for four unique quanta numbers for each uniquely considerable quantum.

620.06 With three positive edges and three negative edges, the tetrahedron provides a vectorial quantum model in conceptual array in which the right helix corresponds to the proton set (with electron and antineutrino) and the left helix corresponds to the neutron set (with positron and neutrino). The neutron group has a fundamental leftness and the proton group has a fundamental rightness. They are not mirror images. In the tetrahedron, the two groups interact integrally. The tetrahedron is a form of energy package.

620.07 The tetrahedron is transformable, but its topological and quantum identity persists in whole units throughout all experiments with physical Universe. All of the definable structuring of Universe is tetrahedrally coordinate in rational number increments of the tetrahedron.

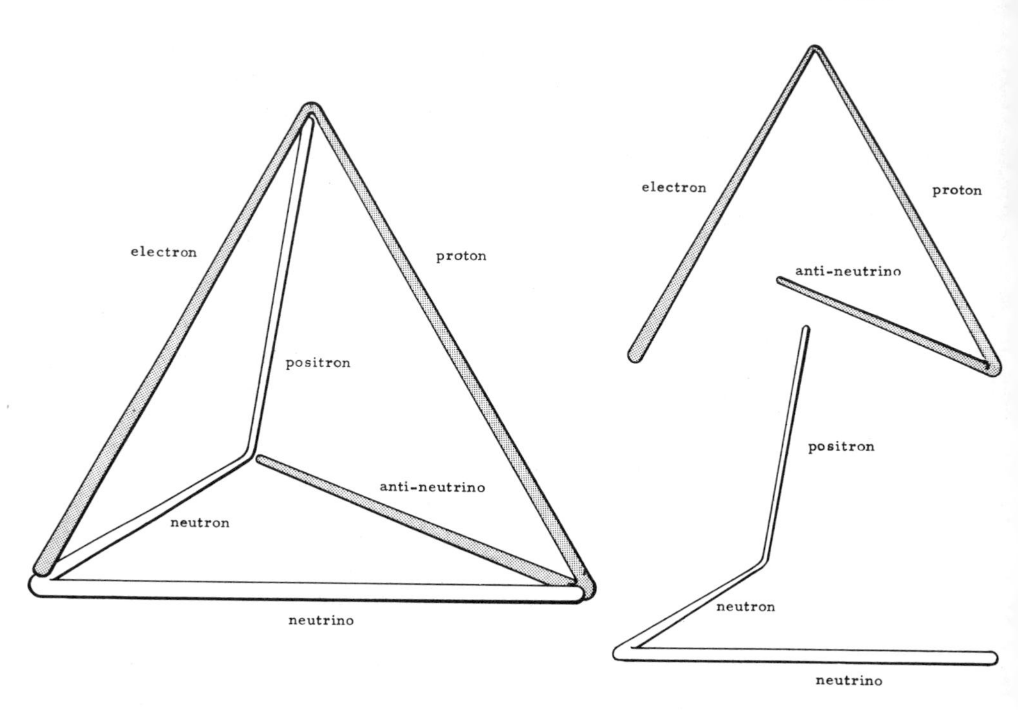

Fig. 620.06 *Tetrahedron as Vectorial Model of Quantum:* The tetrahedron as a basic vectorial model is the fundamental structural system of the Universe. The open-ended triangular spiral as action, reaction, and resultant (proton, electron, and anti-neutrino; or neutron, positron, and neutrino) becomes half quantum. An association of positive and negative half-quantum units identifies the tetrahedron as one quantum.

620.08 Organic chemistry and inorganic chemistry are both tetrahedrally-coordinate. This relates to the thinking process where the fundamental configuration came out a tetrahedron. Nature's formulations here are a very, very high frequency. Nature makes viruses in split seconds. Whatever she does has very high frequency. We come to tetrahedron as the first spontaneous aggregate of the experiences. We discover that nature is using tetrahedron in her fundamental formulation of the organic and inorganic chemistry. All structures are tetrahedrally based, and we find our thoughts resolving themselves spontaneously into the tetrahedron as it comes to the generalization of the special cases that are the physics or the chemistry.

620.09 We are at all times seeking how it can be that nature can develop viruses or billions of beautiful bubbles in the wake of a ship. How does she formulate these lovely geometries so rapidly? She must have some fundamentally pure and simple way of developing these extraordinary life cells at the rate she develops them. When we get to something as simple as finding that the tetrahedron is the minimum thinkable set that subdivides Universe and has relatedness, and that the chemist found all the structuring of nature to be tetrahedral, in some cases vertex to vertex, in others interlinked edge to edge, we find, as our thoughts go this way, that it is a very satisfying experience.

620.10 All polyhedra may be subdivided into component tetrahedra, but no tetrahedron may be subdivided into component polyhedra of less than the tetrahedron's four faces.

620.11 The triangle is the minimum polygon and the tetrahedron is the minimum structural system, for we cannot find an enclosure of less than four sides, that is to say, of less than 720 degrees of interior- (or exterior-) angle interaction. The tetrahedron is a tetrahedron independent of its edge lengths or its relative volume. In tetrahedra of any size, the angles are always sum-totally 720 degrees.

620.12 Substituting the word *tetrahedron* for the number two completes my long attempt to convert all the previously unidentifiable integers of topology into geometrical conceptuality. Thus we see both the rational energy quantum of physics and the topological tetrahedron of the isotropic vector matrix rationally accounting all physical and metaphysical systems. (See Secs. 221.01 and 424.02.)

621.00 *Constant Properties of the Tetrahedron*

621.01 Evaluated in conventional terms of cubical unity, the volume of a tetrahedron is one-third the base area times the altitude; in synergetics, however, the volume of the tetrahedron is unity and the cube is threefold unity. Any asymmetric tetrahedron will have a volume equal to any other tetrahedron so long as they have common base areas and common altitudes. (See Sec. 923.20.)

621.02 Among geometrical systems, a tetrahedron encloses the minimum volume with the most surface, and a sphere encloses the most volume with the least surface.

621.03 A cone is simply a tetrahedron being rotated. Omnidirectional growth—which means all life—can be accommodated only by tetrahedron.

621.04 There is a minimum of four unique planes nonparallel to one another. The four planes of the tetrahedron can never be parallel to one another. So there are four unique perpendiculars to the tetrahedron's four unique faces, and they make up a four-dimensional system.

621.05 Sixth-powering is all the perpendiculars to the 12 faces of the rhombic dodecahedron.

621.06 When we try to fill all space with *regular* tetrahedra, we are frus-

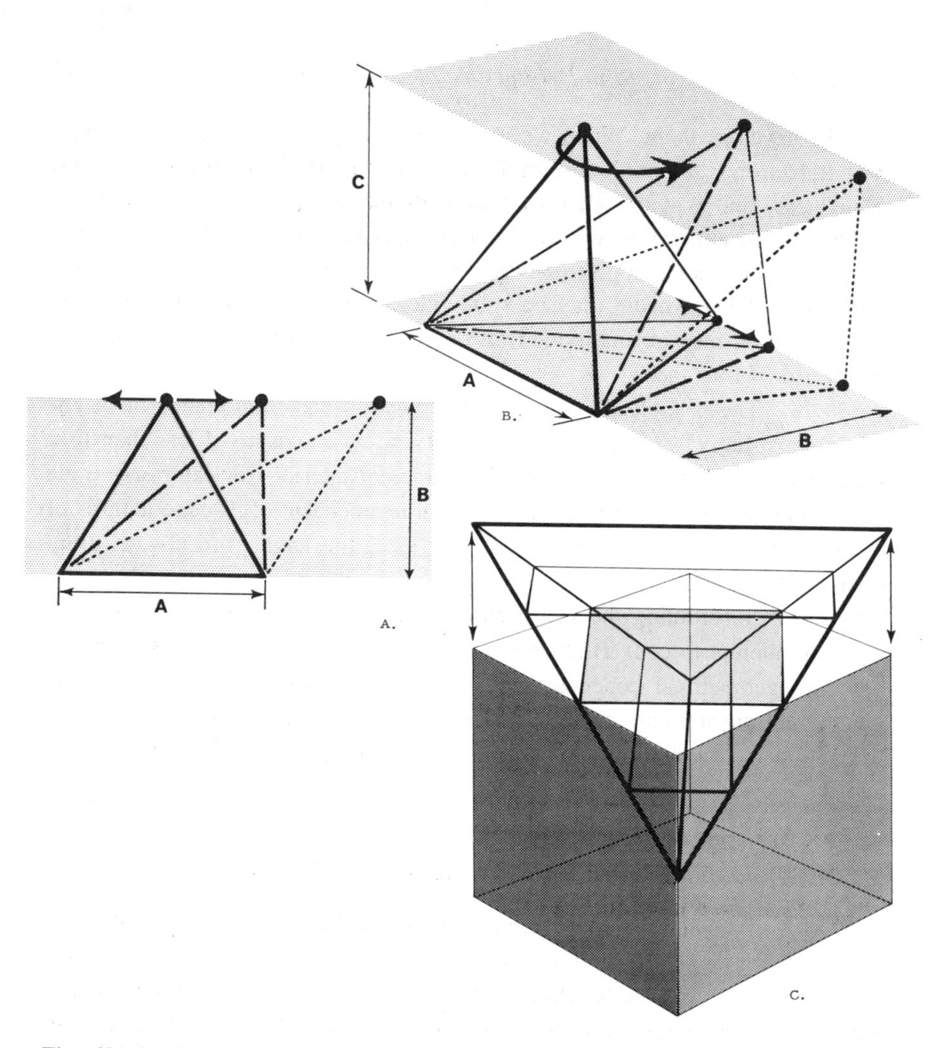

Fig. 621.01 *Constant Properties of the Tetrahedron:*

A. The area of a triangle is one-half the base times the altitude. Any arbitrary triangle will have the same area as any other triangle so long as they have a common base and altitude. Here is shown a system with two constants, *A* and *B*, and two variables—the edges of the triangle excepting *A*.

B. The volume of a tetrahedron is one-third the base area times the altitude. Any arbitrary tetrahedron will have a volume equal to any other tetrahedron so long as they have common base areas and common altitudes. Here is shown a system in which there are three constants, *A, B, C,* and five variables—all the tetrahedron edges excluding *A*.

C. As the tetrahedron is pulled out from the cube, the circumference around the tetrahedron remains equal when taken at the points where cube and tetrahedron edges cross; i.e. any *rectangular* plane taken through the regular tetrahedron will have a circumference equal to any other rectangular plane taken through the same tetrahedron, and this circumference will be twice the length of the tetrahedron edge.

trated because the tetrahedra will not fill in the voids above the triangular-based grid pattern. But the regular tetrahedron is a complementary space filler with the octahedron. Sec. 951 describes irregular tetrahedral allspace fillers.

621.07 The tetrahedron and octahedron can be produced by multilayered closest packing of spheres. The surface shell of the icosahedron can be made of any one layer—but only one layer—of closest-packed spheres; the icosahedron refuses radial closest packing.

621.10 Six Vectors Provide Minimum Stability: If we have one stick standing alone on a table, it may be balanced to stand alone, but it is free to fall in any direction. The same is true of two or three such sticks. Even if the two or three sticks are connected at the top in an interference, they are only immobilized for the moment, as their feet can slide out from under them. Four or five sticks propped up as triangles are free to collapse as a hinge action. Six members are required to complete multidimensional stability—our friend tetrahedron and the six positive, six negative degrees of freedom showing up again.

621.20 Tepee-Tripod: The tepee-tripod affords the best picture of what happens locally to an assemblage of six vectors or less. The three sides of a tepee-tripod are composed first of three vertical triangles rising from a fourth ground triangle and subsequently rocking toward one another until their respective apexes and edges are congruent. The three triangles plus the one on the ground constitute a minimum system, for they have minimum "withinness." Any one edge of our tepee acting alone, as a pole with a universal joint base, would fall over into a horizontal position. Two edges of the tepee acting alone form a triangle with the ground and act as a hinge, with no way to oppose rotation toward horizontal position except when prevented from falling by interference with a third edge pole, falling toward and into congruence with the other two poles' common vertex. The three base feet of the three poles of the tepee-tripod would slide away outwardly from one another were it not for the ground, whose structural integrity coheres the three feet and produces three invisible chords preventing the three feet from spreading. This makes the six edges of the tetrahedron. (See Secs. 521.32 and 1012.37.)

622.00 Polarization of Tetrahedron

622.01 The notion that tetrahedra lack polarity is erroneous. There is a polarization of tetrahedra, but it derives only from considering a *pair* of tetrahedral edge vectors that do not intersect one another. The opposing vector mid-edges have a polar interrelationship.

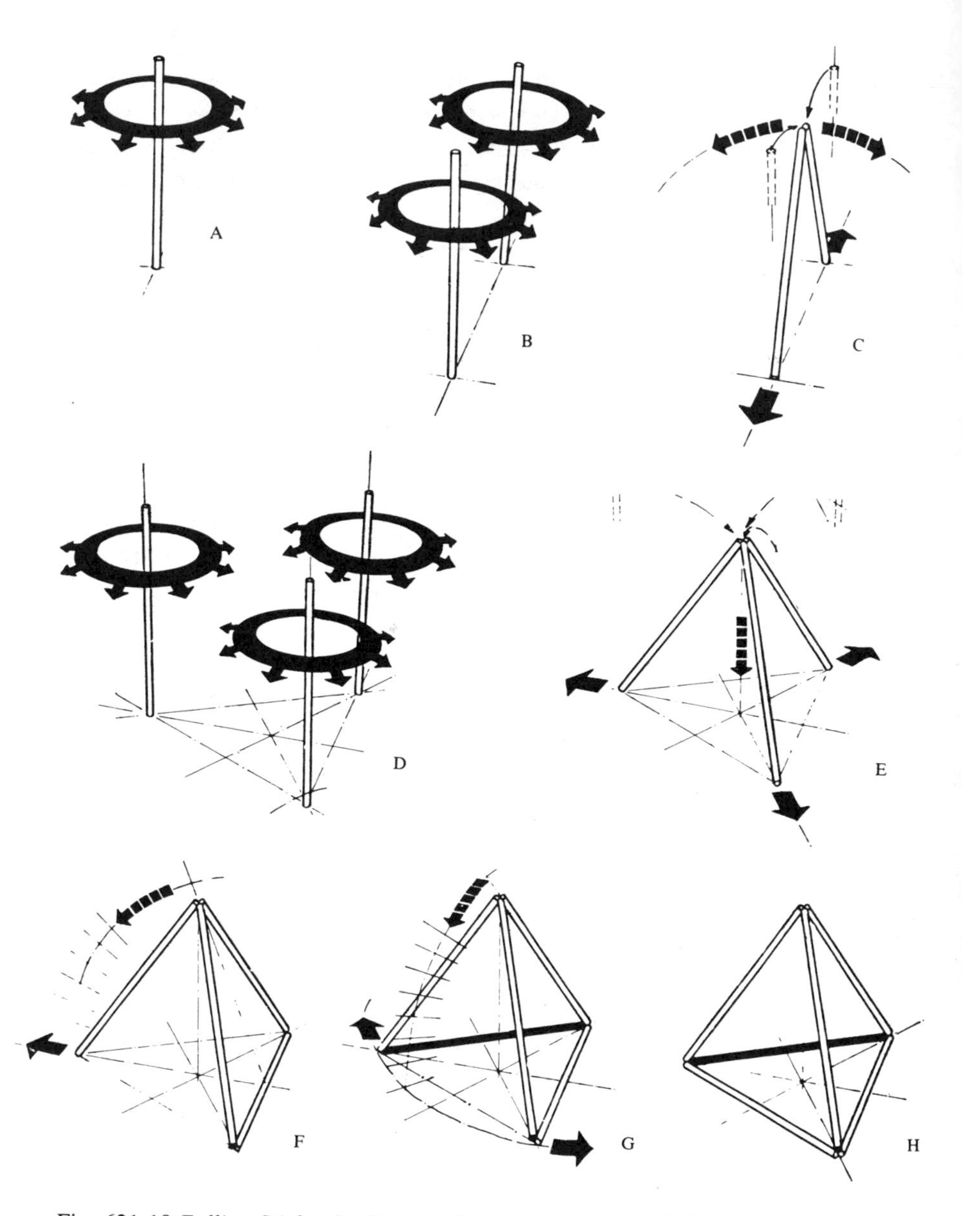

Fig. 621.10 *Falling Sticks: Six Vectors Provide Minimum Stability:*

A. Stick standing alone is free to fall in any direction.
B. Two sticks: free to fall in any direction.
C. Two sticks joined: free to fall in two directions and to slide apart at bases.
D. Three sticks: free to fall in any direction.
E. Three sticks joined: only free to slide apart at bases.
F. Four sticks: a propped-up triangle—the prop is free to slide out.
G. Five members: two triangles may collapse as with a hinge action.
H. Six members: complete multidimensional stability—the tetrahedron.

622.10 **Precessionally Polarized Symmetry**: There is a polarization of tetrahedra, but only by taking a *pair* of opposite edges which are arrayed at 90 degrees (i.e., precessed) to one another in parallelly opposite planes; and only their midpoint edges are axially opposite and do provide a polar axis of spin symmetry of the tetrahedron. There is a fourfold symmetry aspect of the tetrahedron to be viewed as precessionally polarized symmetry. (See Sec. 416.01.)

622.20 **Dynamic Equilibrium of Poles of Tetrahedron**: There is a dynamic symmetry in the relationship between the mid-action, i.e., mid-edge, points of the opposing pair of polar edges of the tetrahedron. The one dot represents the positive pole of the tetrahedron at mid-action point, i.e., action center. The other dot represents the negative pole of the tetrahedron at mid-action point, i.e., at the center of negative energy of the dynamical equilibrium of the tetrahedron.

622.30 **Spin Axis of Tetrahedron**: The tetrahedron can be spun around its negative event axis or around its positive event axis.

623.00 *Coordinate Symmetry*

623.10 **Cheese Tetrahedron**: If we take a symmetrical polyhedron of cheese, such as a cube, and slice parallel to one of its faces, what is left over is no longer symmetrical; it is no longer a cube. Slice one face of a cheese octahedron, and what is left over is no longer symmetrical; it is no longer an octahedron. If you try slicing parallel to one of the faces of all the symmetrical geometries, i.e., all the Platonic and Archimedean "solids," each made of cheese, what is left after the parallel slice is removed is no longer the same symmetrical polyhedron—but with one exception, the tetrahedron.

623.11 Let us take a foam rubber tetrahedron and compress on one of its four faces inward toward its opposite vertex instead of slicing it away. It remains symmetrical, but smaller. If we pull out on a second face at the same rate that we push in on the first face, the tetrahedron will remain the same size. It is still symmetrical, but the pushing of the first face made it get a little smaller, while the pulling of the second face made it get a little larger. By pushing and pulling at the same rate, it remains the same size, but its center of gravity has to move because the whole tetrahedron seems to move. As it moves, it receives one positive alteration and one negative alteration. But in moving it we have acted on only two of the tetrahedron's four faces. We could push in on the third face at a rate different from the first couple, which is already operating; and we could pull out the fourth face at the same rate we are pushing in on the third face. We are introducing two completely different rates of change: one being very fast and the other slow; one being very hard and the other soft. We are introducing two completely different rates of change in physical energy or change in abstract metaphysical conceptuality. These completely different rates are coupled so that the tetrahedron as a medium of exchange remains both symmetrical and the same size, but it has to change its position to accommodate two alterations of the center of gravity positioning but not in the same plane or the same line. So it will be moving in a semihelix. This is another manifestation of precessional resultants.

623.12 The tetrahedron's four faces may be identified as A, B, C, and D. Any two of these four faces can be coupled and can be paired with the other two to provide the dissimilar energy rate-of-exchange accommodation. $\frac{N^2-N}{2}=$ the number of relationships. In this case, $N=4$, therefore, $\frac{16-4}{2}=6$. There are six possible couples: *AB, AC, AD, BC, BD, CD,* and these six couples may be interpaired in $\frac{N^2-N}{2}$ ways; therefore, $\frac{36-6}{2}=15$; which 15 ways are:

(1) *AB-AC*	(6) *AC-AD*	(11) *AD-BD*
(2) *AB-AD*	(7) *AC-BC*	(12) *AD-CD*
(3) *AB-BC*	(8) *AC-BD*	(13) *BC-BD*
(4) *AB-BD*	(9) *AC-CD*	(14) *BD-CD*
(5) *AB-CD*	(10) *AD-BC*	(15) *BD-CD*

Thus any one tetrahedron can accommodate 15 different *amplitude* (A) and, or *frequency* (F) of interexchanging without altering the tetrahedron's size while, however, always changing the tetrahedron's apparent occurrence locale; therefore the number of possible alternative exchanges are three; i.e., *AA, AF, FF;* therefore, $3\times 15=45$ different combinations of *interface couplings* and message contents can be accommodated by the same apparent unit-size tetrahedron, the only resultants of which are the 15 relocations of the tetrahedrons and the 45 different message accommodations.

623.13 Tetrahedron has the extraordinary capability of remaining symmetrically coordinate and entertaining 15 pairs of completely disparate rates of change of three different classes of energy behaviors in respect to the rest of Universe and not changing its size. As such, it becomes a universal joint to couple disparate actions in Universe. So we should not be surprised at all to find nature using such a facility and moving around Universe to accommodate all kinds of local transactions, such as coordination in the organic chemistry or in the metals. The symmetry, the fifteeness, the sixness, the fourness, and the threeness are all constants. This induced "motion," or position displacement, may explain all apparent motion of Universe. The fifteenness is unique to the icosahedron and probably valves the 15 great circles of the icosahedron.

623.14 A tetrahedron has the strange property of *coordinate symmetry,* which permits local alteration without affecting the symmetrical coordination of the whole. This means it is possible to receive changes in respect to one part or direction of Universe and not in the direction of the others and still have the symmetry of the whole. In contradistinction to any other Platonic or Archimedean symmetrical "solid," only the tetrahedron can accommodate local asymmetrical addition or subtraction without losing its cosmic symmetry. Thus the tetrahedron becomes the only exchange agent of Universe that is not itself altered by the exchange accommodation.

623.20 Size Comes to Zero: There are three different aspects of size—linear, areal, and volumetric—and each aspect has a different velocity. As you move one of the tetrahedron's faces toward its opposite vertex, it gets smaller and smaller, with the three different velocities operative. But it always remains a tetrahedron with six edges, four vertexes, and four faces. So the symmetry is not lost and the fundamental topological aspect—its 60-degreeness—never changes. As the faces move in, they finally become congruent to the opposite vertex as all three velocities come to zero at the same time. The 60-degreeness, the six edges, the four faces, and the symmetry were never altered because they were not variables. The only variable was size. Size alone can come to zero. The conceptuality of the other aspects never changes.

624.00 Inside-Outing of Tetrahedron

624.01 The tetrahedron is the only polyhedron, the only structural system that can be turned inside out and vice versa by one energy event.

624.02 You can make a model of a tetrahedron by taking a heavy-steel-rod triangle and running three rubber bands from the three vertexes into the center of gravity of the triangle, where they can be tied together. Hold the three rubber bands where they come together at the center of gravity. The inertia of the steel triangle will make the rubber bands stretch, and the triangle becomes a tetrahedron. Then as the rubber bands contract, the triangle will lift again. With such a triangle dangling in the air by the three stretched rubber bands, you can suddenly and swiftly plunge your hand forth and back through the relatively inert triangle . . . making first a positive and then a negative triangle. (In the example given in Sec. 623.20, the opposite face was pumped through the inert vertex. It can be done either way.) This kind of oscillating pump is typical of some of the atom behaviors. An atomic clock is just such an oscillation between a positive and a negative tetrahedron.

624.03 Both the positive and negative tetrahedra can locally accommodate the 45 different energy exchange couplings and message contents, making 90 such accommodations all told. These accommodations would produce 30 different "apparent" tetrahedron position shifts, whose successive movements would always involve an angular change of direction producing a helical trajectory.

624.04 The extensions of tetrahedral edges through any vertex form positive-negative tetrahedra and demonstrate the essential twoness of a system.

624.05 The tetrahedron is the minimum, convex-concave, omnitriangulated, compound curvature system, ergo, the minimum sphere. We discover that the additive twoness of the two polar (and a priori awareness) spheres at most economical minimum are two tetrahedra and that the insideness and outsideness complementary tetrahedra altogether represent the two invisible complementary twoness that balances the visible twoness of the polar pair.

624.06 When we move one of the tetrahedron's faces beyond congruence with the opposite vertex, the tetrahedron turns inside out. An inside-out tetrahedron is conceptual and of no known size.

624.10 **Inside Out by Moving One Vertex:** The tetrahedron is the only polyhedron that can be turned inside out by moving one vertex within the prescribed linear restraints of the vector interconnecting that vertex with the other vertexes, i.e., without moving any of the other vertexes.

624.11 Moving one vertex of an octahedron within the vectorial-restraint limits connecting that vertex with its immediately adjacent vertexes (i.e., without moving any of the other vertexes), produces a congruence of one-half of the octahedron with the other half of the octahedron.

624.12 Moving one vertex of an icosahedron within the vectorial-constraint limits connecting that vertex with the five immediately adjacent vertexes (i.e., without moving any of the other vertexes), produces a local inward dimpling of the icosahedron. The higher the frequency of submodulating of the system, the more local the dimpling. (See Sec. 618.)

625.00 *Invisible Tetrahedron*

625.01 The Principle of Angular Topology (see Sec. 224) states that the sum of the angles around all the vertexes of a structural system, plus 720 degrees, equals the number of vertexes of the system multiplied by 360 degrees. The tetrahedron may be identified as the 720-degree differential between any definite local geometrical system and finite Universe. Descartes discovered the 720 degrees, but he did not call it the tetrahedron.

625.02 In the systematic accounting of synergetics angular topology, the sum of the angles around each geodesically interrelated vertex of every definite concave-convex local system is always two vertexial unities less than universal, nondefined, finite totality.

625.03 We can say that the difference between any conceptual system and total but nonsimultaneously conceptual—and therefore nonsimultaneously sensorial—scenario Universe, is always one exterior tetrahedron and one interior tetrahedron of whatever sizes may be necessary to account for the balance of all the finite quanta thus far accounted for in scenario Universe outside and inside the conceptual system considered. (See Secs. 345 and 620.12.)

625.04 Inasmuch as the difference between any conceptual system and total Universe is always two weightless, invisible tetrahedra, if our physical conceptual system is a regular equiedged tetrahedron, then its complementation may be a weightless, metaphysical tetrahedron of various edge lengths—ergo, non-mirror-imaged—yet with both the visible and the invisible tetrahedra's corner angles each adding up to 720 degrees, respectively, though one be equiedged and the other variedged.

625.05 The two invisible and *n*-sized tetrahedra that complement all systems to aggregate sum totally as finite but nonsimultaneously conceptual scenario Universe are mathematically analogous to the "annihilated" left-hand phase of the rubber glove during the right hand's occupation of the glove. The difference between the sensorial, special-case, conceptually measurable, finite, separately experienced system and the balance of nonconceptual scenario Universe is two finitely conceptual but nonsensorial tetrahedra. We can say that scenario Universe is finite because (though nonsimultaneously conceptual and considerable) it is the sum of the conceptually finite, after-image-furnished thoughts of our experience systems plus two finite but invisible, *n*-sized tetrahedra.

625.06 The tetrahedron can be turned inside out; it can become invisible. It can be considered as antitetrahedron. The exterior invisible complementary tetrahedron is only concave having only to embrace the convexity of the visible system and the interior invisible complementary tetrahedron is only convex to marry the concave inner surface of the system.

626.00 Operational Aspects of Tetrahedra

626.01 The world military forces use reinforced concrete tetrahedra for military tank impediments. This is because tetrahedra lock into available space by friction and not by fitting. They are used as the least disturbable barrier components in damming rivers temporarily shunted while constructing monolithic hydroelectric dams.

626.02 The tetrahedron's inherent refusal to fit allows it to get ever a little closer; in not fitting additional space, it is always available to accommodate further forced intrusions. The tetrahedron's edges and vertexes scratch and dig in and thus produce the powerfully locking-in-place frictions . . . while stacks of neatly fitting cubes just come apart.

626.03 This is why stone is crushed to make it less spherical and more tetrahedral. This is why beach sand is not used for cement; it is too round. Spheres disassociate; tetrahedra associate spontaneously. The limit conditions involved are the inherent geometrical limit conditions of the sphere enclosing the most volume with the least surface and the fewest angular protrusions, while the tetrahedron encloses the least volume with the most surface and does so with most extreme angular vertex protrusion of any regular geometric forms. The sphere has the least interfriction surface with other spheres and the greatest mass to restrain interfrictionally; while the tetrahedra have the most interfriction, interference surface with the least mass to restrain.

630.00 Antitetrahedron

631.00 Minimum of Four Points

631.01 We cannot produce constructively and operationally a real experience-augmenting, omnidirectional system with less than four points. A fourth point cannot be in the plane approximately located, i.e., described, by the first three points, for the points have no dimension and are unoccupiable as is also the plane they "describe." It takes three points to define a plane. The fourth point, which is not in the plane of the first three, inherently produces a tetrahedron having insideness and outsideness, corresponding with the reality of operational experience.

631.02 The tetrahedron has four unique planes described by the four possible relationships of its four vertexes and the six edges interconnecting them. In a regular tetrahedron, all the faces and all the edges are assumed to be approximately identical.

632.00 Dynamic Symmetry of the Tetrahedron

632.01 There is a symmetry of the tetrahedron, but it is inherently four-dimensional and related to the four planes and the four axes projected perpendicularly to those planes from their respective subtending vertexes. But the tetrahedron lacks three-dimensional symmetry due to the fact that the subtending vertex is only on one side of the triangular plane, and due to the fact that the center of gravity of the tetrahedron is always only one-quarter of its altitude irrespective of the seeming asymmetry of the tetrahedron.

632.02 The dynamic symmetry of the tetrahedron involves the inward projection of four geodesic connectors with the center of area of the triangular face opposite each vertex of the tetrahedron (regular or maxi-asymmetrical); which four vertex-to-opposite-triangle geodesic connectors will all pass through the center of gravity of the tetrahedron—regular, mini- or maxi-asymmetric; and the extension of those geodesics thereafter through the four centers of gravity of those four triangular planes, outwardly from the tetrahedron to four new vertexes equidistant outwardly from the three corners of their respective four basal triangular facet planes of the original tetrahedron. The four exterior vertexes are equidistant outwardly from the original tetrahedron, a distance equal to the interior distances between the centers of gravity of the original tetrahedron's four faces and their inwardly subtending vertexes. This produces four regular tetrahedra outwardly from the four faces of the basic tetrahedron and triple-bonded to the original tetrahedron.

632.03 We have turned the tetrahedron inside out in four different directions and each one of the four are dimensionally similar. This means that each of the four planes of the tetrahedron produces four new points external to the original tetrahedron, and four similar tetrahedra are produced outwardly from the four faces of the original tetrahedron; these four external points, if interconnected, produce one large tetrahedron, whose six edges lie outside the four externalized tetrahedra's 12 external edges.

633.00 *Negative Tetrahedron*

633.01 As we have already discovered in the vector equilibrium (see Sec. 480), each tetrahedron has its negative tetrahedron produced through its interior apex rather than through its outer triangular base. In the vector equilibrium, each tetrahedron has its negative tetrahedron corresponding in dynamic symmetry to its four-triangled, four-vertexed, fourfold symmetry requirement. And all eight (four positive and four negative) tetrahedra are clearly present in the vector equilibrium. Their vertexes are congruent at the center of the vector equilibrium. Each of the tetrahedra has one internal edge circumferentially congruent with the other tetrahedra's edge, and each of the tetrahedra's three internal edges is thus double-bonded circumferentially with three other tetrahedra, making a fourfold cluster in each hemisphere. This exactly balances a similarly bonded fourfold cluster in its opposite hemisphere, which is double-bonded to their hemisphere's fourfold cluster by six circumferentially double-bonded, internal edges. Because there are four equatorial planes of symmetry of the vector equilibrium, there are four different sets of the fourfold tetrahedra clusters that can be differentiated one from the others.

633.02 Each of the eight tetrahedra symmetrically surrounding the nucleus of the vector equilibrium can serve as a nuclear domain energy valve, and each can accommodate 15 alternate intercouplings and three types of message contents; wherefore, the vector equilibrium cosmic nucleus system can accommodate $4 \times 45 = 180$ positive, and $4 \times 45 = 180$ negative, uniquely different energy—or information—transactions at four frequency levels each. We may now identify (a) the four positive-to-negative-to positive, triangular intershuttling transformings within each cube of the eight corner cubes of the two-frequency cube (see Sec. 462 *et seq.*); with (b) the 360 nuclear tetrahedral information valvings as being cooperatively concurrent functions within the same prime nuclear domain of the vector equilibrium; they indicate the means by which the electromagnetic, omniradiant wave propagations are initially articulated.

634.00 Irreversibility of Negative Tetrahedral Growth

634.01 When the dynamic symmetry is inside-outingly developed through the tetrahedron's base to produce the negatively balancing tetrahedron, only the four negative tetrahedra are externally visible, for they hide entirely the four positive triangular faces of the positive tetrahedron's four-base, four-vertex, fourfold symmetry. The positive tetrahedron is internally congruent with the four internally hidden, triangular faces of the four surrounding negative tetrahedra. This is fundamental irreversibility: the outwardly articulated dynamic symmetry is not regeneratively procreative in similar tetrahedral growth. The successive edges of the overall tetrahedron will never be rationally congruent with the edges of the original tetrahedron. This growth of dissimilar edges may bring about all the different frequencies of the different chemical elements.

635.00 Base-Extended Tetrahedron

635.01 The tetrahedron extended through its face is pumpingly or diaphragmatically inside-outable, in contradistinction to the vertexially extended tetrahedron. The latter is single-bonded (univalent); the former is triple-bonded and produces crystal structures. The univalent, single-bonded universal joint produces gases.

636.00 Complementary to Vector Equilibrium

636.01 In the vector equilibrium, we have all the sets of tetrahedra bivalently or edge-joined, i.e., liquidly, as well as centrally univalent. Synergetics calls the basally developed larger tetrahedron the *non-mirror-imaged complementary* of the vector equilibrium.* In vectorial-energy content and dynamic-symmetry content lies the complementarity.

637.00 *Star Tetrahedron*

637.01 The name of this dynamic vector-equilibrium complementary tetrahedron is the *star tetrahedron.* The star tetrahedron is one in which the vectors are no longer equilibrious and no longer omnidirectionally and regeneratively extensible. This star tetrahedron name was given to it by Leonardo da Vinci.

637.02 The star tetrahedron consists of five equal tetrahedra, four external and one internal. Because its external edges are not 180-degree angles, it has 18—instead of six—equi-vector external edges: 12 outwardly extended and six inwardly valleyed; ergo, a total of 18. It is a compound structure. Four of its five tetrahedra, which are nonoutwardly regenerative in unit-length vectors, ergo, non-allspace-filling, are in direct correspondence with the five four-ball tetrahedra which do close-pack to form a large, regular, three-frequency tetrahedron of four-ball edges, having one tetrahedral four-ball group at thc center rather than an octahedral group as is the case with planar and linear topological phenomena. This is not really contradictory because the space inside the four-ball tetrahedron is always a small concave octahedron, wherefore, an octahedron is really at the center, though not an octahedron of six balls as at the center of a four, four-ball tetrahedral "pyramid."

638.00 *Pulsation of Antitetrahedra*

638.01 The star tetrahedron is a structure—but it is a compound structure. The fifth tetrahedron, which is the original one, and only nuclear one accommodates the pulsations of the outer four. Its outward pulsings are broadcast, and its inward pulsings are *re*pulsive—that is why it is a star. The four three-way—12 in total—external pulsations are unrestrained, and the internal pulsations are compressionally repulsed. Leonardo called it the star tetrahedron, not because it has points, but because he sensed intuitively that it gives off radiation like a star. The star tetrahedron is an impulsive-expulsive transceiver whose four, 12-faceted, exterior triangles can either (a) feed in cosmic energy receipts which spontaneously articulate one or another of the 15 interpairings of the six *A, B, C, D,* interior tetrahedron's couplings, or (b) transmit through one of the external tetrahedra whose respective three faces each must be refractively pulsated once more to beam or broadcast the 45 possible *AA, AF, FF* messages.

* The non-mirror-imaged complementary is not a negative vector equilibrium. The vector equilibrium has its own integral negative.

638.02 There is a syntropic pulsation receptivity and an outward pulsation in dynamic symmetry of the star tetrahedron. As an energy radiator, it is entropic. It does not regenerate itself internally, i.e., gravitationally, as does the isotropic vector matrix's vector equilibrium. The star tetrahedron's entropy may be the basis of irreversible radiation, whereas the syntropic vector equilibrium's reversibility—inwardly-outwardly—is the basis for the gravitationally maintained integrity of Universe. The vector equilibrium produces conservation of omnidynamic Universe despite many entropic local energy dissipations of star tetrahedra. The star tetrahedron is in balance with the vector equilibrium—pumpable, irreversible, like the electron in behavior. It has the capability of self-positionability by converting its energy receipts to unique refraction sequences, which could change output actions to other dynamic, distances-keeping orbits, in respect to the—also only remotely existent and operating—icosahedron, and its 15 unique, great-circle self-dichotomizing; which icosahedra can only associate with other icosahedra in either linear-beam export or octahedral orbital hover-arounds in respect to any vector equilibrium nuclear group. (See Sec. 1052.)

638.03 The univalent antitetrahedra twist but do not pump. The single-bonded tetrahedra are also inside-outable, but by torque, by twist, and not by triangular diaphragm pumping. The lines of the univalent antitetrahedron are non-self-interfering. Like the lamp standards at Kennedy International Airport, New York, the three lines twist into plus (+) and minus (−) tetrahedra. *MN* and *OP* are in the same plane, with *A* and *A′* on the opposite sides of the plane. So you have a *vertexial* inside-out twisting and a *basal* inside-out pumping.

638.10 **Three Kinds of Inside-Outing:** Of all the Platonic polyhedra, only the tetrahedron can turn inside out. There are three ways it can do so: by single-, double-, and triple-bonded routes. In double-bonded, edge-to-edge inside-outing, there are pairs of diametric unfoldment of the congruent edges, and the diameter becomes the hinge of reverse positive and negative folding.

639.00 Propagation

639.01 The star tetrahedron is nonreversible. It can only propagate outwardly. (The vector equilibrium can keep on reproducing itself inwardly or outwardly, gravitationally.) The star tetrahedron's four external tetrahedra cannot regenerate themselves; but they are external-energy-receptive, whether that energy be tensive or pressive. The star tetrahedron consists only of *A* Modules; it has no *B* Modules. The star tetrahedron may explain a whole new phase of energetic Universe such as, for instance, Negative Universe.

639.02 The vector equilibrium's closest-packed sphere shell builds outwardly to produce successively the neutron and proton counts of the 92 regenerative chemical elements. The star tetrahedron may build negatives for the post-uraniums. The star tetrahedron's six potential geodesic interconnectors of the star tetrahedron's outermost points are out of vector-length frequency-phase and generate different frequencies each time they regenerate; they expand in size due to the self-bulging effects of the 15 energy message pairings of the central tetrahedron. Because their successive new edges are noncongruent with the edges of the original tetrahedron, the new edge will never be equal to or rational with the original edge. Though they produce a smooth-curve, ascending progression, they will always be shorter—but only a very little bit shorter—than twice the length of the original edge vectors. Perhaps this shortness may equate with the shortening of radial vectors in the transition from the vector equilibrium's diameter to the icosahedron's diameter. (See Sec. 460, Symmetrical Contraction of Vector Equilibrium.) This is at least a contraction of similar magnitude, and mathematical analyses may show that it is indeed the size of the icosahedron's diameter. The new edge of the star tetrahedron may be the same as the reduced radius of the icosahedron. If it is, the star tetrahedron could be the positron, as the icosahedron seems to be the electron. These relationships should be experimentally and trigonometrically explored, as should all the energy-experience inferences of synergetics. The identifications become ever more tantalizingly close.

640.00 Tension and Compression

640.01 One cannot patent geometry per se nor any separately differentiated out, pure principle of nature's operative processes. One can patent, however, the surprise complex behaviors of associated principles where the behavior of the whole is unpredicted by the behavior of the parts, i.e., synergetic phenomena. This is known as invention, a complex arrangement not found in, but permitted by, nature, though it is sometimes superficially akin to a priori natural systems, formulations, and processes. Though superficially similar in patternings to radiolaria and flies' eyes, geodesic structuring is true invention. Radiolaria collapse when taken out of water. Flies' eyes do not provide human-dwelling precedent or man-occupiable, environment-valving structures.

640.02 Until the introduction of geodesic structures, structural analysis and engineering-design strategies regarding clear-span structural enclosures in general, and domical structures in particular, were predicated upon the stress analysis of individual beams, columns, and cantilevers as separate components and thereafter as a solid compressional shell with no one local part receiving much, if any, aid from other parts. Their primarily compressional totality was aided here and there by tensional sinews, but tension was a discontinuous local aid and subordinate. As academically constituted in the middle of this 20th-century, engineering could in no way predict, let alone rely upon, the synergetic behaviors of geodesics in which any one, several, or many of the components could be interchangeably removed without in any way jeopardizing the structural-integrity cohering of the remaining structure. Engineering was, therefore, and as yet is, utterly unable to analyze effectively and correctly tensegrity geodesic structural spheres in which none of the compression members ever touch one another and only the tension is continuous.

640.03 It appeared and as yet appears to follow, in conventional, state-licensed structural engineering, that if tension is secondary and local in all men's structural projections, that tension must also be secondary in man's philosophic reasoning. As a consequence, the popular conception of airplane flight was, at first and for a long time, erroneously explained as a compressional push-up force operating under the plane's wing. It ''apparently'' progressively compressed the air below it, as a ski compresses the snow into a grooved track of icy slidability. The scientific fact remains, as wind-tunnel experiments proved, that three-quarters of the airplane's weight support is furnished by the negative lift of the partial vacuum created atop the airfoil. This is simply because, as Bernoulli showed, it is longer for the air to go around the top of the foil than under the foil, and so the same amount of air in the same amount of time had to be stretched thinner, ergo vacuously, over the top. This stretching thinner of the air, and its concomitant greater effectiveness of interpositioning of bodies (that is, the airplane in respect to Earth), is our same friend, the astro- and nucleic-tensional integrity of dynamic interpatterning causality.

640.10 Slenderness Ratio: Compression members have a limit ratio of length to section: we call it the slenderness ratio. The compression member may very readily break if it is too long. But there is no limit of cross section to length in a tension member; there is no inherent ratio.

640.11 The Greeks, who built entirely in compression, discovered that a stone column's slenderness ratio was approximately 18 to 1, length to diameter. Modern structural-steel columns, with an integral tensional fibering unpossessed by these stone columns, have a limit slenderness ratio of approximately 33 to 1. If we have better metallurgical alloys, we can make longer and longer tension members with less and less section—apparently ad infinitum. But we cannot make longer compression columns ad infinitum.

640.12 If we try to load a slender column axially—for instance, a 36-inch-long by 1/8-inch-diameter steel rod—it tends to bend in any direction away from its neutral axis. If, however, we take a six-inch-diameter bundle of 36-inch-long by 1/8-inch-diameter rods compacted parallel to one another into a closest-packed, hexagonally cross-sectioned bundle, bind them tensionally with circumferential straps in planes at 90 degrees to the axis of the rods, around the bundle's six-inch girth, and then cap both ends of the tightly compacted, hexagonally cross-sectioned bundle with tightly fitting, forged-steel, hexagonal caps, we will have a bundle that will act together as a column. If we now load this 36-inch-high column axially under an hydraulic press, we discover that because each rod could by itself be easily bent, but they cannot bend toward one another because closest packed, they therefore bend away from one another as well as twisting circumferentially into an ever-fattening, twisting cigar that ultimately bursts its girth-tensed bonds. So we discover that our purposeful compressing axially of the bundle column is resulting in tension being created at 90 degrees to our purposeful compressing.

640.20 **Sphere: An Island of Compression:** Aiming of the compressional loading of a short column into the neutral or central-most axis of the column provides the greatest columnar resistance to the compressing because, being the neutral axis, it brings in the most mass coherence to oppose the force. To make a local and symmetrical island of compression from a short column that axial loading has progressively twisted and expanded at girth into a cigar shape, you have to load it additionally along its neutral axis until the ever-fattening cigar shape squashes into a sphere. In the spherical condition, for the first and only time, any axis of the structure is neutral—or in its most effective resistant-to-compression attitude. It is everywhere at highest compression and tension-resisting capability to withstand any forces acting upon it.

640.21 It is not surprising, in view of these properties, that ball bearings prove to be the most efficient compression members known to and ever designedly produced by man. Nor are we surprised to find all the planets and stars to be approximately spherical mass aggregations, as also are the atoms, all of which spherical islands of the macrocosmic and microcosmic aspects of scenario Universe provide the comprehensive, invisible, tensional, gravitational, electromagnetic, and amorphous integrity of Universe with complementarily balancing internality of compressionally most effective, locally and temporarily visible, islanded compressional entities. It is also not surprising, therefore, that Universe islands its spherical compression aggregates and co-

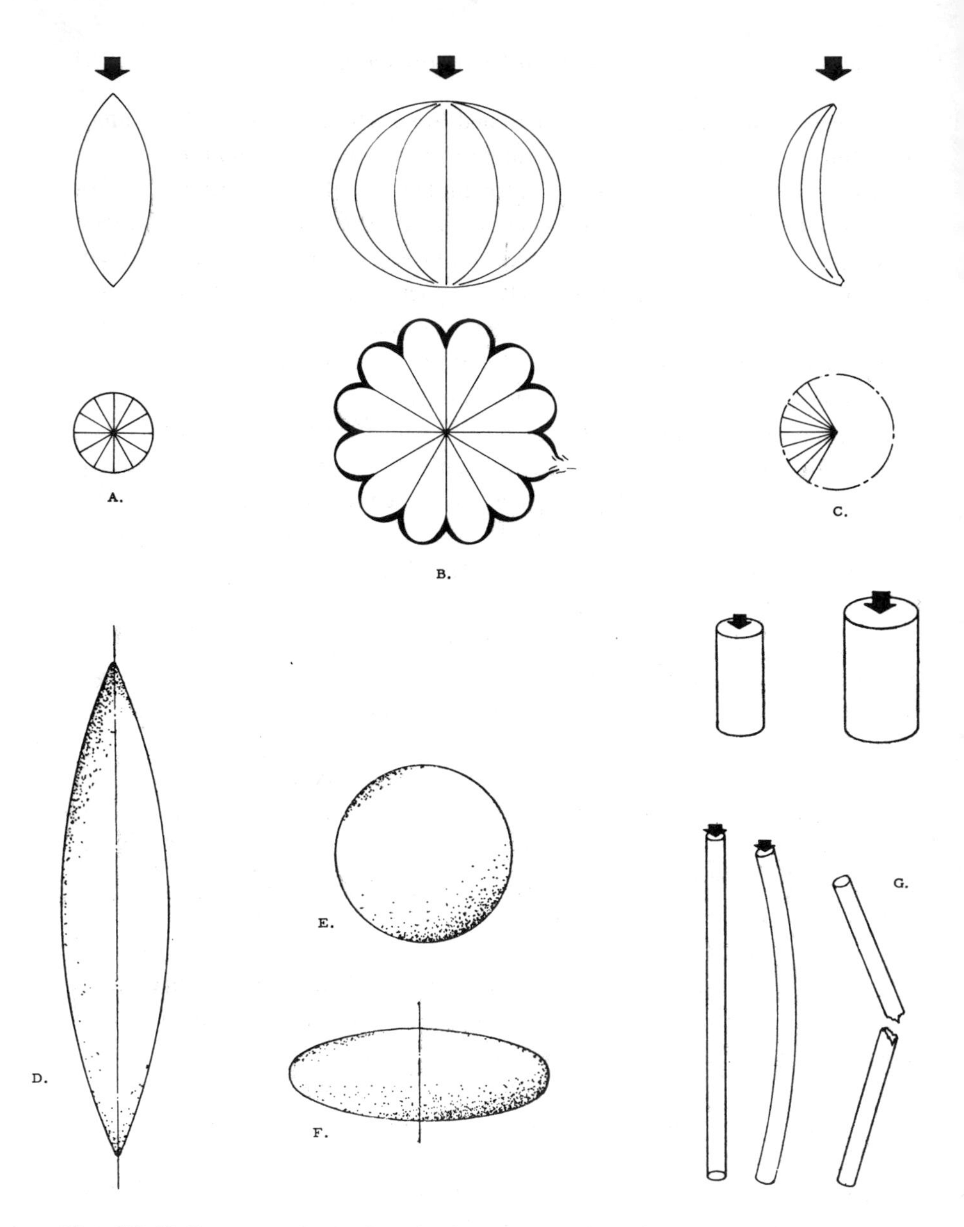

Fig. 640.20 *Compression Members Under Stress:* A cigar shape (A) (with radials short and compact) under pressure in its long axes goes to squash shape (B) (radials long and separating) or to banana shape (C) (radials longer and collecting). Note that on the squash the stretching edge gets thinner and breaks. The cigar (D) has only one neutral axis: axial or polar exaggerated asymmetry. The sphere (E) has an infinity of equi-neutral axes: symmetry. The disc (F) has only one neutral axis: equatorially exaggerated asymmetry. Compression columns (G) tend toward axes of ever lesser radius. As columns become longer in respect to their cross section (slenderness ratio) they tend to flex and break into two shorter columns in an attempt to restore a desirable slenderness ratio.

heres the whole exclusively with tension; discontinuous compression and continuous tension: I call this tensional integrity of Universe *tensegrity*.

640.30 **Precession and Critical Proximity:** Compressions are always local and, when axially increased beyond the column-into-cigar-into-sphere stage of optimum compression-resisting effectiveness, they tend toward edge-sinused, lozenge shapes, then into edge-fractionated discs, and thereafter into a plurality of separately and visibly identifiable entities separating inwardly in a plane at 90 degrees to the compressional forces as the previously neighboring atoms became precessionally separated from one another beyond the critical threshold between the falling-inward, massive integrity coherence proclivities of islanded ''matter''—beyond that proclivity threshold of critical proximity, now to yield precessionally at 90 degrees to participate in the remotely orbiting patterns characterizing 99.99 percent of all the celestially accountable time-distance void of known Universe.

640.40 **Wire Wheel:** In the high- and low-tide cooperative precessional functionings of tension versus compression, I saw that there are times when each are at half tide, or equally prominent in their system relationships. I saw that the exterior of the equatorial compressional island rim-atoll of the wire wheel must be cross-sectionally in tension as also must be its hub-island's girth. I also saw that all these tension-vs-compression patterning relationships are completely reversible, and are entirely reversed, as when we consider the compressively spoked artillery wheel vs. the tensionally spoked wire wheel. I followed through with the consideration of these differentiable, yet complementarily reversible, functions of structural systems as possibly disclosing the minimum or fundamental set of differentiability of nonredundant, precessionally regenerative structural systems. (See Sec. 537.04.)

640.41 As I considered the 12 unique vectors of freedom constantly and nonredundantly operative between the two poles of the wire wheel—its islanded hub and its islanded equatorial rim-atoll, in effect a Milky Way-like ring of a myriad of star islands encircling the hub in a plane perpendicular to the hub axis—I discerned that this most economic arrangement of forces might also be that minimum possible system of nature capable of displaying a stable constellar compressional discontinuity and tensional continuity. A *one*-island system of compression would be an inherently continuous compression system, with tension playing only a redundant and secondary part. Therefore, a one-island system may be considered only as an optically illusory ''unitary'' system, for, of course, at the invisible level of atomic structuring, the coherence of the myriad atomic archipelagos of the ''single'' pebble's compression-island's mass is sum-totally and only provided by comprehensively con-

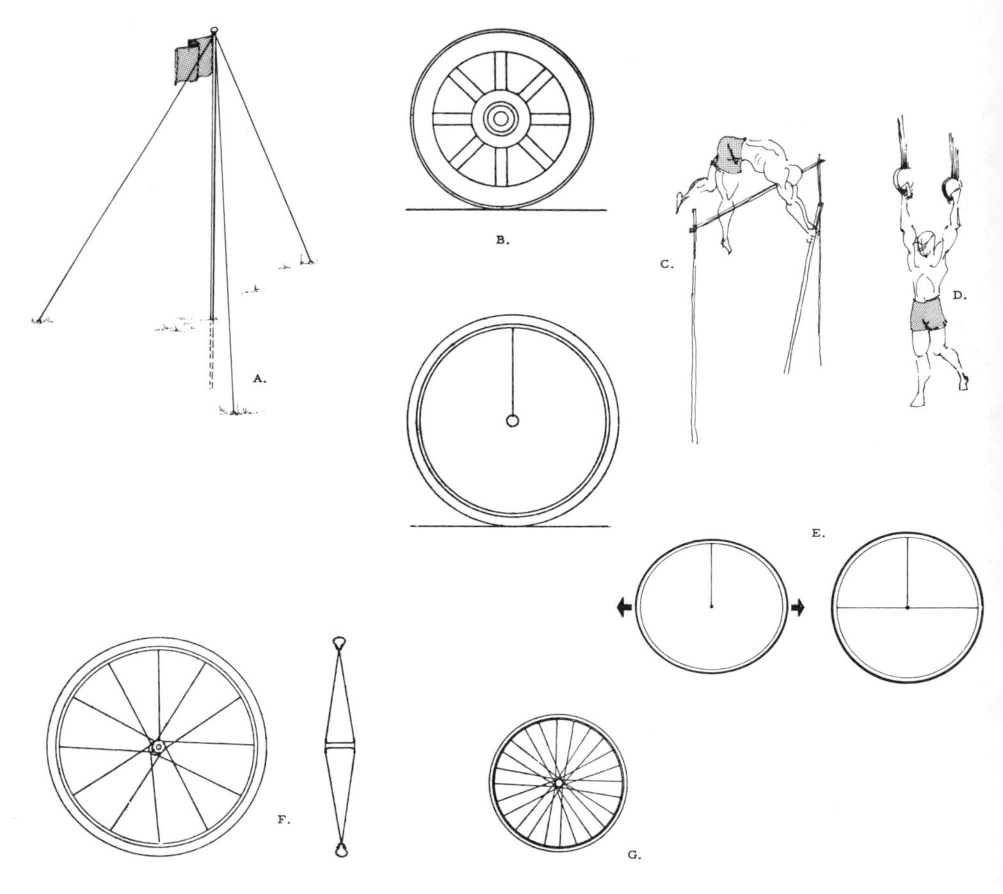

Fig. 640.41A *Stabilization of Tension: Minimum of 12 Spokes:*

A. A solid mast without stays stands erect by itself in ''solid'' earth. Tension stays may be added at end of the lever arm helping against hurricane ''uprooting.'' Men have until now employed a compression continuity as the primary load-carrying structural system with tension employed secondarily to stabilize angular relationships.

B. The old artillery wheel provides a series of vaulting poles.

C. Pole vaulting along, a ''pushing-up'' load.

D. Hanging in tension like the wire wheel.

E. The wire wheel provides a series of tension slings. The axle load of the wire wheel is hung from the top of the wheel, which tries to belly out, so spokes as additional tension members are added horizontally to keep it from bellying.

F. It takes a minimum of 12 spokes to fix the hub position in relation to the rim: six positive diaphragm and six negative diaphragm, of which respectively three each are positively and negatively opposed turbining torque members.

G. Many spokes keep rim from bending outwardly any further while load is suspended by central vertical spokes successively leading from top of wheel to hub and its load.

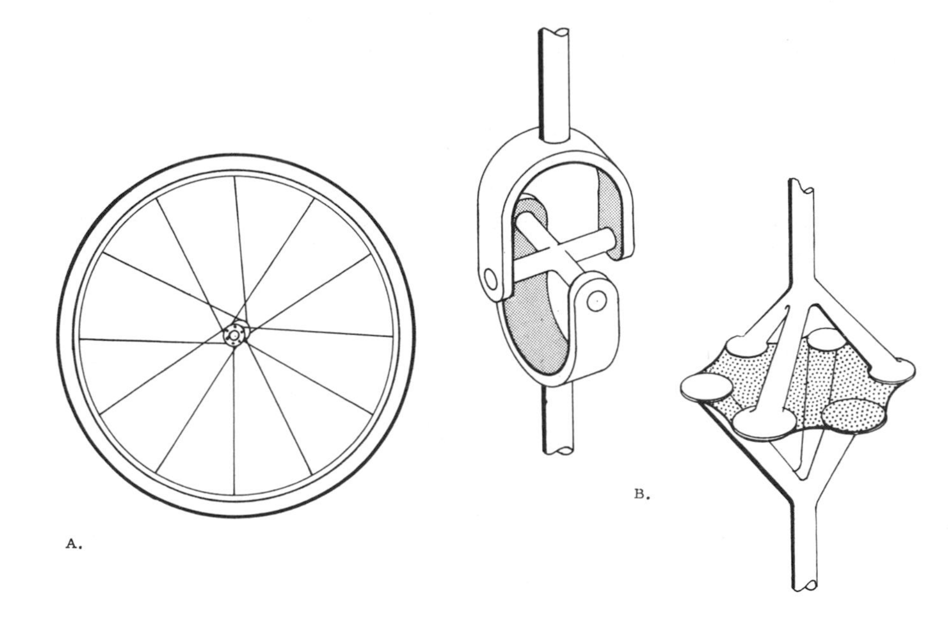

Fig. 640.41B *Minimum of Twelve Spokes Oppose Torque: Universal Joint:*

A. It takes a minimum of 12 spokes to overcome the turbining evident with the minimum four vectors of restraint. This is demonstrated with the 12-spoke wire wheel with its six positive diaphragm and six negative diaphragm of which respectively three each are positively and negatively opposed turbining or torque members.

B. Two-axis and three-axis "universal joints," analogous to the wire wheel as a basic system relying on the differentiation of tension and compression for its effectiveness. These all may be considered basic tensegrity systems.

tinuous tension. This fact was invisible to, and unthought of by, historical man up to yesterday. Before the discovery of this fact in mid-20th-century, there was naught to disturb, challenge, or dissolve his "solid-rock" and other "solid-things" thinking. "Solid thinking" is as yet comprehensively popular and is even dominant over the practical considerations of scientists in general, and even over the everyday logic of many otherwise elegantly self-disciplined nuclear physicists.

640.42 As I wondered whether it was now possible for man to inaugurate an era of thinking and conscious designing in terms of comprehensive tension and discontinuous compression, I saw that his structural conceptioning of the wire wheel documented his intellectual designing breakthrough into such thinking and structuring. The compressional hub of the wire wheel is clearly islanded or isolated from the compressional "atoll" comprising the rim of the wheel. The compressional islands are interpositioned in structural stability only by the tensional spokes. This is clearly a tensional integrity, where tension is primary and comprehensive and compression is secondary and local. This reverses the historical structural strategy of man. His first wire wheel had many and varied numbers of spokes. From mathematical probing of generalized principles and experimentally proven knowledge governing the tensional integrity of the wire wheel, we discover that 12 is the minimum number of spokes necessary for wire wheel stability. (See Sec. 537, Twelve Universal Degrees of Freedom.)

640.50 Mast in the Earth: In his primary regard for compressional structuring, man inserts a solid mast into a hole in the "solid" Earth and rams it in as a solid continuity of the unitary solid Earth. In order to keep the wind from getting hold of the top of the mast and breaking it when the hurricane rages, he puts tension members in the directions of the various winds acting at the ends of the levers to keep it from being pulled over. The set of tension stays is triangulated from the top of the masthead to the ground, thus taking hold of the extreme ends of the potential mast-lever at the point of highest advantage against motion. (See illustration 640.41A.) In this way, tension becomes the helper. But these tensions are secondary structuring actions. They are also secondary adjuncts in man's solidly built, compressional-continuity ships. He puts in a solid mast and then adds tension helpers as shrouds. To man, building, Earth, and ship seemed alike, compressionally continuous. Tension has been secondary in all man's building and compression has been primary, for he has always thought of compression as solid. Compression is that "realistic hard core" that men love to refer to, and its reality was universal, ergo comprehensive. Man must now break out of that habit and learn to play at nature's game where tension is primary and where tension explains the coherence of the whole. Compression is convenient, very convenient, but always secondary and discontinuous.

640.60 Tensed Rope: There is a unique difference in the behaviors of tension and compression. When we take a coil of rope of twisted hemp and pull its ends away from one another, it both uncoils along its whole length and untwists locally in its body. This is to say that a tensed rope or tensed object tends to open its arcs of local curvation into arcs of ever greater radius. But we find that the rope never attains complete straightness either of its whole length or of its separate local fibers or threads. Ropes are complexes of spirals. Tensed mediums tend to a decreasing plurality of arcs, each of the remainder continually tending to greater radius but never attaining absolute straightness, being always affected in their overall length by other forces operating upon them. We see that tension members keep doing bigger and bigger arc tasks. The big patterns of Universe are large-radius patterns, and the small patterns are small-radius patterns. Compressed columns tend to spiral-arc complexes of ever-increasing radius. So we find the compression complexes tend to do the small local structural tasks in Universe, and the tension complexes tend to do the large structural tasks in Universe. As tension accounts for the large patternings and pattern integrities, compression trends into locally small pattern integrities.

641.00 High Tide and Low Tide

641.01 No tension member is innocent of compression, and no compression member is innocent of tension. That is, when we are tensing a rope visibly and axially, its girth contracts as it goes untwistingly into compression, precessionally, compressing in planes at 90 degrees to the axis of our purposeful tensing. We learn experimentally that tension and compression always and only coexist and operate precessionally at right angles to one another, covarying in such a way that one is ebbing toward low tide while the other is flowing toward high tide, or vice versa, in respect to relative human apprehendability, as they fluctuate between visibly obvious and human subvisibility.

641.02 Tension and compression are inseparable and coordinate functions of structural systems, but one may be at its ''high tide'' aspect, i.e., most prominent phase, while the other is at low tide, or least prominent aspect. The *visibly* tensioned rope is compressively contracted in almost *invisible* increments of its girth dimensions everywhere along its length. This low-tide aspect of compression occurs in planes perpendicular to its tensed axis. Columns *visibly* loaded by weights applied only to their top ends are easily seen to have their vertical axes in compression, but *invisibly* the horizontal girths of these columns are also in tension as the result of a cigar-shaped, swelling pattern of forces acting in the column at right angles to its loaded axis, which tends, *invisibly,* to transform toward the shape of a squash or a banana. As a result of the visible, or high-tide, vertical compressioning aspect of such axial loading of the column's system, this swelling force imperceptibly stretches, or tenses, the column's girth as a low-tide reciprocal function of the overall structural-integrity reciprocity.

642.00 Functions

642.01 Functions are never independent of one another. There is a plurality of coexistent behaviors in nature; these are the complementary behaviors.

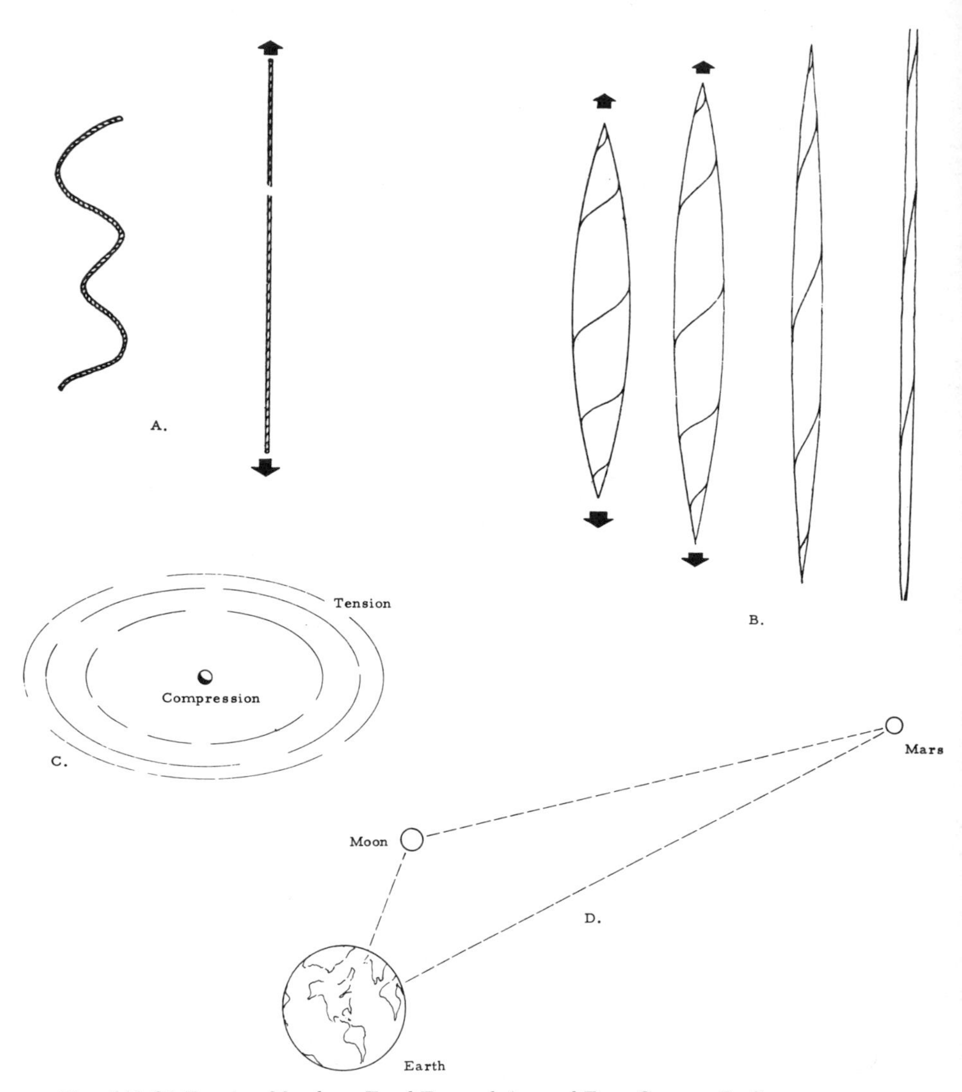

Fig. 641.01 *Tension Members Tend Toward Arcs of Ever Greater Radius:*

A. Slack rope and tensed rope: tensed rope tends toward "straight," i.e. toward arcs of ever greater radius, but never attains complete "straightness."

B. As tension increases: neutral axis lengthens and girth contracts (becomes more compact). Therefore, the long-dimension profile arcs increase in radius and spiral arcs' "radii" increase in dimensions but never attain "straightness" of relation between two "fixed" points, as there are no experiences of fixed points and straight points.

C. Tension goes toward arcing of larger and fewer different radii all ultimately spirally closing back on self. Tension: inherently comprehensive and finite. Compression goes toward relatively smaller radius and toward more of smaller and multiplying microcosmic differentiation. Compression: inherently local and infinite.

D. Tension as gravity: a tension structure is nature's fundamental pattern-cohering principle.

Functions occur only as inherently cooperative and accommodatively varying subaspects of synergetically transforming wholes. Functions are covariants. Wave magnitude and frequency are experimentally interlocked as covarying cofunctions, and both are experimentally gear-locked with energy quanta. The meaning of a function is that it is part of a complementary pattern. No function exists by itself: *X* only in respect to *Y*. Tension and compression are always and only interfunctioning covariables whose seeming relative importance is a consequence of local pattern inspection. Multiplication is accomplished only by division. Universe expands through progressively differentiating out or multiplying discrete considerations.

643.00 Tension and Compression: Summary

Compression is IN.	*Tension* is OUT.
Compression is dispersive both laterally and circumferentially, inherently electrostatic because differentiative, divisive, temporary, and local.	*Tension* is omniradially conversive and is both electromagnetically and gravitationally tensive because eternally and integrally comprehensive.
Compression tends to local dichotomy and multiplication by separation.	*Tension* is unit: universally cohering and comprehensively finite.
Compression is locally expressive in discrete tones and frequencies internal to the octave.	*Tension* is both internal and external to the octave and is harmonic with either the unit octave or octave pluralities.
Compression accumulates potential. As demonstrated in the arch, compression is limited to absolute, local, and within-law relationships of one fixed system.	*Tension* is comprehensive, attractive, and gravitational. Tension is inherently integral and eternally, invisibly, infinitely comprehensive. Tension is comprehensively without-law.
Compression tends toward arcs of decreasing radius.	*Tension* tends toward arcs of increasing radius.
Compressions are plural.	*Tension* is singular.
Compression is time.	*Tension* is eternity.
Compression is specifically directional.	*Tension* is both omni- and supradirectional.
Compression is inherently partial.	*Tension* is inherently total.

644.00 Limitless Ratios of Tension

644.01 I adopted as a working hypothesis that there is a limit to slenderness ratio of the girth diameter of a compression member in respect to its longitudinal axis and that there is no limit to the slenderness ratio of structures dominated by tensional components. Astronomical magnitudes of structural-system coherence are accomplished by tensionally dominated structural functions of zero slenderness ratio, i.e., by gravitational functioning. Compressionally dominated structural components tend toward contour transformation in which the radius of curvature steadily decreases under axial loading, that is, the cigar-shaped column forces tend toward squash- or bananalike bending of their contours. This tending of compressionally loaded systems toward arcs of lessening radius is in direct contrast to the contour transformation tending of tensionally dominated structural components, which always tend toward arcs of ever-increasing radius of axial profile. For instance, the coil of rope tends toward straightening out when terminally tensed, but it never attains absolute straightness; instead, it progresses toward ever-greater radius of locally spiraling but overall orbital arcing, which must eventually cycle back upon itself. Tensionally dominated patterning is inevitably self-closing and finite.

644.02 Compressionally dominated functions of structural systems are inherently self-diminutive in overall aspect. Tensionally dominated functions of structural systems are inherently self-enlarging in overall involvement. The sum of all the interactive-force relationships of Universe must continually accelerate their intertransforming in such a manner as to result in ever more remotely and locally multiplied, islanded, compressional functions—comprehensively cohered by ever-enlarging finite patternings of the tensional functions. Universe must be a comprehensively finite integrity, permitting only a locally islanded infinitude of observer-considered and regenerated-differentialing discovery. We have herein discovered a workable man-awareness of a complete reversal of presently accepted cosmology and of general a priori conceptioning regarding the general patterning scheme of Universe, which has heretofore always conceived only of locally finite experiences as omnidirectionally surrounded by seemingly unthinkable infinity.

645.00 Gravity

645.01 The ratio of length to section in tension appears to be limitless. I once wondered whether it was a nonsensical question that we might be trending toward bridges that have infinite length with no section dimension at all. As a sailor, I looked spontaneously into the sky for indicated clues. I observed that the solar system, which is the most reliable structure that we know of, is so constituted that Earth does not roll around Mars as would ball bearings, which is to say that the compressional components of celestial structures are astro-islands, spatially remote from one another, each shaped in the most ideal conformation for highest compressional-structure effectiveness, which is the approximately spherical shape. All other spheroidal shapes (cigar, turnip, egg, potato, spider) have only one most neutral axis. This is why spherical ball bearings are the ideal compressional-system structures of man's devising, as they continually shift their loads while distributing the energetic effects to the most parts in equal, ergo relatively minuscule, shares in the shortest time. I saw that the astro-islands of compression of the solar system are continuously controlled in their progressive repositioning in respect to one another by comprehensive tension of the system. This is what Newton called *gravity.* The effective coherence between island-components varies in respect to their relative proximities and masses, in ratio gains and losses of the second power in respect to the dimensional distance as stated in terms of the radius of one of the component bodies involved.

645.02 Throughout the Universe, we find that tension and compression are energetically and complementarily interactive. A steel wire of ever stronger metallic alloys can span ever greater distances. In this kind of patterning, we find that the nonsimultaneous structural integrities of Universe are arranged by the tensional coherences. As thinner and thinner wire can span ever greater and greater distances, visible cross section of the tension members trends toward ever lesser and lesser diameters.

645.03 Finally, because there is no limit ratio in tension, may we not get to where we have very great lengths and no section at all? We find this is just the way Universe is playing the game. This is demonstrated astronomically because it is just the way the Earth and the Moon are invisibly cohered . . . remotely cohered and coordinated, noncontiguous, nontangent, physical entities with their respective coherence decreasing at a second-power rate of their relative remoteness multiplied by their combined masses. It is the way the solar system coheres.

645.04 The gravitational, mass-attractive cohering is noncontiguous, and so there *is* no cross-sectional diameter or identifiable local entity. This is why tension and coherence are able to approach larger and larger magnitudes while working toward no cross section at all: for there never was any cross section in the adjacent atoms. You have enormous tension with no section at all. This is also true in the atoms: true in the macrocosm and true in the microcosm.

The same relative distance intervenes as between the Earth and the Moon in respect to their relative masses. The only surprise here is that man has been so superficially misled into ever having thought that there could be solids or continuous compressional or tensional structural members. Only man's mentality has been wrong in trying to organize the idea of structure.

645.05 The trends are to increasing amplification of tension to infinite length with no section. Every use of gravity is a use of such sectionless tensioning. The electric tension first employed by man to pull energy through the nonferrous conductors and later to close the wireless circuit was none other than such universally available sectionless tension. In the phenomenon tension, man is in principle given access to unlimited performance. It seems fantastic, but there it is!

645.10 Tension is shown experientially to be nondimensional, omnipresent, finitely accountable, continuous, comprehensive, ergo timeless, ergo eternal. Comprehensive Universe is amorphous and only locally finite as it transformingly differentiates into serially conceptual pattern integrities, some much larger than humanly apprehendable, some much smaller than humanly apprehendable, ever occurring in nonsimultaneous sets of human observings, most of whose time-canceling, harmonically integrative synchronizations are supra- or subhuman sensibility and longevity experienceability and whose periodicities are therefore so preponderantly unexpected as to induce human reaction of overwhelming disorder, so that . . . suddenly, around comes the comet again, for the first known time in humanly recorded experience, periodically closing the gap and periodically pulsing through eternally normal zero.

645.11 In our old ways of thinking, infinity was expressible numerically as $N + 1$. We tried to get a static picture of a sphere, but we could not understand one more layer beyond it and what was beyond that—in order for it to be something.

645.12 In the nonsimultaneous experiencing of Universe, there is no simultaneous "one frame." We are not faced with that at all. We get to the finite physical world of the physicist and then to the local compressionals and we find that the local is continually subdivisible. We started with a whole that was finite and then began to subdivide it. So there is, in a sense, an infinity of subdivisions locally. This is very much the way the intellectual pattern goes, so that the only thing you might call infinity here is the further subdivision of finity. So it is really never infinite because you are not looking at one part. It is never just Plus One. It is always plus the rest of Universe when you separate that One out. You can separate unity up further and further. You can multiply the subdivisions of unity.

646.00 Chemical Bonds

646.01 While tension and compression always and only coexist, their respective structural behaviors differ greatly. Structural columns function most predominantly in compression of inherent limit of length to cross section, whereas tension cables or rods have no cross section diameter-to-length ratio.

646.02 Mass attraction is always involved in bonding. There may not be atomic bonding without either electromagnetic or mass attraction: either will suffice.

646.03 As man's knowledge of chemical-element interalloying improves, it becomes apparent that critically effective, mass-attractive atomic proximities are intensified by symmetrical congruence. The mass attractions increase as of the second power with each halving of the distance of atomic interstices—the length of structural tensile members, such as those of suspension bridge cables, relative to a given cross section of cable diameter or of any given stress. The overall lengths trend to amplify in every-multiplying degree, thus approaching infinite lengths with no cross section at all. Incredible? No! Look at the Moon and Earth flying coheringly around the Sun. Every use of gravity is a use of such *sectionless* tensioning. The electrical tensioning first employed by man to pull energy through the nonferrous conductors, and later to close the wireless circuit, was none other than such universally available sectionless tension.

646.04 Electromagnetic energy is produced by accelerating the inexhaustible mass attraction into other permitted patterns, as we may stir water in a bathtub to develop cyclic rotation.

647.00 Absolute Velocity: Shunting

647.01 Synergetics discloses that the apparently different velocities, or rates of acceleration, ascribed by humans to environmental events are optical aberrations. The seemingly different velocities are a plurality of angularly precessed—or shunted—energy-action systems regeneratively operated in respect to other systems. Velocity is always 186,000 miles per second. All other relative motion patterns are the result of remotely observed, angularly precessed, 186,000 m.p.s., energy-action shunting. Angularly precessed shunting may divert omnidirectional energy into focused (angularly shunted) actions and reactions of either radial or circumferential patterns, or both.

647.02 Frequency modulation is accomplished through precession-shunted circuit synchronizations. ''Valving'' is angular shunting. Competent design is predicated upon frequency modulation by application of the precessional-shunting principle.

186,000 miles per second

One shunt:

(loop)

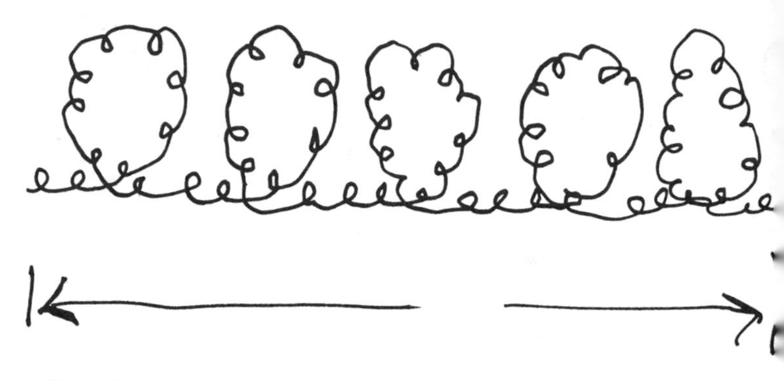

E.g., from here to there, synergetically, relatively to an observer as *10 m.p.h.*

Fig. 647.02.

647.03 Because tension is ever a spiraling arc, it must close back upon itself; it is, therefore, finite and cohesive. Universe is inherently finite and a comprehensive integrity. Compression systems tend, when compression-loaded, to yield to arcs of lesser radius and also, by precessional axial despiraling, tend to unravel and to separate into a plurality of subsystems. Tension systems tend, when axially loaded, to arcs of greater radius. Tension systems tend to greater cohesiveness of precessional inspiraling.

647.04 Discontinuous-compression, continuous-tension structures are finite islands of microcosmic, inwardly precessing, zonal wave-sequence displacements of radial-to-circumferential-to-radial energy knotting regenerations as *nuclear* phenomena—and the whole, which is enclosed in infinite, macrocosmically trending, precessional unravelings, regenerates precessionally as radial-to-circumferential-to-radial *nebular* phenomena—circumferential micro- or macro- being finite, and radial being infinite. Compression is micro and tension is macro.

647.10 In topological systems, vertexes are finite relationships; turbo-systems are in convergence tendencies; and faces are finite sections of infinite open-angle divergent tendencies.

647.20 The equilibriously regenerative octet truss is regenerated as fast and as extensively as man explores and experiences it. As I define Universe as the sum-total aggregate of men's experiences, then we may say that the octet truss-vector equilibrium is universally extensive. *Universally extensive* is a term quite other than *to infinity,* a term the semantic integrity of synergetic geometry may not employ.

647.30 The open end of an angle is infinite, but so too is its convergent end, in that the two actions cannot pass either instantaneously or simultaneously through the same point.

648.00 *Macrocosmic and Microcosmic*

648.01 If we switch our observation from the macrocosmic to the microcosmic, we witness man's probing within the atom, which discloses the same kind of discontinuous-compression, continuous-tension apparently governing the structure of the atom. That is, the islands of energy concentration of the atom and its nucleus are extraordinarily remote from one another in respect to their measurable local-energy-concentration diameters, and all are bounded together by a comprehensive but invisible tensional integrity.

648.02 In the new awareness of synergetics, the remote patternings of Universe are inherently finite, and only the local islands of compression are subdivisible to the degree of infinity projected by the existence of local life and its differential dichotomies of progressive probing. We discover that the more visible, i.e., the more sensorially tunable, the structural functions are, then the more infinitely subdivisible do their potential treatments become. The more invisible the structural functions of Universe, the more comprehensively and comprehendably finite they become.

648.03 As a consequence of these macro-micro structural observations, I also wondered whether man was congenitally limited to his solid structural conceptioning. Man obviously tended to think only of a "solid," brick-on-brick, pile-up law as governing all fundamental forms of structural modifications, i.e., formal local alterations of the "solid" compressional Earth's crust. Could he therefore never participate in the far more efficient structural strategies evidenced in his (only instrumentally harvested) infra- and ultrasensorial data of universal patterning? I saw that man had long known of tensional structures and had experienced and developed tensional capabilities, but apparently only as a secondary accessory of primary compressional structuring.

650.00 *Structural Properties of Octet Truss*

650.01 *Rationale*

650.011 Conventional engineering analysis long ago discovered that a two-way, vertically sectioned beam crossing at 90 degrees, supported from four walls, provided no more strength at the mid-crossing point than could be found in the stronger of the two beams, for they were redundantly acting as hinges, and only one axis of hinging could be articulated at one time.

650.02 In three-way beam crossings, each vertically sectioned beam has a two-way tendency to rock or torque or hinge over from its most favorable aspect of maximum dimension in opposition to gravity into its least favorable aspect, that of least dimension in opposition to gravity. As each beam could hinge from the vertical in two ways, each may be split theoretically into two vertical parts and thus hinge both ways. The three-way beam crossing is thus countered by the simultaneous and symmetrical both-ways split rocking of all three vertical split-beam hinges—as three sets of parallel planes until their edges meet in ridge poles to provide a matrix of tetrahedra, with common lean-to stability and with maximum energy-repose economy, synergetically between a fourth—or horizontal—set of planes.

650.03 While the three beams' sets of uniquely split plus-and-minus vertical planes rotate into three positive-and-negative parallel sets of planes at 35° 16′ off vertical, each of the tilted beam's tops and bottoms is in two parallel and horizontal planes, respectively. This makes a total of four unique and symmetrically oriented planes within the system. Where the four unique sets of planes intercept each other, there is established a system of interconnected lines; as the interconnected lines contain all the stress patternings, struts may be substituted for them and the planar webs may then be eliminated. When struts alone are used for horizontal decking, they are designed to receive loads at their vertexial ends and to send their loads through their neutral axis, whereas beams inefficiently take loads anywhere at 90 degrees to their neutral axes.

650.04 In the octet truss, three planes of beams and their triangularly binding edge patterns rotate tepee-wise * positively and negatively to nonredundant ridge-pole fixity, and with such symmetry as to result in radial distribution of all loads from any one loaded vertex through the neutral axes of all the edges of the system. Loads are precessionally differentiated as either pure-compression or pure-tension stresses. They are metered at even rates because their edge vectors are identical in length. The loads precess further into positive and negative radial and circumferential waves eccentric to the loaded vertex, with the stress distributed positively and negatively throughout those adjacent vertexes surrounding any one loading center, and with the wave distribution in all directions precessing into tensile action the concentric series of rings around the originally loaded vertex. The increasing succession of concentric rings that continually redistribute the received loads act in themselves as unitary systems, with an increasing number of eccentrically dis-

* See Sec. 621.20.

tributive vectors as full-dispersion loads come to symmetric reconcentration at supporting areas in direct pattern reversal. (See Secs. 420 and 825.28.)

650.10 *Inherent Nonredundance*

650.101 The octet truss is synergetic because the four planes comprise a system, and what were previously individual beams, and therefore free systems in themselves, are now fixed components in a larger tetrahedral system, which is inherently nonredundant because it is the minimum fixed system. Ergo, all those previous individual, free-system beams are now converted into one nonredundant complex of basic systems, and all the previous beams' component biological and subchemical structures are systematically refocused in such a manner that all subcomponents are nonredundantly interactive in the second-power rates of effectiveness accruing to the circumferential finiteness of systems in respect to their radial modules.

650.11 The unitary, systematic, nonredundant, octet-truss complex provides a total floor system with higher structural performance abilities than engineers could possibly ascribe to it through conventional structural analysis predicated only upon the behavior of its several parts. It is axiomatic to conventional engineering that if parts are "horizontal," they are beams; and the total floor ability by such conventional engineering could be no stronger than the single strongest beam in the plural group. Thus their prediction falls short of the true behavior of the octet truss by many magnitudes, for in true mathematical fact, no "beams" are left in the complex; that is, there are no members in it loaded at other than polar terminals. Down to the minutest atomic components, the octet truss is therefore proved to be synergetic, and its discovery as a structure—in contradistinction to its aesthetic or superficial appearance—is synergetic in performance; that is, its behavior as a whole is unpredicted by its parts. This makes its discovery as a structure a true surprise, and therefore it is a true invention.

650.12 What is the surprise? It is because we had used three planes of the beams oriented to most favorable ability aspect in respect to gravity, and in so doing we had inadvertently gained a fourth interacting favorable-aspect plane of symmetry not consciously introduced as a previously acquired component of the whole, which thereby made the beams "vanish" into abstract limbo. The fourth plane is strictly the fourth plane of the tetrahedron inadvertently accruing, as does the hinging on of one triangle to two previously hinged equilateral triangles provide inadvertently a fourth triangle: $1 + 2 = 4$. Q.E.D.

650.13 A second derivative surprise is the nonredundance of the larger associated complex of tetrahedra, occasioned by its precessionally induced self-differentiation of functions: when loaded at any one vertex in such a manner that every member acts in axially focused pure tension or pure compression, and with the subsequent loading of any next adjacent vertex, there is inherently induced comprehensive reversal of all the system's pure tension into pure compression functions, and vice versa. That is to say, it is dynamically nonredundant.

700.00 Tensegrity

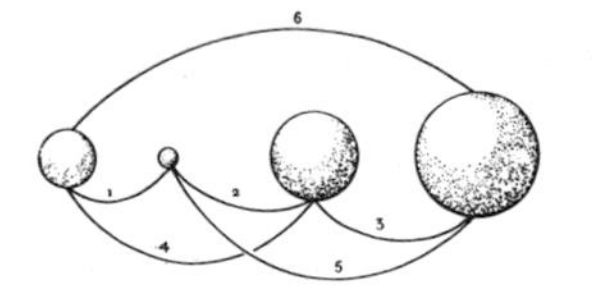

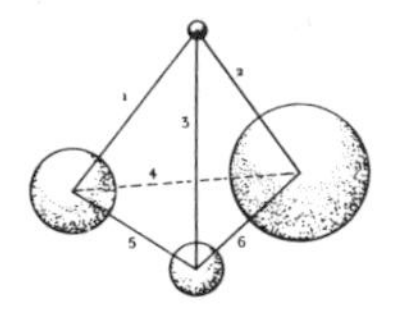

700.01 Definition: Tensegrity

700.011 The word *tensegrity* is an invention: it is a contraction of *tensional integrity*. Tensegrity describes a structural-relationship principle in which structural shape is guaranteed by the finitely closed, comprehensively continuous, tensional behaviors of the system and not by the discontinuous and exclusively local compressional member behaviors. Tensegrity provides the ability to yield increasingly without ultimately breaking or coming asunder.

700.02 The integrity of the whole structure is invested in the finitely closed, tensional-embracement network, and the compressions are local islands. Elongated compression tends to deflect and fail. Compressions are disintegrable because they are not atomically solid and can permit energy penetration between their invisibly amassed separate energy entities. As a compression member tends to buckle, the buckling point becomes a leverage fulcrum and the remainder of the compression member above acts as a lever arm, so that it becomes increasingly effective in accelerating the failure by crushing of its first buckled-in side. The leverage-accelerated penetration brings about precessional dispersal at 90 degrees.

700.03 Tension structures arranged by man depend upon his purest initial volition of interpretation of pure principle. Tension is omnidirectionally coherent. Tensegrity is an inherently nonredundant confluence of optimum structural-effort effectiveness factors.

700.04 All structures, properly understood, from the solar system to the atom, are tensegrity structures. Universe is omnitensional integrity.

701.00 Pneumatic Structures

701.01 Tensegrity structures are pure pneumatic structures and can accomplish visibly differentiated tension-compression interfunctioning in the same manner that it is accomplished by pneumatic structures, at the subvisible level of energy events.

701.02 When we use the six-strut tetrahedron tensegrity with tensegrity octahedra in triple bond, we get an omnidirectional symmetry tensegrity that is as symmetrically compressible, expandable, and local-load-distributing as are gas-filled auto tires.

702.00 Geodesics

702.01 We have a mathematical phenomenon known as a geodesic. A geodesic is the most economical relationship between any two events. It is a special case of geodesics which finds that a seemingly straight line is the shortest distance between two points in a plane. Geodesic lines are also the shortest surface distances between two points on the outside of a sphere. Spherical great circles are geodesics.

703.00 Geodesic-Tensegrity Molecular Kinetics of Pneumatic Systems

703.01 Geodesic domes can be either symmetrically spherical, like a billiard ball, or asymmetrically spherical, like pears, caterpillars, or elephants.

703.02 I prefer to stay with compound curvature because it is structurally stronger than either flat surfaces or simple cylindrical curvature or conical curvature. The new compound-curvature geodesic structures will employ the tensegrity principles. The comparative strength, performance, and weight tables show clearly that the geodesic-dome geometry is the most efficient of all compound-curvatured, omnitriangulated, domical structuring systems.

703.03 All geodesic domes are tensegrity structures whether or not the tension-compression differentiations are visible to the observer. Tensegrity geodesic spheres do what they do because they have the properties of hydraulically or pneumatically inflated structures. Pneumatic structures, such as footballs, provide a firm shape when inflated because the atmospheric molecules inside are impinging outward against the skin, stretching it into accommodating roundness. When more molecules are introduced into enclosures by the air pump, their overcrowding increases the pressure. All the molecules of gas have inherent geometrical domains of activity. The pressurized crowding is dynamic and not static.

703.04 A fleet of ships maneuvering under power needs more room than do the ships of the same fleet when docked side by side. The higher the speed of the individual ships, the greater the sea room required. This means that the enclosed and pressurized molecules in pneumatic structural systems are accelerated in outward-bound paths by the addition of more molecules by the pump and, without additional room, each must move faster to get out of the way of others.

703.05 The pressurized internal liquid or gaseous molecules try to escape from their confining enclosure. The outward-bound molecules impact evenly upon all the inside surface of the enclosure—for instance, upon all of the football's flexible inside skin when it is kicked in one spot from outside. Their many outward-bound impactings force the skin outwardly and firmly in all directions, and the faster they move, the more powerful the impact. This molecular acceleration is misidentified as pressures and firmness of the pneumatic complex. This molecular acceleration distributes the force loads evenly. The outward forces are met by the comprehensive embracement of all the tensile envelope's combined local strengths. All locally impacting external loads, such as the kick given to a point on the football's exterior, are distributed by all the enclosed atmospheric molecules to all of the skin in the innocuously low magnitudes.

703.06 The ability to determine quite accurately what the local loadings of any given pneumatic structure will be under varying conditions and forces is well known and is about as far as the pneumatic sciences have gone in explaining inflated structures. The comfortably equationed state of their art is adequate to their automobile-or-airplane-tire-, balloon-, or submarine-designing needs.

703.07 It is, however, possible to find out experimentally a great deal more about the behavior of those invisible, captive, atmospheric molecules and to arrive at a greater geo-mathematical understanding of the structural relationships between pneumatically inflated bags or vessels and geodesic tensegrity spheres and domes. It is thus possible also to design tensile structures that meet discretely, ergo nonredundantly, the patterns of outwardly impinging forces. It also becomes possible, for the first time, for structural engineers to analyze geodesic domes in a realistic and safe manner. Up to this time, the whole engineering profession has been analyzing geodesics on a strictly *continuous-compression,* crystalline, non-load-distributing, "post-and-lintel" basis. For this reason, the big geodesic domes thus far erected have been way overbuilt by many times their logically desirable two-to-one safety factor.

703.08 While the building business uses safety factors of four, five, or six-to-one, aircraft-building employs only two-to-one or even less because it knows what it is doing. The greater the ignorance in the art, the greater the safety factor that must be applied. And the greater the safety factor, the greater the redundancy and the less the freedom of load distribution.

703.09 First we recall, as has long been known experimentally, that every action has a reaction. For a molecule of gas to be impelled in one direction, it must "shove off from," or be impelled by, another molecule accelerated in an opposite direction. Both of the oppositely paired and impelled action and reaction molecules inside the pneumatically expanded domes will impinge respectively upon two chordally opposed points on the inside of the skin. The middle point of a circular chord is always nearer the center of the circle than are its two ends. For this reason, chords (of arcs of spheres) impinge outwardly against the skin in an acutely glancing angular pattern.

703.10 When two molecules accelerate opposingly from one another at the center of the sphere, their outward trajectories describe a straight line that coincides with the diameter of a sphere. They therefore impinge on the skin perpendicularly, i.e., at 180 degrees, and bounce right back to the sphere center. It is experimentally evidenced that all but two of the myriad molecules of the captive gas do not emanate opposingly from one another at the center of the sphere, for only one pair can occupy one point of tangent bounce-off between any two molecules. If other molecules could occupy the nucleus position simultaneously, they would have to do so implosively by symmetrical self-compression, allowing the sphere to collapse, immediately after which they would all explode simultaneously. No such pulsating implosion-explosion, collapse-and-expand behavior by any pneumatic balls has been witnessed experimentally.

703.11 Molecules of gas accelerating away from one another and trying to proceed in straight trajectories must follow both the shortest-distance geodesic law as well as the angular-reflectance law; they will carom around inside a sphere only in circular paths describing the greatest diameter possible, therefore always in great-circle or geodesic paths.

703.12 For the same reasons, molecules cannot be "stacked up" inside the sphere in parallel or lesser-circle latitude planes. We also found earlier that the molecules could not be exploding simultaneously in all directions from the center of the sphere. If thin, colored vapor streaks are introduced into a transparently skinned, pneumatically pressurized sphere, then only at first superficial observation do the smoke-disclosed molecular motions seem to be demonstrating chaotically random patterns. This is not the case, however, for everything in Universe is in motion and everything in motion is always traveling in the direction of least resistance, wherefore the great circle's inherent polar symmetries of interaction *must impose polar order*—an order that is hidden from the observer only by its articulative velocities, which transcend the human's optically tunable, "velocity-of-motions" spectrum range of apprehending and therefore appear only as clouds of random disorder. Brouwer's theorem shows that when x number of points are stirred randomly on a plane, it can be proved mathematically—when the stirring is stopped—that one of the points was always at the center of the total stirring, and was therefore never disturbed in respect to all the others. It is also demonstrable that any plane surface suitable for stirring things upon, must be part of a system that has an obverse surface polarly opposite to that used for the stirring, and that it too must have its center of stirring; and the two produce poles in any bestirred complex system.

703.13 Every great circle always intercepts any other great circle twice, the interception points always being 180-degree polar opposites. When two force vectors operating in great-circle paths inside a sphere impinge on each other at any happenstance angle, that angle has no amplitude stability. But when a third force vector operating in a great-circle path crosses the other two spherical great circles, a great-circle-edged triangle is formed with its inherently regenerated 180-degree mirror-image polar opposite triangle. With a myriad of successive inside surface caromings and angular intervector impingements, the dynamic symmetry imposed by a sphere tends to equalize the angular interrelationship of all those triangle-forming sets of three great circles which shuntings automatically tend averagingly to reproduce symmetrical systems of omnisimilar spherical triangles always exactly reproduced in their opposite hemispheres, quarterspheres, and octaspheres. This means that if there were only three great circles, they would tend swiftly to interstabilize comprehensively as the spherical octahedron all of whose surface angles and arcs (central angles) average as 90 degrees.

703.14 A vast number of molecules of gas interacting in great circles inside of a sphere will produce a number of great-circle triangles. The velocity of their accomplishment of this structural system of total intertriangulation averaging will seem to be "instantaneous" to the human observer. The triangles, being dynamically resilient, mutably intertransform one another, imposing an averaging of the random-force vectors of the entire system, resulting in angular self-interstabilizing as a pattern of omnispherical symmetry. The aggregate of all the inter-great-circlings resolve themselves typically into a regular pattern of 12 pentagons and 20 triangles; or sometimes more complexedly, into 12 pentagons, 30 hexagons, and 80 triangles described by 240 great-circle chords.

703.15 This is the pattern of the geodesic tensegrity sphere. The numbers of hexagons and triangles and chords can be multiplied in regular arithmetical-geometrical series, but the 12 pentagons, and only 12, will persist as constants; also, the number of triangles will occur in multiples of 20; also, the number of edges will always be multiples of six.

703.16 In the geodesic tensegrity sphere, each of the entirely independent, compressional chord struts represents two oppositely directioned and force-paired molecules. The tensegrity compressional chords do not touch one another. They operate independently, trying to escape outwardly from the sphere, but are held in by the spherical-tensional integrity's closed network system of great-circle connectors, which alone complete the great-circle paths between the ends of the entirely separate, nonintercontacting compressional chords.

704.00 Universal Joints

704.01 The 12-spoke wire wheel exactly opposes all tension, compression, torque, or turbining tendencies amongst its members. Universal joints of two axes or three axes of freedom are analogous to the wire wheel as a basic 12-degrees-of-freedom accommodating, controlling, and employing system whose effectiveness relies upon their discrete mechanical and structural differentiation and disposition of all tension and compression forces. All of these may be considered to be basic tensegrity systems. (See illustration 640.41B.)

704.02 The shafted axis of the two-axis universal joint tends to make it appear as a single-axis system. But it constitutes in actuality an octahedral tensegrity, with its yoke planes symmetrically oriented at 90 degrees to one another. The two-axis tensegrity has been long known and is often successfully employed by mechanics as a flexible-membrane coupling sandwiched between two diametrically opposed yoke-ended shafts, precessionally oriented to one another in a 90-degree star pattern. This only tensionally interlinked, i.e., universally jointed, drive shafting has for centuries been demonstrating the discontinuous compression and only tensionally continuous multiaxial, multidimensional symmetry of tensegrity structuring and energetic work transmission from here to there.

705.00 Simple Curvature: The Barrel

705.01 The barrel represents an advanced phase of the Roman arch principle of stability accomplished by simple (approximately) single-axis curvature. A barrel is comprised of a complete ring around one axis of a number of parallel staves. A cross section cut through the barrel perpendicular to its single axis of curvature shows each of the stave's sections looking like keystones in an arch. Each stave is a truncated section of a triangle whose interior cutaway apex would be at the center of the barrel. The staves employ only the outer trapezoidal wedge-shaped cross section, dispensing with the unnecessary inner part of the triangle. The stave's cross section is wedge-shaped because the outer edge of the stave is longer than the inner edge of the stave. Because the stave's outer-circle chord is longer than its inner-circle chord, it cannot fall inwardly between the other staves and it cannot fall outwardly

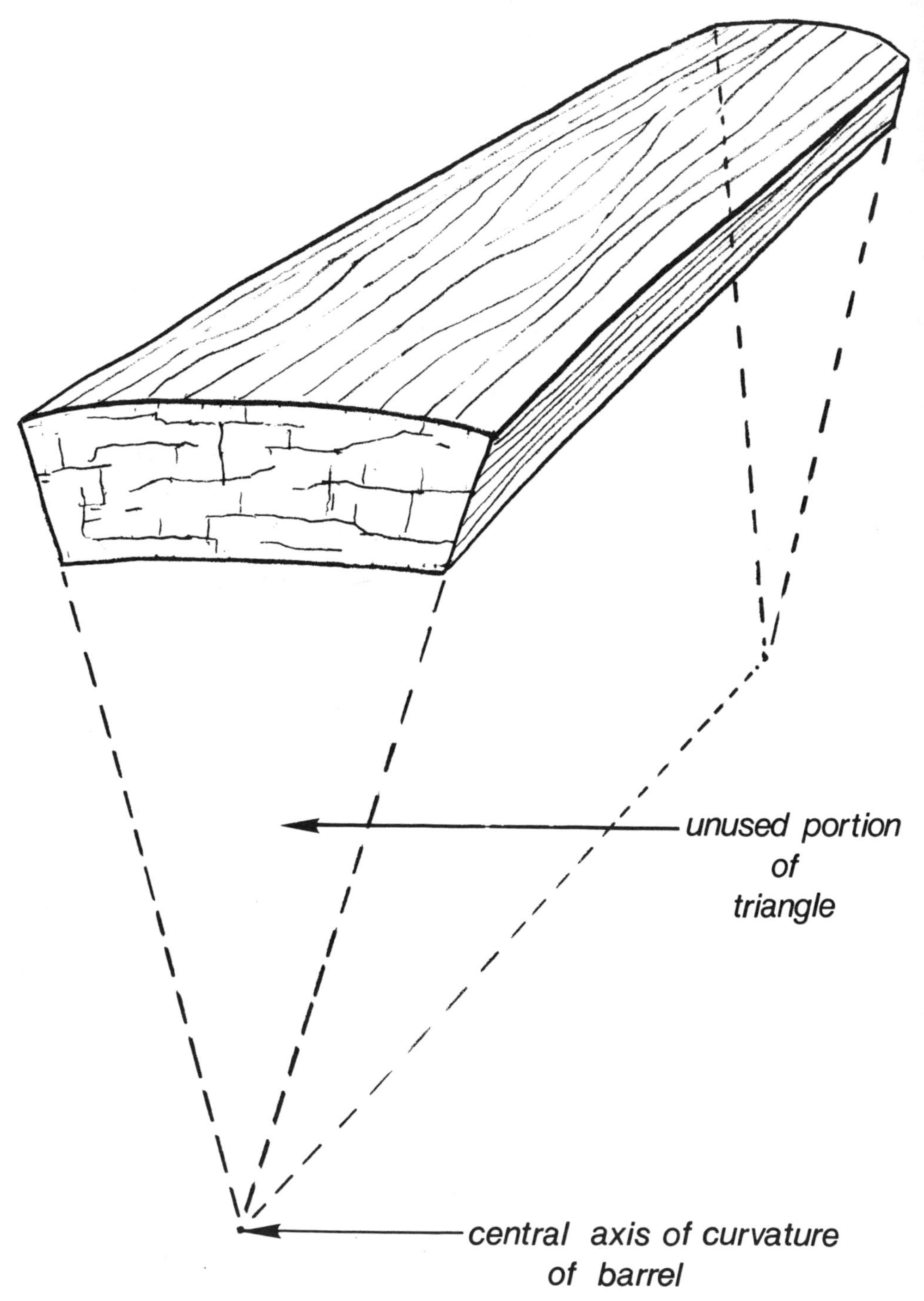
unused portion
of
triangle
central axis of curvature
of barrel

Fig. 705.01.

from close-packed association with the other staves because they are all bound inwardly together by the finitely closed barrel "hoops" of steel.

705.02 All these barrel staves are lined in parallel to one another and are bound cylindrically. They constitute a finite, closed cylinder held together in compression by finitely encompassing tension bands, or hoops, which are parallel to one another and at 90 degrees to the axis of the staves. The staves cannot move outwardly due to the finiteness of the straps closing back upon themselves; they cannot fall inwardly on each other because their external chords are bigger than their internal chords. The tendency of internally loaded cylinders and vertically compressed columns to curve outwardly at their midgirth in their vertical profile is favored by designing and making the barrel staves of greater cross section at their midbarrel portions and the finite, closed-circle bands of lesser diameter near the ends than at the middle. The curving lines of compression thrust back against themselves, while the tension lines tend to pull true and form a finite closure, pressing the short, true chord

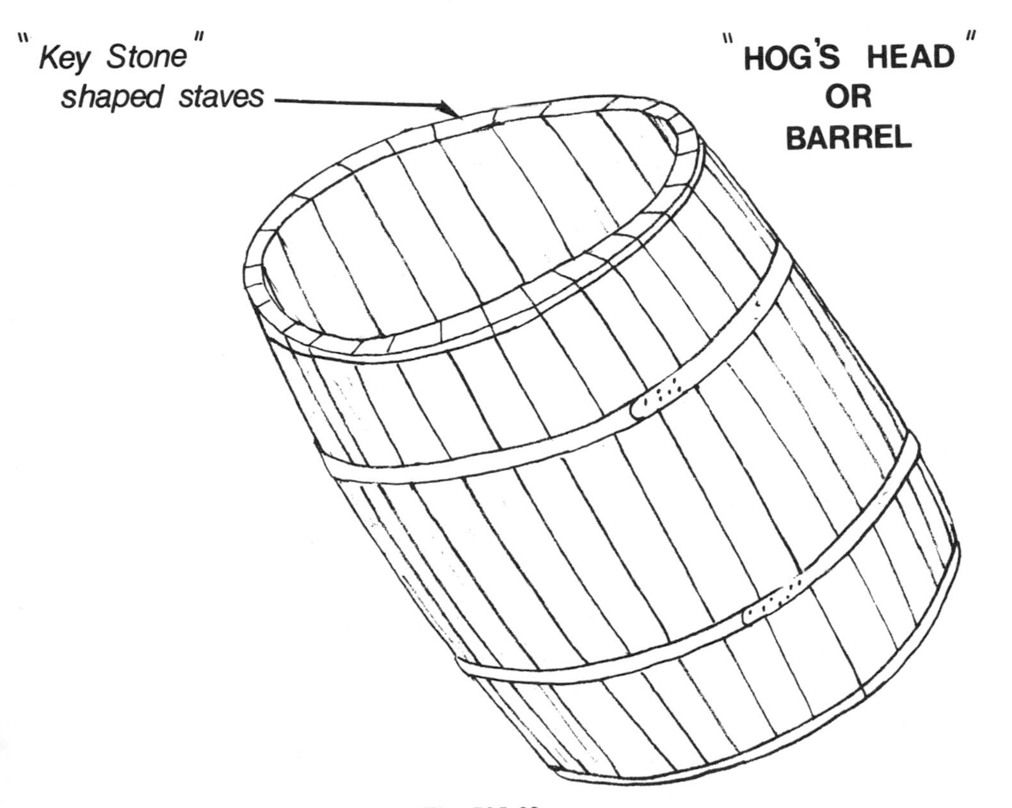

Fig. 705.02.

sections of the staves tightly against one another in a complete circular arch; thus the staves may be flexed, when the barrel is internally filled, without tendency to failure.

705.03 Thus the barrel, when in good material condition, usually proves to be structurally stable and able to withstand the impact of dropping, especially when internally loaded, because the internal load tends to distribute any local shock load to all the enclosing barrel's internal surface and thence to the finitely closed, steel circle bands. Barrels constitute closed circuits of continuous tension finitely restraining discontinuous, though contiguously islanded, staves of compression in dynamic stability. Whether pressure is exerted upon its structure from outside or inside the barrel, the result is always an outward thrust of the staves against the tension members, whose finite closure and cross-sectional strength ultimately absorb all the working or random loads. The vertical forces of gravity in the primary working stresses of internally loaded, simple-curvature structures—such as those of the cylinder, barrel, tree trunk, or Greek column—are translated precessionally into horizontally outward buckling and torque stresses. When, however, such cylinders are not internally loaded and are turned over on their side with their axes horizontal, the stresses are precessed horizontally, outward from the cylinder ends toward the infinite poles of cylindrically paralleled stave lines. Under these conditions, the outer hoops' girth does not aid the structural interstabilization, and the forces of gravity acting vertically against the horizontally paralleled staves develop a lever arm of the topmost staves against the opposite outer staves of the barrel, tending to thrust open the sidemost staves from one another and thus allowing the integrity of the arch to be disintegrated, allowing infinity to enter and disintegrate the system.

705.04 Each of the barrel's tension hoops represents a separately operating, exclusively tensional circle with its plane parallel to, and remote from, the planes of the other, only separately acting, barrel hoops. The tension bands do not touch one another. The tension bands are only parallel to one another and act only at 90 degrees against the staves, which are also only parallel to one another. Neither the staves nor the tension hoops cross one another in such a manner as to provide intertriangulation and its concomitant structural self-stabilization. In fact, they both let infinity into the system to disintegrate it between the only parallel staves and hoops whose separate parts reach forever separately only toward infinity.

705.05 If we take a blowtorch and burn out one of the wooden staves, the whole barrel collapses because infinity floods in to provide enough space between the staves for their arch to be breached and thus collapse disintegratively. What the blowtorch does is to let infinity—or the nothingness of Universe—into the system to intrude between the discontinuous and previously only contiguously crowded together, exclusively compressional members of the system.

705.06 Barrels and casks, which provided great shipping and storage "container advantage" in the past, secured only by finite closure continuities of the only separately acting tension circles, were inherently very limited in structural efficiency due to the infinitely extendable—ergo, infinitely disassociative—staves as well as by the infinity that intruded disintegratively between the barrel's parallel sets of circular bands or hoops.

706.00 Compound Curvature: Spherical Cask

706.01 Engineers and mathematicians both appear presently to be unfamiliar with practical means for discretely analyzing and employing the three-way grids of finitely closed, great-circle triangulations despite the fact that their triangular integrities constitute nature's most powerful and frequently employed structural systems. You can inform yourself experimentally regarding the relative structural effectiveness of flat, simple-cylinder, and compoundly curved sheet material by taking a flat piece of paper, standing it on its edges, and loading that top edge; you will note that it has no structural strength whatsoever—it just crumples. But if we roll-form the same piece of paper into a cylinder, which is what is called simple curvature, we can use the cylinder as a column in which all the compressionally functioning lines are parallel to each other and interact with the closed-circle tensional strength of the paper cylinder's outside surface like the staves and hoops of a barrel.

706.02 But only if we achieve a three-way interaction of great circles can we arrive at the extraordinary stability afforded by the omnitensionally integrated, triangular interstabilization of compound curvature. This we do experimentally with the same sheet of paper, which we now form into a conical shape. Standing the cone on its finitely closed circular base and loading its apex, we find it to be more stable and structurally effective for supporting a concentrated top loading than was either the first sheet or the simply curved cylinder. The top load now thrusts downwardly and outwardly toward the finitely closed, tensionally strong base perimeter, which becomes even stronger if the cone is foldingly converted into a tetrahedron whose insideness concavity and outsideness convexity and omnifinite tensional embracement constitute the prime manifestation of so-called "compound curvature."

706.03 In contrast to our simply curved, cylindrical barrel construction, let us now make a wooden geodesic sphere in which all of the triangular facets are external faces of internally truncated tetrahedra whose interior apexes, had they not been snubly truncated, would each have reached the center of the sphere. Each of the outwardly triangular, internally truncated, tetrahedral cork's edges is covered by finitely closed great-circle tension straps. The steel tension straps are not parallel to each other but are omnitriangularly interconnected to form a spherical barrel. Every great circle of a spherical cask crosses all other great circles of that sphere twice. Any two such—only polarly interconnected—great circles can hinge upon each other like a pair of shears. They are angularly unstable until a third great circle that does not run through the same crossings of the other two inherently crosses both of the first two great circles and, in effect, taking hold of the lever ends of the other two great circles, with the least effort accomplishes stabilization of the oppositely converging angle.

706.04 So we now have an omnitriangulated geodesic sphere of triangulated wooden plugs, or hard wooden corks fashioned of the same barrel stave oak, each one surrounded by and pressed tightly against three other such triangular hardwood corks; each has its exterior triangular facets edged by three great circles whose lengths are greater than the respectively corresponding wooden cork triangle's interior chords so that none of the wooden corks can fall inwardly. The finitely closed great-circle straps are fastened to each other as they cross one another; thus stably interpositioned by triangulation, they cannot slip off the sphere; and none of the wooden triangular corks can fall in on one another, having greater outer edge lengths than those of their inner edges. The whole sphere and its spherical aggregates of omnitriangularly corked surface components are held tightly together in an omnitriangulated comprehensive harness. All of the great circles are intertriangulated in the most comfortable, ergo most economical, interpositions possible.

706.05 If we now take a blowtorch and burn out entirely one of the triangular, truncated wooden corks—just as we burned out one of the barrel's wooden staves—unlike the barrel, our sphere will not collapse. It will not collapse as did the regular barrel when one stave was burnt out. Why does it not collapse? Because the three-way triangular gridding is finitely closed back on itself. Infinity is not let into the system except through the finitely-perimetered triangular hole. The burning out of the triangular, truncated tetrahedron, hardwood cork leaves only a finite triangular opening; and a triangular opening is inherently a stable opening. We can go on to burn out three more of the triangular, truncated wooden corks whose points are adjacent to each other, and while it makes a larger opening, it remains a triangular opening and will still be entirely framed with closed and intertriangulated great circles; hence it will not collapse. In fact, we find that we can burn out very large areas of the geodesic sphere without its collapsing. This three-way finite crossing of most-economical great circles provides a powerful realization of the fundamentals of compound curvature. Compound curvature is inherently self-triangulating and concave-convexing the interaction of those triangles around the exterior vertexes.

706.10 **Sphericity:** Compound curvature, or sphericity, gives the greatest strength with the least material. It is no aesthetic accident that nature encased our brains and regenerative organs in compoundly curvilinear structures. There are no cubical heads, eggs, nuts, or planets.

706.20 **Three-Way Great Circling:** While great circles are the shortest distances around spheres, a *single* great-circle band around a sphere will readily slide off. Every great circle of a sphere must cross other great circles of that sphere twice, with the crossings of any two always 180 degrees apart. Since an infinite number of great circles may run through any two same points on a sphere 180 degrees apart, and since any two great-circle bands are automatically self-interpolarizing, *two* great-circle bands on a sphere can rotate equatorially around their mutual axis and attain congruency, thereafter to act only as one solitary meridian, and therefore also free to slide off the sphere. Not until we have *three* noncommonly polarized, great-circle bands providing omnitriangulation as in a spherical octahedron, do we have the great circles acting structurally to self-interstabilize their respective spherical positionings by finitely intertriangulating fixed points less than 180 degrees apart.

706.21 Since great circles describe the shortest distances between any two spherical points less than 180 degrees apart, they inherently provide the most economical spherical barrel bandings.

706.22 The more minutely the sphere is subtriangulated by great circles, the lesser the local structural-energy requirements and the greater the effectiveness of the mutual-interpositioning integrity. This spontaneous structural self-stabilizing always and only employs the chords of the shortest great-circle arc distances and their respective spherical finiteness tensional integrity.

706.23 When disturbed by energy additions to the system, the triangular plug "corks" can only—and precessionally "prefer"—to be extruded only outwardly from the system, like the resultant of all forces of all the kinetic momentums of gas molecules in a balloon. The omni-outwardly straining forces of all the compressional forces are more than offset by the finitely closed, omni-intertriangulated, great-circle tensions, each of whose interstitial lines, being part of a triangle—or minimum structure—are inherently nonredundant. The resultant of forces of all the omni-intertriangulated great-circle network is always radially, i.e. perpendicularly, inward. The tightening of any one great circle results in an even interdistribution of the greater force of the inward-outward balance of forces.

706.30 **Fail-Safe Advantages:** With each increase of frequency of triangular module subdivisions of the sphere's unitary surface, there is a corresponding increase in the fail-safe advantage of the system's integrity. The failure of a single triangular cork in an omnitriangulated spherical grid leaves a triangular hole, which, as such, is structurally innocuous, whereas the failure of one stave in a simple-curvature barrel admits infinity and causes the whole barrel to collapse. The failure of two adjacent triangular corks in a spherical system leaves a diamond-shaped opening that is structurally stable and innocuous; similarly, the failure of five or six triangles leaves a completely arched, finitely bound, and tensionally closed pent or hex opening that, being circumferentially surrounded by great circles, is structurally innocuous. Failure of a single spherical-tension member likewise leaves an only slightly relaxed, two-way detoured, diamonded relaying of the throughway tensional continuity. Considerable relaxing of the spherical, triangulated-cork barrel system by many local tension failures can occur without freeing the corks to dangerously loosened local rotatability, because the great-circle crossings were interfastened, preventing the tensionally relaxed enlargement of the triangular bonds. The higher the frequency and the deeper the intertrussing, the more fail-safe is this type of spherical structure.

706.31 Structural systems encompassing radial compression and circumferential tension are accomplished uniquely and exclusively through three-way spherical gridding. These radial and circumferential behaviors open a whole new field of structural engineering formulations and an elegance of refinement as the basis for a new tensegrity-enlightened theory of engineering and construction congruent with that of Universe.

707.00 Spherical and Triangular Unity

707.01 Complex Unity and Simplex Unity: The sphere is maximal complex unity and the triangle is minimal simplex unity. This concept defines both the principles and the limits governing finite solution of all structural and general-systems-theory problems.

707.02 Local isolations of "point" fixes, "planes," and "lines" are in reality only dependent aspects of larger, often cosmically vast or micro-, spheric topological systems. When local isolation of infinitely open-ended planes and linear-edged, seemingly flat, and infinite segments are considered apart from their comprehensive spherical contexts, we are confronted with hopelessly special-cased and indeterminate situations.

707.03 Unfortunately, engineering has committed itself in the past exclusively to these locally infinite and inherently indeterminate systems. As a consequence, engineering frequently has had to rely only on such trial-and-error-evolved data regarding local behaviors as the "rate" of instrumentally measurable deflection changes progressively produced in static-load increases, from which data to evolve curves that theoretically predict "failure" points and other critical information regarding small local systems such as columns, beams, levers, and so forth, taken either individually or collectively and opinionatedly fortified with safely "guesstimated" complex predictions. Not until we evolve and spontaneously cultivate a cosmic comprehension deriving from universal, finite, omnitriangulated, nonredundant structural systems can we enjoy the advantage of powerful physical generalizations concisely describing all structural behaviors.

710.00 Vertexial Connections

710.01 When a photograph is made of a plurality of lines crossing through approximately one point, it is seen that there is a blurring or a running together of the lines near the point, creating a weblike shadow between the converging lines—even though the individual lines may have been clearly drawn. This is caused by a refractive bending of the light waves. When the masses of the physically constituted lines converge to critical proximity, the relative impedance of light-wave passage in the neighborhood of the point increases as the second power of the relative proximities as multiplied by a factor of the relative mass density.

710.02 Tensegrity geodesic spherical structures eliminate the heavy sections of compression members in direct contact at their terminals and thus keep the heavy mass of respective compressions beyond critical proximities. As the vertexial connections are entirely tensional, the section mass is reduced to a minimum, and system "frequency" increase provides a cube-root rate of reduction of section in respect to each doubling frequency. In this manner, very large or very small tensegrity geodesic spheroids may be designed with approximate elimination of all microwave interferences without in any way impairing the structural dimensional stability.

710.03 The turbining, tensionally interlaced joints of the tensegrity-geodesic spheroids decrease the starlike vertexial interference patterns.

711.00 Gravity as a Circumferential Force

711.01 Circumference: Circumference $= \pi D = C$. Wherefore, we can take a rope of a given D length and lay it out circumferentially to make it a circle with its ends almost together, but with a tiny gap between them.

711.02 Then we can open out the same rope to form only a half-circle in which the diameter doubles that of the first circle and the gap is wide open.

711.03 Halfway between the two, the gap is partially open.

711.04 As we open gaps, we make the sphere bigger. The comprehensive tension wants to make it smaller. Struts in the gap prevent it from becoming smaller. Struts make big. Tension makes small. The force of the struts is only outward. The force of the tension network is only inward.

711.10 **Circumferential Advantage over Radial:** Gravity is a spherically circumferential, omniembracingly contractive force. The resultant is radially inward, attempting to make the system get smaller. The *circumferential* mass-interattraction effectiveness has a constant coherent advantage ratio of 12 to 1 over the only *radially* effective mass attraction; ergo, the further inward within the embraced sphere, the greater the leverage advantage of the circumferential network over the internal compaction; ergo, the greater the radial depth within, the greater the pressure.

711.20 **Ratio of Tensors:** Locally on a circle, each particle has two sideways tensors for each inward tensor. One great-circle plane section through a circle shows two sideways tensors for one inward vector. But, on the surface of a sphere, each particle has *six* circumferential tensors for each single inward radial vector. When you double the radius, you double the chord.

711.30 **Struts as Chords in a Spherical Network:** When inserting a strut into a tensegrity sphere, we have to pull the tension lines outward from the system's center, in order to insert the strut between the vertexes of those lines. As we pull outward, the chordal distance of the gap between the spheric tension lines increases.

711.31 If we wish to open the slot in the basketball or football's skin through which its pneumatic bladder is to be inserted, we pull it outwardly and apart to make room inside.

711.32 The most outward chord of any given central angle of a circle is the longest. The omnicircumferential, triangularly stabilized, interconnecting tension lines of the spherical-network system cannot get bigger than its discretely designed dimensions and the ultimate tensile strength of the network's tensors, without bursting its integrity. The comprehensive spherical-tensor network can only relax inwardly. When all in place, the tensegrity-compression struts can only prevent the tension network from closing inward toward the sphere's center, which is its comprehensive proclivity.

711.33 The synergetic force of the struts (that is, their total interrelationship tendency) is not predicted by any one strut taken singly. It is entirely omniradially outward. The force of the strut is not a chordal two-way thrust.

711.34 A fully relaxed spherical tensegrity structure may be crumpled together in a tight bundle without hurting it, just as a net shopping bag can be stuffed into a small space. Thereafter, its drooped, untaut tension members can only yield outward radially to the dimensionally predesigned and prefabricated limits of the omniclosed spheric system, which must be progressively opened to accommodate the progressive interconstruction of the predesigned, prefabricated chordal lengths of the only circumferentially arrayed compression struts.

711.35 The compression struts are islanded from one another, that is, in each case, neither of the separate compression strut's ends touches any part of any other compression strut in the spheric system. As struts are inserted into the spheric-tension network, the whole spheric system is seen to be expanding omnioutwardly, as do pneumatic balloons when air is progressively introduced into their previously crumpled skins.

711.36 The comprehensive, finitely closed tension network's integrity is always pulling the islanded compression struts inward; it is never pushing them, nor are they pushing it, any more than a rock lying on Earth's crust thrusts horizontally sidewise. The rock is held where it is by the comprehensively contractive Earth's inter-mass-attraction (gravitational) field, or network. But the more rocks we add, the bigger the sphere held comprehensively together by the omnitensively cohering, gravitational consequences of the omni-interattractive mass aggregate.

712.00 Clothesline

712.01 Surprising behaviors are found in tensegrity structures. The illustration shows a house and a tree and a clothesline. The line hangs low between the house and the tree. To raise the line so that the clothes to be dried

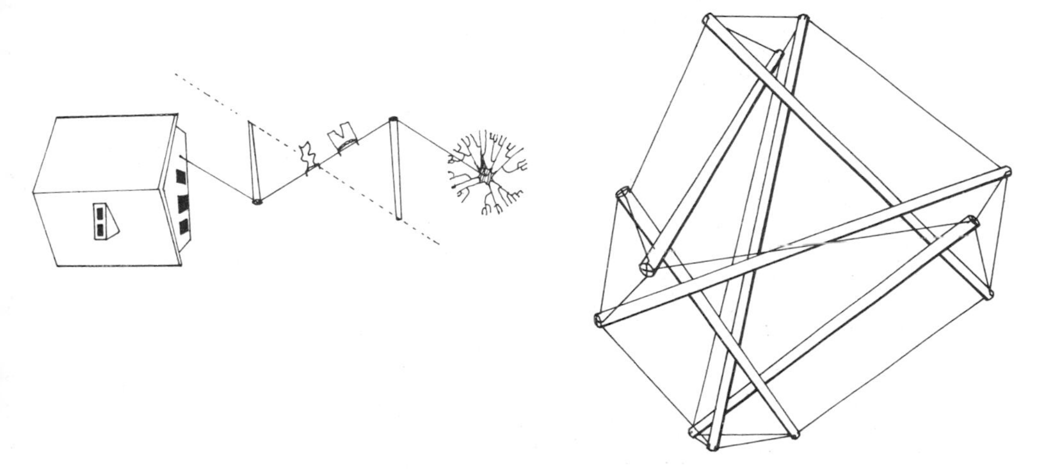

Fig. 712.01 *Tensegrity Behavior.*

will not sweep the ground, the line is elevated by a pole that has one end thrust against the ground and the other end pushed outwardly against the line. The line tightens with the pole's outer end at the vertex of an angle stretched into the line. The line's angle shows that the line is yielding in the direction away from the thrusting pole.

712.02 As the clothesline tightens and bends, it always yields *away* from the pushing strut. In spherical tensegrity structures the islanded compression struts *pull* the tension lines to *angle toward* the strut ends.

712.03 When we release a compression member from a tensegrity sphere, one end does not thrust *by* the tension member to which it was fastened in a circumferential direction. It was not fastened in *thrust* or *sheer*. It was not pushing circumferentially. It was resisting being compressed, and like a cork in a bottle, it was employing its frictional contact with the tension net at both its ends to *resist* its only tendency, which was to exit radially outward from the system's center.

713.00 Discontinuous Compression

713.01 Subvisible Discontinuity: In the Babylonian, Egyptian, and Ionian eras of ways of looking at, thinking about, and formulating, there evolved a concept of a "first family" of geometrical "solids," in which each member was characterized by all of its faces being identical and all of its edges being one length only. Humans were then unaware of what physics was only much later to discover experimentally: that nature discloses no evidence of a continuum. Experiment discloses only aggregates of separate, finitely closed events. Ergo, there are no solids.

713.02 Their optical illusion and stubbornly conditioned reflexes have since motivated one generation after another to go on teaching and accepting the misconception of geometric "solids," "planes," and "straight lines," where physics has discovered only wavilinear trajectories of high-frequency, yes-no event pulsations. With the misconception of straight lines came the misconception of the many lines going through the same point at the same time. Wherefore the 12 edges that define the cube were assumed to be absolute straight lines, and therefore sets of them ran simultaneously into the thus absolutely determined eight corner "points" of the cube.

713.03 Humans were accustomed to the idea that edges come together at one certain point. But we now know operationally that if we look at any of the edges of any item microscopically, there is no such absolute line, and instead there is seen to be an aggregate of atomic events whose appearance as an aggregate is analogous to the roughly rounding, wavilinear profiled, shoulder "edge" of a rock cliff, sand, or earth bluff standing high above the beach of the shore lying below, whose bluff and beach disclose the gradual erosion of the higher land by the sea.

713.04 The corners of the solids are also just like the corners of an ocean-side bluff that happens to have its coastwise direction changed at 90 degrees by large geological events of nature such as an earthquake fault. Such an easterly coastline's bluff casts dark shadows as the Earth rotates; seen from airplanes at great altitudes, long sections of that black coastal shadow may appear illusionarily as "straight."

713.05 We can make Platonic figures in nonsolid tensegrity where none of the lines go through any of the same points at the same time, and we realize that the only seemingly continuous, only mass-interattractively cohered, atomic "Milky Way" tensor strands spanning the gaps between the only seemingly "solid," omni-islanded, vectorially compressioned struts, do altogether permit a systematic, visually informed, and realistically comprehended differentiation between the flexible tensor and inflexible vector energy-event behaviors, all of which are consistent with all the experimental information accruing to the most rigorous scientific discipline.

713.06 The eye can resolve intervals of about 1/100th of an inch or larger. Below that, we do not see the aggregates as points. Thereafter, we see only "solid"-color surfaces. But our color receptivity, which means our only-human-optics-tunable range of electromagnetic radiation frequencies, cannot "bring in," i.e., resonatingly respond to, more than about one-millionth of the now known and only instrumentally tune-in-able overall electromagnetic-wave-frequency range of physical Universe. This is to say that humans can tune in directly to less than one-millionth of physical reality—ergo, cannot "see" basic atomic and molecular-structuring events and behaviors, but our synergetic tensegrity principles of structuring are found instrumentally to be operative to the known limits of both micro- and macro-Universe system relationships as the discontinuous, entropic, radiational, and omnicohering, collecting gravitational syntropics. (See Sec. 302.)

713.07 Convergence: While we cannot see the intervals between atomic-event waves, the tensegrity structuring principles inform our consideration of the invisible events. Every time we instrumentally magnify the illusionarily converging geometrical "lines" defining the edges of "solids," we see them only wavilinearly converging toward critical proximity but never coming completely together; instead, twisting around each other, then slivering again, never having gone through the same "points."

713.08 When we first try to differentiate tension and compression in consciously attempting to think about the behavior of structures in various locals of Universe, it becomes apparent that both macro-Universe and micro-

Universe are only tensionally cohered phenomena. They both obviously manifest discontinuous compression islands. It is evidenced, in cosmically structured systems, both macro and micro, that compression members never touch one another. Earth does not roll "ball bearing" around on the surface of Mars; nor does the Moon roll on Earth, and so forth. This structural scheme of islanded spheres of compression, which are only mass-attractively cohered, also characterizes the atomic nucleus's structural integrities. Tensegrity discoveries introduce new and very different kinds of structural principles which seem to be those governing all structuring of Universe, both macrocosmic and microcosmic.

713.20 Compression Members

713.21 **Behavior of Compression Members in Spherical Tensegrity Structures:** In spherical tensegrity constructions, whenever a tension line interacts with a compression strut, the line does *not* yield in a circumferential direction *away* from the strut. The islanded compression member, combining its two ends' oppositely outward thrust, *pulls* on the omni-integrated tension network only acting as a radially outward force in respect to the sphere's center.

713.22 When we remove a compression member from a tensegrity sphere of more than three struts, the compression member of the original triangular group, when released on one end, does not *shove by* the tension member to which it was fastened. It is not fastened in *shove* or *sheer*. It pulls outwardly of the spherical system, away from the tension members at both of its ends simultaneously; when released, it pops only outwardly from the sphere's center.

713.23 When inserting a strut into a tensegrity sphere, you are pulling the tensional network only outwardly of the system in order to allow the strut to get *into* the system, that is, toward the structure's center. The strut pulls only outward on the two adjacent tension members to which it is fixed, trying to escape only radially outwardly from the system's center.

714.00 Interstabilization of Local Stiffeners

714.01 **Local, Discontinuous, Compressional Strut Waves Interstabilizing Two Concentric, Differentially Radiused Tensegrity Spheres:** Highly stable, nonredundant, rigidly trussed, differently radiused, concentric spherical tensegrity structures of hexagonal-pentagonal, inner *or* outer (but not both) surface dimples, symmetrically interspersing their omnitriangularly interlinked, spherically closed systems, may be constructed with swaged crossings of high-tensile-steel-cabled, spherical nets and locally islanded compressional struts occurring discontinuously as inbound-outbound, triangularly intertrussing, locally islanded compressional struts. The struts may then be either hydraulically actuated to elongate them to designed dimensions, or may be locally jacked in between the comprehensively prefabricated, spherical-system tensional network.

714.02 The local struts are so oriented that they always and only angle inwardly and outwardly between the concentric, differently radiused, comprehensively finite, exterior and interior, tensional spherical nets. The result is an interstabilized dynamic equilibrium of positive and negative waves of action. Such tensegrity sphere structures are limited in size only by the day-to-day limits of industrial production and service-logistics techniques. Large tensegrity spheres can have their lower portions buried in reinforced-concrete as tie-down bases to secure them against hurricane-drag displacement.

715.00 Locked Kiss

715.01 As we increase the frequency of triangular-module subdivisions of a tensegrity geodesic sphere, we thus also increase the number of compression struts, which get progressively halved in length, while their volumes and weights shrink eightfold. At the same time, the arc altitude between the smaller arcs and chords of the sphere decreases, while the compression members get closer and closer to the adjacent compression members they cross. Finally, we reach the condition where the space between the struts is the same dimension as the girth radius of the struts. At this point, we can let them kiss-touch; i.e., with the ends of two converging struts contacting the top middle of the strut running diagonally to those two struts and immediately below their ends. We may then lock the three kissing members tensionally together in their kiss, but when we do so, we must remember that they were not pushing one another when they ''kissed'' and we locked them in that equilibrious, ''most comfortable'' position of contact coincidence. Tensegrity spheres are not fastened in shear, even though their *locked kiss* gives a superficially ''solid'' continuity appearance that is only subvisibly discontinuous at the atomic level.

716.00 Complex Continuity and Discontinuity in Tensegrity Structures

716.01 The terminal junctures of four three-strut tensegrity octahedra are all 180-degree junctures. They appear to be compressionally continuous, while the central coherence of the three struts appears visibly discontinuous. The complex tensegrity presents a visibly deceptive appearance to the unwary

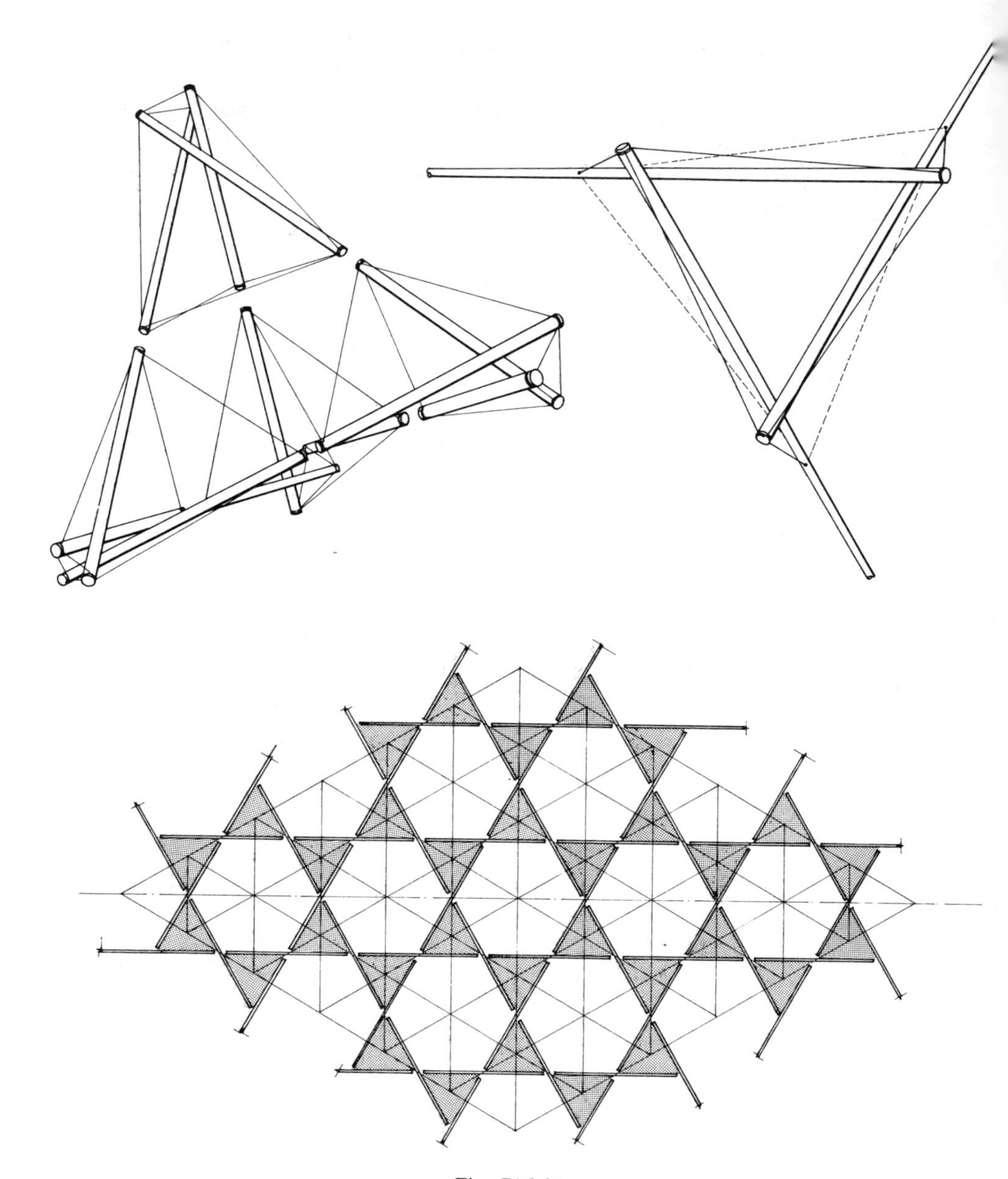

Fig. 716.01.

observer. The two joined legs of the basic units appear as single units; as such, they appear to be primary elements of the complex tensegrity, whereas we learn from construction that our elements are the three-strut octahedra and that the cohering principle of the simplest elements is tensegrity.

716.02 The fundamental, repeatable unit used to form the spherical tensegrity structures is a flattened form of the basic three-strut tensegrity octahedron.

716.03 The basic 12-frequency tensegrity matrix employs collections of the basic three-strut units joined at dead center between single- and double-bonded discontinuity. The shaded triangles in the illustration represent the sites for each of the three-strut units. This matrix is applied to the spherical triacontrahedron—consequently, the large 12-frequency rhombus (illustration 716.01C) is one-thirtieth of the entire sphere.

716.10 Convergence

716.11 Whereas man seems to be blind in employing tensegrity at his level of everyday consciousness, we find that tensegrity structures satisfy our conceptual requirement that we may not have two events passing through the same point at the same time. Vectors converge in tensegrity, but they never actually get together; they only get into critical proximities and twist by each other.

717.00 Single- and Double-Bonding in Tensegrity Spheres

717.01 Basic three-strut tensegrities may be joined in single-bonding or double-bonding to form a complex, 270-strut, isotropic tensegrity geodesic sphere. It can be composited to rotate negatively or positively. A six-frequency triacontrahedron tensegrity is shown in illustration 717.01.

717.02 Complexes of basic three-strut tensegrities are shown with axial alignment of exterior terminals to be joined in single bond as a 90-strut tensegrity.

720.00 Basic Tensegrity Structures

720.01 In basic tensegrity structures, the spheric-tension network system is completely continuous. The ends of each compression member connect only with the tension network at various points on the tensional catenaries

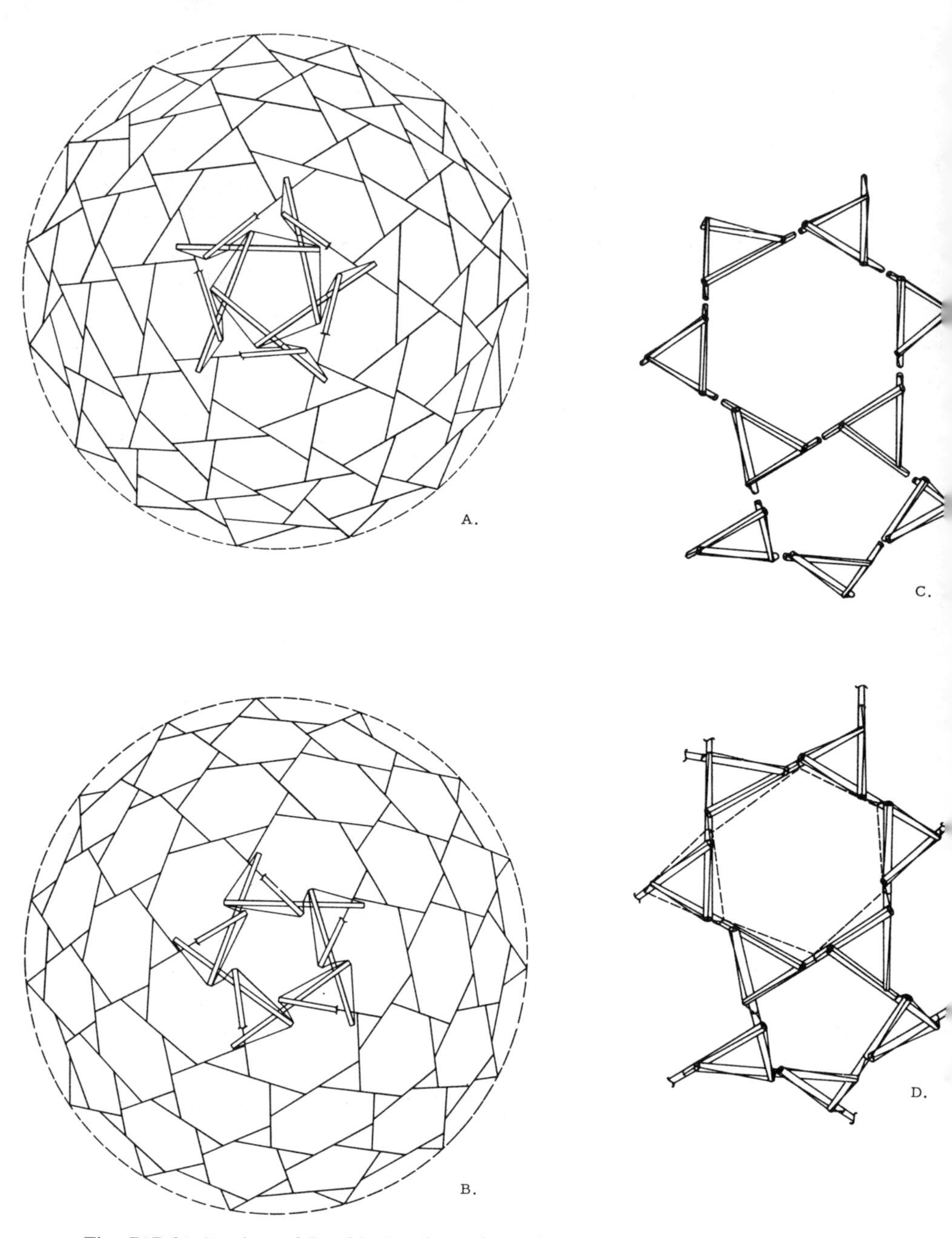

Fig. 717.01 *Single and Double Bonding of Members in Tensegrity Spheres:*

A. Negatively rotating triangles on a 270-strut tensegrity geodesic sphere with double-bonded triangles.

B. A 270-strut isotropic tensegrity geodesic sphere: single bonded turbo triangles forming a complex six-frequency triacontahedron tensegrity.

C. Complex of basic three-strut tensegrities, with axial alignment whose exterior terminals are to be joined in single bond as 90-strut tensegrity.

D. Complex of basic three-strut tensegrity units with exterior terminals now joined.

nearest to the respective ends of the system's omni-islanded compression vector struts. The tension members running between the ends of the struts may be double or single. Double tension members best distribute the loads and most economically and nonredundantly accommodate the omnidistributive stress flows of the system. The catenaries always yield in obtuse or acute "V" shapes at their points of contact with the *strut* islands of compression.

720.02 Conventional building with continuous compression and discontinuous tension is accustomed to fastening compression members to their buildings in shear, that is, in predictably, calculatable, "slide-by" pushing actions, where one force opposes another in parallel but opposite directions.

720.03 But in tensegrity structures, the tension members pull away from the compression strut ends, which the V-shape tension connections demonstrate. If two people take positions on opposite sides of a tensegrity sphere and pull on polarly opposite struts in opposite directions from one another, it will be seen that all around the sphere there is a uniform and symmetrical response to the opposite pulling (or pushing). Pulling on two opposite parts makes the whole sphere grow symmetrically in size, while pushing forces the whole sphere to shrink contractively and symmetrically. Cessation of either the pulling or the pushing causes the sphere to take its size halfway between the largest and smallest conditions, i.e., in its equilibrious size. This phenomenon is a typical four-dimensional behavior of synergetic intertransforming. It explains why it is that all local celestial systems of Universe, being cohered with one another tensionally, pull on one another to bring about an omniexpanding physical Universe.

720.10 **Micro-Macro Structural Model:** If you just tauten one point in a tensegrity system, all the other parts of it tighten evenly. If you twang any tension member anywhere in the structure, it will give the same resonant note as the others. If you tauten any one part, the tuning goes to a higher note everywhere in the structure. Until its tension is altered, each tensegrity structure, as with every chemical element, has its own unique frequency. In a two-sling tensegrity sphere, every part is nonredundant. If tension goes up and the frequency goes up, it goes up uniformly all over. As tensegrity systems are tautened, they approach but never attain rigidity, being nonredundant structures. Anything that we would call rigid, such as one of the atoms of a very high integrity pattern, is explained by this type of tensegrity patterning.

720.11 The kinetically interbalanced behaviors of tensegrity systems manifest discretely and elucidate the energy-interference-event patternings that integrate to form and cohere all atoms. The tensegrity system is always the equilibrious-balance phase, i.e., the omnipotential-energy phase visually articulate of the push-pull, in-out-and-around, pulsating and orbiting, precessionally shunted reangulations, synergetically integrated.

720.12 The circumferentially islanded tensegrity struts are energy vectors in action, and the tension lines are the energy tensors in action. Their omnisystem interpatterning shows how the circumferentially orbiting tensegrity struts' lead ends are pulled by the center of mass of the next adjacent inwardly positioned vector strut. The mass attraction pulls inwardly on the lead ends of the precessionally articulating, self-orbiting, great-circle chord vector struts, thus changing their circumferential direction. They are precessionally and successively deflected from one tangent course to the next, circumferentially inward and onward, tangent vector course. Thus each of the vectors is successively steered to encircle the same tensegrity system center. In this manner, a variety of energy-interference, kinetic-equilibrium patterns results in a variety of cosmically local, self-regenerative, micro-macro structural systems such as atoms or star systems.

721.00 Stability Requires Six Struts

721.01 Stability requires six struts, each of which is a combinedly push-pull structural member. It is a synergetic (Sec. 101) characteristic of minimum structural (Sec. 610) systems (Sec. 402) that the system is not stable until the introduction of the last structural component (Sec. 621.10) essential to completion of minimum omnisymmetric array.

721.02 Redundancy (Sec. 723) can be neither predicted nor predetermined by observation of either the integral constraints or external freedoms of energetic behaviors of single struts, or beams, or columns, or any one chain link of a series that is less than 12 in number, i.e., six positive vectors and six negative tensors. Of these 12, six are open-endedly uncoordinate, disintegrative forces that are always omni-cohered by six integrative forces in finitely closed coordination.

722.00 Push-Pull Members

722.01 Minimum structural-system stability requires six struts, each of which is a push-pull member. Push-pull structural members embody in one superficially solid system both the axial-linear tension and compression functions.

722.02 Tensegrity differentiates out these axial-linears into separately co-functioning compression vectors and tensional tensors. As in many instances of synergetic behavior, these differentiations are sometimes subtle. For instance, there is a subtle difference between Eulerian topology, which is polyhedrally superficial, and synergetic topology, which is nuclear and identifies spheres with vertexes, solids with faces, and struts with edges. The subtlety lies in the topological differentiation of the relative abundance of these three fundamental aspects whereby people do not look at the four closest-packed spheres forming a tetrahedron in the same way that they look at a seemingly solid stone tetrahedron, and quite differently again from their observation of the six strut edges of a tetrahedron, particularly when they do not accredit Earth with providing three of the struts invisibly cohering the base ends of the camera tripod.

723.00 Redundance

723.01 There are metaphysical redundancies, repeating the same thing, saying it in a little different way each time.

723.02 There are physical redundancies when, for instance, we have a mast stepped in a hole in the ground and three tensional stays at 120 degrees. When a fourth tension member is led to an anchor at an equiradiused distance from the mast base and at one degree of circular arc away from one of the original three anchors, we then have two tension stays running side by side. When the two stays are thus approximately parallel, we find it is impossible to equalize the tensions exactly. One or the other will get the load, not both.

723.03 It is structural redundancy when a square knot is tied and an amateur says, "I'm going to make that stronger by tying more square knots on top of it." The secondary knots are completely ineffective because the first square knot will not yield. There is a tendency of the second square knot to "work open" and thus deteriorate the first knot. Structural redundancies tend to deteriorate the effectiveness of the primary members.

723.04 There are two classes of redundant acts:

(1) conscious and knowledgeably competent, and

(2) subconscious and ignorantly fearful cautionaries.

723.05 Building codes of cities, formulated by politicians fearful of the calumny of what may befall them if buildings fall down, ignorantly insist on doubling the thickness of walls. Building codes require a safety factor of usually five, or more, to one.

723.06 Aircraft designing employs a safety factor of two to one—or even no safety factor at all, while cautioning the pilot through instrumental indication of when he is approaching limit condition. The deliberately imposed safety factors of society's building conglomerates introduce redundancy breeding redundancy, wherein—as with nuclear fusion, chain-reacting—the additional weights to carry the additional weights multiply in such a manner as to increase the inefficiency imposed by the redundancy at an exponential rate implicit in Newton's mass-attraction gravitational law: every time we double the safety factor, we fourfold the inefficiency and eightfold the unnecessary weight.

724.00 Three and Only Basic Structures

724.01 The original six vector-edge members of the tensegrity tetrahedron may be transformed through the tensegrity-octahedron phase and finally into the tensegrity-icosahedron phase. The same six members transform their relation to each other through the full range of the three (only) fundamental structures of nature: the tetrahedron, the octahedron, and the icosahedron. (See Secs. 532.40, 610.20, 724, 1010.20, 1011.30 and 1031.13.)

724.02 The same six members transform from containing one volume to containing 18.51 volumes. These are the principles actively operative in atomic-nucleus behavior in visual intertransformations.

724.10 **Tensegrity Octahedron:** The simplest form of tensegrity is the octahedron with three compression members crossing each other. The three compression struts do not touch each other as they pass at the center. They are held together only at their terminals by the comprehensive triangular tension net. The same three-islanded struts of the tensegrity octahedron may be mildly reorganized or asymmetrically transformed.

724.11 The struts may be the same length or of different lengths. Some tensional edges may be lengthened while other tensional edges of the surface triangles are shortened. The compression members still do not touch each other. One figure is a positive and the other a negative tensegrity octahedron. They can be joined together to make a new form: the tensegrity icosahedron.

724.20 **Tensegrity Icosahedron:** The six-islanded-strut icosahedron and its allspace-filling, closest-packing capability provide omni-equi-optimum economy tensegrity Universe structuring.

724.30 **Six-Strut Tensegrities:** Two three-strut tensegrities may be joined together to make the tensegrity icosahedron. This form has six members in three parallel sets with their ends held together in tension. There are 12 terminals of the six struts (the two octahedra—each with three struts of six ends—combined). When you connect up these 12 terminals, you reveal the 12 vertexes of the icosahedron. There are 20 triangles of the icosahedron clearly described by the tension members connecting the 12 points in the most economical omnitriangular pattern.

724.31 In the tensegrity icosahedron, there are six tension members, which join parallel struts to each other. If these tension members are removed

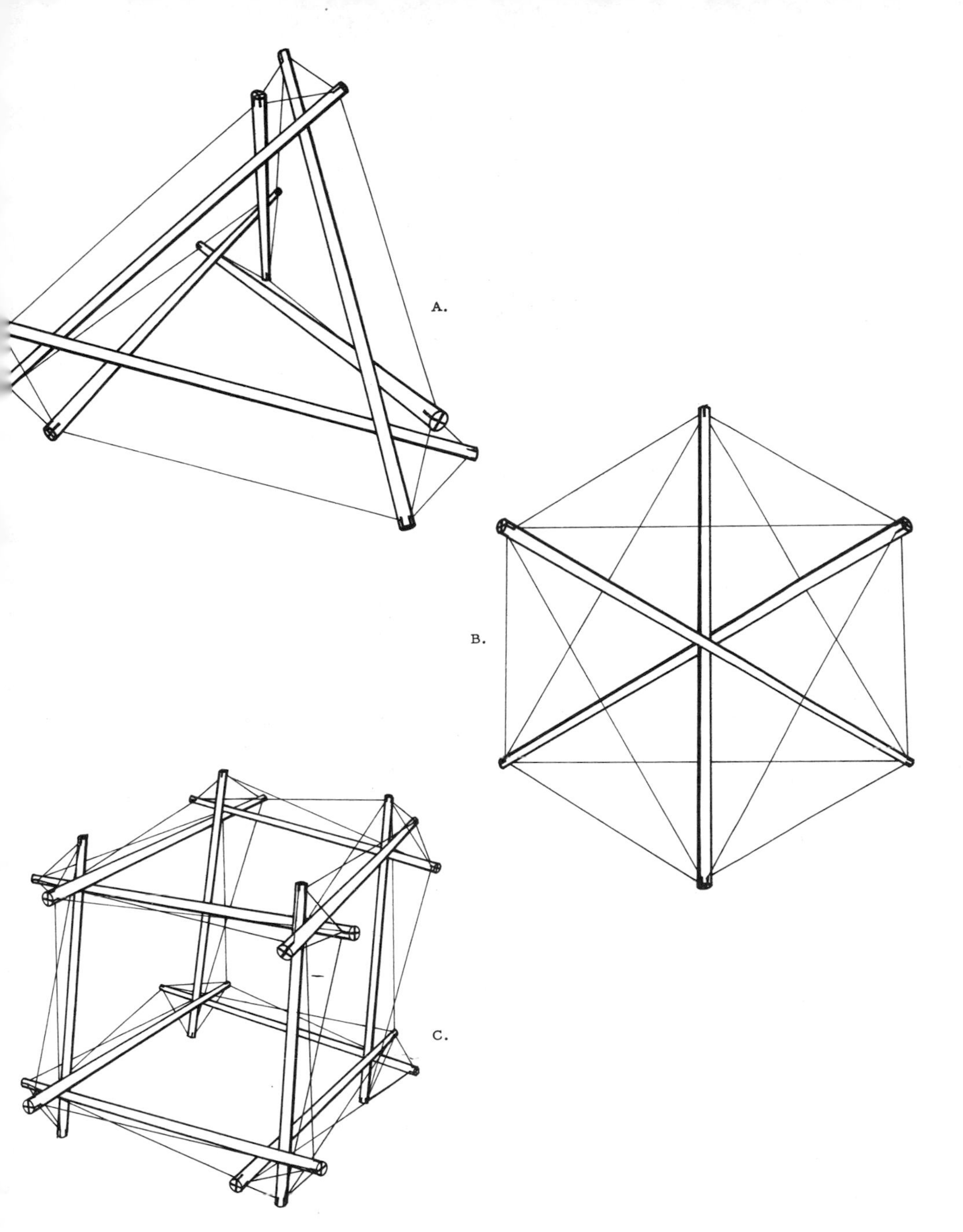

Fig. 724.10:

A. A six-strut tensegrity tetrahedron showing central-angle turbining.

B. The three-strut tensegrity octahedron. The three compression struts do not touch each other as they pass at the center of the octahedron. They are held together only at their terminals by the comprehensive, triangular tension net. It is the simplest form of tensegrity.

C. The 12-strut tensegrity cube, which is unstable.

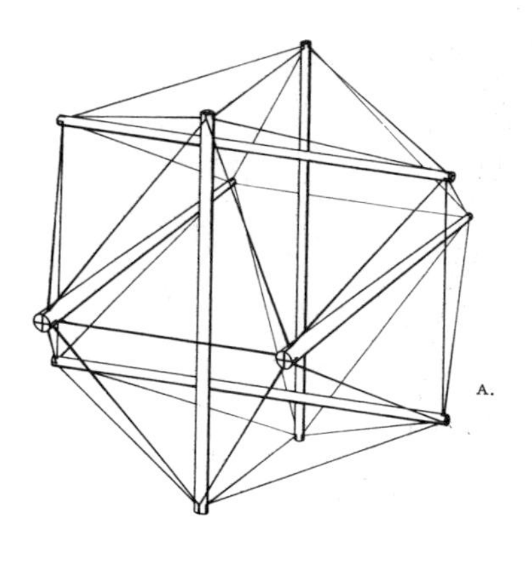

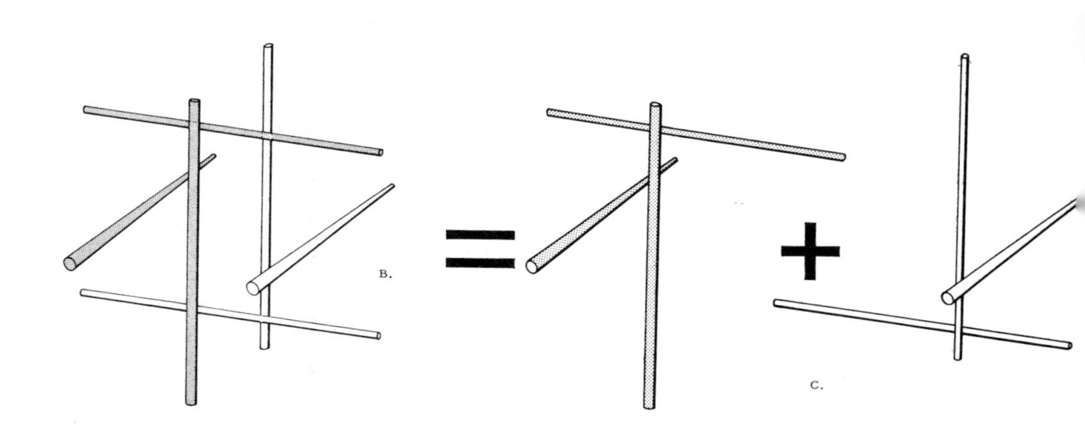

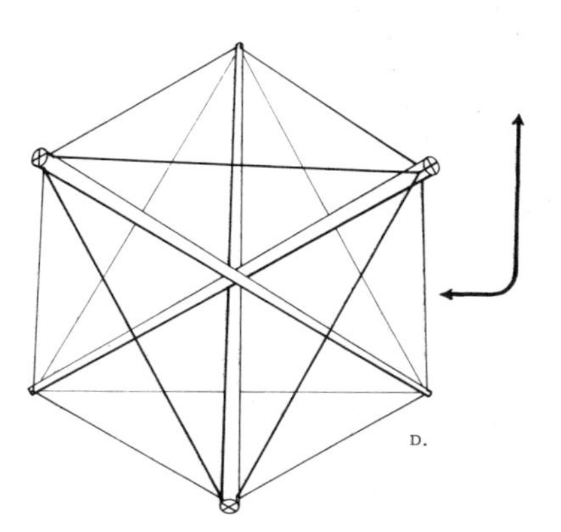

Fig. 724.30 *Behavior of Tensegrity Icosahedron.*

from the icosahedron, only eight triangles remain from the original 20. These eight triangles are the eight transforming triangles of the symmetrical contraction of the vector equilibrium "jitterbug." (See Sec. 460.) Consequently, this "incomplete" icosahedron demonstrates an expansion-contraction behavior similar to the "jitterbug," although pulsing symmetrically inward-outward within more restricted limits.

724.32 If two opposite and parallel struts are pushed or pulled upon, all six members will move inwardly or outwardly, causing the icosahedron to contract or expand in a symmetrical fashion. When this structure is fully expanded, it is the regular icosahedron; in its contracted state, it becomes an icosahedron bounded by eight equilateral triangles and 12 isosceles triangles (when the missing six tension members are replaced). All 12 vertexes may recede from the common center in perfect symmetry of expansion or, if concentrated load is applied from without, the whole system contracts symmetrically, i.e., all the vertexes move toward their common center at the same rate.

724.33 This is not the behavior we are used to in any structures of our previous experiences. These compression members do not behave like conventional engineering beams. Ordinary beams deflect locally or, if fastened terminally in tension to their building, tend to contract the building in axial asymmetry. The tensegrity "beam" does not act independently but acts only in concert with "the whole building," which contracts only symmetrically when the beam is loaded.

724.34 The tensegrity system is synergetic—a behavior of the whole unpredicted by the behavior of the parts. Old stone-age columns and lintels are energetic and only interact locally with whole buildings. The whole tensegrity-icosahedron system, when loaded oppositely at two diametric points, contracts symmetrically, and because it contracts symmetrically, its parts get symmetrically closer to one another; therefore, gravity increases as of the second power, and the whole system gets uniformly stronger. This is the way atoms behave.

725.00 Transformation of Tensegrity Structures

725.01 Six-strut tensegrity tetrahedra can be transformed in a plurality of ways by changing the distribution and relative lengths of its tension members to the six-strut icosahedron.

725.02 A theoretical three-way coordinate expansion can be envisioned, with three parallel pairs of constant-length struts, in which a stretching of tension members is permitted as the struts move outwardly from a common center. Starting with a six-strut octahedron, the structure expands outwardly, going through the icosahedron phase to the vector-equilibrium phase.

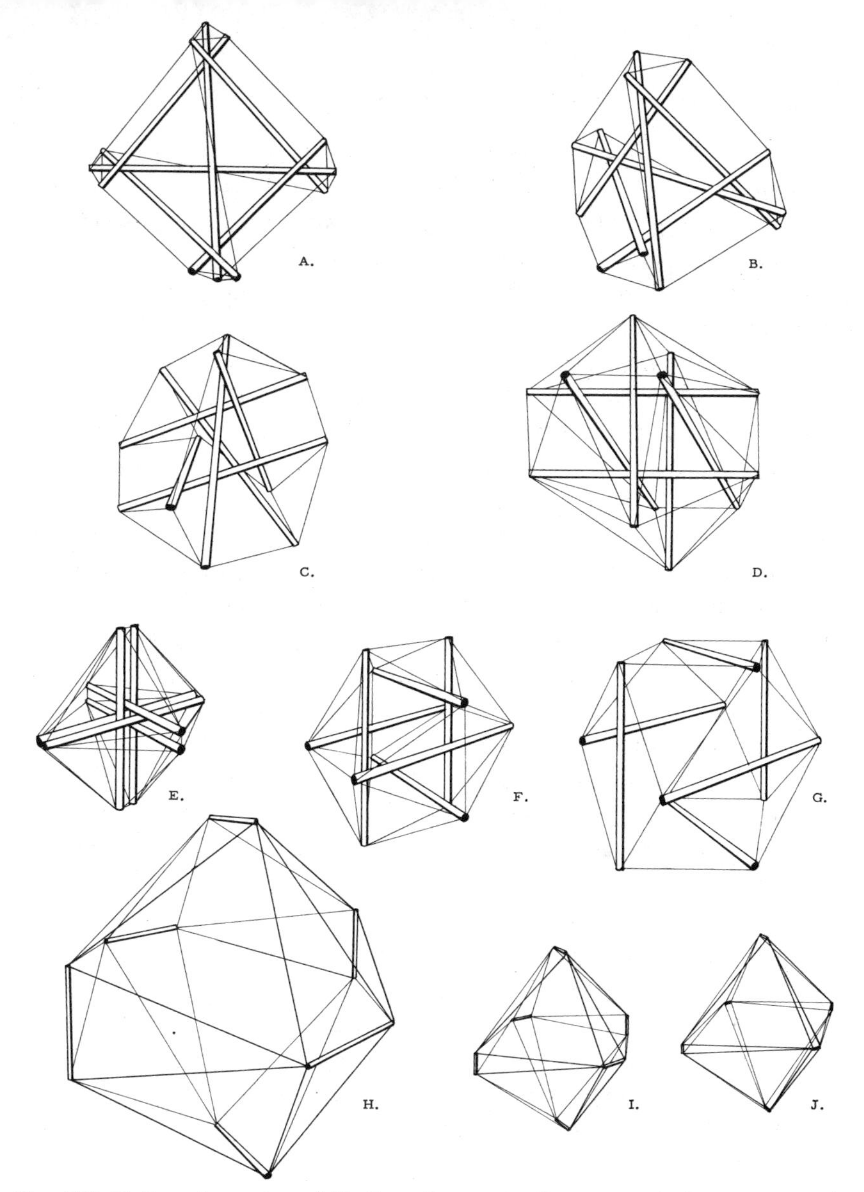

Fig. 725.02 *Transformation of Six-Strut Tensegrity Structures:* A six-strut tensegrity tetrahedron (A) can be transformed by changing the distribution and relative lengths of its tension members (B, C) to the six-strut icosahedron (D). A *theoretical* three-way coordinate expansion can be envisioned with three parallel pairs of constant-length struts in which a stretching of tension members is permitted as the struts move outwardly from a common center. Starting with a six-strut octahedron (E), the structure expands outwardly going through the icosahedron phase (F) to the vector-equilibrium phase (G). When the structure expands beyond the vector equilibrium, the six struts become the edges of figure H. They consequently lose their structural function (assuming the original distribution of tension and compression members remains unchanged). As the tension members become substantially longer than the struts, the struts tend to approach relative zero and the overall shape of the structure approaches a *super* octahedron (I, J).

725.03 When the structure expands beyond the vector equilibrium, the six struts become the edges of the figure; they consequently lose their structural function (assuming that the original distribution of tension and compression members remains unchanged). As the tension members become substantially longer than the struts, the struts tend to approach relative zero, and the overall shape of the structure approaches a super octahedron.

726.00 Six-Pentagonal Tensegrity Sphere

726.01 **The Symmetrical, Six-Great-Circle-Planed, Pentagonally Equatored Tensegrity Sphere:** A basic tensegrity sphere can be constituted of six equatorial-plane pentagons, each of which consists of five independent and nonintertouching compression struts, totaling 30 separate nonintertouching compression struts in all. This six-pentagon-equatored tensegrity sphere interacts in a self-balanced system, resulting in six polar axes that are each perpendicular to one of its six equatorial pentagonal planes. Twelve lesser-circle-planed polar pentagons are found to be arrayed perpendicular to the six polar axes and parallel to the equatorial pentagon planes. It also results in 20 triangular interweavings, which structuring stabilizes the system.

726.02 Instead of having cables connecting the ends of the struts to the ends of the next adjacent struts in the six-axes-of-symmetry tensegrity structure, 60 short cables may be led from the ends of each prestressed strut either to the midpoint of the next adjacent strut or to the midpoint of tension lines running from one end to the other of each compression strut. Each of the two ends of the 30 spherical-chord compression struts emerges as an energy action, out over the center of action-and-reaction-effort vectors of the next adjacent strut, at which midpoint the impinging strut's effort is angularly precessed to its adjacent struts. Thus each strut precessionally transfers its effort and relayed interloadings to the next two adjacent struts. This produces a dynamically regenerative, self-interweaving basketry in which each compression strut is precessed symmetrically outwardly from the others while simultaneously precessing inwardly the force efforts of all the tensional network.

726.03 In this pattern of six separate, five-strut-membered pentagons, the six pentagonal, unsubstanced, but imaginable planes cut across each other equiangularly at the spheric center. In such a structure, we witness the cosmic principles that make possible the recurrence of locally regenerative structural patterns. We are witnessing here the principles cohering and regenerating the atoms. The struts are simple, dynamic, energy-event vectors that derive their regenerative energies from an eternally symmetrical interplay of inbound-outbound forces of systems that interfere with one another to maintain critical fall-in, shunt-out proximities to one another.

730.00 Stabilization of Tension in Tensegrity Columns

730.10 Symmetric Juxtaposition of Tetrahedra

730.11 All polyhedra may be subdivided into component tetrahedra. Every tetrahedron has four vertexes, and every cube has eight vertexes. Every cube contains two tetrahedra (*ABCD* and *WXYZ*). Each of its faces has two diagonals, the positive set and the negative set. These may be called the symmetrically juxtaposed positive and negative tetrahedra, whose centers of volume are congruent with one another as well as congruent with the center of volume of the cube. It is possible to stack cubes into two columns. One col-

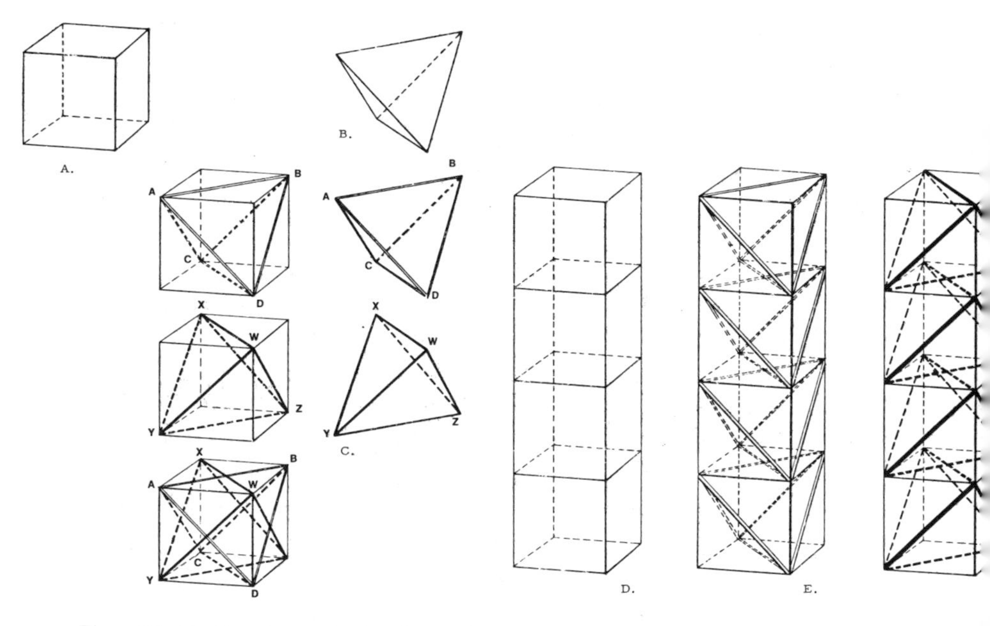

Fig. 730.11 *Functions of Positive and Negative Tetrahedra in Tensegrity Stacked Cubes:* Every cube has six faces (A). Every tetrahedron has six edges (B). Every cube has eight corners and every tetrahedron has four corners. Every cube contains two tetrahedra (*ABCD* and *WXYZ*) because each of its six faces has two diagonals, the positive and negative set. These may be called the symmetrically juxtaposed positive and negative tetrahedra whose centers of gravity are congruent with one another as well as congruent with the center of gravity of the cube (C). It is possible to stack cubes (D) into two columns. One column contains the positive tetrahedra (E) and the other contains the negative tetrahedra (F).

umn can demonstrate the set of positive tetrahedra, and the other column can demonstrate the set of negative tetrahedra.

730.12 In every tetrahedron, there are four radials from the center of volume to the four vertexes. These radials provide a model for the behavior of compression members in a column of tensegrity-stacked cubes. Vertical tension stays connect the ends of the tetrahedral compression members, and they also connect the successive centers of volume of the stacked spheres—the centers of volume being also the junction of the tetrahedral radials. As the two centers of volume are pulled toward one another by the vertical tension stays, the universally jointed radials are thrust outwardly but are finitely restrained

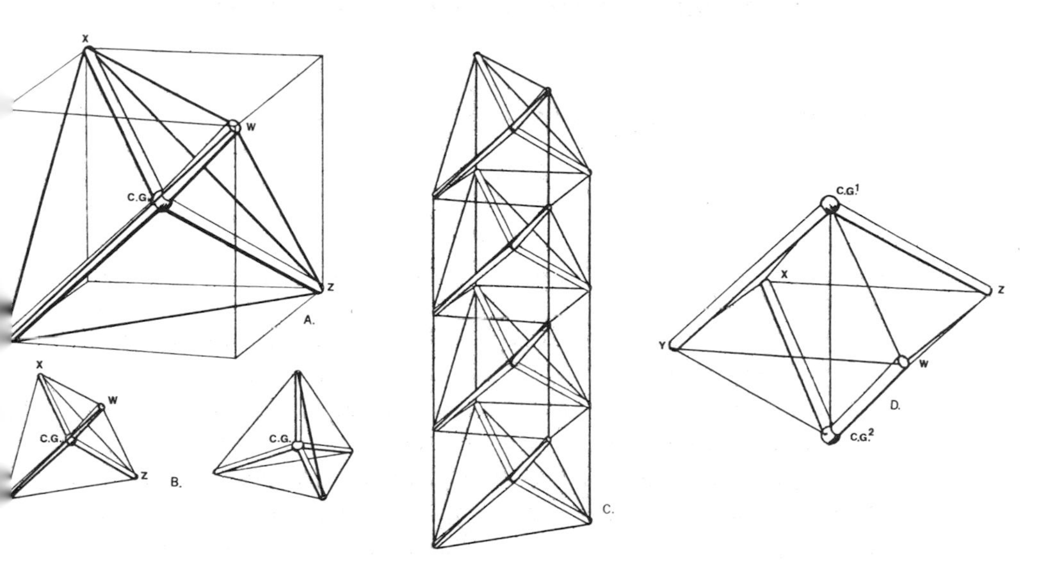

Fig. 730.12 *Stabilization of Tension in Tensegrity Column:* We put a steel sphere at the center of gravity of a cube which is also the center of gravity of tetrahedron and then run steel tubes from the center of gravity to four corners, *W,X,Y,* and *Z,* of negative tetrahedron (A). Every tetrahedron's center of gravity has four radials from the center of gravity to the four vertexes of the tetrahedron (B). In the juncture between the two tetrahedra (D), ball joints at the center of gravity are pulled toward one another by a vertical tension stay, thus thrusting universally jointed legs outwardly, and their outward thrust is stably restrained by finite sling closure *WXYZ.* This system is nonredundant: a basic discontinuous-compression continuous-tension or "tensegrity" structure. It is possible to have a stack (column) of center-of-gravity radial tube tetrahedra struts (C) with horizontal (approximate) tension slings and vertical tension guys and diagonal tension edges of the four superimposed tetrahedra, which, because of the (approximate) horizontal slings, cannot come any closer to one another, and, because of their vertical guys, cannot get any further away from one another, and therefore compose a stable relationship: a structure.

by the sliding closure *XYZW* interlinking the tetrahedral integrities of the successive cubes.

730.13 This system is inherently nonredundant, as are all discontinuous-compression, continuous-tension tensegrity structures. The approximately horizontal slings cannot come any closer to one another, and the approximately vertical stays cannot get any farther from one another; thus they comprise a discrete-pattern, interstabilizing relationship, which is the essential characteristic of a *structure*.

740.00 Tensegrity Masts: Miniaturization

740.10 Positive and Negative

740.11 Stacked columns of "solidly," i.e., compressionally continuous and only compressionally combined, cubes demonstrate the simultaneous employment of both positive and negative tensegrities. Because both the positive and the negative tensegrity mast are independently self-supporting, either one provides the same overall capability. It is a kind of capability heretofore associated only with "solid" compressional struts, masts, beams, and levers—that is, either the positive- or the negative-tensegrity "beam-boom-mast" longitudinal structural integrity has the same capability independently as the two of them have together. When the two are combined, either the positive- or the negative-tensegrity set, whichever is a fraction stronger than the other, it is found experimentally, must be doing all the strut work at any one time. The unemployed set is entirely superfluous, ergo redundant. All "solid" structuring is redundant.

740.12 If the alternate capabilities of the positive and negative sets are approximately equal, they tend to exchange alternately the loading task and thus generate an oscillating interaction of positive vs. negative load transferral. The energies of their respective structural integrities tend to self-interdeterioration of their combined, alternating, strut-functioning longevity of structural capability. The phenomenon eventually approaches crystallization. All the redundant structures inherently accelerate their own destruction in relation to the potential longevity of their nonredundant tensegrity counterparts.

740.20 Miniaturization

740.21 It is obvious that each of the seemingly "solid" compression struts in tensegrity island complexes could be replaced by miniature tensegrity

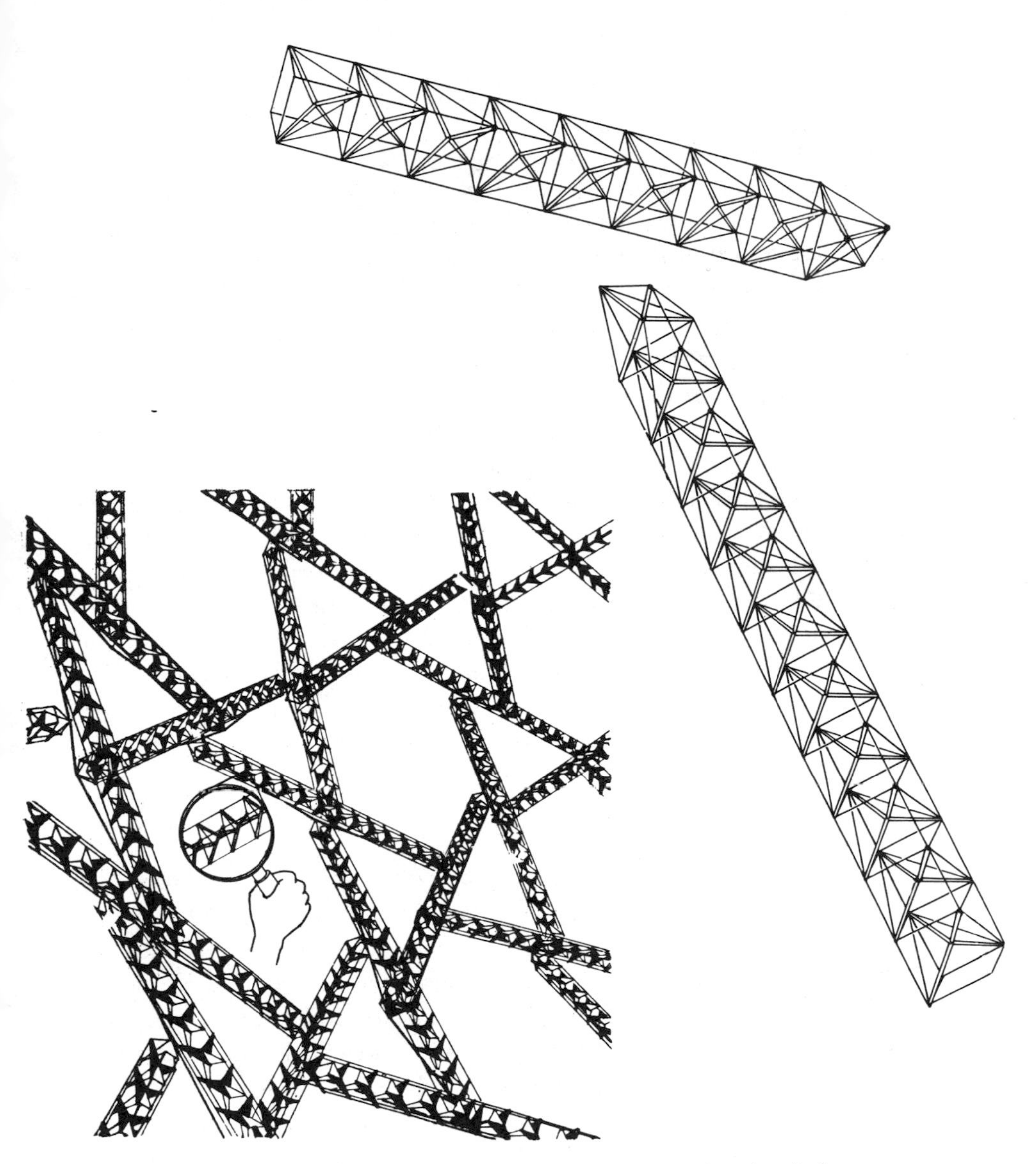

Fig. 740.21 *Tensegrity Masts as Struts: Miniaturization Approaches Atomic Structure:* The tensegrity masts can be substituted for the individual (so-called solid) struts in the tensegrity spheres. In each one of the separate tensegrity masts, acting as struts, in the tensegrity spheres it can be seen that there are little (so-called) *solid struts*. A miniature tensegrity mast may be substituted for each of those *solid struts*. The subminiature tensegrity masts within the tensegrity struts of the tensegrity sphere and a subsubminiature tensegrity mast may be substituted for each of those *solid* struts, and so on to subsubsubminiature tensegrities until we finally get down to the size of the atom and this becomes completely compatible with the atom for the atom is tensegrity and there are no "solids" left in the entire structural system. There are no solids in structures, ergo no solids in Universe. There is nothing incompatible with what we may *see* as solid at the visual level and what we are finding out to be the structural relationships in nuclear physics.

masts. There is nothing to keep us from doing this but technological techniques for operating at microlevels. It is simply that each of the struts gets smaller: as we look at each strut in the tensegrity mast, we see that we could make another much smaller miniature tensegrity mast to replace it. Every time we can see a separate strut and can devise means for making a tensegrity strut of that overall size, we can substitute it for the previously "solid" strut. By such a process of progressive substitutions in diminishing order of sizes, leading eventually via sub-sub-sub-miniaturizing tensegrities to discovery of the last remaining stage of the seemingly "solid" struts, we find that there is a minimum "solid-state" strut's column diameter, which corresponds exactly with two diameters of the atoms of which it is constructed. And this is perfectly compatible, because discontinuity characterizes the structuring of the atoms. The atom is a tensegrity, and there are no "solids" left in the entire structural system. We thus discover that tensegrity structuring and its omnirationally constituted regularities are cosmically a priori, disclosing that Universe is not redundant. It is only humanity's being born ignorant that has delayed all of humanity's escape from the self-annihilating effect of the omniredundance now characterizing most of humanity's activities.

740.30 No Solids in Structures

740.31 There are no solids in structures. Ergo, there are no solids in Universe. There is nothing incompatible with what we may see as "structure" at the superficial level and what we are finding out to be the structural relationships in nuclear physics. It is just that we did not have the information when yesterday we built so solidly. This eliminates any further requirement of the now utterly obsolete conception of "solid" anything as intervening in the man-tuned sensorial ranges between the macro- and micro-world of ultra- and infrasensorial integrity. We have tensegrity constellations of stars and tensegrity constellations of atoms, and they are just Milky Way-like star patterns of relative spaces and critical proximities.

750.00 Unlimited Frequency of Geodesic Tensegrities

750.10 Progressive Subdividing

750.11 The progressive subdivision of a given metal fiber into a plurality of fibers provides tensile capabilities of the smaller fibers at increased magnitudes up to hundreds and thousandsfold that of the originally considered unit section. This is because of the increased surface-to-mass ratios and because all tensile capability of structure is inherently invested in the external beginnings of structural systems, which are polyhedra, with the strength enclosing the microcosm that the structural system inwardly isolates.

750.12 Geodesic tensegrity spheres are capable of mathematical treatment in such a manner as to multiply the frequency of triangular modular subdivision in an orderly second-power progression. As relative polyhedral size is diminished, the surface decreases at a velocity of the second power of the linear-dimension shrinkage, while the system volume decreases at a velocity of the third power. Weight-per-surface area relates directly to the surface-to-volume rate of linear-size decrease or increase.

750.20 Unlimited Subdivisibility of Tensional Components

750.21 The higher the frequency, the greater the proportion of the structure that is invested in tensional components. Tensional components are unlimited in length in proportion to their cross-section diameter-to-length ratios. As we increase the frequency, each tension member is parted into a plurality of fibers, each of whose strength is multiplied many times per unit of weight and section. If we increase the frequency many times, the relative overall weight of structures rapidly diminishes, as ratioed to any linear increase in overall dimension of structure.

750.22 The only limit to frequency increase is the logistic practicality of more functions to be serviced, but the bigger the structure, the easier the local treatability of high-frequency components.

750.23 In contrast to all previous structural experience, the law of diminishing returns is operative in the direction of decreasing size of geodesic tensegrity structures, and increasing return is realized in the direction of their increasing dimensions.

751.00 Pneumatic Model

751.01 If the frequency is high enough, the size of the interstices of the tensegrity net may become so relatively small as to arrest the passage of any phenomena larger than the holes. If the frequency is high enough, neither water nor air molecules can pass through. The geodesic tensegrity may be designed to keep out the weather complex while admitting radar's microwaves and light from the Sun.

751.02 If we raise the structural-system frequency sufficiently, we will decrease the residual compression islands to the microcosmic magnitude of atoms, which only serves to disclose that the atoms and their nuclei are themselves geodesic tensegrity structures, ergo, compatible with this ultimate frequency limit—a fact that is now, in the 1960s and '70s, swiftly looming into the nuclear physicist's ken.

751.03 We now comprehend that geodesic tensegrity structuring provides the first true and visualizable model of pneumatic structures in which the relative thickness of the enclosing films, in proportion to diameter, rapidly decreases with the increasing size of the balloons or spheric networks.

751.04 In the case of geodesic tensegrity structures, no overcrowding of interior gas molecules, imprisoned within a submolecular mesh net, is necessary to thrust the net's structure outward from its spherical geometric center, because the compressional struts, locally islanded, as outward-thrusting struts at both their ends, push the spherical net outwardly at every vertexial advantage of network convergence. Geodesic tensegrities are the "hollowed-out" balloons discarding their redundantly "solid" air core.

751.05 The geodesic tensegrity is a hollowed-out balloon in which those specific molecules of gas that happen to be impinging from within against the skin at any one moment (thus pushing it outwardly) are replaced by the islanded geodesic struts, and all other redundant molecules are discarded. It is possible to sew pockets on the inside surface of a balloon skin corresponding in pattern to the islanded tensegrity geodesic strut-end positions and to insert into those pockets stiff battens that cause the otherwise limp balloon bag to take spherical shape, as it would if filled with a pressured-in gas.

751.06 Local stiffeners of skin suitable to preferred activities, at any structural focus, can be had by increasing the inward-outward angular strut depths and the local-surface-frequency patternings—thus thickening the truss depth without weight penalties. Here we have nature's own trick of local stiffening as accomplished by the higher-frequency, closest-packing pattern of isotropically moduled local cartilages, and even higher-frequency local bone structuring, as ratioed to the frequency of tissue cells of animal flesh.

751.07 If we employ hydraulic pressure within the local islands of compression for dimensional stability, and if we employ gas molecules between the liquid molecules for local shock-load compressibility (ergo, flexibility), we will find that our geodesic tensegrity structures will in every way have taken advantage of the same structural-strategy principles employed by nature in all her sizes of biological formulations.

751.08 Geodesic tensegrities are true pneumatic structures in purest design frequency principle. They obviate the randomness and redundance characterizing the work of designers dealing only with pneumatics who happen to be successful in blowing air into a bladder while being utterly dependent upon the subvisible behaviors of chemical phenomena. Geodesic tensegrity engineering enables discrete separation of all the structural events into two diametrically opposed magnitude classes: all the outward-bound phenomena which are too large to pass through all the interstices of all the inward-bound events in the too-small class. This is the same kind of redundancy that occurs in reinforced concrete which, if drilled out wherever redundant components exist, would disclose an orderly four-prime-magnitude complex octahedron-tetrahedron truss network, disencumbered of more than 50 percent of its weight.

751.09 Tensegrity geodesic spheroids have none of the portal pressure-lock problems of "solid-oozing" pneumatic balloons. The pressure is discretely localized and locked in place by the tension net, and therefore it cannot escape.

751.10 Tensegrity geodesic spheroids may have several frequencies simultaneously—a low-frequency major web and a high-frequency minor local web. If they are of sufficiently high frequency of secondary or minor webbing to exclude atmospheric molecules, they may be partially vacuumized; that is, they may be made air-floatable.

760.00 Balloons

761.00 Net

761.01 People think spontaneously of a balloon as a continuous skin or solidly impervious unitary and spherically enclosed membrane holding the gas. They say that because the gas cannot get out and because it is under pressure, the pressure makes the balloon spheroidal. This means that the gas is pushing the skin outwardly in all directions. People think of a solid mass of air jammed into a pneumatic bag. But if we look at this skin with a microscope, we find that it is not a continuous film at all; it is full of holes. It is made up of molecules that are fairly remote from one another. It is in reality a great energy aggregate of Milky Way-like atomic constellations cohering only gravitationally to act as the invisible, tensional integrities of the fibers with which the webbing of the pneumatic balloon's net is woven.

761.02 In a gas balloon, we do not have a continuous membrane of film. There is no such thing as a continuous "solid" skin or a "solid" or a "continuous" anything in Universe. What we do have is a network pattern, a network of energy actions interspersed with vast spaces or lack of energy events. The mass-interattracted atomic components not only are not touching each other, but they are as remote from one another as are Sun and its planets

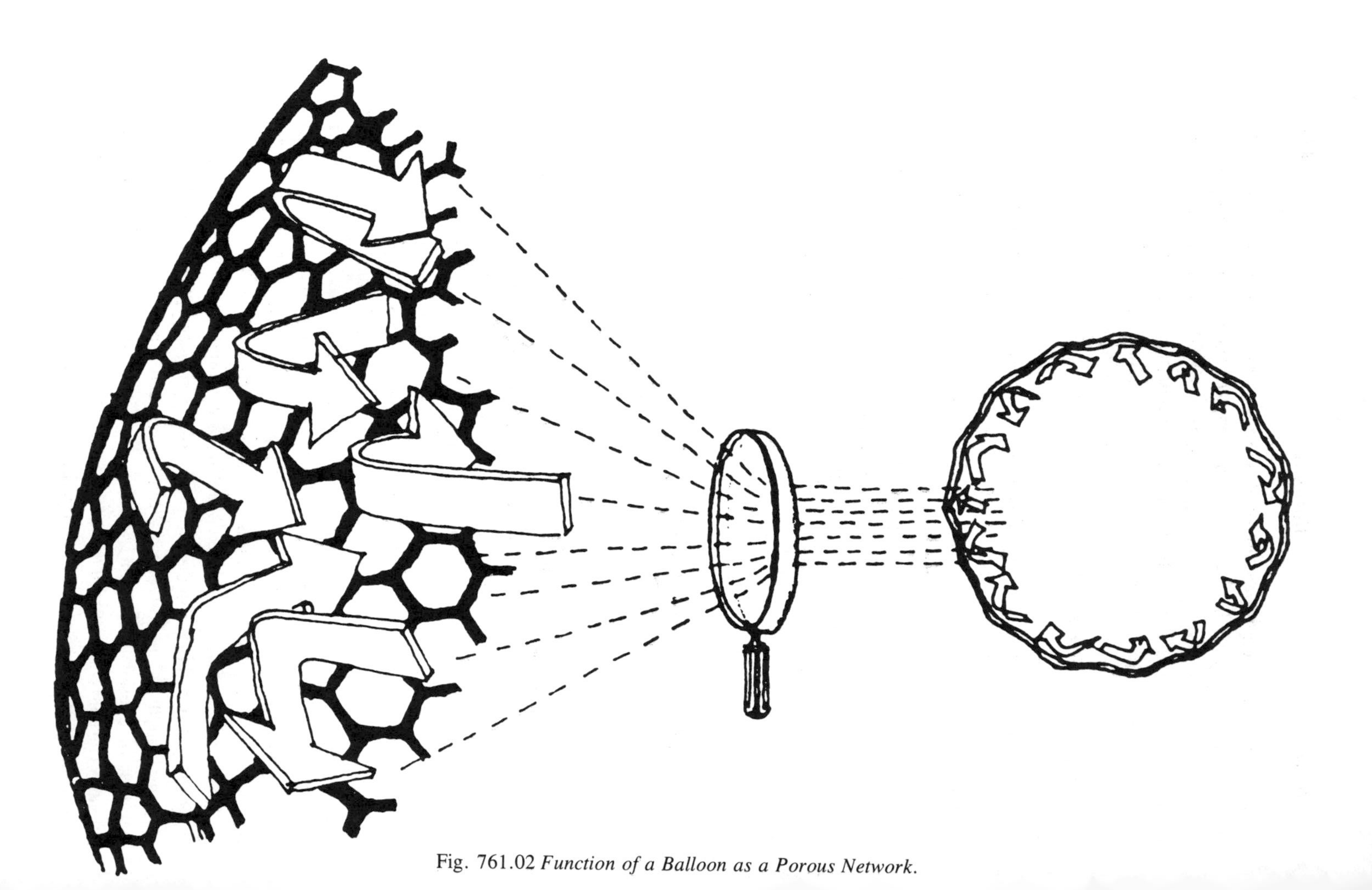

Fig. 761.02 *Function of a Balloon as a Porous Network.*

in the relative terms of respective diameters of each of the phenomena involved.

761.03 The spaces between the energy-action-net components are smaller, however, than are the internally captivated and mutually interrepelled gas molecules; wherefore the gas molecules, which are complex low-frequency energy events, interfere with the higher-frequency, omnienclosing net-webbing energy events. The pattern is similar to that of fish crowded in a spherical net and therefore running tangentially outward into the net in approximately all directions. Fish caught in nets produce an enclosure-frustrated, would-be escape pattern. In the tensegrities, you have gravity or electromagnetism producing the ultimate tension forces, but you don't have any strings or ultimately smallest solid threads. The more we think about it and the more we experiment, the less reliable becomes our concept "solid." The balloon is indeed not only full of holes, but it is in fact utterly discontinuous. It is a net and not a bag. In fact, it is a spherical galaxy of critically neighboring energy events.

761.04 The balloon is a net in which the holes are so small that the molecules are larger than the holes and therefore cannot get out. The molecules are gas, but they have a minimum dimension, and they cannot get out of the holes. The next thing that we discover is the pressure of the gases explained by their kinetics. That is, the molecules are in motion; they are not rigid. There is nothing static at all pushing against the net. They are hitting it like projectiles. All of the molecules of gas are trying to get out of the system: this is what gives it the high pressure. The middle of the chord of an arc is always nearer to the center of the sphere than the ends of the chord. Chord ends are always pushing the net outwardly from the system's spherical center. The molecules are stretching the net outwardly until the skin acts to resist the outward motion and relaxes inwardly. The skin is finite and closes back upon itself in apparently all circumferential directions. The net represents a tensional force with the arrows bound inwardly, balancing all the molecules, hitting them, caroming around, with every molecular action having its chordal reaction. But the molecules do not huddle together at the center and then simultaneously explode outward to hit the balloon skin in one omnidirectionally outbound wave. Not only are there critical proximities that show up physically, but there are critical proximities tensionally and critical proximities compressionally—that is, there are repellings.

761.05 What makes the net take the shape that it does is simply the molecules that happen to hit it. The molecules that are not hitting it have nothing to do with its shape. There is potential that other molecules might hit the network, but that is not what we are talking about. The shape it has is by virtue of the ones that happen to hit it.

761.06 When we crowd the gassy molecules into a container, they manifest action, reaction, and resultant. When one molecule goes out to hit the net, it is also pushing another molecule inwardly or in some other direction. We discover mathematically that it would be impossible to get all of them to go to an absolute common center because that would require a lot more pressure. It would have to be a smaller space so the patterns are not all from the center outwardly against the bag. Each one of the patterns is ricocheting around the bag; some are hitting the net and some are only interfering with and precessing each other and changing angles without hitting the net.

762.00 Paired Swimmers

762.01 The molecules near the surface of the net are coursing in chordally ricocheting great-circle patterns around the net's inner surface. Because every action has its reaction, it would be possible to pair all the molecules so that they would behave as, for instance, two swimmers who dive into a swimming tank from opposite ends, meet in the middle, and then, employing each other's inertia, shove off from each other's feet in opposite directions. We have an acceleration effectiveness equal to what they experience when shoving off from the tank's "solid" wall. When you are swimming, you dive from one end of the tank, which gives you a little acceleration into the water. When you get to the end of the tank, you can put up your feet, double up your body, and shove off from the wall again. Likewise, two swimmers can meet in the middle of the tank, double up their bodies, put the soles of their feet together, and thrust out in opposite directions. The phenomenon is similar to the discontinuous compression and continuous tension of geodesics. The molecules are in motion and have to have some kind of a reaction set; each molecule caroming around, great-circularly hitting glancing blows, then making a chord and then another glancing blow, has to have another molecule to shove off from. They are the ones that are accounting for all the work. Each one would have to be balanced as a balanced pair of forces. We discover that all we are accounting for can be paired. So there is a net of arrows outwardly in the middle of the chord pulling against the net of arrows pointing inwardly.

762.02 The pattern indicates that we could have each and all of the paired molecules bounce off their partners and dart away in opposite directions, with each finally hitting the balloon net and pushing it outwardly as they each angled in glancing blows in new directions, but always toward the net at another point where, in critical repelling proximities, they would all pair off nonsimultaneously but at high frequency of re-repellment shove-offs to ricochet off the net at such a high frequency of events as to keep the net stretched outwardly in all directions. This represents what the molecules of balloon-

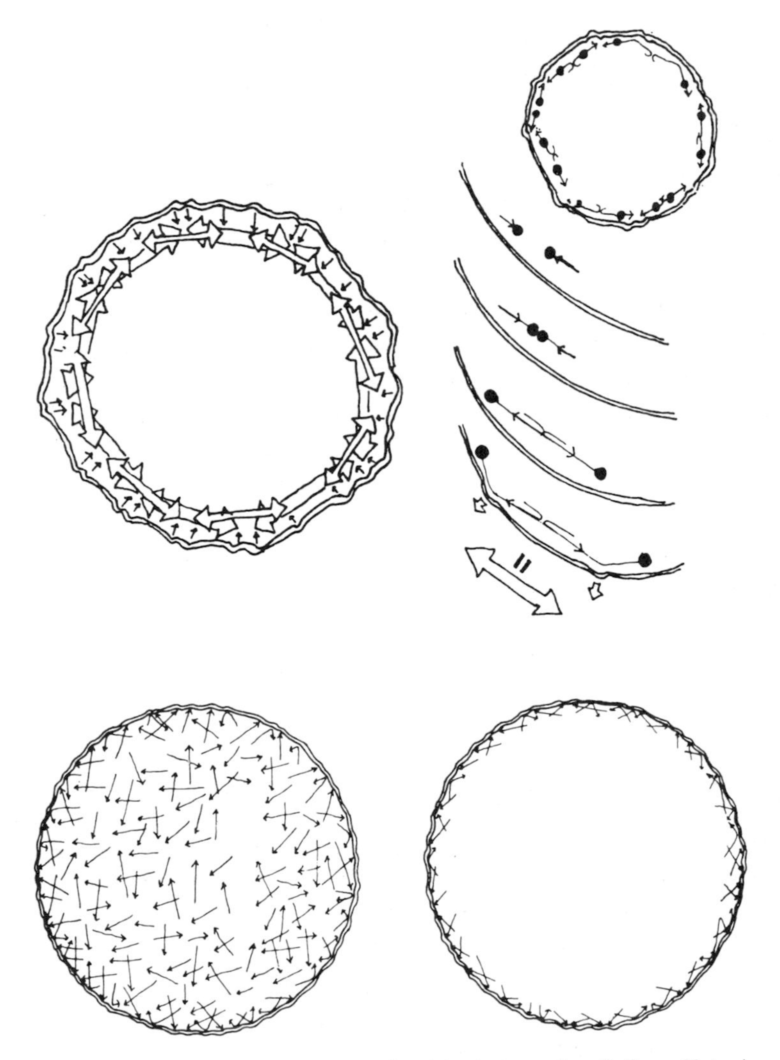

Fig. 762.01 *Chordal Ricochet Pattern in Stretch Action of a Balloon Net:* A gas balloon's exterior tension "net" has the shape that it has because some of the molecules are too large to escape and, crowded by the other molecules, are hitting the balloon. But the molecules do not huddle together at the center and then simultaneously explode outwardly to hit the balloon skin in one omnidirectionally outbound wave. The molecules near the surface are coursing in chordally ricocheting patterns all around the inner net's surface. I therefore saw that—because every action has its reaction—it would be possible to pair all the molecules so that they would behave as can two swimmers who dive into a swimming tank from opposite ends, meet in the middle and then, employing each other's inertia, shove off from each other's feet in opposite directions.

confined gases are doing. With discontinuous compression and continuous tension, we can make geodesic structures function in the same way.

763.00 Speed and Concentration of Airplanes

763.01 As we find out in electromagnetics where there are repellings and domains of actions, the kinetic actions of these gas molecules seem to require certain turning-radius magnitudes. When you pressure too many of these patterns into the same area, there is not enough room to avoid interferences, and they develop a very high speed. Increased speed decreases interference probability caused by increased crowding.

763.02 Airplanes in the sky seem to be great distances apart. But the minute they come in for a landing, they are slowed down and are very much closer to each other. If you have phenomena at very high speeds, their amount of time at any one point is a very short time: the amount of time there would be at a given point for something to hit it would be very much lessened by the speed. The higher the velocity, the lesser the possibility of interference at any one point. So we have the motion patterns of the molecules making themselves more comfortable inside the balloon by increasing their velocity, thereby reducing the interferences that are developing. The velocity then gives us what we call pressure or heat: it can be read either way. If you feel the pneumatic bag, you may find it getting hotter. You can feel an automobile tire getting hotter as it is pumped full.

764.00 Escape from Compression Structuring

764.01 Geodesics introduces tension as the integrity of structure. Geodesics is in fundamental contradistinction to the compressional arches where men made lesser rings of stone and bricks and so forth, like Santa Sophia, fitting them together beautifully and shaping them very mathematically to prevent their slipping or falling inwardly from one another to break the integrity of the compressional rings. In Santa Sophia, they put a chain around the bottom of the dome to take care of the outward thrust of the enormous weights of the aggregate trying to come apart. They could not build an exclusively compressionally composed dome that would not thrust outwardly at the base, so they put the chains around the bases to prevent their collapsing.

764.02 We have seen that in tensional structures there is no limit of length to cross section: you can make as big a pneumatic bag as you want. In the comprehensive, geodesically omnitriangulated, tensegrity structures, we are able to reach unlimited spans because our only limitation is tension, where there is no inherent limit to cross section due to length. We get to where there is no cross section visible at all, as in the pull between the Earth and the Moon. With such structural insights we can comprehend the structure of an apple in terms of noncompressible hydraulic compression and critical proximity cellular wall tensioning. Synergetics identifies tensegrity with high-tensile alloys, pneumatics, hydraulics, and load distribution.

765.00 Snow Mound

765.01 A child playing in sticky snow may make a big mound of snow and hollow it out with his hands or a shovel to make a cave. The snow is fascinating because you can push it together and it will take on shapes. It has coherence. Almost every child with mittens on has built himself a mound and then started chipping away to make a cave. Looking at the hollowed mound from the outside, he may discover that he has made a rough dome. He might then conclude that whatever makes the structure stand up has to do with the circumferential interactions of the snow crystals and their molecules and the latter's atoms. He finds that he can get in it and that the structural integrity has nothing to do with the snow that used to be at the middle. So we may develop a strong intuition about this when we are very young: that it is the circumferential set of molecules that are accounting for the structural integrity of the dome.

765.02 The child may then find by experiment that he might hollow out the pneumatic network and put not only one hole, but many holes, in the snowdome shell, and it continues to stand up. It becomes apparent that it would be possible to take a pneumatic balloon, pair the molecules doing the work, and get rid of all the molecules at the center that were not hitting the balloon—for it is only the molecules that hit the balloon at high frequency of successive bounce-offs that give the balloon its shape.

766.00 Tensegrity Geodesic Three-Way Grid

766.01 What happens in the snow mound is also what happens in the three-way tensegrity geodesic spherical grid. In the balloon, we get paths of these positively and negatively paired kinetic molecules reacting from one another in a random set of directions. If they went into one path only, they would make a single circle, which would push the balloon outwardly only at its equator, making a disc and allowing the poles to collapse. If they made a two-way stack of parallel lesser circles as a cylinder, the cylinder would contract axially into a disc.

766.02 A gas-filled balloon is not stratified. If it were, it would collapse like a Japanese lantern.

766.03 A two-way grid would make only unstable squares and diamonds, which would elongate into a tubular snake.

766.04 Once we have three or more sets of angularly independent, great-circularly continued, push-pull paths, they must inherently triangulate by

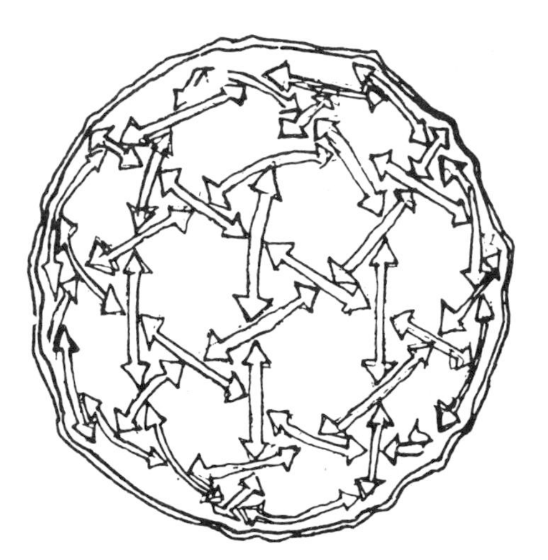

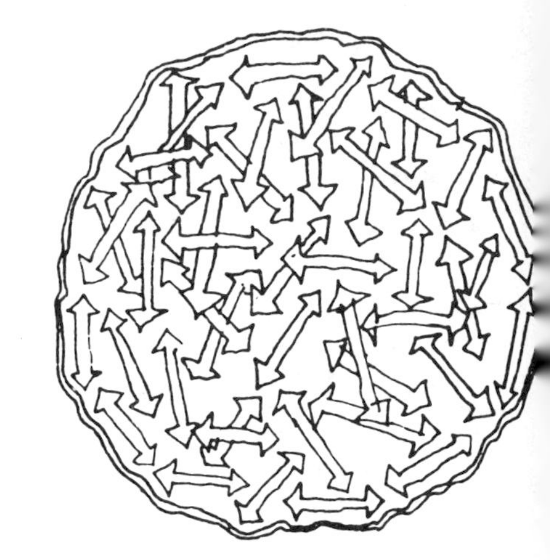

Fig. 765.02 *Stabilization of Three-Way-Grid Tensegrity Sphere:* What happens with the snow mound is also exactly what happens in a three-way-grid tensegrity-geodesic spherical grid. In the balloon we get paths of these positively and negatively paired, kinetic molecules reacting from one another in a random set of directions. If they went into one path only, they would make a single circle which would push the balloon outwardly only at its equator making a disc and allowing the poles to collapse. If they made a two-way stack of parallel lesser circles as a cylinder, the cylinder would contract axially into a disc. A two-way grid would make only unstable squares and diamonds, which would elongate into a tubular snake. But once we have three or more sets of angularly independent circularly continued push-pull paths, they must inherently triangulate by push-pull stabilization of opposite angles. Triangulation means self-stabilizing, which creates omnidirectional symmetry, which makes an inherent three-way spherical symmetry grid, which is the geodesic structure.

push-pull into stabilization of opposite angles. Triangulation means self-stabilizing; which creates omnidirectional symmetry; which makes an inherent three-way spherical symmetry grid; which is the geodesic structure.

770.00 System Turbining in Tensegrity Structures

770.10 Comprehensive System Turbining

770.11 The whole system turbines positively, or the whole system turbines negatively. There are no polar or opposite hemisphere differences of these systems. There are no ''rights'' or ''lefts'' in Universe.

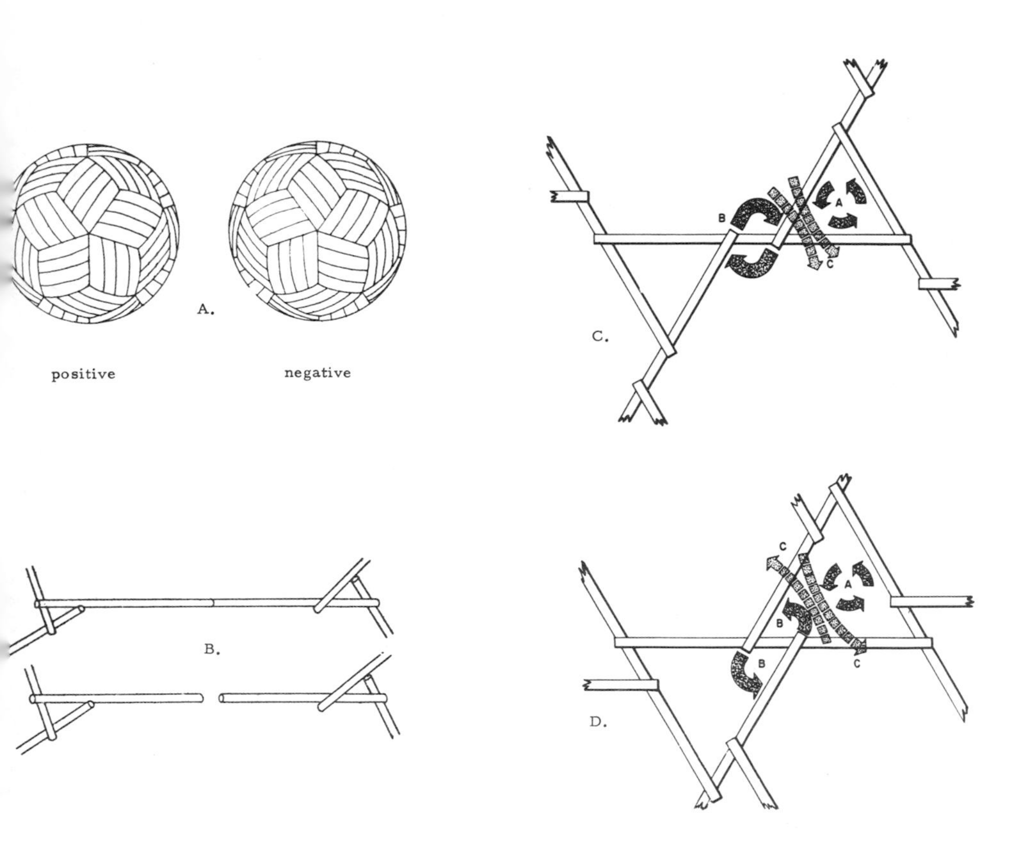

Fig. 770.11 *System Turbining in Tensegrity Structures:*

A. The two above both have six axes of symmetry. It is the patterning of the thirty diamond domains of the icosahedron's thirty edges, the rhombic triacontahedron.
B. The linear-congruence juncture of two positive or two negative turbining-surface three-strut tensegrity units.
C. Single-bonded tensegrity: turbining tendencies of thrusts of *C* about *A* and *B* are additive.
D. Double-bonded tensegrity: turbining tendencies of thrusts of *C* about *A* are opposed to those about *B*.

770.12 Three-strut tensegrity units exhibit either two positive or two negative turbining surfaces at their linear congruence junctures.

770.13 In single-bonded tensegrity structures, turbining tendencies are additive. In double-bonded tensegrity structures, turbining tendencies are opposed.

770.20 Central-Angle and Surface-Angle Turbining

770.21 Turbining in tensegrity systems may derive from either central angles or surface angles. There is inherent comprehensive positive or negative turbining of finite systems in both central and surface angles. Central-angle turbining effects surface-angle turbining.

780.00 Allspace Filling

780.10 **Conceptual Definition of Allspace Filling:** The multiply furnished but thought-integrated complex called space by humans occurs only as a consequence of the imaginatively recallable consideration (see Sec. 509) of an insideness-and-outsideness-defining array of contiguously occurring and consciously experienced time-energy events.

780.11 Unitary conceptuality requires spontaneous aggregating of relevant magnitudes and frequencies of experience recalls.

780.12 Conceptualization is inherently local in time as are the separate frames of scenario Universe's conceptualities nonconceptually identical. Conceptuality is always momentary and local.

780.13 When we speak of allspace filling, we refer only to a conceptual set of in-time local relationships. This is what we mean by tunability.

780.14 The limits of an allspace-filling array are nondefinable. Nondefinable is not the same as infinite.

780.20 **Galactic Orientation:** Apparently simultaneous static-system conceptualization is ''relatively'' misinterpretable as an environmentally experienceable condition of the individual which he reflexively identifies as ''instantaneous''—a word as yet frequently used with omnipopularly misassumed fidelity to reality.

780.21 This instantaneously infinite static Universe misconception is vastly fortified as the living observers go outside the house on a clear, ''still'' night and stand fixedly *under* the stars, gazing fixedly at the fixed stars, and, as we say poetically, ''turn this instant into eternity,'' within which cosmically arrested moment, subconsciously stimulated by the latest newspaper item regarding a way-up-there quasar or other astro discovery, we say in spontaneously expressed curiosity: ''I wonder what's outside the outside of all these omnidirectionally positioned stars?''

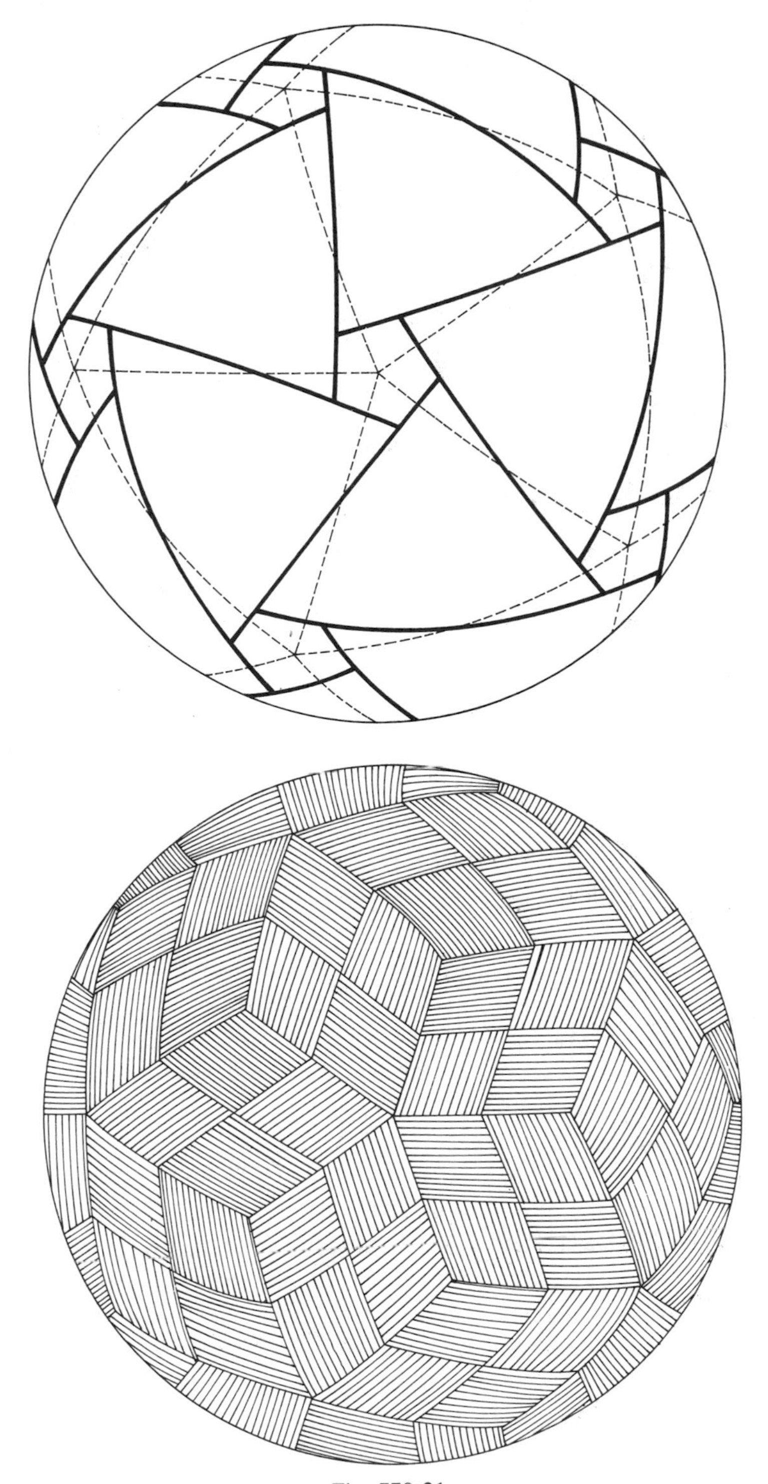

Fig. 770.21.

780.22 This brain-fixing fixity's conceptual interpretation of experience is permitted only by the infinitesimally short life span of humans in the thus-far-discovered historical magnitudes of universal history's events. As we stand "fixedly" in "space" at the terrestrial latitude most occupied by Earthian humans, we are revolving around the Earth's axis at 600 m.p.h. Together with our Moon, we orbit around Sun at 60,000 m.p.h. (which orbital speed is three times as fast as the Earth-Moon-ferrying, Apollo-rocketed space vehicle), and all the while our solar system, situated about three-quarters of the way outward from the center of our celestial galaxy—together with all that inwardly active galaxy's billions of stars—are cosmically merry-go-rounding at approximately a million miles per hour, and all the while we participate in all these motions, our Milky Way merry-go-round galaxy itself may be, and is scientifically thought to be, involved in comprehensive motion at an even higher velocity. Due to the omnieverywhere-expanding Universe (interpretation of observed data), our galaxy and all the other of the billions of galaxies of Universe are alike in traveling outwardly from one another isotropically at additional millions of miles per hour.

780.23 This expanding-Universe concept is easy to phrase in words as reported, but lucid comprehension of its import involves experientially "impossible," three-dimensional, space-motion conceptionalizing, for in order to travel away cosmically from each of all the spherically surrounding galaxies of our Milky Way, any one of the billions of galaxies seemingly would have to go outwardly in all directions in order to go away from each of them simultaneously. Obviously, however, this could not be accomplished by any one of them moving in only one direction—which is humanity's way of thinking of motion—unless there were a center of galaxy of Universe outwardly from which all others move *exactly* and only radially, *or* unless all of Universe and all of the galaxies and each and all phenomena within them, including the smallest nuclear particle, are either *expanding* systematically and simultaneously or are *shrinking* systematically and simultaneously, all changing in size at a rate that is just a bit faster than the speed of light, with either the universal contraction or universal expansion of all points in Universe producing the same effect of uniform withdrawal from one another.

780.24 This may be the universal effect of the *speed* of gravity, whose force (possibly in order to eternally cohere Universe) is, as is often found experimentally, always just a fraction greater than the cosmic speed of inherently disintegrative radiation. (See Sec. 231.) This conceptioning becomes lucid if one is familiar with the vector equilibria and their identity with isotropism, which spontaneously accommodates coexpansion or contraction independent of any Universe center, every nuclear point within the system being a Universe center, with all its 12 most immediate neighbors always being equidistant and bearing at the same total of central-angle magnitudes from one another,* with the circumferentially closed, embracing vector forces always more effective than their equal and opposite radial vectors' non-cooperative, open-ended, disintegrative forces.

780.25 Humans standing on Earth gazing outwardly from Earth at the stars cannot see the stars in the celestial sphere in the direction of their feet. Earth is in the way. Earth is so much in the way that humans at sea on a calm, clear night can see only about half the celestial sphere at any one time. An astronaut out "space-walking" can see approximately all of the stars of the celestial array omnisurrounding him at vast varieties of distances from him, though they all seem to be superficially on the same concave surface of the same black sphere at whose center the astro-observer seems to be. Remembering the difference between the Earth-standing observer's totality of sky and the astronaut's also optically illusioned but far more comprehensively stimulated conjuring of the concept "totality," we can understand why the Earth-standing observer on a completely overcast day cannot see the cloud cover as a dimensionally definable phenomenon, whereas the astronaut seeing the Earth at a distance wrapped in its cloud cover can see Earth and its biosphere as a dimensionally defined entity.

780.26 When we speak of the cosmic limits of seemingly allspace filling, we refer to the totally surrounding, indefinable, extensive allspace-filling effect of fog upon an observer in that dense fog. It seemingly has no shape. Nor has that fog a "shape" even when it lifts into the sky above the observer and fills the whole overhead spherical domain. Observed from outer space at the same moment, however, mantling Earth may seem to have momentarily stable descriptability akin to that of a frozen icefield. Then the same fog or cloud blanket may be viewed at the same time by a third human from a mountaintop just protruding through the cloud. The third observer sees that the clouds are intertransforming in complex, high-speed turbulence, vanishing here in rain and being newly formed elsewhere by Sun-drawn evaporation. Every atom involved in Earth's ocean-atmosphere-intertransforming H_2O cloud-cover phenomena, visible or invisible, has its integrity, and the allspace-filling events become other than visible transformation events, yet may indeed be kept account of by you and me and Universe, with its mathematical integrities of complexedly interaccommodative principles of intertransformative events always occurring interconsiderately.

780.27 Seen from Moon, the total local dimensional involvement of such Earthian atmospheric-oceanic intertransforming events is well within the field of a telephoto-lensed, video-recording camera as well as of a battery of

* I.e., 60 degrees. The nucleus of a square would have a completely different distance to its corners than the corners would have to each other.

frequency sensors "seeing" the humanly invisible events transpire. The intertransformings are finitely packagable and analyzable in conformity to allspace-filling laws. That these same events seem boundless to the Earthian observer uninformed by the celestial-scanning intruments need not obscure our realization that what we mean by allspace-filling regularities are omni-intertransformable—ergo, are scenarios of an aggregate of nonsimultaneously overlapping, energy-transforming events in which one or a few isolated frames of special-case considerations fail to disclose the meaning accruing only to large-continuity consideration of the whole story.

780.28 As a cosmic, generalized, intertransformability system *field,* our allspace-filling synergetics matrix accommodates and equates these behaviors. Allspace filling is a scenario: the eternally self-regenerative scenario of cosmic integrity.

780.30 **Eternality:** "Eternal" identifies only the metaphysical, weightless, abstract principles, which, to hold true in all special-case experiences, are inherently eternal.

780.31 Angles are eternally transcendental to time-size limits. The angle is a subdivision of one cycle quite independent of the length size (time) of the angle-defining radii edges of the angle. One-sixth of unity: the circle is one-sixth independent of time and size.

780.32 Regularity is eternal. But the regularities are eternally omni-interaccommodative, permitting approximately limitless freedoms of selectable alternative developments involving a vast plurality of time-dimensioned frequency involvements.

780.40 **Unitary Conceptuality of Allspace Filling:** Allspace filling means all unitarily conceptual space filling, because Universe, though finite, is an aggregate of nonsimultaneous and only partially overlapping event transformations which, being nonsimultaneous and differentially rate-frequenced, are never momentarily subject to total unitarily synchronized—ergo, simultaneous, apparently static system—conceptualization.

781.00 Accommodation of Aberration

781.01 We can take hold of any two parts of a tensegrity sphere and treat it as an omnidirectional, expansion-contraction accordion. In the same way, we can take hold of any two parts of a rubber-vectored, isotropic-vector matrix and, so long as the contiguous faces of the octahedron-tetrahedron field remain congruent, the matrix can be distorted by angular variation, spin, orbit, inside-outing, expansion, knotting, or torque without losing any of all the fundamental regularities of the omniconsidered, allspace-filling set.

781.02 Activated by tension and compression, *two* remote-from-one-another external triangles of an elongated isotropic rubber-vectored matrix structural system may be congruently associated to close the system's "insideness" back upon itself to form a large, flexible structural-system ring with a circularly closed insideness, like a serpent biting off its own tail and swallowing the "open" end.

781.03 In order to be a *system* definitively—ergo, topologically accommodated throughout all transformative transactions of dividing the insideness from the outsideness—and to be *structural,* the system-dividing medium must be omnitriangulated—ergo, having only triangular openings.

781.04 When the structural system's remote structural parts are joined back on themselves to continue the insideness-integrity's division from the outsideness, the only "holes" in the system (which may be coupled to join the insideness back on itself) are triangular wholes, with their respective three corners identifiable as *A, B, C,* and *A', B', C',* respectively. They could be nontwistingly joined *A* to *A'*, *B* to *B'*, *C* to *C'*, or by twisting the elongated rubber-vectored system's ends 120 degrees, they could be joined as *A* to *B'*, *B* to *C'*, *C* to *A'*, or twisted more to *A* to *C'*, *B* to *A'*, *C* to *B'*—*or* they could be twisted 360 degrees and fastened *A* to *A'*, *B* to *B'*, and *C* to *C'*—or several such always 120-degree-incremented twists and multiples thereof could occur.

782.00 Distortion of Vector-Equilibrium Frame

782.10 **Accommodation of Aberration: Corollary:** An allspace-filling isotropic complex consisting entirely of triple-bonded tetrahedra and octahedra can become nonisotropically distorted yet remain allspace filling, i.e., all six or several edges of the tetrahedron and the correspondingly bonded edges of the octahedra can become coordinatedly dissimilar and yet be allspace filling.

782.11 Throughout the distortions and aberrations of the octahedron-tetrahedron field, the ratio of the octahedral volume as fourfold the tetrahedral volume remains constant.

782.12 The whole synergetic hierarchy of rationally related *A* and *B* Quanta Modules and topological values remains constant.

782.20 **Regularities:** Such potential distortion of the vector-equilibrium frame of reference introduces an almost infinite variety of nuclear sphere's connect-and-disconnect conditions without in any way altering any of the other topological regularities discussed throughout synergetics.

782.21 Infinite variety of local, individual initiations and terminations within eternal cosmic integrity of total order is implicit.

782.22 Regularity is total.
Variability is local.

782.23 Finite—and the concept *finite*'s only-speculatively-inferred impossible condition of conditionless "infinite"—identify only special-case physical experience, ergo, are experientially always finitely terminal. Frequency is of time and is finitely terminated.

782.30 **Variability of Spherical Magnitudes:** All or partial differentialing of the six always-congruent tetrahedron and octahedron edges of allspace filling also introduces variation in the size of the spheres that could surround any one vertex of the system. Whereas the original isotropic vector matrix, with all its vector "lines" the same, provided the set of vertexes that were the centers of unit-radius, omni-closest-packed spheres, we now witness experimentally with a stretchable, rubber-vectored, allspace-filling, originally unit-edged, triple-bonded complex of tetrahedra and octahedra, that the whole system may be stretched, torqued, and angularly wrench-distorted. Ergo, we witness the ways in which the vector equilibrium, or most intereconomical vectorial relationship of 12-around-one sphere centers in closest packing, may be omnidirectionally distorted to accommodate a plurality of spherical magnitudes in an as-yet closest possible interrelated neighborhood array of the respective centers of disparate-size spheres, with some spheres tangent to their neighbors and others disconnected.

782.40 **Isotropic Modular Grid:** In the same experimental model exploring manner, we discover that whereas locally verifiable parallel lines running off to the horizon appear to converge, it becomes a local observational experience reality that what is constructed as a many-miles-wide, -high, and -long isotropic vector matrix of 10-foot modules, with its vertexes occupied by omniuniform radius spheres of 10-foot diameters each, in omni-closest packing, may be photographed or drawn as seen in perspective from one locus outside the system. The sizes of the individual spheres and of the edge lengths and triangular "areas" are experientially witnessable as progressively diminishing in size as they extend, respectively, remote from the observer. The size variations may be measured accurately on a viewer's modular-gridded, hairwire-in-glass screen, mounted vertically, immediately in front of the viewer. It is also experientially observable and documentable that despite these observed progressively diminishing alterations of relative intersystem size, the topological-inventory characteristics of relative abundance of vertexes, faces, edges, *A* and *B* Modules, and the sum total of angles around the vertexes—all remain unaltered.

782.50 **Time as Relative Size Experience:** Local variability within total order synergetically explains and defines the experience "time," which is relative size experience. The magnitude of the event characteristics is always accounted in respect to other time cycles of experiences. The cosmically permitted and experientially accommodated actuality of the individual's unique variety of sensorially differentiated local in time-space experiences also accommodates the experienceability in pure principle of individually unique physical life in concert with the only metaphysically operative, cosmically liaisoned, weightless, abstractly conceptual mind, by means of all of which physically and metaphysically coordinate experienceable principles it is experimentally discoverable how genetic programming accomplishes the "instinctive" conditioning of subconscious, brain-monitored, relative pulsation aberration and transformation controls, which are all reliably referenced entirely subconsciously to the eternally undisturbed, cosmic-coordination regularities unbeknownst to the individual biological organism "experience."

783.00 *Moebius Strip and Klein Bottle*

783.01 Moebius's arbitrarily shadow-edged strip and Klein's rimless bottle are only self-deceptively conceivable as absolute solids or as absolute continuities with inherently absolute edges and lines. The Moebius strip does not have an edge: it is a tube. Lack of any experimental evidence of any such phenomena as absolute solids or absolute continuities with inherently absolute edges and lines induced physicists to abandon the concepts of solids and absolute continuities.

783.02 In their bottle and strip propositions, both Klein and Moebius employ the working assumption of absolute solids and surface continuums. The humanly experienceable surprise qualities of their findings are the same surprise experiences of audiences of expert magicians who seem to produce results by means other than those which they actually employ. The implied significance of the bottle and strip findings vanishes in the presence of the synergetic surprise of the topological constants of the vector equilibrium's hierarchical regularities independent of size, inside-outing, turbining, and so forth. Unlike Moebius's and Klein's experimentally undemonstrable constructed substances, the information input of synergetics and tensegrity are wedded experientially only with the full gamut of the thus far published experimental findings of astrophysics, chemistry, and microbiology.

784.00 Allspace-Filling Tensegrity Arrays

784.10 **Basic Allspace Fillers:** The tensegrity tetrahedron and the tensegrity octahedron are volumetrically complementary, and together they may fill allspace. The tensegrity icosahedron refuses to complement either itself or the tetrahedron or octahedron in filling allspace, but isolates itself in space, or goes on to make up triple-bondedly into larger octahedra, which may then complement tetrahedra to fill allspace.

784.11 Tensegrity icosahedra provide by far the most volume with the least structural effort of the three basic structural systems. The tetrahedron has the least volume with the most surface; the octahedron is in the middle; and the icosahedron gives the most with the least.

784.12 In the icosahedron, five quanta give twenty units of enclosed volume, which means four units of volume for each energy quantum invested in the enclosing structure. Whereas in the tetrahedron one quantum will enclose only one unit of volume. The octahedron gives two units of volume for each quantum. Therefore, the icosahedron gives the most for the least effort.

784.13 The three-islanded tensegrity octahedron, in its positive and negative phases, is fundamental to all tensegrity structures. (See illustration 724.10.)

784.20 **Eight-Icosahedra Tensegrity Array:** The three sets of parallel pairs of struts which form the tensegrity icosahedron may be considered as parallel to the three axes of the *XYZ* coordinate system. The same three sets of parallel pairs of the tensegrity icosahedron may be considered also as two omni-axial sets of tensegrity octahedra. This octahedron-icosahedron parallelism relationship explains why it is possible to collect tensegrity icosahedra in approximately unlimited periodic arrays. A set of eight icosahedra is shown in the illustration.

784.21 Note that the rows of parallel struts can be repeated to infinity and the length of each strut can be infinitely long. The tension net that forms the icosahedron edges stabilizes the array of struts.

784.30 **Tensegrity Icosahedra Surrounding a Nuclear Icosahedron:** Six icosahedra may be arrayed around a nuclear icosahedron in a true *XYZ*-coordinate model.

784.40 **Limitless Array of Tensegrity Icosahedra:** In addition to single-strut tensegrity icosahedral systems, it is possible to organize an only-time-limited, omnidirectionally extensible, uniformly periodic array of ten-

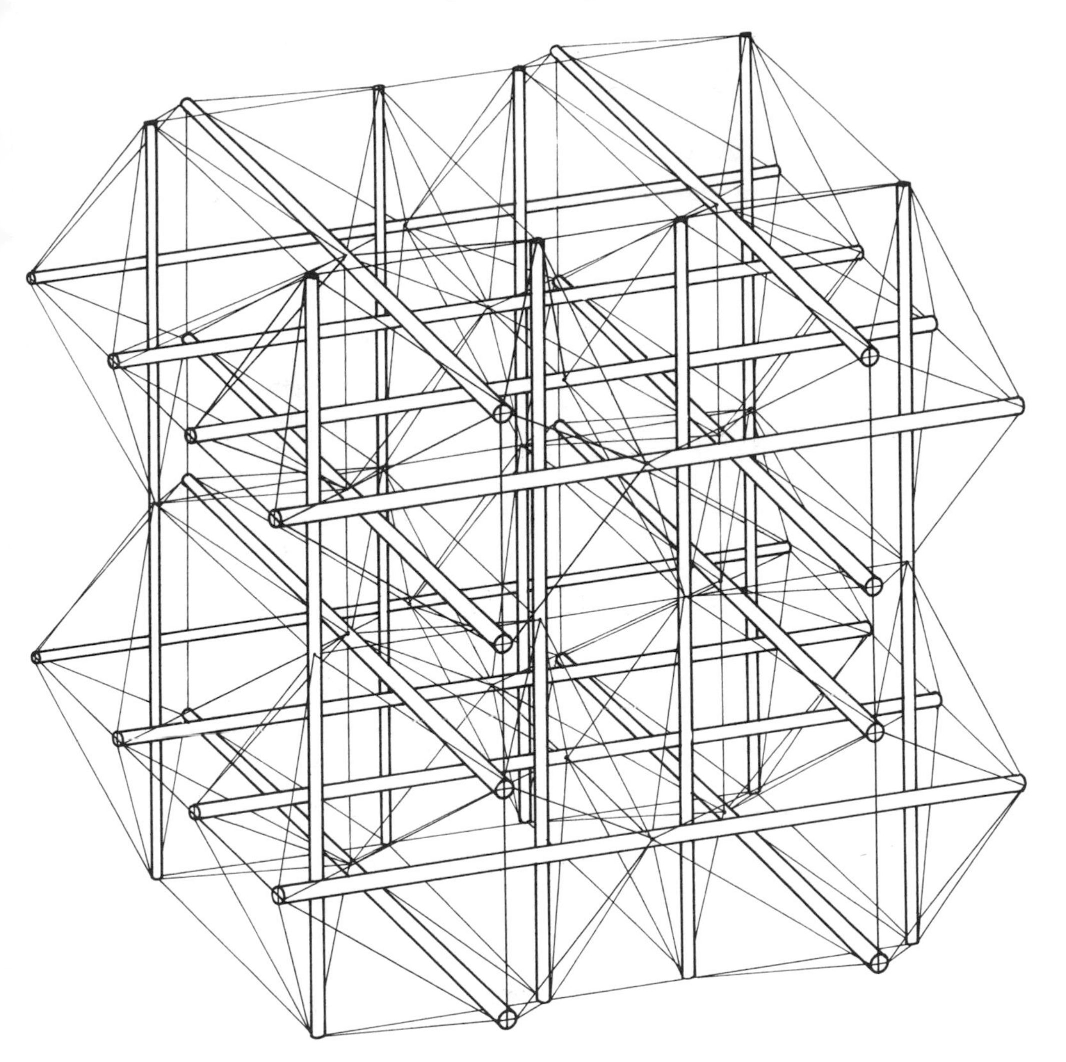

Fig. 784.20.

segrity icosahedra in which each compression member of finite length is common to only two icosahedra.

784.41 This system consists of a series of omnidirectionally staggered layers of icosahedra. A spatial array of six icosahedra is shown both as a tensegrity system and as a collection of "transparent" icosahedra. The lower diagram indicates the method of staggering which results in each compression strut being shared by only two adjacent icosahedra.

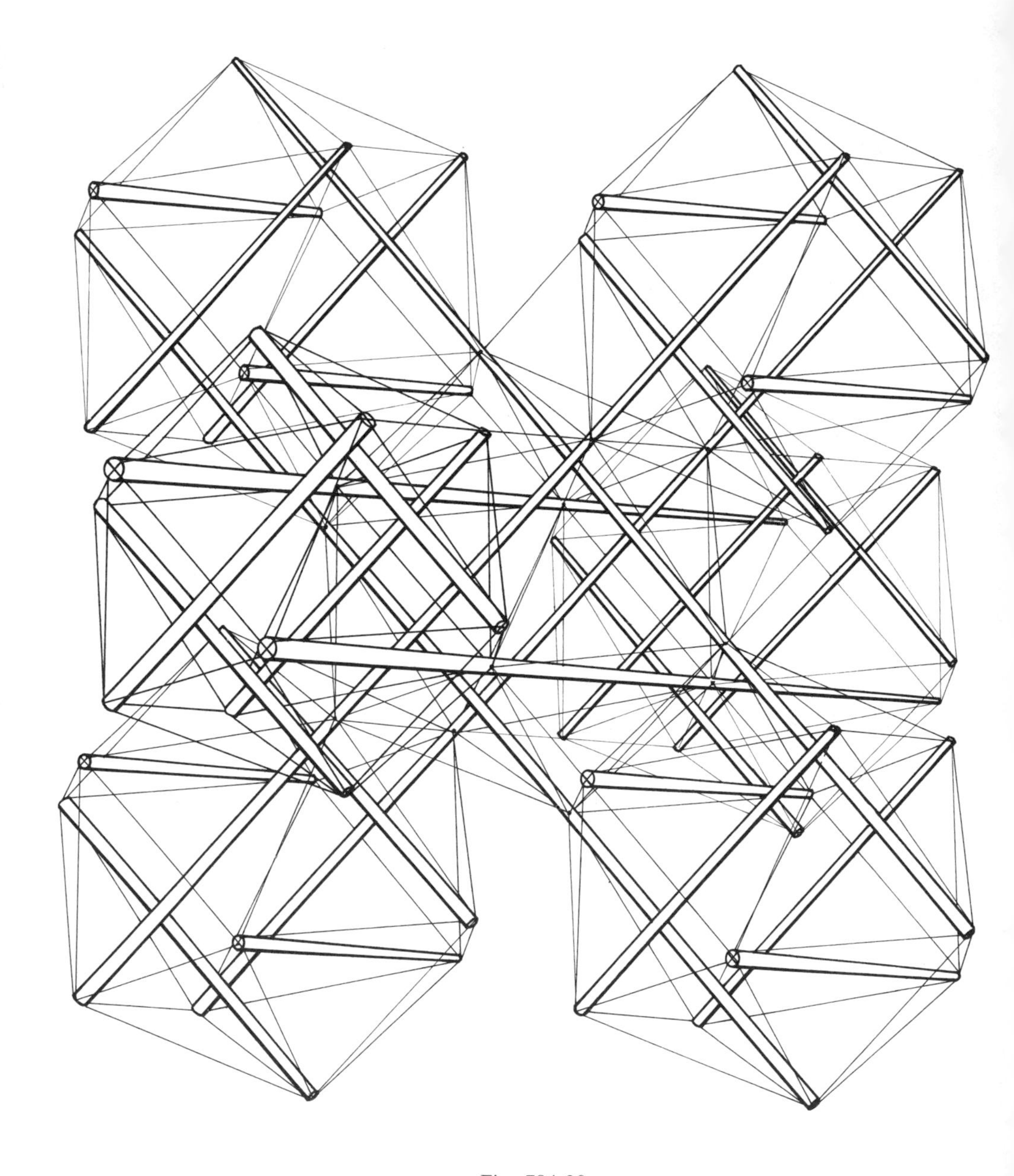

Fig. 784.30.

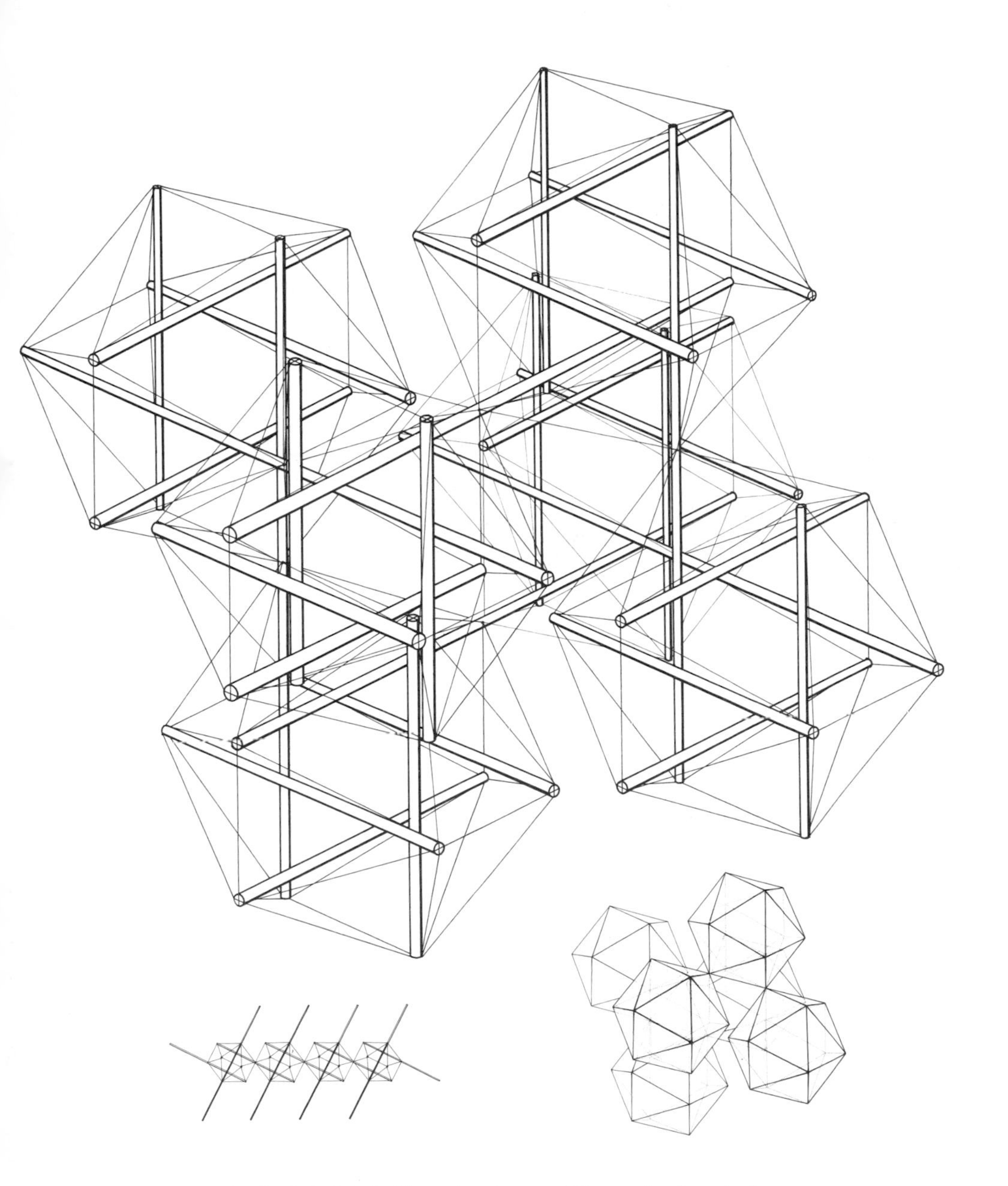

Fig. 784.41 *Indefinitely Extensive Array of Tensegrity Icosahedra.*

800.00 Operational Mathematics

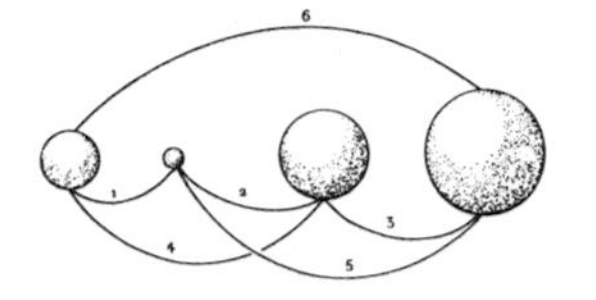

800.00 OPERATIONAL MATHEMATICS

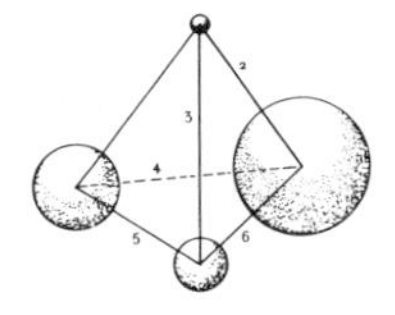

801.00 *Sensoriality: Sweepout*

801.01 *Alternate Faculties of Sensation*

801.02 Information is experience. Experience is information. We have all experienced the information given to us directly through our own sensing faculties or relayed to us by others through our sensing faculties, but as originally sensed directly by others and not by ourselves. The only way that we know that we "are," that we are alive in Universe, is through information apprehended by our own sensorial faculties. We can hear, see, taste, smell, and touch-feel. We have all experienced the information-relaying relationships between the old life and the new life. The old life is excited to see how early the new life develops, coordinates, and responds both consciously to external information and subconsciously to internally programmed instructions of the brain or of the genes. The old life tries to speed the development of the new life's communicated comprehending by pointing to first the child's and then the "old life" speaker's eye and saying, "eye, eye, eye," "mouth, mouth, mouth," and "ear, ear, ear," while pointing to those instruments until the child responds by making a similar sound. However, it is seldom that we observe parents thus engaged with their children refer to their internal organs, such as the endocrine glands. In fact, parents may not even know of these glands, let alone where they are situated. Such word-coaching by oldsters of youngsters relies almost exclusively upon identification of superficial characteristics and comprehends only in superficial degree those organs to which they refer.

801.03 Let us imagine a scientifically conducted experiment designed to disclose the unique behavioral characteristics of each of those four prime sensing faculties without which we could not apprehend Universe and could not have sense of being.

801.04 Let us suppose that you are blindfolded and that your mouth, nostrils, and ears are also simultaneously bound closed. Only your tactile sensing is operative. To find out about yourself and local Universe, you would begin by reaching out around you with your arms-extended hands. You could learn environmental conditions through your hands. You could lean forward, and the sense of balance would tell you how far you can reach without shifting your base position. You discover that you are prospecting with your sensitive skin terminals, as does an insect with all its radially and circumferentially orientable feelers. Your most extreme and mobile skin feelers are your toes and your fingers. You are trying to get terminal reach information before you move on from your safe base. You will not risk shifting your weight until you are certain that you will be supported. You will not move into a place so small you cannot turn around and escape. Without changing your base, and standing with all your weight on your left foot, you learn that you can stretch out and sweep out with your arms while at the same time sweeping space and testing the ground's firmness with your right leg. Thus you learn that there is a maximum range of information gathering, which is the distance between the right foot's big toe and your left hand's middle fingertip. Most of us have a toe-to-fingertip reach of about six or eight feet. In these sense-limited conditions, our only way of finding out about Universe is tactile, through touch alone. Very quickly, we become supersensitive with our feet and hands, particularly with our feet and legs in gravitational balancing. Every child learns this in summer while at camp. At home, his parents won't let him stay up after dark, but maybe at night he is very fond of a path that goes down to the water. He starts going on that path and finds himself running along in the dark. Even though you can't see, you remember well the pattern of turns, depressions, hills, and dales. Your feet feel familiar with the path; the rhythm of steps and heartbeats subconsciously monitors your memory-bank control of your running along that familiar path barefootedly in the dark. We find experimentally that we can remember patterns tactilely and feel very safe following them. We are even able to run back and forth over a local complex of familiar ground and we can run at about 10 miles per hour. Wherefore our static tactile information-gathering, which commands a maximum spherical range of 10 feet in diameter, is augmented by the 10-miles-per-hour dynamic range-minding capability.

801.05 Ecology is the science of cataloguing, ordering, and inspecting patterns of life. Different kinds of life demonstrate different patterns. There is a difference of radius of sweepout of wolves, seagulls, and man. If we humans had only the tactile sense to go by in our ecological patterning, we could only sweep out a fairly small territory, but we could get so used to it that we would probably run around in the known territory. (See Sec. 1005.20.)

801.06 But now suppose that you cover up all your skin and uncover your nostrils and your mouth. Your eyes and ears are still covered and your feet and hands are now tied down so you cannot move. You have only olfactory information. Under these conditions, men's measurements are governed by three factors: (1) the radius of the permeation of gases within gases; (2) the concentration and viscosity of such gases, such as orange groves, pine woods, and so forth; and (3) the wind. Men coming in from months at sea have smelled orange groves and pine trees at somewhere around a mile offshore in still air. Such gases remain sufficiently concentrated to be detectable at a mile. (Of course, dogs can smell at greater distances than our human standing-still olfactory range of about a mile.) If the wind is blowing, the velocity is enhanced so we get smoke from forest fires at great distances. In great, 400-miles-per-hour, high altitude, jet stream winds, the smellable concentration can persist to a range of even 100 miles. Whereas our tactile sense's static range is 10 feet, which equals about 1/500th of a mile; and its dynamic velocity range augmentation is 10 miles per hour, we find our olfactoral static range of information-gathering is 100 miles and its dynamic range is 400 miles per hour.

801.07 If we now shut off the mouth and nostrils—with eyes and skin also blanked out—and we then open up only the ears, we cannot see, smell, or feel; we have only sounds to reckon by. Men have heard sounds at very great distances. Sounds will bounce on the water, into the atmosphere, and back on the water again. Sound is a wave phenomenon that men have heard at ranges up to 100 miles, as in the case of the atomic bomb. The speed of sound in the air is about 700 miles per hour; the static hearing range is about 100 miles, while the dynamic hearing range is 1,100 miles per hour (700 m.p.h. + 400 m.p.h. jet-stream wind = 1,100 m.p.h.).

801.08 We next shut off the tactile, olfactory, and oral sensing, then uncover and open our eyes. Men see stars that are billions of miles away. We know the velocity of light is 186,000 miles per second, or about 700 million miles per hour. We find that the visual sensing is in an entirely different order of magnitude. The tactile, olfactory, and oral faculties as a group are so minuscule as compared to the range of the visual that they cannot even be considered together.

801.09 Human Sense Ranging and Information Gathering

	Radius of Static Ranging:	Dynamic Velocity:
Tactile	1/1,000th of a mile	10 miles per hour
Olfactory	1 mile	400 miles per hour
Aural	100 miles	1,100 miles per hour
Visual	6,000,000,000,000,000,000 miles *	700,000,000 miles per hour

* One light year is six trillion miles, and humans see Andromeda with naked eye one million light years away, which means six quintillion miles.

If we try to plot two curves of these static and dynamic human sensing capabilities on a chart, we will have no trouble in positioning the first three senses; but to reach the point on the chart at which the sight capabilities occur, we will have to take an airplane and fly for many days to reach those positions. It is clear that as we recede from the first three sets of points, they will tend gradually to appear as one. This disparity has not been taught to us. We were told that our senses were approximately equal and alternate capabilities. Court imposed "damage costs" for their respective losses are approximately equal. We found out the disparity ourselves by examining the limit-case conditions, which can only be discovered by physical experience. This method of discovery is called "operational procedure."

801.10 Sense Coordination of the Infant

801.11 One of the most surprising things about a newborn child is that it is already tactilely coordinated. Even in the first day, the baby is so well coordinated tactilely that if you put your finger against its palm, the baby will close its hand firmly and deftly around your finger, although it is not using its ears or eyes at all. If you will now exert a tiny bit of tension effort to remove your finger, the child will respond at once by opening its hand. The infant will repeat the closing and opening response to your initiatives as many times as you may wish to initiate. This should not surprise us if we realize that the baby has been in *tactile* communication with its mother for months before evacuating her womb, within which, however, its visual, olfactory, and aural faculties were muted and inoperative. Not much time after birth the child employs for the first time its olfactory glands and starts searching the mother's breast and the source of milk. Quite a few days later it begins to hear; and very much later, it sees. The sequence in which the child's faculties become employed corresponds to the order of increased range of its respective faculties of information apprehending.

801.12 Thus we find the child successively coordinating the first three faculties: the tactile, the olfactory, and the aural. He begins to learn how they work together and quite rapidly gets to be very skillful in coordinating and handling the information coming to him through these senses. It is only days later that he begins to use his sight. He tries tactilely, olfactorily, and aurally to confirm what he sees to be reality. He cannot do so over any great distance because neither his arms and hands nor his tasting mouth will reach very far. Months later, the child crawls to check tactilely, olfactorily, and aurally on phenomena still further away; and thereby to coordinatingly sort out his information inputs; and thereby develop a scheme of—and a total sense of—reality and repetitive event expectancy. He crawls over to the chair to find that his eyes have reported to him correctly that the chair is indeed there. He begins to check up and coordinate on more distant objects, and he finds his visual ability to be reliable. The child seeing the Fourth of July fireworks for the first time sees a flash and then hears a boom. Maybe that doesn't mean so much to him, because boom (aural) and flash (optical) may be different phenomena; but when he sees a man hammering a fence post, he has by this time been hammering a whole lot and he knows the sound that makes. He may not be very sure of the fireworks in the sky, the flash and the boom, but he is really very confident about the sound of the hammering of the fence post. When he sees the man hammer and then hears the sound a fraction later, he begins to realize that there is some lag in the rates in which he gets information from different faculties. His eye gets it faster than his ear.

801.13 The three postnatal senses the child coordinates are secondary. The first prenatal one, the tactile, is primary. The real emphasis of the judgment of life is on the tactile, the primary, the thing you can touch.* The ranges of the first three senses are so close together, and sight is so different, that we may best rank them as #1, *touch,* being a primary set; with both #2, *olfactoral* coupled with #3, *aural,* as a secondary set; and #4 *sight,* as a tertiary set: wherefore in effect, touch is the *yesterday set;* while the olfactoral and aural (what you are smelling, eating, saying, and hearing) are the *now set;* while sight (what only may be next) is the *future set.* (We can seem to see, but we have not yet come to it.) Whereas reality is *eternally now,* human apprehending demonstrates a large assortment of lags in rates of cognitions whose myriadly multivaried frequencies of myriadly multivaried, positive-negative, omnidirectional aberrations, in multivaried degrees, produce such elusively off-center effects as possibly to result in an illusionary awareness of an approximately unlimited number of individually different awareness patterns, all of whose relative imperfections induce the illusion of a reality in which "life" is terminal, because physically imperfect; as contrasted to mind's discovery of an omni-interaccommodative complex of a variety of different a priori, cosmic, and eternal principles, which can only be intellectually discovered, have no weight, and apparently manifest a perfect, abstract, eternal design, the metaphysical utterly transcendent of the physical.

801.14 The 186,000-miles-per-second speed of light is so fast that it was only just recently measured, and it doesn't really have much meaning to us. You don't have a sense of 700 million miles per hour. If you did get to "see" that way, you would be spontaneously conscious of seeing the Sun eight minutes after the horizon had obscured it; ergo, consciously seeing an arc around the Earth's curvature. We are not seeing that way as yet. To explain

* You can reflect philosophically on some of the things touch does, like making people want to get their hands on the coin, the key, or whatever it may be.

our sight, we call it "instantaneous." We say we can see instantaneously. This fact has misled us very greatly. You insist that you are seeing the black-and-white page of this book, do you not? You're not. You have a brain-centered television set, and the light is bouncing off the page. The resultant comes back through your optical system and is scanned and actually goes back into the brain, and you are seeing the page in your brain. You are not seeing the page out in front of you. We have gotten used to the idea that we see outside of ourselves, but we just don't do so. It only takes about a billionth of a second for the light to bounce off the page and get in the brain to be scanned, so the child is fooled into thinking that he is seeing outside of himself. And we are misinforming ourselves in discounting the lag and assuming that we see it "over there." No one has ever seen outside themselves.

801.20 The Omnidirectional TV Set

801.21 Children looking at TV today look at it quite differently from the way it was to the first generation of TV adults. It begins to be very much a part of the child's life, and he tends to accredit it the way adults accredit what they get from their eyes. When children are looking at a baseball game, they are right there in the field. All of our vision operates as an omnidirectional TV set, and there is no way to escape it. That is all we have ever lived in. We have all been in omnidirectional TV sets all our lives, and we have gotten so accustomed to the reliability of the information that we have, in effect, projected ourselves into the field. We may insist that we see each other out in the field. But all vision actually operates inside the brain in organic, neuron-transistored TV sets.

801.22 We have all heard people describe other people, in a derogatory way, as being "full of imagination." The fact is that if you are not full of imagination, you are not very sane. All we do is deal in brain images. We traffic in the memory sets, the TV sets, the recall sets, and certain incoming sets. When you say that you see me or you say "I see you," or "I touch you," I am confining information about you to the "tactile you." If I had never had a tactile experience (which could easily be if I were paralyzed at conception), "you" might be only where I smell you. "You" would have only the smellable identity that we have for our dogs. You would be as big as you smell. Then, if I had never smelled, tasted, nor experienced tactile sensing, you would be strictly the *hearable you*.

801.23 What is really important, however, about you or me is the *thinkable you* or the *thinkable me,* the abstract metaphysical you or me, what we have done with these images, the relatedness we have found, what communications we have made with one another. We begin to realize that the dimensions of the *thinkable you* are phenomenal, when you hear Mozart on the radio, that is, the metaphysical—only intellectually identifiable—eternal Mozart who will always be there to any who hears his music. When we say "atom" or think "atom" we are intellect-to-intellect with livingly thinkable Democritus, who first conceived and named the invisible phenomenon "atom." Were exclusively tactile Democritus to be sitting next to you, surely you would not recognize him nor accredit him as you do the only-thinkable Democritus and what he thought about the atom. You say to me: "I see you sitting there." And all you see is a little of my pink face and hands and my shoes and clothing, and you can't see *me,* which is entirely the thinking, abstract, metaphysical me. It becomes shocking to think that we recognize one another only as the touchable, nonthinking biological organism and its clothed ensemble.

801.24 Reconsidered in these significant identification terms, there is quite a different significance in what we term "dead" as a strictly tactile "thing," in contrast to the exclusively "thinking" you or me. We can put the touchable things in the ground, but we can't put the thinking and thinkable you in the ground. The fact that I see you only as the touchable you keeps shocking me. The baby's spontaneous touching becomes the dominant sense measure, wherefore we insist on measuring the inches or the feet. We talk this way even though these are not the right increments. My exclusively tactile seeing inadequacy becomes a kind of warning, despite my only theoretical knowledge of the error of seeing you only as the touchable you. I keep spontaneously seeing the tactile living you. The tactile is very unreliable; it has little meaning. Though you know they are gentle, sweet children, when they put on Hallowe'en monster masks they "look" like monsters. It was precisely in this manner that human beings came to err in identifying life only with the touchable physical, which is exactly what life isn't. (See Sec. 531.)

810.00 One Spherical Triangle Considered as Four

811.00 Bias on One Side of the Line

811.01 We have all been brought up with a plane geometry in which a triangle was conceived and defined as an area *bound* by a closed line of three edges and three angles. A circle was an area *bound* by a closed line of unit radius. The area outside the closed boundary line was not only undefinable but was inconceivable and unconsidered.

811.02 In the abstract, ghostly geometry of the Greeks, the triangle and circle were inscribed in a plane that extended laterally to infinity. So tiny is man and so limited was man's experience that at the time of the Greeks, he had no notion that he was living on a planet. Man seemed obviously to be living on an intuitively expansive planar world around and above which passed the Sun and stars, after which they plunged into the sea and arose again in the morning. This cosmological concept of an eternally extended, planar-based Earth sandwiched between heaven above and hell below made infinity obvious, ergo axiomatic, to the Greeks.

811.03 The Greek geometers could not therefore define the planar extensibility that lay outside and beyond the line of known content. Since the surface outside of the line went to infinity, you could not include it in your computation. The Greeks' concept of the geometrical, bound-area of their triangle—or their circle—lay demonstrably on only one bound-area side of the line. As a consequence of such fundamental schooling, world society became historically biased about everything. Continually facing survival strategy choices, society assumed that it must always choose between two or more political or religious "sides." Thus developed the seeming nobility of loyalties. Society has been educated to look for logic and reliability only on one side of a line, hoping that the side chosen, on one hand or the other of indeterminately large lines, may be on the *inside* of the line. This logic is at the head of our reflexively conditioned biases. We are continually being pressed to validate one side of the line or the other.

811.04 You can "draw a line" only on the surface of some system. All systems divide Universe into insideness and outsideness. Systems are finite. Validity favors neither one side of the line nor the other. Every time we draw a line operationally upon a system, it returns upon itself. The line always divides a whole system's unit area surface into two areas, each equally valid as unit areas. Operational geometry invalidates all bias.

812.00 Spherical Triangle

812.01 The shortest distance between any two points on the surface of a sphere is always described by an arc of a great circle. A triangle drawn most economically on the Earth's surface or on the surface of any other sphere is actually always a spherical triangle described by great-circle arcs. The sum of the three angles of a spherical triangle is never 180 degrees. Spherical trigonometry is different from plane trigonometry; in the latter, the sum of any triangle's angles is always 180 degrees. There is no plane flat surface on Earth, wherefore no plane triangles can be demonstrated on its surface. Operationally speaking, we always deal in systems, and all systems are characterized projectionally by spherical triangles, which control all our experimental transformations.

812.02 Drawing or scribing is an operational term. It is impossible to draw without an object upon which to draw. The drawing may be by depositing on or by carving away—that is, by creating a trajectory or tracery of the operational event. All the objects upon which drawing may be operationally accomplished are structural systems having insideness and outsideness. The drawn-upon object may be either symmetrical or asymmetrical. A piece of paper or a blackboard is a system having insideness and outsideness.

812.03 When we draw a triangle on the surface of Earth (which previously unscribed area was unit before the scribing or drawing), we divide Earth's surface into two areas on either side of the line. One may be a little local triangle whose three angles seem to add up to 180 degrees, while the other big spherical triangle complementing the small one to account together for all the Earth's surface has angles adding up to 900 degrees or less. This means that each corner of the big triangle complementing the small local one, with corners seeming to be only 60 degrees each, must be 300 degrees each, for there are approximately 360 degrees around each point on the surface of a sphere. Therefore the sum of all the three angles of the big Earth triangles, which inherently complement the little local 60-degree-per-corner equilateral triangles, must be 900 degrees. The big 900-degree triangle is also an area bounded by three lines and three angles. Our schooled-in bias renders it typical of us to miss the big triangle while being preoccupied only locally with the negligibly sized triangular area.

812.04 If you inscribe one triangle on a spherical system, you inevitably describe four triangles. There is a concave small triangle and a concave big triangle, as viewed from inside, and a convex small triangle and a convex big triangle, as viewed from outside. Concave and convex are not the same, so at minimum there always are inherently four triangles.

813.00 Square or Triangle Becomes Great Circle at Equator

813.01 If we draw a closed line such as a circle around Earth, it must divide its total unit surface into two areas, as does the equator divide Earth into southern and northern hemispheres. If we draw a lesser-sized circle on Earth, such as the circle of North latitude 70°, it divides Earth's total surface into a very large southern area and a relatively small northern area. If we go outdoors and draw a circle on the ground, it will divide the whole area of our planet Earth into two areas—one will be *very* small, the other *very* large.

813.02 If our little circle has an area of one square foot, the big circle has an area of approximately five quadrillion square feet, because our 8,000-mile-

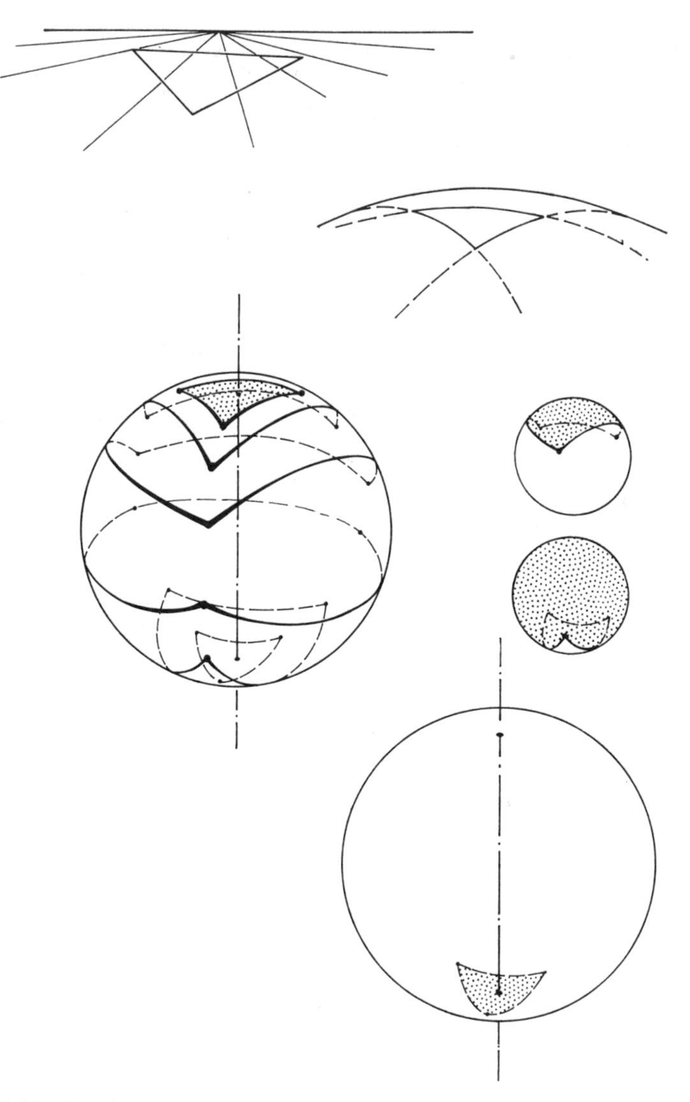

Fig. 812.03: The Greeks defined a triangle as an area bound by a closed line of three edges and three angles. A triangle drawn on the Earth's surface is actually a spherical triangle described by three great-circle arcs. It is evident that the arcs divide the surface of the sphere into two areas, each of which is bound by a closed line consisting of three edges and three angles, ergo dividing the total area of the sphere into two complementary triangles. The area apparently "outside" one triangle is seen to be "inside" the other. Because every spherical surface has two aspects—convex if viewed from outside, concave if viewed from within—each of these triangles is, in itself, two triangles. Thus one triangle becomes four when the total complex is understood. "Drawing" or "scribing" is an operational term. It is impossible to draw without an object upon which to draw. The drawing may be by depositing on or by carving away, that is, by creating a trajectory or tracery of the operational event. All the objects upon which drawing may be operationally accomplished are structural systems having insideness and outsideness. The drawn-upon object may be symmetrical or asymmetrical, a piece of paper or a blackboard system having insideness and outsideness.

diameter Earth has an approximately 200-million-square-mile surface. Each square mile has approximately 25 million square feet, which, multiplied, gives a five followed by fifteen zeros: 5,000,000,000,000,000 square feet. This is written by the scientists as 5×10^{15} square feet; while compact, this tends to disconnect from our senses. Scientists have been forced to disconnect from our senses due to the errors of our senses, which we are now able to rectify. As we reconnect our senses with the reality of Universe, we begin to regain competent thinking by humans, and thereby possibly their continuance in Universe as competently functioning team members—members of the varsity or University team of Universe.

813.03 If, instead of drawing a one-square-foot circle on the ground—which means on the surface of the spherical Earth—we were to draw a square that is one foot on each side, we would have the same size local area as before: one square foot. A square as defined by Euclid is an area bound by a closed line of four equal-length edges and four equal and identical angles. By this definition, our little square, one foot to a side, that we have drawn on the ground is a closed line of four equal edges and equal angles. But this divides all Earth's surface into two areas, both of which are equally bound by four equal-length edges and four equal angles. Therefore, we have two squares: one little local one and one enormous one. And the little one's corners are approximately 90 degrees each, which makes the big square's corners approximately 270 degrees each. While you may not be familiar with such thinking, you are confronted with the results of a physical experiment, which inform you that you have been laboring under many debilitating illusions.

813.04 If you make your small square a little bigger and your bigger one a little smaller by increasing the little one's edges to one mile each, you will have a local one square mile—a customary unit of western United States ranches—and the big square will be approximately 199,999,999 square miles. As you further increase the size of the square, using great-circle lines, which are the shortest distances on a sphere between any two points, to draw the square's edges, you will find the small square's corner angles increasing while the big one's corner angles are decreasing. If you now make your square so that its area is one half that of the Earth, 100 million square miles, in order to have all your edges the same and all your angles the same, you will find that each of the corners of both squares is 180 degrees. That is to say, the edges of both squares lie along Earth's equator so that the areas of both are approximately 10 million square miles.

814.00 Complementarity of System Surfaces

814.01 The progressive enlargement of a triangle, a pentagon, an octagon, or any other equiedged, closed-line figure drawn on any system's surface produces similar results to that of the enlarging square with 180 degrees to each corner at the equator. The closed-line surface figure will always and only divide the whole area into two complementary areas. Each human making this discovery experimentally says spontaneously, "But I didn't mean to make the big triangle," or "the big square," or indeed, the big mess of pollution. This lack of intention in no way alters these truths of Universe. We are all equally responsible. We are responsible not only for the big complementary surface areas we develop on systems by our every act, but also for the finite, complementary outward tetrahedron automatically complementing and enclosing each system we devise. We are inherently responsible for the complementary transformation of Universe, inwardly, outwardly, and all around every system we alter.

820.00 Tools of Geometry

821.01 The Early Greek geometers and their Egyptian and Babylonian predecessors pursued the science of geometry with three basic tools; the dividers, the straightedge, and the scriber. They established the first rule of the game of geometry, that they could not introduce information into their exploration unless it was acquired empirically as constructed by the use of those tools. With the progressive interactive use of these three tools, they produced modular areas, angles, and linear spaces.

821.02 The basic flaw in their game was that they failed to identify and define as a tool the *surface* on which they inscribed. In absolute reality, this surface constituted a fourth tool absolutely essential to their demonstration. The absolute error of this oversight was missed at the time due to the minuscule size of man in relation to his planet Earth. While there were a few who conceived of Earth as a sphere, they assumed that a local planar condition existed—which the vast majority of humans assumed to be extended to infinity, with a four-cornered Earth plane surrounded by the plane of water that went to infinity.

821.03 They assumed the complementary tool to be a plane. Because the plane went to infinity in all planar directions, it could not be defined and therefore was spontaneously overlooked as a *tool* essential to their empirical demonstrating. What they could not define, yet obviously needed, they identified by the ineffable title "axiomatic," meaning "Everybody knows that." Had they recognized the essentiality of defining the fourth tool upon which they inscribed, and had they recognized that our Earth was spherical—ergo, finite; ergo, definite—they could and probably would have employed strategies completely different from that of their initiation of geometry with the exclusive use of the plane. But to the eastern Mediterranean world there lay the flat, infinite plane of the Earth at their feet on which to scratch with a scriber.

821.10 **Dividers:** The ends of two sticks can be bound together to serve as dividers. A straightedge stick could be whittled by a knife and sighted for straightness and improved by more whittling.

821.11 The opening of the dividers could be fixed by binding on a third stick between the other two ends, thus rigidifying by triangulation. Almost anyone at sea or in the desert could start playing this game.

825.00 Greek Scribing of Right-Angle Modularity in a Plane

825.01 It was easy for the Greeks to use their fixed dividers to identify two points on the plane marked by the divider's two ends: A and B, respec-

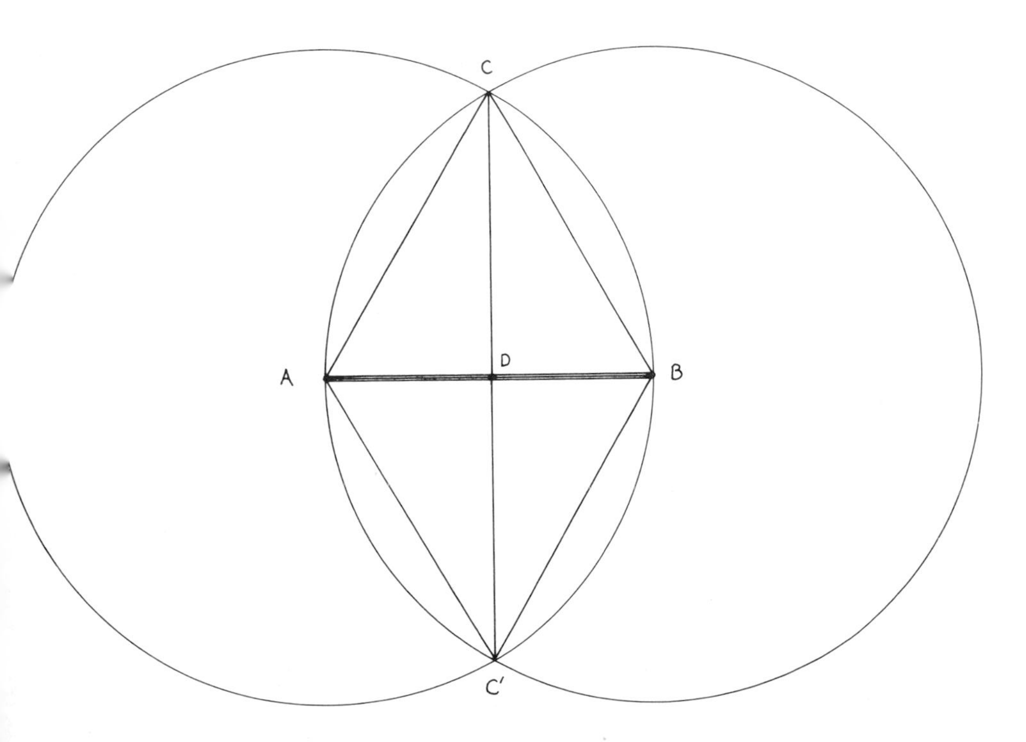

Fig. 825.01.

tively. Employing their straightedge, they could inscribe the line between these two points, the line *AB*. Using one end of the dividers as the pivot point at one end of the line, *A*, a circle can be described around the original line terminal: circle *A*. Using point *B* as a center, a circle can be described around it, which we will call circle *B*. These two circles intersect one another at two points on either side of the line *AB*. We will call the intersection points *C* and *C′*.

825.02 By construction, they demonstrated that points *C* and *C′* were both equidistant from points *A* and *B*. In this process, they have also defined two equilateral triangles *ABC* and *ABC′*, with a congruent edge along the line *AB* and with points *C* and *C′* equidistant on either side from points *A* and *B*, respectively.

825.10 Right Triangle

825.11 They then used a straightedge to connect points *C* and *C′* with a line that they said bisected line *AB* perpendicularly, being generated by equidistance from either point on either side. Thus the Greeks arrived at their right triangle; in fact, their four right triangles. We will designate as point *D* the intersection of the lines *CC′* and *AB*. This gave the Greeks four angles around a common point. The four right triangles *ADC*, *BDC*, *ADC′*, and *BDC′* have hypotenuses and legs that are, as is apparent from even the most casual inspection, of three different lengths. The leg *DB*, for instance, is by equidistance construction exactly one-half of *AB*, since *AB* was the radius of the two original circles whose circumferences ran through one another's centers. By divider inspections, *DB* is less than *CD* and *CD* is less than *CB*. The length of the line *CD* is unknown in respect to the original lines *AB*, *BD*, or *AC*, lines that represented the original opening of the dividers. They have established, however, with satisfaction of the rules of their game, that 360 degrees of circular unity at *D* could be divided into four equal 90-degree angles entirely and evenly surrounding point *D*.

825.20 Hexagonal Construction

825.21 Diameter: The Greeks then started another independent investigation with their three tools on the seemingly flat planar surface of the Earth. Using their dividers to strike a circle and using their straightedge congruent to the center of the circle, they were able with their scriber to strike a seemingly straight line through the center of construction of the circle. As the line passed out of the circle in either direction from the center, it seemingly could go on to infinity, and therefore was of no further interest to them. But inside the circle, as the line crossed the circumference at two points on either side of its center, they had the construction information that the line equated the opening of the dividers in two opposite directions. They called this line the diameter: DIA + METER.

825.22 Now we will call the center of the constructed circle *D* and the two intersections of the line and the circumference *A* and *B*. That $AD = DB$ is proven by construction. They know that any point on the circumference is equidistant from *D*. Using their dividers again and using point *A* as a pivot, they drew a circle around *A;* they drew a second circle using *B* as a pivot. Both of these circles pass through *D*. The circle around *A* intersects the circle around *D* at two points, *C* and *C'*. The circle around *B* intersects the circle around *D* at two points, *E* and *E'*. The circle around *A* and the circle around *B* are tangent to one another at the point *D*.

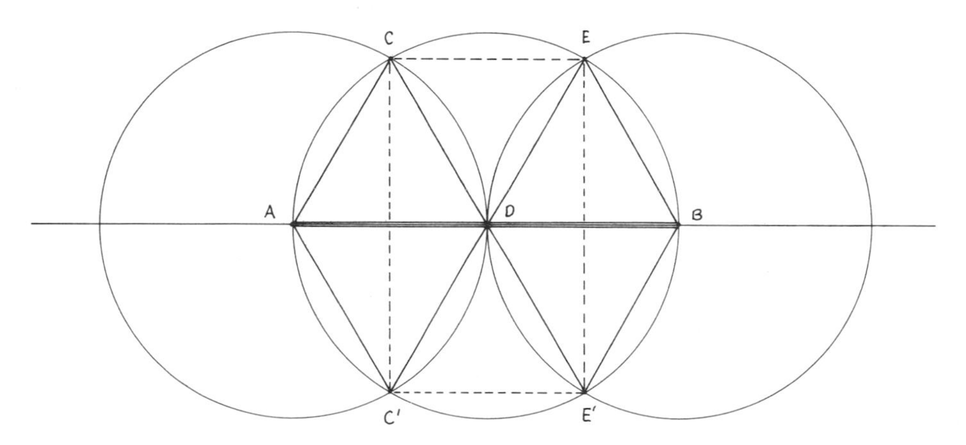

Fig. 825.22.

825.23 They have now constructed four equilateral triangles in two pairs: *ADC* and *ADC'* as the first pair, and *DBE* and *DBE'* as the second pair. They know that the lines *AC, CD, AC'*, and *DC'* are all identical in length, being the fixed opening of the dividers and so produced and proven by construction. The same is true of the lines *DE, EB, DE'*, and *BE'*—they are all the same. The Greeks found it a tantalizing matter that the two lines *CE* and *C'E'*, which lie between the vertexes of the two pairs of equilateral triangles, seemed to be equal, but there was no way for them to prove it by their construction.

825.24 At first it seemed they might be able to prove that the increments *CE* and *C'E'* are not only equal to one another, but are equal to the basic radius of the circle *AD;* therefore, the hexagon *ACEBE'C'* would be an equilateral hexagon; and hexagons would be inherently subdivisible into six 60-degree equilateral triangles around the central point, and all the angles would be of 60 degrees.

825.25 There seemed to be one more chance for them to prove this to be true, which would have provided an equiangular, equiedged, triangularly stable structuring of areal mensuration. This last chance to prove it was by first showing by construction that the line *ADB,* which runs through the point of tangency of the circles *A* and *B,* is a straight line. This was constructed by the straightedge as the diameter of circle *D.* This diameter is divided by four equal half-radii, which are proven to be half-radii by their perpendicular intersection with lines both of whose two ends are equidistant from two points on either side of the intersecting lines. If it could be assumed that: (1) the lines *CE* and *C'E'* were parallel to the straight line *ADB* running through the point of tangency as well as perpendicular to both the lines *CC'* and *EE';* and (2) if it could be proven that when one end between two parallels is perpendicular to one of the parallels, the other end is perpendicular to the other parallel; and (3) if it could be proven also that the perpendicular distances between any two parallels were always the same, they could then have proven *CE*=*CD*= *DE* =*D'E',* and their hexagon would be equilateral and equiradial with radii and chords equal.

825.26 Pythagorean Proof

825.261 All of these steps were eventually taken and proven in a complex of other proofs. In the meantime, they were diverted by the Pythagoreans' construction proof of "the square of the hypotenuse of a right triangle's equatability with the sum of the squares of the other two sides," and the construction proof that any non-right triangle's dimensional values could be obtained by dropping a perpendicular upon one of its sides from one of its vertexes and thus converting it into two right triangles each of which could be solved arithmetically by the Pythagoreans' "squares" without having to labor further with empirical constructs. This arithmetical facility induced a detouring of strictly constructional explorations, hypotheses, and proofs thereof.

825.27 Due to their misassumed necessity to commence their local scientific exploration of geometry only in a supposed plane that extended forever without definable perimeter, that is, to infinity, the Ionians began using their right-triangle exploration before they were able to prove that six equilateral triangles lie in a circle around point *D.* They could divide the arithmetical 360 degrees of circular unity agreed upon into six 60-degree increments. And, as we have already noted, if this had been proven by their early constructions with their three tools, they might then have gone on to divide all planar space with equilateral triangles, which models would have been very convenient in connection with the economically satisfactory point-locating capability of triangulation and trigonometry.

825.28 Euclid was not trying to express forces. We, however—inspired by Avogadro's identical-energy conditions under which different elements disclosed the same number of molecules per given volume—are exploring the possible establishment of an operationally strict *vectorial geometry field,* which is an isotropic (everywhere the same) vector matrix. We abandon the Greek perpendicularity of construction and find ourselves operationally in an omnidirectional, spherically observed, multidimensional, omni-intertransforming Universe. Our first move in spherical reality scribing is to strike a quasi-sphere as the vectorial radius of construction. Our dividers are welded at a fixed angle. The second move is to establish the center. Third move: a surface circle. The radius is uniform and the lesser circle is uniform. From the triangle to the tetrahedron, the dividers go to direct opposites to make two tetrahedra with a common vertex at the center. Two tetrahedra have six internal faces = hexagon = genesis of bow tie = genesis of modelability = vector equilibrium. Only the dividers and straightedge are used. You start with two events—any distance apart: only one module with no subdivision; ergo, timeless; ergo, eternal; ergo, no frequency. Playing the game in a timeless manner. (You have to have division of the line to have frequency, ergo, to have time.) (See Secs. 420 and 650.)

825.29 Commencing proof upon a sphere as representative of energy convergent or divergent, we may construct an equilateral triangle from any point on the surface. If we describe equilateral (equiangular) triangles whose chords are identical to the radii, the same sphere may be intersected alternately by four great-circle planes whose circles intercept each other, respectively, at 12 equidistant points in such a manner that only two circles intersect at any one point. As this system is described, each great circle becomes symmetrically subdivided into six equal-arc segments whose chords are identical to the radii. From this four-dimensional tribisection, any geometrical form may be described in whole fractions.

825.30 Two-Way Rectilinear Grid

825.31 To the Greeks, a two-way, rectilinearly intersecting grid of parallel lines seemed simpler than would a three-way grid of parallel lines. (See Chapter 11, "Projective Transformation.") And the two-way grid was highly compatible with their practical coordinate needs for dealing with an assumedly flat-plane Universe. Thus the Greeks came to employ 90-degreeness and unique perpendicularity to the system as a basic additional dimensional requirement for the exclusive, and consequently unchallenged, three-dimensional geometrical data coordination.

825.32 Their arithmetical operations were coordinated with geometry on the assumption that first-power numbers represented linear module tallies, that second-power N^2 = square increments, and that third-power N^3 = cubical increments of space. First dimension was length expressed with one line. Two dimensions introduced width expressed with a cross of two lines in a plane. Three dimensions introduced height expressed by a third line crossing perpendicularly to the first two at their previous crossing, making a three-way, three-dimensional cross, which they referred to as the *XYZ* coordinate system. The most economical distance measuring between the peripheral points of such *XYZ* systems involved hypotenuses and legs of different lengths. This three-dimensionality dominated the 2,000-year scientific development of the *XYZ*—c.g_t.s. "Comprehensive Coordinate System of Scientific Mensurations." As a consequence, identifications of physical reality have been and as yet are only awkwardly characterized because of the inherent irrationality of the *peripheral* hypotenuse aspects of systems in respect to their *radial XYZ* interrelationships.

825.33 Commanded by their wealth-controlling patrons, pure scientists have had to translate their theoretical calculations of physical-system behaviors into coordinate relationship with physical reality in order to permit applied science to reduce theoretical inventions to physical practice and use. All of the analytic geometers and calculus mathematicians identify their calculus-derived coordinate behaviors of theoretical systems only in terms of linear measurements taken outwardly from central points of reference; they locate the remote event points relative to those centers only by an awkward set of perpendicularities emanating from, and parallel to, the central *XYZ* grid of perpendicular coordinates. The irrationality of this peripheral measuring in respect to complexedly orbited atomic nuclei has occasioned the exclusively mathematical processing of energy data without the use of conceptual models.

826.00 Unity of Peripheral and Radial Modularity

826.01 Had the Greeks originally employed a universal model of *x*-dimensional reality as their first tool upon and within which they could further inscribe and measure with their divider, scriber, and straightedge, they would have been able to arrive at unity of circumferential as well as radial modularity. This would have been very convenient to modern physics because all the accelerations of all the constantly transforming physical events of Universe are distinguished by two fundamentally different forms of acceleration, angular and linear.

826.02 **Hammer Throw:** When a man accelerates a weight on the end of a cord by swinging it around his head, the weight is restrained by the cord and it accumulates the energy of his exertions in the velocity it maintains in a circular pattern. This is angular acceleration, and its velocity rates and angular momentum are calculated in central-angle increments of the circular movement accomplished within given units of time. When the weight's cord is released by its human accelerator, it then goes into linear acceleration and its accomplished distance is measured in time increments following its release and its known release velocity, which calculations are modified by any secondary restraints.

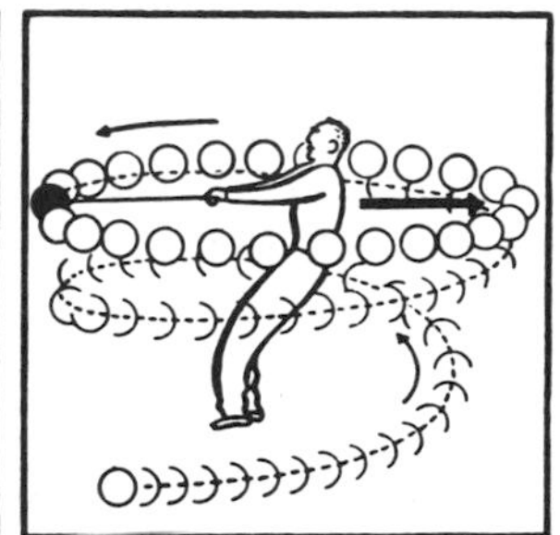
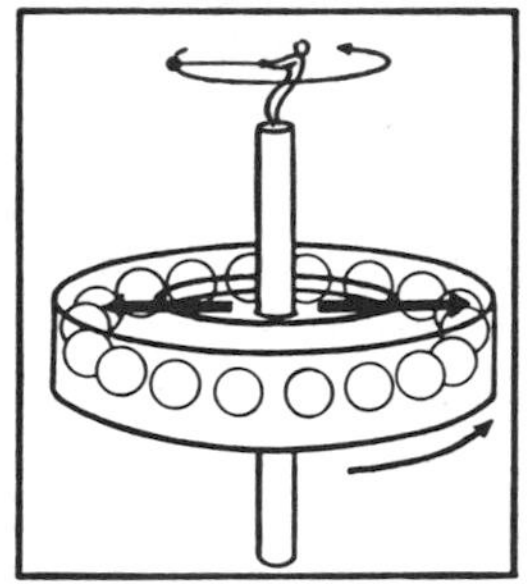
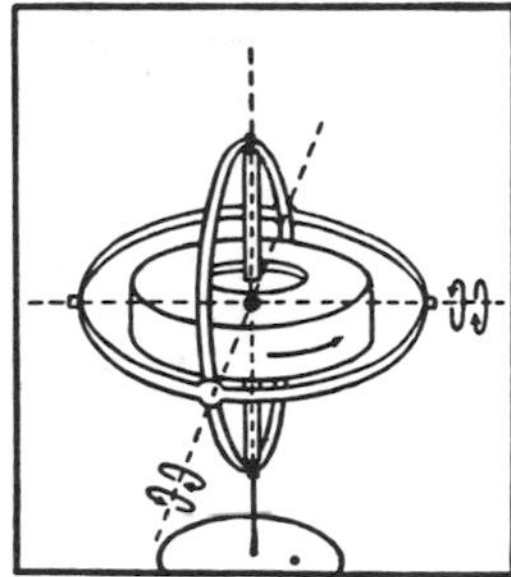

Fig. 826.02.

826.03 The angular accelerations relate then to the myriad of circular or elliptical orbitings of components of systems around their respective centers or focii, and are intimate to original acceleration-generating factors such as the "hammer thrower" himself and his muscle as the metabolic powering by the beef he ate the day before, which gained its energy from vegetation it had eaten, which gained its energy from the Sun's radiation by photosynthesis—all of whose attendant relative efficiencies of energy relaying were consequent upon the relative design efficacies and energy divergence to complementary environment conditions of the total synergetically effective system with the eventually total regenerative Universe itself.*

* This is a typical illustration of total energy accounting, which all society must become conversant with in short order if we are to pass through the crisis and flourish upon our planet. If we do succeed, it will be because, among other planetary events, humans will have come to recognize that the common wealth equating accounting must be one that locks fundamental and central energy incrementations—such as kilowatt hours—to human physical-energy work capability and its augmentation by the mind-comprehending employability of generalized principles of Universe, as these may be realistically appraised in the terms of increasing numbers of days for increasing numbers of lives we are thus far technically organized to cope with, while accommodating increasing hours and distances of increasing freedoms for increasing numbers of human beings. All

826.04 **Science as a Tool:** The linear measurements represent the radial going-away accelerations or resultants of earlier or more remote events as well as of secondary restraints. The rigid rectilinear angularity of the 90-degree-central-angle *XYZ* mensuration instituted by the Greeks made impossible any unit language of direct circumferential or peripheral coordination between angular and linear phenomena. As a consequence, only the radial and linear measurements have been available to physics. For this reason, physics has been unable to make simultaneous identification of both wave and particle aspects of energy events.

826.05 The Greeks' planar inception of geometry and its diversion first into theoretical mathematical calculations and ultimate abandonment of models has occasioned the void of ignorance now existing between the sciences and the humanities imposed by the lack of logical and unitarily moduled conceptual systems. This, in turn, has occasioned complete social blindness to either the facts or the potential benefits of science to humanity. Thus science has now come to represent an invisible monster to vast numbers of society, wherefore society threatens to jettison science and its ''obnoxious'' technology, not realizing that this would lead swiftly to genocide. Central to this crisis of terrestrially situate humans is the necessity for discovering and employing a comprehensively comprehendible universal coordinate system that will make it swiftly lucid to world society that science and technology are only manipulative tools like inanimate and cut-offable hands which may be turned to structuring or destructuring. How they are to be employed is not a function of the tools but of human choice. The crisis is one of the loving and longing impulse to understand and be understood, which results as informed comprehension. It is the will to structure versus ignorant yielding to fear-impulsed reflexive conditioning that results from being born utterly helpless. Intellectual information-accumulating processing and anticipatory faculties are necessary, and are only slowly discovered as exclusively able to overcome the ignorantly feared frustrating experiences of the past. Science must be seen as a tool of fundamental advantage for all, which Universe requires that man understand and use exclusively for the positive advantage of all of humanity, or humanity itself will be discarded by Universe as a viable evolutionary agent.

826.06 It is to this dilemma that we address ourselves; not being interested in palliatives, we backtrack two and a half millennia to the turning of the road where we entered in the hope of regaining the highway of lucid rationality. Using the same Greek tools, but not starting off with a plane or the

---

of this fundamental data can be introduced into world computer memories, which can approximately instantly enlighten world humanity on its increasingly more effective options of evolutionary cooperation and fundamentally spontaneous social commitment.

subsequently substituted blackboard of the pedagogues working indoors and deprived of direct access to the scratchable Earth surface used by the Near Eastern ancients,* we will now institute scientific exploration in the measurement of physical reality.

830.00 Foldability of Great Circles

831.00 Sheet of Paper as a System

831.01 Our steel dividers have sharp, straightedged legs, each tapering into sharp points. We can call these dividers "scissors." Scissors are dividers of either linear or angular, i.e. circular, differentiation. We can even make our explorations with some superficial accommodation of the Greeks' propensity for using a plane. For instance, we can take a finite piece of paper, remembering (operationally), however, that it has "thickness" and "edges," which are in fact small area faces. If it is a rectilinear sheet of typewriter paper, we recognize that it has four minor faces and two major faces. The major faces we call "this side" and "the other side," but we must go operationally further in our consideration of what the "piece of paper" is. Looking at its edges with a magnifying glass, we find that those surfaces round over rather brokenly, like the shoulders of a hillside leading to a plateau. We find the piece of paper to be fundamentally the same kind of entity as that which we have watched the baker make as he concocts, stirs, and thickens his piecrust dough, which, after powdering with flour, can be formed into a spherical mass and set upon a flour-powdered surface to be progressively rolled into a thick sheet that may be cut into separate increments of the same approximate dimensions as the "sheet" of typewriter paper.

831.10 Moebius Strip

831.11 In the same operational piecrust-making strictness of observation, we realize that the phase of topology that Moebius employed in developing his famous strip mistakenly assumed that the strip of paper had two completely nonconnected faces of such thinness as to have no edge dimension whatsoever. When we study the Moebius strip of paper and the method of twisting one of its ends before fastening them together and scribing and cutting the central line of the strip only to find that it is still a single circle of

* With the blackboard the pedagogues were able to bring infinity indoors.

twice the circumference and half the width of the strip, we realize that the strip was just a partially flattened section of our piecrust, which the baker would have produced by making a long hard roll, thinner than a breadstick and flattened out with his wooden roller. What Moebius really did was to take a flattened tube, twist one of its ends 180 degrees, and rejoin the tube ends to one another. The scribed line of cutting would simply be a spiral around the tube, which made it clear that the two alternate ends of the spirals were joined to one another before the knifing commenced.

831.20 *Cutting Out Circular Cookies*

831.21 We can use the leverage of the sheet length of flatness of the paper against the fulcrum of the sheet of paper's thinness to fold it as a relatively flat system, even as the baker could fold over the unbaked piecrust. Or we can scribe upon the paper with our geometrical tools in an approximately accurate measuring manner. What we have done is to flatten our system in a measurefully knowing manner. For operational accommodation, but always by construction, we can for the moment consider the paper's surface as did the Greeks their infinitely extending plane, but we are aware and will always be responsible for "the rest of the system" with which we are working, though we are momentarily preoccupied with only a very local area of the whole.

831.22 We can scribe a circle around the pivotal *A*-end of the dividers, and we can do so in an approximate "plane." We can strike or scribe the approximately straight diameter through the circle's center. We can now use our divider-scissors to divide the finite circle of paper from the finite balance of the paper system lying outside the circle—that is, we can scissor or "cut" out the area contained by the circle from the balance of the paper, as the baker cuts out circular, wafer-thin cookies. We are at all times dividing reality multidimensionally, no matter how relatively diminutive some of its dimensions may be.

831.23 Because we are dealing with multidimensional reality, we must note operationally that in cutting out our circular piece of paper, we are also cutting our original piece of typewriter paper into two pieces, the other piece of which has a circular hole in its overall rectilinear area. We must keep ourselves conscious of this complementary consequence even though we are for the moment interested only in the cut-out circular piece pricked with the original center of the divider-generated circumference. (The Maori, whose prime love was the Pacific Ocean, looked upon islands as holes in their ocean and upon what man calls harbors or bays as protrusions of the ocean inserted into the land.) Now, from our cut-out circle and our inventory of construction-produced information, we learn experimentally that we can lift any point of the perimeter of the circle and fold it over so that the point of the perimeter is congruent with *any* other point on the perimeter; in doing so, we find that we are always folding the circular system of paper into two semicircles whose hinge lines always run through the points of origin.

831.24 By construction, we can demonstrate that the circle of paper may be folded along its constructionally scribed diameter, and because all of its perimeter points are equidistant from the center of the circle, the semicircular edges are everywhere congruent. We find that we can fold the circle along any of its infinite number of diameters and the two half-circle circumferences (or perimeters = run arounds = racetracks) will always be congruent as folded together. The same infinity of diameters could be used to fold the paper-circle diameters in the opposite direction on the underside of the original plane.

831.25 Having deliberately colored our original paper's two opposite major sides with two different colors, red and white, we will see that our set of paper-circle folding along its infinity of diameters resulted in red half-circles, while the folding in the opposite direction produced all white half-circles. We also discover that as we fold from flat whole circle to congruence with the other half-circle, among any of the infinity of diameters along which to fold, the circumference of any one side of the circle moves toward the circumference on the other half, and as it travels 180 degrees around its diameter hinges, its perimeter thus describes a hemisphere of points all equidistant from the same center of all the hinges.

831.26 Having worked from a unitary plane and employing the infinity of diameters to fold in opposite directions, we discover that all the combined red and white opposite semicircular foldings altogether have produced a sphere consisting of two complementary hemispheres, one red and one white, which altogether represent all the rotatings of the equidistant circumferences, always from the same common center of all the diameters, which fact we know by construction of the diameters by our straightedge along which we scribed through the original center mark of our generation of the circle.

831.30 Six Cases of Foldability of Great Circles

831.31 There are six cases of folding employed in the proof of sixthing of the circle—or hexagoning the circle. (See Illus 831.31.) Case 1 is a limit case with congruence of all diameters.

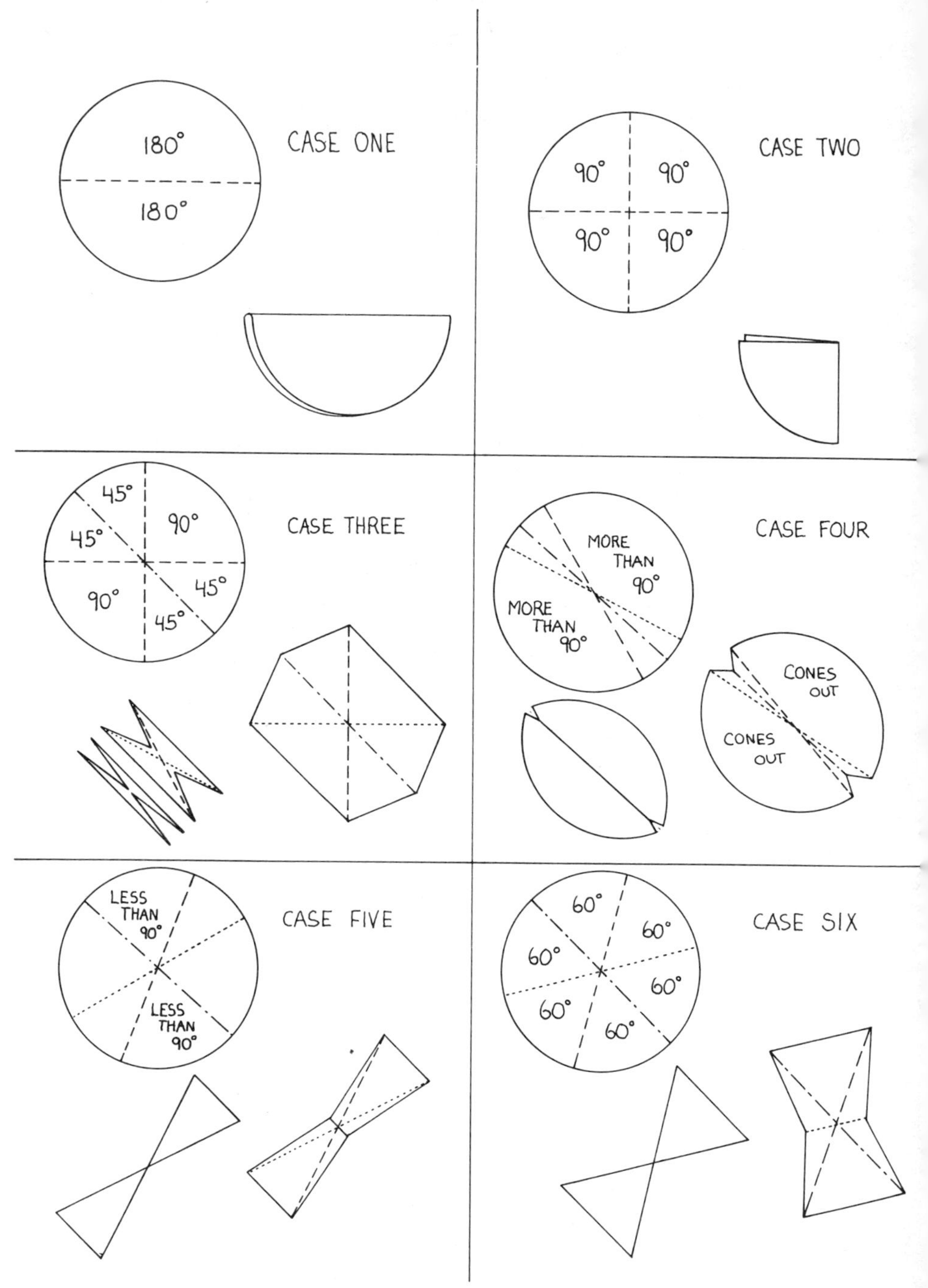

Fig. 831.31 *Foldability of Great Circles.*

835.00 Bow-Tie Construction of Spherical Octahedron

835.01 With one of the sharp points (*A*) of dividers (*AB*) fixed at a point (*X*) on a flat sheet of paper, sharp point *B* is rotated cuttingly around until an equiradius circle of paper is cut out. It is discovered experimentally that if any point on the circular perimeter is folded over to any other point on the circle's perimeter, that the circle of paper always folds in such a manner that one-half of its perimeter—and one-half of its area—is always congruent with the other half; and that the folded edge always runs through the exact center point *X* of the circle and constitutes a diameter line of the circle. This demonstrates that a diameter line always divides both the whole circular area and the circle's perimeter-circumference into two equal halves. If one diameter's end corner *W* of the circle, folded into halves, is folded over once more to congruence with the corner *W'* at the other end of the diameter, once again it will be constructively proved that all of the circle's perimeter is congruent with itself in four folded-together layers, which operational constructing also divides the whole circle into four equal parts, with the second folded diameter *Y-Y'* perpendicular to the first diameter, ergo producing four right-angled corners at the center of the circle as marked by the two diameter fold lines, *W-W'* and *Y-Y'*. If we now open the paper circle and turn it over to its reverse side, we fold in a third diameter line *T-T'* by making circumference point *W* congruent with circumference point *Y* (which inadvertently makes point *W'* congruent with *Y'*), we will find that we have exactly halved the right angles *WXY* and *W'XY'*, so that the perimeter distances *WT* or *TY* are each exactly half the perimeter distance *WY*, and either *W'T'* or *T'Y'* are each one-half the perimeter distances of either *WY*, *YW'*, *W'Y'*, or *Y'W*.

835.02 If we now turn the paper circle over once more we find that the spring in the fold lines of the paper will make point *T* and *T'* approach each other so that the whole circle once again may be folded flat to produce four congruent surfaces of the paper folded into an overall composite quarter circle with the two quarter-circle outer layers, and four one-eighth circle's two inner layers coming to congruent fold-around terminal tangency at the midpoint and center of the folded, right-angle, quarter-circle packet, with *W* congruent with *Y* and *W'* congruent with *Y'* and *T* congruent with *T'*. Thus it is proven that with three diameter foldlines the whole circle can be subdividingly folded into six arc-and-central-angle increments, ergo also unfoldable again into whole-circle flatness. (See Illus. 835.02.)

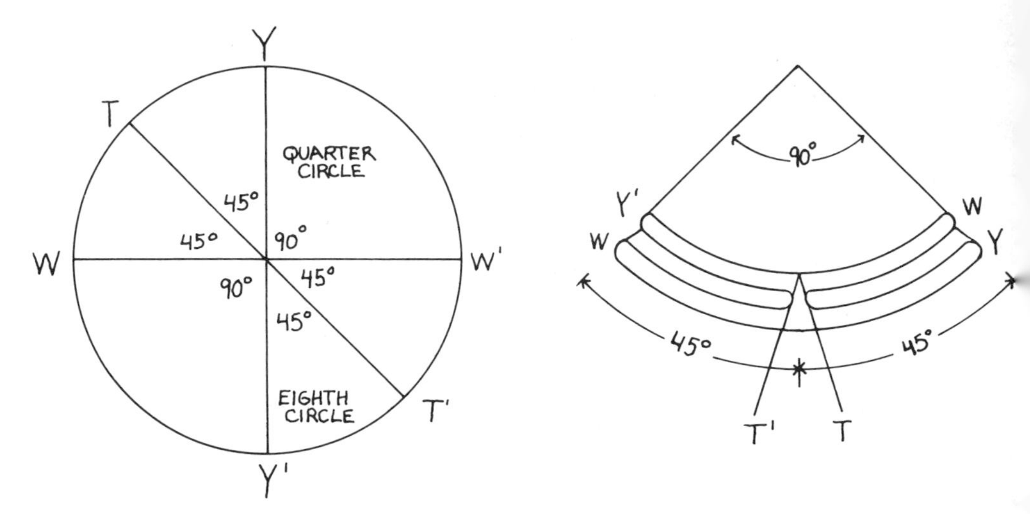

Fig. 835.02.

835.03 We know that every point on the perimeter of the folded semicircle is equidistant from the point of origin. We may now go to one end of the folded-edge diameter and fold the paper in such a manner that two ends of the diameter are congruent. This will fold the paper circle into four quadrants which, by construction congruence, are exactly equal. The legs of the 90-degree angle formed around the origin of the circle by this second folding are the same in length, being the same radius as that of the circle, ergo, of the halved diameter produced by the second folding. The angle edges and the radii are identical. When we open the quarter-circle of four faces folded together into the semicircle, we find that the second fold edge, which produced the 90-degree angle, is the radius of the diameter perpendicular to the first diameter folded upon. The points where this perpendicular diameter's ends intersect the circumference of the circle are equidistant, by construction, from the diameter ends of the first folded-edge diameter of the semicircle. This folded semicircle, with its secondary fold-mark of verticality to its origin, can be partially folded again on that perpendicular radius so that the partially folded semicircle and its partially folded, vertically impinging fold-line constitute an angularly winged unit, with appearance similar to the outer hard covers of a partially opened book standing bottomless with the book's hard covers vertically perpendicular to a table. This flying-winged, vertically hinged pair of double-thickness quarter-circles will be found to be vertically stable when stood upon a table, that is, allowed to be pulled vertically against the table by gravity. In structural effect, this winged quarter-pair of open, standing "book covers" is a tripod because the two diameter ends, *A* and *B*, and the circle's origin point, C, at the middle represent three points, A, B, C, in triangular array touching the table, which act as a triangle base for the tripod whose apex is at the perimeter, T, of the semicircle at the top terminal of the vertical fold. The tripod's legs are uneven, one being the vertical radius of the original circle, TC, and the other two being the equidistance chords, a and b, running from the top of the vertical "book" column's back and leading directly to the two wing terminals, A and B, of the first folded diameter of the original circle. The weight of the paper on either side of the vertical fold extended on only one side of any line produces weight or gravitational effect to keep the vertical edge vertical, not allowing it to lean farther in the direction of the legs due to the relative structural rigidity of the paper itself.

835.04 We will now take five additional pieces of paper, making six in all, producing the circles on each of the same radius with our dividers welded and using the scissor function of the dividers' cuttingly ground straightedges. We cut the circles out and fold them in the manner already described to produce the vertically standing, angular interaction of the four quadrants of paper, standing as a vertically edged tripod with double thicknesses of the paper in arced flanges acting as legs to stabilize their verticality. We now have six such assemblies. We can take any two of them that are standing vertically and bring the vertical edges of their tripods together. (We know that they are the same size and that the vertical hinges are dimensionally congruent because they are all of the same radius length produced by the dividers.) We move two of their vertically folded edges into tangential congruence, i.e., back-to-back. The vertical perimeter terminals of their vertically folded hinges and their circle-center origins at the bottom of the hinges are congruent.

835.05 To hold their vertical hinges together and to free our hands for other work, we slip a bobby pin over their four thicknesses of paper, holding their two angles together in the pattern of a cross as viewed vertically from above or below. This construction produces a *quadripod*. Now I can grasp this cross between two of my fingers inserted into the angles of the cross and lift it from the table, turning it upside down in my hands and finding the other side of the cross, all four lines of which are in the same (approximately flat) plane, in contradistinction to the way the cross looked when those four folded edges sat on the table and had four arcing lines running in four different directions from their vertical congruence. I will insert a bobby pin to hold together the cross at its folded-line intersection. With its flat cross down, it will now stand as the partial profiling of a hemisphere. When I put the arced cross down on the table, it will roll around as would half a wooden ball. Placing it on the table in this roll-around hemisphere attitude, I can stabilize it with underprops so that the plane of four folded edges coming together on top will be approximately horizontal and parallel to the table top.

835.06 I may now take one of the four additional quadri-folded, partially opened, hinged, quarter-circle, double-thickness assemblies first described as able to stand vertically by themselves. Each of the four can be made to stand independently with one of its 90-degree, quarter-circle wings lying horizontally on the table and its other quarter-fold wing standing approximately vertically. The four quarter-circles on the table can be slid together to form a whole circle base; bobby pins can be inserted at their four circumference terminals to lock them together in a circle; and their four approximately vertical flanges can be hinged into true verticality so that they form two half-circle arcs, passing through one another perpendicularly to one another. They will have a common vertical radius (by construction) at the common top terminal, and all of their four vertical hinges' two crossing bobby pins can be inserted to lock this vertex together. This assembly of four of the six units with circular base can now be superimposed upon the first pair of hinges sitting on the table with hemisphere down and its planar cross up. The four cross ends of that first assembly can be hinged around into congruence with the 90-degree circumference points of the top assembled four units, with everything firmly congruent by construction.

835.07 We will now take bobby pins and fasten the folded flanking edge ends of the top-four assembly congruent with each of the four edges of the hemispherical cross group on the table.

835.08 Fastened by bobby pins at the congruent perimeter terminals of the folded cross lines, this top assembly stabilizes the previously unstable angular space between any two of the cross-forming hemispherical groups prop-stabilized, bowl down, with the plane of its four-way hinge cross horizontal and parallel to the table. The angle between any two of the horizontally crossed assembly members is now stabilized at exactly one-quarter of a circle by the integrity of construction procedures of our experiment. This produces one complete horizontal circle with 90-degree triangular webbing of double-folded paper perpendicular to two other perpendicularly intersecting vertical circles, each of which also consist of four 90-degree triangular webs of double-thickness paper, each of all 12 of which 90-degree triangular webs structurally stabilizes the six radial hinges of the three *XYZ* axes of this spherically profiled system assembly, prop-stabilized not to roll on the table.

835.09 In effect, we have the original six circular pieces reassembled with one another as two sets of three circles symmetrically intercepting one another. We know that each of the six quadrantly folded units fit into the remaining angular spaces because, by construction, each of the angles was folded into exactly one-quarter of a circle and folded together exactly to complete their circle. And we know that all the radial hinges fit together because they are constructionally of equal length. We have now a triangularly stabilized structure constituting what is called the *spherical octahedron.* Its vertical axis has polar terminals we call north and south. South is congruent with the table, and north is at the apex of the assembly. It has four equatorial points lying in a plane horizontal to the table. It is called the spherical octahedron because it has an external pattern subdivided exactly, evenly, and symmetrically by eight spherical triangles, four in the northern hemisphere and four in the southern hemisphere.

835.10 We find that the construction has three distinct planes that are all symmetrical and perpendicular to one another; the horizontal equatorial plane and two vertical planes intersect each other on the north-south polar axis perpendicularly to one another, which perpendicularity is constructurally inherent. Each of the perpendicularly intersecting great circles is seen to be of a double thickness due to the folding of the six original paper great circles, which now appear, deceptively, as three, but are not continuous planes, being folded to make their hinges congruent. (See Illus. 835.10.)

835.11 The spherical octahedron provides the basis for the frame of reference of the constructionally proven verticality of its axis in respect to its equa-

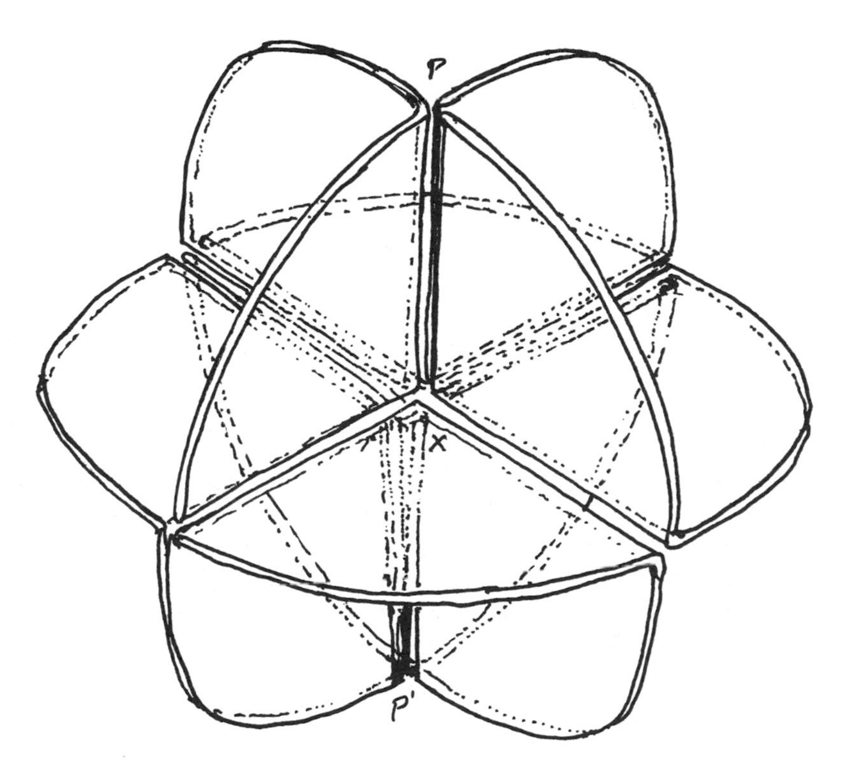

Fig. 835.10 *Six Great Circles Folded to Form Octahedron.*

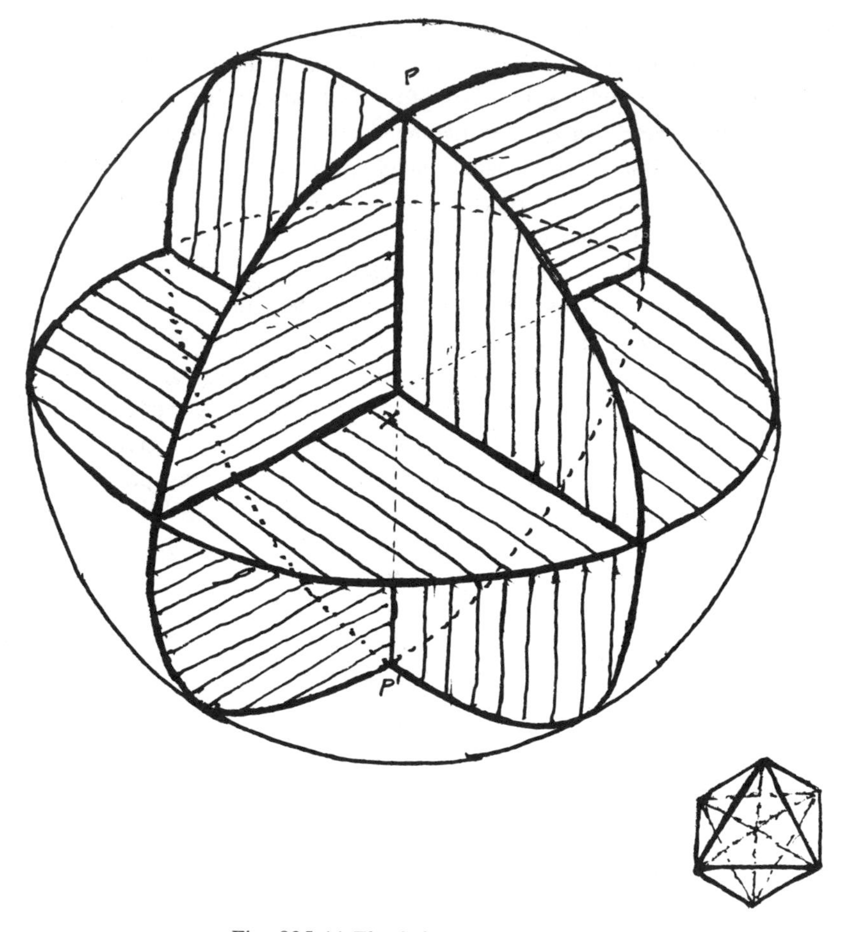

Fig. 835.11 *The Spherical Octahedron.*

torial plane and the equidistance of the poles from all the perimeter points. (See Illus. 835.11.)

835.12 As we rotate this octahedron rapidly on any one of its three axes, the rotated perimeters generate optically what can be called a dynamically generated true sphere. By construction, every point on the sphere's dynamically high-frequency event-occurring is equidistant from the central origin—our initial scribing position of one end of the dividers whose central angle we locked by welding it into unalterability.

836.00 *Spherical Octahedron: Alternate Assembly*

836.01 In addition to the foregoing operational development of the octahedron, we discover that the assembling of the spherical octahedron called for a fundamental asymmetry of procedure. That is, we assemble two of its quadrantly folded great circles to form one hemisphere and four of the quadrantly folded great circles to form the other hemisphere. In this method, the equator has to be included in either the northern or the southern hemisphere.

836.02 Therefore, in attempting to find some other method of assembly, we find that the spherical octahedron can be alternatively assembled in three sets of two quadrantly folded great circles. This is done by following all of the general procedures for developing the six quadrantly folded circles and their stand-up-ability as open book backs, producing a tripoding stability with the angle hinged by the variability of the vertical book-back spine.

836.03 We will now make three pairs of these variantly angled, quadrantly folded circles. We find that instead of standing one of them as a book with its hinge-spine vertical, the book can be laid with one of its faces parallel to the table and the other pointing approximately vertically, outward from the table. Due to the relative inflexibility of the double-folded angle of the great-circle construction paper, the book can be laid on its front face or its back face. We will take two of them sitting on the front face of the book with their backs reaching outwardly, vertically away from the table. We move two in front of us, one right and one left. We will rotate the right-hand one counterclockwise, 90 degrees around its vertical axis. Then we move the quadrant angle of thc right-hand one into congruence with the quadrant angle of the left-hand one. We stabilize the variable angle of the right-hand one between its vertical and horizontal parts by fastening with two bobby pins the constructionally produced stable quadrant of the folded parts of the right-hand unit. This gives us a constructionally proven one-eighth of a sphere in an asymmetrical assembly, having the 180-degree axis of the sphere lying congruent with the table. On one end of the axis, we have the stabilized quadrant angle; on the other end of the axis, we have the open, unstabilized angle.

836.04 With the other four of our six quadrantly folded circles, we make two more paired assemblies in exactly the same manner as that prescribed for the first paired assembly. We now have three of these assemblies with their axes lying on the table; on the left-hand side of all three, there will be found the stabilized, spherical-octant triangle. On the left side, there is a folded quadrant, where the angle between the vertical axis of the spherical octant is approximately 90 degrees from the folded axis lying on the table—but an unstabilized 90 degrees; it can be stabilized into 90-degreeness by virtue of the fact that both of its open folded edges are radii of the sphere by construction and have an accommodating, open hinge-line. We notice then that the three axes lying on the table, as the interior edge of the semicircle of double-ply folded paper, represent the three *XYZ* axes of the octahedron as well as the *XYZ* 90-degree coordinates of the international scientific standards of comprehensive mensuration—as, for instance, the *X* axis represents the height, *Y* the width, and *Z* the breadth. Geographically, this would represent the north and south poles and the four perpendicular quadrants of the equator.

836.05 Our operational-construction method employs the constant radius and identifies every point on the circumference and every point on the internal radii. This is in contradistinction to analytic geometry, in which the identification is only in terms of the *XYZ* coordinates and the perpendiculars to them. Analytic geometry disregards circumferential construction, ergo, is unable to provide for direct identification of angular accelerations.

836.06 These three subassemblies of the six folded quadrants are inherently asymmetrical. It was the fundamental asymmetry that made it possible to make the spherical octahedron with only three whole circles of paper, but we found it could only be accomplished symmetrically with six quadrantly folded great circles, with the symmetry being provided by the duality, by the *twoness*.

836.07 All three assemblies are identically asymmetric. The *loaded XYZ* axes hold the *Y* axis vertically. Pick up the *Y* axis and turn it 90 degrees to the *X* axis. This brings one of the stabilized quadrants of the *Y* axis into congruence with one of the nonstabilized quadrants of the *X* axis—to stabilize it. With the *Y* axis now at 90 degrees to the *X* axis, we can fasten the two assemblies into place with bobby pins.

836.08 Take the *Z* axis assembly and hold it so that it is perpendicular to both the *X* and *Y* axes; this will bring the three constructionally proven folded quadrangles into congruence with the three folded, as yet unstabilized, 90-degree sinuses of the *X* and *Y* axis assemblies.

840.00 Foldability of Four Great Circles of Vector Equilibrium

841.00 Foldability Sequence

841.11 Using the method of establishing perpendiculars produced by the overlapping of unit-radius circles in the first instance of the Greeks' exclusively one-planar initiation of their geometry (see Illus. 455.11), a diameter *PP′* perpendicular to the first straightedge constructed diameter *DD′* can be constructed. If we now fold the paper circles around *DD′* and *PP′*, it will be found that every time the circles are folded, the points where the perpendicular to that diameter intercept the perimeter are inherently congruent with the same perpendicular's diametrically opposite end.

841.12 The succession of positive and negative foldings in respect to the original plane folded around a plurality of diameters of that plane will define a sphere with inherent poles *P* and *P'*, which occur at the point of crossing of the rotated perpendiculars to the folded-upon diameters, the *PP'* points being commonly equidistant from the first prime, as yet unfolded circle cut out from the first piece of paper. This constructional development gives us a sphere with a polar axis *PP'* perpendicular to the original plane's circle at the center of that circle. We can also fold six great circles of unit radius, first into half-circle, 180-degree-arc units, and then halve-fold those six into 90-degree "bookends," and assemble them into a spherical octahedron with three axes, and we can rotate the octahedron around axis *PP'* and thus generate a spherical surface of uniform radii.

841.13 We could also have constructed the same sphere by keeping point *A* of the dividers at one locus in Universe and swinging point *B* in a multiplicity of directions around *A* (see Illus. 841.15). We now know that every point on the surface of an approximate sphere is equidistant from the same center. We can now move point *A* of the dividers from the center of the constructed sphere to any point on the surface of the sphere, but preferably to point *P* perpendicular to an equatorially described plane as in 841.11 and 841.12. And we can swing the free point *B* to strike a circle on the surface of the sphere around point *P*. Every point in the spherical surface circle scribed by *B* is equidistant chordally from *A*, which is pivotally located at *P*, that is, as an apparently straight line from *A* passing into and through the inside of the spherical surface to emerge again exactly in the surface circle struck by *B*, which unitary chordal distance is, by construction, the same length as the radius of the sphere, for the opening of our divider's ends with which we constructed the sphere was the same when striking the surface circle around surface point *A*.

841.14 We now select any point on the spherical surface circle scribed by point *B* of the dividers welded at its original radius-generating distance with which we are conducting all our exploration of the spheres and circles of this operational geometry. With point *A* of the dividers at the north-polar apex, *P'*, of the spherical octahedron's surface, which was generated by rotating the symmetrical assembly of six 90-degree, quadrangularly folded paper circles. Axis *PP'* is one of its three rectilinearly interacting axes as already constructively described.

841.15 We now take any point, *J*, on the spherical surface circle scribed by the divider's point *B* around its rotated point at *P*. We now know that *K* is equidistant chordally from *P* and from the center of the sphere. With point *A* of our dividers on *J*, we strike point *K* on the same surface circle as *J*, which makes *J* equidistant from *K*, *P*, and *X*, the center of the sphere. Now we know by construction integrity that the spherical radii *XJ, XK,* and *XP* are the same length as one another and as the spherical chords *PK, JK,* and *JP*. These six equilength lines interlink the four points *X, P, J,* and *K* to form the regular equiedged tetrahedron. We now take our straightedge and run it chordally from point *J* to another point on the same surface circle on which *JK* and *K* are situated, but diametrically opposite *K*. This diametric positioning is attained by having the chord-describing straightedge run inwardly of the sphere and pass through the axis *PP'*, emerging from the sphere at the surface-great-circle point *R*. With point *A* of the dividers on point *R* of the surface circle—on which also lies diametrically point *K*—we swing point *B* of the dividers to strike point *S* also on the same spherical surface circle around *P*, on which now lie also the points *J, K* and *R*, with points diametrically opposite *J*, as is known by construction derived information. Points *R, S, P,* and *X* now describe another regular tetrahedron equiedged with tetrahedron *JKPX;* there is one common edge, *PX,* of both tetrahedra. *PX* is the radius of the spherical, octahedrally constructed sphere on whose surface the circle was struck around one of its three perpendicularly intersectioned axes, and the three planes

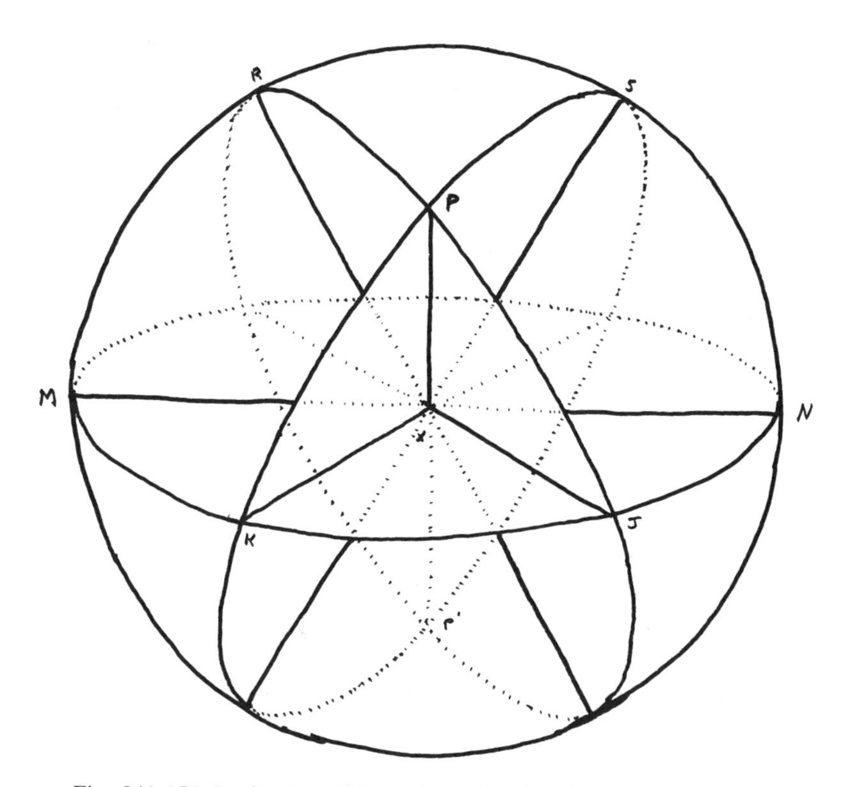

Fig. 841.15A *Realization of Four Great Circles of Vector Equilibrium.*

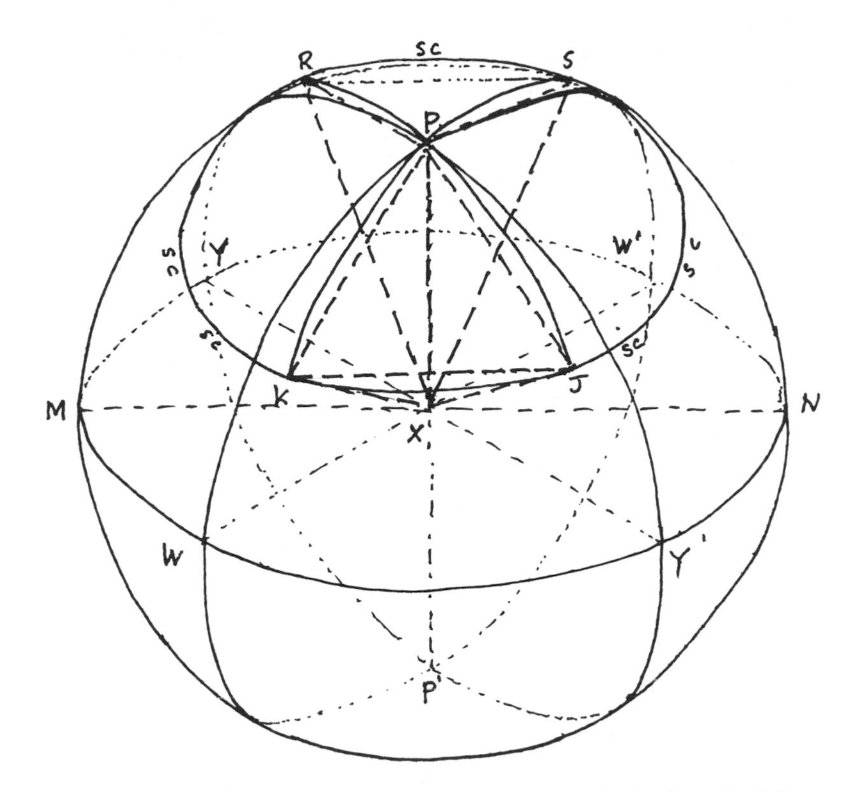

Fig. 841.15B *Fixed Radius Striking a Small Circle on the Surface of a Sphere.*

through them intersect congruently with the three axes by construction. *PX* is perpendicular to the equatorial plane passing through *W, Y, W', Y'* of the spherical octahedron's three axes *PP', WW',* and *YY'*.

841.16 We may now take the congruent radius edge *PX* of the two tetrahedra and separate it into the two radii $(PX)^1$ and $(PX)^2$ and rotate their two *P* ends (*PM*) and (*PN*), away from one another around the sphere's center, *X*, until (*PM*) and (*PX*) are diametrically opposite one another. Therefore, points (*PM*) and (*PN*) are now lying in the octahedron's equatorial plane *WXW'Y'*. We may now rotate points *J, K, R,* and *S* around the (*PM*), *X* (*PN*) axis until points *J, K, R,* and *S* all lie in the octahedral plane *WY W'X'*, which converts the opened unitary construction first into a semifolded circle and then into a circle congruent with the octahedron's equatorial plane, all of which six-hinged transformation was permitted as all the seven points—(*PM*), *J, K,* (*PN*), *S, R, X*—were at all times equidistant from one another, with no restraints placed on the motion. We now have the hexagonally divided circle as a constructionally proven geometrical relationship; and therefore we have what the Greeks could not acquire: i.e., a trisected 180-degree angle; ergo a six-equiangular subdivision of spherical unity's 360 degrees into 60-degree omniequiangularity; ergo a geometrically proven isotropic vector matrix operational evolvement field.

841.17 With our operationally considerate four tools of divider, straightedge and scriber and measurably manipulable scribable system as a material object (in this case, a sheet of paper; later, four sheets of paper), and with our constructionally proven symmetrical subdivision of a circle into six equilateral triangles and their six chord-enclosed segments, we now know that all the angles of the six equilateral triangles around center *X* are of 60 degrees; ergo, the six triangles are also equiangular. We know that the six circumferential chords are equal in length to the six radii. This makes it possible to equate rationally *angular* and *linear* accelerations, using the unit-radius chord length as the energy-vector module of all physical-energy accelerations. We know that any one of the 12 lines of the equilaterally triangled circle are always either in 180-degree extension of, or are parallel to, three other lines. We may now take four of these hexagonally divided circles of paper. All four circular pieces of paper are colored differently and have different colors on their opposite faces; wherefore, there are eight circular faces in eight colors paired in opposite faces, e.g., red and orange, yellow and green, blue and violet, black and white.

841.18 We will now take the red-orange opposite-faced construction-paper circle. We fold it first on its (*PM*)-(*PN*) axis so that the red is hidden inside and we see only an orange half-circle's two-ply surface. We next unfold it again, leaving the first fold as a crease. Next we fold the circle on its *RX* axis so that the orange face is inside and the red is outside the two-ply, half-circled foldup. We unfold again, leaving two crossing, axially folded creases in the paper. We next fold the same paper circle once more, this time along its *JS* axis in such a manner that the orange is inside and once again only the red surface is visible, which is the two-ply, half-circle folded condition.

841.19 We now unfold the red-orange opposite-faced colored-paper circle, leaving two positive and one negative creases in it. We will find that the circle of paper is now inclined by its creases to take the shape of a double tetrahedron *bow tie,* as seen from its openings end with the orange on the inside and the red on the outside. We may now insert a bobby pin between points (*PM*) and (*PN*), converting this hexagonally subdivided and positively–negatively folded circle back into the mutually congruent *PX* edge, two (hinge-bonded, bivalent) tetrahedra: *JKPX* and *RSPX*.

841.20 We may now fold the other three circles into similar, edge-bonded, tetrahedral bow-tie constructions in such a manner that number two is yellow outside and green inside; number three of the 60-degree-folded bow ties is blue outside and violet inside; and the fourth bow tie, identical to the other three bow ties' geometrical aspects of 60-degree equiangularity and equiradius chord edges, is black outside and white inside.

841.21 We may take any two of these bow ties—say, the orange inside and the green inside—and fasten each of their outside corners with bobby pins, all of their radii being equal and their hinges accommodating the interlinkage.

841.22 Each of these paired bow-tie assemblies, the orange-green insiders and the violet-white insiders, may now be fastened bottom-to-bottom to each other at the four external fold ends of the fold cross on their bottoms, with those radial crosses inherently congruent. This will reestablish and manifest each of the four original circles of paper, for when assembled symmetrically around their common center, they will be seen to be constituted of four great circles intersecting each other through a common center in such a manner that only two circular planes come together at any other than their common center

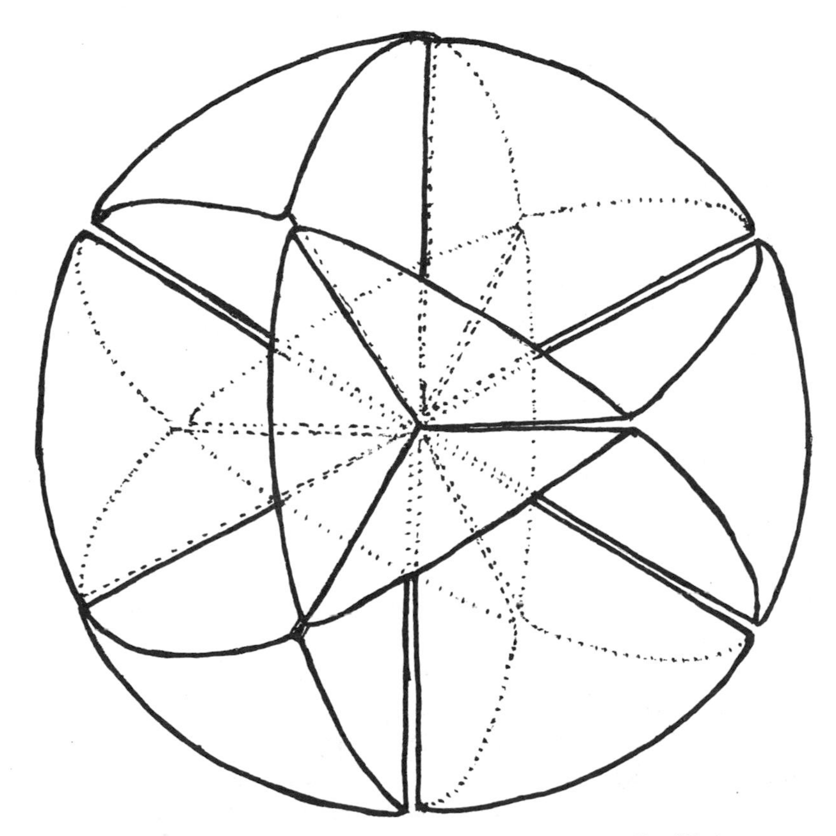

Fig. 841.22 *Foldability of Four Great Circles of Vector Equilibrium.*

point and in such a manner that each great circle is divided entirely into six equilateral triangular areas, with all of the 12 radii of the system equilengthed to the 24 circumferential chords of the assembly. Inasmuch as each of the 12 radii is shared by two great circle planes, but their 24 external chords are independent of the others, the seeming loss of 12 radii of the original 24 is accounted for by the 12 sets of congruent pairs of radii of the respective four hexagonally subdivided great circles. This omniequal line and angle assembly, which is called the vector equilibrium, and its radii-chord vectors accommodate rationally and simultaneously all the angular and linear acceleration forces of physical Universe experiences.

842.00 Generation of Bow Ties

842.01 When we consider the "jitterbug" vector equilibrium contracting into the icosahedron, bearing in mind that it is all double-bonded, we discover that, when the jitterbug gets to the octahedron phase, there really are two octahedra there. . . . Just as when you get three great circles, each one is doubled so that there are really six. . . . In making my tests, taking whole great circles of paper, doing my spherical trigonometry, learning the central angles, making those bow ties as a complex, which really amounts to tetrahedra bonded edge-to-edge with a common center, they link up as a chain and finally come together to make the icosahedron in a very asymmetrical manner. The 10, 12, and 15 great circles re-establish themselves, and every one of them can be folded.

842.02 You cannot make a spherical octahedron or a spherical tetrahedron by itself. You can make a spherical cube with two spherical tetrahedra in the pattern of the six great circles of the vector equilibrium. It becomes a symmetrically triangulated cube. In fact, the cube is not structurally stabilized until each of its six unstable, square-based, pyramidal half-octahedra are subdivided respectively into two tetrahedra, because one tetrahedron takes care of only four of the eight vertexes. For a cube to be triangulated, it has to have two tetrahedra.

842.03 There is no way to make a single spherical tetrahedron: its 109° 28′ of angle cannot be broken up into 360-degree-totaling spherical increments. The tetrahedron, like the octahedron, can be done only with two tetrahedra in conjunction with the spherical cube in the pattern of the six great circles of the vector equilibrium.

842.04 Nor can we project the spherical octahedron by folding three whole great circles. The only way you can make the spherical octahedron is by making the six great circles with all the edges double—exactly as you have them in the vector equilibrium—as a strutted edge and then it contracts and becomes the octahedron.

842.05 There is a basic cosmic sixness of the two sets of tetrahedra in the vector equilibrium. There is a basic cosmic sixness also in an octahedron minimally-great-circle-produced of six great circles; you can see only three because they are doubled up. And there are also the six great circles occurring in the icosahedron. All these are foldable of six great circles which can be made out of foldable disks.

842.06 This sixness corresponds to our six quanta: our six vectors that make one quantum.

842.07 There are any number of ways in which the energy can go into the figure-eight bow ties or around the great circle. The foldability reveals holdings patterns of energy where the energy can go into local circuits or go through the points of contact.

900.00 Modelability

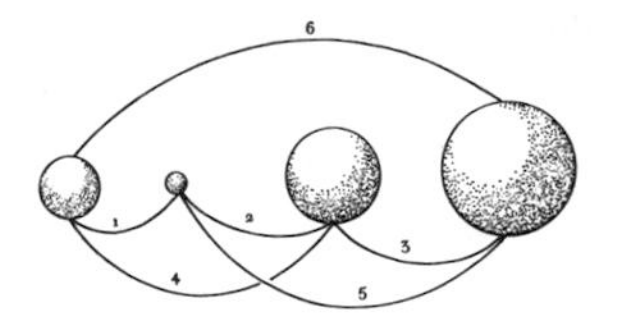

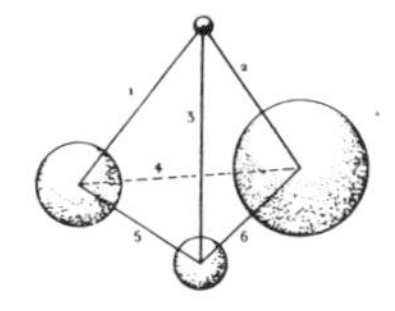

901.00 Basic Disequilibrium LCD Triangle

901.01 Definition

901.02 The Basic Disequilibrium 120 LCD Spherical Triangle of synergetics is derived from the 15-great-circle, symmetric, three-way grid of the spherical icosahedron. It is the lowest common denominator of a sphere's surface, being precisely $^{1}/_{120}$th of that surface as described by the icosahedron's 15 great circles. The trigonometric data for the Basic Disequilibrium LCD Triangle includes the data for the entire sphere and is the basis of all geodesic dome calculations. (See Sec. 612.)

901.03 As seen in Sec. 610.20 there are only three basic structural systems in Universe: the tetrahedron, octahedron, and icosahedron. The largest number of equilateral triangles in a sphere is 20: the spherical icosahedron. Each of those 20 equiangular spherical triangles may be subdivided equally into six right triangles by the perpendicular bisectors of those equiangular triangles. The utmost number of geometrically similar subdivisions is 120 triangles, because further spherical-triangular subdivisions are no longer similar. The largest number of similar triangles in a sphere that spheric unity will accommodate is 120: 60 positive and 60 negative. Being spherical, they are positive and negative, having only common arc edges which, being curved, cannot hinge with one another; when their corresponding angle-and-edge patterns are vertex-mated, one bellies away from the other: concave or convex. When one is concave, the other is convex. (See Illus. 901.03 and drawings section.)

901.04 We cannot further subdivide the spherical icosahedron's equiangular triangles into similar, half-size, equiangular triangles, but we can in the planar icosahedron. When the sides of the triangle in the planar icosahedron are bisected, four similar half-size triangles result, and the process can be continued indefinitely. But in the spherical icosahedron, the smaller the tri-

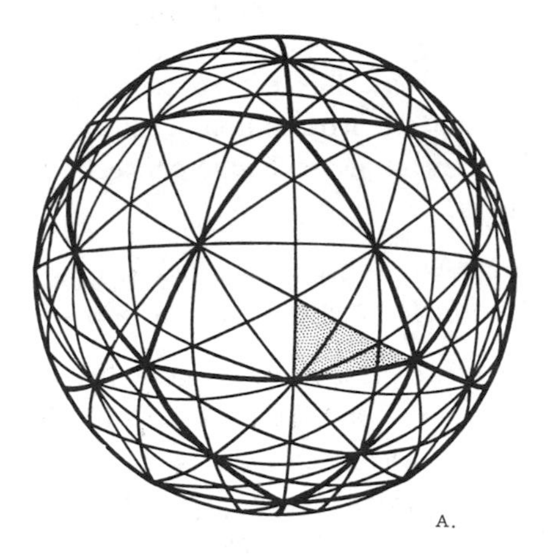

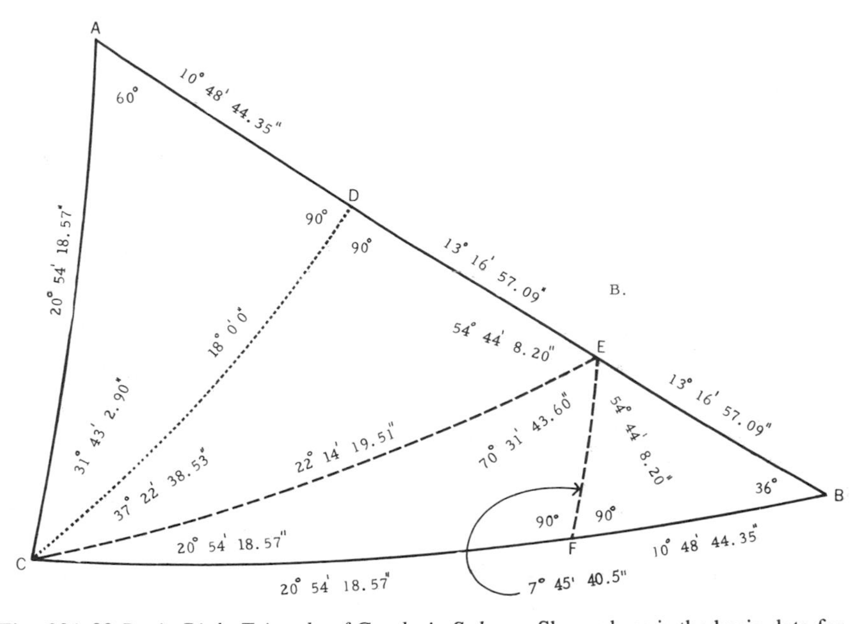

Fig. 901.03 *Basic Right Triangle of Geodesic Sphere:* Shown here is the basic data for the 31 great circles of the spherical icosahedron, which is the basis for all geodesic dome calculations. The basic right triangle as the lowest common denominator of a sphere's surface includes all the data for the entire sphere. It is precisely $^1/_{120}$th of the sphere's surface and is shown as shaded on the 31-great-circle-sphere (A). An enlarged view of the same triangle is shown (B) with all of the basic data denoted. There are three different external edges and three different internal edges for a total of six different edges. There are six different internal angles other than 60° or 90°. Note that all data given is spherical data, i.e. edges are given as central angles and face angles are for spherical triangles. No chord factors are shown. Those not already indicated elsewhere are given by the equation $2 \sin \theta/2$, where θ is the central angle. Solid lines denote the set of 15 great circles. Dashed lines denote the set of 10 great circles. Dotted lines denote the set of 6 great circles.

angle, the less the spherical excess; so the series of triangles will not be similar. Each corner of the icosahedron's equiangular triangles is 72 degrees; whereas the corners of its mid-edge-connecting triangle are each approximately 63 degrees.

901.10 Geodesic Dome Calculations

901.11 When two great-circle geodesic lines cross, they form two sets of similar angles, any one of which, paired with the other, will always add to 180°. (This we also learned in plane geometry.) When any one great circle enters into—or exits from—a spherical triangle, it will form the two sets of similar angles as it crosses the enclosing great-circle-edge-lines of that triangle.

901.12 As in billiards or in electromagnetics, when a ball or a photon caroms off a wall it bounces off at an angle similar to that at which it impinged.

901.13 If a great-circle-describing, inexhaustibly re-energized, satellite ball that was sufficiently resilient to remain corporeally integral, were suddenly to encounter a vertical, great-circle wall just newly mounted from its parent planet's sphere, it would bounce inwardly off that wall at the same angle that it would have traversed the same great-circle line had the wall not been there. And had two other great-circle walls forming a right spherical triangle with the first wall been erected just as the resilient ball satellite was hitting the first great-circle wall, then the satellite ball would be trapped inside the spherical-triangle-walled enclosure, and it would bounce angularly off the successively encountered walls in the similar-triangle manner unless it became aimed either at a corner vertex of the triangular wall trap, or exactly perpendicularly to the wall, in either of which cases it would be able to escape into the next spherical area lying 180° ahead outside the first triangle's walls.

901.14 If, before the satellite bouncingly earned either a vertexial or perpendicular exit from the first-described spherical triangle (which happened to be dimensioned as one of the 120 LCD right triangles of the spherical icosahedron) great-circle walls representing the icosahedron's 15 complete great circles, were erect—thus constructing a uniform, spherical, wall patterning of 120 (60 positive, 60 negative) similar spherical, right triangles—we would find the satellite sphere bouncing around within one such spherical triangle at exactly the same interior or exiting angles as those at which it would have crossed, entered into, and exited, each of those great-circle boundaries of those 120 triangles had the wall not been so suddenly erected.

901.15 For this reason the great-circle interior mapping of the symmetrically superimposed other sets of 10 and 6 great circles, each of which—

together with the 15 original great circles of the icosahedron—produces the 31 great circles of the spherical icosahedron's total number of symmetrical *spinnabilities* in respect to its 30 mid-edge, 20 mid-face, and 12 vertexial poles of half-as-many-each axes of spin. (See Sec. 457.) These symmetrically superimposed, 10- and 6-great-circles subdivide each of the disequilibrious 120 LCD triangles into four lesser right spherical triangles. The exact trigonometric patterning of any other great circles orbiting the 120-LCD-triangled sphere may thus be exactly plotted within any one of these triangles.

901.16 It was for this reason, plus the discovery of the fact that the icosahedron—among all the three-and-only prime structural systems of Universe (see Sec. 610.20)—required the least energetic, vectorial, structural investment per volume of enclosed local Universe, that led to the development of the Basic Disequilibrium 120 LCD Spherical Triangle and its multifrequenced triangular subdivisioning as the basis for calculating all high-frequency, triangulated, spherical structures and structural subportions of spheres; for within only one disequilibrious LCD triangle were to be found all the spherical chord-factor constants for any desired radius of omnisubtriangulated spherical structure.

901.17 In the same way it was discovered that local, chord-compression struts could be islanded from one another, and could be only tensionally and non-inter-shearingly connected to produce stable and predictably efficient enclosures for any local energetic environment valving uses whatsoever by virtue of the approximately unlimited range of frequency-and-angle, subtriangle-structuring modulatability.

901.18 Because the 120 basic disequilibrious LCD triangles of the icosahedron have $2^1/_2$ times less spherical excess than do the 48 basic equilibrious LCD triangles of the vector equilibrium, and because all physical realizations are always disequilibrious, the Basic Disequilibrium 120 LCD Spherical Triangles become most realizably *basic* of all general systems' mathematical control matrixes.

902.00 *Properties of Basic Triangle*

902.01 Subdivision of Equilateral Triangle: Both the spherical and planar equilateral triangles may be subdivided into six equal and congruent parts by describing perpendiculars from each vertex of the opposite face. This is demonstrated in Fig.902.01, where one of the six equal triangles is labeled to correspond with the Basic Triangle in the planar condition.

902.10 Positive and Negative Alternation: The six equal subdivision triangles of the planar equilateral triangle are hingeable on all of their adjacent

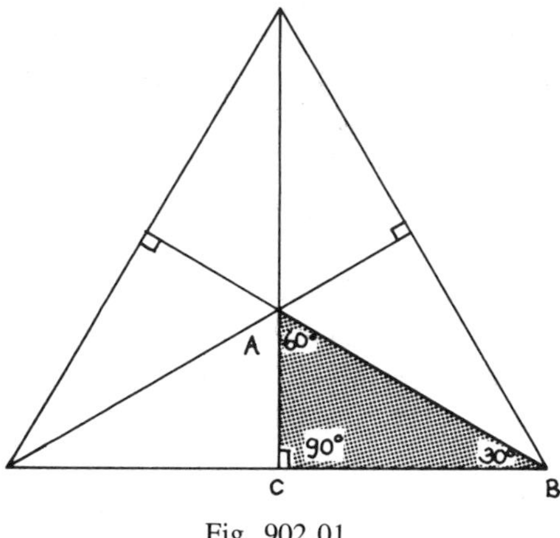

Fig. 902.01.

lines and foldable into congruent overlays. Although they are all the same, their dispositions alternate in a positive and negative manner, either clockwise or counterclockwise.

902.20 **Spherical Right Triangles:** The edges of all spherical triangles are arcs of great circles of a sphere, and those arc edges are measured in terms of their central angles (i.e., from the center of the sphere). But plane surface triangles have no inherent central angles, and their edges are measured in relative lengths of one of themselves or in special-case linear increments. Spherical triangles have three surface (corner) angles and three central (edge) an-

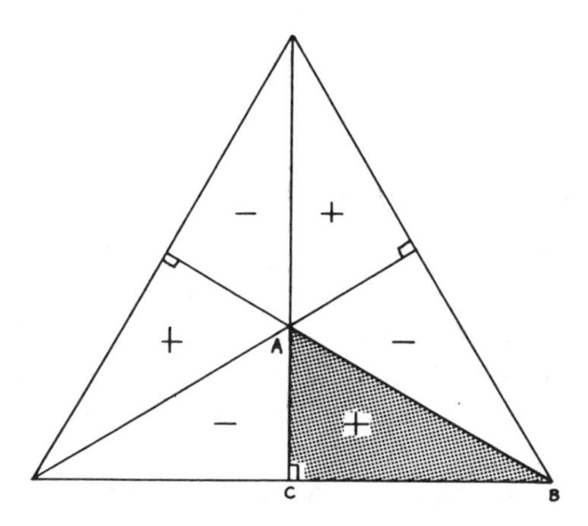

Fig. 902.10.

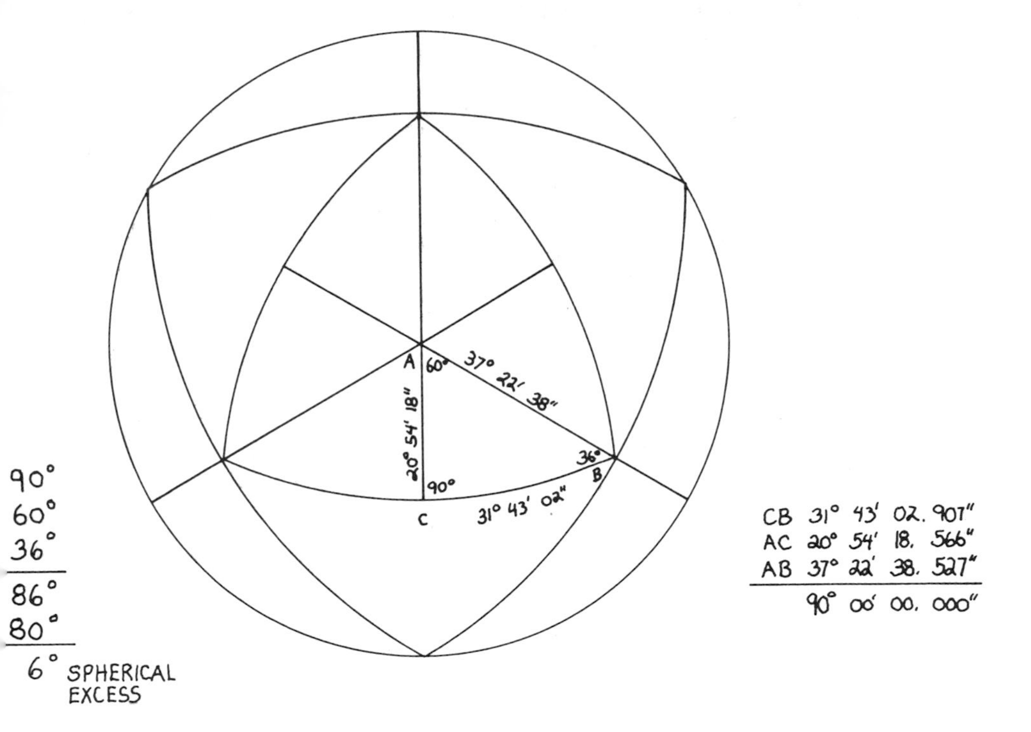

Fig. 902.20.

gles. The basic data for the central angles provided below are accurate to 1/1,000 of a second of arc. On Earth

1 nautical mile	= 1 minute of arc
1 nautical mile	= approximately 6,000 feet
1 second of arc	= approximately 100 feet
$^1/_{1,000}$ second of arc	= approximately $^1/_{10}$ foot
$^1/_{1,000}$ second of arc	= approximately 1 inch

These calculations are therefore accurate to one inch of Earth's arc.

902.21 The arc edges of the Basic Disequilibrium 120 LCD Triangle as measured by their central angles add up to 90° as do also three internal surface angles of the triangle's *ACB* corner:

$BCE = 20°\ 54'\ 18.57'' = ECF$
$ECD = 37°\ 22'\ 38.53'' = DCE$
$DCA = 31°\ 43'\ 02.9'' = ACD$
$\overline{90°\ 00'\ 00''}$

902.22 The spherical surface angle *BCE* is exactly equal to two of the arc edges of the Basic Disequilibrium 120 LCD Triangle measured by their central angle. *BCE* = arc *AC* = arc *CF* = 20° 54′ 18.57″.

902.30 **Surface Angles and Central Angles:** The Basic Triangle *ACB* can be folded on the lines *CD* and *CE* and *EF*. We may then bring *AC* to coincide with *CF* and fold *BEF* down to close the tetrahedron, with *B* congruent with *D* because the arc *DE* = arc *EB* and arc *BF* = arc *AD*. Then the tetrahedron's corner *C* will fit exactly down into the central angles *AOC, COB,* and *AOB*. (See Illus. 901.03 and 902.30.)

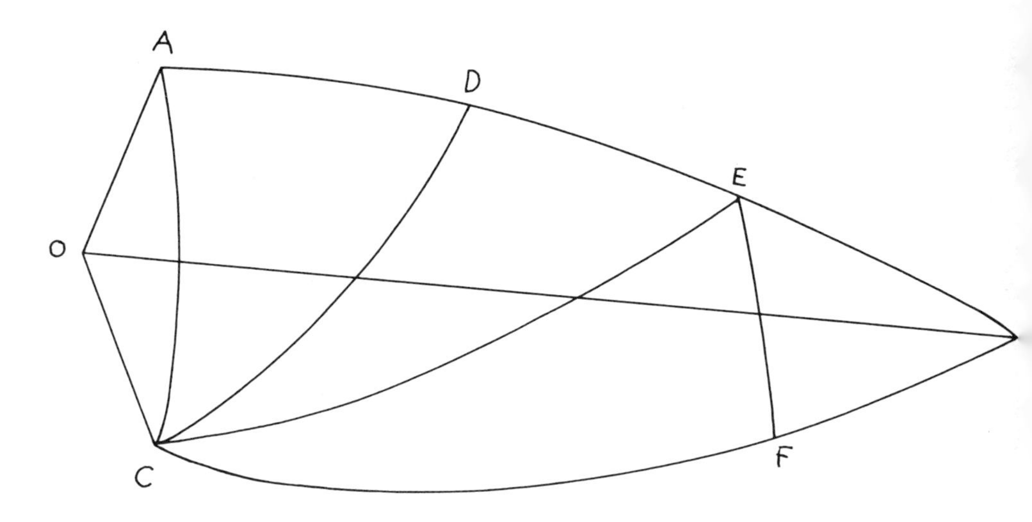

Fig. 902.30.

902.31 As you go from one sphere-foldable great-circle set to another in the hierarchy of spinnable symmetries (the 3-, 4-, 6-, 12-sets of the vector equilibrium's 25-great-circle group and the 6-, 10-, 15-sets of the icosahedron's 31-great-circle group), the central angles of one often become the surface angles of the next-higher-numbered, more complex, great-circle set while simultaneously some (but not all) of the surface angles become the respective next sphere's central angles. A triangle on the surface of the icosahedron folds itself up, becomes a tetrahedron, and plunges deeply down into the congruent central angles' void of the icosahedron (see Sec. 905.47).

902.32 There is only one noncongruence: the last would-be hinge, *EF* is an external *arc* and cannot fold as a straight line; and the spherical surface angle *EBF* is 36 degrees whereas a planar 30 degrees is called for if the surface is cast off or the arc subsides chordally to fit the 90-60-30 right plane triangle.

902.33 The 6 degrees of spherical excess is a beautiful whole, rational number excess. The 90-degree and 60-degree corners seem to force all the excess into one corner, which is not the way spherical triangles subside. All the angles lose excess in proportion to their interfunctional values. This particular condition means that the 90 degrees would shrink and the 60 degrees would shrink. I converted all the three corners into seconds and began a proportional decrease study, and it was there that I began to encounter a ratio that seemed rational and had the number 31 in one corner. This seemed valid as all the conditions were adding up to 180 degrees or 90 degrees as rational wholes even in both spherical and planar conditions despite certain complementary transformations. This led to the intuitive identification of the Basic Disequilibrium 120 LCD Triangle's foldability (and its fall-in-ability into its own tetra-void) with the *A* Quanta Module, as discussed in Sec. 910, which follows.

905.00 Equilibrium and Disequilibrium Modelability

905.01 **Tetrahedron as Model:** Synergetics is the geometry of thinking. How we think is epistemology, and epistemology is modelable; which is to say that knowledge organizes itself geometrically, i.e., with models.

905.02 Unity as two is inherent in life and the resulting model is tetrahedral, the conceptuality of which derives as follows:

—life's inherent unity is two;
—no otherness = no awareness;
—life's awareness begins with otherness;
—otherness is twoness;
—*this* moment's awareness is different from previous awareness;
—differentiations of time are observed directionally;
—directions introduce vectors (lines);
—two time lines demonstrate the observer and the observed;
—the interconnection of two lines results in a tetrahedron;
—sixfold interrelatedness is conceptual:

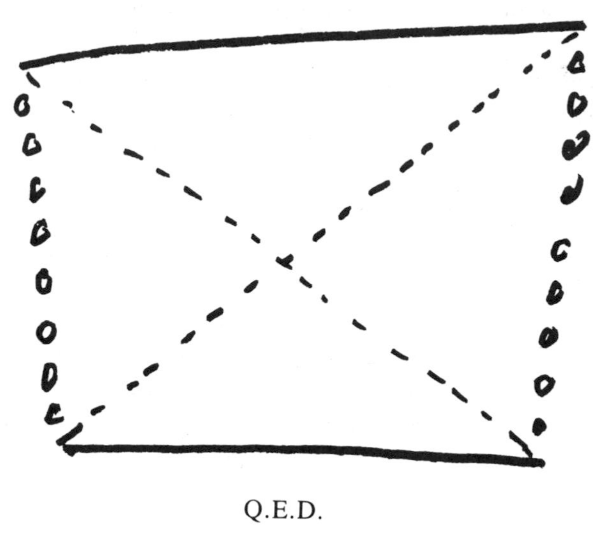

Q.E.D.

Fig. 905.02.

905.10 Doubleness of Unity

905.11 The prime number twoness of the octahedron always occurs in structuring doubled together as four—i.e., 2^2—a fourness which is also doubleness of unity. Unity is plural and, at minimum, is two. The unity volume 1 of the tetrahedron is, in structural verity, two, being both the outwardly displayed convex tetrahedron and the inwardly contained concave tetrahedron. (See Chart 223.64, columns 2, 12, and 15)

905.12 The three-great-circle model of the spherical octahedron only "seems" to be three; it is in fact "double"; it is only foldably produceable in unbroken (whole) great-circle sheets by edge-combining *six* hemicircularly folded whole great circles (see Sec. 840). Thus it is seen that the octahedron—as in Iceland spar crystals—occurs only doubly, i.e., omnicongruent with itself, which is "quadrivalent."

905.13 Among the three possible omnisymmetrical prime structural systems—the tetrahedron, octahedron, and icosahedron—only the tetrahedron has each of its vertexes diametrically opposite a triangular opening. (See Illus. 610.20.) In the octahedron and icosahedron, each vertex is opposite another vertex; and each of their vertexes is diametrically blocked against articulating a self-inside-outing transformation. In both the octahedron and the icosahedron, each of the vertexes is tense-vector-restrained from escaping *outwardly* by the convergent vectorial strength of the system's other immediately surrounding—at minimum three—vertexial event neighbors. But contrariwise, each of the octahedron's and icosahedron's vertex events are constrainingly impulsed *inwardly* in an exact central-system direction and thence impelled toward diametric exit and inside-outing transformation; and their vertex events would do so were it not for their diametrically opposed vertexes, which are surroundingly tense-vector-restrained from permitting such outward egress.

905.14 As a consequence of its uniquely unopposed diametric vertexing—ergo permitted—diametric exit, only the tetrahedron among all the symmetric polyhedra can turn itself pulsatingly inside-out, and can do so in eight different ways (see Sec. 624); and in each instance, as it does so, one-half of its combined concave-convex unity ''twoness'' is always inherently invisible.

905.15 The octahedron, however, restrainingly vector-blocked as described, can only infold itself pulsatingly to a condition of hemispherical congruence like a deflated basketball. Thus the octahedron's concave-convex, unity-twoness state remains plurally obvious. You can see the concave infolded hemisphere nested into the as-yet outfolded convex hemisphere. Verifying the octahedron's *fourness* as being an evolutionary transformation of the tetrahedron's unity-twoness, we may take the four triangles of the tetrahedron which were edge-hinged together (bivalently) and reassemble them univalently (that is, corner-to-corner) and produce the octahedron, four of whose faces are triangular (ergo structurally stable) voids. This, incidentally, introduces the structural stability of the *triangle* as a visualizable yet physical nothingness.

905.16 The triangle is structure. Structure is spontaneous pattern stabilization of a complex of six individual events. Structure is an integral of six events. Structure is a pattern integrity. Pattern integrity is conceptual relationship independent of size. The integrity of the nuclear structuring of the atoms is conceptually thinkable, as are the associability and disassociability proclivities of chemistry, virology, biology, and all nonbiological structuring and mechanics.

905.17 Any and all of the icosahedron's vertexes pulsate individually and independently from the convex to concave state only in the form of local dimpling, because each only-from-outward-motion-restrained vertex—being free to articulate inwardly toward its system center, and having done so—becomes abruptly five-vector-restrained by its immediate neighboring vertexial event convergences; and the abrupt halting of its inward travel occurs before it reaches the system center. This means that one vertex cannot pulse inwardly more deeply than a local dimple similar to the popping in of a derby hat. (See Sec. 618.30.)

905.18 Both the coexisting concave and convex aspects of the icosahedron—like those of the octahedron, but unlike those of the unique case of the tetrahedron—are always visually obvious on the inside and outside of the only locally dimpled-in, or nested-in, vertex. In both the octahedron and the icosahedron, the concave-convex, only inwardly pulsative self-transforming always produces visually asymmetrical transforming; whereas the tetrahedron's permitted inside-outing pulsatively results only in a visible symmetry, the quasi-asymmetry being invisibly polarized with the remainder of Universe outside the tetrahedron, which, being omniradially outward, is inferentially—but not visually—symmetrical; the only asymmetrical consideration of the tetrahedron's inside-outing being that of an initial direction of vertexial exiting. Once exited, the visible remaining symmetrical tetrahedron is in verity the inside-outness of its previously visible aspects. (See Sec. 232.01.)

905.19 In either of the two sets of four each as alternatively described, one of the polar states is always visible and the other complementarily invisible. This is a dynamic relationship. Dynamically, all four of each of the two sets of the tetrahedral potential are co-occurrently permitted and are required by omni-action-reaction-resultant synergetics. The seeming significance of the separately considered asymmetries are cancelled by the omnidirectional symmetry.

905.20 The vertexes are the unique, individual, ergo in-time events; and the nonvertex voids are the outdividual, ergo out, timeless, sizeless nonevents. The both outwardly and inwardly escaping nonevents complement the embryo, local-in-time, special-case, convergent-event, systemic pattern fixation of individual intercomplementary event identities. (See Sec. 524.)

905.21 *In* is unidirectional, pointable. *Out* is omnidirectional, unpointable—go out, to-go-out, or go-in-to-go-out on the other side. Any direction from *here* is out; only one direction from *here* is in. Go either temporarily in to go diametrically out on the other side of the individually identical local *in,* or go anydirectionally out . . . to the complete, eternal, unidentifiable, nonness, noneness of the a priori mysterious, integrally regenerative, inherently complex Universe.

905.22 So-called edges and vectors are inherently only convergent or divergent interrelationships between multiply-identifiable, point-to-able, vertex fixes.

905.23 Because each tetrahedron has both four vertexes and four subtending nonvertex voids, we can identify those four diametrically complementary sets of all minimal cosmic structural systems as the four visible vertexes and the four nonvisible nonvertexes, i.e., the triangularly symmetrical, peripheral voids. The tetrahedron thus introduces experientially the cosmic principle of the visible and invisible pairs or couples; with the nonvisible vertex as the inside-out vertex, which nonvertex is a nonconvergence of events; whereas the vertexes are visible event convergences.

905.30 Hierarchy of Pulsating Tetrahedral Arrays

905.31 Among the exclusively, three and only, prime cosmic structural systems—the tetra, octa, and icosa—only the tetrahedron's pulsative transforming does not alter its overall, visually witnessable symmetry, as in the case of the "cheese tetrahedron" (see Sec. 623). It is important to comprehend that any *one* of the two sets of four each potential vertexial inside-outing pulsatabilities of the tetrahedron—considered only by themselves—constitutes polarized, but only invisible, asymmetry. In one of the two sets of four each potential inside-outings we have three-of-each-to-three-of-the-other (i.e., trivalent, triangular, base-to-base) vertexial bonding together of the visible and invisible, polarized pair of tetrahedra. In the other of the two sets of four each potential inside-outings we have one-vertex-to-one-vertex (i.e., univalent, apex-to-apex) interbonding of the visible and invisible polarized pair of tetrahedra.

905.32 Because each simplest, ergo prime, structural system tetrahedron has at minimum four vertexes (point-to-able, systemic, event-patterned fixes), and their four complementary system exit-outs, are symmetrically identified at mid-void equidistance between the three other convergent event identity vertexes; and because each of the two sets of these four half-visible/half-invisible, polar-paired tetrahedra have both three-vertex-to-three-vertex as well as single-vertex-to-single-vertex inside-out pulsatabilities; there are eight possible inside-outing pulsatabilities. We have learned (see Sec. 440) that the vector equilibrium is the nuclear-embracing phase of all eight "empty state" tetrahedra, all with common, central, single-vertex-to-single-vertex congruency, as well as with their mutual outward-radius-ends' vertexial congruency; ergo the vector equilibrium is bivalent.

905.33 The same vector equilibrium's eight, nuclear-embracing, bivalent tetrahedra's eight nuclear congruent vertexes may be simultaneously outwardly pulsed through their radially-opposite, outward, triangular exits to form eight externally pointing tetrahedra, which thus become only univalently, i.e., only-single-vertex interlinked, and altogether symmetrically arrayed around the vector equilibrium's eight outward "faces." The thus-formed, eight-pointed star system consisting of the vector equilibrium's volume of 20 (tetrahedral unity), plus the eight star-point-arrayed tetrahedra, total volumetrically to 28. This number, 28, introduces the prime number *seven* factored exclusively with the prime number *two,* as already discovered in the unity-twoness of the tetrahedron's always and only, co-occurring, concave-convex inherently disparate, behavioral duality. This phenomenon may be compared with the 28-ness in the Coupler accounting (see Sec. 954.72).

905.34 We have also learned in the vector equilibrium jitterbugging that the vector equilibrium contracts symmetrically into the octahedral state, and we thus witness in the octahedron the eight tetrahedra—three-vertex-to-three-vertex (face-to-face, trivalent, triple-interbonded)—which condition elucidates the octahedron's having a volumetric *four* in respect to the tetrahedron's dual unity. Whereas the octahedron's prime number is *two* in respect to the tetrahedron's prime number *one,* it is experientially evidenced that the octahedron always occurs as both the double phase and the fourfold phase of the tetrahedron; i.e., as (a) the tetrahedral invisible/visible, (No-Yes), concave/convex; as well as (b) the octahedral visible/visible, (Yes-Yes), concave/convex: two different twoness manifestations. The tetrahedron has a unity-two duality in both its generalized dynamic potential and kinetic states, having always both the cosmic macro-tetrahedron and the cosmic micro-tetrahedron, both embracingly and inclusively defined by the four convergent event fixes of the minimum structural system of Universe. There is also the fundamental twoness of the tetrahedron's three sets of *two*-each, opposed, 90-degree-oriented edge-vectors whose respective four ends are always most economically omni-interconnected by the four other vectors of the tetrahedron's total of six edge-vectors.

905.35 The jitterbug shows that the bivalent vector equilibrium contracts to the octahedral trivalent phase, going from a twentyness of volume to a fourness of volume, 20→4, i.e., a 5:1 contraction, which introduces the prime number *five* into the exclusively tetrahedrally evolved prime structural system intertransformabilities. We also witness that the octahedron state of the jitterbug transforms contractively even further with the 60-degree rotation of one of its triangular faces in respect to its nonrotating opposite triangular face—wherewith the octahedron collapses into one, flattened-out, two-vector-length, equiedged triangle, which in turn consists of four one-vector-edged, equiangled triangles, each of which in turn consists of two congruent, one-vector-long, equiedged triangles. All eight triangles lie together as two congruent sets of four small, one-vector-long, equiedged triangles. This centrally congruent axial force in turn plunges the two centrally congruent triangles through the inertia of the three sets of two congruent, edge-hinged triangles on the three sides of the congruent pair of central triangles which fold the big triangle's corners around the central triangle in the manner of the three petals folding into edge congruence with one another to produce a tetrahedrally shaped flower bud. Thus is produced one tetrahedron consisting of four quadrivalently congruent tetrahedra, with each of its six edges consisting of four congruent vectors. The tetrahedron thus formed, pulsatively reacts by turning itself inside-out to produce, in turn, another quadrivalent, four-tetrahedra congruence; which visible-to-visible, quadrivalent tetrahedral inside-outing/outside-inning is pulsatively regenerative. (See Illus. 461.08.)

905.36 Herewith we witness both visible and heretofore invisible phases of each of the single tetrahedra thus pulsatively involved. The univalent, apex-to-apex-bonded, four tetrahedra and the three-point-to-three-point, trivalent, base-bonded, four tetrahedra are both now made visible, because what was visible to the point-to-point four was invisible to the three-point-to-three-point four, and vice versa.

905.37 In the two extreme limit cases of jitterbug contraction—both the positive-negative and the negative-positive phases—the two cases become alternately visible, which results in the invisible phase of either case becoming congruent with the other's invisible phase: ergo rendering both states *visible*.

905.38 This pulsating congruence of both the alternately quadrivalent visible phases of the limit case contractions of the vector equilibrium results in an octavalent tetrahedron; i.e., with all the tetrahedron's eight pulsative intertransformabilities simultaneously realized and congruently oriented.

905.39 This hierarchy of events represents a 28-fold volumetric contraction from the extreme limit of univalently coherent expandability of the ever-integrally-unit system of the eight potential pulsative phases of self-intertransformability of the tetrahedron as the minimum structural system of all Universe. In summary we have:

—the 28-volume univalent;
—the 20-volume bivalent;
—the 8-volume quadrivalent;
—the two sets of 1-volume quadrivalent; and finally,
—the complex limit congruence of the 1-volume octavalent tetrahedron.

905.40 As we jitterbuggingly transform contractively and symmetrically from the 20-volume bivalent vector equilibrium phase to the 8-volume quadrivalent octahedral phase, we pass through the icosahedral phase, which is non-selfstabilizing and may be stabilized only by the insertion of six additional external vector connectors between the 12 external vertexes of the vector equilibrium travelling toward convergence as the six vertexes of the trivalent 4-volume octahedron. These six vectors represent the edge-vectors of one tetrahedron.

905.41 The 28-volume, univalent, nucleus-embracing, tetrahedral array extends its outer vertexes beyond the bounds of the nucleus-embracing, closest-packed, omnisymmetrical domain of the 24-volume cube formed by superimposing eight Eighth-Octahedra, asymmetrical, equiangle-based, three-convergent-90-degree-angle-apexed tetrahedra upon the eight outward equiangular triangle facets of the vector equilibrium. We find that the 28-ness of free-space expandability of the univalent, octahedral, nucleus embracement must lose a volume of 4 (i.e., four tetrahedra) when subjected to omni-closest-packing conditions. This means that the dynamic potential of omni-interconnected tetrahedral pulsation system's volumetric embracement capability of 28, upon being subjected to closest-packed domain conditions, will release an elsewhere-structurally-investable volume of 4. Ergo, under closest-packed conditions, each nuclear array of tetrahedra (each of which is identifiable energetically with one energy quantum) may lend out four quanta of energy for whatever tasks may employ them.

905.42 The dynamic vs. kinetic difference is the same difference as that of the generalized, sizeless, metaphysically abstract, eternal, constant sixness-of-edge, fourness-of-vertex, and fourness-of-void of the only-by-mind-conceptual tetrahedron vs. the only-in-time-sized, special-case, brain-sensed tetrahedron. This generalized quality of being dynamic, as being one of a plurality of inherent systemic conditions and potentials, parts of a whole set of eternally co-occurring, complex interaccommodations in pure, weightless, mathematical principle spontaneously producing the minimum structural systems, is indeed the prime governing epistemology of wave quantum physics.

905.43 In consideration of the tetrahedron's quantum intertransformabilities, we have thus far observed only the expandable-contractable, variable-bonding-permitted consequences. We will now consider other dynamical potentials, such as, for instance, the axial rotatabilities of the respective tetras, octas, and icosas.

905.44 By internally interconnecting its six vertexes with three polar axes: *X, Y,* and *Z,* and rotating the octahedron successively upon those three axes, three planes are internally generated that symmetrically subdivide the octahedron into eight uniformly equal, equiangle-triangle-based, asymmetrical tetrahedra, with three convergent, 90-degree-angle-surrounded apexes, each of whose volume is one-eighth of the volume of one octahedron: this is called the Eighth-Octahedron. (See also Sec. 912.) The octahedron, having a volume of four tetrahedra, allows each Eighth-Octahedron to have a volume of one-half of one tetrahedron. If we apply the equiangled-triangular base of one each of these eight Eighth-Octahedra to each of the vector equilibrium's eight equiangle-triangle facets, with the Eighth-Octahedra's three-90-degree-angle-surrounded vertexes pointing outwardly, they will exactly and symmetrically produce the 24-volume, nucleus-embracing cube symmetrically surrounding the 20-volume vector equilibrium; thus with $8 \times 1/2 = 4$ being added to the 20-volume vector equilibrium producing a 24-volume total.

905.45 A non-nucleus-embracing 3-volume cube may be produced by applying four of the Eighth-Octahedra to the four equiangled triangular facets of the tetrahedron. (See Illus. 950.30.) Thus we find the tetrahedral evolvement of the prime number *three* as identified with the cube. Ergo all the prime numbers—1, 2, 3, 5, 7—of the octave wave enumeration system, with its zero-nineness, are now clearly demonstrated as evolutionarily consequent upon tetrahedral intertransformabilities.

905.46 Since the tetrahedron becomes systematically subdivided into 24 uniformly dimensioned *A* Quanta Modules (one half of which are positive and the other half of which are negatively inside-out of the other); and since both the positive and negative *A* Quanta Modules may be folded from one whole triangle; and since, as will be shown in Sec. 905.62, the flattened-out triangle of the *A* Quanta Module corresponds with each of the 120 disequilibrious LCD triangles, it is evidenced that five tetrahedra of 24 *A* Quanta Modules each, may have their sum-total of 120 *A* Modules all unfolded, and that they may be edge-bonded to produce an icosahedral spherical array; and that $2^1/_2$ tetrahedra's 60 *A* Quanta Modules could be unfolded and univalently (single-bondedly) arrayed to produce the same spheric icosahedral polyhedron with 60 visible triangles and 60 invisible triangular voids of identical dimension.

905.47 Conversely, 60 positive and 60 negative *A* Quanta Modules could be folded from the 120 *A* Module triangles and, with their "sharpest" point pointed inward, could be admitted radially into the 60-positive-60-negative tetrahedral voids of the icosahedron. Thus we discover that the icosahedron, consisting of 120 *A* Quanta Modules (each of which is $^1/_{24}$th of a tetrahedron) has a volume of $^{120}/_{24} = 5$. The icosahedron volume is 5 when the tetrahedron is 1; the octahedron 2^2; the cube 3; and the star-pointed, univalent, eight-tetrahedra nuclear embracement is 28, which is 4×7; 28 also being the maximum number of interrelationships of eight entities: $\frac{N^2 - N}{2} = \frac{8^2 - 8}{2} = 28$.

905.48 The three surrounding angles of the interior sharpest point of the *A* Quanta Module tetrahedron are each a fraction less than the three corresponding central angles of the icosahedron: being approximately one-half of a degree in the first case; one whole degree in the second case; and one and three-quarters of a degree in the third case. This loose-fit, volumetric-debit differential of the *A* Quanta Module volume is offset by its being slightly longer in radius than that of the icosahedron, the *A* Module's radial depth being that of the vector equilibrium's, which is greater than that of the icosahedron, as caused by the reduction in the radius of the 12 balls closest-packed around one nuclear ball of the vector equilibrium (which is eliminated from within the same closest-radially-packed 12 balls to reduce them to closest surface-packing, as well as by eliminating the nuclear ball and thereby mildly reducing the system radius). The plus volume of the fractionally protruded portion of the *A* Quanta Module beyond the icosahedron's surface may exactly equal the interior minus volume difference. The balancing out of the small plus and minus volumes is suggested as a possibility in view of the exact congruence of 15 of the 120 spherical icosahedra triangles with each of the spherical octahedron's eight spherical equiangle faces, as well as by the exact congruence of the octahedron and the vector equilibrium themselves. As the icosahedron's radius shortens, the central angles become enlarged.

905.49 This completes the polyhedral progression of the omni-phase-bond-integrated hierarchies of—1-2-3-4, 8—symmetrically expanded and symmetrically subdivided tetrahedra; from the $^1/_{24}$th-tetrahedron (12 positive and 12 negative *A* Quanta Modules); through its octavalent 8-in-1 superficial volume-1; expanded progressively through the quadrivalent tetrahedron; to the quadrivalent octahedron; to the bivalent vector equilibrium; to the univalent, 28-volume, radiant, symmetrical, nucleus-embracing stage; and thence exploded through the volumeless, flatout-outfolded, double-bonded (edge-bonded), 120-*A*-Quanta-Module-triangular array remotely and symmetrically surrounding the nuclear volumetric group; to final dichotomizing into two such flatout half (positive triangular) film and half (negative triangular) void arrays, single-bonded (corner-bonded), icosahedrally shaped, symmetrically nuclear-surrounding systems.

905.50 Rotatability and Split Personality of Tetrahedron

905.51 Having completed the expansive-contractive, could-be, quantum jumps, we will now consider the rotatability of the tetrahedron's six-edge axes generation of both the two spherical tetrahedra and the spherical cube whose "split personality's" four-triangle-defining edges also perpendicularly bisect all of both of the spherical tetrahedron's four equiangled, equiedged triangles in a three-way grid, which converts each of the four equiangled triangles into six right-angle spherical triangles—for a total of 24, which are split again by the spherical octahedron's three great circles to produce 48 spherical triangles, which constitute the 48 equilibrious LCD Basic Triangles of omni-equilibrious eventless eternity (see Sec. 453).

905.52 The spherical octahedron's eight faces become skew-subdivided by the icosahedron's 15 great circles' self-splitting of its 20 equiangular faces into six-each, right spherical triangles, for an LCD spherical triangle total of 120, of which 15 such right triangles occupy each of the spherical octahedron's eight equiangular faces—for a total of 120—which are the same 120 as the icosahedron's 15 great circles.

905.53 The disequilibrious 120 LCD triangle = the equilibrious 48 LCD triangle $\times 2^1/_2$. This $2^1/_2 + 2^1/_2 = 5$; which represents the icosahedron's basic *fiveness* as split-generated into $2^1/_2$ by their perpendicular, mid-edge-bisecting 15 great circles. Recalling the six edge vectors of the tetrahedron as one quantum, we note that $6 + 6 + {}^6/_2$ is $1 + 1 + {}^1/_2 = 2^1/_2$; and that $2^1/_2 \times 6 = 15$ great circles. (This half-positive and half negative dichotomization of systems is discussed further at Sec. 1053.30ff.)

905.54 We find that the split personality of the icosahedron's 15-great-circle splittings of its own 20 triangles into 120, discloses a basic asymmetry caused by the incompleteness of the $2^1/_2$, where it is to be seen in the superimposition of the spherical icosahedron upon the spherical vector equilibrium. In this arrangement the fundamental prime number *fiveness* of the icosahedron is always split two ways: $2^1/_2$ positive phase and $2^1/_2$ negative phase. This half-fiving induces an alternate combining of the half quantum on one side or the other: going to first *three* on one side and *two* on the other, and vice versa.

905.55 This half-one-side/half-on-the-other induces an oscillatory alternating 120-degree-arc, partial rotation of eight of the spherical tetrahedron's 20 equiangled triangles within the spherical octahedron's eight triangles: $8 \times 2^1/_2 = 20$. We also recall that the vector equilibrium has 24 internal radii (doubled together as 12 radii by the congruence of the four-great-circle's hexagonal radii) and 24 separate internal vector chords. These 24 external vector chords represent four quanta of six vectors each. When the vector equilibrium jitterbuggingly contracts toward the octahedral edge-vector doubling stage, it passes through the unstable icosahedral stage, which is unstable because it requires six more edge-vectors to hold fixed the short diagonal of the six diamond-shaped openings between the eight triangles. These six equilength vectors necessary to stabilize the icosahedron constitute one additional quantum which, when provided, adds 1 to the 4 of the vector equilibrium to equal 5, the basic quantum number of the icosahedron.

905.60 The Disequilibrium 120 LCD Triangle

905.61 The icosahedral spherical great-circle system displays:

12 vertexes surrounded by 10 converging angles;
20 vertexes surrounded by 6 converging angles;
30 vertexes surrounded by 4 converging angles

$$\begin{aligned} 12 \times 10 &= 120 \\ 20 \times 6 &= 120 \\ 30 \times 4 &= 120 \\ \hline &\ 360 \text{ converging angle sinuses.} \end{aligned}$$

905.62 According to the Principle of Angular Topology (see Sec. 224), the 360 converging angle sinuses must share a 720-degree reduction from an absolute sphere to a chorded sphere: $^{720°}/_{360°} = 2°$. An average of 2 degrees angular reduction for each corner means a 6 degrees angular reduction for each triangle. Therefore, as we see in each of the icosahedron's disequilibrious 120 LCD triangles, the well-known architects and engineers' 30°-60°-90° triangle has been spherically *opened* to 36°-60°-90°—a "spherical excess,'' as the Geodetic Survey calls it, of 6 degrees. All this spherical excess of 6 degrees has been massaged by the irreducibility of the 90-degree and 60-degree corners into the littlest corner. Therefore, 30 → 36 in each of the spherical icosahedron's 120 surface triangles.

905.63 In subsiding the 120 spherical triangles generated by the 15 great circles of the icosahedron from an omnispherical condition to a neospheric 120-planar-faceted polyhedron, we produce a condition where all the vertexes are equidistant from the same center and all of the edges are chords of the same spherical triangle, each edge having been shrunk from its previous arc length to the chord lengths without changing the central angles. In this condition the spherical excess of 6 degrees could be shared proportionately by the 90°, 60°, 30° flat triangle relationship which factors exactly to 3:2:1. Since $6° = {}^{1}/_{30}$ of 180°, the 30 quanta of six each in flatout triangles or in the 120 LCD spherical triangles' 186 degrees, means one additional quantum crowded in, producing 31 quanta.

905.64 Alternatively, the spherical excess of 6 degrees (one quantum) may be apportioned totally to the biggest and littlest corners of the triangle, leaving the 60-degree, vector equilibrium, neutral corner undisturbed. As we have discovered in the isotropic vector matrix nature coordinates crystallographically in 60 degrees and not in 90 degrees. Sixty degrees is the vector equilibrium neutral angle relative to which life-in-time aberrates.

Flatout *A* Quanta *Module Triangle* *			Basic Draftsman's *Triangle (Flat)*
35° 16′	(minus 5° 16′)	=	30° 00′
60° 00′	(unchanged)	=	60° 00′
84° 44′	(plus 5° 16′)	=	90° 00′
180° 00°			180° 00′

905.65 By freezing the 60-degree center of the icosahedral triangle, and by sharing the 6-degree, spherical-planar, excess reduction between the 36-degree and 90-degree corners, we will find that the *A* Quanta Modules are exactly congruent with the 120 internal angles of the icosahedron. The minus 5° 16′ closely approximates the one quantum 6+ of spherical excess apparent at the surface, with a comparable nuclear deficiency of 5° 16′. (See Table 905.65.)

* These are the interior corner angles of the flatout triangle which is foldable to form one positive (and reverse foldable to produce one negative) *A* Quanta Module. The 5° 16′ is called the ''twinkle angle'' (see Sec. 915).

Table 905.65

Decimal Magnitudes of VE 10-ness (Equil.) + Icosa 5-ness (Disequil.)	Angles around Sharp Vertex of *A* Quanta Module Tetrahedron		Differential		Central Angles of the Spherical Icosahedron's Disequilibrium 120 LCD Triangles	Differential	Decimal Magnitudes of VE 10-ness (Equil.) + Icosa 5-ness (Disequil.)
20°	19° 28′		26′ 18.5 ″		20° 54′ 18.57″	−00° 54′ 18.57″	20°
30°	30°		1° 43′ 02.9 ″		31° 43′ 02.9 ″	− 1° 43′ 02.9 ″	30°
						− 2° 37′ 21.47″	
40°	35° 16′		1° 06′ 38.53″		37° 22′ 38.53″	+ 2° 37′ 21.47″	40°
90°	84° 44′	+	5° 16′	=	90° 00′ 00.00″	5° 14′ 43.34″	90°
			This 5° 16′ is one whole quantum—44′			This 2° 37′ 21.47″ is one quantum split in two	

(There is a basic difference between 5° 16′ and approx. 5° 15′. It is obviously the same "twinkle angle" with residual calculation error of trigonometric irrational inexactitude.)

905.66 The Earth crust-fault angles, steel plate fractionation angles, and ship's bow waves all are roughly the same, reading approximately 70-degree and 110-degree complementation.

Dihedral angle of octahedron = 109° 28′ = 2 × 54° 44′
Dihedral angle of tetrahedron = 70° 32′

$$\frac{\begin{array}{r}109^\circ\ 28' \\ 70^\circ\ 32'\end{array}}{180^\circ\ 00'}$$

$$\begin{array}{r}54^\circ\ 44' \\ +\ 54^\circ\ 44' \\ \hline 109^\circ\ 28'\end{array} \qquad \begin{array}{r}60^\circ\ 00' \\ -\ 54^\circ\ 44' \\ \hline 5^\circ\ 16'\end{array} \qquad \begin{array}{r}5^\circ\ 16' \\ \times\ 2 \\ \hline 10^\circ\ 32'\end{array} \qquad \begin{array}{r}70^\circ\ 32' \\ -\ 60^\circ\ 00' \\ \hline 10^\circ\ 32'\end{array}$$

—If 5° 16′ = unity; 54° 44′ = 60° − 1 quantum; and 70° 32′ = 60° + 2 quanta.

—Obviously, the 70° 32′ and 109° 28′ relate to the "twinkle angle" differential from 60° (cosmic neutral) and to the 109° 28′ central angle of the spherical tetrahedron. (See also Sec. 1051.20.)

905.70 Summary: Wave Propagation Model

905.71 Both in the spherical vector equilibrium and in the disequilibrious icosahedral spherical system, the prime number *five* is produced by the fundamental allspace-filling complementarity of the 1-volume tetrahedron and the 4-volume octahedron.

—Symmetrical: $1 + 4 = 5$

—Asymmetrical: $4 + 1 = 5$

The effect is *symmetrical* when the tetrahedron's four vertexes simultaneously pulse outwardly through their opposite void triangles to produce the "star tetrahedron," one outwardly-pointing tetrahedron superimposed on each of the four faces of a nuclear tetrahedron: i.e., $1 + 4 = 5$. The effect is *asymmetrical* when one outwardly-pointing tetrahedron is superimposed on one face of one octahedron: i.e., $4 + 1 = 5$.

905.72 We now understand how the equilibrious 48 basic triangles transform into the 120 disequilibrious basic triangles. The 120 (60 positive and 60 negative) LCD spherical triangles' central (or nuclear) angles are unaltered as we transform their eternal systemic patterning symmetry from (a) the octahedral form of $^{120}/_{8} = 15$ *A* Quanta Modules per each octa triangle; to (b) the icosahedron's $^{120}/_{20} = 6$ *A* Quanta Modules per each icosa triangle; to (c) the dodecahedron's $^{120}/_{12} = 10$ *A* Quanta Modules per each pentagon. This transformational progression demonstrates the experientially witnessable, wave-producing surface-askewing caused by the 120-degree, alternating rotation of the icosahedron's triangles inside of the octahedron's triangles. Concomitant with this alternating rotation we witness the shuttling of the spherical vector equilibrium's 12 vertexial positions in a successive shifting-back-and-forth between the spherical icosahedron's 12 vertexial positions; as well as the wave-propagating symmetrical, polyhedral alterations of the inward-outward pulsations which generate surface undulations consequent to the radial contractions, at any one time, of only a fractional number of all the exterior vertexes, while a symmetrical set of vertexes remains unaltered.

905.73 This elucidates the fundamental, electromagnetic, inward-outward, and complex great-circling-around type of wave propagation, as does also the model of spheres becoming voids and all the voids becoming spheres, and their omniradiant wave propagation (see Sec. 1032).

905.74 There are also the approximately unlimited ranges of frequency modulatabilities occasioned by the symmetrical subdivisioning of all the prime, equiangled, surface triangles of the tetrahedron, octahedron, and icosahedron. This additionally permitted wave undulation of surface pattern shifting is directly identified with the appearance of photons as spherically clustered and radiantly emittable tetrahedra (see Sec. 541.30).

910.00 A and B Quanta Modules

910.01 All omni-closest-packed, complex, structural phenomena are omnisymmetrically componented only by tetrahedra and octahedra. Icosahedra, though symmetrical in themselves, will not close-pack with one another or with any other symmetrical polyhedra; icosahedra will, however, face-bond together to form open-network octahedra.

910.02 In an isotropic vector matrix, it will be discovered that there are only two omnisymmetrical polyhedra universally described by the configuration of the interacting vector lines: these two polyhedra are the regular tetrahedron and the regular octahedron.

910.10 Rational Fraction Elements

910.11 All other regular, omnisymmetric, uniform-edged, -angled, and -faceted, as well as several semisymmetric, and all other asymmetric polyhedra other than the icosahedron and the pentagonal dodecahedron, are described repetitiously by compounding rational fraction elements of the tetrahedron and octahedron. These elements are known in synergetics as the *A* and *B* Quanta Modules. They each have a volume of $^1/_{24}$th of a tetrahedron.

911.00 Division of Tetrahedron

911.01 The regular tetrahedron may be divided volumetrically into four identical Quarter-Tetrahedra, with all their respective apexes at the center of volume of the regular unit tetrahedron. (See Illus. 913.01.) The Quarter-Tetrahedra are irregular pyramids formed upon each of the four triangular faces of the original unit tetrahedra, with their four interior apexes congruent at the regular tetrahedron's volumetric center; and they each have a volume of one quarter of the regular tetrahedron's volume-1.

911.02 Any of the Quarter-Tetrahedra may be further uniformly subdivided into six identical irregular tetrahedra by describing lines that are perpendicular bisectors from each vertex to their opposite edge of the Quarter-Tetrahedron. The three perpendicular bisectors cut each Quarter-Tetrahedron into six similar tetrahedral pieces of pie. Each one of the six uniformly symmetrical components must be $^1/_6$th of One Quarter, which is $^1/_{24}$th of a regular tetrahedron, which is the volume and description of the *A* Quanta Module. (See Illus. 913.01B.)

912.00 Division of Octahedron

912.01 The regular octahedron has a volume equivalent to that of four unit tetrahedra. The octahedron may be subdivided symmetrically into eight equal parts, as Eighth-Octahedra, by planes going through the three axes connecting its six vertexes. (See Illus. 916.01.)

912.02 The Quarter-Tetrahedron and the Eighth-Octahedron each have an equilateral triangular base, and each of the base edges is identical in length. With their equiangular-triangle bases congruent we can superimpose the Eighth-Octahedron over the Quarter-Tetrahedron because the volume of the Eighth-Octahedron is $^1/_2$ and the volume of the Quarter-Tetrahedron is $^1/_4$. The volume of the Eighth-Octahedron is twice that of the Quarter-Tetrahedron; therefore, the Eighth-Octahedron must have twice the altitude because it has the same base and its volume is twice as great.

913.00 A Quanta Module

913.01 The *A* Quanta Module is $^1/_6$th of a Quarter-Tetrahedron. The six asymmetrical components of the Quarter-Tetrahedron each have a volume of $^1/_{24}$th of the unit tetrahedron. They are identical in volume and dimension, but three of them are positive and three of them are negative. (See Illus. 913.01.)

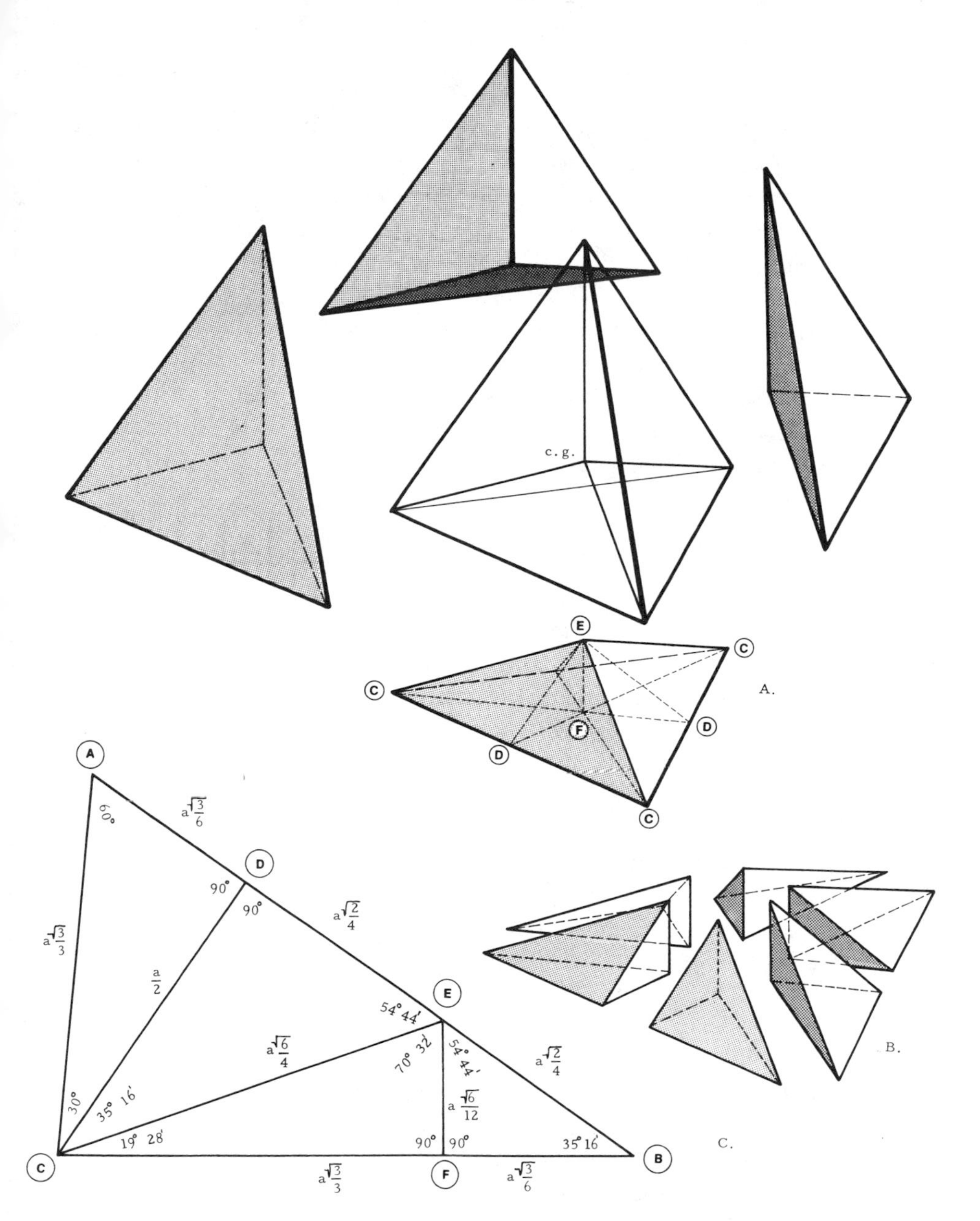

Fig. 913.01 *Division of the Quarter-Tetrahedron into Six Parts:* A *Quanta Module:*

A. The regular tetrahedron is divided volumetrically into four identical quarters.

B. The quarter-tetrahedron is divided ino six identical irregular tetrahedra, which appear as three right-hand and three left-hand volumetric units each equal in volume to $^{1}/_{24}$th of the original tetrahedron. This unit is called the *A* Module.

C. The plane net which will fold into either left or right *A* modules is shown. Vertex *C* is at the vertex of the regular tetrahedron. Vertex *E* is at the center of gravity of the tetrahedron. Vertex *D* is at the mid-edge of the tetrahedron. Vertex *F* is at the center of the tetrahedron face. Note that $AD=FB$, $DE=EB$, and $AC=CF$.

913.10 **Positive and Negative:** The positive and negative *A* Quanta Modules (the + and the −) will not nest in one another congruently despite identical angles, edges, and faces. The pluses are inside-out minuses, which can be shown by opening three of their six edges and folding the three triangles' hinged edges in the opposite direction until their edges come together again.

913.11 The *A* Quanta Module triangle is possibly a unique scalene in that neither of its two perpendiculars bisect the edges that they intersect. It has three internal foldables and no "internally contained" triangle. It drops its perpendiculars in such a manner that there are only three external edge increments, which divide the perimeter into six increments of three pairs.

914.00 *A Quanta Module: Foldability*

914.01 The *A* Quanta Module can be unfolded into a planar triangle, an asymmetrical triangle with three different edge sizes, yet with the rare property of folding up into a whole irregular tetrahedron.

914.02 An equilateral planar triangle *AAA* may be bisected in each edge by points *BBB*. The triangle *AAA* may be folded on lines *BB*, *BB*, *BB*, and points *A*, *A*, *A* will coincide to form the regular tetrahedron. This is very well known.

914.10 **Four Right Angles:** In respect to the *A* Quanta Module flatout triangle or infolded to form the irregular tetrahedron, we find by the method of the module's construction (by perpendicular planes carving apart) that there are four right angles (see Illus. 913.01C):

EFB	*EDC*
EFC	*ADC*

914.20 **Unfolding into a Flat Triangle:** If we go to the vertex at *E* and break open the edges *ED* and *AD*, we can hinge open triangle *EBF* on hinge line *EF*. We can then break open the edge *AC* and fold triangle *ADC*, as well as folding out the two triangles *DEC* and *CEF*, which are connected by the hinge *EC*, so that now the whole asymmetric *A* Quanta Module is stretched out flat as a triangle.

914.21 The *A* Quanta Module unfolds into a scalene triangle; that is, all of its non-90-degree angles are different, and all are less than 90 degrees. Two of the folds are perpendicular to the triangle's sides, thus producing the four right angles. The *A* Quanta Module triangle may be the only triangle fulfilling all the above stated conditions.

914.30 **Spiral Foldability:** The foldability of the *A* Quanta Module planar triangle differs from the inter-mid-edge foldability of the equilateral or isosceles triangle. All the mid-edge-foldable equilateral or isosceles triangles can all form tetrahedra, regular or irregular. In the case of the folded equilateral or isosceles triangle, the three triangle corners meet together at one vertex: like petals of a flower. In the case of the inter-mid-edge-folding scalene triangle, the three corners fail to meet at one vertex to form a tetrahedron.

915.00 Twinkle Angle

915.01 The faces of an *A* Quanta Module unfold to form a triangle with 84° 44′ (30° 00′ + 35° 16′ + 19° 28′) as its largest angle. This is 5° 16′ less than a right angle, and is known as the *twinkle angle* in synergetics (see Illus. 913.01C).

915.02 There is a unique 5° 16′-ness relationship of the *A* Quanta Module to the symmetry of the tetrahedron-octahedron allspace-filling complementation and other aspects of the vector equilibrium that is seemingly out of gear with the disequilibrious icosahedron. It has a plus-or-minus incrementation quality in relation to the angular laws common to the vector equilibrium.

915.10 **A Quanta Module Triangle and Basic Disequilibrium 120 LCD Triangle:** The angles of fold of the *A* Quanta Module triangle correspond in patterning to the angles of fold of the Basic Disequilibrium 120 LCD Triangle, the $^1/_{120}$th of a sphere whose fundamental great circles of basic symmetry subdivide it in the same way. The angles are the same proportionally when the spherical excess subsides proportionally in all three corners. For instance, the angle *ACB* in Illus. 913.01C is not 90 degrees, but a little less.

915.11 It is probable that these two closely akin triangles and their respective folded tetrahedra, whose *A* Module Quantum phase is a rational subdivider function of all the hierarchy of atomic triangulated substructuring, the 120 Basic Disequilibrium LCD triangles and the *A* Module triangles, are the *same quanta* reoccurrent in their most powerful wave-angle oscillating, intertransformable extremes.

915.20 **Probability of Equimagnitude Phases:** The 6° spherical excess of the Basic Disequilibrium 120 LCD Triangle, the 5° 16′ "twinkle angle" of the *A* Quanta Module triangle, and the 7° 20′ "unzipping angle" of birth, as in the DNA tetrahelix, together may in time disclose many equimagnitude phases occurring between complementary intertransforming structures.

916.00 B *Quanta Module*

916.01 The *B* Quanta Module is $^1/_6$th of the fractional unit described by subtracting a Quarter-Tetrahedron from an Eighth-Octahedron. The six asymmetrical components of the fractional unit so described each have a volume of $^1/_{24}$th of the unit tetrahedron. They are identical in volume and dimensioning, but three of them are positive and three of them are negative. (See Illus. 916.01.)

916.02 When the Eighth-Octahedron is superimposed on the Quarter-Tetrahedron, the top half of the Eighth-Octahedron is a fractional unit, like a concave lid, with a volume and weight equal to that of the Quarter-Tetrahedron below it. We can slice the fractional unit by three planes perpendicular to its equilateral triangular base and passing through the apex of the Quarter-Tetrahedron, through the vertexes of the triangular base, and through the mid-points of their respective opposite edges, separating it into six equidimensional, equivolume parts. These are *B* Quanta Modules.

916.03 *B* Quanta Modules are identical irregular tetrahedra that appear as three positive (outside-out) and three negative (outside-in) units. Each of the *B* Quanta Modules can be unfolded into a planar, multitriangled polygon. (See Illus. 916.01F.)

920.00 *Functions of* A *and* B *Modules*

920.01 The *A* and *B* Quanta Modules may possibly quantize our total experience. It is a phenomenal matter to discover asymmetrical polyhedral units of geometry that are reorientably compositable to occupy one asymmetrical polyhedral space; it is equally unique that, despite disparate asymmetric polyhedral form, both have the same volume; and both associate in different kinds of simplex and complex, symmetrical and asymmetrical, coherent systems. While they consist, in their positive and negative aspects, of four different asymmetrical shapes, their unit volume and energy quanta values provide a geometry elucidating both fundamental structuring and fundamental and complex intertransformings, both gravitational and radiational.

921.00 *Energy Deployment in* A *and* B *Quanta Modules*

921.01 By virtue of their properties as described in Secs. 920, 921.20, and 921.30, the centers of energy in the *A* and *B* Quanta Modules can be

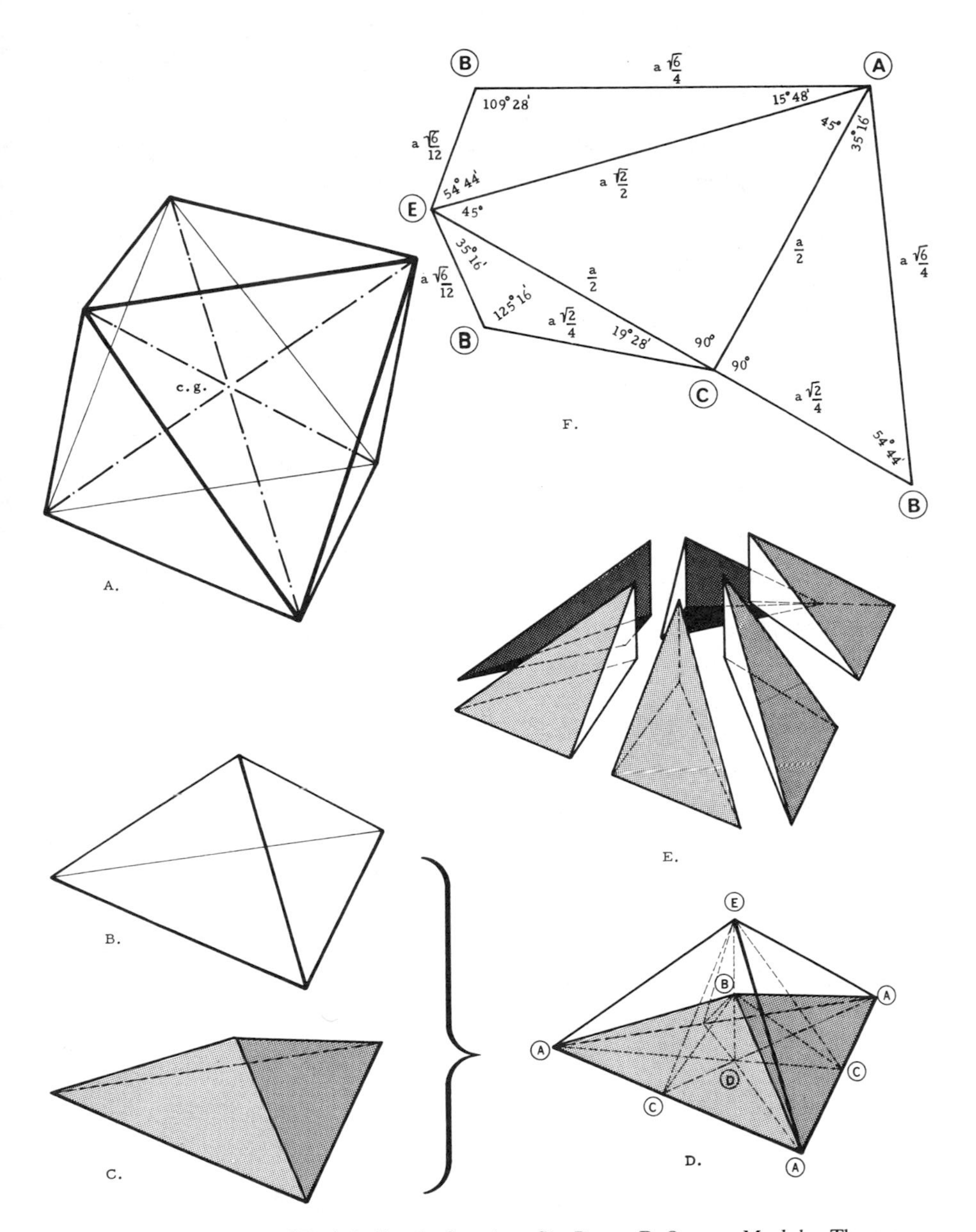

Fig. 916.01 *Division of Eighth-Octahedron into Six Parts:* B *Quanta Module:* The regular octahedron (A) is divided into eight identical units (B) equaling $^1/_8$ of the volume of the octahedron. The quarter tetrahedron as defined by six *A* Modules (C) is subtracted from the $^1/_8$-octahedron (D). This fractional unit is then subdivided into six identical irregular tetrahedra that appear as three right-hand and three left-hand units and are referred to as *B* Modules. They are equal in volume to the *A* Modules and are consequently also $^1/_{24}$th of the regular tetrahedron. In (F) is shown the plane net which will fold into either the right or left *B* Module. Vertex *A* is at the vertex of the octahedron. Vertex *C* is at the mid-edge of the octahedron. Vertex *E* is at the center of gravity of the octahedron.

locally reoriented within the same space without disturbing contiguously surrounding configurations of closest-packed geometry; these local reorientations can either concentrate and hold or deploy and distribute the energies of the respective *A* and *B* Quanta Modules, in the first case concentrating the centers of energy inwardly, and in the second case deploying the centers of energy outwardly.

921.02 In X-ray diffraction, you can see just such alternate energy concentrations of omnideployed patterns in successive heat treatments of metals. You can hit a piece of metal and you will find by X-ray diffraction that a previously concentrated array of centers of energy have been elegantly deployed. When you take the temper out of the metal, the energy centers will again change their positions. The metal's coherence strength is lessened as the energy centers are outwardly deployed into diffused remoteness from one another. When the centers of energy are arranged closer to one another, they either attract or repulse one another at the exponentially increasing rates of gravitational and radiational law. When we heat-treat or anneal metals and alloys, they transform in correspondence with the reorientabilities of the *A* and *B* Quanta Modules.

921.03 The identical volumes and the uniquely different energy-transforming capabilities of the *A* and *B* Quanta Modules and their mathematically describable behaviors ($10F^2+2$) hint at correspondence with the behaviors of neutrons and protons. They are not mirror images of one another, yet, like the proton and neutron, they are energetically intertransformable and, due to difference of interpatternability, they have difference in mass relationship. Whether they tend to conserve or dissipate energy might impose a behavioral difference in the processes of measuring their respective masses. A behavioral proclivity must impose effects upon the measuring process.

921.04 The exact energy-volume relationship of the *A* and *B* Quanta Modules and their probable volumetric equivalence with the only meager dimensional transformations of the 120 LCD tetrahedral voids of the icosahedron (see Sec. 905.60) may prove to have important physical behavior kinships.

921.10 **Energy Behavior in Tetrahedra:** A tetrahedron that can be folded out of a single foldable triangle has the strange property of holding energy in varying degrees. Energy will bounce around inside the tetrahedron's four internal triangles as we described its bouncing within one triangle (see Sec. 901). Many bounce patterns are cyclically accomplished without tendency to bounce out of tetrahedrons, whether regular or irregular, symmetrical or asymmetrical.

921.11 The equiangled, omni-sixty-degreed, regular tetrahedron can be opened along any three edges converging at any one of its vertexes with its edge-separated vertexial group of three triangles appearing as a three-petaled flower bud about to open. By deliberately opening the three triangular petals, by rotating them outward from one another around their three unsevered base-edge hinges, all three may be laid out flat around the central base triangle to appear as a two-frequency, edge-moduled, equiangular triangle consisting of four internal triangles. Energy tends by geodesical economy and angular law to be bounce-confined by the tetrahedron.

921.12 The irregular, asymmetrical, tetrahedral *A* Quanta Module's four triangular facets unfold spirally into one asymmetrical triangle.

921.13 But the triangular facets of the *B* Quanta Module unfold inherently into four mutually dissimilar but interhinged 90-degree triangles.

921.14 All the interior edges of the triangles, like the edges of a triangular billiard table, will provide unique internal, bouncing, corner-pocket-seeking patterns. An equilateral, equiangled triangle will hold the bouncing with the least tendency to exit at the pocketed corners. The more asymmetrical the triangular billiard table, the more swiftly the angular progression to exit it at a corner pocket. The various bounce patterns prior to exit induce time-differentiated lags in the rate of energy release from one tetrahedron into the other tetrahedron.

921.15 Energy bounces around in triangles working toward the narrowest vertex, where the impossibility of more than one line going through any one point at any one time imposes a twist vertex exit at the corners of all polyhedra. Therefore, all triangles and tetrahedra ''leak'' energy, but when doing so between two similar corresponding vertexes-interconnected tetrahedra, the leaks from one become the filling of the other.

921.20 Energy Characteristics of *A* Quanta Module: The *A* Quanta Modules can hold energy and tend to conserve it. They do so by combining with one another in three unique ways, each of which combine as one regular tetrahedron; the regular tetrahedron being a fundamental energy-holding form—the energy being held bounce-describing the internal octahedron of every tetrahedron.

921.21 The *A* Quanta Modules can also combine with the *B* Quanta Modules in seven ways, each of which result in single whole tetrahedra, which, as noted, hold their energy within their inherent octahedral centers.

921.30 Energy Characteristics of *B* Quanta Module: The *B* Quanta Modules can vertex-combinedly hold energy but tend to release it.

921.31 While all the single triangles will hold swift-motion energies only for relatively short periods of time, the four very asymmetrical and dissimilar triangles of the *B* Quanta Module will release energy four times faster than any one of their asymmetrical tetrahedral kin.

921.32 The *B* Quanta Modules do not retain energy, and they cannot combine with one another to form a single tetrahedron with energy-introverting and -conserving proclivities.

921.40 Summary: Though of equal energy potential or latent content, the *A*s and the *B*s are two different systems of unique energy-behavior containment. One is circumferentially embracing, energy-impounding, integratively finite, and nucleation-conserving. The other is definitively distintegrative and nuclearly exportive. *A* is outside-inwardly introvertive. *B* is outside-outwardly extrovertive. (See Illus. 924.20.)

922.00 Conceptual Description and Contrast

922.01 The *A* Quanta Module is all of the nonconsidered, nonconceptual, finite, equilibrious, not-now-tuned-in Universe.

922.02 The *B* Quanta Module is the only momentarily extant considered subdivision of disequilibrious Universe, i.e., the attention-preoccupying, special-case local system. The *B* Quanta Module is always the real live ''baby''; it is most asymmetrical.

923.00 Constant Volume

923.10 Precession of Two Module Edges: There are six edges of a tetrahedron, and each edge precesses the opposite edge toward a 90-degrees-maximum of attitudinal difference of orientation. Any two discrete, opposite edges can be represented by two aluminum tubes, *X* and *Y* (see Illus. 923.10 D), which can move longitudinally anywhere along their respective axes while the volume of the irregular tetrahedra remains constant. They may shuttle along on these lines and produce all kinds of asymmetrical tetrahedra, whose volumes will always remain unit by virtue of their developed tetrahedra's constant base areas and identical altitudes. The two tubes' four ends produce the other four interconnecting edges of the tetrahedron, which vary as required without altering the constantly uniform volume.

923.20 Constant Volume: A comparison of the end views of the *A* and *B* Quanta Modules shows that they have equal volumes as a result of their equal base areas and identical altitudes. (See Sec. 621.)

923.21 A line can be projected from its origin at the center of area of the triangular base of a regular tetrahedron, outward through the opposite apex of

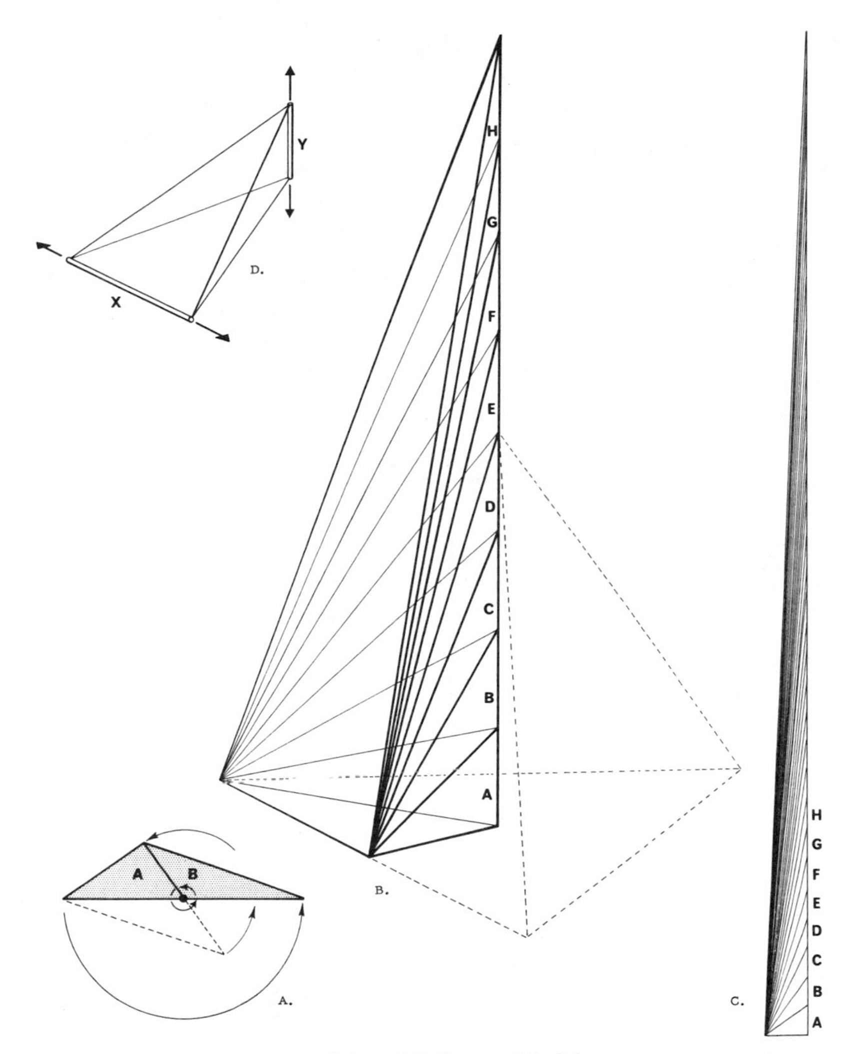

Fig. 923.10 *Constant Volume of* A *and* B *Quanta Modules:*

A. A comparison of the end views of the *A* and *B* Quanta Modules shows that they have equal volumes by virtue of the fact that they have equal base areas and identical altitudes.

B. It follows from this that if a line, originating at the center of area of the triangular base of a regular tetrahedron, is projected through the apex of the tetrahedron to infinity, is subdivided into equal increments, it will give rise to additional Modules to infinity. Each additional Module will have the same volume as the original *A* or *B* Module, and as the incremental line approaches infinity the Modules will tend to become lines, but lines still having the same volume as the original *A* or *B* Module.

C. End view shows Modules beyond the *H* Module shown in (B).

D. The two discrete members *X* and *Y* can move anywhere along their respective axes and the volume of the irregular tetrahedron remains constant. The other four edges vary as required.

the tetrahedron to any desired distance. When subdivided into increments equal to the distance between its triangular-base center and its apex, and when each of these equilinear increments outward beyond the apex is interconnected by three lines leading to each of the three corners of the base triangle, then each of the successive volumetric additions will be of identical volume to that of the original tetrahedra, and the overall form will be that of a tetrahedron which become progressively longer and sharp-pointed with each addition. (See Illus. 923.10 A, B, and C.) As the ever-sharpening and elongating tetrahedron approaches infinity, the three elongating edges tend to parallelism; i.e., toward what is known as parallax in astronomy. The modules will tend to congruence with the parallaxing lines. Each full-line-long length model of these congruent lines will have the same volume as the original module.

923.30 Energy Accommodation: The *A* and *B* Quanta Modules start with unit base and add unit altitude, C, D, E, F, and so forth, but as each additional altitude is superimposed, the volume remains the same: a volume of one. We find these linear incrementation assemblies getting longer, with their additional volumes always one. Suppose we think about this progression as forming an electric-wire conductor and divide its circular base into three 120-degree angles. Its progressive conic increments could grow and operate in the same manner as our constant-volume, tetrahedral modules.

923.31 We will inherently superimpose progressive base-to-apex attenuating sections. In the electric conductor wire, this means that whatever energy increment is fed into the first base module will tend to be conducted at various unit frequencies along the line. Each unique frequency introduced at the base will create its unique conic altitude incrementation. The outermost, line-long cone's energy quantum will always be the same as that of the initial base cone. Finally, the last and outermost cone is just as long as the wire itself—so there is an outside charge on the wire tending to fluoresce a precessional broadcasting of the initial inputs at 90 degrees; i.e., perpendicularly away from the wire. This may elucidate antenna behaviors as well as long-distance, high-voltage, electric energy conductions which tend to broadcast their conducted energy. (For further elaboration of the constant-volume, tetrahedral models, see Secs. 961.10, 20, 30 and 40.)

924.00 Congruence of Centers

924.10 Congruence of *A* and *B* Quanta Module Centers: Within either the *A* or *B* Quanta Modules the

centers of effort;
centers of energy;

centers of gravity;
centers of radiation;
centers of volume; and
centers of field

are congruent, i.e., identical. The same centers are involved. We will call their six congruent centers their synergetic centers.

924.11 But the *A* (+) and *A* (−), and *B* (+) and *B* (−) respective volumetric centers are never congruent. However, the positive or the negative *AAB* aggregates (these are the "Mites." See Sec. 953.10) have identical volumetric centers.

924.20 Table of Tetrahedral Functions of *A* and *B* Quanta Modules: See page 514.

930.00 Tetrahelix: Unzipping Angle

930.10 Continuous Pattern Strip: "Come and Go"

930.11 Exploring the multiramifications of spontaneously regenerative reangulations and triangulations, we introduce upon a continuous ribbon a 60-degree-patterned, progressively alternating, angular bounce-off inwards from first one side and then the other side of the ribbon, which produces a wave pattern whose length is the interval along any one side between successive bounce-offs which, being at 60 degrees in this case, produces a series of equiangular triangles along the strip. As seen from one side, the equiangular triangles are alternately oriented as *peak away,* then *base away,* then *peak away* again, etc. This is the patterning of the only equilibrious, never realized, angular field state, in contradistinction to its sine-curve wave, periodic realizations of progressively accumulative, disequilibrious aberrations, whose peaks and valleys may also be patterned between the same length wave intervals along the sides of the ribbon as that of the equilibrious periodicity. (See Illus. 930.11.)

930.20 **Pattern Strips Aggregate Wrapabilities:** The equilibrious state's 60-degree rise-and-fall lines may also become successive cross-ribbon fold-lines, which, when successively partially folded, will produce alternatively a tetrahedral- or an octahedral- or an icosahedral-shaped spool or reel upon which to roll-mount itself repeatedly: the tetrahedral spool having four

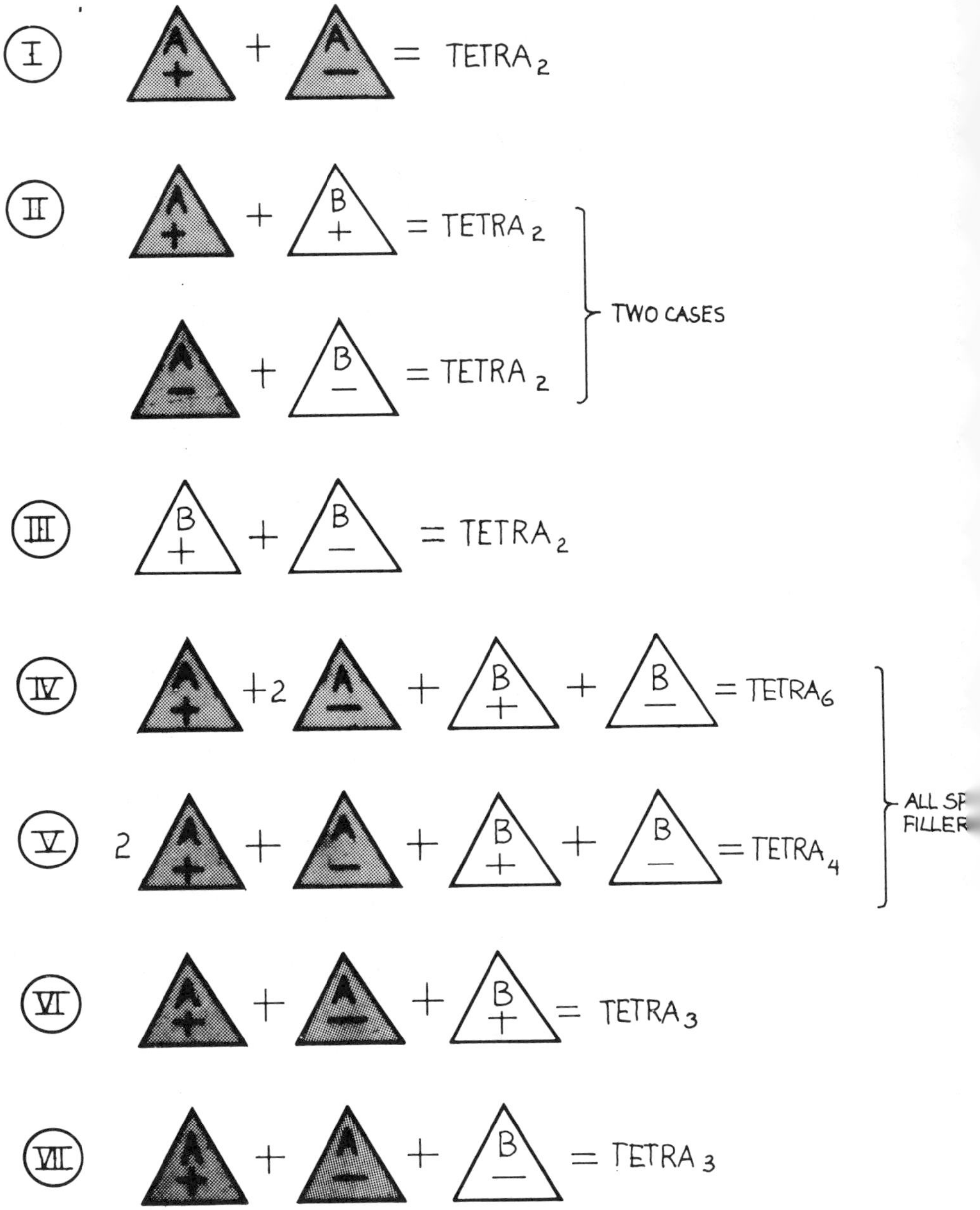

I. A+ + A− = TETRA $_2$

II. A+ + B+ = TETRA $_2$
A− + B− = TETRA $_2$
TWO CASES

III. B+ + B− = TETRA $_2$

IV. A+ + 2 A− + B+ + B− = TETRA $_6$

V. 2 A+ + A− + B+ + B− = TETRA $_4$

VI. A+ + A− + B+ = TETRA $_3$

VII. A+ + A− + B− = TETRA $_3$

Table 924.20 *Tetrahedral Functions of* A *and* B *Quanta Modules.*

Fig. 930.11: This continuous triangulation pattern strip is a 60°, angular, "come and go" alternation of very-high-frequency energy events of unit wavelength. This strip folded back on itself becomes a series of octahedra. The octahedra strips then combine to form a space-filling array of octahedra and tetrahedra, with all lines or vectors being of identical length and all the triangles equilateral and all the vertexes being omnidirectionally evenly spaced from one another. This is the pattern of "closest packing" of spheres.

successive equiangular triangular facets around its equatorial girth, with no additional triangles at its polar extremities; while in the case of the octahedral reel, it wraps closed only six of the eight triangular facets of the octahedron, which six lie around the octahedron's equatorial girth with two additional triangles left unwrapped, one each triangularly surrounding each of its poles; while in the case of the icosahedron, the equiangle-triangulated and folded ribbon wraps up only 10 of the icosahedron's 20 triangles, those 10 being the 10 that lie around the icosahedron's equatorial girth, leaving five triangles uncovered around each of its polar vertexes. (See Illus. 930.20.)

930.21 The two uncovered triangles of the octahedron may be covered by wrapping only one more triangularly folded ribbon whose axis of wraparound is one of the *XYZ* symmetrical axes of the octahedron.

930.22 Complete wrap-up of the two sets of five triangles occurring around each of the two polar zones of the icosahedron, after its equatorial zone triangles are completely enclosed by one ribbon-wrapping, can be accomplished by employing only two more such alternating, triangulated ribbon-wrappings.

930.23 The tetrahedron requires only *one* wrap-up ribbon; the octahedron *two;* and the icosahedron *three,* to cover all their respective numbers of triangular facets. Though all their faces are covered, there are, however, alternate and asymmetrically arrayed, open and closed edges of the tetra, octa, and icosa, to close all of which in an even-number of layers of ribbon coverage per each facet and per each edge of the three-and-only prime structural systems of Universe, requires *three,* triangulated, ribbon-strip wrappings for the tetrahedron; *six* for the octahedron; and *nine* for the icosahedron.

930.24 If each of the ribbon-strips used to wrap-up, completely and symmetrically, the tetra, octa, and icosa, consists of transparent tape; and those tapes have been divided by a set of equidistantly interspaced lines running parallel to the ribbon's edges; and three of these ribbons wrap the tetrahedron, six wrap the octahedron, and nine the icosahedron; then all the four equiangular triangular facets of the tetrahedron, eight of the octahedron, and 20 of the icosahedron, will be seen to be symmetrically subdivided into smaller equiangle triangles whose total number will be N^2, the second power of the number of spaces between the ribbon's parallel lines.

930.25 All of the vertexes of the intercrossings of the three-, six-, nine-ribbons' internal parallel lines and edges identify the centers of spheres closest-packed into tetrahedra, octahedra, and icosahedra of a frequency corresponding to the number of parallel intervals of the ribbons. These numbers (as we know from Sec. 223.21) are:

$2F^2 + 2$ for the tetrahedron;

$4F^2 + 2$ for the octahedron; and

$10F^2 + 2$ for the icosahedron (or vector equilibrium).

930.26 Thus we learn sum-totally how a ribbon (band) wave, a wave-band, can self-interfere periodically to produce in-shuntingly all the three prime structures of Universe and a complex isotropic vector matrix of successively shuttle-woven tetrahedra and octahedra. It also illustrates how energy may be wave-shuntingly self-knotted or self-interfered with (see Sec. 506), and their energies impounded in local, high-frequency systems which we misidentify as only-seemingly-static matter.

931.00 Chemical Bonds

931.10 Omnicongruence: When two or more systems are joined vertex to vertex, edge to edge, or in omnicongruence—in single, double, triple, or quadruple bonding, then the topological accounting must take cognizance of the congruent vectorial build in growth. (See Illus. 931.10.)

931.20 Single Bond: In a single-bonded or univalent aggregate, all the tetrahedra are joined to one another by only one vertex. The connection is like an electromagnetic universal joint or like a structural engineering pin joint; it can rotate in any direction around the joint. The mutability of behavior of single bonds elucidates the compressible and load-distributing behavior of gases.

931.30 Double Bond: If two vertexes of the tetrahedra touch one another, it is called double-bonding. The systems are joined like an engineering *hinge;* it can rotate only perpendicularly about an axis. Double-bonding characterizes the load-distributing but noncompressible behavior of liquids. This is edge-bonding.

931.40 Triple Bond: When three vertexes come together, it is called a *fixed* bond, a three-point landing. It is like an engineering fixed joint; it is *rigid.* Triple-bonding elucidates both the formational and continuing behaviors of crystalline substances. This also is face-bonding.

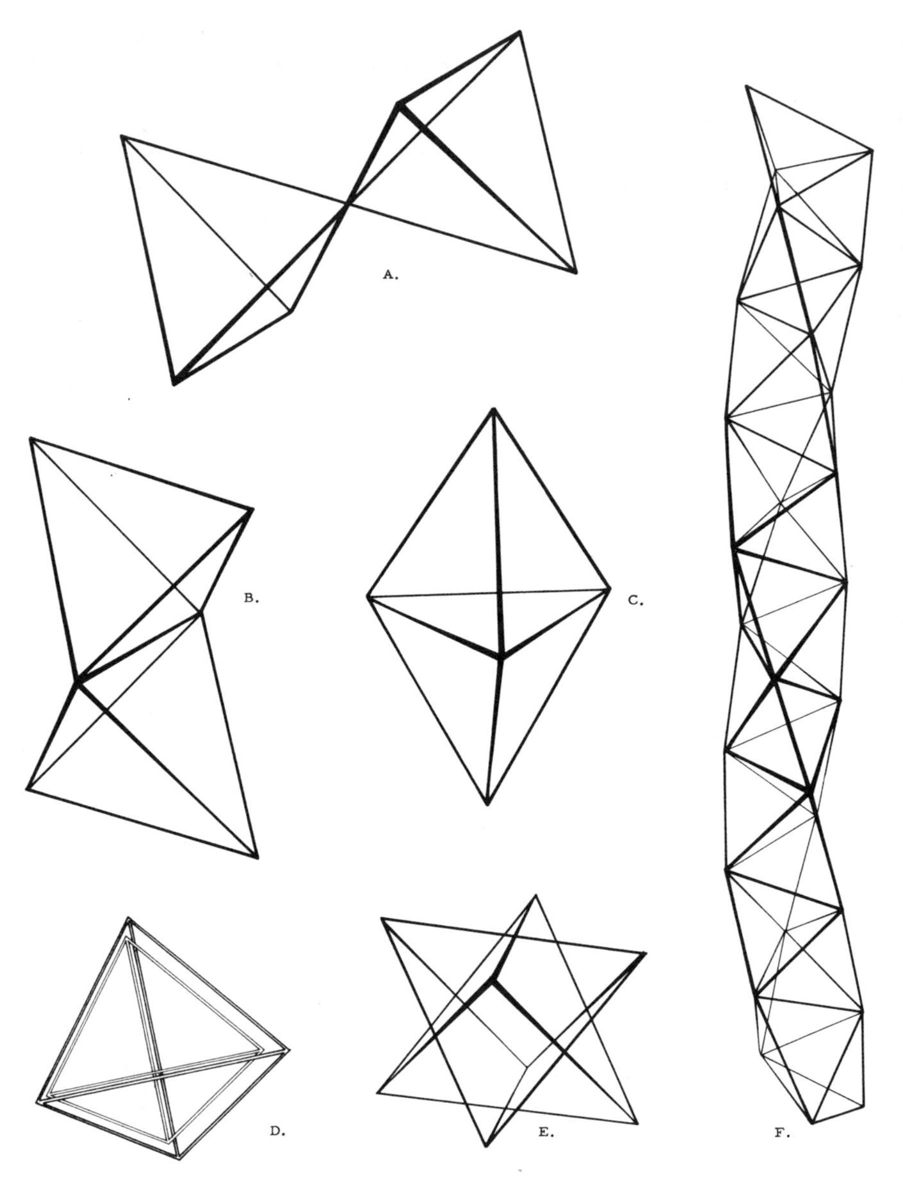

Fig. 931.10 *Tetrahedral Characteristics of Chemical Bonding: Tetrahelix:* Chemical bonds as demonstrated by arrangements of tetrahedra:

A. The single-bonded tetrahedron system is like an engineering pin joint: it can move in any direction. It characterizes the behavior of gases.

B. The double-bonded tetrahedron system is like an engineering hinge joint: it can rotate about an axis. It characterizes the behavior of liquids.

C. The triple-bonded tetrahedron system is like an engineering fixed joint: it is rigid. It demonstrates the behavior of crystalline substances.

D & E. The quadri-bond and mid-edge-coordinate tetrahedron systems demonstrate the super strength of substances such as diamond and the metals.

F. The tetrahelix: a helical array of triple-bonded tetrahedra.

931.50 **Quadruple Bond:** When four vertexes are congruent, we have quadruple-bonded *densification*. The relationship is quadrivalent. Quadri-bond and mid-edge coordinate tetrahedron systems demonstrate the super-strengths of substances such as diamonds and metals. This is the way carbon suddenly becomes very dense, as in a diamond. This is multiple self-congruence.

931.51 The behavioral hierarchy of bondings is integrated four-dimensionally with the synergies of mass-interattractions and precession.

931.60 **Quadrivalence of Energy Structures Closer-Than-Sphere Packing:** In 1885, van't Hoff showed that all organic chemical structuring is tetrahedrally configured and interaccounted in vertexial linkage. A constellation of tetrahedra linked together entirely by such single-bonded universal jointing uses lots of space, which is the openmost condition of flexibility and mutability characterizing the behavior of gases. The medium-packed condition of a double-bonded, hinged arrangement is still flexible, but sum-totally as an aggregate, allspace-filling complex is noncompressible—as are liquids. The closest-packing, triple-bonded, fixed-end arrangement corresponds with rigid-structure molecular compounds.

931.61 The closest-packing concept was developed in respect to spherical aggregates with the convex and concave octahedra and vector equilibria spaces between the spheres. Spherical closest packing overlooks a much closer packed condition of energy structures, which, however, had been comprehended by organic chemistry—that of quadrivalent and fourfold bonding, which corresponds to outright congruence of the octahedra or tetrahedra themselves. When carbon transforms from its soft, pressed-cake, carbon black powder (or charcoal) arrangement to its diamond arrangement, it converts from the so-called closest arrangement of triple bonding to quadrivalence. We call this self-congruence packing, as a single tetrahedron arrangement in contradistinction to closest packing as a neighboring-group arrangement of spheres.

931.62 Linus Pauling's X-ray diffraction analyses revealed that all metals are tetrahedrally organized in configurations interlinking the gravitational centers of the compounded atoms. It is characteristic of metals that an alloy is stronger when the different metals' unique, atomic, constellation symmetries have congruent centers of gravity, providing mid-edge, mid-face, and other coordinate, interspatial accommodation of the elements' various symmetric systems.

931.63 In omnitetrahedral structuring, a triple-bonded linear, tetrahedral array may coincide, probably significantly, with the DNA helix. The four unique quanta corners of the tetrahedron may explain DNA's unzipping dichotomy as well as—*T-A; G-C*—patterning control of all reproductions of all biological species.

932.00 *Viral Steerability*

932.01 The four chemical compounds guanine, cytosine, thymine, and adenine, whose first letters are GCTA, and of which DNA always consists in various paired code pattern sequences, such as GC, GC, CG, AT, TA, GC, in which A and T are always paired as are G and C. The pattern controls effected by DNA in all biological structures can be demonstrated by equivalent variations of the four individually unique spherical radii of two unique pairs of spheres which may be centered in any variation of series that will result in the viral steerability of the shaping of the DNA tetrahelix prototypes. (See Sec. 1050.00 et. seq.)

932.02 One of the main characteristics of DNA is that we have in its helix a structural patterning instruction, all four-dimensional patterning being controlled only by frequency and angle modulatability. The coding of the four principal chemical compounds, GCTA, contains all the instructions for the designing of all the patterns known to biological life. These four letters govern the coding of the life structures. With new life, there is a parent-child code controls unzipping. There is a dichotomy and the new life breaks off from the old with a perfect imprint and control, wherewith in turn to produce and design others.

933.00 *Tetrahelix*

933.01 The tetrahelix is a helical array of triple-bonded tetrahedra. (See Illus. 933.01.) We have a column of tetrahedra with straight edges, but when face-bonded to one another, and the tetrahedra's edges are interconnected, they altogether form a hyperbolic-parabolic, helical column. The column spirals around to make the helix, and it takes just ten tetrahedra to complete one cycle of the helix.

933.02 This tetrahelix column can be equiangle-triangular, triple-ribbon-wave produced as in the methodology of Secs. 930.10 and 930.20 by taking a ribbon three-panels wide instead of one-panel wide as in Sec. 930.10. With this triple panel folded along both of its interior lines running parallel to the three-band-wide ribbon's outer edges, and with each of the three bands interiorly scribed and folded on the lines of the equiangle-triangular wave pattern, it will be found that what might at first seem to promise to be a straight, prismatic, three-edged, triangular-based column—upon matching the next-nearest above, wave interval, outer edges of the three panels together (and taping them together)—will form the same tetrahelix column as that which is produced by taking separate equiedged tetrahedra and face-bonding them together. There is no distinguishable difference, as shown in the illustration.

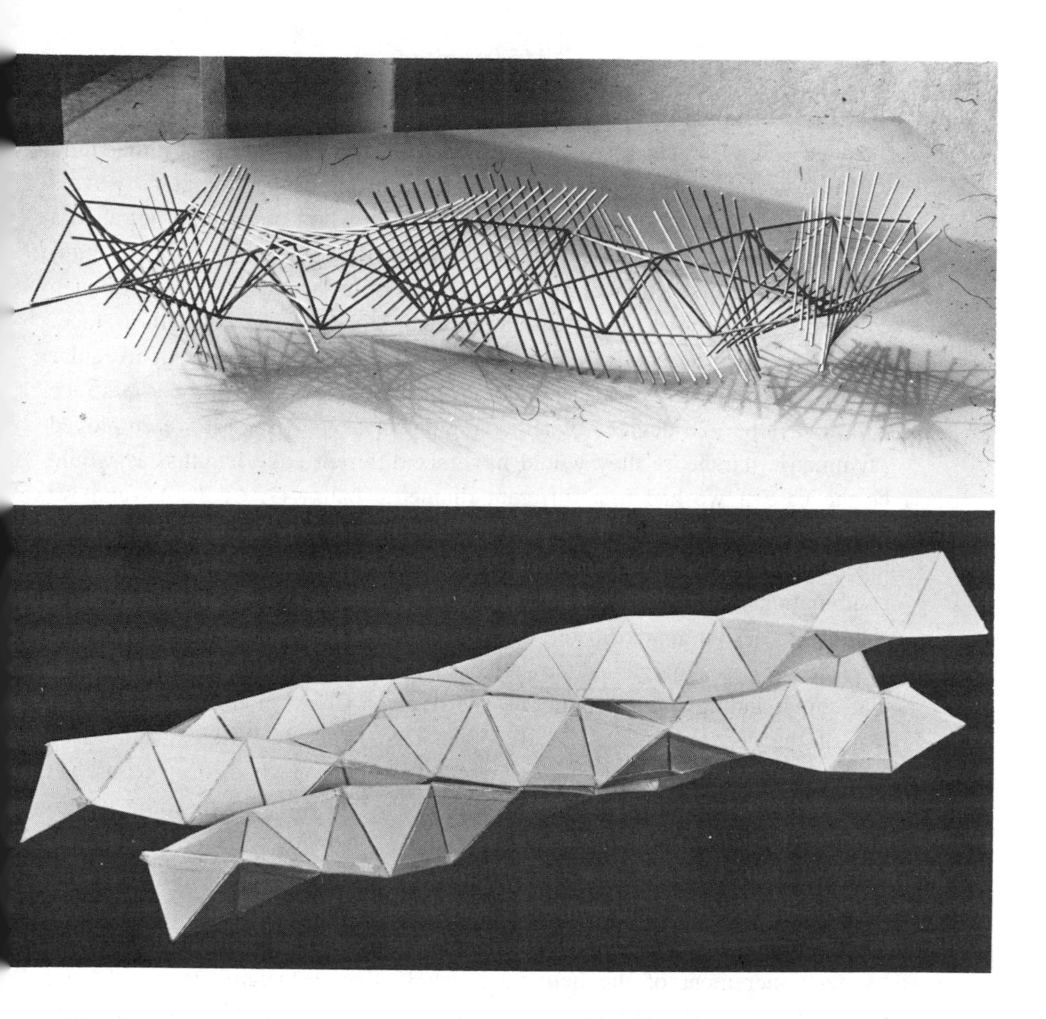

Fig. 933.01: These helical columns of tetrahedra, which we call the tetrahelix, explain the structuring of DNA models of the control of the fundamental patterning of nature's biological structuring as contained within the virus nucleus. It takes just 10 triple-bonded tetrahedra to make a helix cycle, which is a molecular compounding characteristic also of the Watson-Crick model of the DNA. When we address two or more positive (or two or more negative) tetrahelixes together, they nestle their angling forms into one another. When so nestled the tetrahedra are grouped in local clusters of five tetrahedra around a transverse axis in the tetrahclix nestling columns. Because the dihedral angles of five tetrahedra are 7° 20′ short of 360°, this 7° 20′ is sprung-closed by the helix structure's spring contraction. This backed-up spring tries constantly to unzip one nestling tetrahedron from the other, or others, of which it is a true replica. These are direct (theoretical) explanations of otherwise as yet unexplained behavior of the DNA.

933.03 The tetrahelix column may be made positive (like the right-hand-threaded screw) or negative (like the left-hand-threaded screw) by matching the next-nearest-below wave interval of the triple-band, triangular wave's outer edges together, or by starting the triple-bonding of separate tetrahedra by bonding in the only alternate manner provided by the two possible triangular faces of the first tetrahedron furthest away from the starting edge; for such columns always start and end with a tetrahedron's edge and not with its face.

933.04 Such tetrahelical columns may be made with regular or irregular tetrahedral components because the sum of the angles of a tetrahedron's face will always be 720 degrees, whether regular or asymmetric. If we employed asymmetric tetrahedra they would have six different edge lengths, as would be the case if we had four different diametric balls—*G, C, T, A*—and we paired them tangentially, *G* with *C,* and *T* with *A,* and we then nested them together (as in Sec. 623.12), and by continuing the columns in any different combinations of these pairs we would be able to modulate the rate of angular changes to design approximately any form.

933.05 This synergetics' tetrahelix is capable of demonstrating the molecular-compounding characteristic of the Watson-Crick model of the DNA, that of the deoxyribonucleic acid. When Drs. Watson, Wilkins, and Crick made their famous model of the DNA, they made a chemist's reconstruct from the information they were receiving, but not as a microscopic photograph taken through a camera. It was simply a schematic reconstruction of the data they were receiving regarding the relevant chemical associating and the disassociating. They found that a helix was developing.

933.06 They found there were 36 rotational degrees of arc accomplished by each increment of the helix and the 36 degrees aggregated as 10 arc increments in every complete helical cycle of 360 degrees. Although there has been no identification of the tetrahelix column of synergetics with the Watson-Crick model, the numbers of the increments are the same. Other molecular biologists also have found a correspondence of the tetrahelix with the structure used by some of the humans' muscle fibers.

933.07 When we address two or more positive or two or more negative tetrahelixes together, the positives nestle their angling forms into one another, as the negatives nestle likewise into one another's forms.

934.00 Unzipping Angle

934.01 If we take three columns of tetrahelixes and nest them into one another, we see that they also apparently internest neatly as with a three-part rope twist; but upon pressing them together to close the last narrow gap between them we discover that they are stubbornly resisting the final closure because the core pattern they make is one in which five tetrahedra are triple-bonded around a common edge axis—which angular gap is impossible to close.

934.02 Five tetrahedra triple-bonded to one another around a common edge axis leave an angular sinus * of 7° 20′ as the *birth unzipping* angle of DNA-RNA behaviors. This gap could be shared 10 ways, i.e., by two faces each of the five circle-closing tetrahedra, and only 44 minutes of circular arc per each tetra face, each of whose two faces might be only alternatingly edge-bonded, or hinged, to the next, which almost-closed, face-toward-face, hinge condition would mechanically accommodate the spanned coherence of this humanly-invisible, 44-minutes-of-circular-arc, distance of interadherence.

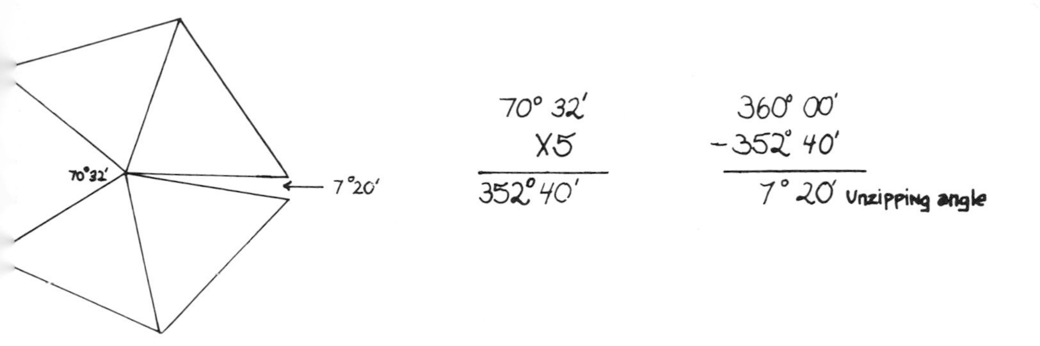

Fig. 934.02.

Making such a tetrahelix column could be exactly accomplished by only hinging one edge of each tetrahedron to the next, always making the next hinge with one of the two-out-of-three edges not employed in the previous hinge. Whatever the method of interlinkage, this birth dichotomy is apparently both *accommodated by* and *caused by* this invisible, molecular biologist's 1° 28′ per tetra and 7° 20′ per helical-cycle hinge opening.

934.03 Unzipping occurs as the birth dichotomy and the new life breaks off from the old pattern with a perfect imprint and repeats the other's growth pattern. These helixes have the ability to nest by virtue of the hinge-spring linkage by which one is being imprinted on the other. Positive columns nest with and imprint only upon positive helix columns and negative helix columns nest with and imprint their code pattern only with and upon negative helix columns. Therefore, when a column comes off, i.e., *unzips,* it is a replica of the original column.

* Sinus means *hollow* or *without* in Latin.

934.04 We know that the edge angle of a tetrahedron is 70° 32′, and five times that is 352° 40′, which is 7° 20′ less than 360°. In other words, five tetrahedra around a common edge axis do not close up and make 360 degrees, because the dihedral angles are 7° 20′ short. But when they are brought together in a helix—due to the fact that a hinged helix is a coil spring—the columns will twist enough to permit the progressive gaps to be closed. No matter how long the tetrahelix columns are, their sets of coil springs will contract enough to bring them together. The backed-up spring tries constantly to *unzip* one nesting tetrahedron from the others of which it is a true replica. These are only synergetical conjectures as to the theoretical explanations of otherwise as yet unexplained behaviors of the DNA.

940.00 Hierarchy of Quanta Module Orientations

940.10 Blue A Modules and Red B Modules

940.11 A *Modules:* We color them *blue* because the *A*s are energy conservers, being folded out of only one triangle.

940.12 B *Modules:* We color them *red* because the *B*s are energy distributors, not being foldable out of only one triangle.

940.13 This coloring will provide quick comprehension of the energy behaviors unique to the various geometrical systems and their transformations—for instance, in the outermost module layer shell of the vector equilibrium, all the triangular faces will be blue and all the square faces will be red, indicating that the eight tetrahedra of the vector equilibrium are conserving the system's structural integrity and will permit export of energy from the square faces of the system without jeopardizing the system's structural integrity.

941.00 Relation of Quanta Modules to Closest-Packed Sphere Centers

942.01 Illustrations of the *A* and *B* Quanta Modules may be made with spherical segment arcs of unit radius scribed on each of their three triangular faces having a common vertex at the sphere's center. The common center of those circular arcs lies in their respectively most acute angle vertexes; thus, when assembled, those vertexes will lie in the centers of the closest-packed spheres of which each *A* and *B* Quanta Module embraces a part, $^1/_{144}$th of a sphere, as well as its proportional part of the space between the closest-packed spheres.

942.00 Progression of Geometries in Closest Packing

942.01 Two balls of equal radius are closest packed when tangent to one another, forming a linear array with no ball at its center. Three balls are closest packed when a third ball is nested into the valley of tangency of the first two, whereby each becomes tangent to both of the other two, thus forming a triangle with no ball at its center. Four balls are closest packed when a fourth ball is nested in the triangular valley formed atop the closest-packed first three; this fourth-ball addition occasions each of the four balls becoming tangent to all three of the other balls, as altogether they form a *tetrahedron,* which is an omnidirectional, symmetrical array with no ball at its center but with one ball at each of its four corners. (See Sec. 411.)

942.02 Four additional balls can be symmetrically closest packed into the four nests of the closest-packed tetrahedral group, making eight balls altogether and forming the *star tetrahedron,* with no ball at its center.

942.03 Five balls are closest packed when a fifth ball is nested into the triangular valley on the reverse side of the original triangular group's side within whose triangular valley the fourth ball had been nested. The five form a polar-symmetry system with no ball at its center.

942.04 Six balls are closest packed when two closest-packed triangular groups are joined in such a manner that the three balls of one triangular group are nested in the three perimeter valleys of the other triangular group, and vice versa. This group of six balls is symmetrically associated, and it constitutes the six corners of the regular *octahedron,* with no ball at its center.

942.05 Eight additional balls can be mounted in the eight triangular nests of the octahedron's eight triangular faces to produce the *star octahedron,* a symmetrical group of 14 balls with no ball at the group's center.

942.10 Tetrahedron: The tetrahedron is composed exclusively of *A* Modules (blue), 24 in all, of which 12 are positive and 12 are negative. All 24 are asymmetrical, tetrahedral energy conservers.* All the tetrahedron's 24 blue *A* Modules are situate in its only one-module-deep outer layer. The tetrahedron is all blue: all energy-conserving.

942.11 Since a tetrahedron is formed by four mutually tangent spheres with no sphere at its center, the *A* Modules each contain a portion of that sphere whose center is congruent with the *A* Module's most acute corner.

* For discussion of the self-containing energy-reflecting patterns of single triangles that fold into the tetrahedron—symmetrical or asymmetrical—see Secs. 914 and 921.

942.12 The tetrahedron is defined by the lines connecting the centers of the tetrahedron's four corner spheres. The leak in the tetrahedron's corners elucidates entropy as occasioned by the only-critical-proximity but non-touching of the tetrahedron's corners-defining lines. We always have the twisting—the vectorial near-miss—at the corners of the tetrahedron because not more than one line can go through the same point at the same time. The construction lines with which geometrical entities are structured come into the critical structural proximity only, but do not yield to spontaneous mass attraction, having relative Moon-Earth-like gaps between their energy-event-defining entities of realization. (See Sec. 921.15.)

942.13 The tetrahedron has the minimum leak, but it does leak. That is one reason why Universe will never be confined within one tetrahedron, or one anything.

942.15 Quarter-Tetrahedra: Quarter-Tetrahedra have vector-edged, equiangled, triangular bases that are congruent with the faces of the regular tetrahedron. But the apex of the Quarter-Tetrahedron occurs at the center of volume of the regular tetrahedron.

942.16 The Quarter-Tetrahedra are composed of three positive *A* Quanta Modules and three negative *A* Quanta Modules, all of which are asymmetrical tetrahedra. We identify them as six energy quanta modules. These six energy quanta modules result when vertical planes running from the three vertexes to their three opposite mid-edges cut the Quarter-Tetrahedron into six parts, three of which are positive and three of which are negative.

942.17 The triangular conformation of the Quarter-Tetrahedron can be produced by nesting one uniradius ball in the center valley of a five-ball-edged, closest-packed, uniradius ball triangle. (See Illus. 415.55C.) The four vertexes of the Quarter-Tetrahedron are congruent with the volumetric centers of four uniradius balls, three of which are at the corners and one of which is nested in the valley at the center of area of a five-ball-edged, equiangle triangle.

942.18 The Quarter-Tetrahedron's six edges are congruent with the six lines of sight connecting the volumetric centers of the base triangle's three uniradius corner balls, with one uniradius ball nested atop at the triangle's center of area serving as the apex of the Quarter-Tetrahedron.

942.20 Isosceles Dodecahedron: The isosceles dodecahedron consists of the regular tetrahedron with four Quarter-Tetrahedra extroverted on each of the regular tetrahedron's four triangular faces, with the extroverted Quarter-Tetrahedra's volumetric centers occurring outside the regular tetrahedron's four triangular faces, whereas the central nuclear tetrahedron's four Quarter-

Tetrahedra are introverted with their volumetric centers situate inwardly of its four outer, regular, equiangled, triangular faces.

942.21 The isosceles dodecahedron is composed of 48 blue *A* Modules, 24 of which are *introverted;* that is, they have their centers of volume inside the faces of the central, regular tetrahedron and constitute the nuclear layer of the isosceles dodecahedron. An additional 24 *extroverted A* Modules, with their volumetric centers occurring outside the four triangular faces of the central tetrahedron, form the outermost shell of the isosceles dodecahedron. The isosceles dodecahedron is all blue both inside and outside.

942.30 Octahedron: The octahedron or "Octa" is composed of 96 energy quanta modules of which 48 are red *B* Quanta Modules and 48 blue *A* Quanta Modules. It has two module layers, with the inner, or nuclear, aggregate being the 48 red *B*s and the outer layer comprised of the 48 blue *A*s. The octahedron is all blue outside with a red nucleus.

942.31 The octahedron has distributive energies occurring at its nucleus, but they are locked up by the outer layer of *A* Modules. Thus the tendency of the 48 red *B* Module energy distributors is effectively contained and conserved by the 48 blue *A* Module conservators.

942.40 Cube: The cube is composed of a total of 72 energy quanta modules, of which there are 48 blue *A* Modules and 24 red *B* Modules. The cube is produced by superimposing four Eighth-Octahedra upon the four equiangle triangular faces of the regular tetrahedron.

942.41 The cube is three module layers deep, and the layering occurs around each of its eight corners. All of the cube's nuclear and outer-shell-modules three-layer edges are seen to surface congruently along the six diagonal seams of the cube's six faces. The inner nucleus of the cube consists of the blue introverted tetrahedron with its 24 *A* Modules. This introverted tetrahedron is next enshelled by the 24 blue *A* Modules extroverted on the introvert nuclear tetrahedron's four faces to form the isosceles dodecahedron. The third and outer layer of the cube consists of the 24 red *B* Modules mounted outward of the isosceles dodecahedron's 24 extroverted *A* Modules.

942.42 Thus, as it is seen from outside, the cube is an all-red tetrahedron, but its energy-distributive surface layer of 24 red *B* Modules is tensively overpowered two-to-one and cohered as a cube by its 48 nuclear modules. The distributors are on the outside. This may elucidate the usual occurrence of cubes in crystals with one or more of their corners truncated.

942.43 The minimum cube that can be formed by closest packing of spheres (which are inherently stable, structurally speaking) is produced by nesting four balls in the triangular mid-face nests of the four faces of a three-layer, ten-ball tetrahedron, with no ball at its volumetric center. This produces an eight-ball-cornered symmetry, which consists of 14 balls in all, with no ball at its center. This complex cube has a total of 576 *A* and *B* Modules, in contradistinction to the simplest tetra-octa-produced cube constituted of 72 *A* and *B* Modules.

942.50 **Rhombic Dodecahedron:** The rhombic dodecahedron is composed of 144 energy quanta modules. Like the cube, the rhombic dodecahedron is a three-module layered nuclear assembly, with the two-layered octahedron and its exclusively red *B* Moduled nucleus (of 48 *B*s) enveloped with 48 exclusively blue *A* Modules, which in turn are now enclosed in a third shell of 48 blue *A* Modules. Thus we find the rhombic dodecahedron and the cube co-occurring as the first three-layered, nuclearly centered symmetries—with the cube having its one layer of 24 red *B* Modules on the outside of its two blue layers of 24 *A* Modules each; conversely, the rhombic dodecahedron has its two blue layers of 48 *A* Modules each on the outside enclosing its one nuclear layer of 48 red *B* Modules.

942.51 The most simply logical arrangement of the blue *A* and red *B* Modules is one wherein their $^1/_{144}$th-sphere-containing, most acute corners are all pointed inward and join to form one whole sphere completely contained within the rhombic dodecahedron, with the contained-sphere's surface symmetrically tangent to the 12 mid-diamond facets of the rhombic dodecahedron, those 12 tangent points exactly coinciding with the points of tangency of the 12 spheres closest-packed around the one sphere. (For a discussion of the rhombic dodecahedron at the heart of the vector equilibrium, see Sec. 955.50.)

942.60 **Vector Equilibrium:** The vector equilibrium is composed of 336 blue *A* Modules and 144 red *B* Modules for a total of 480 energy quanta modules: $480 = 2^5 \times 5 \times 3$. The eight tetrahedra of the vector equilibrium consist entirely of blue *A* Modules, with a total of 48 such blue *A* Modules lying in the exterior shell. The six square faces of the vector equilibrium are the six half-octahedra, each composed of 24 blue *A*s and 24 red *B*s, from which inventory the six squares expose 48 red *B* Modules on the exterior shell. An even number of 48 *A*s and 48 *B*s provide an equilibrious exterior shell for the vector equilibrium: what an elegance! The distributors and the conservators balance. The six square areas' energies of the vector equilibrium equal the triangles' areas' energies. The distributors evacuate the half-octahedra faces and the basic triangular structure survives.

942.61 The vector equilibrium's inherently symmetrical, closest-packed-sphere aggregate has one complete sphere occurring at its volumetric center for the first time in the hierarchy of completely symmetrical, closest-packed sets. In our multilayered, omniunique patterning of symmetrical nuclear assemblies, the vector equilibrium's inner layer has four energy quanta modules in both its eight tetrahedral domains and its six half-octahedra domains, each of which domains constitutes exactly one volumetric twentieth of the vector equilibrium's total volume.

942.62 The blue *A* Modules and the red *B* Modules of the vector equilibrium are distributed in four layers as follows:

	Tetrahedral *A*s	Octahedral *A*s	Octahedral *B*s	Layer Total
1st innermost layer	48	48	48	144
2nd middle layer	48	48	48	144
3rd middle layer	48	48	0	96
4th outermost layer	48	0	48	96
	192	144	144	480
	144			
Total:	336		144	480
	A Modules		*B* Modules	Quanta Modules

942.63 In both of the innermost layers of the vector equilibrium, the energy-conserving introvert *A* Modules outnumber the *B* Modules by a ratio of two-to-one. In the third layer, the ratio is two-to-zero. In the fourth layer, the ratio of *A*s to *B*s is in exact balance.

942.64 Atoms borrow electrons when they combine. The open and unstable square faces of the vector equilibrium provide a model for the lending and borrowing operations. When the frequency is three, we can lend four balls from each square. Four is the greatest number of electrons that can be lent: here is a limit condition with the three-frequency and the four-ball edge. All the borrowing and lending operates in the squares. The triangles do not get jeopardized by virtue of lending. A lending and borrowing vector equilibrium is maintained without losing the structural integrity of Universe.

942.70 Tetrakaidecahedron: The tetrakaidecahedron—Lord Kelvin's "Solid"—is the most nearly spherical of the regular conventional polyhedra; ergo, it provides the most volume for the least surface and the most unobstructed surface for the rollability of least effort into the shallowest nests of closest-packed, most securely self-cohering, allspace-filling, symmetrical, nuclear system agglomerations with the minimum complexity of inherently concentric shell layers around a nuclear center. The more evenly faceted and the more uniform the radii of the respective polygonal members of the hierarchy of symmetrical polyhedra, the more closely they approach rollable sphericity. The four-facet tetrahedron, the six-faceted cube, and the eight-faceted octahedron are not very rollable, but the 12-faceted, one-sphere-containing rhombic dodecahedron, the 14-faceted vector equilibrium, and the 14-faceted tetrakaidecahedron are easily rollable.

942.71 The tetrakaidecahedron develops from a progression of closest-sphere-packing symmetric morphations at the exact maximum limit of one nuclear sphere center's unique influence, just before another nuclear center develops an equal magnitude inventory of originally unique local behaviors to that of the earliest nuclear agglomeration.

942.72 The first possible closest-packed formulation of a tetrakaidecahedron occurs with a three-frequency vector equilibrium as its core, with an additional six truncated, square-bottomed, and three-frequency-based and two-frequency-plateaued units superimposed on the six square faces of the three-frequency, vector-equilibrium nuclear core. The three-frequency vector equilibrium consists of a shell of 92 unit radius spheres closest packed symmetrically around 42 spheres of the same unit radius, which in turn closest-pack enclose 12 spheres of the same unit radius, which are closest packed around one nuclear sphere of the same unit radius, with each closest-packed-sphere shell enclosure producing a 14-faceted, symmetrical polyhedron of eight triangular and six square equiedged facets. The tetrakaidecahedron's six additional square nodes are produced by adding nine spheres to each of the six square faces of the three-frequency vector equilibrium's outermost 92-sphere layer. Each of these additional new spheres is placed on each of the six square facets of the vector equilibrium by nesting nine balls in closest packing in the nine possible ball matrix nests of the three-frequency vector equilibrium's square facets; which adds 54 balls to the 1

12

42

92

---

146 surrounding the nuclear ball to produce a grand total of 200 balls symmetrically surrounding one ball in an all-closest-packed, omnidirectional matrix.

942.73 The tetrakaidecahedron consists of 18,432 energy quanta modules, of which 12,672 are *A*s and 5,760 are *B*s; there are 1,008 *A*s and only 192 *B*s in the outermost layer, which ratio of conservancy dominance of *A*s over distributive *B*s is approximately two-to-one interiorly and better than five-to-one in the outermost layer.

943.00 Table: Synergetics Quanta Module Hierarchy (see also drawings section)

943.01 The orderly elegance of progressive numbers of concentric shells, starting with one as a discrete arithmetical progression, as well as the pattern of energy quanta modules growth rate, and their respective layer-transformation pairings of positive and negative arrangements of *A* and *B* Quanta Modules, of which there are always an even number of (+) or (−) *A*s or *B*s, is revealed in the synergetic quanta module hierarchy of topological characteristics.

950.00 Allspace Filling

950.01 The regular tetrahedron will not associate with other regular tetrahedra to fill allspace. (See Sec. 780.10 for a conceptual definition of allspace.) If we try to fill allspace with tetrahedra, we are frustrated because the tetrahedron will not fill all the voids above the triangular-based grid pattern. (See Illus. 950.31.) If we take an equilateral triangle and bisect its edges and interconnect the mid-points, we will have a "chessboard" of four equiangular triangles. If we then put three tetrahedra chessmen on the three corner triangles of the original triangle, and put a fourth tetrahedron chessman in the center triangle, we find that there is not enough room for other regular tetrahedra to be inserted in the too-steep valleys lying between the peaks of the tetrahedra.

950.02 If we remove the one tetrahedral chessman from the center triangle of the four-triangle chessboard and leave the three tetra-chessmen standing on the three corner triangles, we will find that one octahedral chessman (of edges equal to the tetra) exactly fits into the valley lying between the first three tetrahedra; but this is not allspace-filling exclusively with tetrahedra.

950.10 Self-Packing Allspace-Filling Geometries

950.11 There are a variety of self-packing allspace-filling geometries. Any one of them can be amplified upon in unlimited degree by high-frequency permitted aberrations. For instance, the cube can reoccur in high-

Table 943.00 Synergetics Quanta Module Hierarchy

Whole Balls at Center:	System:	Volume	Layers	Interior *A*s Implosive Conserver	Interior *B*s Explosive Exportive	Exterior Shells: *A*s Implosive Conserver	Exterior Shells: *B*s Explosive Exportive	Total Quanta Modules	Total Tetrahedral Quanta Modules (24)
0	Tetrahedron	1	1			24		24	1
0	Isosceles Dodecahedron	2	2	24		24		48	2
0	Octahedron	4	2		48	48		96	4
0	Cube	3	3	48			24	72	3
1 *	Rhombic Dodecahedron #1	6	1 (triple deep)					144	6
0	Rhombic Dodecahedron #2	6	3		48	96		144	6
1 **	Vector Equilibrium	20	4	288	96	48	48	480	20
	Tetrakaidecahedron			11,664	5,568	1,008	192	18,432	768

* Sun only; no satellites

** Sun+12 partial sphere satellites

frequency multiples with fundamental rectilinear aspects—with a cubical node on the positive face and a corresponding cubical void dimple on the negative face—which will fill allspace simply because it is a complex of cubes.

950.12 There are eight familiar self-packing allspace-fillers:

(1) The *cube*. (6 faces) Discoverer unknown.
(2) The *rhombic dodecahedron*. (12 faces) Discoverer unknown. This allspace filler is the one that occurs most frequently in nature. Rhombic dodecahedron crystals are frequently found in the floor of mineral-rich deserts.
(3) Lord Kelvin's *tetrakaidecahedron*. (14 faces)
(4) Keith Critchlow's *snub-cornered tetrahedron*. (16 faces)
(5) The *truncated octahedron*. (14 faces)
(6) The *trirectangular tetrahedron*. (4 faces) Described by Coxeter, "Regular Polytopes," p. 71. (See Illus. 950.12 B.)
(7) The *tetragonal disphenoid*. (4 faces) Described by Coxeter, "Regular Polytopes," p. 71. (See Illus. 950.12 C.)
(8) The *irregular tetrahedron* (*Mite*). (4 faces) Discovered and described by Fuller. (See Illus. 950.12 A.)

950.20 Cubical Coordination

950.21 Because the cube is the basic, prime-number-three-elucidating volume, and because the cube's prime volume is three, if we assess space volumetrically in terms of the cube as volumetric unity, we will exploit three times as much space as would be required by the tetrahedron employed as volumetric unity. Employing the extreme, minimum, limit case, ergo the prime structural system of Universe, the tetrahedron (see Sec. 610.20), as prime measure of efficiency in allspace filling, the arithmetical-geometrical volume assessment of relative space occupancy of the whole hierarchy of geometrical phenomena evaluated in terms of cubes is threefold inefficient, for we are always dealing with physical experience and structural systems whose edges consist of events whose actions, reactions, and resultants comprise one basic energy vector. The cube, therefore, requires threefold the energy to structure it as compared with the tetrahedron. We thus understand why nature uses the tetrahedron as the prime unit of energy, as its energy quantum, because it is three times as efficient in every energetic aspect as its nearest symmetrical, volumetric competitor, the cube. All the physicists' experiments show that nature always employs the most energy-economical strategies.

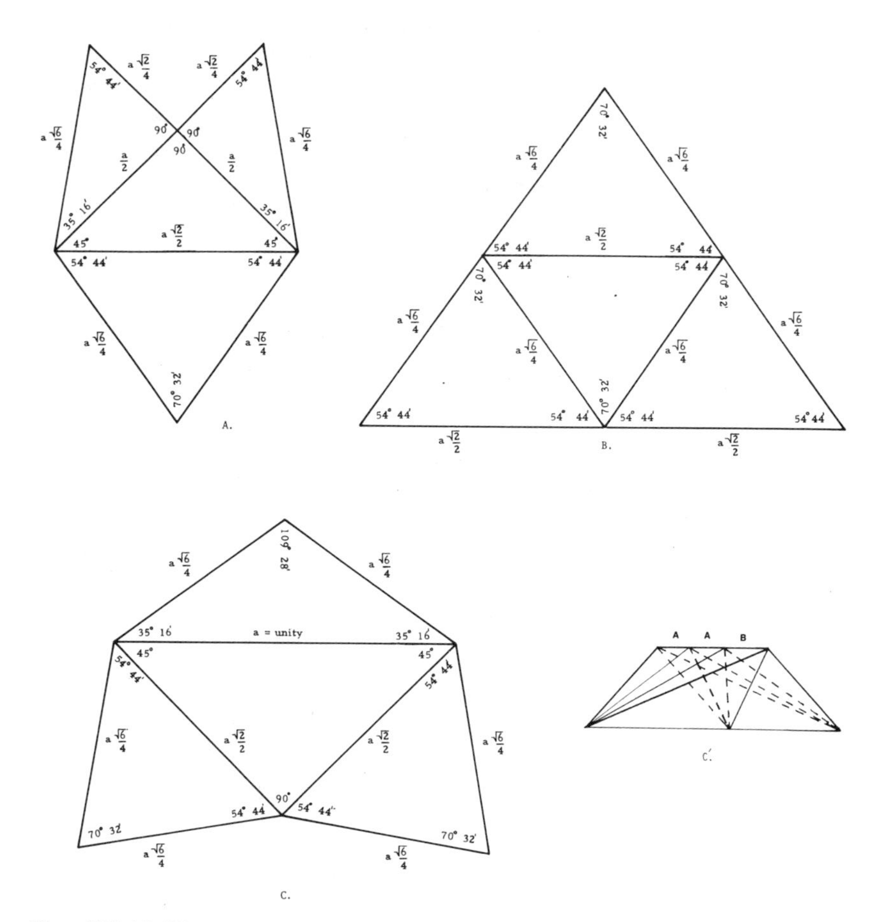

Fig. 950.12 *Three Self-Packing, Allspace-Filling Irregular Tetrahedra:* There are three self-packing irregular tetrahedra that will fill allspace without need of any complementary shape (not even with the need of right- and left-hand versions of themselves). One, the Mite (A), has been proposed by Fuller and described by Coxeter as a tri-rectangular tetrahedron in his book *Regular Polytopes,* p. 71. By joining together two Mites, two varieties of irregular tetrahedra, both called Sytes, can be formed. The tetragonal disphenoid (B), described by Coxeter, is also called the isosceles tetrahedron because it is bounded by four congruent isosceles triangles. The other Syte is formed by joining two Mites by their right-triangle faces (C). It was discovered by Fuller that the Mite has a population of two A quanta modules and one B quanta module (not noted by Coxeter). It is of interest to note that the B quanta module of the Mite may be either right- or left-handed (see the remarks of Arthur L. Loeb). Either of the other two self-packing irregular tetrahedra (Sytes) have a population of four A quanta modules and two B quanta modules, since each Syte consists of two Mites.

Since the Mites are the limit case all space-filling system, Mites may have some relationship to quarks. The A quanta module can be folded out of one planar triangle, suggesting that it may be an energy conserver, while the B quanta module can not, suggesting that it may be an energy dissipator. This gives the Mite a population of two energy conservers (A quanta module) and one energy dissipator (B quanta module).

950.30 Tetrahedron and Octahedron as Complementary Allspace Fillers: A *and* B *Quanta Modules*

950.31 We may ask: What can we do to negotiate allspace filling with tetrahedra? In an isotropic vector matrix, it will be discovered that there are only two polyhedra described internally by the configuration of the interacting lines: the regular tetrahedron and the regular octahedron. (See Illus. 950.31.)

950.32 All the other regular symmetric polyhedra known are also describable repetitiously by compounding rational fraction elements of the tetrahedron and octahedron: the *A* and *B* Quanta Modules, each having the volume of $^1/_{24}$th of a tetrahedron.

950.33 It will be discovered also that all the polygons formed by the interacting vectors consist entirely of equilateral triangles and squares, the latter occurring as the cross sections of the octahedra, and the triangles occurring as the external facets of both the tetrahedra and octahedra.

950.34 The tetrahedra and octahedra complement one another as space fillers. This is not very satisfactory if you are looking for a monological explanation: the "building block" of the Universe, the "key," the ego's wished-for monopolizer. But if you are willing to go along with the physicists, recognizing complementarity, then you will see that tetrahedra plus octahedra—and their common constituents, the unit-volume, *A* and *B* Quanta Modules—provide a satisfactory way for both physical and metaphysical, generalized cosmic accounting of all human experience. Everything comes out rationally.

951.00 Allspace-Filling Tetrahedra

951.01 The tetrahedra that fill allspace by themselves are all asymmetrical. They are dynamic reality only-for-each-moment. Reality is always asymmetrical.

951.10 **Synergetic Allspace-Filling Tetrahedron:** Synergetic geometry has one cosmically minimal, allspace-filling tetrahedron consisting of only four *A* Quanta Modules and two *B* Quanta Modules—six modules in all—whereas the regular tetrahedron consists of 24 such modules and the cube consists of 72. (See Illus. 950.12.)

953.00 Mites and Sytes: Minimum Tetrahedra as Three-Module Allspace Fillers

953.10 **Minimum Tetrahedron: Mite:** Two *A* Quanta Modules and one *B* Quanta Module may be associated to define the allspace-filling positive and

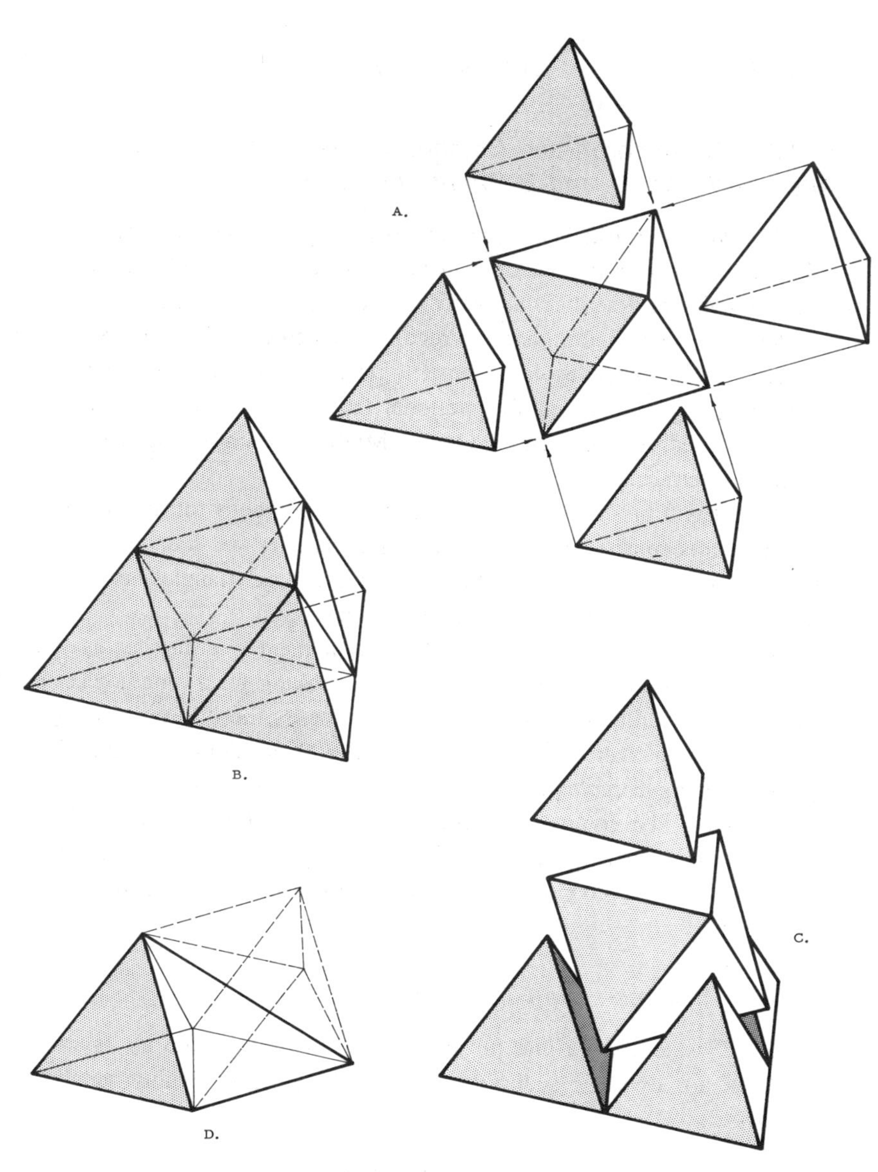

Fig. 950.31 *Tetrahedra and Octahedra Combine to Fill Space:* Regular tetrahedra alone will not fill space, but when four tetrahedra (A) are grouped to define a larger tetrahedron (B), the resulting central space is an octahedron (C). Therefore tetrahedra and octahedra will combine to fill all the space. If the volume of the smaller tetrahedron is equal to one then the volume of the larger tetrahedron is eight, i.e. edge length two to the third power ($2 \times 2 \times 2$). (When we double the linear dimension of a figure we always increase its volume eight-fold.) If the volume of the large tetrahedron is eight the central octahedron must have a volume of exactly four, while the small tetrahedra each equal one. The volume of a pyramid is $^1/_3$ the base area times the height. Therefore the $^1/_4$-octahedron (D) has exactly the same volume as its corresponding tetrahedron, further proof that the regular octahedron has exactly four times the volume of a regular tetrahedron of the same edge length.

negative sets of three geometrically dissimilar, asymmetric, but *unit volume* energy quanta modules which join the volumetric center hearts of the octahedron and tetrahedron. For economy of discourse, we will give this minimum allspace-filling *AAB* complex three-quanta module's asymmetrical tetrahedron the name of Mite (as a contraction of Minimum Tetrahedron, allspace filler). (See drawings section.)

953.20 **Positive or Negative:** Mites can fill allspace. They can be either positive (+) or negative (−), affording a beautiful confirmation of negative Universe. Each one can fill allspace, but with quite different energy consequences. Both the positive and negative Mite Tetrahedra are comprised, respectively, of two *A* Quanta Modules and one *B* Quanta Module. In each Mite, one of the two *A*s is positive and one is negative; the *B* must be positive when the Mite is positive and negative when the Mite is negative. The middle *A* Quanta Module of the *AAB* wedge-shaped sandwich is positive when the Mite and its *B* Quanta Module are negative. The Mite and its *B* Quanta Module have like signs. The Mite and its middle *A* Quanta Module have unlike signs.

953.21 If there were only positive Universe, there would be only Sytes (see Sec. 953.40). But Mites can be either plus or minus; they accommodate both Universes, the positive and the negative, as well as the half-positive and half-negative, as manifestations of fundamental complementarity. They are true rights and lefts, not mirror images; they are inside out and asymmetrical.

953.22 There is a noncongruent, ergo mutually exclusive, tripartiteness (i.e., two *A*s and one *B* in a wedge sandwich) respectively unique to either the positive or the negative world. The positive model provides for the interchange between the spheres and the spaces.* But the Mite permits the same kind of exchange in negative Universe.

953.23 The cube as an allspace filler requires only a positive world. The inside-out cube is congruent with the outside-out cube. Whereas the inside-out and outside-out Mites are not congruent and refuse congruency.

953.24 Neither the tetrahedron nor the octahedron can be put together with Mites. But the allspace-filling rhombic dodecahedron and the allspace-filling tetrakaidecahedron can be exactly assembled with Mites. Their entire componentation exclusively by Mites tells us that either or both the rhombic dodecahedron and the tetrakaidecahedron can function in either the positive or the negative Universe.

953.25 The allspace-filling functions of the (+) or (−) *AAB* three-module Mite combines can operate either positively or negatively. We can

* See Sec. 1032.10.

take a collection of the positives or a collection of the negatives. If there were only positive outside-out Universe, it would require only one of the three alternate six-module, allspace-filling tetrahedra (see Sec. 953.40) combined of two *A* (+), two *A* (−), one *B* (+), and one *B* (−) to fill allspace symmetrically and complementarily. But with both inside-out and outside-out worlds, we can fill all the outside-out world's space positively and all the inside-out world's space negatively, accommodating the inherent complementarity symmetry requirements of the macro-micro cosmic law of convex world and concave world, while remembering all the time that among all polyhedra only the tetrahedron can turn itself inside out.

953.30 **Tetrahedron as Three-Petaled Flower Bud:** Positive or negative means that one is the inside-out of the other. To understand the inside-outing of tetrahedra, think of the tetrahedron's four outside faces as being blue and the four inside faces as being red. If we split open any three edges leading to any one of the tetrahedron's vertexes, the tetrahedron will appear as a three-petaled flower bud, just opening, with the triangular petals hinging open around the common triangular base. The opening of the outside-blue–inside-red tetrahedron and the hinging of all its blue bud's petals outwardly and downwardly until they meet one another's edges below the base, will result in the whole tetrahedron's appearing to be red while its hidden interior is blue. All the other geometrical characteristics remain the same. If it is a regular tetrahedron, all the parts of the outside-red or the outside-blue regular tetrahedron will register in absolute congruence.

953.40 **Symmetrical Tetrahedron: Syte:** Two of the *AAB* allspace-filling, three-quanta module, asymmetric tetrahedra, the Mites—one positive and one negative—may be joined together to form the six-quanta-module, semisymmetrical, allspace-filling Sytes. The Mites can be assembled in three different ways to produce three morphologically different, allspace-filling, asymmetrical tetrahedra: the *Kites, Lites,* and *Bites,* but all of the same six-module volume. This is done in each by making congruent matching sets of their three, alternately matchable, right-triangle facets, one of which is dissimilar to the other two, while those other two are both positive-negative mirror images of one another. Each of the three pairings produces one six-quanta module consisting of two *A*(+), two *A*(−), one *B*(+), and one *B*(−).

953.50 **Geometrical Combinations:** All of the well-known Platonic, Archimedean, Keplerian, and Coxeter types of radially symmetric polyhedra may be directly produced or indirectly transformed from the whole unitary combining of Mites without any fractionation and in whole, rational number increments of the *A* or *B* Quanta Module volumes. This prospect may bring us within sight of a plenitudinous complex of conceptually discrete, energy-importing, -retaining, and -exporting capabilities of nuclear assemblage components, which has great significance as a specific closed-system complex with unique energy-behavior-elucidating phenomena. In due course, its unique behaviors may be identified with, and explain discretely, the inventory of high-energy physics' present prolific production of an equal variety of strange small-energy "particles," which are being brought into split-second existence and observation by the ultrahigh-voltage accelerator's bombardments.

953.60 **Prime Minimum System:** Since the asymmetrical tetrahedron formed by compounding two *A* Quanta Modules and one *B* Quanta Module, the Mite, will compound with multiples of itself to fill allspace and may be turned inside out to form its noncongruent negative complement, which may also be compounded with multiples of itself to fill allspace, this minimum asymmetric system—which accommodates both positive or negative space and whose volume is exactly $^1/_8$th that of the tetrahedron, exactly $^1/_{32}$nd that of the octahedron, exactly $^1/_{160}$th that of the vector equilibrium of zero frequency, and exactly $^1/_{1280}$th of the vector equilibrium of initial frequency $(=2)$, $1280 = 2^8 \times 5$—this Mite constitutes the generalized nuclear geometric limit of rational differentiation and is most suitably to be identified as the *prime minimum system;* it may also be identified as the prime, minimum, rationally volumed and rationally associable, structural system.

954.00 *Mite as the Coupler's Asymmetrical Octant Zone Filler* (*see drawings section*)

954.01 The Coupler is the asymmetric octahedron to be elucidated in Secs. 954.20 through 954.70. The Coupler has one of the most profound integral functionings in metaphysical Universe, and probably so in physical Universe, because its integral complexities consist entirely of integral rearrangeability within the same space of the same plus and/or minus Mites. We will now inspect the characteristics and properties of those Mites as they function in the Coupler. Three disparately conformed, nonequitriangular, polarized half-octahedra, each consisting of the same four equivolumetric octant zones occur around the three half-octants' common volumetric center. These eight octant zones are all occupied, in three possible different system arrangements, by identical asymmetrical tetrahedra, which are Mites, each consisting of the three *AAB* Modules.

954.02 Each of these $^1/_8$ octant-zone-filling tetrahedral Mite's respective surfaces consists of four triangles, *CAA, DEE,* EFG^1, and EFG^2, two of which, *CAA* and *DBB*, are dissimilar isosceles triangles and two of which, EFG^1 and EFG^2, are right triangles. (See Illus. 953. 10.) Each of the dissimilar isosceles triangles have one mutual edge, *AA* and *BB*, which is the base respectively of both the isosceles triangles whose respective symmetrical apexes, *C* and *D*, are at different distances from that mutual baseline.

954.03 The smaller of the mutually based isosceles triangle's apex is a right angle, *D*. If we consider the right-angle-apexed isosceles triangle *DBB* to be the horizontal base of a unique octant-zone-filling tetrahedron, we find the sixth edge of the tetrahedron rising perpendicularly from the right-angle apex, *D*, of the base to *C* (*FF*), which perpendicular produces two additional right triangles, FGE^1 and FGE^2, vertically adjoining and thus surrounding the isosceles base triangle's right-angled apex, *D*. This perpendicular *D* (*FF*) connects at its top with the apex *C* of the larger isosceles triangle whose baseline, *AA*, is symmetrically opposite that *C* apex and congruent with the baseline, *BB*, of the right-angle-apexed isosceles base triangle, *BBD*, of our unique octant-filling tetrahedral Mite, *AACD*.

954.04 The two vertical right triangles running between the equilateral edges of the large and small isosceles triangles are identical right triangles, EFG^1 and EFG^2, whose largest (top) angles are each 54° 44′ and whose smaller angles are 35° 16′ each.

$$\begin{array}{r} 90^\circ\ 00' \\ 54^\circ\ 44' \\ 35^\circ\ 16' \\ \hline 180^\circ\ 00' \end{array}$$

954.05 As a tetrahedron, the Mite has four triangular faces: *BBD, AAC*, EFG^1, and EFG^2. Two of the faces are dissimilar isosceles triangles, *BBD* and *AAC;* ergo, they have only two sets of two different face angles each—*B, D, A*, and *C*—one of which, *D*, is a right angle.

954.06 The other two tetrahedral faces of the Mites are similar right triangles, *EFG*, which introduce only two more unique angles, *E* and *F*, to the Mite's surface inventory of unique angles.

954.07 The inventory of the Mite's twelve corner angles reveals only five different angles. There are two *A*s and two *F*s, all of which are 54° 44′ each, while there are three right angles consisting of one *D* and two *G*s. There are two *B*s of 45° 00′ each, two *E*s of 35° 16′ each, and one *C* of 70° 32′. (See drawings section.)

954.08 Any of these eight interior octant, double-isosceles, three-right-angled-tetrahedral domains—Mites—(which are so arrayed around the center of volume of the asymmetrical octahedron) can be either a positively or a negatively composited allspace-filling tetrahedron.

954.09 We find the Mite tetrahedron, *AACD*, to be the smallest, simplest, geometrically possible (volume, field, or charge), allspace-filling module of the isotropic vector matrix of Universe. Because it is a tetrahedron, it also qualifies as a structural system. Its volume is exactly $^1/_8$th that of its regular *tetrahedral* counterpart in their common magnitude isotropic vector matrix; within this matrix, it is also only $^1/_{24}$th the volume of its corresponding allspace-filling *cube,* $^1/_{48}$th the volume of its corresponding allspace-filling *rhombic dodecahedron,* and $^1/_{6144}$th the volume of its one other known unique, omnidirectional, symmetrically aggregatable, nonpolarized-assemblage, unit-magnitude, isotropic-vector-matrix counterpart, the allspace-filling *tetrakaidecahedron.*

954.10 Allspace-Filling Hierarchy as Rationally Quantifiable in Whole Volume Units of *A* or *B* Quanta Modules

Synergetics' Name	*Quanta Module Volume*	*Type Polyhedron*	*Symmetrical or Asymmetrical*
Mite	3	Tetrahedron	Asymmetrical
Syte (3 types)	6	Tetrahedron	Asymmetrical
Kites			
Lites			
Bites			
Coupler	24	Octahedron	Asymmetrical
Cube	72	Cube	Simple Symmetrical
Rhombic Dodecahedron	144	Rhombic Dodecahedron	Complex Symmetrical
Tetrakaidecahedron	18,432	Tetrakaidecahedron	Complex Symmetrical

954.20 **Coupler:** The basic complementarity of our octahedron and tetrahedron, which always share the disparate numbers 1 and 4 in our topological analysis (despite its being double or 4 in relation to tetra = 1), is explained by the uniquely asymmetrical octahedron, the Coupler, that is always constituted by the many different admixtures of *AAB* Quanta Modules; the Mites, the Sytes, the cube (72 *A*s and *B*s), and the rhombic dodecahedron (144 *A*s and *B*s).

954.21 There are always 24 *A*s or *B*s in our uniquely asymmetrical octahedron (the same as one tetra), which we will name the Coupler because it occurs between the respective volumetric centers of any two of the adjacently matching diamond faces of all the symmetrical, allspace-filling rhombic dodecahedra (or 144 *A*s and *B*s). The rhombic dodecahedron is the most-faceted, identical-faceted (diamond) polyhedron and accounts, congruently and symmetrically, for all the unique domains of all the isotropic-vector-matrix vertexes. (Each of the isotropic-vector-matrix vertexes is surrounded symmetrically either by the spheres or the intervening spaces-between-spheres of the closest-packed sphere aggregates.) Each rhombic dodecahedron's diamond face is at the long-axis center of each *Coupler* (vol. = 1) asymmetric octahedron. Each of the 12 rhombic dodecahedra is completely and symmetrically omnisurrounded by—and diamond-face-bonded with—12 other such rhombic dodecahedra, each representing one closest-packed sphere and that sphere's unique, cosmic, intersphere-space domain lying exactly between the center of the nuclear rhombic dodecahedron and the centers of their 12 surrounding rhombic dodecahedra—the *Couplers* of those closest-packed-sphere domains having obviously unique cosmic functioning.

954.22 A variety of energy effects of the *A* and *B* Quanta Module associabilities are contained uniquely and are properties of the *Couplers,* one of whose unique characteristics is that the Coupler's topological volume is the exact prime number one of our synergetics' tetrahedron (24 *A*s) accounting system. It is the asymmetry of the *B*s (of identical volume to the *A*s) that provides the variety of other than plusness and minusness of the all-*A*-constellated tetrahedra. Now we see the octahedra that are allspace filling and of the same volume as the *A*s in complementation. We see proton and neutron complementation and non-mirror-imaging interchangeability and intertransformability with 24 subparticle differentiabilities and 2, 3, 4, 6, combinations—enough to account for all the isotopal variations and all the nuclear substructurings in omnirational quantation.

954.30 **Nuclear Asymmetric Octahedra:** There are eight additional asymmetric octahedra Couplers surrounding each face of each *Coupler*. It is probable that these eight asymmetric nuclear octahedra and the large variety of each of their respective constituent plus and minus Mite mix may account for all the varieties of intercomplex complexity required for the permutations of the 92 regenerative chemical elements. These eight variables alone provide for a fantastic number of rearrangements and reorientations of the *A* and *B* Quanta Modules within exactly the same geometric domain volume.

954.31 It is possible that there are no other fundamental complex varieties than those accounted for by the eight nuclear Coupler-surrounding asymmetrical octahedra. There is a mathematical limit of variation—with our friend octave coming in as before. The Coupler may well be what we have been looking for when we have been talking about "number one." It is quite possibly one nucleon, which can be either neutron or proton, depending on how you rearrange the modules in the same space.

954.32 There are enough coincidences of data to suggest that the bombardment-produced energy entities may be identified with the three energy quanta modules—two *A* Quanta Modules and one *B* Quanta Module—allspace-filler complexities of associability, all occurring entirely within one uniquely proportioned, polarized, asymmetrical, nonequilateral, eight-triangle-faceted polyhedron—the Coupler—within whose interior only they may be allspace-fillingly rearranged in a large variety of ways without altering the external conformation of the asymmetrical, octahedral container.

954.40 **Functions of the Coupler:** In their cosmic roles as the basic allspace-filling complementarity pair, our regular tetrahedron and regular octahedron are also always identified respectively by the disparate numbers 1 and 4 in the column of relative volumes on our comprehensive chart of the topological hierarchies. (See Chart 223.64.) The volume value 4—being 2^2—also identifies the prime number 2 as always being topologically unique to the symmetrical octahedron while, on the same topological hierarchy chart, the *uniquely asymmetrical* allspace-filling *octahedron,* the Coupler, has a volume of 1, which volume-1-identity is otherwise, topologically, uniquely identified only with the non-allspace-filling regular symmetrical tetrahedron.

954.41 The uniquely asymmetrical octahedron has three *XYZ* axes and a center of volume, *K*. Its *X* and *Y* axes are equal in length, while the *Z* axis is shorter than the other two. The uniquely asymmetrical octahedron is always polarly symmetrical around its short *Z* axis, whose spin equatorial plane is a square whose diagonals are the equilengthed *X* and *Y* axes. The equatorially spun planes of both the *X* and *Y* axes are similar diamonds, the short diagonal of each of these diamonds being the *Z* axis of the uniquely asymmetrical octahedron, while the long diagonal of the two similar diamonds are the *X* and *Y* axes, respectively, of the uniquely asymmetrical octahedron.

954.42 The uniquely asymmetrical octahedron could also be named the polarly symmetrical octahedron. There is much that is unique about it. To begin with the "heart," or center of volume of the asymmetrical octahedron (knowable also as the polarly symmetrical octahedron, of geometrical volume 1), is identified by the capital letter *K* because *K* is always the kissing or tangency point between each and every sphere in all closest-packed unit radius sphere aggregates; *and it is only through* those 12 kissing (tangency) points symmetrically embracing every closest-packed sphere that each and all of the 25 unique great circles of fundamental crystallographic symmetry must pass—those 25 great circles being generated as the 3, 4, 6, 12 = 25 great circle equators of spin of the only-four-possible axes of symmetry of the vector equilibrium. Therefore it is only through those volumetric heart *K* points of the uniquely asymmetrical octahedra that energy can travel electromagnetically, wavelinearly, from here to there throughout Universe over the shortest convex paths which they always follow.

954.43 The uniquely asymmetrical octahedron is always uniformly composed of exactly eight asymmetrical, allspace-filling, double-isosceles tetrahedra, the Mites, which in turn consist of *AAB* three-quanta modules each. Though outwardly conformed identically with one another, the Mites are always either positively or negatively biased internally in respect to their energy valving (amplifying, choking, cutting off, and holding) proclivities, which are only "potential" when separately considered, but operationally effective as interassociated within the allspace-filling, uniquely asymmetrical octahedron, and even then muted (i.e., with action suspended as in a holding pattern) until complexes of such allspace-filling and regeneratively circuited energy transactions are initiated.

954.44 The cosmically minimal, allspace-filling Mites' inherent bias results from their having always one $A+$ and one $A-$ triple-bonded (i.e., face-bonded) to constitute a symmetrical isosceles (two-module) but non-allspace-filling tetrahedron to either one of the two external faces, of which either one $B+$ or one $B-$ can be added to provide the allspace-filling, semisymmetrical double-isosceles, triple right-angled, three-moduled Mite, with its positive and negative bias sublimatingly obscured by the fact that either the positive or the negative quantum biasing add together to produce the same overall geometrical space-filling tetrahedral form, despite its quanta-biased composition. This obscurity accounts for its heretofore unbeknownstness to science and with that unbeknownstness its significance as the conceptual link between the heretofore remote humanists and the scientists' cerebrating, while with its discovery comes lucidly conceptual comprehension of the arithmetical and geometrical formings of the whole inventory of the isotopes of all the atoms as explained by the allspace-filling variety of internal and external associabilities and reorientings permitted within and without the respective local octant-filling of the, also in-turn, omni-space-filling, uniquely asymmetrical octahedron, the Coupler.

954.45 As learned in Sections 953 and 954, one plus-biased Mite and one minus-biased Mite can be face-bonded with one another in three different allspace-filling ways, yet always producing one energy-proclivity-balanced, six-quanta-moduled, double-isosceles, allspace-filling, asymmetrical tetrahedron: the Syte. The asymmetric octahedron can also be composed of four such balanced-bias Sytes (4 *A*s—2 +, 2 −—and 2 *B*s—1 +, 1 −). Since there are eight always one-way-or-the-other-biased Mites in each uniquely asymmetrical octahedron, the latter could consist of eight positively biased or eight negatively biased Mites, or any omnigeometrically permitted mixed combination of those 16 (2^4) cases.

954.46 There are always 24 modules (16 *A*s and 8 *B*s—of which eight *A*s are always positive and the eight other *A*s are *always* negative, while the eight *B*s consist of any of the eight possible combinations of positives and negatives) * in our uniquely asymmetrical octahedron. It is important to note that this 24 is the same 24-module count as that of the 24-*A*-moduled regular tetrahedron. We have named the uniquely asymmetrical octahedron the *Coupler*.

954.47 We give it the name the Coupler because it always occurs between the adjacently matching diamond faces of all the symmetrical allspace-filling rhombic dodecahedra, the "spherics" (of 96 *A*s and 48 *B*s). The rhombic dodecahedron has the maximum number (12) of identical (diamond) faces of all the allspace-filling, unit edge length, symmetrical polyhedra. That is, it most nearly approaches sphericity, i.e., the shortest-radiused, symmetrical, structural, polyhedral system. And each rhombic dodecahedron exactly embraces within its own sphere each of all the closest-packed unit radius spheres of Universe, and each rhombic dodecahedron's volumetric center is congruent with the volumetric center of its enclosed sphere, while the rhombic dodecahedron also embracingly accounts, both congruently and symmetrically, for all the isotropic-vector-matrix vertexes in closest-packed and all their "between spaces." The rhombic dodecahedra are the unique cosmic domains of their respectively embraced unit radius closest-packed spheres. The center of area, *K*, of each of the 12 external diamond faces of each rhombic dodecahedron is always congruent with the internal center of volume (tangent sphere's kissing points), *K*, of all the allspace-filling uniquely asymmetrical octahedra.

954.48 Thus the uniquely asymmetrical octahedra serve most economically to join, or couple, the centers of volume of each of the 12 unit radius spheres tangentially closest packed around every closest packed sphere in Universe, with the center of volume of that omnisurrounded, ergo nuclear, sphere. However the asymmetrical, octahedral coupler has three axes (*X*, *Y*, *M*), and only its *X* axis is involved in the most economical intercoupling of the energy potentials centered within all the closest-packed unit radius spheres. The *Y* and *M* axes also couple two alternative sets of isotropic-

* 8 all plus 0 minus
7 plus 1 minus
6 plus 2 minus
5 plus 3 minus
4 plus 4 minus
3 plus 5 minus
2 plus 6 minus
1 plus 7 minus
0 plus 8 minus

These combinations accommodate the same bow-tie wave patterns of the Indigs (see Sec. 1223). This eight-digited manifold is congruent with the Indig bow-tie wave—another instance of the congruence of number and geometry in synergetics. Because of the prime quanta functioning of the allspace-filling Mites, we observe an elegant confirmation of the omniembracing and omnipermeative pattern integrities of synergetics.

vector-matrix centers. The *M* axis coupling the centers of volume of the concave vector equilibria shaped between closest-packed sphere spaces, and the *Y* axis interconnecting all the concave octahedral between spaces of unit-radius closest-packed sphere aggregates, both of which concave between-sphere spaces become spheres as all the spheres—as convex vector equilibria or convex octahedra—transform uniformly, sumtotally, and coincidentally into concave-between-unit-radius-sphere spaces. The alternate energy transmitting orientations of the locally contained *A* and *B* Quanta Modules contained within the 12 couplers of each nuclear set accommodate all the atomic isotope formulations and all their concomitant side effects.

954.49 We also call it the Coupler because its volume = 1 regular tetrahedron = 24 modules. The Couplers *uniquely bind together* each rhombic dodecahedron's center of volume with the centers of volume of all its 12 omniadjacent, omniembracing, rhombic dodecahedral "spherics."

954.50 But it must be remembered that the centers of volume of the rhombic dodecahedral spherics are also the centers of each of all the closest-packed spheres of unit radius, and their volumetric centers are also omnicongruent with all the vertexes of all isotropic vector matrixes. The Couplers literally couple "everything," while alternatively permitting all the varieties of realizable events experienced by humans as the sensation of "free will."

954.51 We see that the full variety of energy effects made by the variety of uniquely permitted *A*-and-*B*-Module rearrangeabilities and reassociabilities within the unique volumetric domain of the Coupler manifest a startling uniqueness in the properties of the Coupler. One of the Coupler's other unique characteristics is that its volume is also the exact prime number 1, which volumetric oneness characterizes only one other polyhedron in the isotropic-vector-matrix hierarchy, and that one other prime-number-one-volumed polyhedron of our quantum system is the symmetric, initial-and-minimal-structural system of Universe: the 24-module regular tetrahedron. Here we may be identifying the cosmic bridge between the equilibrious prime number one of metaphysics and the disequilibrious prime number one of realizable physical reality.

954.52 It is also evidenced that the half-population ratio asymmetry of the *B* Modules (of identical volume to the *A* Modules) in respect to the population of the *A* Modules, provides the intramural variety of rearrangements—other than the $^{1}/_{1}$ plus-and-minusness—of the all-*A*-Module-constellated regular tetrahedron.

954.53 The Coupler octahedron is allspace-filling and of the same 24-module volume as the regular tetrahedron, which is not allspace-filling. We go on to identify them with the proton's and neutron's non-mirror-imaged complementation and intertransformability, because one consists of 24 blue *A* Modules while the other consists of sixteen blue *A*s and eight red *B*s, which renders them not only dissimilar in fundamental geometric conformation, but behaviorally *different* in that the *A*s are energy-inhibiting and the *B*s are either energy-inhibiting or energy-dissipating in respect to their intramural rearrangeabilities, which latter can accommodate the many isotopal differentiations while staying strictly within the same quanta magnitude units.

954.54 When we consider that each of the eight couplers which surround each nuclear coupler may consist of any of 36 different *AAB* intramural orientations, we comprehend that the number of potentially unique nucleus and nuclear-shell interpatternings is adequate to account for all chemical element isotopal variations, as well as accommodation in situ for all the nuclear substructurings, while doing so by omnirational quantation and without any external manifestation of the internal energy kinetics. All that can be observed is a superficially static, omniequivectorial and omnidirectional geometric matrix.

954.55 Again reviewing for recall momentum, we note that the unique asymmetrical Coupler octahedron nests elegantly into the diamond-faceted valley on each of the 12 sides of the rhombic dodecahedron (called spheric because each rhombic dodecahedron constitutes the unique allspace-filling domain of each and every unit radius sphere of all closest-packed, unit-radius sphere aggregates of Universe, the sphere centers of which, as well as the congruent rhombic dodecahedra centers of which, are also congruent with all the vertexes of all isotropic vector matrixes of Universe).

954.56 Neatly seated in the diamond-rimmed valley of the rhombic dodecahedron, the unique asymmetrical octahedron's *Z* axis is congruent with the short diagonal, and its *Y* axis is congruent with the long diagonal of the diamond-rimmed valley in the rhombic dodecahedron's face into which it is seated. This leaves the *X* axis of the uniquely asymmetrical octahedron running perpendicular to the diamond face of the diamond-rimmed valley in which it so neatly sits; and its *X* axis runs perpendicularly through the *K* point, to join together most economically and directly the adjacent hearts (volumetric centers) of all adjacently closest-packed, unit radius spheres of Universe. That is, the *X* axes connect each nuclear sphere heart with the hearts of the 12 spheres closest-packed around it, while the *Y* axis, running perpendicularly to the *X* axis, most economically joins the hearts (volumetric centers) of the only circumferentially adjacent spheres surrounding the nuclear sphere at the heart of the rhombic dodecahedron, but not interconnecting with those nuclear spheres' hearts. Thus the *Y* axes interlink an omnisymmetrical network of tangential, unit-radius spheres in such a manner that each sphere's heart is interconnected with the hearts of only six symmetrically interarrayed tangentially adjacent spheres. This alternate interlinkage package of each-with-six, instead of six-with-twelve, other adjacent spheres, leaves every other space in a closest-packed, isotropic-vector-matrixed Universe centrally unconnected from its heart with adjacent hearts, a condition which, discussed elsewhere, operates in Universe in such a way as to permit two of the very important phenomena of Universe to occur: (1) electromagnetic wave propagations, and (2) the ability of objects to move through or penetrate inherently noncompressible fluid mediums. This phenomenon also operates in such a manner that, in respect to the vertexes of isotropic vector matrixes, only every other one becomes the center of a sphere, and every other vertex becomes the center of a nonsphere of the space interspersing the spheres in closest packing, whereby those spaces resolve themselves into two types—concave vector equilibria and concave octahedra. *And,* whenever a force is applied to such a matrix every sphere becomes a space and every space becomes a sphere, which swift intertransforming repeats itself as the force encounters another sphere, whereby the sphere vanishes and the resulting space is penetrated.

954.57 We now understand why the *K* points are the *kinetic* switch-off-and-on points of Universe.

954.58 When we discover the many rearrangements within the uniquely asymmetric Coupler octahedra of volume one permitted by the unique self-interorientability of the *A* and *B* Modules without any manifest of external conformation alteration, we find that under some arrangements they are abetting the *X* axis interconnectings between nuclear spheres and their 12 closest-packed, adjacently-surrounding spheres, or the *Y* axis interconnectings between only every other sphere of closest-packed systems.

954.59 We also find that the *A* and *B* Module rearrangeabilities can vary the intensity of interconnecting in four magnitudes of intensity or of zero intensity, and can also interconnect the three *X* and *Y* and *M* systems simultaneously in either balanced or unbalanced manners. The unique asymmetric octahedra are in fact so unique as to constitute the actual visual spin variable mechanisms of Dirac's quantum mechanics, which have heretofore been considered utterly abstract and nonvisualizable.

954.70 **The Coupler: Illustrations:** The following paragraphs illustrate, inventory, sort out, and enumerate the systematic complex parameters of interior and exterior relationships of the 12 Couplers that surround every unit-radius sphere and every vertexial point fix in omni-closest-packed Universe, i.e., every vertexial point in isotropic vector matrixes.

954.71 Since the Coupler is an asymmetric octahedron, its eight positive or negative Mite (*AAB* module), filled-octant domains introduce both a positive and a negative set of fundamental relationships in unique system sets of eight as always predicted by the number-of-system-relationships formula: $\frac{N^2-N}{2}$, which with the system number eight has 28 relationships.

954.72 There being three axes—the X, Y, and M sets of obverse-reverse, polar-viewed systems of eight—each eight has 28 relationships, which makes a total of three times $28=84$ integral axially regenerated, and 8 face-to-face regenerated K-to-K couplings, for a total of 92 relationships per Coupler. However, as the inspection and enumeration shows, each of the three sets of 28, and one set of 8 unique, hold-or-transmit potentials subgroup themselves into geometrical conditions in which some provide energy intertransmitting facilities at four different capacity (quantum) magnitudes: 0, 1, 2, 4 (note: $4=2^2$), and in three axial directions. The $X-X'$ axis transmits between—or interconnects—every spheric center with one of its 12 tangentially adjacent closest-packed spheres.

954.73 The $Y-Y'$ axis transmits between—or interconnects—any two adjacent of the six octahedrally and symmetrically interarrayed, concave vector equilibria conformed, 'tween-space, volumetric centers symmetrically surrounding every unit-radius, closest-packed sphere.

954.74 The $M-M'$ axis interlinks, but does not transmit between, any two of the cubically and symmetrically interarrayed eight concave octahedra conformed sets of 'tween-space, concave, empty, volumetric centers symmetrically surrounding every unit-radius, closest-packed sphere in every isotropic vector matrix of Universe.

954.75 The eight K-to-K, face-to-face, couplings are energizingly interconnected by one Mite each, for a total of eight additional interconnections of the Coupler.

954.76 These interconnections are significant because of the fact that the six concave vector equilibria, $Y-Y'$ axis-connected 'tween-spaces, together with the eight concave octahedral 'tween-spaces interconnected by the $M-M'$ axis, are precisely the set of spaces that transform into spheres (or convex vector equilibria) as every sphere in closest-packed, unit-radius, sphere aggregates transforms concurrently into either concave vector equilibria 'tween-spaces or concave octahedra 'tween-sphere spaces.

954.77 This omni-intertransformation of spheres into spaces and spaces into spheres occurs when any single force impinges upon any closest-packed liquid, gaseous, or plasmically closest-packed sphere aggregations.

954.78 The further subdivision of the A Modules into two subtetrahedra and the subdividing of the B Modules into three subtetrahedra provide every positive Mite and every negative Mite with seven plus-or-minus subtetrahedra of five different varieties. Ergo $92\times7=644$ possible combinations, suggesting their identification with the chemical element isotopes.

955.00 Modular Nuclear Development of Allspace-Filling Spherical Domains

955.01 The 144 *A* and *B* Quanta Modules of the rhombic dodecahedron exactly embrace one whole sphere, and only one whole sphere of closest-packed spheres as well as all the unique closest-packed spatial domains of that one sphere. The universal versatility of the *A* and *B* Quanta Modules permits the omni-invertibility of those same 144 Modules within the exact same polyhedral shell space of the same size rhombic dodecahedron, with the omni-inversion resulting in six $^1/_6$th spheres symmetrically and intertangentially deployed around one concave, octahedral space center.

955.02 On the other hand, the vector equilibrium is the one and only unique symmetric polyhedron inherently recurring as a uniformly angled, centrially triangulated, complex collection of tetrahedra and half-octahedra, while also constituting the simplest and first order of nuclear, isotropically defined, uniformly modulated, inward-outward-and-around, vector-tensor structuring, whereby the vector equilibrium of initial frequency, i.e., "plus and minus one" equilibrium, is sometimes identified only as "potential," whose uniform-length 24 external chords and 12 internal radii, together with its 12 external vertexes and one central vertex, accommodates a galaxy of 12 equiradiused spheres closest packed around one nuclear sphere, with the 13 spheres' respective centers omnicongruent with the vector equilibrium's 12 external and one internal vertex.

955.03 Twelve rhombic dodecahedra close-pack symmetrically around one rhombic dodecahedron, with each embracing exactly one whole sphere and the respective total domains uniquely surrounding each of those 13 spheres. Such a 12-around-one, closest symmetrical packing of rhombic dodecahedra produces a 12-knobbed, 14-valleyed complex polyhedral aggregate and not a single simplex polyhedron.

955.04 Since each rhombic dodecahedron consists of 144 modules, $13 \times 144 = 1{,}872$ modules.

955.05 Each of the 12 knobs consists of 116 extra modules added to the initial frequency vector equilibrium's 12 corners. Only 28 of each of the 12 spheres' respective 144 modules are contained inside the initial frequency vector equilibrium, and 12 sets of 28 modules each are $^7/_{36}$ths embracements of the full 12 spheres closest packed around the nuclear sphere.

955.06 In this arrangement, all of the 12 external surrounding spheres have a major portion, i.e., $^{29}/_{36}$ths, of their geometrical domain volumes protruding outside the surface of the vector equilibrium, while the one complete nuclear sphere is entirely contained inside the initial frequency vector equilibrium, and each of its 12 tangent spheres have $^{7}/_{36}$ths of one spherical domain inside the initial frequency vector equilibrium. For example, $12 \times 7 = {}^{84}/_{36} = 2^{1}/_{3} + 1 = 3^{1}/_{3}$ spheric domains inside the vector equilibrium of 480 quanta modules, compared with 144×3.333 rhombic dodecahedron spherics $= 479.5+$ modules, which approaches 480 modules.

955.07 The vector equilibrium, unlike the rhombic dodecahedron or the cube or the tetrakaidecahedron, does *not* fill allspace. In order to use the vector equilibrium in filling allspace, it must be complemented by eight Eighth-Octahedra, with the latter's single, equiangular, triangular faces situated congruently with the eight external triangular facets of the vector equilibrium.

955.08 Each eighth-octahedron consists of six *A* and six *B* Quanta Modules. Applying the eight 12-moduled, 90-degree-apexed, or "cornered," eighth-octahedra to the vector equilibrium's eight triangular facets produces an allspace-filling cube consisting of 576 modules: one octahedron $= 8 \times 12$ modules $= 96$ modules. $96 + 480$ modules $= 576$ modules. With the 576-module cube completed, the 12 (potential) vertexial spheres of the vector equilibrium are, as yet, only partially enclosed.

955.09 If, instead of applying the eight eighth-octahedra with 90-degree corners to the vector equilibrium's eight triangular facets, we had added six half-octahedra "pyramids" to the vector equilibrium's six square faces, it would have produced a two-frequency octahedron with a volume of 768 modules: $6 \times 48 = 288 + 480 =$ an octahedron of 768 modules.

955.10 **Mexican Star:** If we add both of the set of six half-octahedra made up out of 48 modules each to the vector equilibrium's six square faces, and then add the set of eight Eighth-Octahedra consisting of 12 modules each to the vector equilibrium's eight triangular facets, we have not yet completely enclosed the 12 spheres occurring at the vector equilibrium's 12 vertexes. The form we have developed, known as the "Mexican 14-Pointed Star," has six square-based points and eight triangular-based points. The volume of the Mexican 14-Pointed Star is $96 + 288 + 480 = 864$ modules.

955.11 Not until we complete the two-frequency vector equilibrium have we finally enclosed all the original 12 spheres surrounding the single-sphere nucleus in one single polyhedral system. However, this second vector-equilibrium shell also encloses the inward portions of 42 more embryo spheres tangentially surrounding and constituting a second closest-packed concentric sphere shell embracing the first 12, which in turn embrace the nuclear sphere; and because all but the corner 12 of this second closest-packed sphere shell nest mildly into the outer interstices of the inner sphere shell's 12 spheres, we cannot intrude external planes parallel to the vector equilibrium's 14 faces without cutting away the internesting portions of the sphere shells.

955.12 On the other hand, when we complete the second vector equilibrium shell, we add 3,360 modules to the vector equilibrium's initial integral inventory of 480 modules, which makes a total of 3,840 modules present. This means that whereas only 1,872 modules are necessary to entirely enclose 12 spheres closest packed around one sphere, by using 12 rhombic dodecahedra closest packed around one rhombic dodecahedron, these 13 rhombic dodecahedra altogether produce a knobby, 14-valleyed, polyhedral star complex.

955.13 The 3,840 modules of the two-frequency vector equilibrium entirely enclosing 13 whole nuclear spheres, plus fractions of the 42 embryo spheres of the next concentric sphere shell, minus the rhombic dodecahedron's 1,872 modules, equals 1,968 extra modules distributable to the 42 embryo spheres of the two-frequency vector equilibrium's outer shell's 42 fractional sphere aggregates omnioutwardly tangent to the first 12 spheres tangentially surrounding the nuclear sphere. Thus we learn that $1{,}968 - 1{,}872 = 96 = 1$ octahedron.

955.14 Each symmetrical increase of the vector-equilibrium system "frequency" produces a shell that contains further fractional spheres of the next enclosing shell. Fortunately, our *A* and *B* Quanta Modules make possible an *exact* domain accounting, in whole rational numbers—as, for instance, with the addition of the first extra shell of the two-frequency vector equilibrium we have the 3,360 additional modules, of which only 1,872 are necessary to complete the first 12 spheres, symmetrically and embryonically arrayed around the originally exclusively enclosed nucleus. Of the vector equilibrium's 480 modules, 144 modules went into the nuclear sphere set and 336 modules are left over.

955.20 **Modular Development of Omnisymmetric, Spherical Growth Rate Around One Nuclear Sphere of Closest-Packed, Uniradius Spheres:** The subtraction of the 144 modules of the nuclear sphere set from the 480-module inventory of the vector equilibrium at initial frequency, leaves 336 additional modules, which can only compound as sphere fractions. Since there are 12 equal fractional spheres around each corner, with 336 modules we have $^{336}/_{12}$ths. $^{336}/_{12}\text{ths} = 28$ modules at each corner out of the 144 modules needed at each corner to complete the first shell of nuclear self-embracement by additional closest-packed spheres and their space-sharing domains.

955.21 The above produces $^{28}/_{144}\text{ths} = {}^{7}/_{36}$ths present, and $^{116}/_{144}$ths $= {}^{29}/_{36}$ths per each needed.

955.30 **Possible Relevance to Periodic Table of the Elements:** These are interesting numbers because the $^{28}/_{144}$ths and the $^{116}/_{144}$ths, reduced to their least common denominator, disclose two prime numbers, i.e., *seven* and *twenty-nine,* which, together with the prime numbers 1, 2, 3, 5, and 13, are already manifest in the rational structural evolvement with the modules' discovered relationships of unique nuclear events. This rational emergence of the prime numbers 1, 3, 5, 7, 13, and 29 by whole structural increments of whole unit volume modules has interesting synergetic relevance to the rational interaccommodation of all the interrelationship permutation possibilities involved in the periodic table of the 92 regenerative chemical elements, as well as in all the number evolvements of all the spherical trigonometric function intercalculations necessary to define rationally all the unique nuclear vector-equilibrium intertransformabilities and their intersymmetric-phase maximum aberration and asymmetric pulsations. (See Sec. 1238 for the Scheherazade Number accommodating these permutations.)

955.40 Table: Hierarchy of *A* and *B* Quanta Module Development of Omni-Closest-Packed, Symmetric, Spherical, and Polyhedral, Common Concentric Growth Rates Around One Nuclear Sphere, and Those Spheres' Respective Polyhedral, Allspace-Filling, Unique Geometrical Domains (Short Title: *Concentric Domain Growth Rates*)

		A *and* B *Quanta Module Inventory*		*Spherical Domains*
Rhombic Dodecahedron	=	144 modules	=	1
Initial-Frequency Vector Equilibrium	=	480 modules	=	$3^{1}/_{3}$
Octahedron	=	96 modules	=	$^{2}/_{3}$
Cube	=	72 modules	=	$^{1}/_{2}$
Tetrahedron	=	24 modules	=	$^{1}/_{6}$

955.41 Table: Spherical Growth Rate Sequence

(1) Modular Development of Omnisymmetric Spherical Growth Rate Around One Nuclear Sphere.

(2) Nuclear Set of Rhombic Dodecahedron:
144 modules 1 sphere

(3) Vector Equilibrium, Initial Frequency:
480 modules—Itself and $2^{1}/_{3}$ additional spheres

(4) Cube—Initial Frequency:
576 modules 4 spheres

(5) Octahedron, Two-Frequency:
768 modules $5^{1}/_{3}$ spheres

(6) Mexican 14-Point Star:
864 modules 6 spheres
(7) Rhombic Dodecahedron, 12-Knobbed Star:
1,872 modules 13 spheres
(8) Vector Equilibrium, Two-Frequency:
3,840 modules $26^1/_9$ spheres

955.50 Rhombic Dodecahedron at Heart of Vector Equilibrium: Nature always starts every ever freshly with the equilibrious isotropic-vector-matrix field. Energy is not lost; it is just not yet realized. It can be realized only disequilibriously.

955.51 At the heart of the vector equilibrium is the ball in the center of the rhombic dodecahedron.

955.52 Look at the picture which shows one-half of the rhombic dodecahedron. (See Illus. 955.52.) Of all the polyhedra, nothing falls so readily into a closest-packed group of its own kind as does the rhombic dodecahedron, the most common polyhedron found in nature.

960.00 Powers and Dimensions

960.01 The Coordination of Number Powers and Geometrical Dimensions

960.02 Powering means the multiplication of a number by itself.

960.03 Number powers refer to the *numbers of times* any given number is multiplied by itself. While empty set numbers may be theorized as multipliable by themselves, so long as there is time to do so, all experimental demonstrability of science is inherently time limited. Time is size and size is time. Time is the only dimension. In synergetics time-size is expressible as *frequency*.

960.04 Recalling our discovery that angles, tetrahedra, and topological characteristics are system constants independent of size, the limit of experimentally demonstrable powering involves a constant vector equilibrium and an isotropic vector matrix whose omnisymmetrically interparalleled planes and electable omniuniform frequency reoccurrences accommodate in time-sizing everywhere and anywhere regenerative (symmetrically indestruct, tetrahedral, four-dimensional, zerophase, i.e., the vector equilibrium) rebirths of a constant, unit-angle, structural system of convergent gravitation and divergent

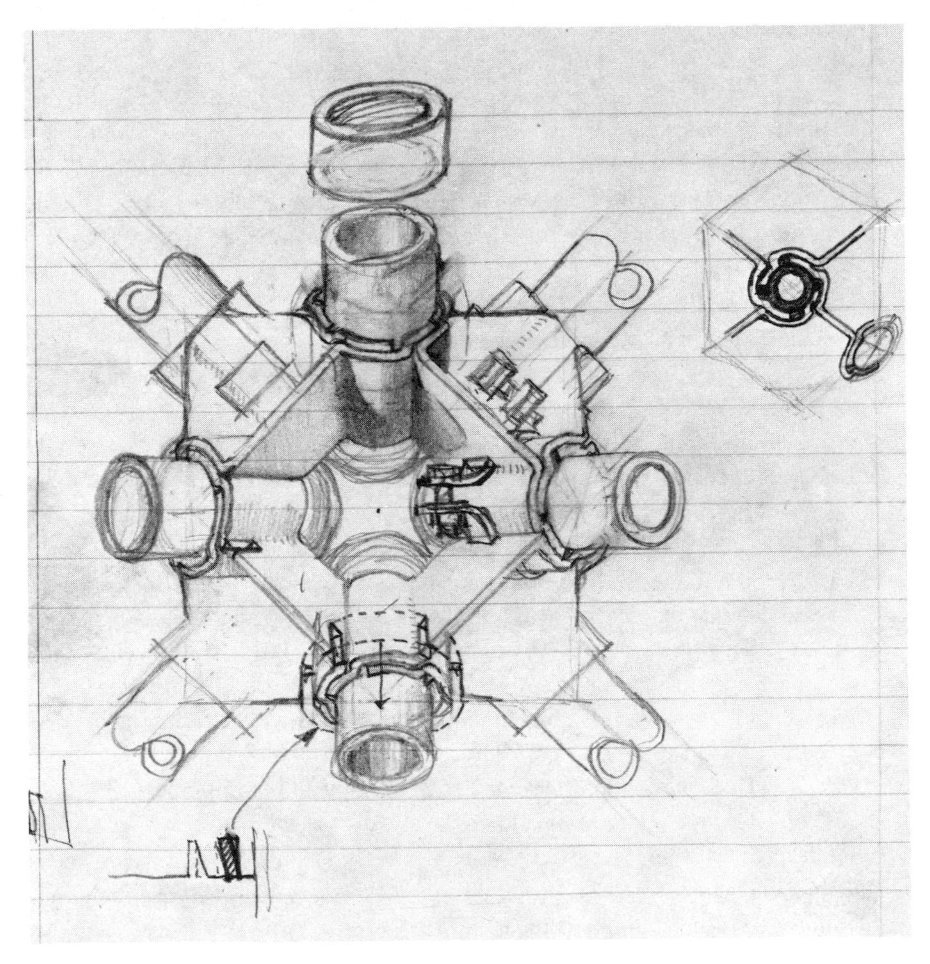

Fig. 955.52 *Basic Joint for Isotropic Vector Equilibrium Matrix.*

radiation resonatability, whose developed frequencies are the specific, special-case, time-size dimensionings.

960.05 Dimensional growth is not occasioned by an increase in exponential powers. It is brought about by increasing subdivision of the constant whole of Universe to isolate a locally considerable increment. For instance, $E = Mc^2$ says that the amount of energy involved in the isolated "mass" as a local event complex of Universe under consideration in this particular instance is to be determined by reference to the constant amount of cosmic energy involved in the constant rate of growth of a spherical, electromagnetic, wave surface, which constant is c^2. Because the potential energy is in vector equilibrium packages, the centers of energy rebirth are accommodated by the isotropic vector matrix. The constant power is the frequency $10F^2 + 2$, which accommodates all the exportive-importive, entropic-syntropic, regeneration patterning of Universe.

960.06 The only dimension is time, the time dimension being the radial dimension outward from or inward toward any regenerative center, which may always be anywhere, yet characterized by always being at the center of system regeneration.

960.07 The time dimension is frequency.

960.08 Any point can tune in any other point in Universe. All that is necessary is that they both employ the same frequency, the same resonance, the same system, center to center.

960.09 The total nothingness involved is accounted by $20F^3$. The third power accounts both the untuned nothingness and the finitely tuned somethingness. The 20 is both Einstein's empty set *M* and all the other untuned non-*M* of Universe. The $20F^3$ is the total Universe momentarily all at one time or timeless center. Eternity is 1 = No frequency $1^3 = 1 \times 20$. The 20 is eternally constant. The rate of wave growth corresponding to Einstein's $c^2 = 10F^2 + 2$.

960.10 Thus the isotropic vector matrix of synergetics' convergence and divergence accommodates elegantly and exactly both Einstein's and Newton's radiation and gravitation formulations, both of which are adequately accounted only in second-powered terms.

960.11 Distance is time. Distance is only frequency-accountable.

960.12 Newton's intermass attraction increases at the second power as the time-distance between is halved. Newton and Einstein deal only with mass and frequency to the second power. Their masses are relatively variable. In one, mass is acceleratingly expended; in the other, mass is acceleratingly collected. (See Sec. 1052.21.)

960.13 In synergetics, the total mass somethingness to be acceleratingly expended is $10F^2$, with always a bonus spin-aroundable-polar-axis 2: Me and the Otherness. In synergetics, the total nothingness and somethingness involved in both inbound and outbound field is $20F^3$. (Nothing = 10. Something = 10. Both = 20.) The multiplicative twoness of me and the otherness. The vector equilibrium and the icosahedron are the prime number *five* polyhedra; the multiplicative, concave-convex twoness: $2 \times 5 = 10$. F^3 = Unexpected nothingness $F^1 \times$ Expected somethingness $F^2 = F^3$.

961.00 Unitary Quantation of Tetrahedron

961.01 The area of a triangle is arrived at by multiplying the length of the baseline by one-half of the triangle's apex altitude.

961.02 The volume of a tetrahedron is the product of the area of the base and one-third of its altitude.

961.03 A minimum garland of "granteds" combines only synergetically to disclose the following:

961.10 **Granted: A Slidable Model of Constant Volume:** Granted any point *A* that is movable limitlessly anywhere within one of two planes parallel to one another at a given perpendicular distance *X* from one another, and, cogliding anywhere within the other parallel plane, two parallel lines lying at a given perpendicular distance *Y* from one another, and a point *B* that is slidable anywhere along one of the parallel lines, along the other line of which (two parallel lines) is a slidable pair of points, *C* and *D*, always slidable only at a constant and given distance *Z* from one another; it will be found that the vast variety of tetrahedra to be formed by interconnecting these four points (two independently variable and two only covariable) will *always enclose the same volume*. (See Sec. 923 and Illus. 923.10D.)

961.11 Provided the relationship between *X, Y,* and *Z* remain constant as described, and the distances *X* and *Y* in respect to the "constructed" distance are always such that

$$Y = \sqrt{Z^2 - (^1/_2\, Z)^2}$$

and

$$X = \sqrt{Z^2 - (^2/_3\, Y)^2}$$

then, by varying *Z* to correspond with the distance between two experiential event foci, all the other vertexial positions of all the tetrahedra of equal volume can be described by revolution of the constantly cohered tetrahedral system around the axial line running through the two points *C* and *D*. This axial line may itself be angularly reoriented to aim the tetrahedral system by combining and interconnecting circuitry closing in any direction, thereby to reach any other two points in Universe to be tetrahedrally interjoined in unitary quantation.

961.12 With the observer-articulator's experientially initiated and interpositioned two control points *C* and *D*, these uniformly quantated observer-articulator variable initial "tunings," accomplished exclusively by frequency and angle modulations, may "bring in" subjectively-objectively, receivingly-and-transmittingly, omnicosmic events occurring remotely in nonsimultaneously evolving and only otherness-generated self-awareness and deliberately thinkable self-conceptioning of progressively omnicomprehending embracements and penetrations. This in turn enables the conscious designing capabilities to be realized by these omnicosmic reaching tetrahedral coordinations (which are resolvable into generalized, quantum-regularized sets, consisting of only two human individual mentalities' predeterminable variables consisting exclusively of frequency and angle modulations identified only with the self-and-other, *C*- and *D*-defined, selectable wavelength *Z* and the *Z* axis' angular aiming and reorientation regulatability of the *Z* axis' ever-constant reorientations of its *X*- and *Y*-dependent coordinates in exclusively angle- and frequency-determined invariant relationship), all of whose synergetic integrities' intersignificance realizations are eternally interaccommodated by the tetrahedral structural system's prime conceptual initiations.

961.20 **Granted: A Model of Comprehensive Covariability:** Granted that a tetrahedron of given altitude *X*, with a base triangle of given altitude *Y* and given baseline edge length *Z*, is volumetrically constant independent of the omnivariable interangling of its four vertexes, and five variable-length edges, and four variable triangular faces, whose comprehensive covariability can altogether accommodate any symmetric or asymmetric aspect transformability to correspond exactly in all its interangular face relationships and relative edge lengths with any tetrahedron to be formed by interconnecting any four points in Universe, provided the relative values of *X* and *Y* in respect to *Z* (which is the only experientially known distance) are always such that:

$$Y = \sqrt{Z^2 - (^1/_2\, Z)^2}$$

and,

$$X = \sqrt{Z^2 - (^2/_3\, Y)^2}$$

As the values only of *Z* are altered, the respective value of the uniformly volumed tetrahedra will vary at a rate of the third power of *Z*'s linear change.

961.30 **Granted: A Model for Third-Power Rate of Variation:** Granted that there is then in respect to any two points in Universe a tetrahedron that can be given any symmetrical or asymmetrical tetrahedral shape, any of whose volumes will remain uniform or will vary uniformly at a third-power rate in respect to any alteration of the distance between the two initial control points on the axial control line; then, any four points in Universe, provided one is not in the plane of the other three, can be interconnected by varying the angular orientation of the control-line axis and the distance between the two central control points.

961.31 Being generalized, these three relative distance-control coordinates *X*, *Y*, and *Z* are, of course, also present in the special-case, omnirectilinear $XYZ - c.g_t.s.$ coordinate system. That the most economical time distances between the two parallel planes and two parallel lines are coincident with perpendiculars to those parallel planes and lines does not impose any rectilinear profiling or structuring of the tetrahedron, which is a unique, four-planes-of-symmetry, self-structuring system, as the three-plane-defined cube of basic reference is not.

961.40 **Granted: A Model for Six Degrees of Freedom:** Granted the area of a triangle is base times one-half the altitude, with one given length of line *AB* marked on a flat plane and another infinitely extensible line number two lying in the same plane as short line *AB*, with line two parallel to *AB;* then connecting any point *C* on line two with both *A* and *B* will produce a constant-area triangle *ABC*. Holding *AB* fixed and moving only *C* in any direction on line two, the shape of triangle *ABC* will change, but its area will be constant. If we move *C* along line two in one direction the three edges will approach congruence with one another, appearing only as a line but being, in fact, a constant-area triangle.

961.41 Granted the volume of a tetrahedron is its base area times one-third of its altitude, we can now take the permitted, special condition discussed in Sec. 961.40 whereby *C* on line two is equidistant from both of line one's terminal-defining points *A* and *B*. We may next take a fourth point *D*, lying in an infinitely extensible second plane which is parallel to the first infinitely extensible plane defined by points *ABC*. With *D* equidistant from *A, B,* and *C,* the volume of the regular tetrahedron *ABCD* will not be altered by letting *D* travel to any point in plane two while point *C* travels to any point on line two. Thus we learn that constant-volume tetrahedron *ABCD* might become so distended as to appear to be a line of no volume. Since there could be no volumeless line produced operationally, we may assume that all visible lines must be at minimum extended tetrahedra.

961.42 These variabilities of the constant-volume tetrahedron and its constant-area faces will permit congruence of the four vertexes of the tetrahedron with any four points of Universe by simply taking the initial distance *AB* to suit the task. This unit linear adjustment is a familiar wavelength tuning function. Here we have the six cosmic degrees of freedom (see Sec. 537.10); whereby we are free to choose the length of only one line to be held constant, while allowing the other five edge-lines of the tetrahedron to take any size. We can connect any four points in Universe and produce a tetrahedron that is matchable with whole, unit, rational-number, volume increments of the *A* and *B* Quanta Modules.

961.43 With large, clear plexiglass Models of the *A* and *B* Quanta Modules, we can easily see their clearly defined centers of volume. The centers of area of the triangular faces are arrived at by bisecting the edges and connecting the opposite angle. The center of volume of the tetrahedron is arrived at by interconnecting the four centers of triangular area with their opposite vertexes. These four lines constructed with fine, taut wires will converge to tangency at—and then diverge away from—the tetrahedron's center of volume.

961.44 The lines defining the center of four triangles and the center of volume inherently divide the modules into 24 equal parts. The same progressive subdivisions of the last 24 can be continued indefinitely, but each time we do so the rational bits become more and more asymmetrical. They get thinner and thinner and become more and more like glass splinters. By varying the frequency we can make any shape tetrahedron from the regular to the most asymmetric.

961.45 *The modules make all the geometries—all the crystallography. Any probabilities can be dealt with. With the* two *of Euler: and Gibbs—the Me-and-Other-Awareness—the beginning of time, if there is time. . . . It starts testing the special cases that have time. They are absolutely quantized. The* As *are blue and the* Bs *are red. The blues and reds intertransform. Every sphere becomes a space, and every space becomes a sphere, palpitating in the wire model of electromagnetic wave action.*

961.46 The *A* and *B* Quanta Modules become linear, as did the progression of concentric, common-base, uniform, linear, frequencied, electric-impulse conductors (see Sec. 923.21); and as also did the concentric, annually-frequencied, common-base-into-cone-rotated tetrahedra (see Sec. 541.30); the free energy put in at the base electronically, when you close the circuit at the beginning of the wire—you get the same package out at the other end, the same quanta. The longer the wire gets—or the tree grows—as it approaches parallelism, the more the energy packages begin to precess and to branch out at right angles.

961.47 Fluorescing occurs until all the juice is finally dissipated off the wire—or until all of this year's additional frequency's growth is realized in new branches, twigs, leaves and tetrahedrally-precessed buds. Birth: buds: *A* and *B* Modules; three-, four-, five-, and six-petallings: tetra, octa, icosa, rhombic-dodeca bud petals. The original input—the six *A* Quanta Modules of the original base tetrahedron—becomes distributive at 90 degrees. Coaxial cables tend to divert the precessional distributives inwardly to reduce the loss.

961.48 When great electrostatic charges built into clouds become dischargingly grounded (to Earth) by the excellently-conducting water of rain, and lightning occurs; we see the Earthward, precessionally-branching lightning. In grounding with Earth, lightning often closes its circuit through the tree's branches, whose liquid, water-filled, cell fibers are the most efficient conductors available in conducting the great electric charges inward to Earth through the trunk and the precessionally-distributive roots' branchings. Lines are tetrahedra. Lines can wave-bounce in ribbons and beams: tetra, octa, and icosa energy lock-up systems. $E = Mc^2$. All tetra and only tetra are volumetric, i.e., quanta-immune to any and all transformation.

962.00 *Powering in the Synergetics Coordinate System*

962.01 In the operational conventions of the *XYZ*—$c.g_t.s.$ coordinate system of mathematics, physics, and chemistry, exponential powering meant the development of dimensions that require the introduction of successively new perpendiculars to planes not yet acquired by the system.

962.02 In synergetics, powering means only the frequency modulation of the system; i.e., subdivision of the system. In synergetics, we have only two directions: radial and circumferential.

962.03 In the *XYZ* system, three planes interact at 90 degrees (three dimensions). In synergetics, four planes interact at 60 degrees (four dimensions).

962.04 In synergetics there are four axial systems: *ABCD*. There is a maximum set of four planes nonparallel to one another but omnisymmetrically mutually intercepting. These are the four sets of the unique planes always comprising the isotropic vector matrix. The four planes of the tetrahedron can never be parallel to one another. The synergetics *ABCD*-four-dimensional and the conventional *XYZ*-three-dimensional systems are symmetrically intercoordinate. *XYZ* coordinate systems cannot rationally accommodate and directly articulate angular acceleration; and they can only awkwardly, rectilinearly articulate linear acceleration events.

962.05 Synergetic geometry discloses the rational fourth- and fifth-powering modelability of nature's coordinate transformings as referenced to the 60-degree equiangular isotropic-vector equilibrium.

962.06 *XYZ* volumetric coordination requires three times more volume to accommodate its dimensional results than does the 60-degree coordination calculating; therefore, *XYZ* 90-degree coordination cannot accommodate the fourth and fifth powers in its experimental demonstrability, i.e., modelability.

962.07 In the coordinate vectorial topology of synergetics, exponential powers and physical model dimensioning are identified with the number of vectors that may intercept the system at a constant angle, while avoiding parallelism or congruence with any other of thc uniquely convergent vectors of the system.

962.10 **Angular and Linear Accelerations:** Synergetics accommodates the direct expression of both angular and linear accelerations of physical Universe. The frequency of the synergetics coordinate system, *synchrosystem,* simultaneously and directly expresses both the angular and linear accelerations of nature.

962.11 The Mass is the consequence of the angular accelerations. c^2 or G^2 of linear acceleration of the same unit inventory of forever regeneratively finite physical Universe, ever intertransforming and transacting in association (angular) or disassociation (linear) interaccelerations.

962.12 The "three-dimensional" *XYZ*—$c.g_t.s.$ system of coordination presently employed by world-around science can only express directly the linear accelerations and evolve therefrom its angular accelerations in awkward mathematics involving irrational, non-exactly-resolvable constants. $c.g_t.s.$ per second, $M \times F^2$ is cubistically awkwardized into calculatively tattering irrationality.

962.20 **Convergence:** In the topology of synergetics, powering is identifiable only with the uniangular vectorial convergences. The number of superficial, radiantly regenerated, vertex convergences of the system are identified with second powering, and not with anything we call "areas," that is, not with surfaces or with any experimentally demonstrable continuums.

962.30 **Calculation of Local Events:** All local events of Universe may be calculatively anticipated in synergetics by inaugurating calculation with a local vector-equilibrium frame and identifying the disturbance initiating point, direction, and energy of relative asymmetric pulsing of the introduced resonance and intertransformative event. (Synergetics Corollary, see Sec. 240.39.)

962.40 **Time and Dimension:** Synergetic geometry embraces all the qualities of experience, all aspects of being. Measurements of width, breadth, and height are awkward, inadequate descriptions that are only parts of the picture. Without weight, you do not exist physically; nor do you exist without a specific temperature. You can convert the velocity-times-mass into heat. Vectors are not abstractions, they are resolutions. Time and heat and length and weight are inherent in every dimension. Ergo, time is no more the fourth dimension than it is the first, second, or third dimension.

962.41 No time: No dimension. Time is dimension.

962.42 Time is in synergetic dimensioning because our geometry is vectorial. Every vector = mass × velocity, and time is a function of velocity. The velocity can be inward, outward, or around, and the arounding will always be chordal and exactly equated with the inwardness and outwardness time expendabilities. The Euclidian-derived *XYZ* coordinate geometry cannot express time equi-economically around, but only time in and time out. Synergetics inherently has time equanimity: it deals with anything that exists always in 1 × 1 time coordination.

962.50 **Omnidirectional Regeneration:** The coordinate systems of synergetics are omnidirectionally regenerative by both lines and planes parallel to the original converging set. The omnidirectional regeneration of synergetic coordination may always be expressed in always balanced equivalence terms either of radial or circumferential frequency increments.

963.00 First Power: One Dimension

963.01 In conventional *XYZ* coordination, one-dimensionality is identified geometrically with linear pointal frequency. The linear measure is the first power, or the edge of the square face of a cube.

963.02 In synergetics, the first-power linear measure is the radius of the sphere.

963.10 **Synergetics Constant:** The synergetics constant was evolved to convert third-power, volumetric evaluation from a cubical to a tetrahedral base and to employ the *ABCD*-four-dimensional system's vector as the linear computational input. In the case of the cube this is the diagonal of the cube's square face. Other power values are shown in Table 963.10. We have to find the total vector powers involved in the calculation. In synergetics we are always dealing in energy content: when vector edges double together in quadrivalence or octavalence, the energy content doubles and fourfolds, respectively. When the vector edges are half-doubled together, as in the icosahedron phase of the jitterbug—halfway between the vector equilibrium 20 and the octahedron compression—to fourfold and fivefold contraction with the vectors only doubled, we can understand that the volume of energy in the icosahedron (which is probably the same 20 as that of the vector equilibrium) is just compressed. (See Secs. 982.45 and 982.54.)

963.11 In Einstein's $E=Mc^2$, M is volume-to-spherical-wave ratio of the system considered. Mass is the integration of relative weight and volume. What Einstein saw was that the weight in the weight-to-volume ratio, i.e., the Mass, could be reduced and still be interpreted as the latent energy-per-volume ratio. Einstein's M is partly identified with volume and partly with relative energy compactment within that spherical wave's volume. There are then relative energy-of-reality concentration-modifiers of the volumes arrived at by third powering.

963.12 All of the frozen volumetric and superficial area mensuration of the past has been derived exclusively from the external linear dimensions. Synergetics starts system mensuration at the system center and, employing omni-60-degree angular coordinates, expresses the omni-equal, radial and chordal, modular linear subdivisions in "frequency" of module subdivisioning of those radii and chords, which method of mensuration exactly accommodates both gravitational (coherence) and radiational (expansion) calculations. As the length of the vectors represents given mass-times-velocity, the energy involvements are inherent in the isotropic vector matrix.

SYNERGETICS CONSTANT

Dymaxion Hierarchy of Vector Generated Field-, Volume-, Mass-, Charge-, Potential of Geometric Forms, i.e. Potentials of Basic Energetic Transformations Where 3-fold axii and 4-fold axii rotate on 6 axii	*With energy potential in equilibrium dymaxion vector $= 2^2\sqrt{9/8}$ $= 2 \times 1.0198255$ $= 2.039651 = V$ Volume=*	(2) *480×Col. 1. Where 'A' particle =480 makes rational whole sub-particle fractionation by interaction of planes of 25 great circles*	(3) *Where edges of cube and all other planar bound forms=2 as per common Greek 3 Dimension. 'Coord.' System 'regular solids'. Euclidian platonic*	(4) *Where edges of cube and all other planar bound forms=2 as per common Greek 3 Dimension. 'Coord.' System 'regular solids'*	(5) *Edges of tet, octa, icosa, dymax, rombid, dodeca, tetraxidec, all=2. but diagonal of cube and rombidec '2' =2, radius of spheres=2, or arc of spheres=2 as marked*	(6) *Special formula*	(7) *Ratios*	
'A' PARTICLE 1/6 of ¼ of regular Tet. 1/6 of Tet. formed on 4 faces of regular Tet. with apex at C. at G. of Tet.	½V=edge (outer) .0416666 =1/24 of unity	Rational						
ICOSACENTET Each of 20 tets. Formed on 20 faces of Icosa with apex at C. of G. of Icosa.	V=outer edge .9255		.10908	.8726	.8726			
TETRAHEDRON (Regular Tet.) +equal triangular faces.	V=edge 1.0000	< Rational 11.52000	.1179	.9432	.9428	$\frac{\left(\sqrt{\frac{V^2}{2}}\right)^3}{3}$		
CUBE (1) Edge of Cube $3\sqrt{3}=1.4422$. Cube= Tet.+4| (⅛ Octa) on its faces fills all space. If edge of Cube=V, Vol.= 8.4904.	V=diagonal face 3.0000	<Rational	1.0000	8.0000	2.828428	$Vol = \left(\sqrt{\frac{V^2}{2}}\right)^3$		
OCTAHEDRON (Regular Octa) 8 equal triangular faces.	V=edge 4.0000	<Rational	.4714	3.7712	3.7712			
RHOMBICDODECAHEDRON (Rombidec. 1) Fills all space. 12 equilateral rhomboid faces=Octa and 8| (¼ Tet.). Radius Tet.=V.	V=long diag. face 6.0000	<Rational			5.6576			
CUBE (2) Where edge of Cube is Vector= 2.039651.	V=edge 8.4900	Complementary Rational (with Icosahedron)						
ICOSAHEDRON (Icosa) 20 triangular faces. Radius=1.93909. Perpendicular from C.G. Icosa to C.G. triangular face=1.574.	V=edge 18.5100	Complementary Rational (with Cube 2)	2.1817	17.4536	17.4526	$\frac{20}{3}\sqrt{\frac{V2^3}{2}}$	Vol. icosa.: icosasphere 'R' = 1: 1.61725	
DYMAXION (Dymax) 6 square and 8 triangular faces. All edges and radii identical and are identical vectors in omnidirectional equilibrium.	V=edge and radius 20.0000	<Rational	2.3574	18.8592	18.85618		Vol. dymax.: dymaxisphere 'A' = 1: 1.54753, factor = 1/20 vol. dymaxisphere	dymaxion: dymaxisphere 'R' = 1.77715 Note $\sqrt{\pi}$ = 1.772454
RHOMBICDODECAHEDRON (Rombidec. 2) Fills space. 12 rhomboid faces where	V=edge							

Solid	V						
TETRAXIDECAHEDRON (Tetraxidec. 1) Lord Kelvin's all-space solid. 6 square, 8 hexagonal faces. Dymaxion + 1⅔ Octa = total 1⅓ Dymaxion.	V=edge 96.0000	<Rational					
ICOSASPHERE (A) Where arc 63° 26′ = arc edge of spherical triangle of 20 Spherical triangles of Sphere = arc = 2.039651.	V=arc 63° 26′ 27.788	Spheres			arc 63° 26′=2 26.1989		icosasphere 'A': dymaxisphere 'A' = 1: 1.1813
ICOSASPHERE (R) Where radius = V = 2.039651.	V=radius 30.570	Spheres			radius=2 28.8216		icosasphere 'R': dymaxisphere 'R' = 1: 1.16268; icosasphere 'R': dymaxisphere 'A' = 1.012452
DYMAXISPHERE (A) Where arc 60° = V = 2.039651.	V=arc 60° 30.950	Spheres			arc 60°=2 29.18051; radius=1.9098	$\frac{288}{\pi}$ = *Vol. SP.* $\pi R = 6$	dymaxisphere 'A': dymaxisphere 'R' = 1.148379
DYMAXISPHERE (R) Where arc 60° = V = 2.039651.	V=radius 35.540	Spheres			radius=2 33.51029 arc=2.0944		
DODECAHEDRON 12 hexagonal faces.	V=edge 65.018	<Rational	7.6631	61.3048			
TETRAXIDECAHEDRON (Tetraxidec. 2)	V=edge 136.0000	<Rational	16.0242	128.1936			
	∧ Note that 10 out of the 13 planar bound solids are rational and 2 others are complimentary rational		∧ Of above only cube is rational	∧ Of above only cube is rational	∧ Of above none is rational. To convert above solids to values in column (1) multiply by 1.06066 = $\sqrt{9/8}$		

$$2\sqrt[6]{9/8} = \text{dymaxion vector constant}$$

$1/2\,V = 1.0198255$
$(1/2\,V)^3 = 1.0400440504$
$(1/2\,V)^3 = 1.0606605$
$(1/2\,V)^6 = 1.12500000$
$= 1\,1/8$
$= 9/8$

```
  1   2   3   4   5   6   7   8
o - o - o - o - o - o - o - o - o
1   2   3   4   5   6   7   8   9
```

Unique maxima of dymax nucleus employs above: 9 balls, 8 spaces.

This ratio has significant implications as in natural number behavior as indicated by basic sphere and dymaxion and octa-tets. It is seen that number integers take octave—nine modular congruence.

Table 963.10 *Dymaxion Energetic Geometry*, 1950

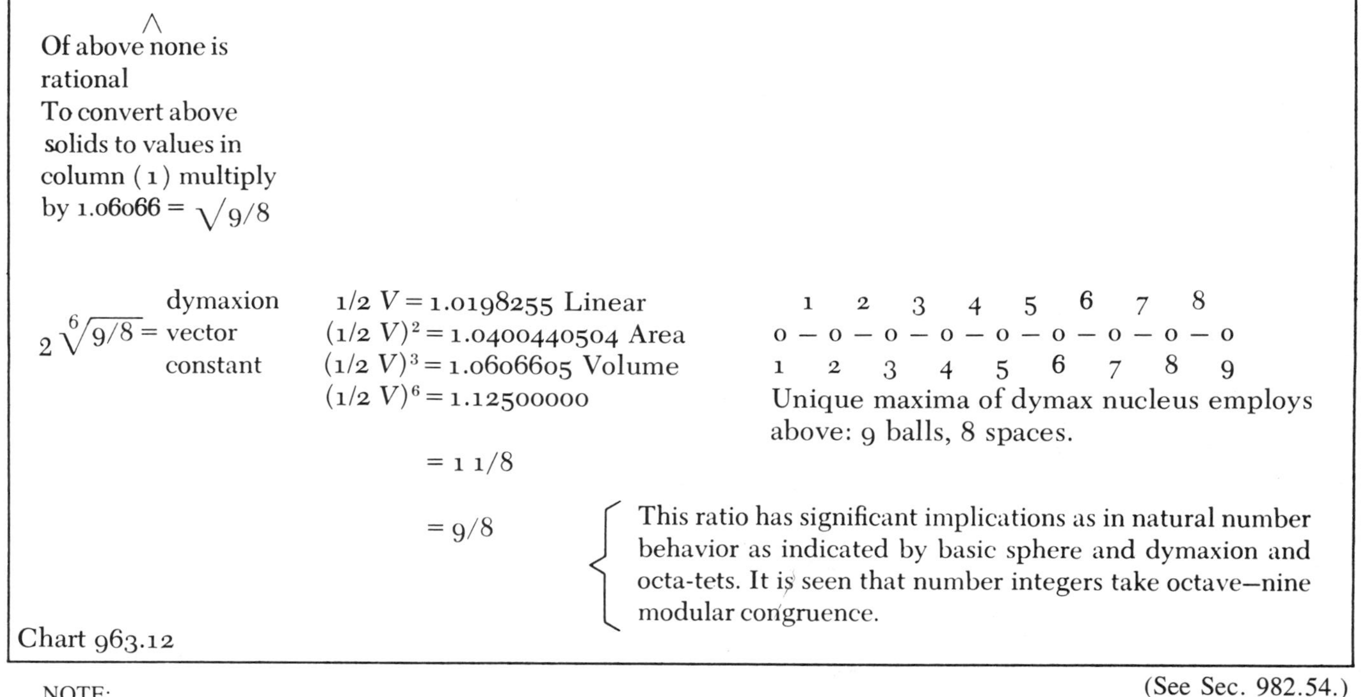

Of above none is rational

To convert above solids to values in column (1) multiply by $1.06066 = \sqrt{9/8}$

$2\sqrt[6]{9/8} =$ dymaxion vector constant

$1/2\ V = 1.0198255$ Linear

$(1/2\ V)^2 = 1.0400440504$ Area

$(1/2\ V)^3 = 1.0606605$ Volume

$(1/2\ V)^6 = 1.12500000$

$= 1\ 1/8$

$= 9/8$

```
  1   2   3   4   5   6   7   8
o — o — o — o — o — o — o — o — o
1   2   3   4   5   6   7   8   9
```

Unique maxima of dymax nucleus employs above: 9 balls, 8 spaces.

This ratio has significant implications as in natural number behavior as indicated by basic sphere and dymaxion and octa-tets. It is seen that number integers take octave—nine modular congruence.

Chart 963.12

(See Sec. 982.54.)

NOTE:

It may readily be seen from the table why the preoccupation of mathematicians and laymen alike with the XYZ rectilinear coordinate system and its cube has obscured the existence of the synergetics hierarchy. It would most naturally be discovered only by the pursuit of the concept of vectorial omnidirectional equilibrium. Rational mensuration is not possible with the cube edge as mensural unity. In synergetics mensural unity derives from the diagonal of the cube as a radial line from sphere center to sphere center. The original formula, first published in 1950, is recapitulated above.

963.13 Synergetics is a priori nuclear; it begins at the center, the center of the always centrally observing observer. The centrally observing observer asks progressively, "What goes on around here?"

964.00 Second Power: Two Dimensions

964.01 In conventional *XYZ* coordination, two-dimensionality is identified with areal pointal frequency.

964.02 In synergetics, second powering = point aggregate quanta = area. In synergetics, second powering represents the rate of system surface growth.

964.10 **Spherical Growth Rate:** In a radiational or gravitational wave system, second powering is identified with the point population of the concentrically embracing arrays of any given radius, stated in terms of frequency of modular subdivisions of either the radial or chordal circumference of the system. (From Synergetics Corollary, see Sec. 240.44.)

964.20 **Vertexial Topology:** Second powering does not refer to "squaring" or to surface amplification, but to the number of the system's external vertexes in which equating the second power and the radial or circumferential modular subdivisions of the system (multiplied by the prime number *one,* if a tetrahedral system; by the prime number *two,* if an octahedral system; by the prime number *three,* if a triangulated cubical system; and by the prime number *five,* if an icosahedral system), each multiplied by two, and added to by two, will accurately predict the number of superficial points of the system.

964.30 **Shell Accounting:** Second power has been identified uniquely with surface area, and it is still the "surface," or *shell.* But what physics shows is very interesting: there are no continuous shells, there are only energy-event foci and quanta. They can be considered as points or "little spheres." The second-power numbers represent the number of energy packages or points in the outer shell of the system. The second-power number is derived by multiplying the frequency of wave divisions of the radius of the system, i.e., F^2 = frequency to the second power.

964.31 In the quantum and wave phenomena, we deal with individual packages. We do not have continuous surfaces. In synergetics, we find the familiar practice of second powering displaying a congruence with the points, or separate little energy packages of the shell arrays. Electromagnetic frequencies of systems are sometimes complex, but they always exist in complementation of gravitational forces and together with them provide prime rational integer characteristics in all physical systems. Little energy actions, little separate stars: this is what we mean by quantum. Synergetics provides geometrical conceptuality in respect to energy quanta.

965.00 *Third Power: Three Dimensions*

965.01 In a radiational or gravitational wave system, third powering is synergetically identified with the total point population involvement of all the successively propagated, successively outward bound in omniradial direction, wave layers of the system. Since the original point was a tetrahedron and already a priori volumetric, the third powering is in fact sixth powering: $N^3 \times N^3 = N^6$.

965.02 Third powering = total volumetric involvement—as, for instance, total molecular population of a body of water through which successive waves pass outwardly from a splash-propagated initial circle. As the circle grows larger, the number of molecules being locally displaced grows exponentially.

965.03 Third powering identifies with a symmetric swarm of points around, and in addition to, the neutral axial line of points. To find the total number of points collectively in all of a system's layers, it is necessary to multiply an initial quantity of one of the first four prime numbers (times two) by the third power of the wave frequency.

965.04 Perpendicularity (90-degreeness) uniquely characterizes the limit of three-dimensionality. Equiangularity (60-degreeness) uniquely characterizes the limits of four-dimensional systems.

966.00 *Fourth Power: Four Dimensions*

966.01 In a radiational or gravitational wave system, fourth powering is identified with the interpointal domain volumes.

966.02 It is not possible to demonstrate the fourth dimension with 90-degree models. The regular tetrahedron has four unique, omnisymmetrically interacting face planes—ergo, four unique perpendiculars to the four planes.

966.03 Four-dimensionality evolves in omnisymmetric equality of radial and chordal rates of convergence and divergence, as well as in all symmetrically interparalleled dimensions. All of synergetics' isotropic-vector-matrix field lines are geodesic and weave both four-dimensionally and omnisymmetrically amongst one another, for all available cosmic time, without anywhere touching one another.

966.04 The vector-equilibrium model displays four-dimensional hexagonal central cross section.

966.05 Arithmetical fourth-power energy evolution order has been manifest time and again in experimental physics, but could not be modelably ac-

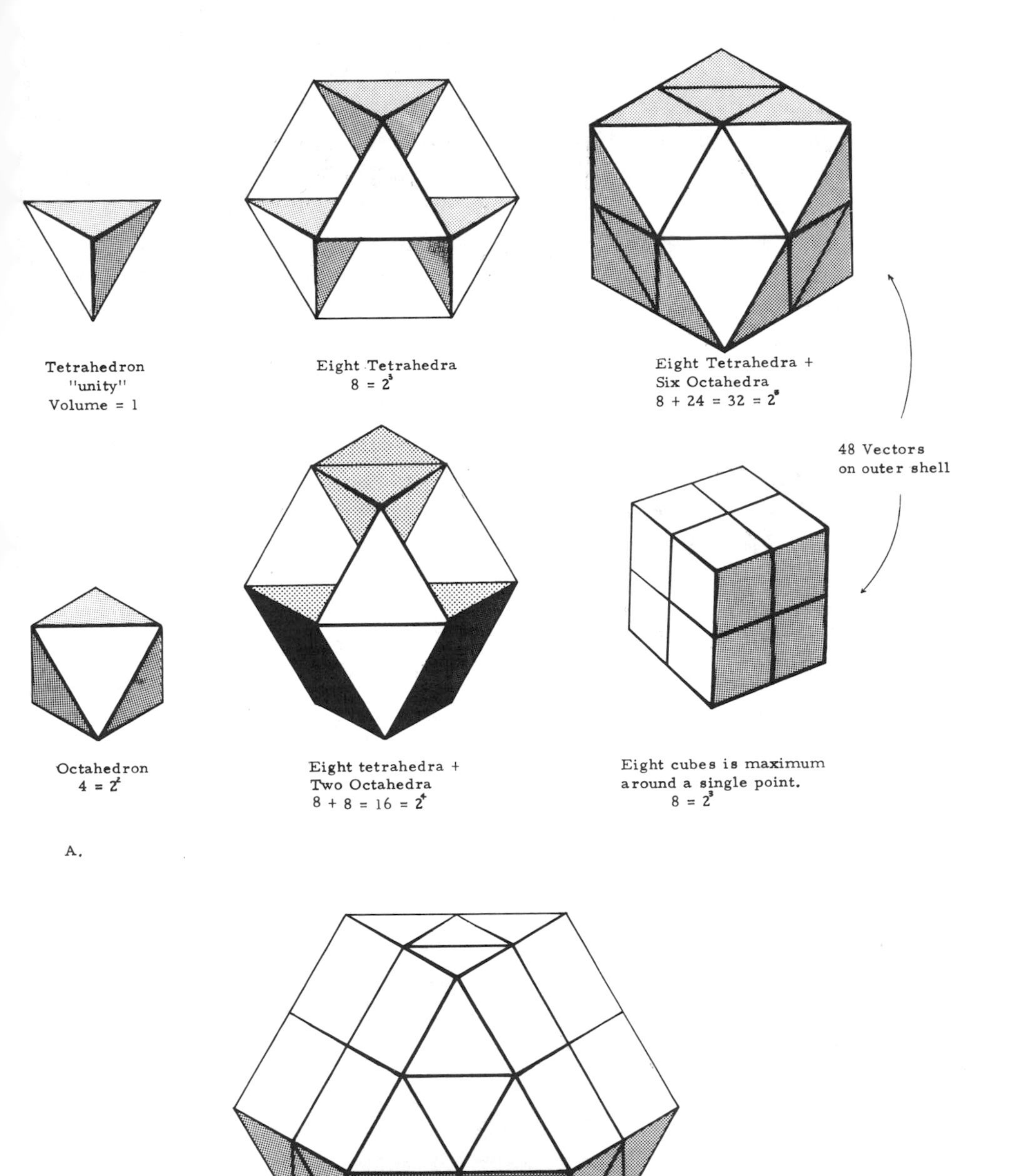

Fig. 966.05 *Tetrahedral Modelability of 2nd, 3rd, 4th, and 5th Power Relationships:*

A. Polyhedral assemblies around a single point having volumes that are integral powers of two when referred to tetrahedron as unity volume.

B. Two-frequency vector equilibrium: $5 \cdot 2^5 = 160$.

commodated by the $XYZ - c.g_t.s.$ system. That the fourth dimension can be modelably accommodated by synergetics is the result of complex local intertransformabilities because the vector equilibrium has, at initial frequency zero, an inherent volume of 20. Only eight cubes can be closest packed in omnidirectional embracement of any one point in the *XYZ* system: in the third powering of two, which is eight, all point-surrounding space has been occupied. In synergetics, third powering is allspace-fillingly accounted in tetrahedral volume increments; 20 unit volume tetrahedra close-pack around one point, which point surrounding reoccurs isotropically in the centers of the vector equilibria. When the volume around one is 20, the frequency of the system is at one. When the *XYZ* system modular frequency is at one, the cube volume is one, while in the vector-equilibrium synergetic system, the initial volume is 20. When the frequency of modular subdivision of *XYZ* cubes reads two, the volume is eight. When the vector equilibria's module reads two, the volume is $20F^3 = 20 \times 8 = 160$ tetrahedral volumes—$160 = 2^5 \times 5$—thus demonstrating the use of conceptual models for fourth- and fifth-powering volumetric growth rates. With the initial frequency of one and the volume of the vector equilibrium at 20, it also has 24×20 *A* and *B* Quanta Modules; ergo is inherently initially 480 quanta modules. $480 = 2^5 \times 5 \times 3$. With frequency of two the vector equilibrium is $160 \times 24 = 3840$ quanta modules. $3840 = 2^8 \times 3 \times 5$. (See Illus. 966.05.)

966.06 Because the volume of one cube equals the volume of three regular tetrahedra, it is now clear that it was only the threefold overstuffing which precluded its capability of providing conceptual modelability of fourth powering. It was the failure of the exclusively three-dimensional *XYZ* coordination that gave rise to the concept that fourth-dimensionality is experimentally undemonstrable—ergo, its arithmetical manifestation even in physics must be a mysterious, because nonconceivable, state that might be spoken of casually as the "time dimension."

966.07 In an omnimotional Universe, it is possible to join or lock together two previously independently moving parts of the system without immobilizing the remainder of the system, because four-dimensionality allows local fixities without in any way locking or blocking the rest of the system's omnimotioning or intertransforming. This independence of local formulation corresponds exactly with life experiences in Universe. This omnifreedom is calculatively accommodated by synergetics' fourth- and fifth-power transformabilities. (See Sec. 465, "Rotation of Wheels or Cams in Vector Equilibrium.") (See Illus. 465.01.)

966.08 In three-dimensional, omni-intermeshed, unclutchable, mechanical systems, if any gear is blocked, the whole gear train is locked. In a four-dimensional unclutchable gear system, a plurality of local gears may be locked, while the remainder of the system interarticulates freely. Odd numbers of individual gears (not gear teeth) lock and block while even numbers reciprocate freely in mechanical gear trains.

966.10 **Fourth Power in Physical Universe:** While nature oscillates and palpitates asymmetrically in respect to the omnirational vector-equilibrium field, the plus and minus magnitudes of asymmetry are rational fractions of the omnirationality of the equilibrious state, ergo, omnirationally commensurable to the fourth power, volumetrically, which order of powering embraces all experimentally disclosed physical volumetric behavior.

966.11 The minimum set of events providing macro-micro differentiation of Universe is a set of four local event foci. These four "stars" have an inherent *sixness* of relationship. This four-foci, six-relationship set is definable as the tetrahedron and coincides with quantum mechanics' requirements of four unique quanta per each considerable "particle."

966.12 In synergetics, all experience is identified as, a priori, unalterably four-dimensional. We do not have to explain how Universe began converting chaos to a "building block" and therefrom simplex to complex. In synergetics Universe is eternal. Universe is a complex of omni-interaccommodative principles. Universe is a priori orderly and complexedly integral. We do not need imaginary, nonexistent, inconceivable points, lines, and planes, out of which non-sensible nothingness to inventively build reality. Reality is a priori Universe. What we speak of geometrically as having been vaguely identified in early experience as "specks" or dots or points has no reality. A point in synergetics is a tetrahedron in its vector-equilibrium, zero-volume state, but too small for visible recognition of its conformation. A line is a tetrahedron of macro altitude and micro base. A plane is a tetrahedron of macro base and micro altitude. Points are real, conceptual, experienceable visually and mentally, as are lines and planes.

966.20 **Tetrahedron as Fourth-Dimension Model:** Since the outset of humanity's preoccupation exclusively with the *XYZ* coordinate system, mathematicians have been accustomed to figuring the area of a triangle as a product of the base and one-half its perpendicular altitude. And the volume of the tetrahedron is arrived at by multiplying the area of the base triangle by one-third of its perpendicular altitude. But the tetrahedron has four uniquely symmetrical enclosing planes, and its dimensions may be arrived at by the use of perpendicular heights above any one of its four possible bases. That's what the fourth-dimension system is: it is produced by the angular and size data arrived at by measuring the four perpendicular distances between the tetrahedral centers of volume and the centers of area of the four faces of the tetrahedron.

966.21 As in the calculation of the area of a triangle, its altitude is taken as that of the triangle's apex above the triangular baseline (or its extensions); so with the tetrahedron, its altitude is taken as that of the perpendicular height of the tetrahedron's vertex above the plane of its base triangle (or that plane's extension outside the tetrahedron's triangular base). The four obtuse central angles of convergence of the four perpendiculars to the four triangular mid-faces of the regular tetrahedron pass convergently through the center of tetrahedral volume at 109° 28′.

970.00 First- and Third-Power Progressions of Vector Equilibria

970.01 Operational Note: In making models or drawing the concentric growth of closest-packed sphere-shells, we are illustrating with great-circle cross sections through the center of the vector equilibrium; i.e., on one of its symmetrically oriented four planes of tetrahedral symmetry; i.e., with the hexagonally cross-section, concentric shells of half-*VE*s.

970.02 Your eye tends quickly to wander as you try to draw the closest-packed spheres' equatorial circles. You have to keep your eye fixed on the mid-points of the intertriangulated vectorial lines in the matrix, the mid-points where the half-radiuses meet tangentially.

970.03 In the model of $10F^2+2$, the green area, the space occupied by the sphere per se, is really two adjacent shells that contain the *insideness of the outer shell* and the *outsideness of the inner shell.* These combine to produce tangentially paired shells—ergo, two layers.

970.10 Rationality of Planar Domains and Interstices: There is a $12F^2+2$ omniplanar-bound, volumetric-domain marriage with the $10F^2+2$ strictly spherical shell accounting. (See tables at Sec. 955.40 and at Sec. 971.00.)

970.11 Both the total inventories of spheres and their planar-bound domains of closest-packed sphere *VE* shells, along with their interstitial, "concave" faceted, exclusively vector equilibrium or octahedral spaces, are rationally accountable in nonfractional numbers.

970.12 Synergetics' isotropic-vector-matrix, omnisymmetric, radiantly expansive or contractive growth rate of interstices that are congruent with closest-packed uniradius spheres or points, is also rational. There is elegant, omniuniversal, metaphysical, rational, whole number equating of both the planar-bound polyhedral volumes and the spheres, which relationships can all be discretely expressed without use of the irrational number *pi* (π), 3.14159, always required for such mathematical expression in strictly *XYZ* coordinate mathematics.

970.13 A sphere is a convexly expanded vector equilibrium, and all inter-closest-packed sphere spaces are concavely contracted vector equilibria or octahedra at their most disequilibrious pulsative moments.

970.20 Spheres and Spaces: The successive $(20F^3) - 20(F-1)^3$ layer-shell, planar-bound, tetrahedral volumes embrace only the tangential inner and outer portions of the concentrically closest-packed spheres, each of whose respective complete concentric shell layers always number $10F^2 + 2$. The volume of each concentric vector-equilibrium layer is defined and structured by the isotropic vector matrix, or octet truss, occurring between the spherical centers of any two concentric-sphere layers of the vector equilibrium, the inner part of one sphere layer and the outer part of the other, with only the center or nuclear ball being both its inner and outer parts (see drawings section).

970.21 There is realized herewith a philosophical synergetic sublimity of omnirational, universal, holistic, geometrical accounting of spheres *and spaces* without recourse to the transcendentally irrational *pi* (π). (See drawings section.) (See Secs. 954.56 and 1032.)

971.00 Table of Basic Vector Equilibrium Shell Volumes

971.01 Relationships Between First and Third Powers of *F* Correlated to Closest-Packed Triangular Number Progression and Closest-Packed Tetrahedral Number Progression, Modified Both Additively and Multiplicatively in Whole Rhythmically Occurring Increments of Zero, One, Two, Three, Four, Five, Six, Ten, and Twelve, All as Related to the Arithmetical and Geometrical Progressions, Respectively, of Triangularly and Tetrahedrally Closest-Packed Sphere *Numbers* and Their Successive Respective Volumetric Domains, All Correlated with the Respective Sphere Numbers and Overall Volumetric Domains of Progressively Embracing Concentric Shells of Vector Equilibria: Short Title: *Concentric Sphere Shell Growth Rates* (see drawings section).

971.02 The red zigzag between Columns 2 and 3 shows the progressive, additive, triangular-sphere layers accumulating progressively to produce the regular tetrahedra.

971.03 Column 4 demonstrates the waves of SIX integer additions to the closest-packed tetrahedral progression. The first SIX zeros accumulate until we get a new nucleus. The first two of the zero series are in fact one invisible zero: the positive zero plus its negative phase. Every six layers we gain one new, additional nucleus.

971.04 Column 5 is the tetrahedral number with the new nucleus.

971.05 In Column 6, the integer SIX functions as zero in the same manner in which NINE functions innocuously as zero in all arithmetical operations.

971.06 In Column 6, we multiply Column 5 by a constant SIX, to the product of which we add the six-stage 0, 1, 2, 3, 4, 5 wave-factor growth crest and break of Column 7.

971.07 Column 7's SIXness wave synchronizes elegantly the third-power arithmetical progression of N, i.e., with the integer-metered volumetric growth of N. Column 7's SIXness identifies uniquely with the rhombic dodecahedron's volume-quantum number. Column 7 tells us that the third powers are most fundamentally identified with the one central, holistic, nuclear-sphere-containing, or six-tangented-together, one-sixth sphere of the six vertexes of the 144 A and B Moduled rhombic dodecahedra.

971.08 Columns 6 and 7 show the *five-sixths* cosmic geometry's sphere/space relationship, which is also relevant to:

—120 icosa's basic sphere surface triangles as the outer faces of the icosahedron's 120 centrally convergent similar tetrahedra, which 120 modules of icosahedral unity correspond in respect to the radially centralized, or circumferentially embracing 144 modules uniquely constituting and exclusively defining the rhombic dodecahedron sphere;

—as 120 is to 144;

—the icosahedron is to spherical unity as 5 is to 6;

—as is also any one shell of the vector equilibrium's concentric closest-packed sphere count to its corresponding concentric omnispace volume count, i.e., as 10 is to 12.

971.09 Column 10 lists the cumulative, planar-bound, tetrahedral volumes of the arithmetical progression of third powers of the successive frequencies of whole vector equilibria. The vector equilibrium's initial non-frequencied tetra-volume, i.e., its quantum value, is 20. The formula for obtaining the frequency-progressed volumes of vector equilibrium is:

Volume of $VE = 20F^3$.

971.10 In Column 11, we subtract the previous frequency-vector equilibrium's cumulative volume from the new one-frequency-greater vector equilibrium's cumulative volume, which yields the tetrahedral volume of the outermost shell. The outer vector equilibrium's volume is found always to be:

$$10\ (2+12\ \frac{R^2-R}{2})$$

971.11 Incidentally, the $\frac{R^2-R}{2}$ part of the formula is inherent in the formula $\frac{N^2-N}{2}$, which determines the exact number of unique relationships always existing between any number of items.

971.20 Pulsation Between Icosahedron and Vector Equilibrium: There is manifest in the icosahedral fiveness, in contradistinction to the vector equilibrium's sixness, the seemingly ever annihilatable and ever-re-creatable integer, eternally propagating the complex of unique frequencies of the 92 inherently regenerative chemical elements as well as all the other unique resonances and frequencies of the electromagnetic, protoplasmic, pneumatic-hydraulic, and crystallographic spectrums, whose omnidirectional yes-no pulsativeness occasions the omniexperienceable, exclusively wavilinear, optically or instrumentally tunable, allness of time-accommodated human experience.

972.00 Universal Integrity Model

972.01 Gravitational-Radiational: In its introvert mode, the rhombic dodecahedron interconnects six 1/6th spheres and manifests gravity. In its extrovert mode, the rhombic dodecahedron comprises one whole sphere at its center with no other spheres implied; these are the spheres that together fill allspace. The extrovert mode of the rhombic dodecahedron manifests radiation.

972.02 The rhombic dodecahedron is the integrated sphere (syntropic) OR the disintegrated sphere (entropic).

973.00 Basic Tetrahedra as Volumetric Modules

973.01 Basic Tetrahedron: Each Basic Tetrahedron (Syte) * is semi-symmetric, four of its six edges consisting always of two pairs of equal-length edges and only two being of odd lengths. The Syte itself consists of six entirely asymmetric Modules, four of which are dissimilar to the other two:

2 $A(+)$ positive Modules
2 $A(-)$ negative Modules
1 $B(+)$ positive Module
1 $B(-)$ negative Module

* See Sec. 953.40.

973.10 **Regular Tetrahedron:** The Plato-identified "regular" (i.e., omnisymmetric) tetrahedron is comprised entirely of *A* Modules: 12 positive *A*s and 12 negative *A*s, but the symmetric, "regular," Platonic, equiangled, equi-edged tetrahedron cannot by itself fill allspace as could the cube, the rhombic dodecahedron, and the tetrakaidecahedron—none of which allspace-filling forms have self-stabilizing structural-conformation integrity, not being comprised of triangles, which alone can stabilize pattern.

973.11 The "regular" Platonic tetrahedron may be combined with the octahedron to fill allspace.

973.12 The volumes of all the symmetrical Platonic polyhedra, except the icosahedron and its pentagonal dodecahedron, are whole, low-order-number multiples of the "regular" Platonic tetrahedron, consisting itself of 24 modules, making that tetrahedron seemingly the "basic unit of measure" of all polyhedra. The inability of that "regular" tetrahedron to fill all cosmic space turns our comprehensive, cosmic-coordinate-system exploration to the consideration of the least common divisor aggregates of the *A* and *B* Modules.

973.20 **Functions of *A* and *B* Quanta Modules and Sytes:** The *A* and *B* Quanta Modules are omnitriangulated and individually asymmetric but not maximally asymmetric. The *A* Quanta Module has three of its 12 total angles at 90 degrees, and the *B* Quanta Module has two 90-degree angles. The *A* Quanta Module has a 30-degree and a 60-degree angle. The *B* Quanta Module has two 45-degree angles.

973.21 These angles all represent low-order whole fractions of unity: $^1/_4$, $^1/_6$, $^1/_8$, $^1/_{12}$, and $^1/_{16}$ of unity in a planar circle; all the other angles of the *A* and *B* Quanta Modules are unit and symmetric central angles of the tetrahedron and octahedron.

973.22 The variety of their mixability produces what need be only momentary bewilderment and only an illusion of "disorder" occasioned initially by the subtlety and muchness of the unfamiliar.

973.23 This brings us to consider the only superficially irregular, only semiasymmetric Syte as possibly being the most separately universal structural-system entity.

973.24 The Syte, consisting of only six modules and filling allspace in a threefold intertransformable manner, is found to be far more universal and "primitive" than the regular tetrahedron. The Mite is the single most universal and versatile structural component—save for its own subcomponents, the *A* and *B* Quanta Modules, which of unit volume and non-mirror-imaged complementation do indeed initially provide, singly or in complementation, the beginnings of all cosmic structuring.

973.25 The Syte's six Modules are always subdivided into two sets of three Modules each—two *A*s, one *B*—of which two sets of the three Modules each are identically dimensioned both angularly and linearly, but one is inside out of the other. Therefore, *one set* of three Modules—two *A*s, one *B*—is positively *outside-outed,* and the other is negatively *inside-outed.*

973.30 **Particle and Wave Involvement:** Particle Quanta Equation: (Prime numbers 2 and 5)

$$10F^2 + 2$$

973.31 Wave Quanta Equation: (Prime numbers 2, 5, *and 3*)

$$10\ (2 + 12\ \frac{F^2 - F}{2})$$

973.32 One is particle involvement; the other is total involvement. Inadvertently, they correlate the sphere and all the other polyhedra rationally.

973.33 The difference is the difference between using the tetrahedron as volumetric unity, while the physicist has always been using the cubic centimeter of water—and then only lifting it in one direction, against gravity, against the imagined plane of the world. But, synergetics moves omnidirectionally, inwardly, outwardly, and aroundly. (See Secs. 505.40, 1009.36 and 1012.37.)

974.00 Initial Frequency

974.01 The initially potential-only frequency ($F = +1, -1$) vector equilibrium has a volume of 20 regular tetrahedra, each of which consists of 24 *A* modules. $20 \times 24 = 480$ modules = initial vector equilibrium.

974.02 The initial-frequency vector equilibrium has alternatively either a *radiant* rhombic dodecahedron core or a *gravitational* rhombic dodecahedron core, either of these alternates being of identical overall size and shape. Both consist of 144 modules. From the 480 modules of the vector equilibrium, we subtract the 144 modules of the rhombic dodecahedron, which leaves 366 modules surrounding either the *radiant* or *gravitational* rhombic dodecahedron nuclear-sphere-enveloping core. Each module = $^1/_{144}$th of our spherical domain. $336/144 = 2^1/_3$; and $480/144 = 3^1/_3$; therefore, one nuclear-sphere domain surrounded by the parts of exactly $2^1/_3$ additional spherical-domain-producing modules, distributed symmetrically around the nucleus in exactly 12 groups of 28 modules each. $2^1/_3$ spheres divided by $12 = {}^7/_{36}$ths of one spherical domain. $2^1/_3 = {}^7/_3 = {}^{84}/_{36}$ spheres $= {}^7/_{36}$ths of a nuclear sphere. We do not produce any complete regular polyhedron by adding 28 modules to each of the 12 rhombic dodecahedron faces. While 28 modules, i.e., $^7/_{36}$ths of one spherical domain, may be added to each corner vertex of the vector equilibrium, they do not produce any complete regular polyhedron at *initial frequency* where F = both $+1$ and -1. Each nucleated vector equilibrium = $3^1/_3$ spheres exactly, and not 13 whole spheres as do 12 closest packed around one. They have the centers of 12 fractional spheres ($^7/_{36}$ each) close packed around one whole sphere.

974.03 Each cube = 3 tetrahedra × 24 modules = 72 modules. 144 modules = 1 sphere. Each initial-frequency cube = $^1/_2$ a spherical domain. Eight cubes in a F^2 cube = four spherical domains. Eight cubes have one whole central nuclear sphere and eight $^1/_8$th spheres on the eight outer corners. The eight $F = 1$ cubes combined to = $1F = 2$ cubes have $8 \times 72 = 576$ modules.

974.04 One octahedron has $4 \times 24 = 96$ modules = $^{96}/_{144} = ^2/_3$ spherical domain. One vector equilibrium = $3^1/_3$ spheres. Therefore, one vector equilibrium plus one octahedron = four spheres = one tetrahedron of four closest-packed spheres. One Eighth-Octahedron = $^{96}/_8 = 12$ modules. If we add eight Eighth-Octahedra to each of the vector equilibrium's eight triangular faces, we produce a cube of $480 + 96$ modules = 576 modules, which is the same as the eight-cube $F = 2$ cube.

980.00 Pi *and Synergetics Constant*

980.01 Relative Superficial and Volumetric Magnitudes

980.02 **Starting With Just Twoness:** Granted a beyond-touch-reach apartness between two initially inter-self-identifying cosmic events of the basic otherness generating the awareness called "life," the approximate distance between the volumetric centers of the respective event complexes can only be guessed at as observingly informed by a sequence of angular-differentialing of any two fixed-distance-apart, "range-finder" optics, integral to the observer in respect to which some feature of the remote pattern characteristics of the otherness correspond with some self-sensible features integral to the observing selfness.

980.03 Self has no clue to what the overall size of the away-from-self otherness may be until the otherness is in tactile contact with integral self, whereby component parts of both self and otherness are contactingly compared, e.g., "palm-to-palm." Lacking such tactile comparing, self has no clue to the distance the other may be away from self. The principle is manifest by the Moon, whose diameter is approximately one-million times the height of the average human. The Moon often appears to humans as a disc no bigger than their fingernails.

980.04 Without direct contact knowledge, curiosity-provoked assumptions regarding the approximate distance *T* can be only schematically guessed at relativistically from a series of observationally measured angular relationship changes in the appearance of the observed otherness's features in respect to experienced time-measured intervals of evolutionary transformation stages of self; such as, for instance, self-contained rhythmic frequencies or self-conceptualizability of angular-integrity relationships independent of size. Relative macro-micro system differentialing of direct-experience-stimulated cosmic conceptuality initiates progressive self-informing effectiveness relative to covarying values integral to any and all self-evolutionarily developing observational history.

980.05 For instance, it is discoverable that with linear size increase of the tetrahedral structural systems (see Sec. 623.10), the tetrahedral surface enclosure increases as the second power of the linear growth rate, while the volumetric content coincreases at a third-power rate of the linear rate of size increase. Ergo, with a given tensile strength of cross section of material (itself consisting of nebular aggregates of critically proximate, mass-interattracted, behavioral-event integrities), which material is completely invested in the tetrahedral envelope stretched around four events, with one of the events not being in the plane of the other three. The envelope of a given amount of material must be stretched thinner and thinner as the tetrahedron's four vertexes recede from one another linearly, the rate of the skin material thinning being a second power of the rate of linear retreat from one another of the four vertexial events. All the while the interior volume of the tetrahedron is increasing at a third-power rate and is being fed through one of its vertexes with an aggregate of fluid matter whose atomic population is also increasing at a third-power volumetric rate in respect to the rate of linear gain by symmetrical recession from one another of the four vertexial points.

980.06 A child of eight years jumping barefootedly from rock to rock feels no pain, whereas a grownup experiences not only pain but often punctures the skin of the bottom of the feet because the weight per square inch of skin has been increased three- or fourfold. If humans have not learned by experience that the surface-to-volume relationships are not constant, they may conclude erroneously that they have just grown softer and weaker than they were in childhood, or that they have lost some mystical faculty of childhood. Realizing intuitively or subconsciously with self-evolution-gained information and without direct knowledge regarding the internal kinetics of atoms and molecules in the combined fluid-gaseous aggregate of organisms, we can intuit cogently that naturally interrepellent action-reaction forces are causing the interior gaseous molecules to accelerate only outwardly from one another because the closest-packed limits will not accommodate inwardly, while expansion is ever less opposed and approaches entirely unlimited, entirely unpacked condition. Sensing such relationships without knowing the names of the principles involved, humans can comprehend in principle that being confined only by the ever-thinning films of matter stretched about them on, for instance, a tetrahedron's surface, the third-power rate of increase of the bursting force of the contained volume of gases against the second-power growth rate of the ever more thinly stretched film, in respect to the first-power growth rate of the system, swiftly approaches parting of the enclosing film without knowing that the subvisible energy events have receded beyond the critical proximity limits of their mutual mass-interattractiveness and its inter-fall-in-ness proclivities, instead of which they interprecess to operate as individually remote cosmic orbitings. All of these principles are comprehensible in effective degree by individuals informed only by repeated self-observation of human saliva's surface tension behaviors of their lung-expelled, tongue-formed, mouth-blowable air bubbles as they swiftly approach the critical proximity surface-tension conditions and burst.

980.07 Such information explains to self that the critical dimensional interrelationships are to be expected regarding which their own and others' experimental measurements may lead them to comprehend in useful degree the complex subvisible organisms existing between the energy states of electromagnetics, crystals, hydraulics, pneumatics, and plasmics. Thus they might learn that the smaller the system, the higher the surface-tension effectiveness in respect to total volumetric-force enclosure and interimpact effects of locally separate system events; if so, they will understand why a grasshopper can spring outward against a system's gravity to distances many times the greatest height of the grasshopper standing on the system and do so without damaging either its mechanical or structural members; on the other hand, humans are unable to jump or spring outwardly from Earth's surface more than one module of their own height, and if they were dropped toward the system from many times that height, it would result in the volumetric-content-mass-concentration acceleration bursting their mass intertensioning's critical limits.

980.08 Thus locally informed of relative magnitude-event behaviors, the individual could make working assumptions regarding the approximate distances as though, informed of the observed presence of enough event details of the otherness corresponding with those of the within-self-complex, as provided by the relative electromagnetic-frequency color effects that identify substances and their arrangement in the otherness corresponding to the observer's integral-event complex, the individual has never heard of or thought of the fact that he is not "seeing things" but is tuning in electromagnetic wave programs. The foregoing embraces all the parameters of the generalized principles governing always and only self-inaugurated education and its only secondary augmentability by others.

980.09 Flying-boat aviators landing in barren-rock- and ice-rimmed waters within whose horizons no living organism may be observed are completely unable to judge the heights of cliffs or valleys and must come in for a landing at a highly controllable glide angle and speed suitable for safe touch-in landing.

980.10 Once there has been contact of the observer with the otherness, then the approximate *T* distance estimation can be improved by modular approximations, the modules being predicated on heartbeat intervals, linear pacings, or whatever. These do, however, require time intervals. No otherness: No time: No distance. The specific within-self rhythm criterion spontaneously employed for time-distance-interval measurements is inconsequential. Any cycle tunable with the specific-event frequencies will do.

981.00 Self and Otherness Sequence

981.01 Coincidentally synchronized with the discovery of self through the discovery of otherness and otherness's and self's mutual inter-rolling-around (see Sec. 411), we have self-discovery of the outside me and the inside me, and the self-discovery of the insideness and outsideness of the otherness. The inside me in my tummy is directionally approachable when I stick my finger in my mouth.

981.02 Now we have the complete coordinate system of self-polarizing in-out-and-aroundness apprehending and comprehending of self experience, which initiates life awareness and regeneratively processes the evoluting agglomeration of individual experiences. Individual experiences are always and only special-case physical manifestations of utterly abstract, cosmically eternal, generalized principles observable at remotely large and small as well as at everyday local middling time ranges, all of which accumulate progressively to provide potential convertibility of the experience inventory from energetic apprehensions into synergetically discovered comprehensions of a slowly increasing inventory of recognized, inherently and eternally a priori generalized principles from which gradually derive the inventory of human advantages gainable through the useful employment of the generalized principles in special-case artifacts and inventions, which are realized and accumulate only through mutually acknowledged self-and-other individual's omnidirectional observations of the multioverlapped relay of only discontinuously living consciousness's apprehension-comprehension-awareness evolution of the totally communicated and ever-increasing special-case information and synergetically generalized knowledge environment, all of which integrally evolving overlapped and nonsimultaneously interspliced finite experience awareness aggregate is experientially identified as nonunitarily conceptual, but finitely equatable, Universe.

981.03 Going beyond the original formulation of the four-sphere-vertexed minimum structural system (Sec. 411), we observe that the addition of a fifth spherical otherness to the four-ball structural system's symmetry brings about a polarized-system condition. The fifth ball cannot repeat the total mutually intertouching experienced by each of the first four as they joined successively together. The fifth sphere is an oddball, triangularly nested diametrically opposite one of the other four and forming the apex of a second structural-system tetrahedron commonly based by the same three equatorially triangulated spheres. This brings about a condition of two polar-apex spheres and an equatorial set of three. Each of the three at the middle touches not only each other but each of the two poles. While each of the equatorial three touches four others of the fivefold system, each of the two polar spheres touches only three others. Due to this inherent individual differentiability, the fivefoldedness constitutes a self-exciting, pulsation-propagating system. (Compare the atomic time clock, which is just such a fivefolded, atomic-structured, mutually based tetrahedral configuration.)

981.04 This is a second-degree polarization. The first polarization was subsystem when the selfness discovered the otherness and the interrelatedness became an axis of cospinnability, only unobservably accomplished and only intuitively theorized during the initial consciousness of inter-rolling-around anywhere upon one another of the mutually interattracted tangency of self and first otherness, which simultaneous and only theoretically conceivable axial-rotation potential of the *self* and *one other* tangential pairing could only be witnessingly apprehended by a secondly-to-be-discovered otherness, as it is mass-attractively drawn toward the first two from the unthinkable nowhere into the somewhere.

981.05 The whole associated self-and-otherness discloses both *in-outing* (A) and *arounding* (B), which are of two subclasses, respectively: (1) the individually coordinate, and (2) the mutually coordinate.

A. (1) is individually considered, radial or diametric, inward and outward exploration of self by self;
(2) is comprehensive expansion or contraction only mutually and systematically accomplishable;

B. (1) is individual spinnability;
(2) is orbiting of one by another, which is only mutually accomplishable.

981.06 The couple may rotate axially, but it has no surrounding environment otherness in respect to which it can observe that it is rotating axially (or be mistaken and egotistically persuaded that the entire Universe is revolving axially around "self").

981.07 Not until a sixth otherness appears remotely, approaches, and associates with the fivefold system can the latter learn from the newcomer of its remote witnessing that the fivefold system had indeed been rotating axially. Before that sixth otherness appears, the two polar balls of the fivefold polarized system symmetry attract each other through the hole in their common base—the triangular three-ball equator—and their approach-accelerated, second-power rate of interattraction increases momentum, which wedge-spreads open the equatorial triangle with the three equatorial spheres centrifugally separated by the axial spin, precessionally arranged by dynamic symmetry into a three-ball equatorial array, with the three spheres spaced 120 degrees apart and forming the outer apexes of three mutually edged triangles with the two axially tangent polar spheres constituting the common edge of the three longitudinally arrayed triangles.

981.08 Then along comes a sixth ball, and once more momentum-produced dynamic symmetry rearranges all six with three uniradius spheres in the northern hemisphere and three in the southern hemisphere: i.e., they form the octahedron, spinning on an axis between the face centers of two of its eight triangular faces, with the other six triangles symmetrically arrayed around its equator. Dynamic symmetry nests the next ball to arrive at the axial and volumetric center occurring between the north and south polar triangular groups, making two tetrahedra joined together with their respective apexes congruent in the center ball and their respective triangular base centers congruent with the north and south poles. Now the mass-interattracted, dynamically symmetried group of seven spheres is centered by their common mid-tetra apex; since the sevenness is greater than the combined mass of the next six arriving spheres, the latter are dynamically arranged around the system equator and thus complete the vector equilibrium's 12-around-one, isotropic, closest-packed, omnicontiguous-embracing, nuclear containment.

981.09 As awareness begins only with awareness of otherness, the mass-interattracted accelerating acceleration—at a second-power rate of gain as proximity is progressively halved by the self and otherness interapproach—both generates and locally impounds the peak energy combining at tangency, now articulated only as round-and-about one another's surfaces rolling.

981.10 Self has been attracted by the other as much as the other has been attracted by self. This initial manifest of interacceleration force must be continually satisfied. This accumulative force is implicit and is continually accountable either as motion or as structural-system coherence. Four balls manifest structural interstabilization, which combiningly multiplied energy is locked up as potential energy, cohering and stabilizing the structural system, as is manifest in the explosive release of the enormous potential energy locked into the structural binding together of atoms.

981.11 With all the 12 spherical othernesses around the initial self-oneness sphericity apparently uniformly diametered with self, the positive-negative vectorial *relativity* of nuclear equilibrium is operationally established. The pattern of this nuclear equilibrium discloses four hexagonal planes symmetrically interacting and symmetrically arrayed (see Sec. 415) around the nuclear center.

981.12 Awareness of otherness involves mutually intertuned event frequencies. The 12 othernesses around the initially conceiving self-oneness establish both an inward and an outward synchroresonance. Circuit frequency involves a minimum twoness. This initial frequency's inherent twoness is totally invested as *one inward* plus *one outward* wave—two waves appearing superficially as one, or none.

981.13 The self extension of the central sphere reproduces itself outwardly around itself until it is completely embraced by self-reproduced otherness, of which there are exactly 12, exact-replica, exactly spherical domains symmetrically filling all the encompassing space outside of the initial sphere's unique closest-packed cosmic domain, which includes each sphere's exact portion of space occurring outside and around their 12 points of intertangency. The portion of the intervening space belonging to each closest-packed sphere is that portion of the space nearest to each of the spheres as defined by planes halfway between any two most closely adjacent spheres. There are 12 of these tangent planes symmetrically surrounding each sphere whose 12 similar planes are the 12 diamond-shaped facets of the rhombic dodecahedron. The rhombic dodecahedra are allspace filling. Their allspace-filling centers are exactly congruent with the vertexes of the isotropic vector matrix—ergo, with the centers of all closest-packed unit radius sphere complexes.

981.14 This is the self-defining evolution of the sphere and spherical domain as omnisymmetrically surrounded by identical othernesses, with the self-regenerative surroundment radially continuous.

981.15 We now reencounter the self-frequency-multipliable vector equilibrium regeneratively defined by the volumetric centers of the 12 closest-packed rhombic dodecahedral spheric domains exactly and completely surrounding one such initial and nuclear spheric domain.

981.16 What seemed to humans to constitute initiation and evolution of dimensional connection seemed to start with his scratching a line on a plane of a flat Earth, in which two sets of parallel lines crossed each other perpendicularly to produce squares on the seemingly flat Earth, from the corners of which, four perpendiculars arose to intersect with a plane parallel to the base plane on flat Earth occurring at a perpendicular distance above Earth equal to the perpendicular distance between the original parallel lines intersecting to form the base square. This defined the cube, which seemed to satisfy humanity's common conception of dimensional coordination defined by width, breadth, and height. Not knowing that we are on a sphere—the sphere, even a round pebble, seems too foreign to the obvious planar simplicity seemingly accommodated by the environment to deserve consideration. But we have learned that Universe consists primarily of spherically generated events. Universe is a priori spherically islanded. The star-energy aggregates are all spherical.

981.17 If we insist (as humans have) on initiating mensuration of reality with a cube, yet recognize that we are not living on an infinite plane, and that reality requires recognition of the a priori sphericity of our planet, we must commence mensuration with consideration of Earth as a spherical cube. We observe that where three great-circle lines come together at each of the eight corners of the spherical cube, the angles so produced are all 120 degrees—and not 90 degrees. If we make concentric squares within squares on each of the six spherical-surface squares symmetrically subdividing our planet, we find by spherical trigonometry that the four corner angles of each of the successively smaller squares are progressively diminishing. When we finally come to the little local square on Earth within which you stand, we find the corner angles reduced from 120 degrees almost to 90 degrees, but never quite reaching true 90-degree corners.

981.18 Because man is so tiny, he has for all of history deceived himself into popular thinking that all square corners of any size are exactly 90 degrees.

981.19 Instead of initiating universal mensuration with assumedly straight-lined, square-based cubes firmly packed together on a world plane, we should initiate with operationally verified reality; for instance, the first geometrical forms known to humans, the hemispherical breasts of mother against which the small human spheroidal observatory is nestled. The synergetic initiation of mensuration must start with a sphere directly representing the inherent omnidirectionality of observed experiences. Thus we also start synergetically with wholes instead of parts. Remembering that we have verified the Greek definition of a sphere as experimentally invalidated, we start with a spheric array of events. And the "sphere" has definable insideness and seemingly undefinable outsideness volume. But going on operationally, we find that the sphere becomes operationally omni-intercontiguously embraced by other spheres of the same diameter, and that ever more sphere layers may symmetrically surround each layer by everywhere closest packings of spheres, which altogether always and only produces the isotropic vector matrix. This demonstrates not only the uniformly diametered domains of closest-packed spheres, but also that the domains' vertexially identified points of the system are the centers of closest-packed spheres, and that the universal symmetric domain of each of the points and spheres of all uniformly frequencied systems is always and only the rhombic dodecahedron. (See Sec. 1022.11.)

981.20 All the well-known Platonic polyhedra, as well as all the symmetrically referenced crystallographic aberrations, are symmetrically generated in respect to the centers of the spheric domains of the isotropic vector matrix and its inherently nucleating radiational and gravitational behavior accommodating by concentrically regenerative, omnirational, frequency and quanta coordination of vector equilibria, which may operate propagatively and coheringly in respect to any special-case event fix in energetically identifiable Universe.

981.21 The vector equilibrium always and only represents the first omnisymmetric embracement and nucleation of the first-self-discovered-by-otherness sphere by the completely self-embracing, twelvefold, isotropic, continuous otherness.

981.22 Sphere is prime awareness.

981.23 Spheric domain is prime volume.

981.24 Only self-discoverable spheric-system awareness generates all inwardness, outwardness, and aroundness dimensionality.

982.00 Cubes, Tetrahedra, and Sphere Centers

982.01 Spheric Domain: As the domain hierarchy chart shows (see "Concentric Domain Growth Rates," Sec. 955.40), the inherent volume of one prime spheric domain, in relation to the other rational low order number geometric volumes, is exactly sixfold the smallest omnisymmetrical structural system polyhedron: the tetrahedron.

982.02 The spheric domain consists of 144 modules, while the tetrahedron consists of 24 modules. $^{24}/_{144} = {}^{1}/_{6}$.

982.03 The vector equilibrium consists of 480 modules. $^{24}/_{480} = {}^{1}/_{20}$.

982.04 Within the geometries thus defined, the volume of the cube = 3. The cube consists of 72 modules. $^{72}/_{24} = {}^{3}/_{1}$. The initial cube could not contain one sphere because the minimum spheric domain has a volume of six.

982.05 The initial generated minimum cube is defined by four $^{1}/_{8}$th spheres occurring close-packingly and symmetrically only at four of the cube's eight corners; these four corners are congruent with the four corners of the prime tetrahedron, which is also the prime structural system of Universe.

982.06 We thus discover that the tetrahedron's six edges are congruent with the six lines connecting the four fractional spheres occurring at four of the eight alternate corners of the cube.

982.10 Noncongruence of Cube and Sphere Centers: The centers of cubes are not congruent with the sphere centers of the isotropic vector matrix. All the vectors of the isotropic vector matrix define all the centers of the omni-interconnections of self and otherness of omnicontiguously embracing othernesses around the concentrically regenerating, observing self sphere.

982.11 The vector equilibrium represents self's initial realization of self both outwardly and inwardly from the beginning of being between-ness.

982.12 The cube does occur regularly in the isotropic vector matrix, but none of the cubes has more than four of their eight corners occurring in the centers of spheres. The other four corners always occur at the volumetric centers of octahedra, while only the octahedra's and tetrahedra's vertexes always occur at the volumetric centers of spheres, which centers are all congruent with all the vertexes of the isotropic vector matrix. None of the always co-occurring cube's edges is congruent with the vectorial lines (edges) of the isotropic vector matrix. Thus we witness that while the cubes always and only co-occur in the eternal cosmic vector field and are symmetrically oriented within the field, none of the cubes' edge lines is ever congruent or rationally equatable with the most economical energetic vector formulating, which is always rational of low number or simplicity as manifest in chemistry. Wherefore humanity's adoption of the cube's edges as its dimensional coordinate frame of scientific-event reference gave it need to employ a family of irrational constants with which to translate its findings into its unrecognized isotropic-vector-matrix relationships, where all nature's events are most economically and rationally intercoordinated with omni-sixty-degree, one-, two-, three-, four-, and five-dimensional omnirational frequency modulatability.

982.13 The most economical force lines (geodesics) in Universe are those connecting the centers of closest-packed unit radius spheres. These geodesics interconnecting the closest-packed unit radius sphere centers constitute the vectors of the isotropic (everywhere the same) vector matrix. The instant cosmic Universe insinuatability of the isotropic vector matrix, with all its lines and angles identical, all and everywhere equiangularly triangulated—ergo, with omnistructural integrity but always everywhere structurally double- or hinge-bonded, ergo, everywhere nonredundant and force-fluid—is obviously the idealized eternal coordinate economy of nature that operates with such a human-mind-transcending elegance and bounty of omnirational, eternal, optional, freedom-producing resources as to accomplish the eternal regenerative integrity of comprehensively synergetic, nonsimultaneous Universe.

982.14 The edges of the tetrahedra and octahedra of the isotropic vector matrix are always congruent with one another and with all the vectors of the system's network of closest-packed unit radius spheres.

982.15 The whole hierarchy of rationally relative omnisymmetrical geometries' interdimensional definability is topologically oriented exactly in conformance with the ever cosmically idealizable isotropic vector matrix.

982.16 Though symmetrically coordinate with the isotropic vector matrix, none of the co-occurring cube's edges is congruent with the most economical energy-event lines of the isotropic vector matrix; that is, the cube is constantly askew to the most economic energy-control lines of the cosmic-event matrix.

982.20 **Starting With Parts: The Nonradial Line:** Since humanity started with parallel lines, planes, and cubes, it also adopted the edge line of the square and cube as the prime unit of mensuration. This inaugurated geomathematical exploration and analysis with a part of the whole, in contradistinction to synergetics' inauguration of exploration and analysis with total Universe, within which it discovers whole conceptual systems, within which it identifies subentities always dealing with experimentally discovered and experimentally verifiable information. Though life started with whole Universe, humans happened to pick one part—the line, which was so short a section of Earth arc (and the Earth's diameter so relatively great) that they assumed the Earth-scratched-surface line to be straight. The particular line of geometrical reference humans picked happened *not* to be the line of most economical interattractive integrity. It was neither the radial line of radiation nor the radial line of gravity of spherical Earth. From this nonradial line of nature's event field, humans developed their formulas for calculating areas and volumes of the circle and the sphere only in relation to the cube-edge lines, developing empirically the "transcendentally irrational," ergo incommensurable, number *pi* (π), 3.14159 . . . ad infinitum, which provided practically tolerable approximations of the dimensions of circles and spheres.

982.21 Synergetics has discovered that the vectorially most economical control line of nature is in the diagonal of the cube's face and not in its edge; that this diagonal connects two spheres of the isotropic-vector-matrix field; and that those spherical centers are congruent with the two only-diagonally-interconnected corners of the cube. Recognizing that those cube-diagonal-connected spheres are members of the closest packed, allspace-coordinating, unit radius spheres field, whose radii = 1 (unity), we *see* that the isotropic-vector-matrix's field-occurring-cube's diagonal edge has the value of 2, being the line interconnecting the centers of the *two* spheres, with each half of the line being the radius of one sphere, and each of the whole radii perpendicular to the same points of intersphere tangency.

982.30 **Diagonal of Cube as Control Length:** We have learned elsewhere that the sum of the second powers of the two edges of a right triangle equals the second power of the right triangle's hypotenuse; and since the hypotenuse of the two similar equiedged right triangles formed on the square face of the cube by the sphere-center-connecting diagonal has a value of two, its second power is four; therefore, half of that four is the *second power* of *each* of the equi-edges of the right triangle of the cube's diagonaled face: half of four is two.

982.31 The square root of $2 = 1.414214$, ergo, the length of each of the cube's edges is 1.414214. The $\sqrt{2}$ happens to be one of those extraordinary relationships of Universe discovered by mathematics. The relationship is: the number one is to the second root of two as the second root of two is to two: $1 : \sqrt{2} = \sqrt{2} : 2$, which, solved, reads out as $1 : 1.414214 = 1.414214 : 2$

982.32 The cube formed by a uniform width, breadth, and height of $\sqrt{2}$ is $\sqrt{2^3}$, which $= 2.828428$. Therefore, the cube occurring in nature with the isotropic vector matrix, when conventionally calculated, has a volume of 2.828428.

982.33 This is exploratorily noteworthy because this cube, when calculated in terms of man's conventional mensuration techniques, would have had a volume of one, being the first cube to appear in the omni-geometry-coordinate isotropic vector matrix; its edge length would have been identified as the prime dimensional input with an obvious length value of *one*—ergo, its volume would be one: $1 \times 1 \times 1 = 1$. Conventionally calculated, this cube with a volume of one, and an edge length of one, would have had a face diagonal length of $\sqrt{2}$, which equals 1.414214. Obviously, the use of the diagonal of the cube's face as the control length results in a much higher volume than when conventionally evaluated.

982.40 Tetrahedron and Synergetics Constant: And now comes the big *surprise,* for we find that the cube as coordinately reoccurring in the isotropic vector matrix—as most economically structured by nature—has a volume of *three* in synergetics' vector-edged, structural-system-evaluated geometry, wherein the basic structural system of Universe, the tetrahedron, has a volume of *one*.

982.41 A necklace-edged cube has no structural integrity. A tension-linked, edge-strutted cube collapses.

982.42 To have its cubical conformation structurally (triangulated) guaranteed (see Secs. 615 and 740), the regular equiangled tetrahedron must be inserted into the cube, with the tetrahedron's six edges congruent with each of the six vacant but omnitriangulatable diagonals of the cube's six square faces.

982.43 As we learn elsewhere (Secs. 415.22 and 990), the tetrahedron is not only the basic structural system of Universe, ergo, of synergetic geometry, but it is also *the* quantum of nuclear physics and is, ipso facto, exclusively identifiable as *the* unit of volume; ergo, tetrahedron volume equals one. We also learned in the sections referred to above that the volume of the octahedron is exactly *four* when the volume of the tetrahedron of the unit-vector edges of the isotropic-vector-matrix edge is *one,* and that four Eighth-Octahedra are asymmetrical tetrahedra with an equiangular triangular base, three apex angles of 90 degrees, and six lower-corner angles of 45 degrees each; each of the $^1/_8$th octahedron's asymmetric tetrahedra has a volumetric value of one-half unity (the regular tetrahedron). When four of the Eighth-Octahedrons are equiangle-face added to the equiangled, equiedged faces of the tetrahedra, they produce the minimum cube, which, having the tetrahedron at its heart with a volume of one, has in addition four one-half unity volumed Eighth-Octahedra, which add two volumetric units on its corners. Therefore, $2 + 1 = 3$ = the volume of the cube. The cube is volume three where the tetrahedron's volume is one, and the octahedron's volume is four, *and the cube's diagonally structured faces have a diagonal length of one basic system vector of the isotropic vector matrix.* (See Illus. 463.01.)

982.44 Therefore the edge of the cube = $\sqrt{^1/_2}$.

982.45 Humanity's conventional mensuration cube with a volume of one turns out in energetic reality to have a conventionally calculated volume of 2.828428, but this same cube in the relative-energy volume hierarchy of synergetics has a volume of 3.

$$\frac{3}{2.828428} = 1.06066$$

982.46 To correct 2.828428 to read 3, we multiply 2.828428 by the *synergetics conversion constant* 1.06066. (See Chart 963.10.)

982.47 Next we discover, as the charts at Secs. 963.10 and 223.64 show, that of the inventory of well-known symmetrical polyhedra of geometry, all but the cube have irrational values as calculated in the *XYZ* rectilinear-coordinate system—"cubism" is a convenient term—in which the cube's edge and volume are both given the prime mensuration initiating value of *one.* When, however, we multiply all these irrational values of the Platonic polyhedra by the synergetic conversion constant, 1.06066, all these values become unitarily or combinedly rational, and their low first-four-prime-number-accommodation values correspond exactly with those of the synergetic hierarchy of geometric polyhedra, based on the tetrahedron as constituting volumetric unity.

982.48 All but the icosahedron and its "wife," the pentagonal dodecahedron, prove to be volumetrically rational. However, as the tables show, the icosahedron and the vector-edged cube are combiningly rational and together have the rational value of three to the third power, i.e., 27. We speak of the pentagonal dodecahedron as the icosahedron's wife because it simply outlines the surface-area domains of the 12 vertexes of the icosahedron by joining together the centers of area of the icosahedron's 20 faces. When the pentagonal dodecahedron is vectorially constructed with flexible tendon joints connecting its 30 edge struts, it collapses, for, having no triangles, it has no structural integrity. This is the same behavior as that of a cube constructed in the same flexible-tendon-vertex manner. Neither the cube nor the pentagonal dodecahedron is scientifically classifiable as a structure or as a structural system (see Sec. 604).

982.50 **Initial Four-Dimensional Modelability:** The modelability of the *XYZ* coordinate system is limited to rectilinear-frame-of-reference definition of all special-case experience patternings, and it is dimensionally sized by arbitrary, e.g., $c.g_t.s.$-system, subdivisioning increments. The initial increments are taken locally along infinitely extensible lines always parallel to the three sets of rectilinearly interrelated edges of the cube. Any one of the cube's edges may become the one-dimensional module starting reference for initiating the mensuration of experience in the conventional, elementary, energetical * school curriculum.

982.51 The *XYZ* cube has no initially moduled, vertex-defined nucleus; nor has it any inherent, common, most-economically-distanced, uniform, in-out-and-circumferentially-around, corner-cutting operational interlinkage, uniformly moduled coordinatability. Nor has it any initial, ergo inherent, time-weight-energy-(as mass charge or EMF) expressibility. Nor has it any omni-intertransformability other than that of vari-sized cubism. The *XYZ* exploratory coordination inherently commences differentially, i.e., with partial system consideration. Consider the three-dimensional, weightless, timeless, temperatureless volume often manifest in irrational fraction increments, the general reality impoverishments of which required the marriage of the *XYZ* system with the $c.g_t.s.$ system in what resembles more of an added partnership than an integration of the two.

982.52 The synergetics coordinate system's initial modelability accommodates four dimensions and is operationally developable by frequency modulation to accommodate fifth- and sixth-dimensional conceptual-model accountability. Synergetics is initially nuclear-vertexed by the vector equilibrium and has initial in-out-and-around, diagonaling, and diametrically opposite, omni-shortest-distance interconnections that accommodate commonly uniform wavilinear vectors. The synergetics system expresses divergent radiational and convergent gravitational, omnidirectional wavelength and frequency propagation in one operational field. As an initial operational vector system, its (mass × velocity) vectors possess all the unique, special-

* *Energetical* is in contradistinction to *synergetical*. Energetics employs isolation of special cases of our total experience, the better to discern unique behaviors of parts undiscernible and unmeasurable in total experience.

case, time, weight, energy (as mass charge or EMF) expressibilities. Synergetics' isotropic vector matrix inherently accommodates maximally economic, omniuniform intertransformability.

982.53 In the synergetics' four-, five-, and six-dimensionally coordinate system's operational field the linear increment modulatability and modelability is the isotropic vector matrix's vector, with which the edges of the co-occurring tetrahedra and octahedra are omnicongruent; while only the face diagonals—and not the edges—of the inherently co-occurring cubes are congruent with the matrix vectors. Synergetics' exploratory coordination inherently commences integrally, i.e., with whole-systems consideration. Consider the one-dimensional linear values derived from the initially stated whole system, six-dimensional, omnirational unity; any linear value therefrom derived can be holistically attuned by unlimited frequency and one-to-one, coordinated, wavelength modulatability. To convert the *XYZ* system's cubical values to the synergetics' values, the mathematical constants are linearly derived from the mathematical ratios existing between the tetrahedron's edges and the cube's corner-to-opposite-corner distance relationships; while the planar area relationships are derived from the mathematical ratios existing between cubical-edged square areas and cubical-face-diagonaled-edged triangular areas; and the volumetric value mathematical relationships are derived from ratios existing between (a) the cube-edge-referenced third power of the—often odd-fractioned—edge measurements (metric or inches) of cubically shaped volumes and (b) the cube-face-diagonal-vector-referenced third power of exclusively whole number vector, frequency modulated, tetrahedrally shaped volumes. (See Sec. 463 and 464 for exposition of the diagonal of the cube as a wave-propagation model.)

982.54 The mathematical constants for conversion of the linear, areal, and volumetric values of the *XYZ* system to those of the synergetics system derive from the synergetics constant (1.060660). (See Sec. 963.10 and Chart 963.12.) The conversion constants are as follows:

(a) *First Dimension:* The first dimensional cube-edge-to-cube-face-diagonal vector conversion constant from *XYZ* to synergetics is as 1:1.060660.

(b) *Second Dimension:* The two-dimensional linear input of vector vs. cube-edged referenced, triangular vs. square area product identity is $1.060660^2 = 1.125 = 1\,^1/_8\text{th} = ^9/_8\text{ths}$. The second-power value of the vector, $^9/_8$, is in one-to-one correspondence with "congruence in modulo nine" arithmetic (see Secs. 1221.18 and 1221.20); ergo is congruent with wave-quanta modulation (see Secs. 1222 and 1223).

(c) *Third Dimension:* The three-dimensional of the cube-edge vs. vector-edged tetrahedron vs. cube volumetric identity is $1.060660^3 = 1.192$.

982.55 To establish a numerical value for the sphere, we must employ the synergetics constant for cubical third-power volumetric value conversion of the vector equilibrium with the sphere of radius 1. Taking the vector equilibrium at the initial phase (zero frequency, which is unity-two diameter: ergo unity-one radius) with the sphere of radius 1; i.e., with the external vertexes of the vector equilibrium congruent with the surface of the sphere $= 4/3$ *pi* (π) multiplied by the third power of the radius. Radius $= 1$. $1^3 = 1$. $1 \times 1.333 \times 3.14159 = 4.188$. 4.188 times synergetics third-power constant $1.192 = 5$ = volume of the sphere. The volume of the radius 1 vector equilibrium = 2.5. VE sphere = 2 VE.

982.56 We can assume that when the sphere radius is 1 (the same as the nuclear vector equilibrium) the Basic Disequilibrium 120 LCD tetrahedral components of mild off-sizing are also truly of the same volumetric quanta value as the *A* and *B* Quanta Modules; they would be shortened in overall greatest length while being fractionally fattened at their smallest-triangular-face end, i.e., at the outer spherical surface end of the 120 LCD asymmetric tetrahedra. This uniform volume can be maintained (as we have seen in Sec. 961.40).

982.57 Because of the fundamental 120-module identity of the nuclear sphere of radius 1 ($F = 0$), we may now identify the spherical icosahedron of radius 1 as *five;* or as 40 when frequency is $2F^2$. Since 40 is also the volume of the F^2 vector-equilibrium-vertexes-congruent sphere, the unaberrated vector equilibrium $F^2 = 20$ (i.e., $8 \times 2^1/_2$ nuclear-sphere's inscribed vector equilibrium). We may thus assume that the spherical icosahedron also subsides by loss of half its volume to a size at which its volume is also 20, as has been manifested by its prime number *five,* indistinguishable from the vector equilibrium in all of its topological hierarchies characteristics.

982.58 Neither the planar-faceted exterior edges of the icosahedron nor its radius remain the same as that of the vector equilibrium, which, in transforming from the vector equilibrium conformation to the icosahedral state—as witnessed in the jitterbugging (see Sec. 465)—did so by transforming its outer edge lengths as well as its radius. This phenomenon could be analagous the disappearance of the nuclear sphere, which is apparently permitted by the export of its volume equally to the 12 surrounding spheres whose increased diameters would occasion the increased sizing of the icosahedron to maintain the volume 20-ness of the vector equilibrium. This supports the working assumption that the 120 LCD asymmetric tetrahedral volumes are quantitatively equal to the *A* or *B* Quanta Modules, being only a mild variation of shape. This effect is confirmed by the discovery that 15 of the 120 LCD Spherical Triangles equally and interiorly subdivide each of the eight spherical octahedron's triangular surfaces, which spherical octahedron is described by the three-great-circle set of the 25 great circles of the spherical vector equilibrium. (See drawings section.)

982.59 We may also assume that the pentagonal-faced dodecahedron, which is developed on exactly the same spherical icosahedron, is also another transformation of the same module quantation as that of the icosahedron's and the vector equilibrium's prime number *five* topological identity.

982.60 Without any further developmental use of *pi* (π) we may now state in relation to the isotropic vector matrix synergetic system, that:

The volume of the sphere is a priori always quantitatively:

—$5F^3$ as volumetrically referenced to the regular tetrahedron (as volume = 1); or

—$120F^3$ as referenced to the *A* and *B* Quanta Modules.

982.61 There is realized herewith a succession of concentric, 12-around-one, closest-packed spheres, each of a tetra volume of *five;* i.e., of 120 *A* and *B* Quanta Modules omniembracing our hierarchy of nuclear event patternings. See Illus. 982.61 in the color section, which depicts the synergetics isometric of the isotropic vector matrix and its omnirational, low-order whole number, equilibrious state of the micro-macro cosmic limits of nuclearly unique, symmetrical morphological relativity in their interquantation, intertransformative, intertransactive, expansive-contractive, axially-rotative, operational field. This may come to be identified as the unified field, which, as an operationally transformable complex, is conceptualizable only in its equilibrious state.

982.62 Table of Concentric, 12-Around-One, Closest-Packed Spheres, Each of a Tetra Volume of Five, i.e., 120 A *and* B *Quanta Modules, Omniembracing Our Hierarchy of Nuclear Event Patternings.* (See also Illus. 982.61 in drawings section.)

Symmetrical Form:	*Tetra Volumes*	A *and* B *Quanta Modules*
F^2 Sphere	40	960
F^2 Cube	24	576
F^2 Vector equilibrium	20	480
F^0 Rhombic dodecahedron	6	144
F^0 Sphere (nuclear)	5	120
F^0 Octahedron	4	96
F^0 Cube	3	72
F^0 Vector equilibrium	$2^1/_2$	60
F^0 Tetrahedron	1	24
F^0 Skew-aberrated, disequilibrious icosahedron	← 5 →	120
F^2 Skew-aberrated, disequilibrious icosahedron	← 40 →	960

982.63 Sphere and Vector Equilibrium: Sphere = vector equilibrium in combined four-dimensional orbit and axial spin. Its 12 vertexes describing six great circles and six axes. All 25 great circles circling while spinning on one axis produce a spin-profiling of a superficially perfect sphere.

982.64 The vector equilibrium also has 25 great circles (see Sec. 450.10), of which 12 circles have 12 axes of spin, four great circles have four axes of spin, six great circles have six axes of spin, and three great circles have three axes of spin. (12 + 4 + 6 + 3 = 25)

982.65 Vector equilibrium = sphere at equilibrious, ergo zero energized, ergo unorbited and unspun state.

982.70 **Hierarchy of Concentric Symmetrical Geometries:** It being experimentally demonstrable that the number of *A* and *B* Quanta Modules per tetrahedron is 24 (see Sec. 942.10); that the number of quanta modules of all the symmetric polyhedra congruently co-occurring within the isotropic vector matrix is always 24 times their whole regular-tetrahedral-volume values; that we find the volume of the nuclear sphere to be *five* (it has a volumetric equivalence of 120 *A* and *B* Quanta Modules); that the common prime number *five* topological and quanta-module value identifies both the vector equilibrium and icosahedron (despite their exclusively unique morphologies—see Sec. 905, especially 905.55); that the icosahedron is one of the three-and-only prime structural systems of Universe (see Secs. 610.20 and 1011.30) while the vector equilibrium is unstable—because equilibrious—and is not a structure; that their quanta modules are of equal value though dissimilar in shape; and that though the vector equilibrium may be allspace-fillingly associated with tetrahedra and octahedra, the icosahedron can never be allspace-fillingly compounded either with itself nor with any other polyhedron: these considerations all suggest the relationship of the neutron and the proton for, as with the latter, the icosahedron and vector equilibrium are interexchangingly transformable through their common spherical-state omnicongruence, quantitatively as well as morphologically.

982.71 The significance of this unified field as defining and embracing the minimum-maximum limits of the inherent nuclear domain limits is demonstrated by the nucleus-concentric, symmetrical, geometrical hierarchy wherein the rhombic dodecahedron represents the *smallest,* omnisymmetrical, self-packing, allspace-filling, six-tetra-volume, uniquely exclusive, cosmic domain of each and every closest-packed, unit-radius sphere. Any of the closest-packed, unit-radius spheres, when surrounded in closest packing by 12 other such spheres, becomes the nuclear sphere, to become uniquely embraced by four successive layers of surrounding, closest-packed, unit-radius spheres—each of which four layers is uniquely related to that nucleus—with each additional layer beyond four becoming duplicatingly repetitive of the pattern of unique surroundment of the originally unique, first four, concentric-layered, nuclear set. It is impressive that the unique nuclear domain of the rhombic dodecahedron with a volume of *six* contains within itself and in nuclear concentric array:

—the unity-one-radiused sphere of volume *five;*

—the octahedron of volume *four;*

—the cube of volume *three;*

—the prime vector equilibrium of volume $2^1/_2$; and

—the two regular (positive and negative) tetrahedra of volume *one* each.

This succession of 1, 2, 3, 4, 5, 6 rational volume relationships embraces the first four prime numbers 1, 2, 3, and 5. (See Illus. 982.61 in color section.) The volume-24 (tetra) cube is the *largest* omnisymmetrical self-packing, all-space-filling polyhedron that exactly identifies the unique domain of the original 12-around-one, nuclear-initiating, closest packing of unit-radius spheres. The unit quantum leap of 1—going to 2—going to 3—going to 4—going to 5—going to 6, with no step greater than 1, suggests a unique relationship of this set of six with the sixness of degrees of freedom.*

982.72 The domain limits of the hierarchy of concentric, symmetrical geometries also suggests the synergetic surprise of two balls having only one interrelationship; while three balls have three—easily predictable —relationships; whereas the simplest, ergo prime, structural system of Universe defined exclusively by four balls has an unpredictable (based on previous experience) sixness of fundamental interrelationships represented by the six edge vectors of the tetrahedron.

982.73 The one-quantum "leap" is also manifest when one vector edge of the volume 4 octahedron is rotated 90 degrees by disconnecting two of its ends and reconnecting them with the next set of vertexes occurring at 90 degrees from the previously interconnected-with vertexes, transforming the same unit-length, 12-vector structuring from the octahedron to the first three-triple-bonded-together (face-to-face) tetrahedra of the tetrahelix of the DNA-RNA formulation. One 90-degree vector reorientation in the complex alters the volume from exactly 4 to exactly 3. This relationship of one quantum disappearance coincident to the transformation of the nuclear symmetrical octahedron into the asymmetrical initiation of the DNA-RNA helix is a reminder of the disappearing-quanta behavior of the always integrally end-cohered jitterbugging transformational stages from the 20 tetrahedral volumes of the vector equilibrium to the octahedron's 4 and thence to the tetrahedron's 1 volume. All of these stages are rationally concentric in our unified operational

* For further suggestions of the relationship between the rhombic dodecahedron and the degrees of freedom see Secs. 426, 537.10, and 954.47.

field of 12-around-one closest-packed spheres that is only conceptual as equilibrious. We note also that per each sphere space between closest-packed spheres is a volume of exactly one tetrahedron: $6 - 5 = 1$.

982.80 **Closest Packing of Circles:** Because we may now give the dimensions of any sphere as $5F^n$, we have no need for *pi* in developing spheres holistically. According to our exploratory strategy, however, we may devise one great circle of one sphere of unit rational value, and, assuming our circle also to be rational and a whole number, we may learn what the mathematical relationship to *pi* may be—lengthwise—of our a priori circle as a whole part of a whole sphere. We know that *pi* is the length of a circle as expressed in the diameters of the circle, a relationship that holds always to the transcendentally irrational number 3.14159. But the relationship of volume 5 to the radius of one of our spheres is not altered by the circumference-to-diameter relationship because we commence with the omnidimensional wholeness of reality.

982.81 We recall also that both Newton and Leibnitz in evolving the calculus thought in terms of a circle as consisting of an infinite number of short chords. We are therefore only modifying their thinking to accommodate the manifest discontinuity of all physical phenomena as described by modern physics when we explore the concept of a circle as an aggregate of short event-vectors—tangents (instead of Newtonian short chords) whose tangential overall length must be greater than that of the circumference of the theoretical circle inscribed within those tangent event-vectors—just as Newton's chords were shorter than the circle encompassing them.

982.82 If this is logical, experimentally informed thinking, we can also consider the closest-tangential-packing of circles on a plane that produces a non-all-area-filling pattern with concave triangles occurring between the circles. Supposing we allowed the perimeters of the circles to yield bendingly outward from the circular centers and we crowded the circles together while keeping themselves as omni-integrally, symmetrically, and aggregatedly together, interpatterned on the plane with their areal centers always equidistantly apart; we would find then—as floor-tile makers learned long ago—that when closest packed with perimeters congruent, they would take on any one of three and only three possible polygonal shapes: the hexagon, the square, or the triangle—closest-packed hexagons, whose perimeters are exactly three times their diameters. Hexagons are, of course, cross sections through the vector equilibrium. The hexagon's six radial vectors exactly equal the six chordal sections of its perimeter.

982.83 Assuming the vector equilibrium hexagon to be the relaxed, cosmic, neutral, zero energy-events state, we will have the flexible but not stretchable hexagonal perimeter spun rapidly so that all of its chords are centrifugally expelled into arcs and the whole perimeter becomes a circle with its radius necessarily contracted to allow for the bending of the chords. It is this circle with its perimeter equalling six that we will now convert, first into a square of perimeter six and then into a triangle of perimeter six with the following results:

Circle	radius 0.954930	perimeter 6
Hexagon	radius 1.000000	perimeter 6 (neutral)
Square	radius 1.060660	perimeter 6
Triangle	radius 1.1546	perimeter 6

(In the case of the square, the radius is taken from the center to the corner, not the edge. In the triangle the radius is taken to the corner, not the edge.) We take particular note that the radius of the square phase of the closest-packed circle is 1.060660, the synergetics constant.

982.84 In accomplishing these transformations of the uniformly-perimetered symmetrical shapes, it is also of significance that the area of six equiangular, uniform-edged triangles is reduced to four such triangles. Therefore, it would take more equiperimeter triangular tiles or squares to pave a given large floor area than it would using equiperimetered hexagons. We thus discover that the hexagon becomes in fact the densest-packed patterning of the circles; as did the rhombic dodecahedron become the minimal limit case of self-packing allspace-filling in isometric domain form in the synergetical from-whole-to-particular strategy of discovery; while the rhombic dodecahedron is the six-dimensional state of omni-densest-packed, nuclear field domains; as did the two-frequency cube become the maximum subfrequency self-packing, allspace-filling symmetrical domain, nuclear-uniqueness, expandability and omni-intertransformable, intersymmetrical, polyhedral evolvement field; as did the limit-of-nuclear-uniqueness, minimally at three-frequency complexity, self-packing, allspace-filling, semi-asymmetric octahedron of Critchlow; and maximally by the three-frequency, four-dimensional, self-packing, allspace-filling tetrakaidecahedron: these two, together with the cube and the rhombic dodecahedron constitute the only-four-is-the-limit-system set of self-packing, allspace-filling, symmetrical polyhedra. These symmetrical realizations approach a neatness of cosmic order.

983.00 Spheres and Interstitial Spaces

983.01 Frequency: In synergetics, F=

either, frequency of modular subdivision of one radius;

or, frequency of modular subdivision of one outer chord of a hexagonal equator plane of the vector equilibrium. Thus, $F = r$, *r*adius; or $F = Ch$, *Ch*ord.

983.02 **Sphere Layers:** The numbers of separate spheres in each outer layer of concentric spherical layers of the vector equilibrium grows at a rate: $= 10r^2 + 2$, or $10F^2 + 2$.

983.03 Whereas the space between any two concentrically parallel vector equilibria whose concentric outer planar surfaces are defined by the spheric centers of any two concentric sphere layers, is always $10\ (2 + 12\ \frac{r^2 - r}{2})$, or $10\ (2 + 12\ \frac{F^2 - F}{2})$.

983.04 The difference is the nonsphere interstitial space occurring uniformly between the closest-packed spheres, which is always $6 - 5 = 1$ tetrahedron.

984.00 Rhombic Dodecahedron

984.10 The rhombic dodecahedron is symmetrically at the heart of the vector equilibrium. The vector equilibrium is the ever-regenerative, palpitatable heart of all the omniresonant physical-energy hearts of Universe.

985.00 Synergetics Rational Constant Formulas for Area of a Circle and Area and Volume of a Sphere

985.01 We employ the synergetics constant, "*S*," for correcting the cubical *XYZ* coordinate inputs to the tetrahedral inputs of synergetics:

$S^1 = 1.060660$
$S^2 = 1.12487$
$S^3 = 1.1931750$

We learn that the sphere of radius 1 has a "cubical" volume of 4.188; corrected for tetrahedral value we have $4.188 \times 1.193 = 4.996 = 5$ tetrahedra $= 1$ sphere. Applying the S^2 to the area of a circle of radius 1, ($pi = 3.14159$) $3.14159 \times 1.125 = 3.534$ for the corrected "square" area.

985.02 We may also employ the *XYZ* to synergetics conversion factors between commonly based squares and equiangled triangles: from a square to a triangle the factor is 2.3094; from a triangle to a square the factor is .433. The constant *pi* $3.14159 \times 2.3094 = 7.254 = 7^1/_4$; thus $7^1/_4$ triangles equal the area of a circle of radius 1. Since the circle of a sphere equals exactly four circular areas of the same radius, $7^1/_4 \times 4 = 29 =$ area of the surface of a sphere of radius 1.

985.03 The area of a hexagon of radius 1 shows the hexagon with its vertexes lying equidistantly from one another in the circle of radius 1 and since the radii and chords of a hexagon are equal, then the six equilateral triangles in the hexagon plus $1^1/_4$ such triangles in the arc-chord zones equal the area of the circle: $1.25/6 = .208$ zone arc-chord area. Wherefore the area of a circle of frequency 2 = 29 triangles and the surface of a sphere of radius 2 = 116 equilateral triangles.

985.04 For the 120 LCD spherical triangles $S = 4$; $S = 4$ for four great-circle areas of the surface of a sphere; therefore S for one great-circle area equals exactly one spherical triangle, since $^{120}/_{4} = 30$ spherical triangles vs. $^{116}/_{4} = 29$ equilateral triangles. The S disparity of 1 is between a right spherical triangle and a planar equiangular triangle. Each of the 120 spherical LCD triangles has exactly six degrees of spherical excess, their three corners being 90 degrees, 60 degrees and 36 degrees vs. 90 degrees, 60 degrees, 30 degrees of their corresponding planar triangle. Therefore, 6 degrees per each spherical triangle times 120 spherical triangles amounts to a total of 720 degrees spherical excess, which equals exactly one tetrahedron, which exact excessiveness elucidates and elegantly agrees with previous discoveries (see Secs. 224.07, 224.10, and 224.20).

985.05 The synergetical definition of an operational sphere (vs. that of the Greeks) finds the spheric experience to be operationally always a star-point-vertexed polyhedron, and there is always a 720 degree (one tetrahedron) excess of the Greek's sphere's assumption of 360 degrees around each vertex vs. the operational sum of the external angles of any system, whether it be the very highest frequency (seemingly "pure" spherical) regular polyhedral system experience of the high-frequency geodesic spheres, or irregular giraffe's or crocodile's chordally-interconnected, outermost-skin-points-defined, polyhedral, surface facets' corner-angle summation.

985.06 Thus it becomes clear that $S = 1$ is the difference between the infinite frequency series' perfect nuclear sphere of volume 5 and 120 quanta modules, and the four-whole-great-circle surface area of 116 equilateral triangles, which has an exact spherical excess of 720 degrees = one tetrahedron, the difference between the 120 spherical triangles and the 120 equilateral triangles of the 120-equiplanar-faceted polyhedron.

985.07 This is one more case of the *one tetrahedron : one quantum jump* involved between various stages of nuclear domain intertransformations, all the way from the difference between integral-finite, nonsimultaneous, scenario Universe, which is inherently nonunitarily conceptual, and the maximum-minimum, conceptually thinkable, systemic subdivision of Universe into an omnirelevantly frequenced, tunable set which is always one positive tetrahedron (macro) and one negative tetrahedron (micro) less than Universe: the definitive conceptual vs. finite nonunitarily conceptual Universe (see Secs. 501.10 and 620.01).

985.08 The difference of *one* between the spheric domain of the rhombic dodecahedron's *six* and the nuclear sphere's *five*—or between the tetra volume of the octahedron and the three-tetra sections of the tetrahelix—these are the *prime wave pulsation propagating quanta phenomena* that account for local aberrations, twinkle angles, and unzipping angles manifest elsewhere and frequently in this book.

985.10 Table: Triangular Area of a Circle of Radius 1

F^1 = Zero-one frequency = $7^1/_4$.

Table of whole triangles only with F = Even N, which is because Even N = closed wave circuit.

	F^N	F^2		Triangular Areas of Circle of Radius 1:		
Open	1	1	× $7^1/_4$		$7^1/_4$	
Close	2	4	× $7^1/_4$		29	
Open	3	9	× $7^1/_4$	(63 + $2^1/_4$).	$65^1/_4$	
Close	4	16	× $7^1/_4$	(112 + 4)	116	(also surface of one sphere)
Open	5	25	× $7^1/_4$	(175 + $6^1/_4$)	$181^1/_4$	
Close	6	36	× $7^1/_4$	(252 + 9)	261	

985.20 Spheric Experience: Experientially defined, the spheric experience, i.e., a sphere, is an aggregate of critical-proximity event "points." Points are a multidimensional set of crossings of orbits: traceries, foci, fixes, vertexes coming cometlike almost within intertouchability and vertexing within cosmically remote regions. Each point consists of three or more vectorially convergent events approximately equidistant from one approximately locatable and as yet nondifferentially resolved, point; i.e., three or more visualizable, four-dimensional vectors' most critical proximity, convergently-divergently interpassing region, local, locus, terminal and macrocosmically the most complex of such point events are the celestial stars; i.e., the highest-speed, high-frequency energy event, importing-exporting exchange centers. Microcosmically the atoms are the inbound terminals of such omniorderly exchange systems.

985.21 Spheres are further cognizable as vertexial, star-point-defined, polyhedral, constellar systems structurally and locally subdividing Universe into insideness and outsideness, microcosm-macrocosm.

985.22 Physically, spheres are high-frequency event arrays whose spheric complexity and polyhedral system unity consist structurally of discontinuously islanded, critical-proximity-event huddles, compressionally convergent events, only tensionally and omni-interattractively cohered. The pattern integrities of all spheres are the high-frequency, traffic-described subdivisionings of either tetrahedral, octahedral, or icosahedral angular interference, intertriangulating structures profiling one, many, or all of their respective great-circle orbiting and spinning event characteristics. All spheres are high-frequency geodesic spheres, i.e., triangular-faceted polyhedra, most frequently icosahedral because the icosasphere is the structurally most economical.

990.00 Triangular and Tetrahedral Accounting

990.01 All scientists as yet say "X squared," when they encounter the expression "X^2," and "X cubed," when they encounter "X^3." But the number of squares enclosed by equimodule-edged subdivisions of large gridded squares is the same as the number of triangles enclosed by equimodule-edged subdivisions of large gridded triangles. This remains true regardless of the grid frequency, except that the triangular grids take up less space. Thus we may say "triangling" instead of "squaring" and arrive at identical arithmetic results, but with more economical geometrical and spatial results. (See Illus. 990.01 and also 415.23.)

990.02 Corresponding large, symmetrical agglomerations of cubes or tetrahedra of equimodular subdivisions of their edges or faces demonstrate the same rate of third-power progression in their symmetrical growth (1, 8, 27, 64, etc.). This is also true when divided into small tetrahedral components for each large tetrahedron or in terms of small cubical components of each large cube. So we may also say "tetrahedroning" instead of "cubing" with the same arithmetical but more economical geometrical and spatial results.

990.03 We may now say "one to the second power equals one," and identify that arithmetic with the triangle as the geometrical unit. Two to the second power equals four: four triangles. And nine triangles and 16 triangles, and so forth. Nature needs only triangles to identify arithmetical "powering" for the self-multiplication of numbers. Every square consists of two triangles. Therefore, "triangling" is twice as efficient as "squaring." This is what nature does because the triangle is the only structure. If we wish to learn how nature always operates in the most economical ways, we must give up "squaring" and learn to say "triangling," or use the more generalized "powering."

990.04 There is another very trustworthy characteristic of synergetic accounting. If we prospectively look at any quadrilateral figure that does not have equal edges, and if we bisect and interconnect those mid-edges, we

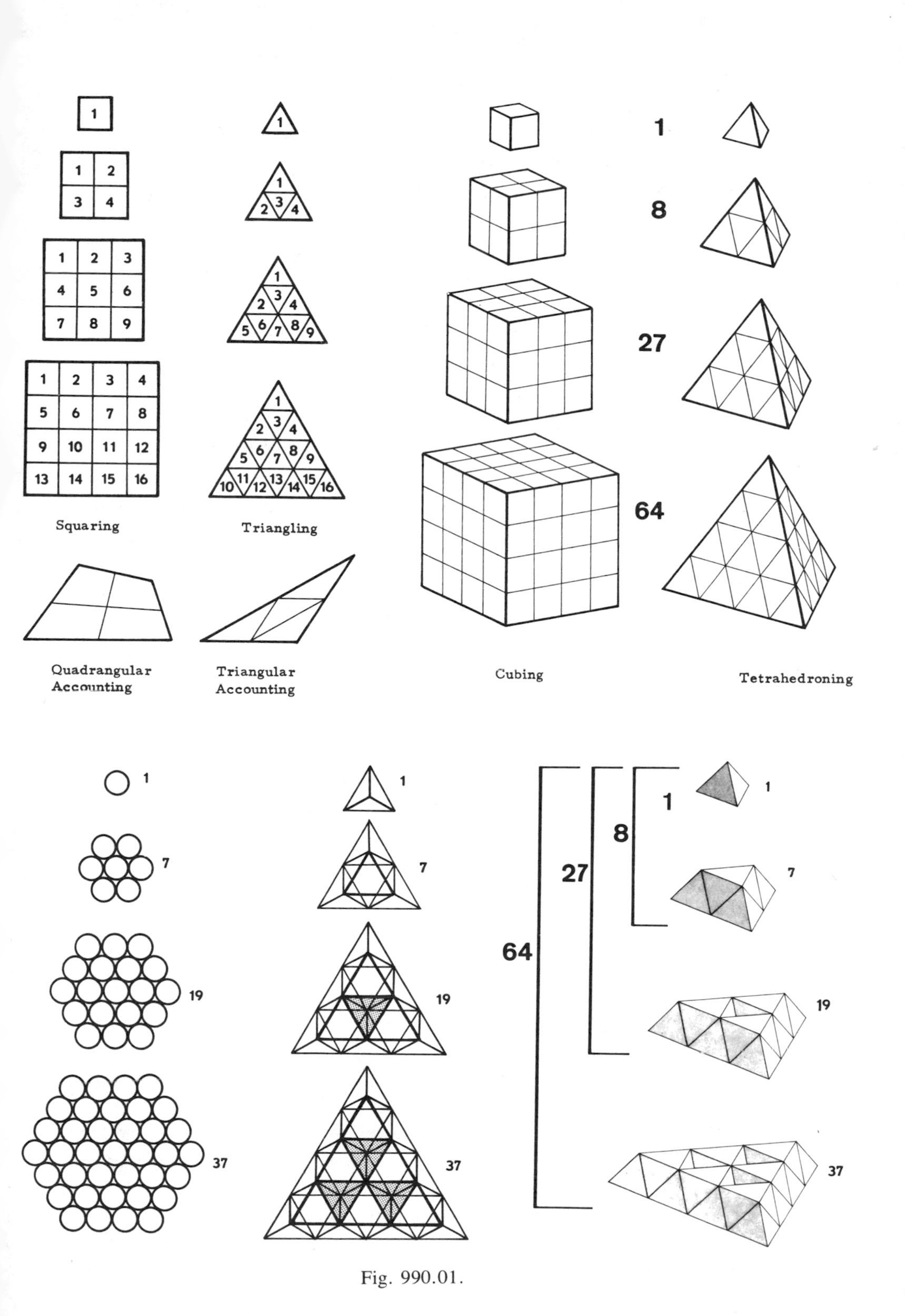

1
1 2
3 4
1 2 3
4 5 6
7 8 9
1 2 3 4
5 6 7 8
9 10 11 12
13 14 15 16
Squaring
1
1 3 2 4
1 3 2 4 5 6 7 8 9
1 3 2 4 5 6 7 8 9 10 11 12 13 14 15 16
Triangling
1
8
27
64
Quadrangular Accounting
Triangular Accounting
Cubing
Tetrahedroning
1
7
19
37
1
7
19
37
64
27
8
1
1
7
19
37

Fig. 990.01.

always produce four dissimilar quadrangles. But when we bisect and interconnect the mid-edges of any arbitrary triangle—equilateral, isosceles, or scalene—four smaller similar and equisized triangles will always result. There is no way we can either bisect or uniformly subdivide and then interconnect all the edge division points of any symmetrical or asymmetrical triangle and not come out with omni-identical triangular subdivisions. There is no way we can uniformly subdivide and interconnect the edge division points of any asymmetrical quadrangle (or any other different-edge-length polygons) and produce omnisimilar polygonal subdivisions. Triangling is not only more economical; it is always reliable. These characteristics are not available in quadrangular or orthogonal accounting.

990.05 The increasingly vast, comprehensive, and rational order of arithmetical, geometrical, and vectorial coordination that we recognize as synergetics can reduce the dichotomy, the chasm between the sciences and the humanities, which occurred in the mid-nineteenth century when science gave up models because the generalized case of exclusively three-dimensional models did not seem to accommodate the scientists' energy-experiment discoveries. Now we suddenly find elegant field modelability and conceptuality returning. We have learned that all local systems are conceptual. Because science had a fixation on the "square," the "cube," and the 90-degree angle as the exclusive forms of "unity," most of its constants are *irrational*. This is only because they entered nature's structural system by the wrong portal. If we use the cube as volumetric unity, the tetrahedron and octahedron have irrational number volumes.

995.00 Vector Models of Magic Numbers

995.01 Magic Numbers

995.02 The magic numbers are the high abundance points in the atomic-isotope occurrences. They are 2, 8, 20, 50, 82, 126, . . . , ! For every nonpolar vertex, there are three vector edges in every triangulated structural system. The Magic Numbers are the nonpolar vertexes. (See Illus. 995.31.)

995.03 In the structure of atomic nuclei, the Magic Numbers of neutrons and protons correspond to the states of increased stability. Synergetics provides a symmetrical, vector-model system to account for the Magic Numbers based on combinations of the three omnitriangulated structures: tetrahedron, octahedron, and icosahedron. In this model system, all the vectors have the value of one-third. The Magic Numbers of the atomic nuclei are accounted for

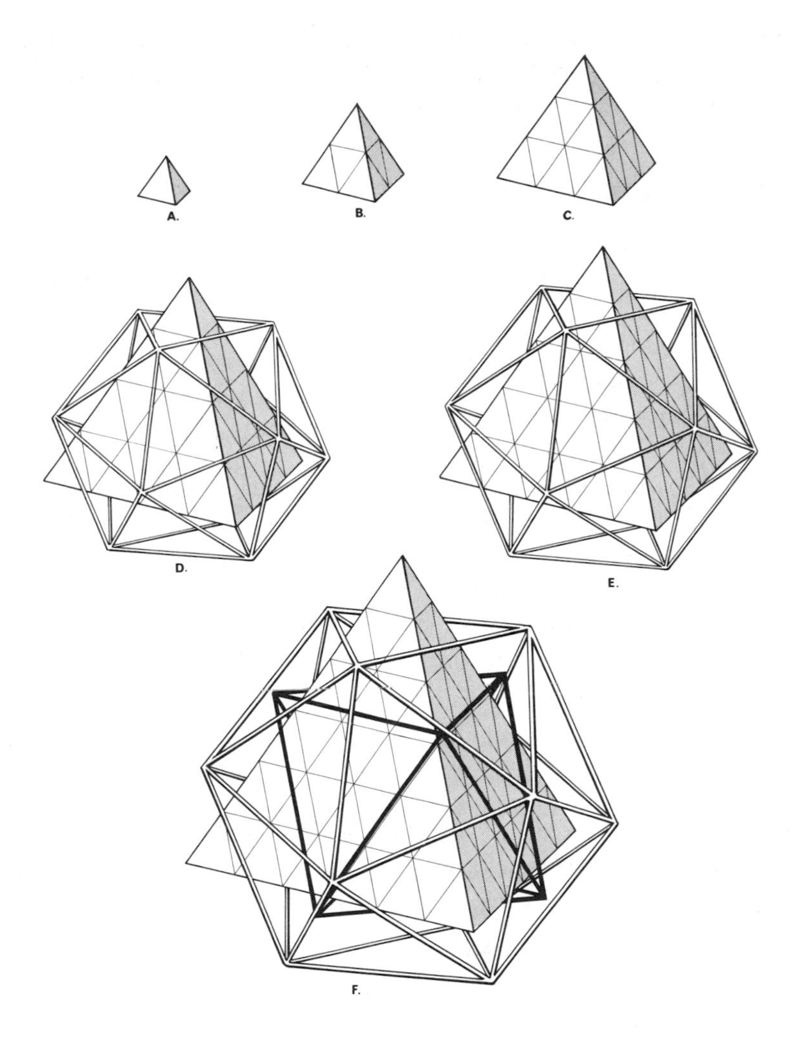

Fig. 995.03 *Vector Models of Atomic Nuclei: Magic Numbers:* In the structure of atomic nuclei there are certain numbers of neutrons and protons which correspond to states of increased stability. These numbers are known as the magic numbers and have the following values: 2, 8, 20, 50, 82, and 126. A vector model is proposed to account for these numbers based on combinations of the three fundamental omnitriangulated structures: the tetrahedron, octahedron, and icosahedron. In this system all vectors have a value of one-third. The magic numbers are accounted for by summing the total number of vectors in each set and multiplying the total by $^{1}/_{3}$. Note that although the tetrahedra are shown as opaque, nevertheless all the internal vectors defined by the isotropic vector matrix are counted in addition to the vectors visible on the faces of the tetrahedra.

by summing up the total number of external and internal vectors in each set of successive frequency models, then dividing the total by three, there being three vectors in Universe for every nonpolar vertex.

995.10 Sequence

995.11 The sequence is as follows:

	(Magic Number:)
One-frequency tetrahedron:	
6 vectors times $^1/_3$	$=2$
Two-frequency tetrahedron:	
24 vectors times $^1/_3$	$=8$
Three-frequency tetrahedron:	
60 vectors times $^1/_3$	$=20$
Four-frequency tetrahedron + One-frequency icosahedron:	
120 vectors + 30 vectors times $^1/_3$	$=50$
Five-frequency tetrahedron + One-frequency icosahedron:	
216 + 30 vectors times $^1/_3$	$=82$
Six-frequency tetrahedron + One-frequency octahedron + One-frequency icosahedron:	
336 + 12 + 30 vectors times $^1/_3$	$=126$

995.20 Counting

995.21 In the illustration, the tetrahedra are shown as opaque. Nevertheless, all the internal vectors defined by the isotropic vector matrix are counted in addition to the vectors visible on the external faces of the tetrahedra.

995.30 Reverse Peaks in Descending Isotope Curve

995.31 There emerges an impressive pattern of regularly positioned behaviors of the relative abundances of isotopes of all the known atoms of the known Universe. Looking like a picture of a mountainside ski run in which there are a series of ski-jump upturns of the run, there is a series of sharp upward-pointing peaks in the overall descent of this relative abundance of isotopes curve, which originates at its highest abundance in the lowest-atomic-numbered elemental isotopes.

995.32 The Magic Number peaks are approximately congruent with the atoms of highest structural stability. Since the lowest order of number of isotopes are the most abundant, the inventory reveals a reverse peak in the otherwise descending curve of relative abundance.

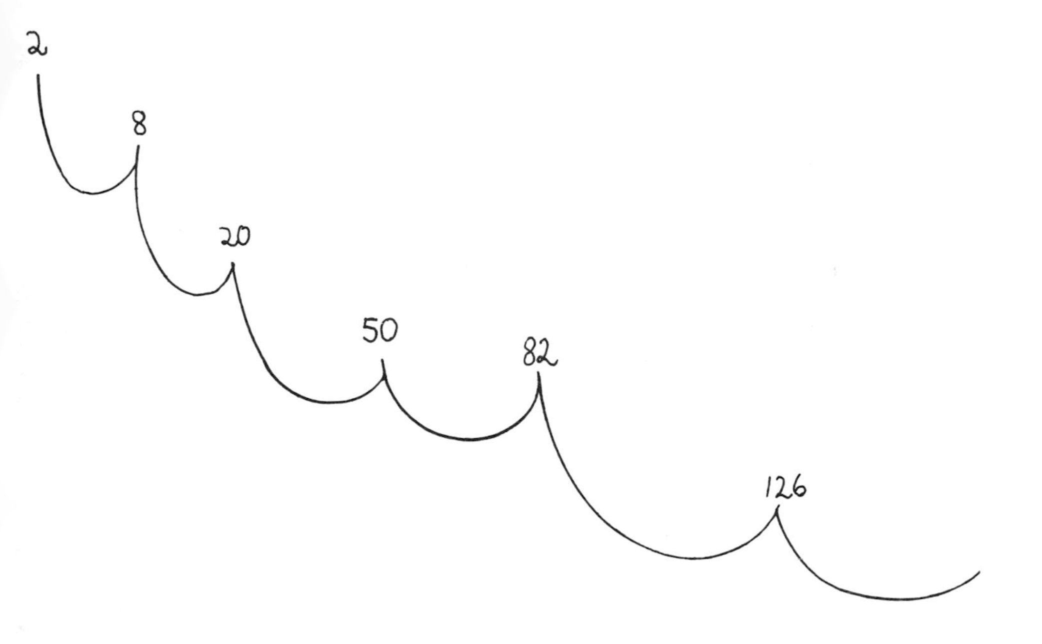

Fig. 995.31

995.33 The vectorial modeling of synergetics demonstrates nuclear physics with lucid comprehension and insight into what had been heretofore only instrumentally apprehended phenomena. In the post-fission decades of the atomic-nucleus explorations, with the giant atom smashers and the ever more powerful instrumental differentiation and quantation of stellar physics by astrophysicists, the confirming evidence accumulates.

995.34 Dr. Linus Pauling has found and published his spheroid clusters designed to accommodate the Magic Number series in a logical system. We find him—although without powerful synergetic tools—in the vicinity of the answer. But we can now identify these numbers in an absolute synergetic hierarchy, which must transcend any derogatory suggestion of pure coincidence alone, for the coincidence occurs with mathematical regularity, symmetry, and a structural logic that identifies it elegantly as the model for the Magic Numbers.

1000.00 Omnitopology

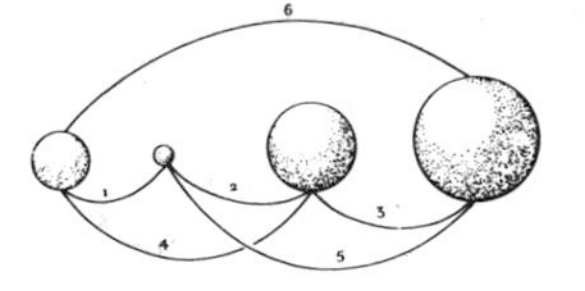

1000.00 OMNITOPOLOGY

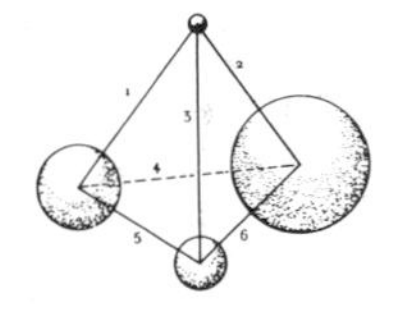

1001.00 *Inherent Rationality of Omnidirectional Epistemology*

1001.10 Spherical Reference: Operationally speaking, the word *omnidirectional* involves a speaker who is observing from some viewing point. He says, "People and things are going every which way around me." It seems chaotic to him at first, but on further consideration he finds the opposite to be true, that only inherent order is being manifest. First, we observe that we do not and cannot live and experience in either a one-dimensional linear world nor in a two-dimensional infinitely extended planar world.

1001.11 Omnidirectional means that a center of a movable sphere of observation has been established a priori by Universe for each individual life's inescapably mobile viewpoint; like shadows, these move everywhere silently with people. These physical-existence-environment surrounds of life events spontaneously resolve into two classes:

(1) those events that are to pass *tangentially* by the observer; and

(2) those event entities other than self that are moving *radially* either toward or away from the observer.

1001.12 The tangentially passing energy events are always and only moving in lines that are at nearest moment perpendicular to the radii of the observer, which means that the multiplicity of his real events does not produce chaos: it produces discretely apprehendable experience increments, all of which can be chartingly identified by angle and frequency data therewith to permit predictable reinterpositioning events and environmental transformations.

1001.13 The observer's unfamiliarity with the phenomena he is observing, the multiplicity of items of interaction and their velocity of transformations, and their omniengulfing occurrences tend to dismay the observer's hope of immediate or reasonable comprehension. Therefore, observers are often induced to discontinue their attempts at technical comprehension of their experience, in a surrender of the drive to comprehend. This fills the potential comprehension void of the observer with a sense of chaos, which sensation he then subconsciously converts into a false rationale by explaining to himself that the environment is inherently chaotic, ergo, inherently incomprehensible. Thus he satisfies himself that he is being super-reasonably ''realistic'' and that Universe is just annoyingly disorderly—ergo, frequently dismissible—which seemingly warrants his invention of whatever kind of make-believe Universe seems momentarily most satisfying to him.

1001.14 The more humanity probes and verifies experimentally by reducing its theories to demonstrable practice in order to learn whether or not the theories are valid, the more clearly does Universe reveal itself as being generated and regenerated only upon a complex of entirely orderly relationships. The inherent spherical center viewpoint with which each individual is endowed generates its own orderly radii of observation in a closed finite system of event observations that are subject to orderly angular subdividing, recording, and interrelating in spherical trigonometric computational relationship to the observer's inherently orderly sphere of reference.

1001.15 The expression ''frame'' of reference is not only ''square'' as imputed by the two-dimensional language of youth, but also by its exclusive three-dimensional axes of reference. Such *XYZ* coordinates impose inept, exclusively rectilinear definings, which are uncharacteristic of the omniwavilinear orbiting Universe reality. Science has not found any continuous surfaces, solids, straight lines, or infinitely extensible, nonclosed-system planes. The only infinity humanity has discovered experimentally is that of the whole-fraction subdivisibility of wholes into parts, as for instance by the progressive halvings that divide the finitely closed circle into ever smaller, central-angle-expressed, arc increments. The spherical dimensions of tangent and angle frequencied intervals can always be searchlighted ''right on'' all actual event tracery.

1001.16 Because spherical trigonometry sounded so formidable, it was omitted from primary education. Humans preferred to rationalize their observed experience exclusively in terms of nonexistent straight lines and planes, and thus they evolved illogical linear and square models of Universe such as the four corners of the wide, wide world with its nonexistent fixed *up* and *down* coordinates. Employment of the ''square'' *XYZ* coordinate frame of mensural reference in all present scientific exploration is similar to going to Washington from Boston only via Chicago because that pattern conformed to the scientists' only right-angled-expressibility of relationships. Of course, if you know calculus, you could evolve a curve plotted on the *XYZ* gridding which may shorten your course; but if you don't know calculus, you have to go via Chicago.

1002.10 *Omnidirectional Nucleus*

1002.11 Omnidirectional invokes a nucleus. Omnidirectional consideration as generalized conceptual pattern integrity requires an inherently regenerative nucleus of conceptual observation reference. Because of omni-closest-packing of 12 spheres triangularly surrounding one, inwardly-outwardly precessed pulsations cannot distribute energy further inwardly than the nuclear sphere's prime volume, ergo nucleus-free, and only geometrically approximatable center of volume; whereafter it can only be distributed outwardly.

1002.12 With 12 omnidirectional, equally-most-economical, alternative-move options accommodating each event, each multiplied in optional diversity by myriads of alternate frequencies-of-occurrence rates, it is inherent to the "game" of Universe that complex redistribution of event identities swiftly ensues, as with a vast omnidirectionally observed kaleidoscope in ever-accelerating acceleration of pulsatively intertransformed pattern continuities.

1002.13 Because there are spaces between closest-packed spheres, energy can be imported syntropically all the way inward to the prime nuclear domain, which thereafter can only be articulated outwardly—ergo, as entropy. The omnidirectional grid of the isotropic vector matrix, whose vertexes always coincide with the sphere centers of all closest-sphere packings, always provides the new spherical reference system that spontaneously accommodates the observer's omnirational accounting of all Universe relations by providing an omnidirectionally observing observer's nuclear-sphere viewpoint; and all the other relevantly-to-be-identified nuclear (star) sphere centers all inherently interpositioned in omnispherical, uniradius, isotropic matrix array with omnivectorially accommodated, omnidirectionally permitted intertransformability, apprehendibility, and discrete vectorially quantated and angularly identified comprehensibility of all intertransformative transactions.

1003.10 *Isotropic-Vector-Matrix Reference*

1003.11 *Isotropic* means everywhere the same, which also means omnidirectionally the same. The isotropic vector matrix provides the actual and only systematic scheme of reference that agrees with all the experimentally disclosed behaviors of nature, while also disclosing only whole-number increments of nature's and individual's special-case objectifications of the often only subjectively apprehended information regarding the generalized principles being employed by nature. All the isotropic-vector-matrix identifications of experience are expressible in terms of angle and frequency. The angles are independent of size and absolutely generalized. The frequencies are all special-case, time-space-limited specifics and identify relative sizes and magnitudes of eternally conceptual generalizations.

1004.10 An Omnisynergetic Coordinate System

1004.11 The omnirational, omnidirectional, comprehensive coordinate system of Universe is omnisynergetic. The name *synergetic* refers specifically to the cosmically rational, most omnieconomic coordinate system with which nature interaccommodates the whole family of eternal generalized principles that are forever omni-interaccommodatively operative. This coordinate system is ever regenerative in respect to the nuclear centers, all of which are rationally accounted for by synergetics.

1005.10 Inventory of Omnidirectional Intersystem Precessional Effects

1005.11 Precession has been thought of only as an angularly reoriented, single-plane resultant of orbiting forces, as expounded, for instance, in the author's 1940 article on the gyroscope (see footnote at Sec. 1009.60). Sun's planets are precessed to orbit in a plane generated at 90 degrees to the axis of its poles. In synergetics, we discover omnidirectional precession as in tensegrity geodesic spheres. When we push inwardly on any two diametrically opposite points of a tensegrity geodesic sphere, the whole sphere contracts symmetrically; when we pull outwardly from one another on any two diametrically opposite and islanded compression members of a geodesic tensegrity sphere, the whole sphere is precessionally and symmetrically expanded. Precession is not an exclusively single-plane, 90-degree reorientation, for it also operates omnidirectionally, as do all electromagnetic wave phenomena, which can, however, be reflectively concentrated and unidirectionally beamed. The fact that waves can be reflectively and refractively focused does not alter the fact that they are inherently omnidirectional.

1005.12 While all great circles of a sphere always cross each other twice, any two such orbits precess one another into 90-degree-polar crossings, while three-way great-circling interprecesses to equiangularly intertriangulate and thus interstabilize each other.

1005.13 Today, society is preoccupied with exclusively linear information inputs.

1005.14 Pushing on one individual pole of a tensegrity geodesic sphere is the same as pushing on two poles, because you only have to push at one point for the inertia of the system to react against your pushing. This point produces a spherical wave set that if uninterfered with, will travel encirclingly around the sphere from any one starting point to its 180-degree antipodes. It is like dropping a pebble into the water: the crest is the expanded phase of Universe, and the trough is the contracted phase of Universe. Looking at the ripples, we see that they are the locally initiated expanding-contracting of whole Universe as a consequence of local energy-event inputs. This is why tensegrity and pneumatic balls bounce. Contracting as they contact, their equally violent expansion impels them away from the—relative to them—inert body of contact.

1005.20 **Biospherical Patterns:** Here we see the interplay of all the biological systems wherein all the "life"-accommodating organisms of Earth's biosphere are exclusively regenerated by energy sent to Earth by radio from the energy-broadcasting stars, but most importantly from the star Sun, by which design-science system the terrestrial vegetation and algae are the only energy radio-receiving sets.

1005.21 You and I and all the other mammals cannot by sunbathing convert Sun's energy to direct life support. In the initial energy impoundment of the powerful Sun-energy radiation's exposure of its leaves and photosynthesis, the vegetation would be swiftly dehydrated were it not water-cooled. This is accomplished by the vegetation putting its roots into the ground and drawing the water by osmosis from the ground and throughout its whole system, finally to atomize it and send it into the atmosphere again to rain down upon the land and become available once more at the roots.

1005.22 Because the rooted vegetation cannot get from one place to another to procreate, all the insects, birds, and other creatures are given drives to cross-circulate amongst the vegetation; for instance, as the bee goes after honey, it inadvertently cross-pollinates and interfertilizes the vegetation. And all the mammals take on all the gases given off by the vegetation and convert them back to the gases essential for the vegetation. All this complex recirculatory system combined with, and utterly dependent upon, all the waters, rocks, soils, air, winds, Sun's radiation, and Earth's gravitational pull are what we have come to call *ecology*.

1005.23 As specialists, we have thought of all these design programmings only separately as "species" and as independent linear drives, some pleasing and to be cultivated, and some displeasing and to be disposed of by humans. But the results are multiorbitally regenerative and embrace the whole planet, as the wind blows the seeds and insects completely around Earth.

1005.24 Seen in their sky-returning functioning as recirculators of water, the ecological patterning of the trees is very much like a slow-motion tornado: an evoluting-involuting pattern fountaining into the sky, while the roots reverse-fountain reaching outwardly, downwardly, and inwardly into the Earth again once more to recirculate and once more again—like the pattern of atomic bombs or electromagnetic lines of force. The magnetic fields relate to this polarization as visually witnessed in the Aurora Borealis. (Illus. 505.41)

1005.30 **Poisson Effect:** Pulling on a rope makes it precess by taut contracting at 90 degrees to the line of pulling, thus going into transverse compression. That's all the Poisson Effect is—a 90-degree resultant rather than a 180-degree resultant; and it's all precession, whether operative hydraulically, pneumatically, crystallographically, or electromagnetically.

1005.31 The intereffect of Sun and planets is precessional. The intereffect of the atom and the electrons is precessional. They can both be complex and elliptical because of the variability in the masses of the satellites or within the nuclear mass. Planar ellipses have two foci, but "to comprehend what goes on in general" we have to amplify the twofold planar elliptical restraints' behavior of precession into the more generalized four-dimensional functions of radiation and gravitation.

1005.32 All observability is inherently nuclear because the observer is a nucleus. From nucleus to circle to sphere, they all have radii and become omniintertriangulated polyhedrally arrayed, interprecessing event "stars."

1005.40 **Genetic Intercomplexity:** DNA-RNA genetics programming is precessionally helical with only a net axial linear resultant. The atoms and molecules are all always polarized, and their total interprecessional effects often produce overall linear resultants such as the stem of a plant. All the genetic drives of all the creatures on our Earth all interact through chemistry, which, as with DNA-RNA, is linearly programmable as a code, all of which is characterized by sequence and intervals that altogether are realized at various morphologically symmetrical and closely intercomplementary levels of close proximity intercomplexity. On the scale of complexity of ecology, for instance, we observe spherically orbiting relay systems of local discontinuities as one takes the pattern of regenerativity from the other to produce an omniembracing, symmetrically interfunctioning, synergetic order. The basic nuclear symmetries and intertransformabilities of synergetics omniaccommodates the omnidirectional, omnifrequencied, precessional integrity.

1005.50 **Truth and Love: Linear and Embracing:** Metaphysically speaking, systems are conceptually independent of size. Their special-case realizations are expressible mathematically in linear equations, although they are only realizable physically as functions of comprehensive-integrity, interprecessionally complex systems. And the tetrahedron remains as the minimum spheric-experience system.

1005.51 The very word *comprehending* is omni-interprecessionally synergetic.

1005.52 The eternal is omniembracing and permeative; and the temporal is linear. This opens up a very high order of generalizations of generalizations. The truth *could not be more omni-important,* although it is often manifestly operative only as a linear identification of a special-case experience on a specialized subject. Verities are semi-special-case. The metaphor is linear. (See Secs. 217.03 and 529.07.)

1005.53 And all the categories of creatures act individually as special-case and may be linearly analyzed; retrospectively, it is discoverable that inadvertently they are all interaffecting one another synergetically as a spherical, interprecessionally regenerative, tensegrity spherical integrity. Geodesic spheres demonstrate the compressionally discontinuous—tensionally continuous integrity. Ecology is tensegrity geodesic spherical programming.

1005.54 Truth is cosmically total: synergetic. Verities are generalized principles stated in semimetaphorical terms. Verities are differentiable. But love is omniembracing, omnicoherent, and omni-inclusive, *with no exceptions.* Love, like synergetics, is nondifferentiable, i.e., is integral. Differential means locally-discontinuously linear. Integration means omnispherical. And the intereffects are precessional.

1005.55 The dictionary-label, special cases seem to go racing by because we are now having in a brief lifetime experiences that took aeons to be differentially recognized in the past.

1005.56 The highest of generalizations is the synergetic integration of truth and love.

1005.60 **Generalization and Polarization:** In cosmic structuring, the general case is tensegrity: three-way great-circling of islands of compression. Polarized precession is special-case. Omnidirectional precession is generalized.

1005.61 It is notable that the hard sciences and mathematics have discovered ever-experimentally-reverifiable generalizations. But the social scientists and the behaviorists have not yet discovered any anywhere-and-everywhere, experimentally-reverifiable generalizations. Only economics can be regarded as other than special-case: that of the utterly uninhibited viewpoint of the individual. Nature's own simplest instructional trick in its economic programming is to give us something we call "hunger" so that we will eat, take in regenerative energy. Arbitrarily contrived "scarcity" is the only kind of behavioral valving that the economists understand. There is no other way the economists know how to cope. Selfishness is a drive so that we'll be sure to regenerate. It has nothing to do with morals. These are organic chemical compounds at work. Stones do not have hunger.

1005.62 Because man is so tiny and Earth is so great, we only can see gravity operating in the perpendicular. We think of ourselves as individuals with gravity pulling us Earthward individually in perpendiculars parallel to one another. But we know that in actuality, radii converge. We do not realize that you and I are convergently interattracted because gravity is so big. The interattraction is there, but it seems so minor we dismiss it as something we call ''aesthetics'' or a ''love affair.'' Gravity seems so vertical.

1005.63 Initial comprehension is holistic. The second stage is detailing differentiation. In the next stage the edges of the tetrahedron converge like petals through the vector-equilibrium stage. The transition stage of the icosahedron alone permits individuality in progression to the omni-intertriangulated spherical phase.

1006.10 Omnitopology Defined

1006.11 Omnitopology is accessory to the conceptual aspects of Euler's superficial topology in that it extends its concerns to the angular relationships as well as to the topological domains of nonnuclear, closest-packed spherical arrays and to the domains of the nonnuclear-containing polyhedra thus formed. Omnitopology is concerned, for instance, with the individually unself-identifying concave octahedra and concave vector-equilibria volumetric space domains betweeningly defined within the closest-packed sphere complexes, as well as with the individually self-identifying convex octahedra and convex vector equilibria, which latter are spontaneously singled out by the observer's optical comprehensibility as the finite integrities and entities of the locally and individual-spherically closed systems that divide all Universe into all the macrocosmic outsideness and all the microcosmic insideness of the observably closed, finite, local systems—in contradistinction to the indefinability of the omnidirectional space nothingness frequently confronting the observer.

1006.12 The closest-packed symmetry of uniradius spheres is the mathematical *limit case* that inadvertently ''captures'' all the previously unidentifiable otherness of Universe whose inscrutability we call ''space.'' The closest-packed symmetry of uniradius spheres permits the symmetrically discrete differentiation into the individually isolated domains as sensorially comprehensible concave octahedra and concave vector equilibria, which exactly and complementingly intersperse eternally the convex ''individualizable phase'' of comprehensibility as closest-packed spheres and their exact, individually proportioned, *concave-in-betweenness* domains as both closest packed around a nuclear uniradius sphere or as closest packed around a nucleus-free prime volume domain. (See illustrations 1032.30 and 1032.31.)

1006.13 Systems are individually conceptual polyhedral integrities. Human awareness's concession of ''space'' acknowledges a nonconceptually defined experience. The omniorderly integrity of omnidirectionally and infinitely extensible, fundamentally coordinating, closest packing of uniradius spheres and their ever coordinately uniform radial expandibility accommodates seemingly remote spherical nucleations that expand radially into omni-intertangency. Omni-intertangency evidences closest sphere packing and its inherent isotropic vector matrix, which clearly and finitely defines the omnirational volumetric ratios of the only concave octahedra and concave vector equilibria discretely domaining all the in-betweenness of closest-packed-sphere interspace. The closest-packed-sphere interspace had been inscrutable a priori to the limit phase of omni-intertangencies; this limit phase is, was, and always will be omnipotential of experimental verification of the orderly integrity of omni-intercomplementarity of the space-time, special-case, local conceptualizing and the momentarily unconsidered, seeming nothingness of all otherness.

1006.14 Human awareness is conceptually initiated by special-case otherness observability. Humans conceptualize, i.e., image-ize or image-in, i.e., bring-in, i.e., capture conceptually, i.e., in-dividualize, i.e., systemize by differentiating local integrities from *out* of the total, nonunitarily conceptualizable integrity of generalized Universe.

1006.20 **Omnitopological Domains:** In omnitopology, spheres represent the omnidirectional domains of points, whereas Eulerian topology differentiates and is concerned exclusively with the numerical equatability of only optically apprehended inventories of superficial vertexes, faces, and lines of whole polyhedra or of their local superficial subfacetings: $(V+F=L+2)$ when comprehensive; $(V+F=L+1)$ when local.

1006.21 In omnitopology, the domains of volumes are the volumes topologically described. In omnitopology, the domain of an external face is the volume defined by that external face and the center of volume of the system.

1006.22 All surface areas may be subdivided into triangles. All domains of external facets of omnitopological systems may be reduced to tetrahedra. The respective domains of each of the external triangles of a system are those tetrahedra formed by the most economical lines interconnecting their external apexes with the center of volume of the system.

1006.23 In omnitopology, each of the lines and vertexes of polyhedrally defined conceptual systems have their respective unique areal domains and volumetric domains. (See Sec. 536.)

1006.24 The respective volumetric domains of a system's vertexes are embracingly defined by the facets of the unique polyhedra totally subdividing the system as formed by the set of planes interconnecting the center of volume of the system and each of the centers, respectively, of all those surface areas of the system immediately surrounding the vertex considered.

1006.25 The exclusively surface domains of a system's vertexes are uniquely defined by the closed perimeter of surface lines occurring as the intersection of the internal planes of the system which define the volumetric domains of the system's respective vertexes with the system's surface.

1006.26 The respective areal domains of external polyhedral lines are defined as all the area on either surface side of the lines lying within perimeters formed by most economically interconnecting the centers of area of the polyhedron's facets and the ends of all the lines dividing those facets from one another. Surface domains of external lines of polyhedra are inherently four-sided.

1006.27 The respective volumetric domains of all the lines—internal or external—of all polyhedra are defined by the most economical interconnectings of all adjacent centers of volume and centers of area with both ends of all their respectively adjacent lines.

1007.10 Omnitopology Compared with Euler's Topology

1007.11 While Euler discovered and developed topology and went on to develop the structural analysis now employed by engineers, he did not integrate in full potential his structural concepts with his topological concepts. This is not surprising as his contributions were as multitudinous as they were magnificent, and each human's work must terminate. As we find more of Euler's fields staked out but as yet unworked, we are ever increasingly inspired by his genius.

1007.12 In the topological past, we have been considering domains only as surface areas and not as uniquely contained volumes. Speaking in strict concern for always omnidirectionally conformed experience, however, we come upon the primacy of topological domains of systems. Apparently, this significance was not considered by Euler. Euler treated with the surface aspects of forms rather than with their structural integrities, which would have required his triangular subdividing of all polygonal facets other than triangles in order to qualify the polyhedra for generalized consideration as structurally eternal. Euler would have eventually discovered this had he brought to bear upon topology the same structural prescience with which he apprehended and isolated the generalized principles governing structural analysis of all symmetric and asymmetric structural components.

1007.13 Euler did not treat with the inherent and noninherent nuclear system concept, nor did he treat with total-system angle inventory equating, either on the surfaces or internally, which latter have provided powerful insights for further scientific exploration by synergetical analysis. These are some of the differences between synergetics and Euler's generalizations.

1007.14 Euler did formulate the precepts of structural analysis for engineering and the concept of neutral axes and their relation to axial rotation. He failed, however, to identify the structural axes of his engineering formulations with the "excess twoness" of his generalized identification of the inventory of visual aspects of all experience as the polyhedral vertex, face, and line equating: $V + F = L + 2$. Synergetics identifies the twoness of the poles of the axis of rotation of all systems and differentiates between polar and nonpolar vertexes. Euler's work, however, provided many of the clues to synergetics' exploration and discovery.

1007.15 In contradistinction to, and in complementation of, Eulerian topology, omnitopology deals with the generalized equatabilities of a priori generalized omnidirectional domains of vectorially articulated linear interrelationships, their vertexial interference loci, and consequent uniquely differentiated areal and volumetric domains, angles, frequencies, symmetries, asymmetries, polarizations, structural-pattern integrities, associative interbondabilities, intertransformabilities, and transformative-system limits, simplexes, complexes, nucleations, exportabilities, and omni-interaccommodations. (See Sec. 905.16.)

1008.10 Geodesic Spheres in Closest Packing

1008.11 What we call *spheres* are always geodesics. While they may superficially appear to be spherical, they are always high-frequency geodesic *embracements*.

1008.12 In the closest packing of omnitriangulated geodesic spheres, the closest the spheres can come to each other is as triangular face-bonding, which is of course triple-bonded, or trivalent. In such cohering tangency, the closest-packed geodesic "spherical" polyhedra would constitute crystalline arrangements and would take up the least amount of space, because the midfaces are radially closer to the center of the sphere than are the midedges (midchords) of the omnifaceted "spheric" polyhedra, while the vertexes are at greatest radius.

1008.13 Taking up a little more room would be closest packing of geodesic spheres by edge-bonding, which is double-bonded, or bivalent. Bivalently tangential spherical polyhedra, being hinged edge-to-edge, may have characteristics similar to liquid or gelatinous aggregates.

1008.14 Single-bonded geodesic spherical polyhedra closest packed point-to-point are univalent. This point-to-point arrangement takes up the most space of all closest-packed spherical tangency agglomerates and may illustrate the behavior of gases.

1008.15 These nuances in closest-packing differentiations may explain many different unexpected and hitherto unexplained behaviors of Universe.

1009.00 Critical Proximity

1009.10 Interference: You Really Can't Get There from Here: Omnitopology recognizes the experimentally demonstrable fact that two energy-event traceries (lines) cannot pass through the same point at the same time. It follows that no event vectors of Universe ever pass through any of the same points at the same time. Wherefore, it is also operationally evidenced that the conceptual-system geometries of omnitopology are defined only by the system withinness and withoutness differentiating a plurality of loci occurring approximately midway between the most intimate proximity moments of the respectively convergent-divergent wavilinear vectors, orbits, and spin equators of the system. (See Sec. 517, "Interference.") The best you can do is to get almost there; this is evidenced by physical discontinuity. Zeno's paradox thus loses its paradoxical aspects.

1009.11 In omnitopology, a vertex (point) is the only-approximate, amorphous, omnidirectional region occurring mid-spatially between the most intimate proximity of two almost-but-never-quite, yet critically intertransformatively, interfering vectors. (See Sec. 518.)

1009.20 Magnitude of Independent Orbiting: Most impressively illustrative of what this means is evidenced by the mass-attractively occasioned falling in toward Earth of all relatively small objects traveling around Sun at the same rate as Earth, Earth itself being only an aggregate of all the atoms that are cotravelers around Sun at the same velocity, while each atom's nucleus is only one-ten-thousandth the diameter of its outer electron shell. There is as much space between the atom's nucleus and its electron-orbit-produced shell as proportionately exists between Sun and its planet Pluto.

1009.21 The tendency to fall in to Earth or any other celestial body will be reduced as a cotraveling object increases its distance away from Earth or any other relatively large body as a consequence of its being given acceleration into orbital speed greater than Earth's Sun-orbiting speed. In 99.9999999999 percent of Universe no body tends to fall in to any other; 99.999999×10^{30} of all known Universe bodies are independently orbiting.

1009.30 Symmetrical Conformation of Flying-Star Teams: We have terms such as "boundary layer" that have to be recognized in hard technology where we find that despite the accurate machining to fine tolerance of such things as steel bearings, there is always a dimensional aberration that is unaccounted for in man's eyes but, when measured instrumentally in nuclear-diameter magnitudes, is as relatively great as that between the stars of the Milky Way. Men think superficially only of lubricants and mechanically-fitting-bearings tolerances whereas—focused at the proper magnitude of conceptuality—what goes on in the affairs of lubricants and bearings discloses discrete geometrical relationships where no event ever makes absolute contact with another. There are simply orbital interferences, where the mass attractions will always be just a little more powerful than the fundamental disintegrative tendencies.

1009.31 The relative frequency timing of orbits is such that as one complex energy event (a body) approaches critical proximity between any two other equal mass bodies to that of the intruder, the group mass interattraction fourfolds. We get to a condition where the approaching body is suspended between two others like landing on an invisible trampoline. Similarly, in man-made machinery as the teeth of gears enter into the matching gears' valleys, the mass-attraction forces finally provide an invisible suspension field whereby none of the atoms ever touches another. (See Sec. 1052.21.)

1009.32 When metallic alloys are produced, we have such conditions, for instance, as four symmetrically orbiting stars producing a tetrahedral flying formation, each trying to orbit away from the other but inter-mass-attractively cohered. When this flying team of four stars in tetrahedral conformation joins together with a second team of four stars in tetrahedral conformation, they take position symmetrically with each member star of the two sets of four becoming congruent with the eight corners of a cube.

1009.33 Now each of the stars in the flying teams has nearer neighbors than it had before, and this mass interattractiveness is multiplied as the second power of relative proximity. Their initial acceleration of 186,000 miles per second keeps their orbits always intact. Each of the flying formations is made up of other flying teams of atoms with a central commander nucleus and a fleet of electrons buzzing around it at 186,000 miles per second; being interfrequenced, the four nucleated team members synchronously interact as the orbits of their electrons in closest proximity are intervally geared in second-power accelerations of intertenuousness, producing an omnicoordinate condition akin to the mid-gear-tooth trampoline (an invisible muscular field).

1009.34 Next, a six-member flying team (octahedron) heaves into critical proximity with the original two teams now flying a group formation in the form of the eight corner positions of the cube. The acceleration stability of each of the flying teams is such that they join with the new six-star team taking symmetrical positions in the middle of the six square faces of the eight-star-team cube. The mass interattraction of the 14 now becomes vastly greater, and the electron-orbit-gear-trampolines of each of the 14 nuclear-flow spherical ships are now in very much greater second-power increase of interattractiveness.

1009.35 This cubical flying team of 14 ships now sights another flying team of 12 ships, and the team of 14 and the team of 12 are flown into group formation with the 12 ships taking station at the midpoints of the 12 edges of the 14-star-team cube. Thus the mass attraction is ever more vastly increased, yet the integrity of their interpositioning and their non-falling-into-one-anotherness is guaranteed by the centrifugal forces of the orbiting superbly balanced by the second-power increase of the gravitational buildup already noted.

1009.36 Thus are planets cohered, and thus are metallic alloys on planets even more powerfully cohered—all within the rules of never-quite-touching; all within the rules of interval; all within the rules of no actual particulate "solids." They may fly wavilinear patterns, but the atoms are found to be as discontinuous as the wavilinear sky trails of the jet airplane. While physics is as yet formally puzzling over the paradox of the wave and the particle, the apparent contradiction is occasioned only by the superficial misconception of a particle where none exists. We deal only with events in pure principle. The sense of physical, textural reality, of awareness itself, which uniquely identifies life and time (in contradistinction to eternal, weightless metaphysics), is inherent to the plurality of frequencies and degrees of freedom that in pure principle theoretically provide different interpositionings within given amounts of time. The plurality of principles, which themselves are interaccommodative, inherently generates awareness differentiability. The exquisite perfection of the total interaccommodation and the limited local set of the tunabilities of the terrestrial living organisms, such as the human instrument vehicle, are all permitted in the general complexity and permit local-focus, limited awareness as individual-seeming perceptivity. (See Sec. 973.30.)

1009.37 What I am saying is that we have only eternity and integrity. Unity is plural in pure principle. The awareness we speak of as life is inherently immortal and equieternal.

1009.40 Models and Divinity: Because of indeterminism, the exclusive tenuous nature of integrity—discontinuity—means that no exact hard particulate models may ever be fashioned by man. The conscientious and competent modelmaker undertaking to make a beautiful tetrahedron suddenly becomes aware that it is impossible to make a perfect corner at a point. There is always both a terminal and a radius and an askew convergence-divergence at noncontacting critical proximity. When he magnifies the edges which look sharp to the naked eye, he sees they are never sharp. The more powerful the magnification he brings to bear on his work, the more he becomes aware of the lumpy radii of the micropatterning of the stuff with which he works. Finally, the electron microscope tells him that the point of a needle is a pile of oranges and that the blade of the razor is a randomly dumped breakwater of spherical rubble. When further meticulously studied and magnified, this superficial seeming randomness proves to be our flying squadrons, earlier described, enjoying a vast number of intricately orderly team maneuvers but with never a pilot in sight. The whole is flown by remote control with fantastic feedback and local automation, all governed by an eternally complex integrity of complementary, interaccommodative principles.

1009.41 Little man on little planet Earth evoking words to describe his experiences, intuiting ever and anon the greater integrity, struggles to form a word to manifest his awareness of the greater integrity. His lips can express, his throat and lungs can produce, in the limited atmosphere of planet Earth, he may make a sound like g o d . . . which is obviously inadequate to identify his inherent attunement to eternal complex integrity. The little humans on little Earth, overwhelmed these millions of years with the power of the bigger over the lesser (muscles), have spontaneously identified the cosmic integrity with the local terrestrial experience. The conditioned-reflex feedbacks have introduced enormous confusion of approximate identification, fusing the local physical muscular authority with the eternal complex integrity, whose absolute generalizability can never be locked into or described as a special case.

1009.50 Acceleration: Physics does not speak of motion; it speaks of acceleration. And physics has identified only two kinds of acceleration, linear and angular. We are informed experientially that this is a misinterpretation of the data.

1009.51 There are indeed two kinds of acceleration, but they are both angular. All accelerations are angular and cyclically complete. There are no open endings in Universe. Physics has discovered only waves, no straight lines.

1009.52 The angular accelerations, however, manifest a vast variety of radii. The differentiation of physics into linear and angular occurred when the humans involved failed to realize that the diameter of the little circle is always a small arc of a vastly greater circle passing through it. The greater the radius, the slower the total cyclic realization. There are no straight lines or "linear infinities." Realization of this is what Einstein spoke of as "curved space." (See Sec. 522.21.)

1009.53 Einstein was up against trying to communicate with the mathematicians in terms of their adopted mathematical models, all of which were—and still are—straight-line, *XYZ* models on a linear frame and with linear coordinates going outward from the model to infinity. So "field" was always a little set of local perpendicular crossings of straight lines each outward bound to an infinity of infinities.

1009.54 All the experimentally harvested information says that the "field" must now be recognized as a complex of never-straight lines that, at their simplest, always will be very short arcs of very great circular orbits. And the orbits are all elliptical due to the fact that unity is plural and at minimum two. There will always be at least one other critical proximity-imposing aberration restraint focus.

1009.55 A single ellipse is a wave system with *two diametric peak phases*—a gear with only two teeth—at 180 degrees from one another. All other gears are multitoothed, high-frequency waves. All is wavilinear.

1009.56 Critical proximity crimping-in is realized by local wave-coil-spring contractions of the little system's diameter by the big system, but local radius is always a wavilinear, short-section arc of a greater system passing through it in pure generalized eternal principle. (See Sec. 541.04.)

1009.60 **Hammer-Thrower:** The effect of bodies in acceleration upon other bodies in acceleration is always precessional, and the resultant is always at an angle other than 180 degrees. Even today, the physicists consider precession to be only mathematically treatable by quantum mechanics because they have failed to realize that the complex intereffects are conceptually comprehendible.*

1009.61 The model of the man fastening a weight to a string and swinging it around his head is the familiar one of the hammer-thrower. With each accelerating cycle the object swung around his head accumulates in its velocity the progressive energy imports metabolically exported through the action of the human's muscles. (See Sec. 826.03 for description of the metabolic energy accumulation of the hammer-thrower.) Men have not accurately interpreted their instinctively articulated performance of slinging, hammer-throwing, baseball pitching, and other angularly accelerated hurlings. When a human picks up a stone and throws it, he thinks of it as a different kind of a sport from the hammer-thrower's activity. But the only difference is that with throwing the stone, his arm is the rod of the hammer and instead of accumulating velocity by many cycles of acceleration, he operates through only one-third of a circle in which he can accumulate a certain amount of metabolic muscle energy to transform into acceleration. Substituting the athlete's

* The author established this fact with the authority of the great specialists in applied precession (i.e., gyroscoping), the chief engineer and vice president for research of the Sperry Gyroscope Company. See the author's article on the gyroscope in *Fortune,* Vol. XXI, No. 5, May 1940.

"hammer" for the stone or baseball, the hammer-thrower is able to build much more of his metabolically generated energy into muscular acceleration, which accumulates to produce very great force. The baseball-throw and the hammer-throw utilize the same principle, except that the rate of accumulation is one-third cycle for the former while it is a plurality of cycles for the latter, thus permitting the introduction of larger amounts of time-of-effort application.

1009.62 A man with a weight on a string swings it above his head and lets go of it, but the man is in such close proximity to Earth that the attraction of Earth takes over and pulls the weight in toward Earth. This tends to misinform the observer, who may lose sight of the fact that the man and Earth and the weight on the string are all going together around Sun at 60,000 mph.

1009.63 Despite the overwhelming power of the attraction of Earth, we must continue to keep in mind the critical-proximity concept. For instance, let us consider two steel magnets lying on a table and apparently not attracting one another simply because Earth-pull against the table and the friction of the table prevent them from indicating their pull for one another. But as they are given a series of pushes toward one another there comes a point when Earth's gravity-induced friction is overcome by the local magnetic interattraction which increases as the second power of the relative interdistance increase; and there comes a moment when friction is overcome and the two magnets start moving toward one another and accelerate to a fast, final-snap closure. It is when such other forces are overcome that the two magnets articulate their interattraction independent of all other forces: this is the point we call critical proximity. (See Sec. 518.)

1009.64 Earth and Moon were, still are, and always will be pulling on the two magnets to some extent—as are all the other galaxies of Universe. The critical-proximity moment is when all the other pulls are overcome by the pull between the two magnets and "falling-in" occurs; and the falling-in is always of the lesser toward the greater.

1009.65 The astronaut can go out space-walking because he and his space vehicle are in the same Universe orbit at the same rate, as would be any other object the space-walker had in his hand. Here is an opportunity for the mutual mass attractions to articulate themselves, except that in this situation, the prime force is the acceleration itself. What the physicists have failed to elucidate to society, and possibly to themselves, as well, is that *linear acceleration is also orbital* but constitutes release from co-orbiting (or critical-proximity orbiting) into the generalized orbiting of all Universe.

1009.66 All the creatures on board planet Earth are in such critical proximity that the falling-in effect of the apple hitting the grass, the rain dropping on the sidewalk, the hammer falling to the floor, or the child bottoming to the deck of the crib are all typical of the critical-proximity programmability of a design integrity, which programmability is employable by humans in design science. All of the creatures of planet Earth are in a "fall-in" programmed by a critical-proximity guarantee.

1009.67 The bee goes after his honey and, inadvertently, at 90 degrees to his honey-seeking plunge, his tail takes on pollen and knocks off pollen to produce a large, slowly orbiting interfertilization of the vegetation's prime-energy impoundment of photosynthesis from the stars—particularly the Sun star—of all the radio-transmitted energies to Earth. Photosynthesis impounds energy, and by orderly molecular formation and crystal building, the synergetic intertransformabilities and the associabilities and disassociabilities of the isotropic-vector-matrix field accommodation occur. What is spoken of as ecology is slowly orbiting local interaction of mutual intersupport within unpremeditatedly accomplished tuning of the prime drive programming of the spontaneous fall-in-ability of the creatures within the critical-proximity conditions: the sugar on the table, the naked girl on the bed.

1009.68 All special-case events are generated in critical proximity. Critical proximity is inherent to all intertransformability and interaccounting of eternally regenerative Universe—as, for instance, in the myriad varieties of frequencies ranging from eons to split-seconds. When Earth's orbit passes through a comet's stardust plume, we witness some of the comet's stardust falling in to Earth captivity, some of it igniting as it enters the atmospheric gases, some falling into Earth, and some with such acceleration as only to pass through the atmosphere leaving meager entropic dust to fall to Earth.

1009.70 **Orbital Escape from Earth's Critical-Proximity Programmability:** Human mind, while discovering generalized principles, eternally persisted in special-case experience sequences, but has gradually developed the capability to employ those principles to put vehicles and then self into such acceleration as to escape the fall-back-in proclivity and to escape the general ecological fall-in program of invisible interorbiting regeneration.

1009.71 As each human being discovers self and others and employs more principles more and more consciously to the advantage of others, the more effectively does the individual retain the integrity of his own unique orbiting in Universe, local though it may seem aboard our planet. His unique orbiting brings him into a vast variety of critical-proximity fall-ins. Man has progressively acquired enough knowledge to raise his vision from the horizontal to the vertical, to stay first atop the watery ocean and next atop the air-ocean heights, and most recently to orbit beyond the biosphere with ever greater independence, with ever greater competence, and with ever greater familiarity with the reliability of the generalized principles.

1009.72 Little individuals in orbit around little berry patches, fruit trees, nut piles, and fishing holes are instinctively programmed to pick up rocks and pile up walls around the patches, orchards, and gathering places. Some men floating on the waters and blown by the wind were challenged to respond to the accelerating frequence of stress and high-energy impacts, and they went into vastly longer orbital voyages. Others went into lesser and slower orbits on camels and horses, or even slower orbits on their own legs. The effect of human beings on other human beings is always precessional. All of us orbit around one another in ever greater acceleration, finally going into greater orbits. The local critical-proximity fall-in and its 99.9999 percent designed-in programming becomes no longer in critical-proximity evidence, while all the time the apprehending and comprehending of the generalized principles elucidates their eternal integrity in contrast to the complex inscrutability of the local critical-proximity aberrations permitted and effected in pure principle whenever the frictional effect on the two stones lying before us overcomes their tendency to fall in to one another—with naught else in Universe but two stones—which statement in itself discloses our proclivity for forgetting all the billions of atoms involved in the two stones, and their great electron orbits around their nuclei, guaranteeing the omniacceleration, yet synergetically and totally cohered by the mass-interattractiveness, which is always more effective (because of its finite closures) than any of the centrifugal disintegrative effects of the acceleration. All the interaberrations imposed on all the orbits bring about all the wave-frequency phenomena of our Universe. The unique wave frequencies of the unique 92 chemical elements are unique to the local critical-proximity event frequency of the elemental event patternings locally and precessionally regenerated. Finally, we must recall that what man has been calling "linear" is simply *big orbit arc* seemingly attained by escaping at 90 degrees from local orbit. There are only two kinds of acceleration, greater and lesser, with the lesser being like the radius of the nucleus of an atom in respect to the diameter of its electron shell.

1009.73 Humanity at this present moment is breaking the critical-proximity barrier that has programmed him to operate almost entirely as a part of the ecological organisms growing within the planet Earth's biosphere. His visit to Moon is only symptomatic of his total, local, social breakout from a land-possessing, fearful barnacle into a world-around-swimming salmon. Some have reached deep-water fish state, some have become world-around-migrating birds, and some have gone out beyond the biosphere. Long ago, man's mind went into orbit to understand a little about the stars. And little man on little Earth has now accumulated in the light emanating from all the stars a cosmic inventory of the relative abundance of each of the 92 regenerative chemical elements present in our thus-far-discovered billion galaxies of approximately a hundred billion stars each, omnidirectionally observed around us at a radius of 11 billion light years. Man can always go into infinitely great, eternal orbit. Mind always has and always will.

1009.80 **Pea-Shooter, Sling-Thrower, and Gyroscope: Gravity and Mass-Attraction:** Highly specializing, formula-preoccupied, conventional academic science of the late twentieth century seems to have lost epistemological sight of the operationally derived mathematics identifying Galileo's accelerating-acceleration of free-falling bodies as being simply R^2, where R is the relative proximity of any two bodies whose mutual interattraction is isolatingly considered. R^2 says that every time the proximity is halved, the mass-interattraction of the two bodies will be increased as 2^2, i.e., fourfolded.

1009.81 Isaac Newton did comprehend this. Newton was inspired by the early Greeks and by Copernicus, Kepler, and Galileo. Newton compounded Kepler's discovery of the mathematical regularities manifest by the differently sized solar system's interattractions with Galileo's discoveries—which information Newton's own intuition then further integrated with ancient experience of the sling-throwers, which showed that the more the sling-thrower converted his mucle power into increasing the speed of the sling orbiting around his head before freeing one end of it to release his stone pellet, the faster and farther would the impelled stone travel horizontally before another more powerful force pulled it inwardly toward Earth's center.

1009.82 The gravitational constant is expressed as a second power. Second-powering means that the number is multiplied by itself. Thus the forces of the accelerating-acceleration of gravity can be calculated, provided the masses of the two interapproaching bodies are multiplied and their relative proximity is expressed in the terms of the relative radii magnitudes of the bodies.

1009.83 When we see the pea-shooter blowing peas out in a trajectory, we see that if it is blown harder, the impelled peas may attain a longer trajectory before they curve down and toward Earth as they yield to gravity. Assuming no wind, the gradual curvature from approximately horizontal to vertical of the peas' trajectory all occurs in the same single vertical plane. When you insert your finger into the blown pea's trajectory, you interfere with the pea and deflect it angularly. This means deflecting the plane with which the pea's horizontal course is translated toward the vertical from below, or sideways, or from any direction. This trajectory altering is a phenomenon described by the physicists as angular deflection.

1009.84 In the same way, you can also deflect the plane of travel of water coming out of a garden hose—to aim the stream of water in any direction you want before gravity overcomes the initial force impelling the water. We cannot see the individual molecules of water we are deflecting one by one when our finger angularly modifies the stream of water at the hose nozzle, but we can see the individual shooter-blown peas that we can deflect individually, thus aiming them to hit various targets. The vertical plane of the pea is deflected sideways by you, and its falling within the plane is directed by gravity. There is a vertical integrity of the trajectory plane. The finger only deflects the horizontal orientation plane. The pea does not have a memory and after initial deflection by your finger does not try to resume the vertical plane of its previous travel.

1009.85 Two forces have operated to determine the pea's trajectory:

(1) gravity, which continues to operate, deflecting it progressively; and

(2) the finger, which momentarily deflected it but is no longer doing so.

When we come back to the spinning Olympic hammer-thrower this time rotating vertically between head-and-foot-clamped ball-bearing turntables which in turn are mounted in gimbaled rings, to whose belt is hooked a complete, 360-degree, ball-to-ball, "grass-skirt-like" ring of horizontally revolving steel balls on the outer ends of steel rods, on the inner ends of which are pairs of triangular steel handles now hooked to the hammer-thrower's belt after his successive angular acceleration of each hammer into the horizontal spinning ring of his "grass skirt." His separately accelerated and horizontally traveling balls are each similar to each of the peas as first blown horizontally out of the pea-shooter tube. Both the peas and the steel balls are being affected by two forces: the peas by gravity pulling upon them and by the force with which they were originally propelled in horizontal trajectory; the spinning steel balls have their original horizontal acceleration, which was so great as to overcome gravity's Earthward-pulling effect, plus their second restraint, that of the steel rods successfully restraining and countering the centrifugal force that seeks to release the balls into *tangential,* not radial trajectory.

1009.86 Because centrifuges separate "heaviers" from "lighters" by expelling the heavy from the light—such as milk centrifuges out of cream—people have mistakenly thought of the expelling as radial rather than tangential. Make yourself a diagram of your own spinning of a weight around your head—you tend to think of it as being released in a horizontal plane at a point on the spun circle directly in the line running between your eyes and the target direction in which you wish the hammer to travel, that is leading perpendicularly outward from the circle in the direction in which the released pellet travels. The fact is that if it were released at that point, it would travel at a direction 90 degrees, or sidewise from your desired trajectory, from that actually realized. Studying the action of an Olympic hammer-thrower, you will find that the spherical hammer and its rod are released at a point facing away

90 degrees from the direction in which the released hammer travels: i.e., the hammer always goes off tangentially from the circle of acceleration. This contradicts the popular conception of a centrifugal force as being radial rather than tangential.

1009.87 Returning to our Olympic hammer-thrower's steel-ball, flying skirt, if you touch evenly their successively passing tops, thus downwardly deflecting each ball of the full circle of ball hammers spinning around, each is discretely deflected, say 30-degrees downwardly, which changes the plane of its individual orbital spinning. Each ''peels off,'' like an airplane flying formation and obeying a command to break company and go into a descending path followed exactly by each successive ball coming into touch-contact with your deflecting finger held rigidly at the same point.

1009.88 If your rigidly held finger is lowered further to another discrete point in the line of travel of the successively revolving balls, and if it is held rigidly at this new point, each of the circle of revolving spherical hammers will again be discretely deflected into an additionally tilted plane (with the hammer-thrower himself as axis of rotation always maintaining perpendicularity to the plane of the hammers' revolution, his axial tilting being accommodated by the three-dimensionally oriented axles of the two gimbal rings within and to which his ball-bearing foot-and-head clamps are firmly attached).

1009.89 We had learned earlier about fixed or progressive-horizontal reangling of the plane of the peas' coincidentally yielding to gravity (Sec. 1009.83), as we tried discrete deflecting of the successive peas shot from the pea-shooter. By experimenting, moving our finger progressively deeper, in deliberately distanced stages, into the peas' profile-described ''tubular'' space-path of travel, we found that the nearer our finger came to the center line of travel within that ''tubular'' space-path, the wider the resulting angle of deflection of the peas' trajectory. When finally our finger crossed the tube's center line, the angular deflection ceased and direct 180-degree opposition to the line of pea travel occurred, whereat all the horizontal force originally imparted to the peas by the pea-shooter's pneumatic pressure-blowing is almost absorbed by impact with the finger. The pea bounces back horizontally for a usually imperceptibly meager distance before yielding entirely to gravity and traveling Earthward at 90 degrees to its original horizontal trajectory.

1009.90 What we also learned observationally before and after deflecting the peas experimentally was that gravity went to work on the peas as soon as they left the tube, and that as the peas were decelerated by air resistance below the rate of acceleration that rendered them approximately immune to the pull of gravity, that latter force became ever more effective as the air resistance took its toll of energy from the peas, and the peas were deflected progressively Earthward. We also observed that no wind was blowing, and when we did not deflect the peas with our finger, they all followed a progressively descending path in exactly the same plane until they hit the ground. Next we learned that if we intruded our finger horizontally a discrete distance into the tubular space-path of the peas, they were deflected at some discrete angle (less than 90 degrees, diametrically away from the point of entry of our finger into the peas' tubular space-path), and that if we did not move our finger further into the tubular space-path, each traveling pea thus interfered with deflected the same angular amount horizontally away from our intruding finger and held that newly angled direction, yielding further only to air resistance and gravity, with the result that each successive pea thus discretely deflected proceeded in a progressively curved trajectory, but always within the same vertical plane. In other words, successively separate and discretely distanced progressive intrusions of our finger into the tubular space-path of travel deflected the vertical plane of the trajectory of the peas into a new but again sustained vertical plane of travel, that new vertical plane occurring each time at a more abrupt angle from the original nonintruded vertical plane of the stream of traveling peas. Thus we learned that we could deliberately aim the peas to hit targets within the range of such traveling. (We have all succeeded in deflecting the trajectory of a pressured stream of water in just such a manner, but we cannot see the individual molecules of water thus deflected and think of it as a continuous stream.) The discretely modified behavior of our pea-shooter's individual peas and the individual steel-ball "hammers" of the Olympic hammer-thrower altogether permit our comprehension of the parts played by individual, but invisible-to-human-eyes energy quanta in bringing about only superficially witnessed motion phenomena that most often appear deceivingly as motionless solids or as swiftly rotating solid flywheels such as those of gyroscopes.

1009.91 Thus we now can understand that our touching the rim of a flywheel of an *XYZ*-axialed and gimbaled gyroscope causes each of the successively and discretely top-touched quanta to be deflected downward into a new plane of travel, accompanied always by the coincident tilting of the axle of the flywheel, which always maintains its perpendicularity to the plane of spin of the flywheel. The tilting of the plane of spin of the flywheel, caused by our finger touching the rim of the spinning wheel, tilts that wheel around an axis of tilt, which axis is the line diametrically crossing the circular plane of spin from the rim point that you touched. This diametric line is the tilting-hinge line. It runs directly away from you across the wheel. This means that as the wheel's extended axle perpendicular to the flywheel plane tilts with the wheel, as permitted by the three-axial degrees of freedom of the gimbaled gyroscope, then the axle tilts in a plane at right angles to the tilting-hinge line in the flywheel. Because the steel wheel and its axle are integral, it would be in exactly the same plane of force in which you applied your touch to the flywheel's rim, if, instead, you took hold of the top bearing housing the flywheel's top axle extremity and pulled that gimbal-freed bearing toward the rim point at which your finger had applied its initial touch, the bearing housing and the axle of the flywheel will rotate exactly sidewise from the direction in which you are pulling on it because that force makes the flywheel tilt hingingly around the line running diametrically across the wheel from your rim-touching point.

1009.92 Thus we learn that pulling the axle bearing atop the gyroscope toward the rim-touching point, which is also incidentally pulling the top axle bearing in the gimbaled system toward yourself, results in the wheel plane tilting around the described hinging line, and the axle and its bearing are thus forced to move coincidentally in a plane perpendicular to that hinge line and in the direction which is tangential to the wheel spinning at the initial touching point. This means that pulling on top of the gyroscope does not result in its yielding toward you, as you might have expected from its three-axial degrees of gimbaled freedom, as it would have done had the wheel not been spinning. Instead, it seemingly travels rotatingly in a plane at 90 degrees to your effort and continues to do so so long as you apply the force, and does so ever more speedily if you increase the force. This yielding at a plane angled at 90 degrees to your (or anyone's) applied effort is *precession,* which is the effect of a body in motion on any other body in motion; the resulting angles of precession are never in a plane congruent with the precessionally actuating force.

1009.93 Since all Universe is in motion, all the intereffects of its energy concentrations as ''matter'' are always intereffecting one another precessionally. The pull of Sun on Earth results in Earth orbiting around Sun at 90 degrees to the line of Sun's mass attraction of Earth. Bodies ''fall'' toward Earth only when their relatively small size and the critical proximity of their respectively mutual orbiting of Sun at 60,000 mph allows their progressive orbital convergence; the lesser body is only negligibly affected by the precessional forces of other astro bodies because of the second-power rate of diminution of intermass-attraction occurring as the intervening distances are increased.

1009.94 All the foregoing illustrates the integration of (1) Newton's mass-attraction law, (2) precession, and (3) synergy. They are all coming together here: Kepler, Galileo, and Newton. The earliest sling-thrower revolving the sling around his head (angular acceleration, as it is called by the physicists) demonstrates the added energy of the sling-thrower extending the trajectory. The pea-shooter does the same thing in linear acceleration. It can extend its trajectories with greater energy, but its pellets, too, yield to the gravity of Earth. Earth is very powerful, but the pea-shooter or the sling-thrower discover that the harder they swing or blow—i.e., the more energy they put into accelerating their pellets—the farther the pellets go horizontally before gravity deflects them at 90 degrees.

1009.95 But there is an integration of the horizontal and vertical planes of applied forces, between the horizontal plane of your varying effort and the vertical plane of the constant Earth's pull. Realization of this integration may be what inspired Newton. Galileo used the phrase "accelerating-acceleration," which means that the velocity is continually increasing. But the sling-thrower's force was discontinued, and the air resistance decelerated its missile until gravity's force at 90 degrees became greater. If the sling-thrower propelled his missile outside the atmosphere of Earth and beyond the critical-proximity limits within which falling in occurs, his missile would keep on traveling ad infinitum in an astro-wandering orbit.

1009.96 The logic of sensorially satisfactory experience acquired in the foregoing elucidation of precession—and the discovery of our self-deceivingly-conditioned reflex in respect to assuming 180 degrees to be the normal angular direction of spin-off instead of reality's 90-degreeness—not only renders precession comprehendible, but can make its 90-degree spin-off and other effects understandably normal and can explain much that has heretofore seemed inexplicable and abnormal. The two angular-acceleration planes become very important devices of comprehension. In our generalization of generalizations, we find that synergy, as "the behavior of whole systems unpredicted by any of the systems' parts taken separately," embraces both the generalized mass attraction and the precessional laws. Apparently, synergy embraces our definition of Universe and is therefore probably the most generalized definition of Universe.

1009.97 The generalizations are of the mind and are omniembracing and omnipermeative. Like the rays of Sun, radiations are radii and are focusable. Gravity cannot be focused; it is circumferentially embracing. Radiation has shadows; gravity has none. Radiation produces the phenomenon known to Einstein as the bending of space, the gravitational field.

1009.98 Gravitation is omniembracing. In the barrel hoops (see Sec. 705), gravity operates only in single and parallel, separate planes. Omni-triangulated geodesic spheres consisting exclusively of three-way interacting great circles are realizations of gravitational-field patterns. Events are forced to bounce in spherically contained circles because they seek the largest possible interior circumference patterns. All great circles cross each other twice. Three or more noncongruent great circles are automatically inter-self-triangulating in their repetitive searching for the "most comfortable" interactions, which always resolve their three-way-great-circle patterning into regular spherical icosahedra, octahedra, or tetrahedra. The gravitational field will ultimately be disclosed as ultra-high-frequency tensegrity geodesic spheres. Nothing else.

1010.00 Prime Volumes

1010.01 A prime volume has unique domains but does not have a nucleus.

1010.02 A prime volume is different from a generalized regenerative system. Generalized regenerative systems have nuclei; generalized prime volumes do not.

1010.03 There are only three prime volumes: tetrahedron, octahedron, and icosahedron. Prime volumes are characterized exclusively by external structural stability.

1010.10 **Domain and Quantum:** The unique insideness domain of a prime system is, in turn, a prime volumetric domain, which is always conceptually defined by the system's topological vertex-interconnecting lines and the areas finitely enclosed by those lines. ($V+F=L+2$.) Prime volumetric domain provides space definition independent of size.

1010.11 Prime volumetric domain and prime areal domain together provide space conceptuality independent of size, just as the tetrahedron provides prime structural system conceptuality independent of size.

1010.12 Complex bubble aggregates are partitioned into prime volumetric domains by interiorly subdividing prime areal domains as flat drawn membranes.

1010.13 A prime volumetric domain has no volumetric nucleus. A prime areal domain has no planar nucleus. So we have *prime system volumetric domains* and *prime system areal domains* and linear interconnections of all vertexes—all with complete topological conceptual interpatterning integrity utterly independent of size.

1010.14 This frees conceptual-integrity comprehending and all the prime constituents of prime-pattern integrity, such as ''volume,'' ''area,'' and ''line,'' from any special-case quantation. All the prime conceptuality of omnitopology is manifest as being a priori and eternally generalized phenomena. Thus *quantum* as prime-structural-system volume is eternally generalized, ergo, transcends any particulate, special-case, physical-energy quantation. Generalized quanta are finitely independent because their prime volumetric-domain-defining lines do not intertouch.

1010.20 **Nonnuclear Prime Structural Systems:** The domain of the tetrahedron is the tetrahedron as defined by four spheres in a tetrahedral, omniembracing, closest-packed tangency network. The domain of an octahedron is vertexially defined by six spheres closest packed in omnitriangular symmetry. The domain of an icosahedron is vertexially defined by 12 spheres omnicircumferentially intertriangulated and only circumferentially symmetrically triangulated in closest packing without a nucleus (in contradistinction to the center sphere of the vector equilibrium, whose 12 outer sphere centers define the vector equilibrium's 12 vertexes; all 13 of the vector equilibrium's spheres are intersymmetrically closest packed both radially and circumferentially).

1010.21 All of the three foregoing non-nuclear-containing domains of the tetrahedron, octahedron, and icosahedron are defined by the four spheres, six spheres, and 12 spheres, respectively, which we have defined elsewhere (see Sec. 610.20, "Omnitriangular Symmetry: Three Prime Structural Systems") as omnitriangulated systems or as prime structural systems and as prime volumetric domains. There are no other symmetrical, non-nuclear-containing domains of closest-packed, volume-embracing, unit-radius sphere agglomerations.

1010.22 While other total closest-packed-sphere embracements, or agglomerations, may be symmetrical or superficially asymmetrical in the form of crocodiles, alligators, pears, or billiard balls, they constitute complexedly bonded associations of prime structural systems. Only the tetrahedral, octahedral, and icosahedral domains are basic structural systems without nuclei. All the Platonic polyhedra and many other more complex, multidimensional symmetries of sphere groupings can occur. None other than the three-and-only prime structural systems, the tetrahedron, octahedron, and icosahedron, can be symmetrically produced by closest-packed spheres without any interioral, i.e., nuclear, sphere. (See Secs. 532.40, 610.20, 1010.20, 1011.30 and 1031.)

1011.00 Omnitopology of Prime Volumes

1011.10 **Prime Enclosure:** Omnitopology describes prime volumes. Prime volume domains are described by Euler's minimum set of visually unique topological aspects of polyhedral systems. Systems divide Universe into all Universe occurring outside the system, all Universe occurring inside the system, and the remainder of Universe constituting the system itself. Any point or locus inherently lacks insideness. Two event points cannot provide enclosure. Two points have betweenness but not insideness. Three points cannot enclose. Three points describe a volumeless plane. Three points have betweenness but no insideness. A three-point array plus a fourth point that is not in the plane described by the first three points constitutes *prime enclosure*.

It requires a minimum of four points to definitively differentiate cosmic insideness and outsideness, i.e., to differentiate macrocosm from microcosm, and to differentiate both of them from *here* and *now*.

1011.11 Systems are domains of volumes. One difference between a domain and a volume is that a domain cannot have an interior point, because if it did, it would be subject to more economical subdivision. For instance, the vector equilibrium is a system and has a volume, but it consists of 20 domains. A vector equilibrium is not a prime domain or a prime volume, because it has a nucleus and consists of a plurality of definitive volumetric domains. The vector equilibrium is inherently subdivisible as defined by most economical triangulation of all its 12 vertexes into eight tetrahedra and 12 quarter-octahedra, constituting 20 identically volumed, minimum prime domains.

1011.20 **Hierarchy of Nuclear Aggregations:** The prime nuclear aggregation of spheres around one sphere is the vector equilibrium. Vector equilibrium constitutes the prime nuclear group because it consists of the least number of spheres that can be closest packed omnitangentially around one nuclear sphere. The vector equilibrium provides the most volumetrically economical pattern of aggregation of 12 balls around a nuclear ball of the same diameter as the 12 surrounding balls; the 13th ball is the center. In other words, 13 is the lowest possible number connected with a structurally stable triangulated nucleus, being omnitriangularly interconnected both radially and circumferentially.

1011.21 An octahedron is at minimum a prime system. Prime systems are generalized. To be realized experimentally in special-case time-space, the octahedron must consist of a high-frequency aggregate of octahedral and tetrahedral components. An octahedral system gains a nucleus with 19 balls, i.e., with 18 uniradius balls around one, as against the minimum nucleated (four-frequency) tetrahedral array of 35 balls, i.e., with 34 balls symmetrically around one. So the octahedron gains a nucleus at a lower frequency than does the tetrahedron.

1011.22 Whether at zero-frequency or multifrequency state, the icosahedron cannot have a tangentially contiguous, ergo statically structural, nuclear sphere of the same radius as those of its closest-packed, single, outer-layer array. It can only have a dynamically structured nucleus whose mass is great enough to impose critical-proximity central dominance over its orbitally icosahedrally arrayed, remotely co-orbiting constellation of concentrated energy events.

1011.23 The vector equilibrium has four hexagonally perimetered planes intersecting each other symmetrically at its center; while the octahedron has only three square-perimetered planes symmetrically intersecting one another at its center. The hexagon has room at its center for a uniradius circle tangent to each of the six circles tangent to one another around it; whereas the square does not have room for such a uniradius circle. Wherefore the minimal four-dimensional coordinate system of the vector equilibrium is the minimum inherently nucleated system. (This is why mathematical physics employing three-dimensional, *XYZ* coordination can only accommodate its experimental evidence of the atomic nucleus by amorphous mathematics.) Like the octahedron, the vector equilibrium also has eight triangular facets; while also explosively extroverting the octahedron's three square central planes, in two ways, to each of its six square external facets, thus providing seven unique planes, i.e. seven-dimensionality. And while the octahedron develops a nucleus at a lower number than does the tetrahedron—or more economically than a tetrahedron—it is indicated that the nuclear arrays are symmetrical and play very great parts in compound chemistry. (The cube develops a nucleus only at a relatively high frequency.) In each one of these, there may be hierarchies that identify the difference between organic and inorganic chemistries. Due to the fact that there are nuclear aggregations in symmetry to which all of our chemistries relate, we may find an organic and inorganic identification of the tetrahedral and octahedral nucleations. The nonnuclear, exclusively volumetric, single-layer, closest-packed, icosahedral aggregate may be identified with the electron "shells" of the compounding atoms.

1011.30 **Prime Tetra, Octa, and Icosa:** Prime means the first possible realization. It does not have frequency. It is subfrequency. One or zero are subfrequency. Interval and differentiation are introduced with two. Frequency begins with three—with triangle, which is the minimum cyclic closed circuitry.

1011.31 Three *linear* events have two intervals, which is the minimum set to invoke the definition frequency. But it is an "open" circuit. The circuit is closed and operative when the triangle is closed and the same three events produce three equi-intervals, rather than two. Equi-interval = "tuned." This is why wave-frequency relationships have a minimum limit and not an infinite series behavior.

1011.32 Frequency and size are the same phenomena. Subfrequency prime tetra, prime octa, and prime icosa are each constituted of only one edge module per triangular facet. While generalizably conceptual, the prime structural systems and their prime domains—linear, areal, and volumetric—are inherently subfrequency, ergo, independent of time and size.

1011.33 *Special case* always has frequency and size-time.

1011.34 *Generalization* is independent of size and time, but the generalization principle must be present in every special case of whatever magnitude of size or time.

1011.35 Prime tetrahedra and octahedra do not have nuclei. In contradistinction to prime tetrahedra and prime octahedra, some complex tetrahedra, complex octahedra, and complex cubes do have a nucleus. They do not develop structurally in strict conformity to closest packing to contain an internal or nuclear ball until additional closest-packed, uniradius sphere layers are added. For instance, the cubical array produced by nesting eight uniradius spheres in the center of the eight triangulated sphere arrays of the nuclear-balled vector equilibrium produces eight tetrahedra single-bondedly arrayed around a nuclear ball. Additional, and symmetrically partial, layers require identification as frequency of reoccurrence of concentric shell embracement. In contradistinction to the other two prime system domains, however, the icosahedron does not accommodate additional closest-packed sphere layers and never develops a static structural nucleus. The icosahedron's closest-packing capability is that of circumferential propagation of only one omni-intertriangulated uniradius sphere and can increase its frequency only as one shell and not as a nucleus.

1011.36 If the icosahedron does develop a further outward shell, it will have to discard its internal shell because the central angles of the icosahedron will not allow room for unit-radius spheres of two or more closest-packed omnitriangulated concentric shells to be constructed. Only one closest-packed shell at a time is permitted. Considered internally, the icosahedron cannot accommodate even one uniradius, tangentially contiguous, interior or nuclear sphere of equal radius to those of its closest-packed, uniradius, outer shell.

1011.37 Speaking externally, either ''prime'' or complex ''frequency'' tetrahedra and octahedra may interagglomerate with one another close-packingly to fill allspace, while icosahedron may never do so. The icosahedra may be face-associated to constitute an ultimately large octahedral structure. Icosahedra may also symmetrically build independent, closest-packed, tetrahedral arrays outwardly on each of their multi-frequenced, 20 triangular facets. Thus it is seen that the icosahedral closest packing can only grow inside-outedly, as does the vector equilibrium grow internally, i.e. inside-inwardly.

1011.38 While the regular icosahedron's radius is shorter in length than its external edge chords, the vector equilibrium has the same radius as each of its edge chords; which explains the vector equilibrium's tolerance of a nucleus and the icosahedron's intolerance of a nucleus.

1011.40 Congruence of Vectors: All vector equilibria of any frequency reveal vectorially that their radially disassociative forces always exactly and balancingly contain their circumferentially integrated—and therefore more embracing than internally disintegrating—forces as manifest by their vectorial edge chords. The vector equilibrium consists of four symmetrically interacting hexagonal planes. Each hexagon displays six radially disintegrative, independently operative, therefore uncompounded, central vectors and their equal-magnitude six, always cooperatively organized and compounded, circumferential chord vectors. Sum-totally, the four hexagons have 24 radial disintegrative vectors and 24 chordally integrative vectors, with the chordals occurring as four closed sets of six vectors each and the radials as four open sets of six vectors each. The planes of any two hexagons of the set of four intersect one another in such a manner that the radii of any two intersecting planes are *congruent,* while the chords are not. This paired congruency of the 24 radial-disintegrative vectors of the four hexagons reduces their visible number to 12. The 24 chordally integrative vectors remain separate and visible as 24 finitely closed in four embracing sets of six each.

1011.41 The phenomenon ''congruence of vectors'' occurs many times in nature's coordinate structuring, destructuring, and other intertransformings, doubling again sometimes with four vectors congruent, and even doubling the latter once again to produce eight congruent vectors in limit-transformation cases, as when all eight tetrahedra of the vector equilibrium become congruent with one another. (See Sec. 461.08.) This phenomenon often misleads the uninformed observer.

1011.50 **Instability of Vector Equilibrium:** If we remove the 12 internal, congruently paired sets of 24 individual radii and leave only the 24 external chords, there will remain the eight corner-interlinked, externally embracing triangles, each of which (being a triangle) is a structure. Between the eight triangular external facets of the vector equilibrium, there also occur six squares, which are not structures. The six square untriangulated faces are the external facets of six nonstructurally stabilized half-octahedra, each of whose four central triangular faces had been previously defined by the now removed 24 radially paired vectors of the vector equilibrium. A half-octahedron, to be stable, has to be complementingly square-face-bonded with its other half. The prime vector equilibrium has only these six half-octahedra, wherefore the circumferential instability of its six square faces invites structural instability. Thus deprived of its internal triangular structuring by removal of all its radial vectors, the vector equilibrium becomes disequilibrious.

1011.51 The prime vector equilibrium has a nucleus surrounded, close-packingly and symmetrically, by 12 uniradius spheres. (See Illus. 222.01.) As we add unit radius sphere layers to the prime vector equilibrium, the 12 balls of the first, or prime, outer layer become symmetrically enclosed by a second closest-packed, unit radius layer of 42 balls circumferentially closest packed. This initiates a vector equilibrium with modular edge and radius intervals that introduce system frequency at its minimum of two.*

1011.52 The edge frequency of two intervals between three balls of each of the vector equilibrium's 24 outer edges identifies the edges of the eight outer facet triangles of the vector equilibrium's eight edge-bonded (i.e., double-bonded) tetrahedra, whose common internal vertex is congruent with the vector equilibrium's nuclear sphere. In each of the vector equilibrium's square faces, you will see nine spheres in planar arrays, having one ball at the center of the eight (see Illus. 222.01), each of whose eight edge spheres belong equally to the adjacent tetrahedra's outwardly displayed triangular faces. This single ball at the center of each of the six square faces provides the sixth sphere to stabilize each of the original six half-octahedra formed by the nuclear ball of the vector equilibrium common with the six half-octahedra's common central vertex around the six four-ball square groups showing on the prime vector equilibrium's surface. This second layer of 42 spheres thus provides the sixth and outermost ball to complete the six-ball group of a prime octahedron, thus introducing structural stability increasing at a fourth-power rate to the vector equilibrium.

1011.53 With the 42-ball layer added to the vector equilibrium, there is no ball showing at the center of any of the triangular faces of the vector equilibrium. The three-ball edges of the 42-ball vector equilibrium provide a frequency of two. Three spheres in a row have two spaces between them. These interconnecting spaces between the centers of area of the adjacent spheres constitute the vectorial interconnections that provide the energetic, or force, frequency of the described systems.

1011.54 Then we come to the next concentric sphere layer, which has 92 balls; its frequency is three, but there are four balls to any one edge. The edges are all common to the next facet, so we only have to credit the balls to one facet or another at any one time; however, we have to do it in total overall accounting, i.e., in terms of how many balls are sum-totally involved in each of the concentrically embracing layers.

1011.55 With the four-ball edge F^3, for the first time, a ball appears in the center of each of the eight triangular facets. These central balls are *potential* nuclei. They will not become new vector-equilibrium nuclei until each potential nuclear sphere is itself surrounded by a minimum of two completely encompassing layers. These potential new nuclei (potentially additional to the as yet only one nucleated sphere at the center of the prime vector equilibrium) occur in the planar triangular facets of the vector equilibrium's eight tetrahe-

* The number of balls in the outer shell of the vector equilibrium = $10F^2 + 2$. The number 42, i.e., F^2, i.e., $2^2 = 4$, multiplied by 10 with the additive $2 = 42$.

dra, which, being tetrahedra, are structural-system integrities (in contradistinction to its six half-octahedra, which—until fortified by their sixth outer-vertex balls of the two-frequency vector equilibrium—were structurally unstable).

1011.56 Though there is one ball in the center of each of the eight triangular facets of the F^3 vector equilibrium, those balls are exposed on the outer surface of their respective tetrahedra and are not omnidirectionally and omnitangentially enclosed, as they would have to be to constitute a fully developed regenerative-system nucleus. Though outer-facetly centered (i.e., planarly central), those eight F^3, triangularly centered balls are not nuclei. To become nuclei, they must await further symmetrically complete, concentric, closest-packed, vector-equilibrium shell embracements which bring about a condition wherein each of the eight new potential nuclei are embraced omnidirectionally and omnitangentially in closest-packed triangulation by a minimum of two shells exclusively unique to themselves, i.e., not shared by any neighboring nuclei. The F^3 vector equilibrium's triangular facets' central surface-area balls are, however, the initial appearance in symmetrical, concentric, vector-equilibrium shell frequency growth of such potentially developing embryo nuclei. They are the first potential nuclei to appear in the progressive closest-packed, symmetrical, concentric layer enclosing of one prime regenerative system's primally nucleated vector equilibrium.

1011.57 But at F^3 we still have only *one true nuclear ball* situated symmetrically at the volumetric center of three layers: the first of 12, the next of 42, and the outer layer of 92 balls. There is only one ball in the symmetrical center of the system. This three-layer aggregate has a total of 146 balls; as noted elsewhere (see Sec. 419.05) this relates to the number of neutrons in Uranium Element #92.

1011.58 Any sphere is in itself a potential nucleus, but it has to have 12 spheres close-packingly and omni-intertangentially embracing it to become a prime nucleated, potentially regenerative system. To stabilize its six half-octahedra requires a second layer of 42 balls. The potentially regenerative prime nucleus can have the first F^0 layer of 12-around-one nucleus, and the next (F^2) layer of 42 around both the nucleus and the first layer, without any new potential nucleus occurring in either of those first two concentric layers. So the vector equilibrium is a *nuclear uniqueness* for the first layer of 12 and the next layer of 42, with no other potential nucleus as yet appearing in its system—in its exterior shell's structural triangular facets—to challenge its nuclear pristinity.

1011.59 While there is a ball in the center of the square faces at the two-frequency, 42-ball level, those square cross sections of half-octahedra are not stable structures. Those square-centered balls are literally structurally superficial, ergo they are *extra balls* that show up but are not structurally stable in any way. They may be released to further re-form themselves into four-ball, prime, tetrahedral, structural systems, or they may be borrowed away from the nuclear system by another nuclear system—as does occur in chemical combines—without damaging the borrowed-from system's structural integrity. The four balls that occur in the core of the square facets of the F^3, 92-ball shell are also borrowable extra balls.

1011.60 In the 92-ball, F^3 third shell, eight potential nuclei occur in the triangular facets. "Four" and "square" do not constitute a structural array. To be structural is to be triangulated. Four balls also occur in each of the square facets, whereas one ball had occurred in the center of each of the six square facets of the previous F^2, 42-ball layer. This means that at the F^3, 92-ball layer, there are five balls in each of the six square-face centers. These five will be complemented by one or more, thus to form six new, detachable, nonnucleated, prime octahedra in the F^4, 162-ball layer by a square group of nine balls in each of the six square facets of the vector equilibrium. The center ball of these nine will now join with the four balls of the F^3 layer and the one ball of the F^2 layer to form altogether a prime, closest-packed octahedron having no nucleus of its own.

1011.61 At the F^4, 162-ball layer, the eight potential nuclei occurring in the mid-triangle faces of the F^3 layer are now omnisurrounded, but as we have seen, this means that each has as yet only the 12 balls around it of the F^0 nuclear-development phase. Not until the F^5, 252-ball layer occurs do the eight potential second-generation nuclei become structurally enclosed by the 42-ball layer, which has as yet no new potential nuclei showing on its surface—ergo, even at the F^5 level, the original prime nucleus considered and enclosingly developed have not become full-fledged, independently qualified, regenerative nuclei. Not until F^6 and the 362-ball layer has been concentrically completed do we now have eight operatively new, regenerative, nuclear systems operating in partnership with the original nucleus. That is, the first generation of omnisymmetrical, concentric, vector equilibrium shells has a total of nine in full, active, operational condition. These nine, $8 + 1$, may have prime identification with the eight operationally intereffective integers of arithmetic and the ninth integer's zero functioning in the prime behaviors of eternally self-regenerative Universe. We may also recall that the full family of Magic Numbers of the atomic isotopes modeled tetrahedrally occurs at the sixth frequency (see Sec. 995).

1011.62 The potential nucleated octahedra that were heralding their eventual development when the six prime (nonnucleated) octahedra occurred at the F^4 level do not develop to full threefold, concentric, shell embracement as operational nuclei for several levels beyond that which had produced the second-generation eight vector-equilibrium nuclear integrities. We become also intrigued to speculate on the possible coincidence of the prime patternings developing here in respect to the 2, 8, 8, 18, 18, etc., sequences . . . of the Periodic Table of the Elements.

1012.00 *Nucleus as Nine = None = Nothing*

1012.01 Nucleus as nine; i.e., non (Latin); i.e., none (English); i.e., nein (German); i.e., neuf (French); i.e., nothing; i.e., interval integrity; i.e., the integrity of absolute generalized octaval cosmic discontinuity accommodating all special-case "space" of space-time reality. (See Secs. 415.43 and 445.10.)

1012.10 **Positive-Negative Wave Pattern:** Both the gravitational and the radiational effects operate exclusively in respect to and through the nucleus, whose unique domains multiply in eighths. Completion of the absolute initial uniqueness of pattern evolution of the nucleus itself brings in the nine as nothingness. How does this happen?

1012.11 Let us take three balls arranged in a triangle. We then take two other uniradius tangent balls lying in the same plane and address them symmetrically to any one corner-ball of the first three so that we have two rows of three balls crossing one another with one ball centrally common to both three-ball lines; so that we have two symmetrically arrayed triangles with one common corner. Obviously the center ball—like a railway switch—has to serve alternately either one three-ball track or the other, but never both at the same time, which would cause a smash-up. If we do the same thing four-dimensionally for the eight tetrahedra of the vector equilibrium, we find that the nuclear center ball is accommodating any one or any pair of the eight tetrahedra and is interconnecting them all. Externally, the eight tetrahedra's 24 vertexes share 12 points; internally, their eight vertexes share one point. The common center ball, being two-in-one (unity two), can be used for a pulse or a space; for an integer or a zero. The one active nucleus is the key to the binary yes-no of the invisible transistor circuitry.

1012.12 As in the 92-ball, three-frequency vector equilibrium, there are four balls to an edge going point to point with a three-space, F^3, in between them. An edge of the four ball could belong either to the adjacent square or to the adjacent triangle. It cannot belong to either exclusively, and it cannot belong to them both simultaneously; it can function for either on modulated-frequency scheduling. It is like our chemical bonding, bivalent, where we get edge-to-common-edgeness.

1012.13 As shown in *Numerology* (Sec. 1223), when we begin to follow through the sequences of wave patterning, we discover this frequency-modulation capability permeating the "Indig's" octave system of four positive, four negative, and zero nine. (See drawings section.)

Indigs of Numerology:

1 = + 1	10 = + 1	19 = + 1
2 = + 2	11 = + 2	20 = + 2
3 = + 3	12 = + 3	21 = + 3
4 = + 4	13 = + 4	22 = + 4
5 = − 4	14 = − 4	23 = − 4
6 = − 3	15 = − 3	24 = − 3
7 = − 2	16 = − 2	25 = − 2
8 = − 1	17 = − 1	26 = − 1
9 = 0	18 = 0	27 = 0 Etc., etc.

1012.14 Applying the Indig-Numerology to the multiplication tables, this wave phenomenon reappears dramatically, with each integer having a unique operational effect on other integers. For instance, you look at the total multiplication patterns of the prime numbers three and five and find that they make a regular X. The fourness (= +4) and the fiveness (= −4) are at the positive-negative oscillation center; they decrease and then increase on the other side where the two triangles come together with a common center in bow-tie form. You find that the sequences of octaves are so arranged that the common ball can be either number eight or it could be zero or it could be one. That is, it makes it possible for waves to run through waves without having interference of waves. (See drawings section.)

1012.15 Each ball can always have a neutral function among these aggregates. It is a nuclear ball whether it is in a planar array or in an omnidirectional array. It has a function in each of the two adjacent systems which performs like bonding. This is the single energy-transformative effect on closest-packed spheres which, with the arhythmical sphere→space→space→sphere→space→space—suggests identity with the neutron-proton interchangeable functioning.

1012.16 The vector equilibrium as the prime convergence-divergence, i.e., gravity-radiation nucleus, provides the nuclear nothingness, the zero point where waves can go through waves without interfering with other waves. The waves are accommodated by the zeroness, by the octave of four positive and four negative phasings, and by a nuclear terminal inside-outing and a unique pattern-limit terminal outside-inning. But there are two kinds of positives and negatives: an inside-outing and an arounding. These are the *additive twoness* and the *multiplicative twoness*. The central ball then is an inside-outness and has its poles so it can accommodate either as a zeroness a wave that might go around it or go through it, without breaking up the fundamental resonance of the octaves.

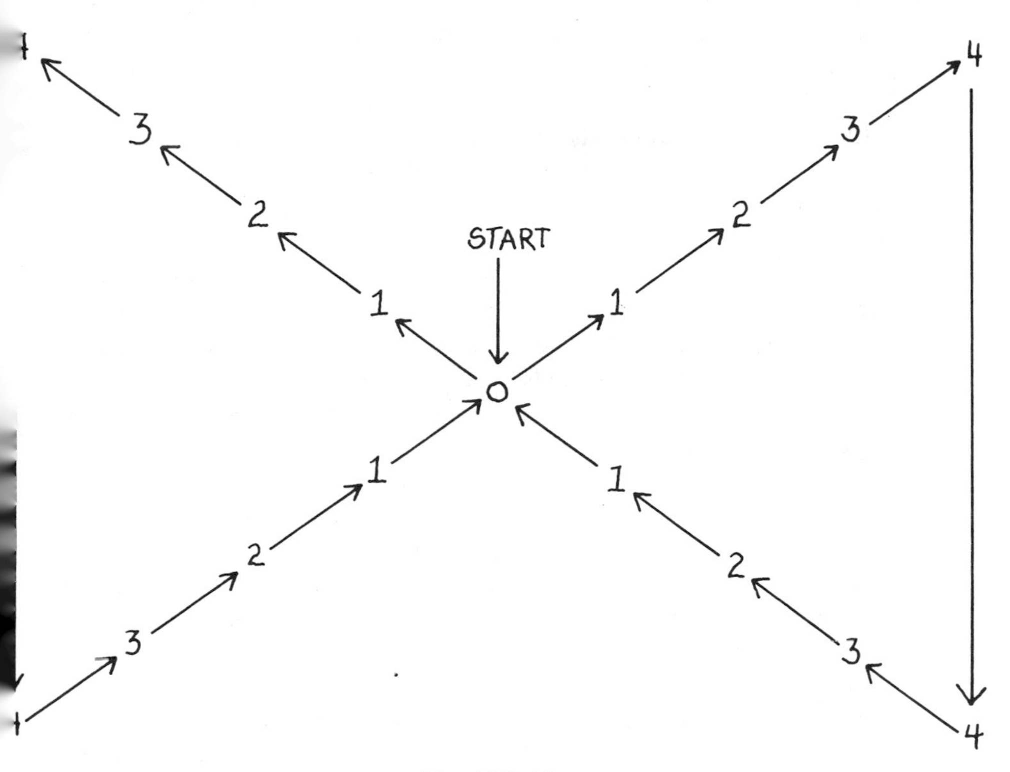

Fig. 1012.15.

1012.20 **Pumping Model:** The center ball of a vector equilibrium is zero. The frequency is zero, just as in the first layer the frequency is one. So zero times ten is zero; to the second power is zero; plus two is two. So the center ball has a value of two. The significance is that it has its concave and its convex. It has both insideness and outsideness congruently. It is as far as you can go. You turn yourself inside out and go in the other direction again. This is a terminal condition.

1012.21 We have then a tetrahedron that has an external and an internal: a terminal condition. Gravity converts to radiation. This is exactly why, in physics, Einstein's supposition is correct regarding the conservation of Universe: it turns around at both the maximum of expansion and the minimum of contraction, because there is clearly provided a limit and its mathematical accommodation at which it turns itself inside out.

1012.22 You get to the outside and you turn yourself outside-in; you come to the center and turn yourself inside-out. This is why radiation does not go to higher velocity. Radiation gets to a maximum and then turns itself inwardly again—it becomes gravity. Then gravity goes to its maximum concentration and turns itself and goes outwardly, becomes radiation. The zero nineness-nucleus provides the means.

1012.30 **Indestructibility of Tetrahedron:** We have here a pumping model of the vector equilibrium. It consists only of the vector lines of the system formed by 12 uniradius spheres closest packed around one sphere of the same radius. The interconnecting lines between those 13 spheres produce the pumping vector equilibrium model's skeleton frame. We have also removed the vector equilibrium's 12 internal double radii to permit the vector equilibrium system to contract; thus we have for the moment removed its nuclear sphere. Every vector equilibrium has eight tetrahedra with 12 common edges, a common central vertex, and 12 common exterior vertexes. Each tetrahedron of the eight has four planes that are parallel to the corresponding four planes of the other seven. Each of the vector equilibrium's eight tetrahedra has an external face perpendicular at its center to a radius developed outwardly from the nucleus. Each of the eight external triangular faces is interconnected flexibly at each of its three corners to one other of the eight triangles. It is found that the whole vector equilibrium external-vector framework can contract symmetrically, with the four pairs of the eight external triangles moving nontorquingly toward one another's opposite triangle, which also means toward their common nucleus. As they do so, each of the four pairs of exterior triangles approaches its opposite. When the eight separate but synchronously contracting tetrahedra diminish in size to no size at all, then all eight planes of the eight triangles pass congruently through the same nuclear center at the same time to form the four planes of the vector equilibrium. (See Sec. 623.)

1012.31 Because each of the eight triangles had converged toward one another as four opposite pairs that became congruent in pairs, we seemingly see only four planes going through the center in the model. There are, however, really eight planes passing through the same vector-equilibrium nuclear point at the same time, i.e., through the empty, sizeless nucleus.

1012.32 As the eight tetrahedra diminished in size synchronously, their edges became uniformly smaller at a velocity of the first power; their areas became smaller at a velocity of the second power; and their volumes became smaller at a velocity of the third power—which are three very different velocities. Finally, they all reached zero velocity and size at the same time. As they became smaller, however, there was no change in their respective fourness of faces; sixness of edges; fourness of vertexes; nor equilateralness; nor equiangularity. These are changeless constants. So what you see in the model is eight sizeless tetrahedra that became one empty, sizeless, congruent set, with all their mathematically constant tetrahedral characteristics unaltered, ergo, conceptually manifest as eternity.

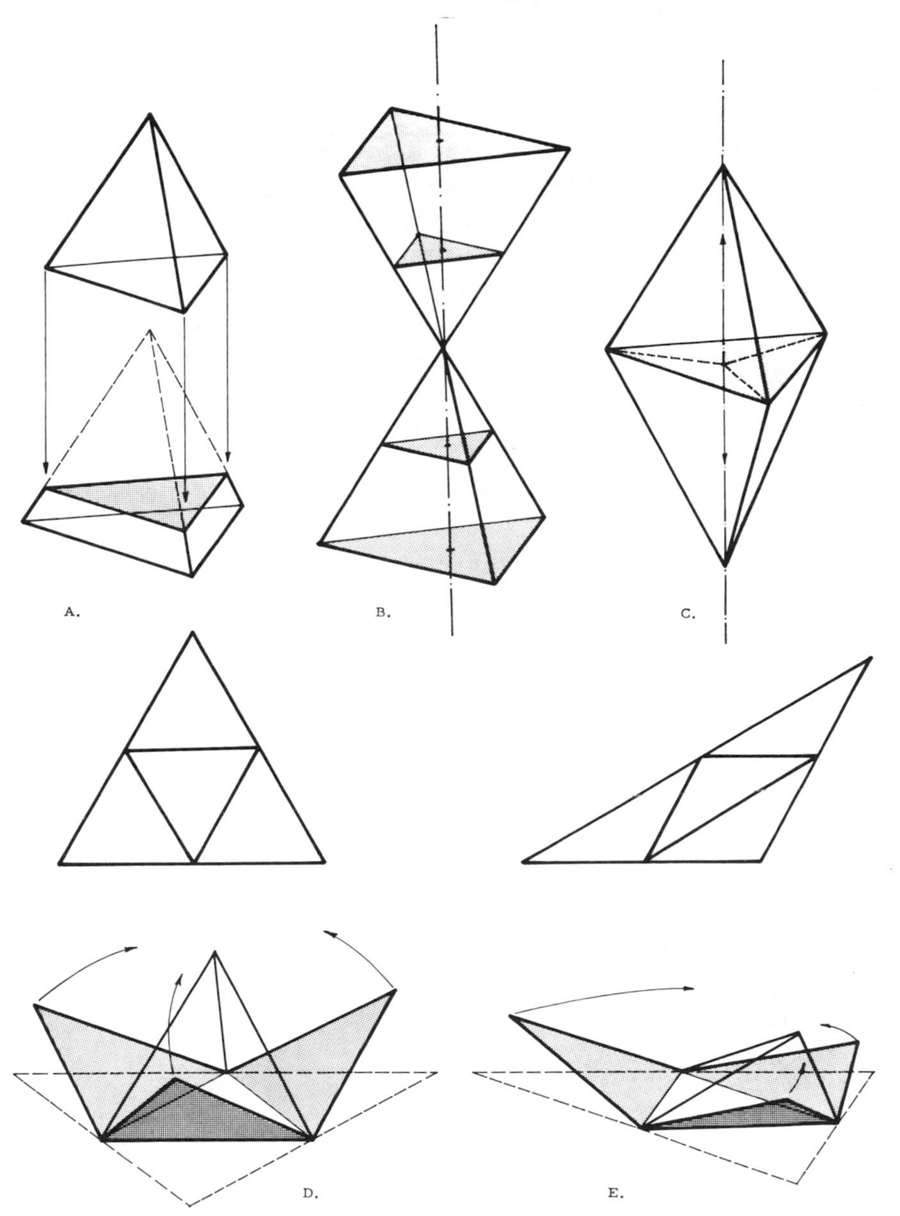

Fig. 1012.30 *Indestructability of the Tetrahedron:*

A. A plane passed through a tetrahedron parallel to a face does not alter its regularity.

B. The extensions of edges through any one vertex to form positive and negative tetrahedra. Another example of the essential twoness of a system.

C. Vertex passed through opposite face to form another version of positive-negative tetrahedra.

D. When an equilateral triangle is divided into four identical smaller triangles it will fold into a tetrahedron.

E. When any arbitrary triangle is divided into four congruent triangles by bisecting its edges and joining them with new edges, it will also fold into a tetrahedron—an irregular tetrahedron bounded by four congruent faces.

1012.33 What we speak of as a point is always eight tetrahedra converged to no size at all. The eight tetrahedra have been brought to zero size and are abstracted from time and special case. They are generalized. Though the empty vector-equilibrium model is now sizeless, we as yet have the planes converging to intercept centrally, indicating the locus of their vanishment. This locus of vanishment is the nearest to what we mean by a point. The point is the macro-micro switchabout between convergence and divergence.

1012.34 We also have learned that a plurality of lines cannot go through the same point at the same time. Therefore, the eight perpendiculars to the centers of area of the triangle faces and the 12 lines that led to their 12 common outer vertexes, like the tetrahedra's volumes and areas, have come to common zero time-space size and can no longer interfere with one another. We find operationally, however, that there never was any paradoxical problem, such as Zeno's ''never completable approach'' concept, for we have learned of the fundamental torque or twist always present in all experientially explored system realization, and we find that as each team of opposite triangles apprehended the other just upon their nearing the center, each is whirled 180 degrees, or is ''half spun'' about, with its three corners never completely converging. Whereafter they diverge.

1012.35 Take three round rods of the same diameter and nest them together in parallel triangulation. They are now closest-possibly-packed together. Now slip a triangular-shaped ring tightly around them and glide it to their midlength point. Now twist the ends of the three in opposite directions; the ends will open outwardly from one another as triangles. Stand the group on one three-point tripod end with wires between those opposite ends to limit their spreading. But they could also have twisted clockwise for the half-spin. They could half-spin alternatively to produce whole-cycle coverage. We find that three lines converge to critical proximity, then twist, and spin around one another. This happens also with all six of the diameters (or all 12 of the radii) of the vector equilibrium. (An articulating model of this can be made with four sets of three stiff brass wires each, laying the four sets in parallel, closest-packed bunches and soldering together the three wire ends at one end of each of the four bunches; it will be found that their total of 12 free ends may be lead through one another's mid-girth in a symmetrically progressive manner, after which these led-through, four sets of ends may be respectively sprung together in sets of three and soldered together; which model then provides a number of very exciting intertransformabilities elucidating the vector equilibrium's significance.) In the vector equilibrium, the ''whole cycles'' are accomplished in four planes corresponding symmetrically to one another as represented by the great-circle planes of the empty-state vector equilibrium or of its eight empty-state tetrahedra.

1012.36 Because the nuclear center of the vector equilibrium is also the generalized volumetric center of the spheres in the closest-packed condition, as well as of the spaces between the spheres, all of which correspond to all the vertexes (or all possible system convergences) of the isotropic vector matrix, we learn that all three vectorial lines of Universe can twist by one another producing half-spin, half-quantum, wave bulgings as they do without frustrating any form of intertransformative event development in Universe, and while also disclosing an absolute compatibility to, and elucidation of, wave-quanta behavior in generalized conceptuality.

1012.37 Reviewing the same phenomenon once again, we make further discovery of the utter interrelatedness of synergetic accommodation, as we find the half-spin "tepee" twist also turning the tetrahedron inside out. (See Sec. 621.20.) Here we find that the vector equilibrium, or the vector equilibrium's eight tetrahedra's external vertexes, all converged toward one another only to suddenly describe four half-great-circle spins as they each turned themselves inside out just before the convergence: thus accomplishing sizeless invisibility without ever coming into contact. Eternal interval is conserved. Thus the paradox of particle discontinuity and wave continuity is conceptually reconciled. (See Sec. 973.30.)

1020.00 Compound Curvature: Chords and Arcs

1021.10 Convexity and Concavity of Tetrahedron

1021.11 The outsides of systems are convex, and their insides are concave. While convexity diffuses radiation impinging upon it, concavity concentrates radiation impinging upon it; ergo, convexity and concavity are not the same.

1021.12 For every tetrahedron, there is one convex and one concave. Because the tetrahedron is inherently the minimum structural system of Universe, it provides the minimum omnicoexisting convexity and concavity condition in Universe.

1021.13 For every tetrahedron, there is an inside tetrahedron and an outside tetrahedron. For every convex spherical polyhedral geodesic system, there is a concave spherical polyhedral geodesic system. One cannot exist without the other either in special case or in sizeless eternal generalization. Spherical arrays and compound curvature begin with the tetrahedron.

1022.10 Minimum Sphere

1022.11 The transcendentally irrational constant *pi* (π) is irrelevant to spherical geodesic polyhedral array calculations because the *minimum sphere* is a tetrahedron. We have learned that a sphere as defined by the Greeks is not experimentally demonstrable because it would divide all Universe into outside and inside and have no traffic between the two. The Greek sphere as defined by them constituted the first and nondemonstrable perpetual-motion machine. Because there could be no holes in it, the Greek sphere would defy entropy. A sphere with no holes would be a *continuum* or a *solid,* which are physical conditions science has not found. We could dispense with all Universe outside the Greek sphere because Universe inside would be utterly conserved and eternally adequate to itself, independent of the rest of Universe outside.

1022.12 What we do have experimentally as a sphere is an aggregate of energy-event foci approximately equidistant in approximately all directions from one approximate energy-event focus. This is a system in which the most economical relationships between embracingly adjacent foci are the great-circle chords, and not the arcs. This is why *pi* (π) is operationally irrelevant. Physics finds that nature always employs the most economical means. Being shorter, chordal distances are more economically traversed than are detouring arcs. All the chords between external points of systems converge with one another concavely and convexly, i.e., with the angles around each external point always adding to less than 360 degrees. They do not come together, as do radii in a plane, with 360 degrees around each point.

1022.13 The chords of an omnidirectional system always come together with concavity on one side and convexity on the other. The angles never add up to 360 degrees, as do those formed on a plane by lines converging radially upon a point. This is why the long-held working assumption of mathematics—that for an infinitesimal moment a sphere is congruent with the plane to which it is tangent—is invalid. Therefore, spherical trigonometry, with its assumption of 360 degrees around a point, is also invalid. Greek spheres cannot be scientifically demonstrated. Almost-spherical polyhedra are the nearest approximation. It can only be treated with as polyhedral—as an aggregate of points in which the most economical relationships are chords; ergo, geodesics.

1022.14 If you find all the connections between all the points, the system is omnitriangulated. A spherical polyhedron is a high-frequency geodesic polyhedron. Its symmetric base may be tetrahedral, octahedral, or icosahedral; but it may not be hexagonal, i.e., with angles adding to 360 degrees around each external point of the system. The sum of all the angles around all the external points of the superficially seeming spherical systems will always add up to 720 degrees less than the number of external vertexes when each is multiplied by 360 degrees.

1022.15 In every geodesic sphere, you can always take out 12 pentagons. These 12 pentagons each drop out one triangle from the hexagonal clusters around all other points. Assuming the dropped-out triangles to be equiangular, i.e., with 60-degree corners, this means that $60 \times 12 = 720°$, which has been eliminated from the total inventory of surface angles. You can always find 12 pentagons on spherically conformed systems such as oranges, which are icosahedrally based; or four triangles with 120-degree corners if the system is tetrahedrally based; or six squares where the system is octahedrally based.

1023.10 Systematic Enclosure

1023.11 If we get too semantically incisive, the reader may lose all connection with anything he has ever thought before. That might not be a great loss. But we assume that the reader can cope with his reflexes and make connections between the old words and new concepts with the new and more apt words. For example, since physics has found no continuums, we have had to clear up what we mean by a sphere. It is not a surface; it is an aggregate of events in close proximity. It isn't just full of holes: it doesn't have any continuum in which to have holes.

1023.12 The word *polyhedron* has to go because it says "many-sided," which implies a continuum. We don't even have the faces. Faces become spaces. They become intervals. They become nothing. The Einsteinian finite Universe—an aggregate nonsimultaneous Universe—is predicated only on the absolute finiteness of each local energy-event package and the logic that an aggregate of finites is itself finite.

1023.13 The spheric experience is simply an ultrahigh frequency of finite event occurrences in respect to the magnitude of the tuning perceptivity of the observer. (High frequency to the human may be low frequency to the mosquito.)

1023.14 If we get rid of the word *polyhedra,* then what word do we have in its place? A high-frequency, omnidirectional, spheric event system. Polyhedra are finite system enclosures. They are topologically describable, finite *system enclosures.* They are Universe dividers. They are not linear dividers, but omnidirectional Universe dividers dividing outside from inside, out from in. A mosquito has macro-micro cosmos system perceptivity at a different level from that of the whale's. Probably each observer organism's stature constitutes its spontaneous observational level of macro-micro subdividing: bigger than me; littler than me; within me; without me.

1023.15 We relinquish the word *polyhedra* to reemploy our new term *systematic enclosure,* which can be generalized to serve creatures of any size—i.e., a tetrahedron big enough for a mosquito or big enough for a whale. Faces are spaces, openings. The four vertexes plus four faces plus six lines of the tetrahedron must become four somethings plus four nothings plus six relations. We add convergence *to something* and divergence *to nothing*—completely independent of size. Since there are no "things," there is no "something." We are talking of an event in pure principle. We have events and no-events. Events: novents: and relationships. Nature employs only one or another of the most equieconomical relationships. The most economical relationships are geodesic, which means most economical relationships. Ergo we have events and novents: geodesics and irrelevance. These are the epistemological stepping-stones.

1023.16 The spheric experience is a high-frequency, omnidirectional complex of events and their relatedness. Since it is concerned with the most economical relatedness, we can also speak of it as a geodesic spherical experience. This is where the importance of chords comes in. A chord is abstract, yet tensive. A chord has pull: we would probably not think about the connections unless there was some pull between them. The function of the chords is to relate. The event is the *vertex*. The reaction is the *chord,* the pulling away. And the resultant is the inadvertent definition of the nothingness of the *areal and volumetric spaces*. The sequence is: Events; chords; no-events. No-events = novents. Areas do not create themselves; as with celestial constellations, they are incidental to the lines between the events. The faces are the bounding of nothingness. Areas and volumes are incidental resultants to finding the connections between events of experience.

1023.17 Not only can there be no awareness until there is otherness to be aware of, but there can be no *magnitude awareness* with only one otherness. You need two otherness experiences with an interval between them in order to have a sense of distance. (Otherwise, you might just be looking at yourself in a mirror.)

1023.18 You can have no sense awareness of shape with just one otherness or two othernesses. *Shape awareness* commences only with three othernesses where the relationship of three as a triangle has finite closure. Shape is what you see areally, and until there is closure, there is no area of otherness.

1023.19 Not until we have four othernesses do we have macrocosmic *volumetric awareness*. Four is required for substantive awareness.

1023.20 *System awareness* begins when we find the otherness surrounding us, when we are omnidirectionally enclosed. The volume sense is only from inside. From outside, four points can look like one point or they can look flat. Not until we turn a tetrahedron inside out do we have microcosmic awareness. Not until we swallow the otherness do we have microcosmic volumetric awareness. We become the outside. At first, we were just the inside. In the womb. In the womb, we had tactile, sensorial awareness of volumetric surroundment by the otherness, but no visual, aural, or olfactoral awareness of the otherness surroundment. The child develops otherness awareness only as outside volumetric surroundment within which he finally discovers *me* the observer, and *me*'s hand.

1024.10 What Is a Bubble?

1024.11 What is a bubble? When oil is spilled on water—unfortunately, an increasing phenomenon—it spreads and spreads as a result of gravity pulling and thinning it out. It thins out because the molecules were piled on top of one another. Gravity is pulling it into single-molecule-thickness array. The individual molecules are mass-interattracted, but the attraction can be focused on the nearest molecules. Molecules can therefore be tensed and will yield in such a manner as to thin out their mass, which can be stretched as a sheet or stretched linearly—for each molecule holds on to only those other molecules within critical proximity.

1024.12 As one floating molecule is surrounded by six other floaters and the six are surrounded by 12, or the next perimeter of 18, pulling on one molecule distributes the pull to six, and the six distribute the pull to 12, and the 12 distribute the pull to 18, and therefore the original pull becomes proportionally reducible and the relative distance between the molecules varies from one surrounding hexagon to the next. This relative proximity brings about varying tension, which brings about varying density. Varying density, we learn in optics, brings about varying refraction of light frequency, ergo, of light as color, which accounts for the rainbow spectrum differentiating witnessed as sunlight strikes oil-covered waters. By passing light through clear plexiglass structural models, the structural strains as distributed throughout the plastic mass are visually witnessable by the red, orange, yellow, green, blue, violet rainbow spectrum.

1024.13 Comprehending the mass-attracted, intertensed integrity of molecules and atoms, witness how the blacksmith can heat his metals in the red-hot condition and hammer the metal into varying shapes, all permitted by the mass-interattraction of the atoms themselves and their geometrical, methodical yielding to rearrangement by forces greater than their local surroundment interattractions. The heating is done to accelerate the atoms' electrons to decrease the relative-proximity interattractiveness and accommodate the geometrical rearranging of the atoms. The cold metals, too, can be hammered, but the energy-as-heat facilitates the rearranging. When metals are reshaped, they do so only as the absolute orderly intertransformative geometry of closest packed atoms permits.

1024.14 Because the atoms and the molecules are subvisible in magnitude to man, he fails to detect the exquisite geometrical orderliness with which they yield to rearrangement while retaining the total interattractiveness occasioned by their initial aggregation within the critical limits of mass-attraction where the attractive force overcomes the individual orbiting integrity. The relative interattraction increases as the second power of the rate at which the interdistances diminish.

1024.15 The atomic proximity within the metals is of such a high order as to give high tensile strength, which is resistance to being pulled or put asunder. Exquisite magnitudes of interattractive proximities have nothing to do with pressure. The phenomenon is coherent density. Density is a pulling together. (The error of reflexing is here comparable to humans' misapprehending the wind's ''blowing'' when we know that it cannot blow; it can only be sucked.) Man is always thinking he can push things when they can only be pulled. Men are pushers. Women are attracters.

1024.16 These principles of interattractive strengths and orderly geometrical yielding to stresses are employed to a high degree in the manufacturing of thin transparent plastic sheets, such as all society is becoming familiar with as a use product.

1024.17 Children are familiar with bubble gum. They are accustomed to seeing the bubble blown until suddenly it becomes transparent. The membrane is yielding circumferentially and tensilely to the pressure differential between the outside atmosphere and the multiplying molecules of gas inside literally hitting the skin, trying to escape. When the molecules of the bubble gum have rearranged themselves in a geometrically orderly manner so as to get the isotropic-vector matrix trussing thinned down finally to a single layer, then it has become transparent. Between the finite Milky Way array, the atoms are in sufficient proximity to hold their single-layer triangulation array of hexagons within hexagons. In this condition, bubbles show the same color differentiation that reflects the tensile variations: what humans have learned to call surface-tension integrity.

1024.18 In the way children blow up small rubber balloons, you can almost see the layers of molecules yielding as if unfolding like an accordion, opening up angle after angle as the balloon yields to stretch. The child witnesses nature yielding to his own internal pressures as nature thins out the atomic and molecular arrangement with the most exquisite delicacy of uniform thickness throughout the stretching. The atoms and molecules distribute the load superbly and open up the many layers to one single layer with a dimensional accuracy inherent in the unique prime geometrical magnitudes of the nucleus-electron orbit frequencies differentiating one chemical element from the other in absolute spectroscopic detectability throughout the so-far-observed Universe. The dimensional integrities are topological and vectorial relative to all the characteristics with which synergetics is concerned.

1024.19 Even as a child blows his bubble gum, the manufacturer of plastic film first extrudes plastic wire; in its most plastic state, its end is conically pierced centrally while a machine blows air into the pierced core (cone) of the wire, which then yields in its absolute geometrical orderliness of intermolecular and interatomic integrities so that the intruded gas stretches the progressively pulled-around and conically intruded wire into a thin, monometrically single-molecule thickness—or a plurality of molecular thicknesses directly and geometrically proportional to the pressure. As the gas is introduced through the apex of the piercer of the wire (like a micro-cratered cone with a compressed air ''volcano'' erupting from within it), it stretches the wire into a bubble expanding at 180 degrees from the gas-introducing point; the now transparently thin-skinned bubble is led into and flatteningly gathered between metal rolls, which progressively close to flatten the bubble into a cylinder form until the whole cylinder of thin film is cut, split, and finally opened up to a single film: the evenness of the bubble stretching has turned the skin of the cylinder into a single sheet. The consistency of the chemical aggregates that nature allows chemists to produce in various chemical situations provides varieties of thicknesses. Mylar polyester, for instance, is inelastic and permits no further yielding; it is not subject to secondary deformation —stretching—such as occurs with rubber. There eventually comes a limit of the orderly rearrangeability of the atomic and molecular structuring beyond which it will no longer flex and at which point it breaks, i.e., disconnects because exceeding its critical-proximity interattraction limits. The relative proximity of the atoms is far more exquisite than that of molecules.

1024.20 Children experience magnets geometrically as metal blocks with thickness, length, and breadth. The magnet blocks can hold together end to end, side to side, or even point to point. You can stand them on their sides as relatively structurally stable, like face-bonding. But they regain flexibility when edge-bonded, or even more so when point-bonded.

1024.21 The bubble gum, the wire film, or the balloon all display invisible pneumatics evenly distributing the tensive energy loads to produce films of uniform thickness. No man could hammer or roll a substance into such exquisite dimensional stability. The popular image has the blacksmith working his will on the semimolten metal, but it is not so. The great armorers and swordmakers found just the opposite; they discovered the way in which nature permits the metals to *yield* and still retain their integrity. Humans cannot see the rearrangements of mountain-reflecting lake waters in atomical and molecular "Between-the-Halves" marching maneuvers to halve at the state of ice; this was arrived at, however, in ever-orderly intertransforming, geometrical integrity, invisible-to-humans magnitude of perception and analysis.

1024.22 Man talks carelessly and ignorantly of such words as *chaos* . . . *turbulence* . . . *turmoil* . . . and (the popular, modern) *pollution* . . . where nothing but absolute order is subvisibly maintained by nature and her transformation arrangements unfamiliar to man. Universe does not have any pollution. All the chemistries of Universe are always essential to the integrity of eternal intertransformation and eternal self-regeneration.

Physicists invent nothing.

Chemists invent nothing.

. . . They find out what nature does from time to time and learn something of what her laws of rearrangement may be, and fortunate humans employ those rules to cooperate consciously with nature's evolution.

1024.23 All humans, endowed at birth with a billion capabilities beyond the knowledge of the parents, evolve in ways that are an utter mystery to them. The exquisite, myriadly endowed child employs that mysterious endowment and intuitionally apprehends itself as inventor of ways of using the orderly laws of Universe to produce tools, substances, and service integrities, to communicate and allow humans to participate in Universe's ever-transforming evolutionary events in an as yet preposterously meager degree, which has given rise to a nature-permitted variety of little humans on tiny planet Earth each becoming Mr. Big, with a suddenly mistaken sense of power over environmental transformations—participation in which permitted him to feel himself as a manager of inventories of logistical multiplicity which, at the most ignorant level, manifests itself as politically assured mandates and political-world gambling = gamboling = ideological warfare = national sovereignties = morally rationalizing public = body politic = individual nations as United Nations.

1024.24 Stress-producing metaphysical gas stretches and strains nature to yield into social-evolution conformations such as the gas-filled plastic tube of Universe. There is an a priori universal law in the controlled complexity that tolerates man's pressurized nonsense, as nature permits each day's seemingly new Universe of semifamiliarities, semiwonders, and semimystery, what humans might think of as history unfolding on this little planet. There is the Game of Cosmic History, in which Universe goes on approximately unaware of human nonsense while accommodating its omnilocal game-playing. Flies have their game. Mosquitoes have their game. Microbes have their game. Lion cubs have their game. Whatever games they may be playing, positive or negative, realistic or make-believe, all the games are fail-safe, alternate circuits, omniconsequential to eternally regenerative Universe integrity. It's all permitted. It all belongs.

1024.25 Only humans play "Deceive yourself and you can fool the world"; or "I know what it's all about"; or "Life is just chemistry"; and "We humans invented and are running the world." Dogs play "Fetch it" to please their masters, not to deceive themselves. The most affectionate of dogs do not play "Burial of our dead"—"Chemistry is for real." Only humans play the game of game of masks and monuments. Fictional history. Historical architecture. Crabs walk sideways; but only human society keeps its eyes on the past as it backs into its future. Madison Avenue aesthetics and ethics. Comic strips and cartoons . . . truth emergent, laughing at self-deception . . . momentary, fleeting glimpses of the glory, inadvertently revealed through faithful accuracy of observation—lucid conceptioning—spoken of as the music of the stars, inadequate to the mystery of integrity . . .

All the poetry,
all the chemistry,
all the stars,
. . . are permitted transformations of all the eternal integrity.
All the constants,
gravitational constant,
radiational constant,
Planck's constant,
. . . above all, mathematics, geometry, physics, are only manifests of the eternal mysteries, love, harmonic integrity beyond further words.

The isotropic vector matrix yields to palm trees and jellyfish as a complex of mathematical integrities. As one will always be to one other. But no other: no one. Other *is* four. No four—but whereas one has no relations; two have only one interrelationship; three have three interrelationships; but four have a minimum of six relationships synergetics. No insideness without four. Without four, no womb: no birth: no life . . . the dawning awareness of the integrity of Universe. For humanity the only permitted infallibly predictable is the eternal cosmic integrity.

1025.10 Closest Packing of Bubbles

1025.11 Isolated bubbles are systematic spheric enclosures. Bubbles are convex and spheric because spheres accommodate the most volume with the least surface, and the pressure differential between inside and outside atmosphere makes them belly out. The enclosing "surfaces" of bubbles are in fact critical-proximity events that produce so-called "surface tension," which is, more accurately, single-molecule-thickness, omnitriangular, mass-interattracted atoms surrounding a gas whose would-be kinetically escaping molecules are larger than the intervals between the spherical membrane's atomic event proximities.

1025.12 Bubbles aggregate in the manner of closest-packed uniradius spheres but behave differently as they aggregate. Only the outer surfaces of the outermost bubbles in the aggregate retain their convex surfaces. Within the aggregate, all the bubbles' pressures become approximately uniform; therefore, relieved of the pneumatic pressure differential between insideness and outsideness, they contract from convex to approximately planar membranes. Here, what would have been spaces between the spheres become planar-bound system enclosures (polyhedra), as do also the corresponding concave octahedra and vector equilibria of hard-shell uniradius spheres in closest packing.

1025.13 Because the bubbles are rarely of unit radius, the closest-packed bubble "polyhedra," corresponding to the closest-packed spheres, disclose only multifrequency-permitted varieties of tensional membrane interfaceting. Yet the fundamental interrelatedness of the seemingly disorderly subdividing of bubble aggregates is elegantly identified with the absolute order of the isotropic vector matrix, in that all the internal polyhedra manifest 14 facets each, though a variety of polygonal shapes and sizes. This 14-ness is also manifest in the closest interpacking of biological cells.

1025.14 The 14 internal facets correspond exactly with the vector equilibrium's 14 faces—eight triangular and six square—which 14-ness, in turn, is directly identifiable with the tetrahedron's sum total of topological aspects: 4 vertexes + 4 faces + 6 edges = 14; as may be experimentally demonstrated with high-frequency tetrahedra, each of whose four vertexes may be truncated, providing four additional triangular facets; and each of whose six edges may be truncated (most crystals have truncated edges), providing six additional rectilinear facets whose terminal ends will now convert the four previous triangular truncated corners into four hexagons. With high-frequency tetrahedra, each of the truncations can be accommodated at different lengths. The truncated tetrahedron's total of 14 facets consisting of eight hexagons and six rectangles may be of a great variety of edge lengths, which variety tends to mislead the observer into thinking of the aggregate as being disorderly.

1030.00 Omniequilibrium

1030.10 **Omniequilibrium of Vector Equilibrium:** I seek a word to express most succinctly the complexedly pulsative, inside-outing, integrative-disintegrative, countervailing behaviors of the vector equilibrium. "Librium" represents the degrees of freedom. Universe is *omnilibrious* because it accommodates all the every-time-recurrent, 12-alternatively-optional degrees of equieconomical freedoms. Omniequilibrious means all the foregoing.

1030.11 The sphere is a convex vector equilibrium, and the spaces between closest-packed uniradius spheres are the concave vector equilibria or, in their contractive form, the concave octahedra. In going contractively from vector equilibrium to equi-vector-edged tetrahedron (see Sec. 460), we go from a volumetric 20-ness to a volumetric oneness, a twentyfold contraction. In the vector-equilibrium jitterbug, the axis does not rotate, but the equator does. On the other hand, if you hold the equator and rotate the axis, the system contracts. Twisting one end of the axis to rotate it terminates the jitterbug's 20-volume to 4-volume octahedral state contraction, whereafter the contraction momentum throws a torque in the system with a leverage force of 20 to 1. It contracts until it becomes a volume of one as a quadrivalent tetrahedron, that is, with the four edges of the tetrahedron congruent. Precessionally aided by other galaxies' mass-attractive tensional forces acting upon them to accelerate their axial, twist-and-torque-imposed contractions, this torque momentum may account for the way stars contract into dwarfs and pulsars, or for the way that galaxies pulsate or contract into the incredibly vast and dense, paradoxically named "black holes."

1030.20 **Gravitational Zone System:** There is no pointal center of gravity. There is a gravitational-zone-system, a zone of concentration with minimum-maximum zone system limits. Vertex is in convergence, and face is in divergence. Synergetics geometry precession explains radial-circumferential accelerational transformations.

1031.10 Dynamic Symmetry

1031.11 When we make the geodesic subdivisions of symmetrically omnitriangulated systems, the three corner angles increase to add up to more than 180 degrees because they are on a sphere. If we deproject them back to the icosahedron, they become symmetrical again, adding to exactly 180 degrees. They are asymmetrical only because they are projected out onto the sphere. We know that each corner of a two-frequency spherical icosahedron has an isosceles triangle with an equilateral triangle in the center. In a four-frequency spherical icosahedron there are also six scalenes: three positive and three negative sets of scalenes, so they balance each other. That is, they are *dynamically symmetrical*. By themselves, the scalenes are asymmetrical. This is synergy. This is the very essence of our Universe. Everything that you and I can observe or sense is an asymmetrical aspect of only sum-totally and non-unitarily-conceptual, omnisymmetrical Universe.

1031.12 Geodesic sphere triangulation is the high-frequency subdivision of the surface of a sphere beyond the icosahedron. You cannot have omnisymmetrical, equiangle and equiedged, triangular, system subdivisioning in greater degree than that of the icosahedron's 20 similar triangles.

1031.13 As we have learned, there are only three prime structural systems of Universe: tetrahedron, octahedron, and icosahedron. When these are projected on to a sphere, they produce the spherical tetrahedron, the spherical octahedron, and the spherical icosahedron, all of whose corner angles are much larger than their chordal, flat-faceted, polyhedral counterpart corners. In all cases, the corners are isosceles triangles, and, in the even frequencies, the central triangles are equilateral, and are surrounded by further symmetrically balanced sets of positive and negative scalenes. The higher the frequency, the more the scalenes. But since the positive and negative scalenes always appear in equal abundance, they always cancel one another out as dynamically complementarily equilateral. This is all due to the fact that they are projections outwardly onto a sphere of the original tetrahedron, octahedron, or icosahedron, which as planar surfaces could be subdivided into high-frequency triangles without losing any of their fundamental similarity and symmetry.

1031.14 In other words, the planar symmetrical is projected outwardly on the sphere. The sphere is simply a palpitation of what was the symmetrical vector equilibrium, an oscillatory pulsation, inwardly and outwardly—an extension onto an asymmetrical surface of what is inherently symmetrical, with the symmetricals going into higher frequency. (See Illus. 1032.12, 1032.30, and 1032.31.)

1031.15 What we are talking about as apparent asymmetry is typical of all life. Nature refuses to stop at the vector-equilibrium phase and always is caught in one of its asymmetric aspects: the positive and negative, inward and outward, or circumferentially askew alterations.

1031.16 Asymmetry is a consequence of the phenomenon time and time a consequence of the phenomenon we call afterimage, or ''double-take,'' or reconsideration, with inherent lags of recallability rates in respect to various types of special-case experiences. Infrequently used names take longer to recall than do familiar actions. So the very consequence of only ''dawning'' and evolving (never instantaneous) awareness is to impose the phenomenon time upon an otherwise timeless, ergo eternal Universe. Awareness itself is in all these asymmetries, and the pulsations are all the consequences of just thought itself: the ability of Universe to consider itself, and to reconsider itself. (See Sec. 529.09.)

1032.00 Convex and Concave Sphere-Packing Intertransformings

1032.10 **Convex and Concave Sphere-Packing Intertransformings as the Energy Patterning Between Spheres and Spaces of Omni-Closest-Packed Spheres and Their Isotropic-Vector-Matrix Field:** When closest-packed uniradius spheres are interspersed with spaces, there are only two kinds of spaces interspersing the closest-packed spheres: the concave octahedron and the concave vector equilibrium. The spheres themselves are convex vector equilibria complementing the concave octahedra and the concave vector equilibria. (See Secs. 970.10 and 970.20.)

1032.11 The spheres and spaces are rationally one-quantum-jump, volumetrically coordinate, as shown by the rhombic dodecahedron's sphere-and-space, and share *sixness* of volume in respect to the same nuclear sphere's own exact *fiveness* of volume (see Secs. 985.07 and 985.08), the morphological dissimilarity of which render them one-quantumly disequilibrious, i.e., asymmetrical phases of the vector equilibrium's complex of both alternate and coincident transformabilities. They are involutionally-evolutionally, inward-outward, twist-around, fold-up and unfold, multifrequencied pulsations of the vector equilibria. By virtue of these transformations and their accommodating volumetric involvement, the spheres and spaces are interchangeably intertransformative. For instance, each one can be either a convex or a concave asymmetry of the vector equilibrium, as the "jitterbug" has demonstrated (Sec. 460). The vector equilibrium contracts from its maximum isotropic-vector-matrix radius in order to become a sphere. That is how it can be accommodated within the total isotropic-vector-matrix field of reference.

1032.12 As the vector equilibrium's radii contract linearly, in the exact manner of a coil spring contracting, the 24 edges of one-half of all the vector equilibria bend outwardly, becoming arcs of spheres. At the same time, the chords of the other half of all the vector equilibria curve inwardly to produce either concave-faced vector-equilibria spaces between the spheres or to form concave octahedra spaces between the spheres, as in the isotropic-vector-matrix field model (see Illus. 1032.12). Both the spheric aspect of the vector equilibrium and the "space" aspect are consequences of the coil-spring-like contraction and consequent chordal "outward" and "inward" arcing complementation of alternately, omnidirectionally adjacent vector equilibria of the isotropic-vector-matrix field.

1032.13 In a tetrahedron composed of four spheres, the central void is an octahedron with four concave spherical triangular faces and four planar triangular faces with concave edges. This can be described as a concave octahe-

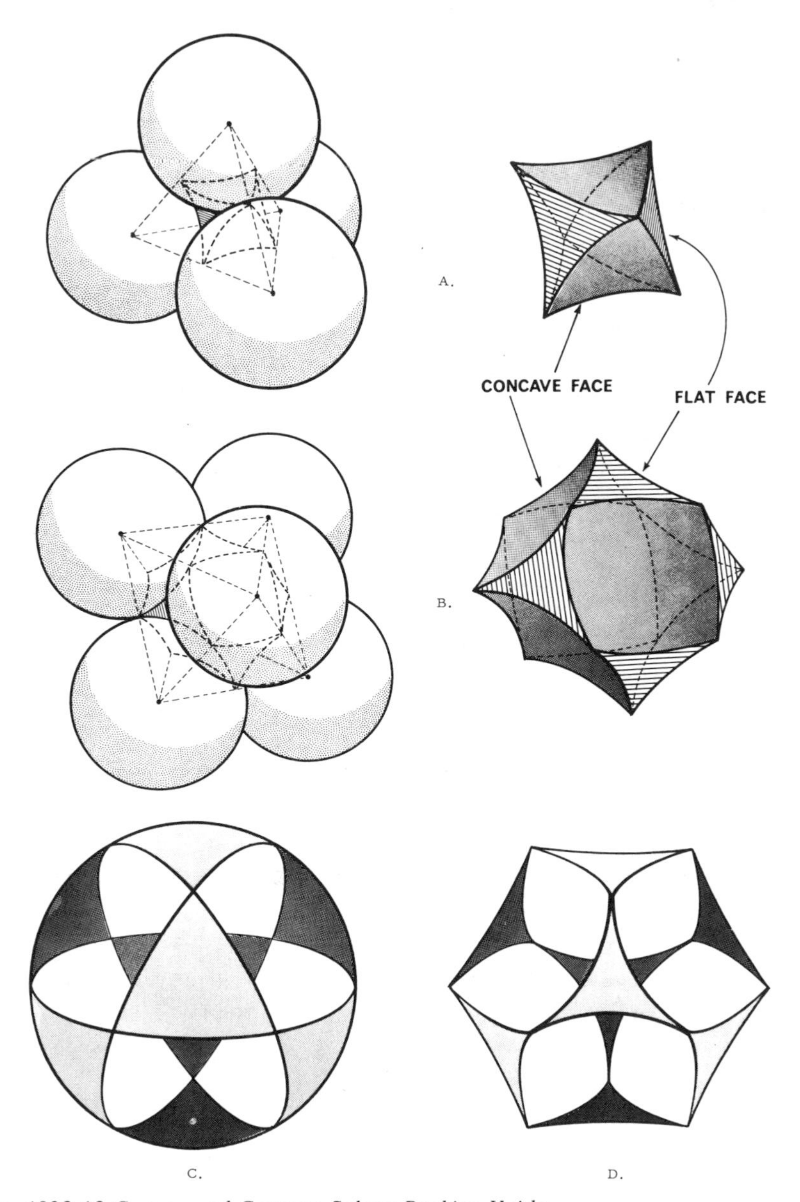

Fig. 1032.12 *Convex and Concave Sphere Packing Voids:*

A. In a tetrahedron composed of four spheres, the central void is an octahedron with four concave spherical triangular faces and four planar triangular faces with concave sides. This can be described as a "concave octahedron."

B. In an octahedron composed of six close-packed spheres, the central void is a vector equilibrium with six concave spherical square faces and eight triangular faces with concave sides: a "concave vector equilibrium."

C. The vector equilibrium with edges arced to form a sphere: a convex vector equilibrium.

D. The vector equilibrium with arcs on the triangular faces defined by spheres tangent at vertexes: a concave vector equilibrium.

dron. In an octahedron composed of six closest-packed spheres, the central void is a vector equilibrium with six concave spherical square faces and eight triangular faces with concave edges: a concave vector equilibrium. The vector equilibrium, with edges arced to form a sphere, may be considered as a convex vector equilibrium. Illus. 1032.12 D shows the vector equilibrium with arcs on the triangular faces defined by spheres tangent at vertexes: a concave vector equilibrium.

1032.20 **Energy Wave Propagation:** The shift between spheres and spaces is accomplished precessionally. You introduce just one energy action—push or pull—into the field, and its inertia provides the reaction to your push or pull; the resultant propagates the everywhere locally sphere-to-space, space-to-sphere omni-intertransformations whose comprehensive synergetic effect in turn propagates an omnidirectional wave. Dropping a stone in the water discloses a planar pattern of precessional wave regeneration. The local unit-energy force articulates an omnidirectional, spherically expanding, four-dimensional counterpart of the planar water waves' circular expansion. The successive waves' curves are seen generating and regenerating and are neither simultaneous nor instantaneous.

1032.21 The only instantaneity is eternity. All temporal (temporary) equilibrium life-time-space phenomena are sequential, complementary, and orderly disequilibrious intertransformations of space-nothingness to time-somethingness, and vice versa. Both space realizations and time realizations are always of orderly asymmetric degrees of discrete magnitudes. The hexagon is an instantaneous, eternal, simultaneous, planar section of equilibrium, wherein all the chords are vectors exactly equal to all the vector radii: six explosively disintegrative, compressively coiled, wavilinear vectors exactly and finitely contained by six chordal, tensively-coil-extended, wavilinear vectors of equal magnitude.

1032.22 Physics thought it had found only two kinds of acceleration: linear and angular. Accelerations are all angular, however, as we have already discovered (Sec. 1009.50). But physics has not been able to coordinate its mathematical models with the omnidirectional complexity of the angular acceleration, so it has used only the linear, three-dimensional, *XYZ,* tic-tac-toe grid in measuring and analyzing its experiments. Trying to analyze the angular accelerations exclusively with straight lines, 90-degree central angles, and no chords involves *pi* (π) and other irrational constants to correct its computations, deprived as they are of conceptual models.

1032.23 Critical-proximity crimping-in of local wave coil-spring contractions of the Little System by the Big System reveals the local radius as always a wavilinear short section of a greater system arc in pure, eternal, generalized principle.

1032.30 **Complementary Allspace Filling of Octahedra and Vector Equilibria:** The closest packing of concave octahedra, concave vector equilibria, and spherical vector equilibria corresponds exactly to the allspace filling of planar octahedra and planar vector equilibria (see Sec. 470). Approximately half of the planar vector equilibria become concave, and the other half become spherical. All of the planar octahedra become concave (see Illus. 1032.30).

1032.31 Concave octahedra and concave vector equilibria close-pack together to define the voids of an array of closest-packed spheres which, in conjunction with the spherical vector equilibria, fill allspace. This array suggests how energy trajectories may be routed over great-circle geodesic arcs from one sphere to another, always passing only through the vertexes of the array—which are the 12 external vertexes of the vector equilibria and the only points where the closest-packed, uniradius spheres touch each other (see Illus. 1032.31).

1040.00 Seven Axes of Symmetry

1041.00 Superficial Poles of Internal Axes

1041.01 There are only three topological axes of crystallography. They are:

Spin of diametrically opposite vertexes Spin of diametrically opposite mid-edges Spin of diametrically opposite centers of face areas	= Three topological types of axes

1042.00 Seven Axes of Symmetry

1042.01 Whatever subdivisions we may make of the tetrahedra, octahedra, and icosahedra, as long as there is cutting on the axes of symmetry, the components always come apart in whole rational numbers, for this is the way in which nature chops herself up.

Fig. 1032.30 *Space Filling of Octahedron and Vector Equilibrium:* The packing of concave octahedra, concave vector equilibria, and spherical vector equilibria corresponds exactly to the space filling of planar octahedra and planar vector equilibria. Exactly half of the planar vector equilibria become convex; the other half and all of the planar octahedra become concave.

Fig. 1032.31 *Concave Octahedra and Concave Vector Equilibria Define Spherical Voids and Energy Trajectories:* "Concave octahedra" and "concave vector equilibria" pack together to define the *voids* of an array of close-packed spheres which in conjunction with the convex spherical vector equilibria fill allspace. This array suggests how energy trajectories may be distributed through great-circle *geodesic* arcs from one sphere to another always passing through the vertexes of the array, which are the vertexes of the vector equilibria and the points where the spheres touch each other.

1042.02 The four sets of unique axes of symmetry of the vector equilibrium, that is, the 12 vertexes with six axes; the 24 mid-edges with 12 axes; and the two different centers of area (a) the eight centers of the eight triangular areas with four axes, and (b) the six centers of the six square areas with three axes—25 axes in all—generate the 25 great circles of the vector equilibrium. These are the first four of the only seven cosmically unique axes of symmetry. All the great circles of rotation of all four of these seven different cosmic axes of symmetry which occur in the vector equilibrium go through all the same 12 vertexes of the vector equilibrium (see Sec. 450).

1042.03 The set of 15 great circles of rotation of the 30 mid-edge-polared axes of the icosahedron, and the set of 10 great circles of rotation of the icosahedron's mid-faces, total 25, which 25 altogether constitute two of the three other cosmic axes of symmetry of the seven-in-all axes of symmetry that go through the 12 vertexes of the icosahedron, which 12 represent the askewedly unique icosahedral rearrangement of the 12 spheres of the vector equilibrium. Only the set of the seventh axis of symmetry, i.e., the 12-vertex-polared set of the icosahedron, go through neither the 12 vertexes of the icosahedron's 12 corner sphere arrangement nor the 12 of the vector equilibrium phase 12-ball arrangement. The set of three axes (that is 12 vertexes, 30 mid-edges, and 20 centers of area) of the icosahedron produce three sets of the total of seven axes of symmetry. They generate the 25 twelve-icosa-vertex-transiting great circles and the six nontransiting great circles for a total of the 31 great circles of the icosahedron. These are the last three of the seven axes of symmetry.

1042.04 We note that the set of four unique axes of symmetry of the vector equilibrium and the fifth and sixth sets of axes of the icosahedron all go through the 12 vertexes representing the 12 spheres either (a) closest-packed around a nuclear sphere in the vector equilibrium, or (b) in their rearrangement without a nuclear sphere in the icosahedron. The six sets of unique cosmic symmetry transit these 12 spherical center corner vertexes of the vector equilibrium and icosahedron; four when the tangential switches of the energy railway tracks of Universe are closed to accommodate that Universe traveling; and two sets of symmetry when the switches are open and the traveling must be confined to cycling the same local icosahedron sphere. This leaves only the seventh symmetry as the one never going through any of those 12 possible sphere-to-sphere tangency railway bridges and can only accommodate local recycling or orbiting of the icosahedron sphere.

1042.05 The seven unique cosmic axes of symmetry describe all of crystallography. They describe the all and only great circles foldable into bow ties, which may be reassembled to produce the seven, great-circle, spherical sets (see Secs. 455 and 457).

Vector Equilibrium		
(squares)	3	
(triangles)	4	
(vertexes)	6	
(midedges)	12	
	25 *	
Icosahedron		
(faces)	10	25 *
(midedges)	15	
(vertexes)	6	
	31	

25
31
56

Axes of Symmetry	
#1	all go through the same 12 vertexes of vector equilibrium and icosahedron
#2	
#3	
#4	
#5	
#6	
#7	go through no vertexes

1050.00 Synergetic Hierarchy

1050.10 Synergy of Synergies: We have the concept of synergy of synergies. Precession is not predicted by mass attraction. Chemical compounds are not prophesied by the atoms. Biological protoplasm is not predicted by the chemical compounds. The design of the elephant or the tree and their unwitting essential respiratory-gas conversion interexchanging is not predicted by the protoplasm. There is nothing about an elephant that predicts islanded star galaxies. As we get into larger and larger systems, the total system is never predicted by its lesser system's components.

1050.11 We know that there is DNA and RNA, any one genetic code of which dictates both a species and within it an individual or special-case formulation. DNA-RNA codes do not explain *why* the protoplasm could produce either an elephant, pine tree, or daisy. They elucidate only *how*. What we call *viral steerability* as produced by the DNA-RNA codes is simply our familiar and generalized *angle-frequency design control*.

1050.12 DNA-RNA angle- and frequency-modulated designs are composed exclusively of four unique chemical constituents that operate as guanine and cytosine; and as thymine and adenine: inseparable but reversible tandem pairs. The first pair occur as *GC* or *CG*. The second pair occur as *TA* or

AT. The DNA-RNA codes may be read in any sequence of those constituents, for instance, as *CG* - *CG* - *CG* - *GC* - *TA* - *AT* - *GC* - *TA* - *TA* - *TA* - *AT* - *CG* - *CG* - *GC*, etc. (See Sec. 932.)

1050.13 We know the codes, but we do not know the "how come" of their producing an elephant. The complementarity of the holisticness of these special-case individuals balances out. An elephant does walk. Elephants are successful designs. We have no evidence of biological species that are inherently incomplete designs. In the hierarchy of hierarchies of synergies, Universe is the unpredicted behavior of any of its sublevel synergetics. We must start our synergetic analysis at the level of Universe and thereafter with the known behavior of the greatest whole and the known behavior of some of its parts, then proceed as permitted mathematically to discover its unknown parts. We have the Greek triangle with its known 180 degrees of angle; which together with the knowledge of the magnitude of any two sides and their included angle, or of any two angles and their included side, etc., permit us to discover the magnitude of the balance of the triangle's six parts. Or, using trigonometry, if we know the magnitude of any two parts, we can ferret out the others.

1050.20 **Trigonometry:** The way we were taught in school about fractions leads to inconsistency. We were taught that fractions can be multiplied, divided, added, and subtracted only when the fractions consisted of identical entities. We could not divide three elephants by four oranges. However, trigonometry introduced functions—which are fractions or ratios, e.g., the sine, cosine, tangent, cotangent, and so forth. Contradicting our earlier lessons about fractions, these trigonometric fractions do mix together angles and edges of spherical triangles. This inconsistency could have been avoided by starting our geometry with spherical trigonometry. We would recognize that what we call a great circle arc or "edge" is indeed a central angle of the sphere. We would learn that we have central and external angles. We would spontaneously see that plane geometry derived from solid geometry and is an oversimplification of localized and superficial aspects of systems. This brings us back to angle and frequency modulation, i.e., outward, inward, and circumferentially around, complementary angle and frequency oscillations and pulsations and the congruence of the linear and angular frequency modulations. By teaching children plane geometry before teaching them spherical trigonometry, society became harnessed with a mathematical contradiction wherein trigonometry deliberately ratioed edge *lines* with *angles*—which clearly seemed to be forbidden by arithmetical fractions' law. Single lines are seemingly very different from angles, because angles involve *two* (convergent) *lines*. If, however, instead of starting elementary education with unrealistic, linear, one-dimensional arithmetic; and then going on to two-dimensional plane geometry; and thence to three-dimensional cubes; and thence to spherical trigonometry . . . if we instead start synergetically with whole systems such as spherical trigonometry, we altogether avoid the concept of an edge and instead learn that the arc-defined edges of spherical triangles are the central angles of the sphere; wherefore both the arc edges and corners are angles, ergo ratioable. Now, having both surface *angles* and central *angles,* we discover that spherical trigonometry is always dealing with whole tetrahedra whose interior apexes are always at the center of the spherical system; and three of whose triangular faces are the great-circle plane triangles hidden within the spheric system; and whose fourth triangular face is always the arc-chord surface triangle of the sphere. These central- and surface-angle understandings are fundamental to transformational thinking, which deals with the falling-inward and precessing-outward proclivities.

1051.00 Circumference and Leverage

1051.10 **Complementarity of Circumferential Oscillations and Inward and Outward Pulsations:** We have demonstrated circumferential complementarity, the circumferential twoness of systems such as the northern and southern hemispheres of our Earth. There is also concave inward and convex outward complementarity, inward and outward twoness. As a consequence, there are also circumferential skew oscillations *and* inward and outward pulsations.

1051.20 **Central and External Angles of Systems:** The tetrahedral integrity of internal (central) angles and external (surface) angles of systems permits the integration of the topological and quantum hierarchies. It is exciting that the three internal radii give us three edges of the tetrahedron's six edges; while the arc chords give us the three other of the tetrahedron's six relationships; and the center of the spheric system and the surface triangle's three corner-vertexes give us the four-vertex-events having the inherent six system relationships; which six are our coincidentally six-positive, six-negative, equieconomical vectorial freedoms (see Sec. 537.10). The central angles gives us what we call the chords of the central-angle arcs. Thus all-system-embracing geodesic lines are expressible in angular fractions of whole circles or cycles.

1051.30 **The Circumferential Field:** The inward-outward complementations of the system are represented by great-circle arcs on the system's surface, whose existence is in reality that of the central angles of the system which subtend those external arcs and create the arc cyclic-duration "lengths." Areal definition of the circumferential—ergo, surface—complementations and their oscillations occur as the surface angles at the vertexes of the system's external mapping.

1051.40 **Angular Functionings of Radiation and Gravity:** The differences between the central angles' and surface angles' functionings are identifiable with radiational and gravitational functionings. Radiation identifies with central angles. Radiation is outwardly divergent. Gravity identifies with the three surface angles' convergent closure into the surface triangle's finite perimeter. Gravity is omniembracing and is not focusable. Gravity is Universe-conservingly effective in its circumferential coherence.

1051.50 **Leverage:** The principle of leverage is employed in shears, nutcrackers, and pliers. The longer the lever arms, the more powerful the pressure applied between the internal central angles of the nutcracker's lever arms. We can make an illuminating model of our planet Earth if we think of it as a spherical bundle of nutcrackers with all their fulcrums at the center of the sphere and all the radii of the sphere acting as the lever arms of the pincers. The whole bunch of pincers have a common universal fulcrum at the common center. The farther out we go on the radial lever arms, the less effort is required to squeeze the ends together to exert nutcracking pressure at the center. If we go around the sphere-embracing circumference progressively tying up the ends of the levers together, we find that it takes very little, local, surface effort tensively between any two surface points to build up excruciatingly powerful, central-compression conditions. The bigger the model, the easier it is to tie it up; ever more delicate an exterior web will hold it together.

1051.51 Look at the relative distance of the atom and its outside electron orbit. The atom's electron field may be equivalent to our magnetic field around this Earth. This elucidates the electromagnetic field of Earth as a world-around, circumferential-embracement field operating ephemerally on the outer ends of 4,000-mile-long levers.

1051.52 Identifying the surface-angle chordings with gravity, we comprehend why it is that as we get deeper and deeper within our Earth, with the pressure continually increasing as we get deeper, we see that the increasing gravitational-compression effect is due to the circumferential containment. The external containment web is always getting hold of the outermost ends of the centrally pinching levers. With this leverage effect, the farther out you go, the more advantage you have and the more powerful work you can do with that lever. Leverage effectiveness increases toward the center, ergo the increasing pressure that we identify with gravity. But it has this circumferential aspect.

1051.53 There is a tendency to misinterpret the increasing pressures occurring inwardly of Earth as "deadweight," i.e., only as a radiationally-inward force, but it must be realized that the "weight" is omnidirectional compression. The gravitational intermass-attraction is progressively augmented, as we go radially outward, by the circumferential mass-interattraction of the relative abundance of elemental atoms, which increases at the second-power rate of the radial-distance outwardly from the Earth's center; and as the pressures bring about ever closer presence of the atoms to one another, there is also an additional second-power exponential gain which results in r^2 varying as $proximity^2 = P^4$, where P = relative compressive force. The surface chordal-angle magnitudes multiplied by radius to the second power produce the relative magnitude of network leverage-advantage resulting in the relative increase in pressure as you go inward toward Earth's center. This is exciting because we now comprehend that gravity is a circumferentially operative force and not a radial force, with precession bringing about the 90-degreeness.

1051.54 Remembering Newton's law of gravity, wherein the relative interattractions are directly proportional to the product of the masses increased by the second power of the distances between the respective mass centers, we realize that doubling the size of a sphere brings about an eightfold multiplication of the circumferential mass-interattraction. In effect, we have a network of chordal cables tensively intertriangulating the progressively outmost ends of the spherical nutcracker bundle with circumferential turnbuckles continually tightening the surface-triangulated tensional embracement network. This means that the pressures being exerted internally are proportional to the fourth power of the relative radial depth inward of Earth's surface.

1051.55 The surface-embracement leverage-advantage of the sphere operating at the fourth power can always overmatch the total volumetric gaining rate as only the third power of radial (frequency), linear gain, as the second-power interproximity attractiveness is further multiplied by the second-power, radial-lever-arm, advantage gains.

1052.00 Universal Integrity

1052.10 Second-Power Congruence of Gravitational and Radiational Constants: The relative mass-energy content magnitude of a polyhedral system is arrived at by multiplying the primitive, frequency-zero, a-priori-state volume (relative to the tetrahedron-equals-one) of the geometric, concentric, structural system's hierarchy, by the second power of the (both minimum and maximum) limit linear velocity of all classes of radiation when unfettered in a vacuum; i.e., multiplying initial volume by the terminal rate at which a spherical wave's outermost, unique-event-distinguishability progressively and omniexpansively occurs, as expressed in terms of the second power of relative frequency of modular subdivision of its initially-occurring, polyhedral system's radius; ergo as manifest in Einstein's equation $E = Mc^2$. Energy equals a given mass with its relative mass-energy compactedness tighteningly modified by the velocity of energy-as-radiation intertransformability potential (not just linearly, but omnidirectionally); ergo as a potentially ever-expansively enlarging spherical wave's outermost-event, one-radial-wavelength-deep surface; ergo second power of system frequency (because wave surfaces grow omni-outwardly as of the second power of the radial, linear frequency) rate of gain. (See Secs. 231.01, 251.05, 529.03 and 541.)

1052.20 **Spherical Field:** As already discovered (see Sec. 964), physics' discovery of universally-multifrequenced, periodic-event-discontinuity *outness* (in complementation to equally frequenced, event-occurrence *in-ness*) is inherent in the always-experientially-verifiable, wave-duration frequency, photon-quantum phenomena; wherefore synergetics had to redefine both volumes and surfaces in terms of *dense* (high-frequency) aggregates of only pointally-positionable, energy events' geometrical formulations, with spherical "surfaces" being in operational reality a dense, outermost, single-photon-thick, "cloud" layer, everywhere approximately equidistant in all directions from one approximately-locatable event center. For this reason the second-power exponential rate of area gain is not to be identified with a continuum, i.e., with a continuous system, but only with the high-frequency outermost layer population aggregate of energy-event points. With numbers of photons and wave frequency per primitive volume, the relative concentration of given masses are determinable.

1052.21 Isaac Newton discovered the celestial gravitation interrelationship and expressed it in terms of the second power of the relative distance between the different masses as determined by reference to the radius of one of the interattracted masses. The gravitational relationship is also synergetically statable in terms of the second power of relative frequency of volumetric quanta concentrations of the respectively interattracted masses. Newton's gravitational constant is a radially (frequency) measured rate of spherical surface contraction, while Einstein's radiational constant is a radial (frequency) rate of spherical expansion. (See Secs. 960.12, 1009.31 and 1052.44.)

1052.30 **Gravitational Constant: Excess of One Great Circle over Edge Vectors in Vector Equilibrium and Icosahedron:** Pondering on Einstein's last problem of the Unified Field Theory, in which he sought to identify and explain the mathematical differentiations between electromagnetics and gravity—the two prime attractive forces of Universe—and recalling in that connection the conclusion of synergetics that gravity operates in spherical embracement, not by direct radial vectors, and recalling that electromagnetics follows the high-tension convex surfaces, possibly the great-circle trunk system of railroad tracks (see Secs. 452 and 458); led to pondering, in surprise, over the fact that the vector equilibrium, which identifies the gravitational behaviors, discloses 25 great circles for the vector equilibrium in respect to its 24 external vector edges, and the icosahedron, which identifies the electron behaviors of electromagnetics, discloses 31 great circles in respect to its 30 external vector edges.

1052.31 In each case, there is an excess of one great circle over the edge vectors. Recalling that the vector edges of the vector equilibrium exactly equal the radial explosive forces, while the icosahedron's 30 external edges are longer and more powerful than its 30 radial vectors, yet each has an excess of one great circle, which great circles must have two polar axes of spin, we encounter once more the *excess two* polar vertexes characterizing all topological systems, and witness the excess of embracingly cohering forces in contradistinction to the explosively disintegrative forces of Universe.

1052.40 **Vector Equilibrium and Icosahedron: Ratio of Gravitational and Electromagnetic Constants:** The vector equilibrium and the icosahedron are the same initial twentyness. But the icosahedron is always in either a positive limit or a negative limit phase of its, only-pulsatingly attained, first-degree structural self-stabilization in the asymmetric transformation of the vector equilibrium, which alternating pulsations are propagated by the eternally opposed, radiant-attractive, always dualistic, inter-self-transformable potential of ideally conceptual unity of Universe.

1052.41 The icosahedral phase of self-structuring is identifiable uniquely with the electron, whose mass relationship to the proton is as 1 : 18.51, whereas the icosahedron's volume is to the vector equilibrium's volume as 20 : 18.51. In this connection it is significant that the vector equilibrium's plural unity is 20, ergo we may say the relationship is as unity : 18.51.

1052.42 The number of icosahedral electrons is always equal to the number of protons that are in the vector equilibrium's idealized form of the same surface layer phenomenon.

1052.43 The nonnucleated icosahedron can and does maintain only one single, one-wave-deep, external layer of omnicircumferentially, omni-intertriangularly tangent, closest-packed, unit-radius, spherically conformed, energy-event packages; while the vector equilibrium is both radially and omnicircumferentially, omnitriangularly closest packed, i.e., in maximum, intertangential, mass-interattractiveness nucleated concentration.

1052.44 Reminiscent of electron proclivities, the icosahedron displays the same surface number of spherically conformed, energy-event packages and its only-one-wavelength-deep, single, outer sphere layer array is omnitriangulated, while the vector equilibrium's surface is arrayed two-fifths in triangulation and three-fifths in open, unstable, square tangency. As spherical agglomerates decrease in radius—as do the vector equilibria's contract to the icosahedral phase—their sphere centers approach one another, and Newton's mass-interattraction law, which shows a second-power gain as the interproximities are halved, imposes an intercoherence condition whereby as their overall system radius decreases, their circumferential mass-interattractions increase exponentially as r^2, where r = radius of the system. (See Sec. 1052.21.)

1053.00 Superficial and Volumetric Hierarchies

1053.10 **Spherical Triangular Grid Tiles:** The interrelationship of the vector equilibrium and the icosahedron when their respective 25- and 31-great-circle grids are superimposed on one another, with the center of area of the vector equilibrium's eight spherical triangles congruent with the areal centers of eight of the icosahedron's 20 spherical triangles, reveals a fundamental, asymmetrical, six-axis, alternative, impulsive-pulsative potential of surface intertransformabilities in respect to which the vector equilibrium serves as the zero between the positive and negative, "relative" asymmetry, deviations.

1053.11 The vector equilibrium's 25 and the icosahedron's 31 spherical-great-circle grids manifest different least-common-denominator, identically angled, spherical triangular "tiles," which together exactly cover and subdivide the spherical surface in whole even numbers of tiles; the vector equilibrium having 48 such LCD triangles and the icosahedron having $2^1/_2$ times as many LCD triangles, i.e., 120.

1053.12 The fundamental fiveness of the icosahedron is split two ways, with $2^1/_2$ going one way (the outside-out way) and $2^1/_2$ going the other way (the inside-out way). The least-common-denominator triangular surface subdivision of the vector equilibrium's sphere provides 48 angularly identical (24 inside-out and 24 outside-out) subdivisions as spherical surface "tiles" that exactly cover one sphere.

1053.13 $^{120}/_{48} = 2^1/_2$; and there are always both the four *positively* skew-rotated and the four *negatively* skew-rotated sets of spherical triangles (two sets of four each), symmetrically borrowed from among the spherical total of 20 equiangled, spherical triangles of each of two spherical icosahedra (each of radius 1)—which four out of 20 ($^{20}/_4 = 5$) spherical icosahedron's triangles' centers of area are exactly concentrically registerable upon every other one of the spherical octahedron's eight triangles, which areal centers of the octahedron's eight triangles are also always concentrically and symmetrically in register with the eight equiangled, spherical triangles of the spherical vector equilibrium when the octahedron and the vector equilibrium spheres are all of the same unity-1 radius. With this registration of four out of eight centers of the icosahedron upon the octahedron–vector equilibrium's eight triangular surfaces each, we find that one icosa set of four skews rotationally positive, while the set of four from another icosahedron phase registers the negative skew rotation, which is a +30 degrees or −30 degrees circumferentially-away-from-zero, rotational askewness for a total of 60 degrees differential between the extremes of both. The remaining 16 out of the total of 20 triangles of each of the two different (plus-or-minus-30-degree) phase icosahedra, subdivide themselves in four sets of four each, each of which sets of four arrange themselves in polarized symmetry upon each of the octahedron's four other spherical triangles which are not concentrically occupied by either the positively- or negatively-skew, concentric sets, of four each, triangles, neither of which four sets of four each non-triangularly-concentric sets repeat the other sets' complementary, asymmetric but polarized, array in superimposition upon the octahedron's four nonconcentrically occupied triangles.

1053.14 It was in discovering this alternate, concentric askewness of icosa-upon-octa, however, that we also learned that the symmetrical, equiangular, spherical triangle areas, filled evenly—but rotationally askew—with sets of 15 of the icosahedron's 120 LCD triangles, exactly registered with the spherical surface area of one of the spherical octahedron's eight triangular faces (each of which are bound by 90-degree corners and 90-degree arc edges). This meant, however, that the 15 LCD icosa triangles' plusly-rotated askew phases are not congruent with one another but are superimposed in alternately askewed arrays, both in the cases of the four concentric triangles and in the cases of the nonconcentrically-registered triangles.

1053.15 Because each of the octahedron's eight faces is subdivided by its respective six sets of spherical "right" triangles (three positive—three negative), whose total of $6 \times 8 = 48$ triangles are the 48 LCD's vector-equilibrium, symmetric-phase triangles, and because $^{120}/_{48} = 2^1/_2$, it means that each of the vector equilibrium's 48 triangles has superimposed upon it $2^1/_2$ positively askew and $2^1/_2$ negatively askew triangles from out of the total inventory of 120 LCD asymmetric triangles of each of the two sets, respectively, of the two alternate phases of the icosahedron's limit of rotational aberrating of the vector equilibrium. This $2^1/_2$ positive superimposed upon the $2^1/_2$ negative, 120-LCD picture is somewhat like a Picasso duo-face painting with half a front view superimposed upon half a side view. It is then in transforming from a positive two-and-one-halfness to a negative two-and-one-halfness that the intertransformable vector-equilibrium-to-icosahedron, icosahedron-to-vector-equilibrium, equilibrious-to-disequilibriousness attains sumtotally and only dynamically a spherical *fiveness* (see Illus. 982.61 in color section).

1053.16 This half-in-the-physical, half-in-the-metaphysical; i.e., half-conceptual, half-nonconceptual; i.e., now you see it, now you don't—and repeat, behavior is characteristic of synergetics with its nuclear sphere being both concave and convex simultaneously, which elucidates the microcosmic, turn-around limit of Universe as does the c^2 the spherical-wave-terminal-limit velocity of outwardness elucidate the turn-around-and-return limit of the macrocosm.

1053.17 This containment of somethingness by uncontained nothingness: this split personality $+2^1/_2$, $-2^1/_2$; $+5$, -5, $+0$, -0; plural unity: this multiplicative twoness and additive twoness of unity; this circumferential-radial; this birth-death, birth-death; physical-metaphysical, physical-metaphysical; yes-no, yes-no-ness; oscillating-pulsating geometrical intertransformability field; Boltzmann importing-exporting elucidates the a priori nature of the associative-disassociative; entropic-syntropic; energetic-synergetic inherency of cosmic discontinuity with its ever locally renewable cyclic continuities, wherewith Universe guarantees the eternally regenerative scenario integrity.

1053.20 **Platonic Polyhedra:** There are 48 spherical triangular tiles of the vector equilibrium nuclear sphere, which 48 triangles' pattern can be symmetrically subdivided into five different sets of symmetrical interpatterning which coincide exactly with the projection outward onto a sphere of the five omnisymmetrical planar-defined Platonic polyhedra, whose linear edges are outlined by the respective chords of the congruent vector equilibrium's symmetrical 25-great-circle grid and the icosahedron's 31-great-circle grid. These equiedged Platonic solids are the icosahedron, the octahedron, the cube, the tetrahedron, and the regular dodecahedron. (The vector equilibrium is one of the Archimedean polyhedra; it was called *cuboctahedron* by the Greeks.)

1053.21 The chords of these five spherical geometric integrities all interact to produce those well-known equiedged polyhedra commonly associated with Plato. The intervolumetric quantation of these five polyhedra is demonstrated as rational when referenced to the tetrahedron as unity. Their surface values can also be rationally quantized in reverse order of magnitude by the 48 spherical triangle tiles in whole, low-order, even numbers. These hierarchies are a discovery of synergetic geometry.

1053.30 **LCD Superficial Quantation of Systems:** Because the icosahedron's 31-great-circle grid discloses 120 least-common-denominator, spherical triangular, whole tiling units, we require a special-case, least-common-surface-denominator identity as a name for the 48 spherical tiles of the vector equilibrium. The 120 spherical surface triangular tiles (60 inside-out and 60 outside-out) do indeed constitute the least-common-spherical and planar polyhedra's whole-surface denominators, ergo LCDs, of all closed systems; for all systems are either simplex (atomic) or complex (molecular) manifests of polyhedra. All systems, symmetrical or asymmetrical, have fundamental insideness (micro) and outsideness (macro) irrelevancies that leave the residual-system relevancies accountable as topological characteristics of the polyhedra.

1053.31 As we have learned elsewhere, the sphere, as demonstrated by the spherical icosahedral subdivisions, discloses a different least-common-denominator spherical subdivision in which there are 120 such tiles (60 positive and 60 negative), which are generalizable mathematically as the least-common surface denominator of surface unity, ergo, of systems in general superficially quantated. Because the icosahedron provides the maximum asymmetries into which the vector equilibrium's universally zero-balanced surface can be transformed, and since the effect of the icosahedron—which introduces the prime number five into Universe systems—is one of transforming, or splitting, equilibrium two ways, we find time after time that the interrelationship of the vector equilibrium and the icosahedron surfaces to be one such elegant manifestation of the number $2^1/_2$—$2^1/_2$ positive and $2^1/_2$ negative, of which the icosahedron's fiveness consists. This half-positive and half-negative dichotomization of systems is the counterpart in pure principle of the nuclear accounting that finds that the innermost ball of the closest-packed symmetrical aggregate always belongs half to a positive world and half to a negative world; that is, the inbound half (implosive) and the outbound half (explosive) altogether make a kinetically regenerative whole centrality that never belongs completely to either world.

1053.32 It is a condition analogous to the sphere with its always and only complementarity of insideness and outsideness, convexity and concavity. A sphere may be thought of as half concave and half convex as well as having two different poles.

1053.33 For the moment, considering particularly spherical-system surfaces, we find the same $2^1/_2$-ness relationship existing between the vector equilibrium and the icosahedron, with their respective least common denominator's surface triangle building tiles (of which the vector equilibrium's 48 LCDs have five of the equiedged Platonic solids and the icosahedron's 120 LCDs have two of the equiedged Platonic solids). The icosahedron-coexistent pentagonal dodecahedron is the special-case system of domains of the icosahedron's 12 vertexes; it is not a structure in its own right. Plato's five omniequifaceted, equiedged and -vertexed, "solids" were the cube, tetrahedron, octahedron, icosahedron, and dodecahedron. All five of these solids are rationally accounted by the LCD spherical surface triangular tilings of the vector equilibrium and the icosahedron.

1053.34 The icosahedron has 120 triangles (60+, 60−), which are the least common denominators of spherical surface unity of Universe; ergo, so important as to have generated, for instance, the ancient Babylonians' adoption of 60 both for increments of time and for circular mensuration. The Babylonians attempted to establish a comprehensive coordinate mensural system that integrated time and matter. Their artifacts show that they had discovered the 60 positive and 60 negative, 120 spherical right triangles of spheres. That their sixtyness did not uncover nature's own rational coordinate system should not be permitted to obscure the fact that the Babylonians were initiating their thinking systematically in polyhedral spherical wholeness and in 60-degree vs. 90-degree coordination, which was not characteristic of the geometrical exploration of a later date by the Egyptians and Greeks, who started very locally with lines, perpendiculars, and planes.

1053.35 The great $2^1/_2$ transformation relations between the vector equilibrium and the icosahedron once again manifest in surface equanimity as the LCD surface triangular tiling, which is *$2^1/_2$ times 48,* or 120.

1053.40 **Superficial Hierarchy:** We have here a total spherical surface subdivisioning hierarchy predicated upon (a) the relative number of LCD ($48/n$) tiles necessary to define each of the following's surface triangles, wherein the tetrahedron requires 12; the octahedron 6; cube 8; and rhombic dodecahedron 4; in contradistinction to (b) their respective volumetric quantations expressed in the terms of the planar-faceted tetrahedron as unity.

1053.41 **Table: Spherical Surface Hierarchy**

Number of Spherical LCD (48 VE) Tiles	*Spherical Conformation*		*Nuclear Sphere's Radius-1 Volumetric Hierarchy*
48	define one Vector Equilibrium sphere	×2 = 96	1
12	define one Tetrahedron face	×2 = 24	3
8	define one Cube face	×2 = 16	4
6	define one Octahedron face	×2 = 12	6
4	define one Rhombic Dodecahedron face	×2 = 8	
4	define one Regular Dodecahedron face	×2 = 8	
$2^1/_2$	define one Icosahedron face	×2 = 5	

1053.50 **Volumetric Hierarchy:** With a nuclear sphere of radius-1, the volumetric hierarchy relationship is in reverse magnitude of the superficial hierarchy. In the surface hierarchy, the order of size reverses the volumetric hierarchy, with the tetrahedron being the largest and the rhombic dodecahedron the smallest.

1053.51 **Table: Volumetric Hierarchy:** The space quantum equals the space domain of each closest-packed nuclear sphere:

Space quantum	= 1
Tetrahedron	= 1
Nuclear vector equilibrium	= 2 1/2
Nuclear icosahedron	= 2 1/2
Cube	= 3
Octahedron	= 4 *
Nuclear sphere	= 5
Rhombic dodecahedron	= 6

1053.60 **Reverse Magnitude of Surface vs. Volume:** Returning to our consideration of the reverse magnitude hierarchy of the surface vs. volume, we find that both embrace the same hierarchical sequence and have the same membership list, with the icosahedron and vector equilibrium on one end of the scale and the tetrahedron on the other. The tetrahedron is the smallest omnisymmetrical structural system in Universe. It is structured with three triangles around each vertex; the octahedron has four, and the icosahedron has five triangles around each vertex. We find the octahedron in between, doubling its prime number twoness into volumetric fourness, as is manifest in the great-circle foldability of the octahedron, which always requires two sets of great circles, whereas all the other icosahedron and vector equilibrium 31 and 25 great circles are foldable from single sets of great circles.

1053.61 The reverse magnitudes of the surface vs. volume hierarchy are completely logical in the case of the total surface subdivision starting with system totality. On the other hand, we begin the volumetric quantation hierarchy with the tetrahedron as the volumetric quantum (unit), and in so doing we build from the most common to the least common omnisymmetrical systems of Universe. In this system of biggest systems built of smaller systems, the tetrahedron is the smallest, ergo, most universal. Speaking holistically, the tetrahedron is predominant; all of this is analogous to the smallest chemical element, hydrogen, being the most universally present and plentiful, constitut-

* The octahedron is always double, ergo, its fourness of volume is its prime number manifest of two, which synergetics finds to be unique to the octahedron.

ing the preponderance of the relative abundance of chemical elements in Universe.

1053.62 The tetrahedron can be considered as a whole system or as a constituent of systems in particular. It is the particulate.

1054.00 Relationship of Gibbs to Euler

1054.10 Synergetic Analysis: Euler's topology and Willard Gibbs' phase rule give us synergetic-analysis capability. Euler differentiated all physical Universe into lines, crossings, and areas: the fundamental visual aspects of our experiences having to do with our eyes, radiation frequencies, and conceptual images. Gibbs' phase rule differentiated the physical Universe into liquid, crystalline, and gaseous phases, which are not so much visual as thermal, which is tactile, and which are always characterized by unique whole-number interattractions, i.e., restraints. Conversely, with successive whole-number degrees of freedom, thermal, sonic, or viscosity frequencies are differentiated in respect to their condition within their respective states as well as between those states.

1054.11 Euler's synergetic differentiation and equatingly accomplished reintegration of Universe deals with energy disassociative as radiation; Gibbs deals with energy associative as matter at various thermal stages. Euler's and Gibbs' are two different system aspects or behaviors of Universe. Euler deals with the static, geometrical field aspects of Universe. Gibbs deals with energy associative as matter, and what the degrees of energetic freedom may be within a local physical complex, and what amounts of energy would have to be added locally to bring about other states.

1054.20 Relationship of Gibbs to Euler

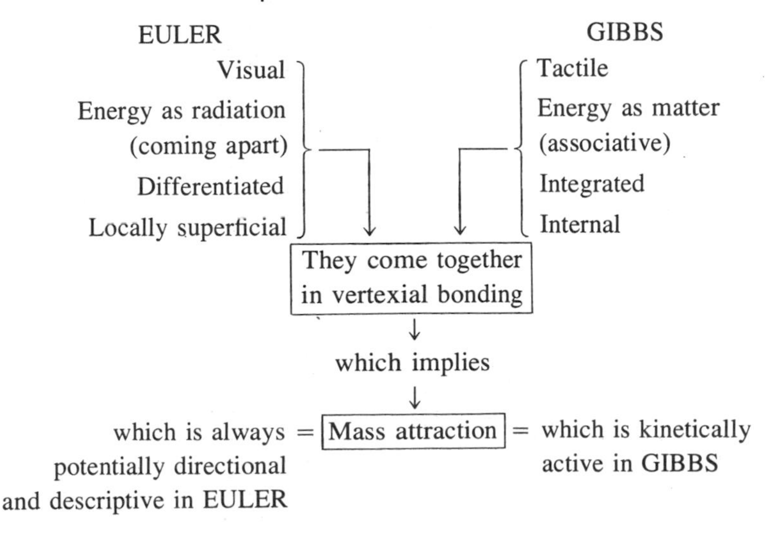

1054.30 **Synergetic Integration of Topology and Quanta:** Synergetics' "breakthrough" integration of Euler's topology and Willard Gibbs' phase rule is explained by the number of intertetrahedral bonds:

Phases:	*Bonds:*		*States:*			
Eccentric	3 bonds	=	Ice →	Face bond tetra	RIGID	
Concentric (medium phase)	2 bonds	=	Water → (medium phase)	Edge bond tetra	FLEX	
Eccentric	1 bond	=	Vapor →	Point bond tetra	FLEX	COMPRESSIVE

Additional bond energies present in the eccentric phases	=	Medium phase + 2
3 + 1, 1 + 3	=	2 + 2

1054.31 The rigid ice stage is characterized by load concentration, no degrees of freedom, and slow creep. The flexible, fluid stage is characterized by hinge-bonding, load distribution, one degree of freedom, and noncompressibility. The flexible, fluid vapor stage is characterized by universal jointing, load distribution, six degrees of freedom, and compressibility.

1054.32 Median unity is two, therefore unity plus two equals four.

Median state = Unity + 2
Frozen state = Median unity − 1 = 1
Vapor state = Median unity + 1 = 3

(3 + 1 = 4; 1 + 3 = 4; 2 + 2 = 4)

Ice = Median freedom minus one freedom
Water = Median freedoms
Vapor = Median freedoms plus one freedom

1054.40 **Topology and Phase** (see Table 1054.40)

1054.50 **Polyhedral Bonding:** Willard Gibbs' phase rule treats with the states of the environment you can sense with your eyes closed: crystallines, liquids, gases, and vapors. Euler's points, lines, and areas are visually described, but they too could be tactilely detected (with or without fingers).

1054.51 The mathematicians get along synergetically using Euler's topology alone. It is the chemists and physicists who cannot predict synergetically without using Gibbs' phase rule.

1054.52 Euler deals with the superficial aspects of polyhedra: of visual conceptuality. He deals only with the convex surfaces of polyhedral systems. Euler deals with unit, integral, single polyhedra, or with their subaspects. He is not concerned with the modus operandi of the associabilities or disassociabilities of a plurality of polyhedra.

1054.53 But Gibbs unknowingly deals with polyhedra that are composited of many polyhedra, i.e., compounds. He does not think or talk about them as polyhedra, but we find the connection between Euler and Gibbs through the polyhedral bonding in respect to Euler's aspects. Euler's lines are double bonds, i.e., hinges. Euler's vertexes are single bonds. Euler's areas are triple bonds. Gibbs accommodates the omnidirectional system complementations of the other senses—thermal, tactile, aural, and olfactory—not just associatively, but radiationally. Gibbs brings in time. Time is tactile. Time is frequency. Our pulses measure its passing.

1054.54 People see things move only relative to other things and feel small vibrations when they cannot see motion. The tactile *feels* angular promontories or sinuses with the fingers or body. *Sinus* means "without"—"nothing," invisible, ergo, nonidentified by Euler. The frequencies we call heat are tactilely sensed. We have radiation-frequency tunability range. Our skin structuring is tuned to frequencies beyond the eye-tunable range, i.e., to ultraviolet and infrared.

1054.55 Euler did not anticipate Gibbs. Gibbs complements Euler—as does synergetics' identification of the two excess vertexes as constituting the axis of conceptual observation in respect to all independent, individual orientations of all systems and subsystems; i.e., quantum mechanics' abstract, nonspinnable "spin."

1054.56 We find Euler and Gibbs coming together in the *vertexial* bonds, or polyhedral "corners," or point convergency of polyhedral lines. The bonds have nothing to do with the "faces" and "edges" they terminally define. Two bonds provide the hinge, which is an edge bonding. One bond gives a universal joint. Triple or areal bonding gives rigidity.

	Inherent Qualities			Old Equation	New Equation	New Equation	New Equation		Prime Numbers	Relative Abundance
	Vertexes	Faces	Edges	V + F = E + 2	✱ + F = E	÷ 2 ●	reduced to common factor			✱ + F = E
Tetrahedron	4	4	6	4 + 4 = 6 + 2	2 + 4 = 6	1 + 2 = 3	1 (1+2=3)		1	
Octahedron	6	8	12	6 + 8 =12 + 2	4 + 8 =12	2 + 4 = 6	2 (1+2=3)		2	
Cube	8	12	18	8 +12 =18 + 2	6 +12 =18	3 + 6 = 9	3 (1+2=3)	ρ^2 or ς^2 ·	3	· ● · [1 + 2 = 3] + ☯
Icosahedron	12	20	30	12+20 =30 + 2	10+20 =30	5 +10 =15	5 (1+2=3)		5	
Vect. Equilib.	12	20	30	12+20 =30 + 2	10+20=30	5 +10 =15	5 (1+2=3)		5	

DEFINITIONS:

✱ Number of points (vertexes) other than those on poles = (V-2=✱) = non-polar vertexes.
☯ Polarity Constant that modifies all systems under consideration, additive twoness.
● Zonality Constant (Zone of Tunability), multiplying twoness.
V Number of vertexes.
F Number of faces.
E Number of edges.
ς Frequency - Modular breakdown.
ρ Wave length.

Gibbs' Phase Rule: F = C + 2 - P

where: F = Degrees of Freedom, i.e. number of variables.
C = Number of Chemical Components.
P = Phases of the System.
2 = Constant.

The phase rule is an equation for determining the number of possible degrees of freedom (variables) that can be given arbitrary values in a system in equilibrium without upsetting the equilibrium. For example in a system consistiing of ice, water. and water vapor, there are three phases: vapor, liquid, and crystalline; and one component: water. Therefore: F = 0. The three phases of water can coexist in equilibrium at a fixed temperature and pressure only, there are no degrees of freedom.

Equivalents	Single-bonded	Double-bonded	Triple-bonded
Phase	gas	liquid	crystalline
Bonds (vertexial)	single	double	triple
Connection	pin	hinge	fix
Inherent Qualities	vertex	edge	face

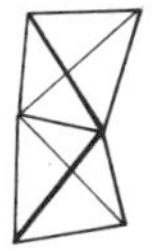

Fig. 1054.40

1054.57 Mass-interattraction is always involved in bonding. You may not have a bond without interattraction, mass or magnetic (integral or induced), all of which are precessional effects. As Sun's pull on Earth produces Earth orbiting, orbiting electrons produce directional field pulls. This was not considered by Euler because he was dealing only with aspects of a single system.

1054.58 Gibbs requires the mass-interattraction without saying so. Mass-interattraction is necessary to produce a bond. Gases may be tetrahedrally bonded singly, corner to corner, or as a universal joint. Gibbs does not say this. But I do.

1054.60 **Orbit as Normal:** Ninety-nine point nine-nine plus percent of the bodies in motion in physical Universe are operating orbitally; therefore interyielding normally; i.e., at 90 degrees to the direction of the applied force.

1054.61 The rare special case of critical proximity, where bodies converge due to the extreme disparity of relative mass magnitude, happens also to be the rare special case in Universe wherein humans happen to exist, being thereby conditioned to think of the special-case exceptional as ''normal,'' thus to misapprehend the normal general behavior. The misapprehension regards the 99.99 percent normal orbital as being strangely perverse. There is much evidence within the critical-proximity environment that demonstrates the normal 90-degree, precessional resultants—as, for instance, when a rope is tensed and reacting at 90 degrees to the direction of the tensing and thus becoming tauter.

1054.70 **Time as Frequency:** The Babylonians tried unsuccessfully to reconcile and coordinate time and space with circular-arc degrees, minutes, and seconds. The *XYZ*, $c.g_t.s.$ metric system accounted time as an exponent. Time was not a unique dimension. It was a uniquely qualifying increment of experience, of obvious existence.

1054.71 Synergetics is the first to introduce the time dimension integrally as the frequency of systems, which initially are metaphysically independent of time and size but, when physically realized, have both time and size, which are identified in synergetics as the frequency of the system: the modular subdividing of the primitive, metaphysical, timeless system.

1054.72 You cannot have time without growthability, which implicitly has a nucleus from which to grow. We would not have discovered the frequency or time dimensions had we not explored the expansiveness-contractiveness and radiational-gravitational behavior of nuclei in pure metaphysical sizeless and timeless principle.

1054.73 It follows that the isotropic-vector-matrix field discovery represents the frame of reference through which all the interpulsating transformations of time realizations transit, but which will never be directly witnessable in the eternally instant static state.

1054.74 Synergetics is an integration of the frequency of Gibbs with the timelessness of Euler. In Table 223.64, Columns 7, 8, and 9 represent the metaphysical timelessness of Euler; Columns 13, 14, and 15 represent the physical-in-time of Gibbs, the thermal, acoustical, sensorial characteristics that are expressible only as frequency.

1055.00 Twentyfoldness of Amino Acid System Indestructibility

1055.01 Return to the Shell of Homogenized Contents of an Egg: There are 20 amino acids, and they can all be made in the laboratory. They always reorganize themselves in geodesic tensegrity patterns. That's why you can pull all of the liquid out of an egg through a tiny needlelike hole, homogenize the contents, and then put it back in the shell, and the embryo will reorganize itself—even after the embryo chick is a week old and has started to form. The amino acids themselves do this.

1055.02 In connection with the 20-amino-acid system's indestructibility, we intuitively sense the necessity to consider the possible interrelations of all of the 20 amino acids' indestructible pattern integrities with other twentyfoldnesses. The number 20 is particularly significant in a plurality of nature's most elementary aspects.

1055.03 Icosahedral Twentyness: There is, for instance, the minimum twentyfoldness of the icosahedron's 20 equiangular, triangular (ergo, structural) facets, which constitute the highest common unit-angle, unit-edge, and unit-vertex structural denominator of universal structural systems. The icosahedron encloses the most volume with the least energy investment as matter or work. Universal limits of eternal abstract principles are indestructible. The discontinuous-compression, continuous-tension, multifrequency geodesic, icosahedral structures are approximately indestructible pattern integrities. They are employed as the protein shells of almost all the viruses. In principle, they are probably involved in the 20 amino acids.

1055.04 Magic Number Twentyness: Then there is the Magic Number twentyness in the relative cosmic abundances of all the atomic-element isotopes, which Magic Numbers we have now identified with mathematical exactitude as constituting a hierarchy of symmetrical, geometrical patterns occurring in mathematical sequence and manifest in the icosahedron-tetrahedron shell-frequency symmetry relationships (see Illus. 995.02).

1055.05 **Vector Equilibrium Twentyness:** Twentyness is significant as the inherent minimum twentyfoldness of the time-space, energy-mass, volume potential of the subfrequency vector equilibrium as quantized by using as unity the geometric volume of the minimum structural system of Universe: the tetrahedron, whose fractional integrity subdivided by the complex of *A* and *B* Module reorientations is in the high order number of magnitude of the aminó acid's interrelationship permutations.

1055.06 **Twentyness in Mass Ratio of Electron and Neutron:** It is relevant in this exploratory speculating to consider that since enzymes are molecular event integrities and involve electron-binding proclivities, this introduces further identification with the fact that the icosahedron's non-closest-packability tends mathematically to be identifiable exclusively with the migrating, trading independence of the electron and its volumetric relationship to the vector equilibrium, i.e., 18.51:20, which is akin to the fractional-number relationship of the electron's mass to the proton's mass.

1055.07 **Twentyness of Maximum Limit Nonnuclear Tetrahedron:** There is another twentyness that seems highly relevant, and that is the twentyness of spherical atoms composing the largest single-shell tetrahedron that can be closest-packingly assembled without a nucleus of its own, which 20-sphered (atomed) tetrahedron has the new potential nucleus to be ''crowned'' when further layers are added; this tetrahedron of 20 occupies each of the eight triangular face regions of the outermost shell of the highest frequency vector equilibrium which is inherently nuclear—that is, it contains only one interior closest-packed sphere. This is its exact volumetric center (see Sec. 414).

1055.08 **Twenty-Sphere Models of DNA-RNA Compounds:** Furthermore, the 20-sphere (atom), closest-packed, non-nucleused tetrahedron consists of five *basic* (because minimum limit) four-ball tetrahedra that, unlike their planar-faceted polyhedral counterpart tetrahedra, can be closest-packingly assembled without octahedral complementation because the octahedra are internal to the four-ball basic tetrahedra. It is further relevant to these considerations that the DNA-RNA code consists always and only of the four chemical compounds—guanine, cytosine, adenine, and thymine—and that the helix that they generate consists entirely of tetrahedra whose four constituents in all vast variety of combinations will always be the same tetrahelixes.

1056.00 *Hierarchy of Generalizations*

1056.01 Epistemology: The more we know the more mysterious it becomes that we can and do know both aught and naught. The number one a priori characteristic of the entirely mysterious *life* is awareness—which develops gradually into comprehension only to become aware of how inherently little we know. But that little we know or may come to know additionally is ever subject to further vast integral exploration, discovery, differentiation, and comprehension.

1056.02 *Nature* is all that we think we do know plus all that we don't know whether or not we know that we don't know. Whatever nature permits is natural. If nature does not permit it, it cannot and does not occur.

1056.03 That there is an a priori unknown is proven by the ever unscheduled, unexpected succession of revelations of additional, theretofore unknown, unconceived-of, generalized principles all of which are discovered and experientially reverifiable as implicit in Universe. It is also retrospectively manifest that this progressively amplifying knowledge, discovered by intuition and mind as constituting eternally operative cosmic relationships, was revealed only because of intuitively pursued, frequent reconsiderations of information complexes redrawn from the ever-recallable special-case experience inventory stored in the humans' brain neuron bank. All that is known emanated exclusively from the previously unknown. (See Sec. 529.21.)

1056.10 Cosmic Hierarchy of Comprehensively Embracing Generalizations

1056.11 = *Integrity:*

The cosmic intellectual integrity manifest by Universe. The orderly interaccommodation of all the generalized principles constitutes a design. Design as a concept of ordered relationships is apprehendable and comprehendable exclusively by intellect. As the human mind progressively draws aside the curtain of unknownness the great design laws of eternally regenerative Universe are disclosed to human intellect. (See Sec. 1056.20, line 38.)

1056.12 = *Synergy:*

The behavior of whole systems unpredicted by behaviors or characteristics of any of the system's parts when assessed separately from the other parts of the system. (See Sec. 1056.20, line 37.)

1056.13 N = Nature:
The totality of both all that is known, *U* (Universe), and all that is unknown, *O*. *N* is the integral of all the integrities always manifest in the progressively discovered generalized eternal principles. (See Sec. 1056.20, line 36.)

1056.14 O = All the Unknown:
The a priori mystery experientially and operationally manifest as a cosmic source by the scientific record of all the *known,* which has always been unpredictedly and successively harvested exclusively from the *a priori unknown,* which nonsimultaneous succession of discoveries thereby discloses that no discovery has as yet exhausted the a priori mysterious exclusive source of all the scientific knowledge—all of which discoveries are always experimentally reverifiable to be forever a priori existent and waiting to be reverified as being eternally coexistent with all the other principles. (See Sec. 1056.20, line 35.)

1056.15 U = Universe: All the Known:
All the thus-far observationally known to exist phenomena. Universe is the aggregate of all of humanity's alltime, consciously apprehended and communicated experiences, including both the explicable and the as-yet unexplained. Communication in this definition can be either self-to-self, or by selves-to-others. It is only by such eternal-generalized-principles-discovering mind's conscious communication to the brain's neuron bank that each generalized-principle-discovering experience becomes an integral special-case asset of humanity's awareness-processing facility. All the foregoing integrate as the *known*. Human awareness first apprehends, then sometimes goes on to comprehend. No guarantees.

1056.20 Cosmic Hierarchy of Comprehensively Embracing and Permeating Generalizations-of-Generalization $= gg^n$

gg	Symbol	Sphere of inclusion		
38.		⌊ COSMIC INTEGRITY		(Sec. 1056.11)
37.		SYNERGY		(Sec. 1056.12)
36.	*N*	NATURE		(Sec. 1056.13)
35.	*O*	UNKNOWN	All that is unknown	(Sec. 1056.14)
34.	*U*	UNIVERSE	All that is known	All known experience (Sec. 1056.15)

gg	Symbol	*Sphere of inclusion.*
33.	*M*	METAPHYSICAL All that is experienceable but weightless, energyless.
32.	*P*	PHYSICAL All the physical is energy. (Note Einstein's $E=Mc^2$ is equivalent to $P=GR^2$)
31.	*G*	SYNTROPY Energy associative as matter precession, gravity, magnetics, interference knotting.
30.	*R*	ENTROPY Energy as radiation, Energy disassociative.
29.	*A*	ASTROPHYSICS The entropic-syntropic, eternally regenerative, synergetical intertransformings of universal evolution.
28.		SOLAR SYSTEM Star systems
27.		EARTH Planetary system in general.
26.		BIOLOGICALS Planetary Biosphere Ecology
25.		HUMANITY Individuals as miniature Universes, each a consequence of unique way of playing the game Universe.
24.		PHILOSOPHY Ideologies, religions, associations.
23.		NATIONS
22.		OTHERS
21.		WE
20.		YOU

gg	Symbol	*Sphere of inclusion*			
19.	Γ̣	THEY			
18.	L̇	ME (intuitive)	Synergetically coordinate sense, intellect. Exploratory sensor, glimpsor, initiator.		
17.	L̇	ME (intellect)	Mathematics, logical conceptioning. Mind discovering and employing eternal principles.		
16.	L̇	ME (sensorial)	Subjective Objective	Brain Neuron	Storing Retrieving Commanding
15.	L̇	ME (memory banked)	Sorted out concepts and data. Booked; libraried, microfiched, computer programmed, interrelated, memory banked around planet, retrieval through satellite relay anywhere.		
14.	L̇	ME (biophysically)	Atomic Physics	Nuclear Structures	
13.	L̇	ME (biochemically)	Exploratory chemistry behavioral proclivities	Atomic compound structures as atomic complexes.	
12.	L̇	ME (scientist)	(Exploratory)	(Science History) Cosmology Cosmogony	
11.	L̇	ME (scientist)	Applied mechanics, structures, electrical and chemical engineering.		
10.	L̇	ME (common sense)	Culture Tribal	Group communication of group sensing, hunting, dancing.	Philosophy needs.
9.	L̇	ME (art)	Individual sense of intuitive communication. Expression of individual philosophy and opinion.		

gg	Symbol	*Sphere of inclusion*			
8.	L̇	ME (incisively disciplined)	Statistics Written Communication Social History	Ideograms Hieroglyphs Phonetics Script.	Accounting Historical Data.
7.	L̇	ME (verbally communicating)			
6.	L̇	ME (gestured communication) human to humans Including smiles, clothing, perfumes, etc.			
5.	L̇	ME (gestured communication) Humans to other creatures articulated			
4.	L̇	ME (gestured communication understood by me)	Animals to humans Yes-no purring—Tail wagging—barking.		
3.	L̇	ME (gestured communication sensed by me)	Nonhuman life to nonhuman life Trees-to-trees Birds-to-trees		
2.	L̇	ME (mute) (communication)	Biologicals to biologicals Thorns—odors —coloring.		
1.	L̇	ME (mute) Gross communication	Stone-to-stone Stone-to-water Thermodynamics, electrolysis, crystallization, erosion.		

L̇ = me ⌉ = others ⌞⌟⌉ = cult

J̇ = you L̇ J̇ = we ⌜⌝⌞⌟ = humanity

Γ̣ = they ⌜⌞⌟ = nation ⨀ = cosmic integrity

⨀ > S > N > O > U > ⌜⌝⌞⌟ > ⌞⌟⌉ > ⌜⌞⌟ > L̇ J̇ > J̇ > L̇

1060.00 Omnisensorial Accommodation

1060.01 The great compressibility of gases is occasioned by the fact that all the tetrahedra are interlinked to one another only by single corners. This is a single bond: it requires the minimum mass-attraction energy of joining. You can fill a very great deal of space with single-bonded tetrahedra; and they are not only highly compressible or infoldable but, being universal-jointed, are most flexible, as are all gases.

1060.02 We will now examine two-bonded associations of tetrahedra. Double-bonding means two mass-attractions. Double-bonds are twice as powerfully cohered and take twice as much energy to disturb their interpatterning. Double-bonding makes a hinge between the tetrahedra. They are, therefore, flexibly interlinked. Forces being applied telegraph throughout the whole system. Both gases and liquids have this property of distributing forces. But whereas single-bonded gases are highly compactible or compressible, double-bonded liquids are noncompressible. If you assemble tetrahedra edge to edge, you cannot compress them any more even though they are flexibly hinged. The coherence of the liquid's viscosity is inherently twice that of the gases.

1060.03 We get even closer inter-mass positioning when there are three-corner bonds (i.e., triangular faces congruent with faces). This produces crystalline rigidity. Crystalline or triple-bonded structuring does not distribute loads as do gases and liquids. Nature designed the triple-bonding to produce the high cohesiveness in tension of crystalline structures. Due to its triple-bonding, the most difficult structure to pull apart is the crystalline.

1061.10 Tree Structure

1061.11 In the structuring of a tree or plant, the crystalline tensions of liquid cell sacs are hydraulically filled in order to distribute the compression and tension loads throughout the whole structure. The hydraulically filled cells of the tree are noncompressible. Thus is the tree capable of holding a five-ton branch out horizontally, due first to the noncompressibility of the liquid content of the cell sacs, and second to the tension being provided at greatest effectiveness by the triple-bond crystalline sac skins. Gases are inserted between the molecules of liquid of the tree's cell sacs. The gases' compressibility provides the compressibility or flexibility of the tree's branches to wave in the wind. If you have ever tried to hold a 25-pound suitcase out horizontally at arm's length, you can appreciate how great a structural task is being performed when a tree's five-ton branches wave yieldingly in the storm without breaking off. You can understand that in an ice storm, the hydraulic content of the tree's cells freezes and can no longer distribute the stresses, and as a consequence during such conditions, many tree branches break off and fall to the ground.

1061.12 We use these combined single-, double-, and triple-bond principles in making the transport airplane's landing gear operate. The pneumatics are in the airplane's rubber tires, and the hydraulics operate as nonfreezing liquids forced through long passageways of the airplane's undercarriage.

1100.00 Triangular Geodesics Transformational Projection

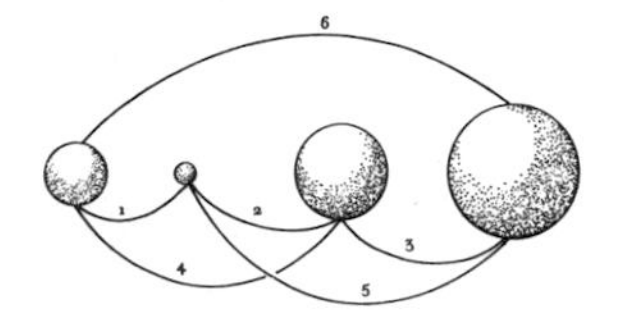

1100.00 TRIANGULAR GEODESICS TRANSFORMATIONAL PROJECTION

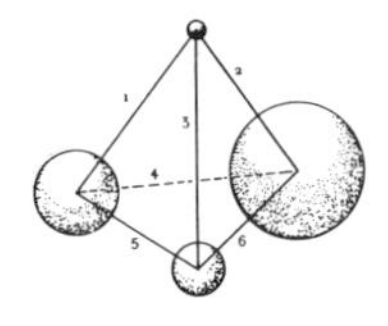

1101.00 Triangular Geodesics Transformational Projection Model

1101.01 Description

1101.02 The transformational projection is contained entirely within a plurality of great-circle-bounded spherical triangles (or quadrangles or multipolygons) of constant, uniform-edge-module (invariant, central-angle-incremented) subdivisioning whose constantly identical edge length permits their hinging into flat mosaic-tile continuities. The planar phase of the transformation permits a variety of hinged-open, completely flat, reorientable, unit-area, world mosaics. The transformational projection model demonstrates how the mosaic tiles migrate zonally. It demonstrates how each tile transforms cooperatively but individually, internally from compound curvature to flat surface without interborder-crossing deformation of the mapping data.

1102.00 Construction of the Model

1102.01 The empirical procedure modeling that demonstrates the transformational projection is constructed as follows:

1102.02 There are three spring-steel straps of equal length, each of which is pierced with rows of holes located at equal intervals, one from the other, along the longitudinal center line of the straps' flat surfaces; the first and last holes are located inward from the ends a distance equal to one-half the width of the steel straps' flat surfaces.

1102.03 Steel rods of equal length are inserted an equal distance through each of the holes in the straps in such a manner that each rod is perpendicular to the parallel surfaces of the strap and therefore parallel to the other rods. Each of the straps and their respective rods form, in effect, a long-toothed comb, with the comb's straight back consisting of the steel strap.

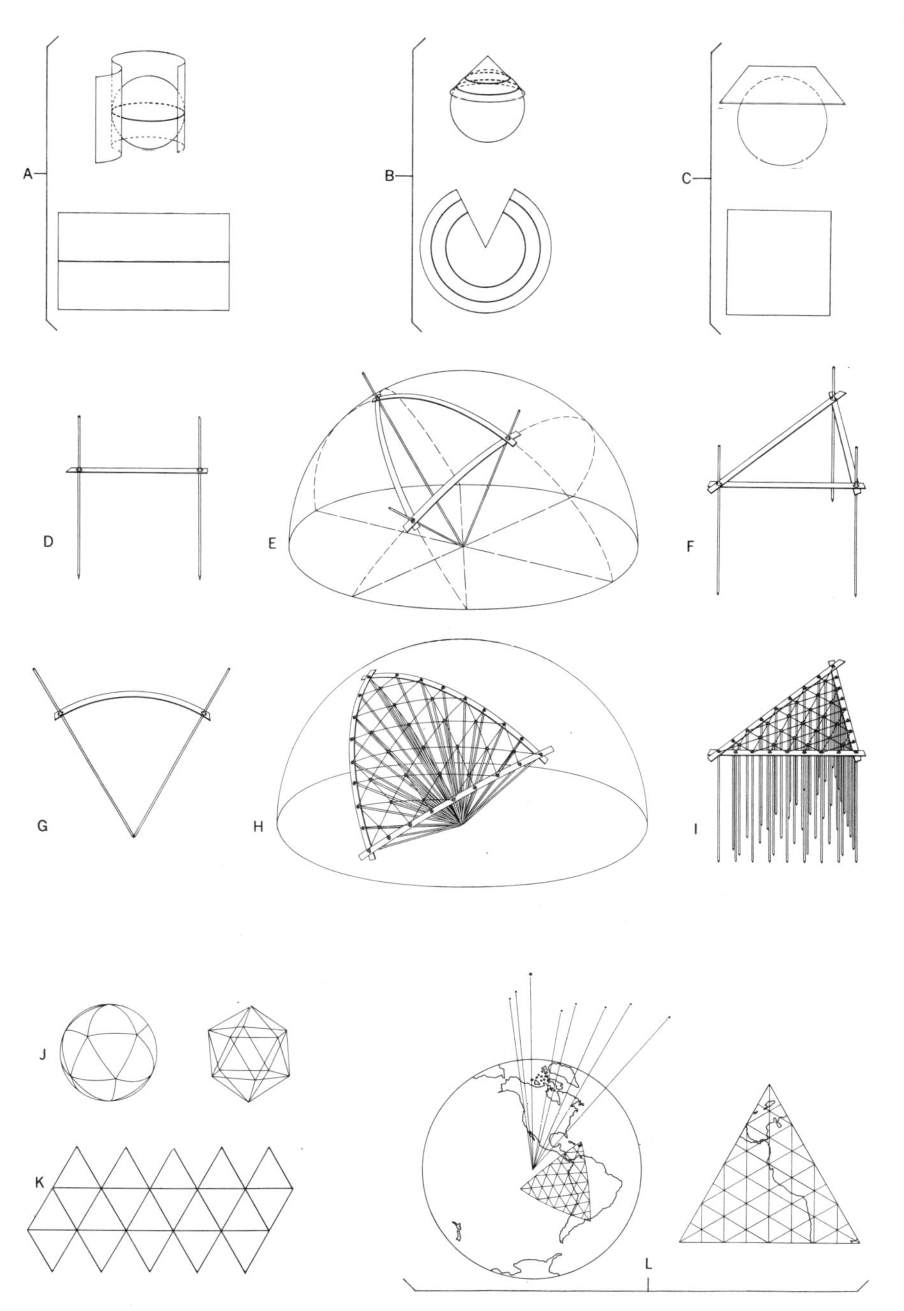

Fig. 1101.02 The projection system of the Dymaxion Airocean World Map divides the sphere into 20 equilateral spherical triangles, which are then flattened to form the icosahedron (J). These 20 triangles are each projected into a flat plane (D, E, F, G, H, I). This method results in a map having less visible distortion than any previously known map projection system to date. To flatten the globe it is simply necessary to "unfold" the icosahedron (K). The more conventional projection systems that are widely used include the Mercator projection (A), the conic projection (B), and the planar projection (C). All three of these systems give rise to considerable visual distortion, unlike the Dymaxion projection.

1102.04 Assuming that the steel straps are flexible, and assuming that each of the rods is absolutely stiff (as when employed as a lever), if the rods have their long lower ends gathered toward one another, at one point the strap will yield by curving of its flat surface to the section of a cylinder whose axis is perpendicular to the plane of the rods and congruent with the rod ends. The strap is bent into an arc of a true circle whose radius is uniformly that of the uniform rod lengths. Each of the rods, as a radius of the circular arc, is perpendicular to the arc. That is, the rods are constantly perpendicular to the strap in either its flat condition or in any of its progressive arcings.

1102.05 Next, one of the two ends of each of the three steel straps is joined to an end of one of the other two straps by means of their end rods being removed and one of the rods being reinserted through their mutual end holes as one strap is superimposed on the other with their respective end holes being brought into register, whereafter, hollow ''stovepipe'' rivets * of complementary inside-outside diameters are fastened through the end holes to provide a journal through which one of the former end rods is now perpendicularly inserted, thus journaled pivotally together like a pair of scissors. The three straps joined through their registered terminal holes form an equilateral triangle of overlapping and rotatably journaled ends. (See Illus. 1101.02 F.)

1102.06 It will next be seen that a set of steel rods of equal length may be inserted an equal distance through each of the holes of each of the straps, including the hollow journaled holes at the ends, in such a manner that each rod is perpendicular to the parallel surfaces of the straps; therefore, each rod is parallel to the others. All of the rods perpendicularly piercing any one of the straps are in a row, and all of their axes are perpendicular to one common plane. The three unique planes of the three rows of rods are perpendicular to each of the straps whose vertical faces form a triangular prism intersecting one another at the central axes of their three corner rods' common hinge extensions. Each of the three planes is parallel to any one rod in each of the other two planes. (See Illus. 1101.02 I.)

1103.00 Flexing of Steel Straps

1103.01 Assuming that each of the rods is absolutely stiff (as when employed as a lever) and that the rotatable journaling in their respective end holes is of such close tolerance that the combined effect of these two qualities of the model is such that any different directions of force applied to any two different rod ends would force the steel straps to yield into circular arc—then it will be clear that the three journaled end rods permit the three corner angles of the triangle to change to satisfy the resulting force or motion differential.

* The rivets resemble hollow, tubelike grommets.

1103.02 If the rods in any one row in any one strap have their ends gathered toward one another, the strap will yield by curving its flat surfaces to the section of a cylinder whose axis is perpendicular to the plane of the rods. If all the rod ends of one strap are pulled together at one point (we refer to the one set of ends on either side of the strap), the strap, being equidistant along its center from that point, will form a segment of a circle, and each of the rods, being radii of that circle, will remain each perpendicular to the strap and all in a single plane perpendicular to the strap throughout the transformation.

1103.03 Now if all the ends of all the rods on one face side or the other of the triangle (since released to its original flat condition of first assembly), and if all of the three rows in the planes perpendicular to each of the three straps forming the triangle are gathered in a common point, then each of the three spring-steel-strap and rod sets will yield in separate arcs, and the three planes of rods perpendicular to them will each rotate around its chordal axis formed between the two outer rivet points of its arc, so that the sections of the planes on the outer side of the chords of the three arcs, forming what is now a constant-length, equiedged (but simultaneously changing from flat to arced equiedged), equiangled (but simultaneously altering corner-angled), spherical triangle, will move toward one another, and the sections of the planes on the inner side of the chords of the three arcs forming the constant, equiedged (but simultaneously changing flat-to-arc equiedged), and equiangled (but simulta-neously altering corner-angled), spherical triangle will rotate away from one another. The point to which all rod ends are gathered will thus become the center of a sphere on the surface of which the three arcs occur, as arcs of great circles—for their planes pass through the center of the same sphere. The sums of the corner angles of the spherical triangles add to more than the 180 degrees of the flat triangle, as do all spherical triangles with the number of degrees and fractions thereof that the spherical triangle is greater than its chorded plane triangle being called the spherical excess, the provision of which excess is shared proportionately in each corner of the spherical triangle; the excess in each corner is provided in our model by the scissorslike angular increase permitted by the pivotal journals at each of the three corners of the steel-strap-edged triangle. (See Illus. 1101.02 H.)

1103.04 The three arcs, therefore, constitute the edges of a spherical equilateral triangle, whose fixed-length steel boundaries are subdivided by the same uniform perimeter scale units of length as when the boundary lines were the "straight" edge components of the flat triangle. Thus we are assured by our model that the original triangle's edge lengths and their submodular divi-sions have not been altered and that the finite closure of the triangle has not been violated despite its transformation from planar to spherical triangles.

1104.00 Constant Zenith of Flat and Spherical Triangles

1104.01 The radii of the sphere also extend outwardly above the surface arcs in equidistance, being perpendicular thereto, and always terminate in zenith points in respect to their respective points of unique penetration through the surface of the sphere.

1104.02 If we now release the rods from their common focus at the center of the sphere and the spring-steel straps return to their normal flatness, all the rods continue in the same perpendicularity to the steel bands throughout the transformation and again become parallel to one another and are grouped in three separate and axially parallel planes. What had been the external spherical zenith points remain in zenith in respect to each rod's point of penetration through the now flat triangle's surface edge. This is an important cartographic property* of the transformational projection, which will become of increasing importance to the future high-speed, world-surface-unified triangulation through aerial and electromagnetic signal mapping, as well as to the spherical world-around data coordination now being harvested through the coordinately ''positioned'' communications satellites ''flying'' in fixed formation with Earth. They and Moon together with Earth co-orbit Sun at 60,000 mph.

1105.00 Minima Transformation

1105.01 If the rods are pushed uniformly through the spring-steel straps so that increasing or decreasing common lengths of rod extend on the side of the triangle where the rods are gathered at a common point, then, as a result, varying ratios of radii length in respect to the fixed steel-strap arc length will occur. The longer the rods, the larger will be the sphere of which they describe a central tetrahedral segment, and the smaller the relative proportional size of the spherical surface triangle bounded by the steel springs—as

* In first contemplating the application of transformational projection for an Earth globe, I realized that the Basic Triangle—120 of which are the lowest common denominator of a sphere—would make a beautiful map. The reason I did not use it was because its sinuses intruded into the continents and there was no possible arrangement to have all the triangles' vertexes occur only in the ocean areas as in the vector equilibrium and icosahedral projection. The Basic LCD Icosahedral Triangle also has a spherical excess of only two degrees per corner, and there would have been no trouble at all to subdivide until the spherical excess for any triangle tile grid was approximately zero. Thus it could have been a 120-Basic-Triangle-grid, at a much higher frequency, but the big detriment was that the spherical trigonometry involved, at that time long before the development of a computer, was so formidable. So the icosahedron was adapted.

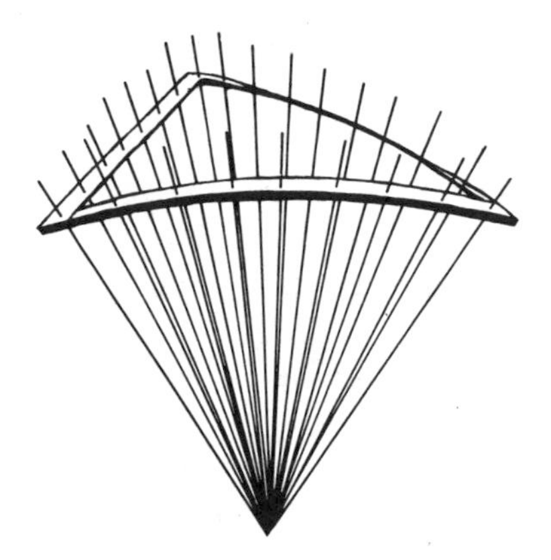

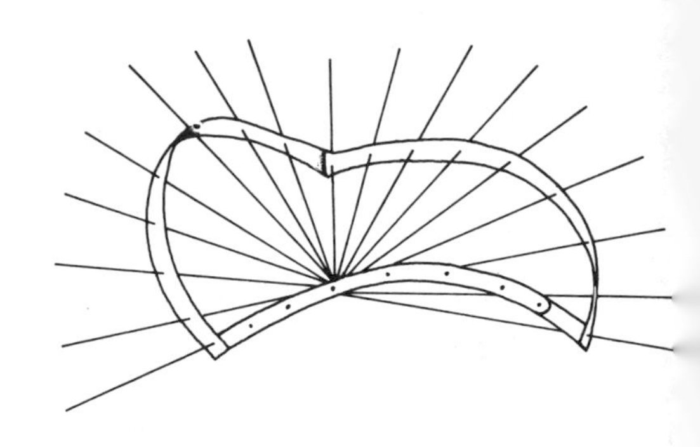

Fig. 1105.01

compared to the whole implicit spherical surface. Because the spherical triangle edge length is not variable, being inherent in the original length of the three identical steel springs, the same overall length can accommodate only an ever smaller spherical surface arc (central-angle subtension) whenever the radii are lengthened to produce a greater sphere.

1105.02 If the ends of the rods gathered together are sufficiently shortened, they will finally attain a minimum length adequate to reach the common point. This minimum is attained when each is the length of the radius of a sphere relative to which the steel spring's length coincides with the length of an arc of 120 degrees. This condition occurs uniquely in a spherical triangle where each of the three vertexes equals 180 degrees and each of the arcs equals 120 degrees, which is of course the description of a single great circle such as the equator.

1105.03 Constituting the *minima transformation* obtainable by this process of gathering of rod ends, it will be seen that the minima is a flat circle with the rods as spokes of its wheel. Obviously, if the spokes are further shortened, they will not reach the hub. Therefore, the minima is not 0—or no sphere at all—but simply the smallest sphere inherent in the original length of the steel springs. At the minima of transformation, the sphere is at its least radius, i.e., smallest volume.

1105.04 As the rods are lengthened again, the implied sphere's radius—ergo, its volume—grows, and, because of the nonyielding length of the outer steel springs, the central angles of the arc decrease, as does also the relative size of the equilateral, equiangular spherical triangle as, with contraction, it approaches one of the poles of the sphere of transformation. The axis running between the two poles of most extreme transformation of the spherical trian-

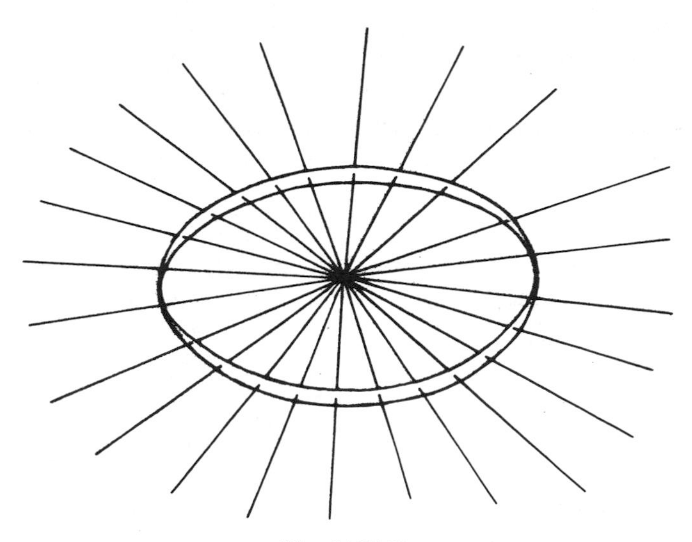

Fig. 1105.03.

gle we are considering runs through all of its transforming triangular centers between its—never attained—minimum-spherical-excess, smallest-conceivable, local, polar triangle on the ever-enlarging sphere, then reversing toward its largest equatorial, three-180-degree-corners, hemisphere—area phase on its smallest sphere, with our triangle thereafter decreasing in relative

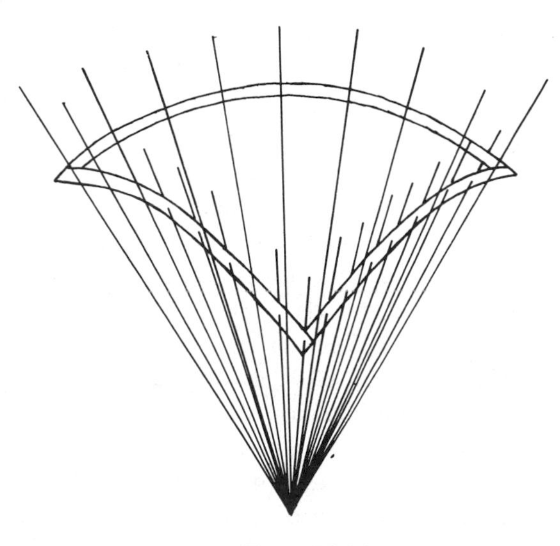

Fig. 1105.04.

spherical surface area as the—never attained—smallest triangle and the sphere itself enlarge toward the—also never attained—cosmically largest sphere. It must be remembered that the triangle gets smaller as it approaches one pole, the complementary triangle around the other pole gets correspondingly larger. It must also be recalled that the surface areas of both the positive and negative complementary spherical triangles together always comprise the whole surface of the sphere on which they co-occur. Both the positive and negative polar-centered triangles are themselves the outer surface triangles of the two complementary tetrahedra whose commonly congruent internal axis is at the center of the same sphere whose total volume is proportionately subdivided between the two tetrahedra.

1106.00 Inside-Outing of Tetrahedron in Transformational Projection Model

1106.10 **Complementary Negative Tetrahedron:** The rod ends can be increased beyond the phase that induced the 180-degree triangle, and the vertexes of the steel-spring surface triangle can go on to be increased beyond 180 degrees each, and thus form a *negative triangle.* This is to say that the original tetrahedron formed between the three vertexes of the spherical triangle on the sphere's surface—with the center of the sphere as the fourth point—will have flattened to one plane when the vertexes are at 180 degrees; at that moment the tetrahedron is a hemisphere. By lengthening the radii again and increasing the triangle's original ''interior'' angles, the tetrahedron will turn itself inside out. In effect, what seems to be a ''small,'' i.e., an only apparently ''plane'' equilateral triangle must always be a small equilateral spherical triangle of a very big sphere, and it is always complemented by the negative triangle completing the balance of the surface of the inherent sphere respective to the three lines and three vertexes of the triangle.

1106.11 No triangular surface is conceivable occurring independently of its inherent sphere, as there is no experimentally demonstrable flat surface plane in Universe reaching outward laterally in all directions to infinity; although this has been illusionarily accepted as ''obvious'' by historical humanity, it is contradictory to experience. The surface of any system must return to itself in all directions and is most economically successful in doing so as an approximate true sphere that contains the most volume with the least surface. Nature always seeks the most economical solutions—ergo, the sphere is normal to all systems experience and to all experiential, i.e., operational consideration and formulation. The construction of a triangle involves a surface, and a curved surface is most economical and experimentally satisfactory. A sphere is a closed surface, a unitary finite surface. Planes are never finite. Once a triangle is constructed on the surface of a sphere—because a triangle is a boundary line closed upon itself—the finitely closed boundary lines of the triangle automatically divide the unit surface of the sphere into two separate surface areas. Both are bounded by the same three great-circle arcs and their three vertexial links: this is the description of a triangle. Therefore, both areas are true triangles, yet with common edge boundaries. It is impossible to construct one triangle alone. In fact, four triangles are inherent to the oversimplified concept of the construction of "one" triangle. In addition to the two complementary convex surface triangles already noted, there must of necessity be two complementary concave triangles appropriate to them and occupying the reverse, or inside, of the spherical surface. Inasmuch as convex and concave are opposites, one reflectively concentrating radiant energy and the other reflectively diffusing such incident radiation; therefore they cannot be the same. Therefore, a minimum of four triangles is always induced when any one triangle is constructed, and which one is the initiator or inducer of the others is irrelevant. The triangle initiator is an inadvertent but inherent tetrahedron producer; it might be on the inside constructing its triangle on some cosmic sphere, or vice versa.

1106.12 It might be argued that inside and outside are the same, but this is not so. While there is an interminable progression of insides within insides in Experience Universe, there is only one outside comprehensive to all insides. So they are not the same, and the mathematical fact remains that four is the minimum of realizable triangles that may be constructed if any are constructed. But that is not all, for it is also experimentally disclosed that not only does the construction of one triangle on the surface of the sphere divide the total surface into two finite areas each of which is bound by three edges and three angles—ergo, by two triangles—but these triangles are on the surface of a system whose unity of volume was thereby divided into two centrally angled tetrahedra, because the shortest lines on sphere surfaces are great circles, and great circles are always formed on the surface of a sphere by planes going through the center of the sphere, which planes of the three-great-circle-arc-edged triangle drawn on the surface automatically divide the whole sphere internally into two spherical tetrahedra, each of which has its four triangles—ergo, inscribing one triangle "gets you eight," like it or not. And each of those eight triangles has its inside and outside, wherefore inscribing one triangle, which is the minimum polygon, like "Open Sesame," inadvertently gets you 16 triangles. And that is not all: the sphere on which you scribed is a system and not the whole Universe, and your scribing a triangle on it to stake out your "little area on Earth" not only became 16 terrestrial triangles but also induced the remainder of Universe outside the system and inside the system to manifest their invisible or nonunitarily conceptual "minimum inventorying'' of ''the rest of Universe other than Earth,'' each of which micro and macro otherness system integrity has induced an external tetrahedron and an internal tetrahedron, each with 16 triangles for a cosmic total of 64 (see Sec. 401.01).

1106.20 **Inside-Outing:** Inside-outing means that any one of the four vertexes of the originally considered tetrahedron formed on the transformational projection model's triangle, with its spherical center, has passed through its opposite face. The minima and the maxima of the spherical equiside and -angle triangle formed by the steel springs is seen to be in negative triangular complement to the smallest 60-degree+ triangle. The vertexes of even the maxima or minima are something greater than 60 degrees each—because no sphere is large enough to be flat—or something less than 300 degrees each.

1106.21 The sphere is at its smallest when the two angles of complement are each 180 degrees on either side of the three-arc boundary, and the minima-maxima of the triangles are halfway out of phase with the occurrence of the minima and maxima of the sphere phases.

1106.22 No sphere large enough for a flat surface to occur is imaginable. This is verified by modern physics' experimentally induced abandonment of the Greeks' definition of a sphere, which absolutely divided Universe into all Universe outside and all Universe inside the sphere, with an absolute surface closure permitting no traffic between the two and making inside self-perpetuating to infinity complex—ergo, the first locally perpetual-motion machine, completely contradicting entropy. Since physics has found no solids or impervious continuums or surfaces, and has found only finitely separate energy quanta, we are compelled operationally to redefine the spheric experience as an aggregate of events approximately equidistant in a high-frequency aggregate in almost all directions from one only approximate event (see Sec. 224.07). Since nature always interrelates in the most economical manner, and since great circles are the shortest distances between points on spheres, and since chords are shorter distances than arcs, then nature must interrelate the spheric aggregated events by the chords, and chords always emerge to converge; ergo, converge convexly around each spheric system vertex; ergo, the sums of the angles around the vertexes of spheric systems never add to 360 degrees. Spheres are high-frequency, geodesic polyhedra (see Sec. 1022.10).

1106.23 Because (a) all radiation has a terminal speed, ergo an inherent limit reach; because (b) the minimum structural system is a tetrahedron; because (c) the unit of energy is the tetrahedron with its six-degrees-of-minimum-freedoms vector edges; because (d) the minimum radiant energy package is one photon; because (e) the minimum polar triangle—and its tetrahedron's contraction—is limited by the maximum reach of its three interior radii edges of its spherical tetrahedron; and because (f) physics discovered experimentally that the photon is the minimum radiation package; therefore we identify the minimum tetrahedron photon as that with radius $= c$, which is the speed of light: the tetrahedron edge of the photon becomes unit radius = frequency limit. (See Sec. 541.30.)

1106.24 The transformational projection model coupled with the spheric experience data prove that a finite minima and a finite maxima do exist, because a flat is exclusively unique to the area confined within a triangle's three points. The *almost flat* occurs at the inflection points between spheric systems' inside-outings and vice versa, as has already been seen at the sphere's minima size; and that at its maxima, the moment of flatness goes beyond approximate flatness as the minima phase satisfies the four-triangle minima momentum of transformation, thus inherently eliminating the paradox of static equilibrium concept of all Universe subdivided into two parts: that inside of a sphere and that outside of it, the first being finite and the latter infinite. The continual transforming from inside out to outside in, finitely, is consistent with dynamic experience.

1106.25 Every great circle plane is inherently two spherical segment tetrahedra of zero altitude, base-to-base.

1106.30 **Inside-Outing of Spheres:** When our model is in its original condition of having its springs all flat (a dynamic approximation) and in one plane, in which condition all the rods are perpendicular to that plane, the rods may be gathered to a point on the opposite side of the spring-steel strap to that of the first gathering, and thus we see the original sphere turned inside out. This occurs as a sphere of second center, which, if time were involved, could be the progressive point of the observer and therefore no "different" point.

1106.31 Considering Universe at minimum unity of two, two spheres could then seem to be inherent in our model. The half-out-of-phaseness of the sphere maxima and minima, with the maxima-minima of the surface triangles, find the second sphere's phase of maxima in coincidence with the first's minima. As the two overlap, the flat phase of the 180-degree triangles of the one sphere's minima phase is the flat phase of the other sphere's maxima. The maxima sphere and the minima sphere, both inside-outing, tend to shuttle on the same polar axis, one of whose smaller polar triangles may become involutional while the other becomes evolutional as the common radii of the two polar tetrahedra refuse convergence at the central sphere. We have learned elsewhere (see Sec. 517) that two or more lines cannot go through the same point at the same time; thus the common radii of the two polar tetrahedra must twistingly avert central convergence, thus accomplishing central core involutional-evolutional, outside-inside-outside, cyclically transformative travel such as is manifest in electromagnetic fields. All of this is implicit in the projection model's transformational phases. There is also disclosed here the possible intertransformative mechanism of the interpulsating binary stars.

1107.00 Transformational Projection Model with Rubber-Band Grid

1107.10 **Construction:** Again returning the model to the condition of approximate dynamic inflection at maxima-minima of the triangle—i.e., to their approximately flat phase of ''one'' most-obvious triangle of flat spring-steel strips—in which condition the rods are all perpendicular to the surface plane of the triangle and are parallel to one another in three vertical planes of rod rows in respect to the triangle's plane. At this phase, we apply a rubber-band grid of three-way crossings. We may consider the rubber bands of ideal uniformity of cross section and chemical composition, in such a manner as to stretch them mildly in leading them across the triangle surface between the points uniformly spaced in rows, along the spring-steel strap's midsurface line through each of which the rods were perpendicularly inserted. The rubber bands are stretched in such a manner that each rubber band leads from a point distant from its respective primary vertex of the triangle to a point on the nearest adjacent edge, that is, the edge diverging from the same nearest vertex, this second point being double the distance along its edge from the vertex that the first taken point is along its first considered edge. Assuming no catenary sag or drift, the ''ideal'' rubber bands of no weight then become the shortest distances between the edge points so described. Every such possible connection is established, and all the tensed, straight rubber bands will lie in one plane because, at the time, the springs are flat—and that one plane is the surface of the main spring-steel triangle of the model.

1107.11 The rubber bands will be strung in such a way that every point along the steel triangle mid-edge line penetrated by the rods shall act as an *origin,* and every second point shall become also the recipient for such a linking as was described above, because each side feeds to the other sides. The ''feeds'' must be shared at a rate of one goes into two. Each recipient point receives two lines and also originates one; therefore, along each edge, *every* point is originating or feeding one vertical connector, while every other, or every *second* point receives two obliquely impinging connector lines in addition to originating one approximately vertically fed line of connection.

1107.12 The edge pattern, then, is one of uniform module divisions separated by points established by alternating convergences with it: first, the convergence of one connector line; then, the convergence of three connector lines; and repeat.

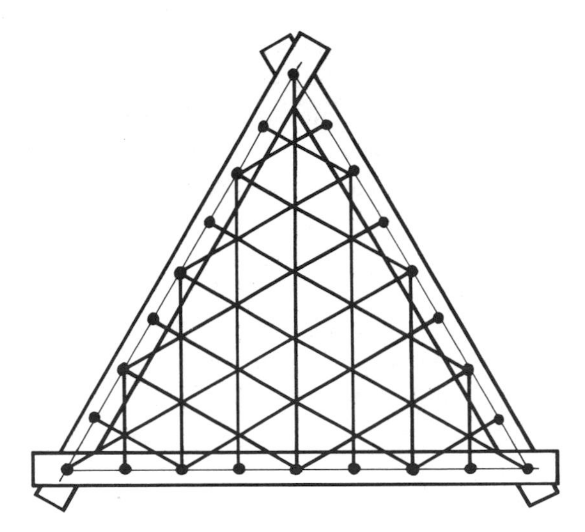

Fig. 1107.12.

1107.13 This linking of the three sides will provide a rubber-band grid of three-way crossings of equi-side and -angle triangular interstices, except along the edges of the main equiangle triangle formed by the spring-steel pieces, where half-equilateral triangles will occur, as the outer steel triangle edges run concurrently through vertexes to and through midpoints of opposite sides, and thence through the next opposite vertexes again of each of the triangular interstices of the rubber-band grid interacting with the steel edges of the main triangle.

1107.14 The rubber-band, three-way, triangular subgridding of the equimodule spring-steel straps can also be accomplished by bands stretched approximately parallel to the steel-strap triangle's edges, connecting the respective modular subdivisions of the main steel triangle. In this case, the rubber-band crossings internal to the steel-band triangle may be treated as is described in respect to the main triangle subtriangular gridding by rubber bands perpendicular to the sides.

1107.20 **Transformation: Aggregation of Additional Rods:** More steel rods (in addition to those inserted perpendicularly through the steel-band edges of the basic triangle model) may now be inserted—also perpendicularly—through a set of steel grommets attached at (and centrally piercing through) each of the points of the three-way crossings of the rubber bands (internal to the big triangle of steel) in such a manner that the additional rods thus inserted through the points of three-way crossings are each perpendicular to the now flat-plane phase of the big basic articulatable steel triangle, and therefore perpendicular to, and coincident with, each of the lines crossing within the big steel triangle face. The whole aggregate of rods, both at edges and at internal intersections, will now be parallel to one another in the three unique sets of parallel planes that intersect each other at 60 degrees of convergence. The lines of the intersecting planes coincide with the axes of the rods; i.e., the planes are perpendicular to the plane of the basic steel triangle and the lines of their mutual intersections are all perpendicular to the basic plane and each corresponds to the axis of one of the rods. The whole forms a pattern of triangularly bundled, equiangular, equilateral-sectioned, parallel-prism-shaped tube spaces.

1107.21 Let us now gather together all the equally down-extending lengths of rod ends to one point. The Greeks defined a sphere as a surface equidistant in all directions from one point. All the points where the rods penetrate the steel triangle edges or the three-way-intersecting elastic rubber-band grid will be equidistant from one common central point to which the rod ends are gathered—and thus they all occur in a spherical, triangular portion of the surface of a common sphere—specifically, within the lesser surfaced of the two spherical triangles upon that sphere described by the steel arcs.

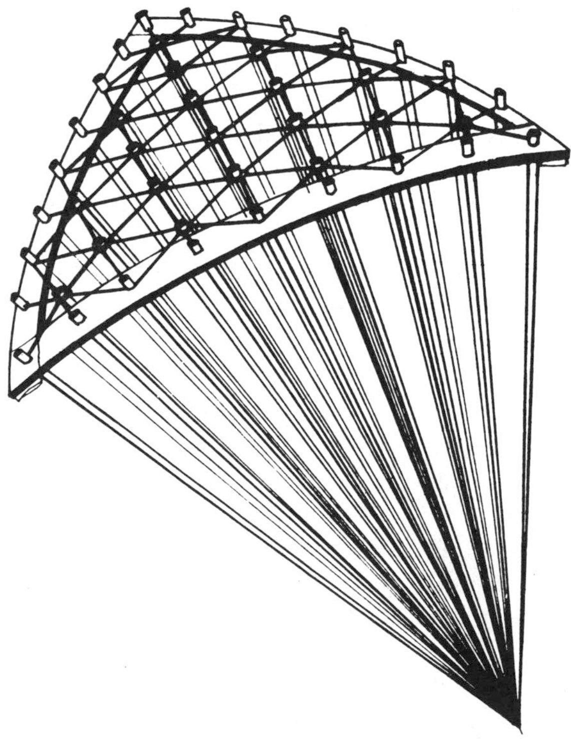

Fig. 1107.21.

Throughout the transformation, all the rods continue their respective perpendicularities to their respective interactions of the three-way crossings of the flexible grid lines of the basic steel triangle's inherently completable surface.

1107.22 If the frequency of uniform spacing of the perpendicularly—and equidistantly—penetrating and extending rods is exquisitely multiplied, and the uniform intervals are thus exquisitely shortened, then when the rod ends are gathered to a common point opposite either end of the basic articulatable steel-band triangle, the gathered ends will be closer together than their previous supposedly infinitely close parallel positioning had permitted, and the opposite ends will be reciprocally thinned out beyond their previous supposedly infinite disposition. Both ends of the rods are in finite condition—beyond infinite—and the parallel phase (often thought of as infinite) is seen to be an *inflection* phase between two phases of the gathering of the ends, alternately, to one or the other of the two spherical centers. The two spherical centers are opposite either the inflection or flat phase of the articulating triangle faces of the basic articulatable triangle of our geodesics transformational projection model.

1107.30 Persistence of Perpendicularity: As the frequency of uniform spacing of the rods increases and as the ends are uniformly and infinitely extended, the distance each end must move to accomplish union is infinitely decreased. As this union takes place and the surface of the original triangle becomes a spherical-surface triangle of infinitely small dimension on an infinitely large sphere, the *chords* between the points of intersection of the rods' triangle's surface, and the *arcs* between the points of intersection, approach infinitely negligible difference. We discern that, while at the inflection point, the rods are at right angles to the chord-arc—"negligible"—mean; that they are not at right angles to the chord at the gathering-of-the-ends phase; and that they are always perpendicular to the mean-arc-chord "infinity"; and that *the condition of perpendicularity is persistent* throughout all the transformation phases. Perpendicularity does not "disappear" at the zero-inflection phase of inside-outing, or positive-to-negative transformation.

1107.31 The above development of our transformational projection model is that of a flexible and two-way steel rod-bristle brush with ends extending evenly—infinitely—in opposite (double infinity) directions and infinitely tightly packed, the bristles being mounted in the steel triangle and its rubber-band-interlaced membrane, which is situated at a central position between two infinite ends and, perpendicular thereto, in both directions.

1107.32 Because the rubber bands seek the shortest distances between their respective points of interaction, and because the steel *arcs* (to which they

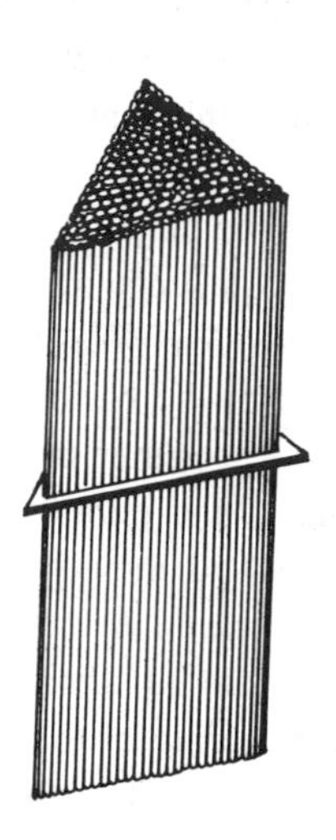
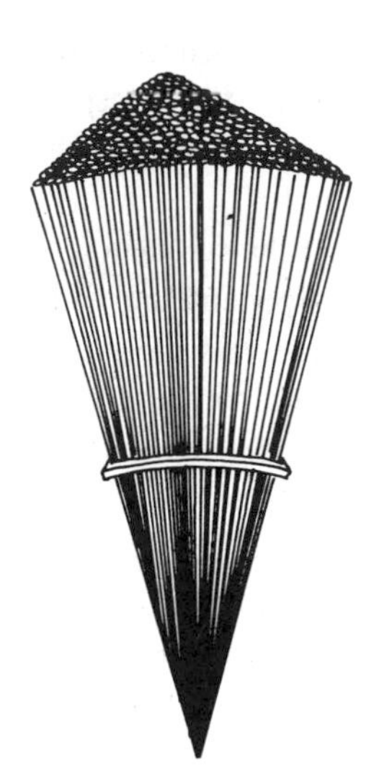

Fig. 1107.31.

are attached at uniform intervals) each *rotate* uniformly (as planes of great circles of the same series of commonly expanding or contracting spheres) *away* from the other *two sides* (of the basic articulatable steel triangle) *toward* one of which two (rotated away from the sides) each rubber band leads (from its own receding position on its awaywardly rotating arc), each band therefore yields elastically, in axial elongation, to permit the continued three-way awayness rotation. Each will persist in finding the progressive set of shortest distances between the points of the spherical triangle's respective perpendicular rod-penetrated surface.

1107.40 **Three-Way Crossings: Zigzags:** Great circles represent the shortest distances between two points on spherical surfaces, and the chords of the arcs between points on spherical surfaces are the even shorter lines of Universe between those points. When the ends of the rods have been gathered together, the rubber bands will be found each to yield complexedly as an integrated resultant of least resistance to the other two bands crossing at each surface point of the grid. They yield respectively each to the other and to the outward thrusting of their rigidly constant steel rods, perpendicularly impinging from within upon the progressively expanding grid. The progressively integrated set of force resultants continuously sorts the rods into sets of rows in the great-circle planes connecting the uniform boundary scale subdivisions of the flexing and outwardly rotating steel-band arcs of the equi-side and -angle articulatable triangle.

1107.41 How do we know empirically that this force-resultant integration is taking place? The stresses pair off into identical zigzags of two-way stress in every chord, in identical magnitude, through the six-functional phases of the six right spherical triangles primarily subdividing the basic equi-side and -angle articulatable steel triangle!

1107.42 How do we know that this is true mathematically? Because the sum-total overall lengths of the vectors in direct opposition are identical, and the sums of their angles are identical!

1110.00 Zenith Constancy of Radial Coordination

1110.01 The *zenith constancy* of the transformational projection's topological trigonometry discretely locates the common zenith points of any commonly centered, concentric-surfaced systems.

1110.02 If camera-equipped telescopes were mounted aboard Earth-dispatched and -controlled satellites that were "locked" in fixed-formation flight positions around Earth, with one such fixed satellite hovering steadily over each vertex of a one-mile-edged world-triangulation grid, and if each telescope was trained so that the eyepiece of its eyepiece-to-optics' axis would be pointed exactly toward the center of Earth and its outer optics' end pointed exactly toward whatever star, if any, may be in exact zenith over the point on the surface of Earth above which the satellite was vertically positioned, a human on Earth at any of those points looking vertically outward into the heavens with a radarscope would discover that satellite as a blip in the middle of his scope-viewing tube's grid.

1110.03 Now let us have an around-the-world simultaneous clicking of the shutters of the cameras attached to each of the telescopes of each of those around-Earth, fixedly hovering photo-satellites with their telescopes pointed to whatever stars may be vertically outward from Earth at their respective omni-Earth-triangulated, one-mile-apart, grid vertexes. Let us assume the photographing telescopes to be very long-barreled to shield those not pointing at Sun from its intense luminosity. A composite mosaic of all those pictures could now be print-mounted spherically on the inside of a translucent 200-foot globe of Earth's conventional geographic data of continents, islands, etc., together with the conventional latitude-longitude grid. Because they were photographed outside Earth's cloud cover, they would present a composite and accurate spherical picture of what the navigators and astronauts call the

FIGURE 8

AUG 15 1950

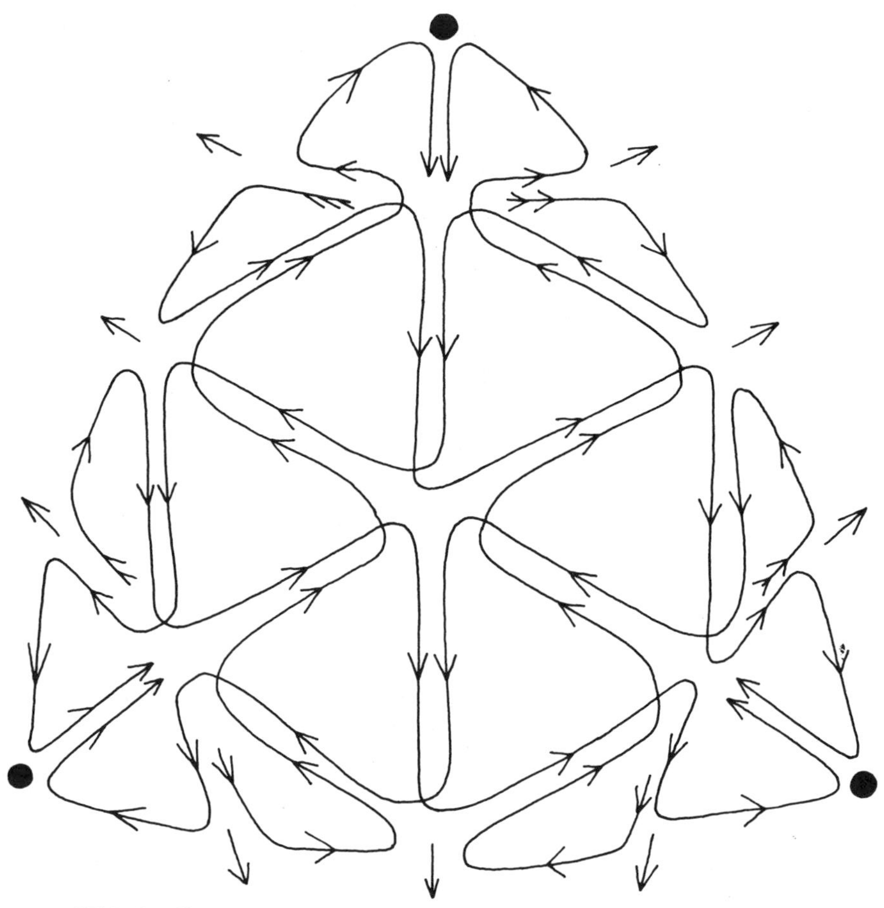

THIS PATTERN SHOWS
2-WAY COMPREHENSIVE
EQUILIBRIUM WITH 3 WAY YIELD AND
BASIC 6-FUNCTION
SATISFACTION WITHOUT INTERNAL
CONFLICT AND SINGLE
EXTERNAL COMPLEMENTARITY NEXT △

Fig. 1107.42.

celestial sphere, with the relative brilliance of the stars in evidence with astronomically calculatable corrections being made in the printing for the Sun's luminosity effects.

1110.04 While this picture was orientationally unique to its one moment in eternity in respect to the Earth-to-celestial-sphere orientation, Earth data per se and the celestial sphere data per se remain constant at their magnitude of scrutability within the lifespan of any human.

1110.05 Because of the accuracy with which this spherical picture was made, it would also be possible to take a transparent-plastic, 20-foot globe of Earth, with the latitude-longitude grid and the continents and islands outlined, together with the marker points identifying the respective positions of the satellite-mounted telescope cameras at the time of the photographing, and to position the 20-foot Earth globe within the 200-foot celestial sphere globe with the miniature Earth's spherical center congruent with the spherical center of the 200-foot celestial sphere.

1110.06 It is then possible to orient the miniature 20-foot-diameter Earth globe so that its polar axis is pointed toward the North Star, making a small correction to correspond with the astronomical correction for the small aberration well known to exist in this respect, which is negligible in this description of the properties of our triangular geodesics transformational projection. We may then rotate the miniature Earth 20-foot globe around its axis until a sighting from its exact center will register each of the satellite camera positions with each of the stars of the 200-foot celestial sphere that the satellites photographed in exact verticality outward from Earth.

1110.07 Earth's highest mountaintop is five miles above sea level, and the ocean's deepest bottom is five miles below sea level. We could now modify the surface of our transparent-plastic, 20-foot model of Earth to show these aberrations, which indicate that some parts of Earth's surface have a differential radial distance from Earth's center; but it would be in evidence that the stars would be in zenith over the same latitude-longitude grid points as would all of the satellite photographic stations.

1110.08 Finding that surface aberrations include only radial-distance variations and changes in the spherical-surface line-of-sight projections from the center, we will now introduce a clear-plastic shell model of a whale and of a crocodile, of such sizes that the crocodile is large enough to omnisurround or swallow the 20-foot miniature Earth globe, and that the whale is large enough to swallow the crocodile yet small enough to be inside the 200-foot-diameter, clear-plastic celestial sphere. With omnidirectional spoke-wires, we will now tensionally position the whale within the 200-foot celestial sphere, and we will tensionally wire-position the crocodile within the whale, and the 20-foot miniature Earth within the crocodile. The miniature Earth is oriented as before, its volumetric center exactly in congruence with the center of volume of the celestial sphere, with all of the stars at the time of the photographing in register with the same satellites that photographed them.

1110.09 Now the whale's and the crocodile's surfaces will be at a great variety of different radii distances from the concentric volumetric centers of the 200-foot and 20-foot spheres. We are going now to coat the surfaces of the transparent whale and transparent crocodile with a photosensitive emulsion. Then we have a high-intensity light source flash at the common volumetric centers of the 20-foot and 200-foot spheres. This process will reproduce on the plastic skin of both the whale and the crocodile—as well as on the celestial 200-foot sphere—the triangular satellite-positioning grid together with the latitude-longitude grid and all Earth's continental and insular outlines. Then, traveling with a pencil-beam strobic light on the outside of the 200-foot celestial sphere, we will point vertically inward against each of the stars, thus projecting their positions radially, i.e., vertically, inwardly to register on the skins of both the whale and the crocodile and on the 20-foot Earth globe. Now, with the human eye at the common concentric centers of volume of the 20-foot and 200-foot spheres, as well as both the whale and the crocodile, we may sight outwardly—which is inherently radially—in all directions, and observe that all the grids and all the geographical and celestial star data appear as one grid, being in exact radial register. We have all the same grids and data on all four of the concentric surfaces: 200-foot celestial sphere, whale, crocodile, and 20-foot Earth globe. That registering of all data is obviously independent of radial distance from the common center; ergo, the only variable in the system is the radius to any given point within the concentric systems.

1110.10 As we have demonstrated with geodesic domes and spheres, what is meant by compound curvature is "omni-intertriangulated structuring (i.e., balanced connectors) of concave-convex surface points." Given a unit radius sphere and the known central angle between any two radii of known length, then the length of the chord running between their outer ends may be calculated trigonometrically by running a line from the sphere center perpendicular to the mid-chord and solving for the right triangle thus formed, whose halved-chord outer edge is the side opposite its central angle, which is half the central angle originally given, and we know that the sine of an angle is the side opposite. When radius is assumed to be *one,* then the well-known sine of one-half the original angle given is the length of that half chord. With the chord length calculatable for a given central angle, it is easy to calculate the length of any line running between the outer end of one of the radii to a position on the other radius at a known distance outward from the spherical center. With this knowledge we can design struts, of suitable structural material—say, aluminum tubes—and we may triangularly interconnect all the vertex points of the triangular grid of the 200-foot sphere. Then we can triangularly interstrut all the grid points on the inside of the whale; then we can-interstrut all the grid vertexes of the crocodile; and finally we can intertriangularly strut the 20-foot Earth globe.

1110.11 Now again, viewing outwardly in all directions from the common volumetric centers of those concentric forms, we will see nothing changed because all the struts will be in register with all the lines of the four separate grids. If we now dissolve the plastic skins from all four shells—the 200-foot celestial sphere, the whale, the crocodile, and the 20-foot globe—we find that all four hold their shapes exactly as before and, being intertrussed (*intertrussed* and *intertriangulated* are the same words: truss: trace: and triangle) between vertexes of the grid, and the grid now being omnitriangularly interstructured, we may again sight outwardly from the volumetric center. A photograph of what we see will reveal only the same lines in exact register that we saw at the time of the original first spherical printing.

1110.12 Since the speed of light permitted astronauts to understand and adopt the light-year in their observational data, we have learned of the great variation of radial distances outwardly to the different stars. In the Big Dipper, one star is 200 light-years farther from Earth than the next one on the handle, which is a distance of 200 quadrillion miles farther away from you and me than is the other. If we ran rods radially from the volumetric center of our model outward perpendicularly through each of the stars shown on the 200-foot celestial sphere to a distance perpendicular outwardly from the 200-footer equal to their distance away in light-years from Earth, with the 200-foot sphere's 100-foot radius equaling that of the nearest star other than Sun, and assume that the camera had photographed only those stars visible to the naked eye, then a few of the rods would reach outwardly ten miles, but most of them would be much nearer in, with one of the Big Dipper's one mile out and another a half-mile out. It would make a vastly varied porcupine if we intertriangularly interconnected the outer terminals of the lines of interconnection, which would as yet be in exact register with the original grid as seen from system center.

1110.13 Now let us separate the four structures by opening up an approximate equator in the outer ones and rejoining the equatorial points. With this celestial porcupine rolled into our deepest ocean and then resting on the bottom, its top would reach outwardly above the ocean surface to the height of Mt. Everest; its densest, most high-frequency-trussed spherical core would be only 200 feet in diameter and would occur at ocean surface. The triangularly trussed 175-foot whale would hold its shape and size, as would the 60-foot crocodile and the little 20-foot miniature Earth. Obviously, they could not appear more differently.

1110.14 Our triangular geodesics transformation projection would show all four of these dissimilar systems in the flat plane in exactly the same manner and in exact register with that of Earth alone as shown later in the icosahedral flat-out of the world map, but with a number (or numbers of different styles) shown at each grid vertex, which number indicates the radius distance of that vertex outwardly from the center point of the system "Earth." Four different colors—blue for the celestial, black for the whale, green for the crocodile, and brown for the 20-foot Earth globe—would identify the relative radius distances outward from the congruent systems' center, which occurs at each vertex of these four utterly different-shaped and -sized systems—all on the same map. This would provide all of the data necessary to reconstruct each of the four systems in exactly the same relative sizes. Every point in the four systems remains in exact perpendicular (zenith), whether in the spherical or planar flat-out phase or any interim transitional phase. This makes possible the design of an airplane or an ocean liner all on one synergetic-geodesic map. And the flat-out map may have its triangular mosaic pieces rearranged in many ways—for instance, to center the oceans or to center the lands. And the building of that airplane or ocean liner, as with the geodesic dome, will generate compound curvature, omnifinite, tensegrity trussing far stronger and lighter than the presently designed and built *XYZ*-parallel coordinate grids and their parallel-plane sectional designing.

1110.15 With omnidirectional, complex, computerized, world-satellite sensing, comprehensive-resources inventorying and interrouting, the triangular geodesics transformational projection can alone bring visual comprehending and schematic-network elucidation.

1110.16 Just as triangular geodesics transformational projection can alone reduce the astronomical to the cosmic middle ground of eye-comprehensible coordination with the mind explorations and formulations in metaphysics in general and mathematics in particular, especially in relation to computer programming, so too may the triangular geodesics transformational projection enlarge the complex invisible microcosmic patterns to eye and sense comprehensibility.

1120.00 Wrapability

1120.01 One roll of paper being unrolled from any one fixed axis wraps up all the faces of a tetrahedron. Two rolls of paper being unrolled from two axes perpendicular to one another wrap up all the faces of the octahedron.

Three rolls of paper being unrolled from three axes * wrap up all the faces of the icosahedron.

1120.02 If the paper were transparent and there were ruled lines on the transparent paper at uniform single intervals, the single lines of the transparent paper wrapping up the tetrahedron will enclose the tetrahedron without any of the lines crossing one another. In wrapping up the octahedron with two rolls of such transparent paper, the lines cross—making a grid of diamonds. Wrapping the icosahedron with the three rolls of transparent, parallel-ruled paper, a three-way grid of omnitriangulation appears.

1120.03 The wrapping of the six-edged tetrahedron with the single roll of paper leaves two opposite edges open, i.e., uncovered by the wrapping-paper roll. The other four opposite edges are closed, i.e., covered by the wrapping-paper roll.

1120.04 The wrapping of the 12-edged octahedron with two rolls of paper leaves two sets of opposite edges open. The other eight opposite edges are closed.

1120.05 The wrapping of the 30-edged icosahedron with three rolls of paper leaves three pairs of opposite edges open or uncovered. To cover those open edges, we need two more rolls. With five wrappings, all 30 edges become enclosed: with five wrappings, 10 faces are double-covered and 10 faces are triple-covered. Only the triple-covered have omnitriangular gridding by the parallel ruled lines. Thus we see that we need a sixth wrapping to make the omnitriangulated three-way grid. At the fifth wrapping, the three-way grid appears about the north and south poles with only a two-way grid on the equatorial triangles. The whole three-way grid six-times rewrapping in omnitriangular gridding at any desired frequency of subdivisioning can thus be accomplished with only one type of continuing, parallel-ruled strip.

1120.06 Wrapping relates to the mid-edges of prime structural systems.

1120.07 It takes three wrappings on three axes to produce the three-way grid on every face of a tetrahedron.

1120.08 Wrapping of the octahedron with two rolls of paper left two opposite edges open. Two strips covered all the faces. Three strips covered all the edges. But a fourth strip is needed to complete the omnitriangulation of each face of the octahedron. (Compare the four axes of the octahedron with the eight faces perpendicular to the center of volume of the octahedron. We are dealing with the axes of the mid-faces.) There are four unique ways to wrap an octahedron from a roll. The three-way grid for each face requires four-way wrappings.

1120.09 If we take a transparent sheet of paper whose width is the alti-

* The six axes of the icosahedron are using the 12 vertexes coming together at 63° 26′ to each other.

tude of the equilateral triangles of the three universal prime structures, both of the edges can be stepped off with vectors of the same length. This produces a series of opposing, regular, uniform, equilateral triangles. The altitude of the equilateral triangle is the width of the transparent paper ruled with parallel lines parallel to the edges of the roll. Along the edge of one side of this roll, we step off increments the same length as the basic vectors of the triangles. We take the midpoint of the first triangle and drop a perpendicular across to the opposite edge of the roll. We step off increments of the same basic vector length. But the step-offs are staggered with the vertex of one triangle opposite the mid-edge of the other.

1120.10 This is how the lines of the tetrahedron keep wrapping up like a spool. That is why in the tetrahedron the axes are all the mid-edges of the poles. One polar pair of opposite edges is left open because the system is *polarized;* therefore, you need the three wrappings—one to cover all the faces, the second to cover all the edges, and the third for omnitriangulation. Three axes = three-way grid = three vectors for every vertex.

1120.11 A single wrapping defines the octahedron even though two faces are left uncovered. It is polarized by the empty opposite triangles. One-half vector lacks rigidity. The interference of two planes is required for the spin. But we have to deal with open edges as well as with open faces. The figure will stand stably because six of the 12 edges are double-spin, with two edges coming together in dihedral angles.

1120.12 Because one preomnitriangulated strip whose width exactly equals the altitude of the tetrahedron can completely spool-wrap all four faces of the tetrahedron, and because a tetrahedron so wrapped has an axis running perpendicular to—and outward through—the two mid-unwrapped-edges of the tetrahedron spool, such a spool may be endlessly wrapped, being a tetrahedron and an omnidirectionally closed system; ergo all the data of evoluting inwardly and outwardly in observable Universe and its scenario of intertransformings could be continuously rephotographed with each cycle and could thus be fed linearly into—and stored in—a computer in the most economical manner to be recalled and rerun, thus coping with all manner of superimpositions and inclusions at recorded dial distances inward and outward as a minimal-simplicity device.

1200.00 Numerology

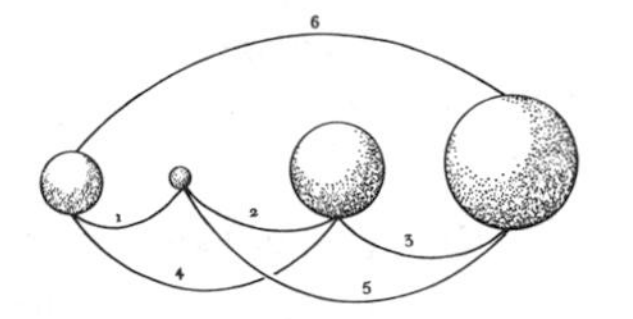

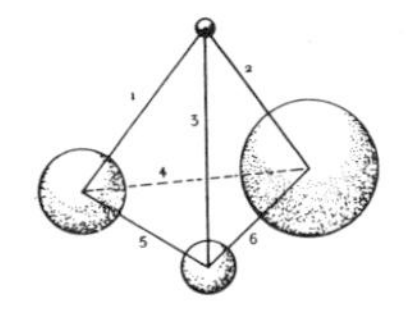

1210.00 Numerology

Historically long perspective
Suggests it as possible
That many of the intriguing
Yet ineffable experiences
Which humanity thus far
Has been unable to explain,
And, therefore, treats with only superstitiously,
May embrace phenomena
Which in due course
Could turn out to be complexes
Of physically demonstrable realities
Which might even manifest
Generalized principles of Universe.

For this and similar reasons
I have paid a lot of attention
To ancient *numerology,*
Thinking that it might contain
Important bases for further understanding
Of the properties of mathematics
And of the intertransformative
Structurings and destructurings
Of the cosmic scenario yclept
''Eternally self-regenerative Universe.''

My intuition does not find it illogical
That humanity has developed and retained

The demisciences of
Astrology and *numerology*—
Demi because they are
Only partially fortified by experimental proofs—
Which nonetheless challenge us tantalizingly
To further explorations
Within which it may be discovered
That generalized scientific laws
Are, indeed, eternally operative.

Our observational awareness
And Newton's proof
Of the mass-attraction law
Governing the Moon's powerful tidal pull
On the Earth's oceans,
In coincidence with our awareness
Of the Moon phase periodicities
Of female humans' menstrual tides,
Gave the Moon's (men's month's) name
To that human blood flow.
Conceivably there could be
Many other effects of celestial bodies
Upon terrestrially dwelling human lives.

In the late 1930s,
When I was science and technology consultant
On its editorial staff, with Russell Davenport
Then managing editor, of *Fortune* magazine,
I found him to be deeply involved with *astrology*.
Russell couldn't understand why
I was not actively excited
By the demiscience—astrology,
Since the prime celestial data derived
From scientific observations.
I was not excited
Because I had no experience data
That taught me incisively
Of any unfailingly predictable influence
Upon myself or other Earthians
Which unerringly corresponded
With the varying positions of the solar planets
At the time of the respective human births.

While the planetary interpositionings
At any given time had been scientifically established,
I had no scientifically cogent means for exploring
Their effect upon terrestrial inhabitants.

On the other hand, I found many cogent clues
For exploring the ancient demiscience of numerology.
Ancient numerologists developed
Many tantalizingly logical theories,
Some of which were
Partially acceptable to formal mathematics,
Such as enumeration by "congruence in modulo eight,"
Or, "congruence in modulo ten,"
Or in increments of twelve.

"Congruence in modulo ten" seemed
Obviously induced by
The convenience of the human's ten fingers
As memory-augmenting,
Sequentially bendable,
Counting devices of serial experiences.
Their common appendages of ten fingers each
Provided humans with "natural" and familiar sets
Of experience aggregates
To match with other newly experienced aggregates
As congruent sets.

There was also the popular enumeration system
Based on modulo twelve.
Human counting systems of twelve were adopted
Because the decimal system
Does not rationally embrace
The prime number three.
Since humanity had so many threefold experiences,
Such as that of the triangle's stability,
Or that of the father-mother-child relationships,
Humanity needed an accounting system
That could be evenly and alternatively subdivided
In increments either
Of one, or two, or three.
Ergo, "congruence in modulo twelve"
Was spontaneously invented.

"Invention" means
To bring into novel special-case use
An eternal and universal principle
Which scientific experiment and comprehension
May attest to be generalized principles.

"Etymology" means
The scientific study of words and their origin.
Through etymology man gave names
To their abstract number set concepts.

English is a crossbred
Worldian language.
It is interwoven with Anglo-Saxon,
Old German, Sanskrit, Latin and Greek roots,
Interspersed with Polynesian, Magyar, Tatar, et al.
The largest proportion of English words
Are derived from India's Sanskrit,
Which itself embraces hundreds
Of lesser known root languages.

There are a few words whose origins
Have thus far defied scientific identification.
There are not many unidentified root words.
Those of unknown origins
Are classified etymologically as "Old Words."
All but one of the world-around
Words or "names" for numbers
Are classified etymologically as "Old Words."
The one exception is the name for "five,"
Whose conceptual derivation comes directly or indirectly
From word roots identifying the human "hand."
None of the other names for numbers
In any of the human languages
Have pragmatic identifiability
With names for any other known
Physical-experience concepts.

To accommodate the cerebrations
Of those who are reflexively conditioned
To recount their experiences

In twelvefold aggregates—
That is, "congruence in modulo twelve"—
Unique names were etymologically evolved
For the numbers *eleven* and *twelve*
As well as for the numbers *one* through *ten*.
In the new world-around-accepted computational system
Of "congruence in modulo ten"—
That is, the *decimal* system—
The numbers zero through ten
Are called "cardinal" numbers.

But the English names "eleven" and "twelve,"
Or French names "onze" and "douze,"
Or the Germans' "elf" and "zwoelf,"
Likewise are cardinal numbers
In the duodecimal system,
And their cardinal names are used
Even when employed in the decimal system.

Following twelvc in the duodecimal system
The number names are no longer *cardinal*.
They are called *ordinal* numbers, which are produced
By combining one, two, or three with ten:
Thir-teen, four-teen, fif-teen, etcetera,
Which are three-ten and four-ten, alliterated
In English, French, and German.
It is not until thirteen is reached
That the process of counting ordinally (three plus ten)
Is employed in the ordinal naming of numbers
Where numbers are communicable by sound.

There are, however, number systems
Based on other pragmatic considerations.
Roman numerals constituted
An exclusively visual method
Of tactilely scoring or scratching
Of a one-by-one exclusively "visual" experience.
When nonliterates were assigned
To counting items such as sheep,
They made one tactile scratch
For one visually experienced sheep,

And a second tactile scratch
As another sheep passed visually,
And another scratch
As the next sheep passed.
The scratch was not a number,
It was only a tactile reaction
To visual experience.
It was a one-by-one,
Tooth-by-tooth intergearing
Of two prime
Sensorially apprehending systems—
Those of touch and sight.

While literate you, in retrospect, could say
That you see *three* scratches,
That is reflexively occasioned
Because you have learned to see groups
And because you have
A sound word for a set of three;
But nonliterate Roman servants who were scoring
Did not have to have number words
To match with tactile one-by-one scratching
Their one-by-one visually experienced,
One-by-one passing-by sheep.
The man doing the scratching
Did not have to have
Any verbal number words or set concepts.
Those landlords, priests, bankers,
Or unsolicited ''protection'' furnishers
Who were interested
In trading, taxing, or extracting
Life-sustaining wealth—
As sheep or wheat productivity—
Alone were concerned
With the specific total numbers of scratches
And of the total sheep or bags of grain
The ignorant servants had scratchingly matched.
From these total numbers
They calculated how many sheep or bags
They could extract for their taxes
Or landlord's tithe,

Or protectionist's fee,
Or banker's "interest"
Without totally discouraging
The sheepherders' or farmers' efforts.

"Pays" means land.
The shepherds and farmers
Were known as pagans
Or paysants, peasants,
I.e., land-working illiterates.

Because the first millennium A.D.
Roman Empire dominating Mediterranean world
Was so pragmatically mastered
By landlords and their calculating priests,
It is in evidence
That the Roman numerals constituted only
A one-by-one scoring system
In which the V for five and X for ten
Were tactilely "sophisticated" supervisor's
Tallying or *totaling* check marks
Which graphically illustrated
Their thumb's angular jutting out
From the four parallel packed fingers
Or digits of the totaler's free hand.
On the other hand, the intellectually conceived Arabic numerals
Were graphic symbols
For the named sets
Of spontaneously perceived number aggregates.
The Arabic numerals
Did not come into use in the Mediterranean world
Until 700 A.D.
This was a thousand years after the Greeks had developed
Their intellectually conceived *geometry*.

The 700 A.D. introduction of Arabic numerals
Into the knowledge-monopolized economic transactions
Of the ignorance-enweakening Roman Empire
And Mediterranean European world in general
Occurred under the so-called "practical" assumption
That the Arabic numerals were only

Economical "shorthand" symbols
For the Roman scratches.
To the nonliterate ninety-nine percent of society,
It was obviously much easier to make a "3" squiggle
Than to make three separate vertical scratch strokes.
But to the illiterate the symbols
Did not conjure forth a number name.

The earliest calculating machine
Is the Chinese-invented abacus.
It is an oblong wooden frame
Which is subdivided
Into a large rectilinear bottom
And small top rectilinear areas
By a horizontal wooden bar
Running parallel to the top of the frame.
The frame's interior space is further subdivided
By a dozen or more
Perpendicularly strung parallel wires
Or thin bamboo rods.
There are four beads
Strung loosely into each of the wires
Below the horizontal crossbar,
And one bead strung loosely
Above the bar on each wire.
Start use of the abacus
With all the beads at bottom
Of their compartments.
In this all-lowered condition,
The columns are all "empty."
To put the number one
Into the first column on the right,
The topmost of the bottom four beads
Is elevated to the horizontal mid-bar.
To put the number two,
Two bottom beads are elevated to this bar.
To put five into the first column,
Lower all four bottom beads
And elevate the top bead.
To enter nine, leave the top bead elevated
And push up four beads

In the bottom section
On the first right-hand wire.
To enter ten,
Lower all beads in the right-hand column
Both above and below the crossbar;
Now elevate one bead
In the bottom section
Of the second column from the right.
The first two right-hand columns read
One and zero, respectively,
Which spells out "ten."
The totaling bead
With a value of five
In the separate compartment
At the head of each column
Permitted the release to *inactive* positioning
At the bottom of their wires
Of the one-by-one elevated bead aggregates.
Lowering of all beads
Permitted "empty columns" to occur.
Moving of the tenness leftward
Permitted progressive positioning,
Which integrated or differentiated out
As multiplication or division.

To those familiar with its use,
The tactile-visual patterns
Of the bead positions of the abacus
Could be mentally re-envisioned, or recalled
And held as afterimage sets
In the *image*-ination,
Which could be mentally manipulated
As columns of so many beads
Which read out progressively
As successively adjacent columns
Of so many beads,
Which, when reaching fiveness,
Called for moving "up" the one bead
Of the totaling head-compartment set,
While releasing the previously aggregated
Lower four beads

To drop into their empty-column condition.
When an additional four beads
Were pushed upwardly in the column,
An additional fiveness accrued.
All the beads in the column were lowered,
And one was entered
On the bottom compartment
Of the next leftward column,
As the two columns now read as "ten."
It was easier to enter
Many columned numbers in the abacus
And to add to them
Multicolumned numbers.
This process then permitted
Multiplication and division as well.

When an abacus was lost overboard or in the sands,
The overseas or over-desert navigator
Could sketch a picture
Of the abacus in the sand
Or on a piece of wood
With its easily remembered columns.
These abacus picturers invented
The "arabic" or abacus numerals
To represent the content
Of the successive columnar content of beads.
Obviously this abacus column imagining
Called also for a symbol
To represent an empty column,
And that symbol became the cyphra—
Or in England, cypher,
Or in American, cipher,
Or what we symbolize as 0,
And much later renamed "zero"
To eliminate the ambiguity
Between the identity of the word cypher
With the word for secret codes
And the word for the empty number,
All of which mathematical abacus elaboration
Became known scoffingly as "abracadabra"

To the 99 percent nonliterate world society,
And to the temporal power leaders
Who feared its portent
As an insidious disrupter
Of their ignorance-fortified authority.

Because of its utterly pragmatic bias,
The Roman culture had no numerical concept
Of "nothing"
That corresponds to the abacus's empty column—
That is, the idea of "no sheep"
Was ridiculous. Humans cannot eat "no sheep."
When the Europeans first adopted the Arabic numerals in 700 A.D.
As "shorthand" for Roman numeral aggregates,
They of course encountered the Arabic cypher,
But they had no thinkably identifiable experiences to associate with it.
"Nothing" obviously lacked "value."
For this reason, the Mediterranean Europeans
Thought of the cypher only as a decoration
Signifying the end of a communication
In the way that we use the word "over"
In contemporary radio communication.
The cypher was just an end *period,*
Just a decorative terminal symbol.

It was not until 1200 A.D.
Or five hundred years later,
That the works of a Persian named Algorismi
Were translated into Latin and introduced into Europe.
Algorismi lived in Carthage, North Africa.
He wrote the first treatise explaining
How the Arabic cypher functioned calculatively
By progressively moving leftward
The newly attained tenness
By elevating one bead at the bottom
Of the bottom section
Of the next leftwardly adjacent column in multiplication
And next rightwardly in division.
Thus complex computation could be effected
Which had been impossible with Roman numerals.

The Arabic cypher had been used
For several millenniums
In the computational manner,
First in the Orient,
Then in Babylon and Egypt.
But such calculations had never before been made
In the Roman Empire's Mediterranean world.
No matter how intuitively
A man might have felt
About the probable significance
Of the principle of leverage
Or about the science of falling bodies,
Previous to the knowledge
Of the cypher's capabilities to position numbers,
He could not compute
Their relative effectiveness values
Without "long" multiplication and division.

The introduction into Europe
Of the computational significance of the cypher
Was an epoch-initiating event
For it made it possible for *anybody* to calculate.
And this was the moment in which
For the first time
The Copernicuses and Tycho Brahes,
The Galileos and Newtons,
The Keplers and Leonardos
Had computational ability.
This broke asunder the Dark Ages
With intellectual enlightenment
Regarding the scientific foundations
And technological responsibilities
Of cosmic miracles,
Now all the more miraculous
As the everyday realizer
Of all humanity's innate capabilities.

When I first went to school in 1899,
The shopkeepers in my Massachusetts town asked me
If I had "learned to do my cyphers"
By which key word—"cypher"—

They as yet identified all mathematics.
Even in 1970
Accountants in India
Are known officially as "cypherists."

Tobias Dantzig, author of *Number: The Language of Science,*
Has traced the etymological history
Of the names for the numbers
In all the known languages of the Earth.
He finds the names for numbers all classifiable
As amongst the "oldest" known words.
Sir James Jeans said
"Science is the attempt
To set in order the facts of experience."
Dantzig, being a good scientist,
Undertook to set in order
The experienced facts of the history
Of the language of number names.
He arranged them experimentally
In their respective ethnic language columns.
Juxtaposed in this way
We are provided with new historical insights.
For instance, we learn
That if we are confronted
With two numbers of different languages,
Words that we have never seen before,
And an authority assures us
That one of these words means "one"
And the other means "two,"
And we are then asked to guess
Which of them means "one"
And which means "two,"
We will be surprised to find
That we can tell easily which is which.
"One" in every language
Starts with a vowel—
Eins, un, odyn, unus, yet, ahed—
And has vowel sound emphasis,
While "two" always has a consonant sound in the front—
Duo, zwei, dva, nee, tnayn, and so forth,
And has a consonant sound emphasis.

For instance, the Irish-Gaelic
Whose ancestors were sea rovers
Say ''an'' for one and ''do'' for two.
These vowel-consonant relations
Hold through into the teens—
Eleven, twelve—in English
Onze, douze—in French
Elf, zwoelf—in German,
With vowels for ''oneness''
And consonants for ''twoness.''

Despite the dissimilarity in different languages
For the names for the same experiences,
And despite the unknown origins of the concepts
From which all numbers but five were derived,
The whole array of names for the numbers
In different languages
Makes it perfectly clear
That the names given the numbers around the world
Grew from the same fundamental
Conceptioning and sound roots.

In view of the foregoing discovery,
We either have to say that some angels
Invented the names for numbers
And the phonetically soundable
Alphabetical letter symbols
With which to spell them
And wrote them on parchments
And air-dropped those number-name leaflets
All around the spherical world,
Thus teaching world-around people the same number names:
Or we have to say that the numbers were invented
By one-world-around-traveling people.

However, if we adopt the latter possibility,
It becomes obvious that no single generation of people
Could, within its lifetime,
Or, in fact, within many lifetimes

Travel all around the world on foot,
For the world's lands are islanded.
But one way humans could get around,
And in a relative hurry,
Was by "high-seas-keeping" sailboats.
It thus becomes intuitively logical
To assume that sailors discovered
And invented the numbers
And inculcated their use
All around the world.

The Polynesians, we know,
Sailed all over the Pacific.
They probably sailed
From there into the Atlantic and Indian oceans
By riding ever-west-toward-east "Roaring Forties"—
The Forty-South latitudes'
Ever-eastward-revolving
Waters and atmospheric winds
Which circle around the vast Antarctic continent.
The "Roaring Forties"
Constitute a gigantic hydraulic-pneumatic merry-go-round,
Which as demonstrated by
World-around single-handing sailors of the 1960s
Enables those who master its ferocious waters
To encircle the world
Within only a year's time.
The Magellans, Cooks, and Slocums
With slower vessels circumnavigated in two years,
In contradistinction to the absolute inability
To go all around the world on foot.
The circumnavigation of the one-ocean world
Which covers three-quarters of our planet
Makes it obvious that the names for numbers
Were conceived by the sailors.
As Magellan, Cook, and later Slocum
Came to the Tierra del Fuego islanders,
They were surrounded by the islanders,
Who lived by pillaging passing ships
And must have been doing so
Profitably for millenniums.

To explain their sustained generations
In an environment approximately devoid
Of favorable human survival,
Except by piracy and salvage
Of the world-around sailing vessels
Funneled through the narrow
And incredibly tumultuous
Waters of the Horn
Running between Antarctica and South America,
With often daily occurring
One-hundred-feet high waves
Cresting at the height
Of ten-story buildings,
Their thousand-ton tops
Tumblingly sheared off to leeward
By hundred-miles-an-hour superhurricanes
Avoidance of whose worst ferocities
Could be accomplished by winding
Through the Strait of Magellan,
Whose fishtrap-like strategic enticement
Often lured Pacific-Atlantic sea traffic
Into those pirates' forlorn domain.

With eighty-five percent of Earth's dry land
And ninety percent of its people
Occupying and dwelling north of the Equator
In the northern, or land-dominant, hemisphere;
And with less than one-tenth of one percent of humanity
Dwelling in the southernmost half
Of the southern, or wave-dominated, Earth hemisphere,
There is more and more scientific evidence accruing
That sailors have been encircling the Earth
South of Good Hope,
North or south of Australia,
And through the Horn
Consciously and competently
For many thousands of millenniums
All unknown to the ninety-nine percent of humanity
That has been "rooted" locally
To their dry-land livelihoods.

The European scholars of the last millennium
Have considered the Polynesians to be illiterate
And therefore intellectually inferior to Europeans
Because the Polynesians didn't have a written history
And used only a binary mathematics,
Or "congruence in modulo two."
The European scholars scoffed,
"The Polynesians can only count to two."

Since the Polynesians lived on the sea
And were naked,
Anything upon which they wrote
Could be washed overboard.
The Polynesians themselves
Often fell overboard.
They had no pockets
Nor any other means
Of retaining reminder devices
Or calculating and scribing instruments
Other than by rings
That could not slip off
From their fingers, ankles, wrists, and necks,
Or by comblike items
That were precariously
Tied into the hair on their heads
Or by rings piercing their ears and noses.
These sea people had to invent ways of calculating and communicating
Principally by brain-rememberable pattern images.
They accomplished their rememberable patterns in sound,
They remembered them in chants.
With day after day of time to spend at sea
They learned to sing and repeat these chants.
Using the successive bow-to-stern,
Canoe and dugout, stiffing ribs and thwarts
Or rafters of their great rafts
As re-minders of successive generations of ancestors,
They methodically and recitationally recalled
The experiences en-chantingly taught to them
As a successive-generation,
Oral relay system

Specifically identified with the paired ancestral parents,
Represented by each pair of ship's ribs or rafters.
When they landed for long periods
They upside-downed their longboats
To provide dry-from-rain habitats.
(The word for "roof" in Japan
Also means "bottom of boat.")
Staying longer than the wood-life of their hulls,
They built long halls patterned after the hulls.
Each successive column and roof rafter
Corresponded with a rib of their long boat.
Gradually they came to carve
Each stout tree column's wood
To represent an ancestor's image.
Each opposing pair of parallel columns
Represented a pair of ancestors:
The male on the one hand
And the female on the other hand.
While most Europeans or Americans can recall
Only ten or less generations of ancestors,
In their chants
The Polynesians can recall
As much as one hundred generations
Of paired ancestors,
And their chants include
The history of their important discoveries
Such as of specific-star-to-specific-star directions to be followed at sea
In order to navigate from here to there.
While many of the words
That their ancestors evolved
To describe their discoveries
Have lost present-day identification,
They continue to sing these words
In faithful confidence
That their significant meaning
Will some day emerge.
Therefore, they teach their children
As they themselves were taught—
To chant successively the special stories
Which include words of lost meaning—
Describing each one of every pair of ancestors.

That is why the Vikings
Had their chants and sagas
And why sailors all around the world
Chant their chanties—"shanties"
As they heave-hoed rhythmically together.

Thus too did the Viking sing their sagas;
And the Japanese and Indian sailors their ragas;
And the Balinese sailors their gagas,
Meaning "tales of the old people,"
Amongst all those high-seas-living world dwellers
Whose single language structure
Served the thirty-million-square-mile living Maoris;
Whereas hundreds of fundamentally different languages
Were of static-existence necessity developed,
For instance, by isolatedly living tribes
Of exclusively inland-dwelling New Guineans.

A nineteenth-century sailor's shanty goes
"One, two, three, four
Sometimes I wish there were more.
Eins, zwei, drei, vier
I love the one that's near.
Yet, nee, same, see
So says the heathen Chinese.
Fair girls bereft
Then will get left
One, two, and three."

As complex twentieth-century,
Electronically actuated computers
Have come into use,
Ever improving methodology
For gaining greater use advantage
Of the computers' capabilities,
As information storing,
Retrieving, and interprocessing devices,
Has induced reassessment
Of relative mathematical systems' efficiencies.
This in turn has induced
Scientific discovery

That binary computation
Or operation by "congruence in modulo two"
Is by far the most efficient and swift system
For dealing universally with complex computation.
In this connection we recall that the Phoenicians
Also as sailor people
Were forced to keep their mercantile records
And recollections in *sound* patterns,
In contradistinction to *tactile* and *visual* scratching—
And that the Phoenicians to implement
Their world-around trading
Invented the Phoenician,
Or Phonetic, or word-sound alphabet,
With which to correlate and record graphically
The various sound patterns and pronunciations
Of the dialects they encountered
In their world-around trading.
And we suddenly realize
How brilliant and conceptually advanced
Were the Phoenicians' high-seas predecessors,
The Polynesians,
For the latter had long centuries earlier
Discovered the binary system of mathematics
Whose "congruence in modulo two"
Provided unambiguous,
Yes–no; go–no go,
Cybernetic controls
Of the electronic circuitry
For the modern computer,
As it had for millenniums earlier
Functioned most efficiently
In storing and retrieving
All the special-case data
In the brains of the Polynesians
By their chanted programming
And their persistent retention
Of the specific but no-longer-comprehended
Sound pattern words and sequences
Taught by their successive
Go–no go, male-female pairs of ancestors.

This realization forces rejection of the European scholars'
Former depreciation of the Polynesian competence,
Which reversal is typical
In both conceptioning and logic
Of the myriad of concept reversals
That are now taking place
And are about to occur
In vastly greater degree
In the late twentieth-century academic world.
The general education system
Has not yet formally acknowledged
The wholesale devaluation
Of their formally held
"Scholarly opinions and hypotheses,"
But that devaluation
Is indeed taking place
And is powerfully manifest
In the students' loss of esteem
For their intellectual wares.

All of the forcgoing
Newly dawning realizations
Point up the significance
Of the world-around physically cross-bred kinship
Of the world's "one-ocean" sailors
Whose Atlantic, Pacific, and Indian waters
Were powerfully interconnected
By the Antarctic-encircling
"Roaring Forties."
Polynesians, Phoenicians, Venetians, Frisians, Vikings
(Pronounced "Veekings" by the Vikings)
All alliterations of the same words.
All evolved from the same ancestors.

The sea was their normal life,
And since three-quarters of the Earth's surface
Is covered with water,
"Normal" life would mean living on the sea.
The Polynesians spontaneously conceive of an island
As a "hole" in the ocean.

Such conceptioning of a negative hole in experience
Brought about their natural invention
Of a symbol for nothing—the zero.
This is negative space conceptioning
And is evident in the Maori paintings.
What is a peninsula to land people
Is a "bay" to them.
The Maori also look at males and females
In the reverse primacy of the land-stranded Western culture.
Seventy-five percent of the planet is covered by the sea.
The sea is normal.
The male is the sailor.
The male is normal.
The penis of the normal sea
Intrudes into the female land.
The bay is a penis of the sea.
The females dwell upon the land.
To the landsman the peninsula or penis
Juts out into the ocean.

On the Indian Ocean side of southeast Africa,
The Zulus are linked with this round-the-world water sailing.
They are probably evolved from the Polynesians of long ago
Swept westward by the monsoons.
I found some of the Zulu chiefs
Wearing discs in their ears
Upon which the cardinal points of the compass
Were clearly marked.
The "Long Ears" of Easter Island
Had their ears pierced and stretched
To accommodate their navigational devices.
Many of the items which European society
Has misidentified in the Fijis as superstitious decoration
Were and as yet are
Navigational information-storing devices,
Being stored, for instance,
As star-pattern combs in their hair,
As rings around their necks,
Or as multiple bracelets
Mounted on their two arms and two legs,
And multiple rings

Upon the four fingers of their hands.
They had thirteen columns of slidable counters,
One neck, eight fingers, two arms, two legs.
Most of the earliest known abacuses
Also have thirteen columns of ring (bead) counters
Which became more convenient to manipulate and retain
When rib-bellied ships
Supplanted the open raft and catamaran.
Once the mathematical conceptioning
Of sliding rings on thirteen columns
Had been evolved by the navigators, traders, magicians,
It was no trick at all
To reproduce the thirteen-column system
In a wooden frame with bamboo slide columns.

By virtue of their ability to go
From the known here to the popularly unknown there,
The navigators were able to psychologically control
Their local island chieftains.
If a chieftain needed a miracle
To offset diminishing credit by his people,
He could confront them with his divine power
By exhibiting some object they had never seen before,
Because it was nonexistent
On their particular island.
All the chieftain had to do
Was to ask the navigators
To exercise their mysterious ability
To disappear at sea
And return days later with an unfamiliar object.
But the navigators kept secret
Their mathematical knowledge
Of offshore celestial navigation
And the lands they thus were able to reach.

To the landed chieftains
The seagoing navigators were mysterious priests.
The South Seas navigators lived and as yet live
Absolutely apart from the chieftains and the tribe
The "priests" taught only their sons about navigation
And they did so only at sea.

A new era dawned
For humanity on our planet
When the Polynesians learned
How to sail zigzaggingly to windward
Into the prevailing west-to-east winds.
Able to sail westward—
Able to follow the Sun—
At far greater sustainable
(All day and all night, day after day)
Sailing speeds than those attainable
By paddling or rowing into head seas;
Having for all time theretofore drifted
In predominantly eastward windblown directions,
Or gone aimlessly where ocean currents bore them,
Yielding to the inevitable
From-west-to-east elements
Bearing them to the American west coasts
And to all the Pacific islands
Throughout the previous x millions of years.

Whereas the Southern Hemisphere ocean
Was dominated by the west-east "Roaring Forties,"
The Polynesians when entering the Northern Hemisphere
Were advantaged not only by their ability
To sail into the wind,
But also by the east-west counter-currents
Of the tropical westward trade winds,
Which they discovered and
Called so because they made it possible
To go back where man had previously been
And thus to integrate world resources.
Thus the secretly held navigational capability
And knowledge of the elemental counting and astronomy
Went westward from Polynesia
Throughout Malaysia and to southern India,
Across the Indian Ocean to Mesopotamia and Egypt
And thence into the Mediterranean.
The powerful priests of Babylon, Egypt, and Crete
Were the progeny of mathematician navigators of the Pacific
Come up upon the land
To guide and miracle-ize the new kings
Of the Western Worlds.

Knowing all about boats,
These navigator priests were the only people
Who knew that the Earth is spherical,
That the Earth is a closed system
With its myriad resources chartable.
But being water people,
They kept their charts in their heads
And relayed the information
To their navigator progeny
Exclusively in esoterical,
Legendary, symbolical codings
Embroidered into their chants.

But some of their numbers
Also sailed deliberately eastward
Carrying their mathematical skills
To west-coast America.

The Mayans used base twenty in their numerical system
By counting with both their fingers and toes.
The number twenty often occurs
In a ''magically'' strategic way.
For an example
We can look at symmetrical aggregates
Of progressively assembled spheres
Closest packed on a plane—a pool table.
First take two balls and make them tangent.
Tangent is the ''closest''
That spheres may come to one another.
We may next nest a third ball
In the valley between the first tangent two.
Now each of the three spheres is tangent to two others
And none can get closer to each other.
These three make a triangle.
There is no ball in the center
Of the triangular group.
We can now add three more balls to the first three
By arranging them tangentially in a row
Along one edge of the first three's triangle.
As yet, all six balls are arranged
As outside edges of the triangle.
Not until we add a fourth row of balls

Nested along one edge of the triangular aggregate
Does a single ball become placed as the nuclear ball
In the center of the triangular "patterned" ball pool-table array.
Ten is the total number of balls
In this first nuclear-ball-containing triangle:
Nine surround the nuclear tenth ball.
And since a triangle is a fundamental structural pattern,
And since the triangular aggregate
Of nine balls around a nuclear one
Is a symmetrical array,
Man's intuitive choice of "congruence in modulo ten"
May have been more subtly conceived
Than simply by coincidence
With the ten digits of his hands.

We will now see what happens experimentally
When sailors stack coconut or orange cargoes
Or when we stack planar groups of triangular aggregates of spheres
On top of one another in such a manner that they will be
Structurally stable without binding agents.

First we will nest six balls
In a closest-packed triangular planar array
On top of the first triangularly arranged ten-ball aggregate.
And on top of those six balls
We can nest three more.
We now have a total of nineteen balls.
We may now nest one more topmost ball
In the one "nest" of the three-ball triangle.
We now have a symmetrical
Tetrahedral aggregate
Consisting of twenty balls
Without any nuclear ball
Occurring in the center
Of the symmetrical tetrahedral pyramid of balls.
We began our vertical stacking
With a symmetrical base triangle of ten balls,
And now we have a tetrahedron composed of twenty balls.
Just as fingers alone may not have been the only reason
For the choice of base ten,
Fingers and toes together may not have been the only reason

That the Mayan priests chose
Congruence in modulo twenty
Or that twenty was considered a magical number.
It might have been the result of an intuitive understanding
Of closest packing of spheres,
Which is something much more fundamental.
For unlike our fingers which lie in a row,
The packing of twenty spheres
That can be grouped symmetrically together without a nucleus
Is a fundamentally significant phenomenon.
In a tetrahedron composed of twenty balls
There is no nucleus.
This may be why twenty appears so abundantly
In the different chemical element isotopes.
And "twenty" is one of the "Magic Numbers"
In the inventory of chemical-element isotopal abundancy in Universe.

In order to position a nuclear ball in the center
Of a symmetrical tetrahedral pyramid of balls,
We need to add another or fifth nested layer of fifteen balls
To one face of the tetrahedron of twenty.
The total number of balls is then thirty-five,
Of which one is the nuclear ball.
If, however, we add four
Progressively larger
Triangular layers of balls
To each of the four triangular faces
Of the twenty-ball, no-nucleus tetrahedron,
It will take exactly one hundred more balls
To enclose the twenty-ball, no-nucleus tetrahedron—
This makes a symmetrical tetrahedron
Of one hundred and twenty balls.
This symmetrical tetrahedron
Is the largest symmetrical assembly
Of closest-packed spheres nowhere containing
Any two-layer-covered nuclear spheres
That is experimentally demonstrable.
In the external affairs of spheres
Such omnidimensional spherical groupings
Of one hundred and twenty same-size balls
Without a nucleus ball

Can be logically identified
With the internal affairs
Of individual spheres,
Wherein we rediscovered
The one hundred and twenty,
Least-common-denominator,
Right spherical triangles of the sphere,
Which are archeologically documented
As having been well known to the Babylonians'
Come-out-upon-the-land-ocean,
Navigator-high-priest mathematicians.

The number 120 also appears as a "Magic Number"
In the relative-abundance hierarchy
Of chemical-element isotopes of Universe.
One hundred and twenty accommodates
Both the decimal and the duodecimal system
(Ten multiplied by twelve).

The Mayans too may have understood
About the tetrahedral closest packing of spheres.
They probably made such tetrahedra
With symmetrically closest-packed stacks of oranges.

The twentieth-century fruit-store man
Spontaneously stacks his spherical fruits
In such closest-packed
Stacking and nesting arrays.
But the physicists didn't pay any attention
To the fruit-store man until 1922.
Then for the first time physicists
Called the tetrahedral stacks of fruit
"Closest packing of spheres."
For centuries past
The numerologists had paid attention
To the closest packing of spheres
In tetrahedral pyramids,
But were given the academic heave-ho
When in the mid-nineteenth century
Physicists abandoned the concept of models.

We have seen
That there are unique or cardinal names
For the concepts one through twelve
In England and Germany,
And for the concepts one through sixteen in France,
But that after that they simply repeat
In whatever congruence modulos
They happen to be working.
The Arabic numerals as well as their names
Are unique and stand alone
Only from zero through nine.
However, eleven is the result of two ones—11,
And twelve is similarly fashioned from two
Previously given symbols,
Namely, one and two—12.

But certain numbers
Such as prime numbers
Have their own cosmic integrity
And therefore ought to be integrally expressed.
What the numerologist does
Is to add numerals horizontally $(120 = 1 + 2 + 0 = 3)$
Until they are all consolidated into one integer.
Numerologists have also assigned
To the letters of the alphabet
Corresponding numbers: *A* is one, *B* is two, *C* is three, etc.
Numerologists wishfully assume
That they can identify
Characteristics of people
By the residual integer
Derived from integrating
All of the integers,
(Which integers
They speak of as digits,
Identifying with the fingers of their hands,
That is, their fingers.)
Corresponding to all the letters
In the individual's complete set of names.
Numerologists do not pretend to be scientific.
They are just fascinated
With correspondence of their key digits

With various happenstances of existence.
They have great fun
Identifying events and things
And assuming significant insights
Which from time to time
Seem well justified,
But what games numerologists
Chose to play with these tools
May or may not have been significant.
Possibly by coincidence, however,
And possibly because of number integrity itself .
Some of the integer intergrating results
Are found to correspond elegantly
With experimentally proven, physical laws
And have subsequently proven to be
Infinitely reliable.
Half a century ago I became interested in seeing
How numerologists played their games.
I found myself increasingly intrigued
And continually integrating digits.

1220.00 Indigs

1220.10 Definition: All numbers have their own integrity.

1220.11 The name *digit* comes from *finger*. A finger is a digit. There are five fingers on each hand. Two sets of five digits give humans a propensity for counting in increments of 10.

1220.12 Curiosity and practical necessity have brought humans to deal with numbers larger than any familiar quantity immediately available with which to make matching comparison. This frequent occurrence induced brain-plus-mind capabilities to inaugurate ingenious human information-apprehending mathematical stratagems in pure principle. If you are looking at all the pebbles on the beach or all the grains of sand, you have no spontaneous way of immediately quantifying such an experience with discrete number magnitude. Quantitative comprehension requires an integrative strategy with which to reduce methodically large unknown numbers to known numbers by use of obviously well-known and spontaneously employed linear-, area-, volume-, and time-measuring tools.

1220.13 **Indig Table A: Comparative Table of Modular Congruences of Cardinal Numbers:** This is a comparative table of the modular congruences of cardinal-number systems as expressed in Arabic numerals with the individual integer symbols integrated as *indigs,* which discloses synergetic wave-module behaviors inherent in nature's a priori, orderly, integrative effects of progressive powers of interactions of number:

Visually	*Nonintegrated*	*Indigs*	*Indig*
1	1	1	
11	11	2	
1 1 1	111	3	
11 11	1111	4	
11 1 11	11111	5	
111 111	111111	6	
111 1 111	1111111	7	
11 11 11 11	11111111	8	
111 111 111	111111111	9	
two hands	1111111111	10	1
too much	11111111111	11	2

1220.14 Man started counting large numbers which he did not recognize as a discrete and frequently experienced pattern by modularly rhythmic repetitive measuring, or matching, with discrete patterns which he did recognize—as, for instance, by matching the items to be counted one for one with the successive fingers of his two hands. This gave him the number of separate items being considered. Heel-to-toe stepping off of the number; or foot-after-foot length dimensions; or progressively and methodically covering areas with square woven floor mats of standard sizes, as the Japanese *tatami* and *tsubo;* or by successive mouthfuls or handfuls or bowls full, counted on the fingers of his hands, then in multiples of hands (i.e., multiples of ten), gave him commonly satisfactory volume measurements.

1220.15 Most readily humans recognized and trusted one and one making two, or one and two making three, or two and two making four. But an unbounded loose set of 10 irregular and dissimilar somethings was not recognizable by numbers in one glance: it was a lot. Nor are five loose, irregular, and dissimilar somethings recognizable in one glance as a number: they are a bunch. But a human hand is boundaried and finitely recognizable at a single glance as a *hand,* but not as a discrete number except by repetitively acquired confirmation and reflexive conditioning. Five is more recognizable as four fingers and a thumb, or even more readily recognizable as two end fingers (the little and the index), two fingers in the middle, and the thumb $(2 + 2 + 1 = 5)$.

1220.16 Symmetrical arrays of identically shaped and sized, integrally symmetric objects evoke spontaneous number identification *from one to six,* but not beyond. Paired sets of identities to six are also spontaneously recognized; hence we have dice and dominoes.

1220.17 Thus humans learned that collections of very large numbers consist of multiples of recognizable numbers, which recognition always goes back sensorially to spontaneously and frequently proven matching correspondence with experientially integrated pattern simplexes. One orange is a point (of focus). Two oranges define a line. Three oranges define an area (a

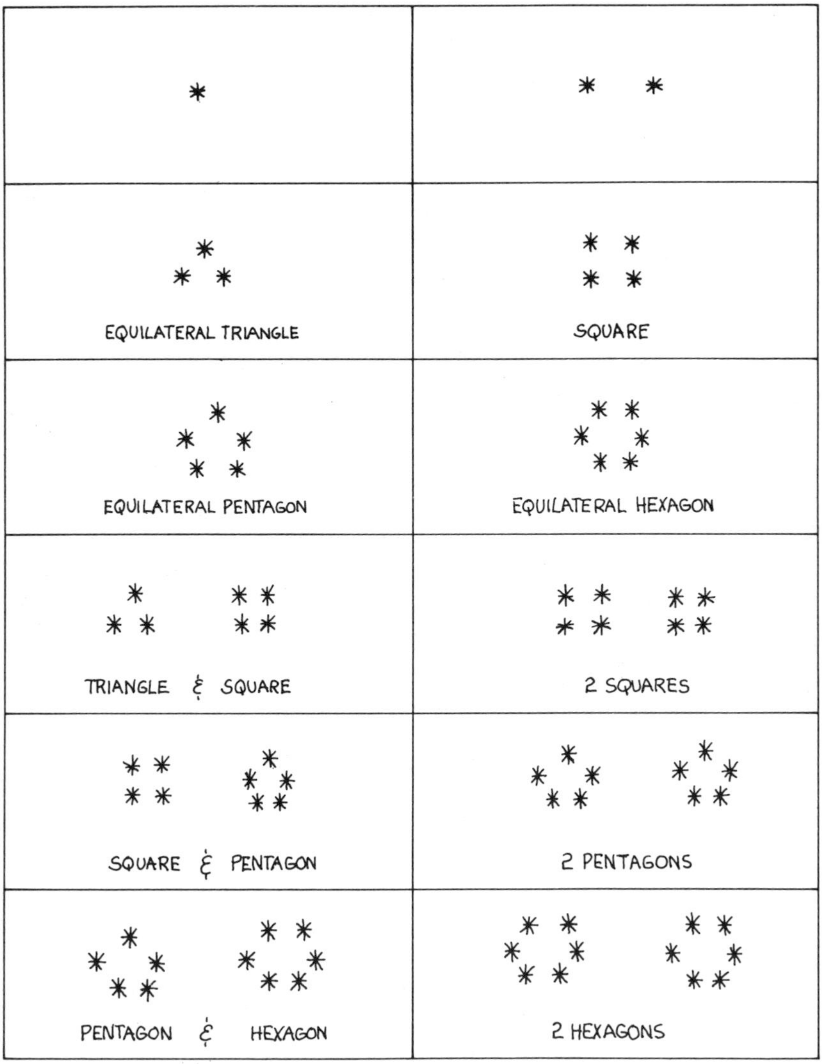

EQUILATERAL TRIANGLE	SQUARE
EQUILATERAL PENTAGON	EQUILATERAL HEXAGON
TRIANGLE & SQUARE	2 SQUARES
SQUARE & PENTAGON	2 PENTAGONS
PENTAGON & HEXAGON	2 HEXAGONS

Fig. 1220.16.

triangle). And four oranges, the fourth nested atop the triangled first three, define a multidimensional volume, a tetrahedron, a scoop, a cup.*

1220.20 Numerological Correspondence: Numerologists do not pretend to be scientific. They are just fascinated with a game of correspondence of their ''key'' digits—finger counts, ergo, 10 digits—with various happenstances of existence. They have great fun identifying the number ''seven'' or the number ''two'' types of people with their own ingeniously classified types of humans and types of events, and thereafter imaginatively developing significant insights which from time to time seem justified by subsequent coincidences with reality. What intrigues them is that the numbers themselves are integratable in a methodically reliable way which, though quite mysterious, gives them faithfully predictable results. They feel intuitively confident and powerful because they know vaguely that scientists also have found number integrity exactly manifest in physical laws.

1220.21 The numerologists have also assigned serial numbers to the letters of the alphabet: *A* is one, *B* is two, *C* is three, etc. Because there are many different alphabets of different languages consisting of various quantities of letters, the number assignments would not correspond to the same interpretations in different languages. Numerologists, however, preoccupied only in their single language, wishfully assumed that they could identify characteristics of people by the residual digits corresponding to all the letters in the individual's complete set of names, somewhat as astrologists identify people by the correspondences of their birth dates with the creative picturing constellations of the Milky Way zoo = Zodiac = Celestial Circus of Animals.

1221.00 Integration of Digits

1221.10 Quantifying by Integration: Early in my life, I became interested in the mathematical potentials latent in the methodology of the numerologists. I found myself increasingly intrigued and continually experimenting with digit integrations. What the numerologist does is to add numbers as expressed horizontally; for instance:

$120 = 1 + 2 + 0 = 3$

Or:

$32986513 = 3 + 2 + 9 + 8 + 6 + 5 + 1 + 3 = 37 = 3 + 7 = 10 = 1 + 0 = 1,$

Numerologically, $32986513 = 1$

* This may have been the genesis of the cube—where all the trouble began. Why? Because man's tetrahedron scoop would not stand on its point, spilled, frustrated counting, and wasted valuable substances. So humans devised the square-based volume: the cube, which itself became an allspace-filling multiple cube building block easily appraised by ''cubing'' arithmetic.

Or:

59865279171 = 5 + 9 = 14 + 8 = 22 + 6 = 28 + 5 = 33 + 2 = 35 + 7 = 42 + 9 = 51 + 1 = 52 + 7 = 59 + 1 = 60 = 6 + 0 = 6,

Numerologically, 59865279171 = 6.

1221.11 Though I was familiar with the methods of the calculus—for instance, quantifying large, irregularly bound areas—explorations in numerology had persuaded me that large numbers themselves, because of the unique intrinsic properties of individual numbers, might be logically integratable to disclose initial simplexes of sensorial interpatterning apprehendibility.

1221.12 Integrating the symbols of the modular increments of counting, in the above case in increments of 10, as expressed in the ten-columnar arrays of progressive residues (less than ten—or less than whatever the module employed may be), until all the columns' separate residues are reduced to one *integral digit,* i.e., an integer that is the ultimate of the numbers that have been integrated. Unity is plural and at minimum two. (See Secs. 240.03; 527.52; and 707.01.)

1221.13 As a measure of communications economy, I soon nicknamed as *indigs* the final unitary reduction of the integrated digits. I use *indig* rather than *integer* to remind us of the process by which ancient mathematicians counting with their fingers (digits) may have come in due course to evolve the term *integer*.

1221.14 I next undertook the indigging of all the successive modular congruence systems ranging from one-by-one, two-by-two pairs to "by the dozens," i.e., from zero through 12. (See modulo-congruence tables, Sec. 1221.20.)

1221.15 The modulo-congruence tables are expressed in both *decimal* and *indig* terms. In each of the 13 tables of the chart, the little superscripts are the indigs of their adjacently below, decimally expressed, corresponding integers.

1221.16 The number of separate columns of the systematically displayed tables corresponds with the modulo-congruence system employed. Inspection of successive horizontal lines discloses the orderly indig amplifying or diminishing effects produced upon arithmetical integer progression. The result is startling.

1221.17 Looking at the chart, we see that when we integrate digits, certain integers invariably produce discretely amplifying or diminishing alterative effects upon other integers.

One produces a plus oneness;

Two produces a plus twoness;

Three produces a plus threeness;

Four produces a plus fourness.

Whereafter we reverse,

Five produces a minus fourness;

Six produces a minus threeness;
Seven produces a minus twoness;
Eight produces a minus oneness.

Nine produces zero plusness or minusness. One and ten are the same. Ten indigs (indig = verb intransitive) as a *one* and produces the same alterative effects as does one. Eleven indigs as *two* and produces the same alterative effects as a two. All the other whole numbers of any size indig to 1, 2, 3, 4, 5, 6, 7, 8, or 9—ergo, have the plus or minus oneness to fourness or zeroness alterative effects on all other integers.

1221.18 Since the Arabic numerals have been employed by the Western world almost exclusively as congruence in modulo ten, and the whole world's scientific, political, and economic bodies have adopted the metric system, and the notation emulating the abacus operation arbitrarily adds an additional symbol column unilaterally (to the left) for each power of ten attained by a given operation, it is reasonable to integrate the separate integers into one integer for each multisymboled number. Thus 12, which consists of 1 + 2, = 3; and speaking numerologically, 3925867 = 4.

1	=	+1	+1	+
2	=	+2	+2	+
3	=	+3	+3	+
4	=	+4	+4	+
5	=	5	−4	+
6	=	6	−3	+
7	=	7	−2	+
8	=	8	−1	+
9	=	0	0	
10	=	1	+1	−
11	=	2	+2	−
12	=	3	+3	−
13	=	4	+4	−
14	=	5	−4	−
15	=	6	−3	−
16	=	7	−2	−
17	=	8	−1	−
18	=	0	0	
19	=	1	+1	
20	=	2	+2	

This provides an octave number system of a plus and minus octave and an (outside-out) and an (inside-out) differentiation, for every system has insideness (concave) and outsideness (convex) as well as two polar hemisystems.

1221.20 **Indig Table B: Modulo-Congruence Tables:** The effects of integers: One is +1. Two is +2. Three is +3. Four is +4. Five is −4. Six is −3. Seven is −2. Eight is −1. Nine is zero; nine is none.

(The superior figures in the Table are the *Indigs*.)

Congruence in Modulo Zero Integrates to Gain or Lose 0:

0 (Like nine) Ø

Congruence in Modulo One Integrates to Gain 1:

1^1
2^2
3^3
4^4
5^5
6^6
7^7
8^8
9^9
10^1
11^2
12^3

(Each row gains 1 in each column)

+1

Congruence in Modulo Two Integrates to Gain 2

1^1	2^2
3^3	4^4
5^5	6^6
7^7	8^8
9^9	10^1
11^2	12^3
13^4	14^5
15^6	16^7

(Each row gains 2 in each column)

+2

Congruence in Modulo Three Integrates to Gain 3:

1^1	2^2	3^3
4^4	5^5	6^6
7^7	8^8	9^9
10^1	11^2	12^3
13^4	14^5	15^6
16^7	17^8	18^9
19^1	20^2	21^3

(Each row gains 3 in each column)

+3

Congruence in Modulo Four Integrates to Gain 4:

1^1	2^2	3^3	4^4
5^5	6^6	7^7	8^8
9^9	10^1	11^2	12^3
13^4	14^5	15^6	16^7
17^8	18^9	19^1	20^2
21^3	22^4	23^5	24^6

(Each row gains 4 in each column)

+4

Congruence in Modulo Five Integrates to Lose 4:

1^1	2^2	3^3	4^4	5^5
6^6	7^7	8^8	9^9	10^1
11^2	12^3	13^4	14^5	15^6
16^7	17^8	18^9	19^1	20^2
21^3	22^4	23^5	24^6	25^7
26^8	27^9	28^1	29^2	30^3

(Each row loses 4 in each column)

−4

Congruence in Modulo Six Integrates to Lose 3:

1^1	2^2	3^3	4^4	5^5	6^6	(Each row loses 3
7^7	8^8	9^9	10^1	11^2	12^3	in each column)
13^4	14^5	15^6	16^7	17^8	18^9	
19^1	20^2	21^3	22^4	23^5	24^6	−3
25^7	26^8	27^9	28^1	29^2	30^3	

Congruence in Modulo Seven Integrates to Lose 2:

1^1	2^2	3^3	4^4	5^5	6^6	7^7	(Each row loses 2
8^8	9^9	10^1	11^2	12^3	13^4	14^5	in each column)
15^6	16^7	17^8	18^9	19^1	20^2	21^3	
22^4	23^5	24^6	25^7	26^8	27^9	28^1	−2

Congruence in Modulo Eight Integrates to Lose 1:

1^1	2^2	3^3	4^4	5^5	6^6	7^7	8^8	(Each row loses 1
9^9	10^1	11^2	12^3	13^4	14^5	15^6	16^7	in each column)
17^8	18^9	19^1	20^2	21^3	22^4	23^5	24^6	
25^7	26^8	27^9	28^1	29^2	30^3	31^4	32^5	−1

Congruence in Modulo Nine Integrates to No Loss or Gain:

1^1	2^2	3^3	4^4	5^5	6^6	7^7	8^8	9^9	(Each row remains
10^1	11^2	12^3	13^4	14^5	15^6	16^7	17^8	18^9	same value in its
19^1	20^2	21^3	22^4	23^5	24^6	25^7	26^8	27^9	column)
28^1	29^2	30^3	31^4	32^5	33^6	34^7	35^8	36^9	Ø

Congruence in Modulo Ten Integrates to Gain 1:

1^1	2^2	3^3	4^4	5^5	6^6	7^7	8^8	9^9	10^1	(Each row gains 1
11^2	12^3	13^4	14^5	15^6	16^7	17^8	18^9	19^1	20^2	in each column)
21^3	22^4	23^5	24^6	25^7	26^8	27^9	28^1	29^2	30^3	
31^4	32^5	33^6	34^7	35^8	36^9	37^1	38^2	39^3	40^4	+1

Congruence in Modulo Eleven Integrates to Gain 2:

1^1	2^2	3^3	4^4	5^5	6^6	7^7	8^8	9^9	10^1	11^2	(Each row gains 2
12^3	13^4	14^5	15^6	16^7	17^8	18^9	19^1	20^2	21^3	22^4	in each column)
23^5	24^6	25^7	26^8	27^9	28^1	29^2	30^3	31^4	32^5	33^6	
34^7	35^8	36^9	37^1	38^2	39^3	40^4	41^5	42^6	43^7	44^8	+2

Congruence in Modulo Twelve Integrates to Gain 3:

1^1	2^2	3^3	4^4	5^5	6^6	7^7	8^8	9^9	10^1	11^2	12^3	(Each row gains 3
13^4	14^5	15^6	16^7	17^8	18^9	19^1	20^2	21^3	22^4	23^5	24^6	in each column)
25^7	26^8	27^9	28^1	29^2	30^3	31^4	32^5	33^6	34^7	35^8	36^9	
37^1	38^2	39^3	40^4	41^5	42^6	43^7	44^8	45^9	46^1	47^2	48^3	+3

1222.00 Absolute Four and Octave Wave

1222.10 Prime Dichotomy: It is found that all decimally expressed whole numbers integrate into only nine digits. Looking at the charts (Indig Table B), we see the nine indigs resultant to the decimal system, or congruence in modulo ten, have integrated further to disclose only nine unique operational effects upon all other integers. These nine interoperational effects in turn reduce into only eight other integer-magnitude-altering effects and one no-magnitude-altering effect. The ''octave'' of eight magnitude-altering sets of indigs in turn disclose primary dichotomy into four positively altering and four negatively altering magnitude operators, with each set arranged in absolute arithmetical sequence of from one to four only.

1222.11 Indig congruences demonstrate that nine is zero and that number system is inherently octave and corresponds to the four positive and four negative octants of the two polar domains (*obverse* and *reverse*) of the octahedron—and of all systems—which systematic polyhedral octantation limits also govern the eight 45-degree-angle constituent limits of 360-degree unity in the trigonometric function calculations.

1222.12 The inherent $+4, -4, 0, +4, -4, 0 \rightarrow$ of number also corresponds (a) to the four varisized spheres integrating tritangentially to form the tetrahedron (see Sec. 1222.20) and (b) to the octantation of the Coupler (see Sec. 954.20) by its eight allspace-filling Mites (*AAB* Modules) which, being inherently plus-or-minus biased, though superficially invariant (i.e., are conformationally identical); altogether provide lucidly synergetic integration (at a kindergarten-comprehendible level) of cosmically basic number behavior, quantum mechanics, synergetics, nuclear physics, wave phenomena in general, and topologically rational accountability of experience in general.

1222.20 Cosmically Absolute Numbers: There are apparently no cosmically absolute numbers other than 1, 2, 3, and 4. This primitive fourness identifies exactly with one quantum of energy and with the fourness of the tetrahedron's primitive structuring as constituting the ''prime structural system of Universe,'' i.e., as the minimum omnitriangulated differentiator of Universe into insideness and outsideness, which alone, of all macro-micro Universe differentiators, pulsates inside-outingly and vice verse as instigated by only one force vector impinging upon it. (See Sec. 624.)

1222.30 Casting Out Nines: We can use any congruence we like, and the pattern will be the same. The wave phenomenon, increasing by four and decreasing by four, is an octave beginning and ending at zero. From this I saw that nine is zero.

1222.31 When I worked for Armour and Company before World War I, I had to add and multiply enormous columns of figures every day. As yet, neither commercially available adding machines nor electric calculators existed. The auditors showed us how to check our multiplications by "casting out nines." This is done by inspecting all the *input* integers of multiplication, first crossing out any nines and then crossing out any combinations of integers that add to nine, exclusively *within* either the (a) multiplicands, (b) multipliers, or (c) products of multiplication, taken separately. This means we do not take combinations of integers occurring in other than their own respective (a), (b), or (c) sets of integers that add up to nine.

1222.32 (a) Multiplier
(b) × Multiplicand
―――――――
(c) Product

Cross out all nines, or any set of integers adding to nine, in any one of either the multiplier (a), the multiplicand (b), or the product (c).

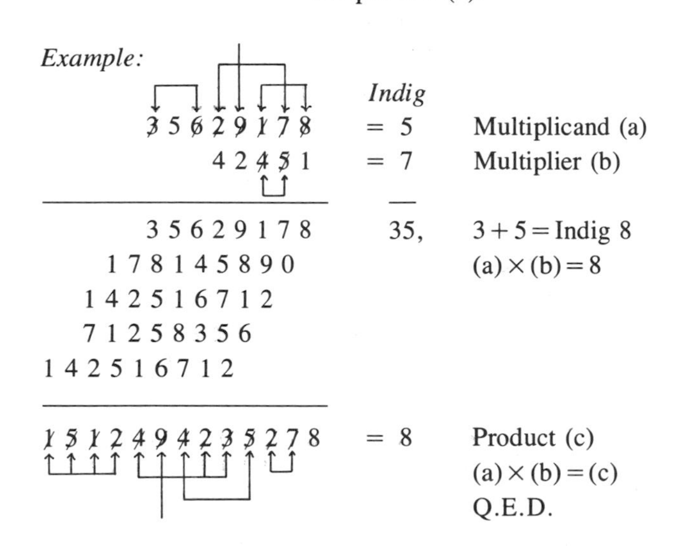

Example:
Indig
3 5 6 2 9 1 7 8 = 5 Multiplicand (a)
4 2 4 5 1 = 7 Multiplier (b)
3 5 6 2 9 1 7 8
1 7 8 1 4 5 8 9 0
1 4 2 5 1 6 7 1 2
7 1 2 5 8 3 5 6
1 4 2 5 1 6 7 1 2
35, 3+5=Indig 8
(a)×(b)=8
1 5 1 2 4 9 4 2 3 5 2 7 8 = 8 Product (c)
(a)×(b)=(c)
Q.E.D.

1223.00 Wave Pulsation of Indigs

1223.10 **Pulsative Octave:** The interaction of all numbers other than nine creates the wave phenomenon described, i.e., the self-invertible, self-inside-outable *octave* increasing and decreasing pulsatively, fourfoldedly, and tetrahedrally. No matter how complex a number-aggregating sequence of events and conditions may be, this same number behavior phenomenon is all that ever happens. There is thus a primitively comprehensive, isotropically distributive, carrier-wave order omniaccommodatively permeating and embracing all phenomena. (See Sec. 1012.10)

1223.11 As the nine columns of Indig Table 2 show, I have integrated the digits of all the different multiplication systems and have always found the positively-negatively pulsative, octave, zero-nine-intervaled, ergo interference-free, carrier-wave pattern to be permeating all of them in four alternative interger-mix sequences; with again, four positively ordered and four negatively ordered sequence sets, all octavely ventilated by zero nines cyclically, ergo inherently, ergo eternally synchronized to non-inter-interferences.

1223.12 As will also be seen in Indig Table 2, the integer carrier waves can pulse in single sets, as in Columns 1 and 8; in double pairs, as in Columns 4 and 5; in triple triplets, as in Columns 3 and 6; and in double quadruplets, as in Columns 2 and 7—always octavely interspersed with zeros and, in the case of Columns 3 and 6, interspersed with zeros triangularly as well as octavely. This also means that the omnidirectional wave interpermutatings are accommodated as points or as lines; or as triangular areas; or as tetrahedral volumes—both positive and negative.

1223.13 Thus we are informed that the carrier waves and their internal-external zero intervalling are congruent with the omnitriangulated, tetraplaned, four-dimensional vector equilibria and the omniregenerative isotropic matrix whose univectorings accommodate any wavelength or frequency multiplying in respect to any convergently-divergently nuclear system loci of Universe.

1223.14 Not only is there an external zero intervalling between all the unique octave-patterning sets in every one of the four positive, four negative systems manifest, but we find also the wave-intermodulating indigs *within* each octave always integrating sum-totally internally to the octaves themselves as *nines,* which is again an internal zero content—this produces in effect a positive zero function vs. a negative zero function, i.e., an inside-out and outside-out zero as the ultracosmic zero-wave pulsativeness.*

1223.15 Thus we discover the modus operandi by which radio waves and other waves pass uninterferingly through seeming solids, which are themselves only wave complexes. The lack of interference is explained by the crossing of the high-frequency waves through the much lower-frequency waves at the

* See Sec. 1012, which describes a closest-sphere-packing model of the same phenomenon. If we make an *X* configuration with one ball in the center common to both triangles of the *X*, the ball at the intersection common to both represents the zero—or the place where the waves can pass through each other. The zero always accommodates when two waves come together. We know that atoms close-pack in this manner, and we know how wave phenomena such as radio waves behave. And now we have a model to explain how they do not interfere.

FUNCTIONAL PROPERTIES OF DIGITS - THERE ARE ONLY 16 FUNCTIONING NUMBERS

First 8 are negative -8 to -1; 9 punctuates without value; next 8 (10-17) are plus +1 to +8.
All higher numbers integrate to the basic number functions of punctuated octave.

[illegible] digits to any number effects the integrated digits by the amount in the second column.	
1	-8
2	-7
3	-6
4	-5
5	-4
6	-3
7	-2
8	-1
9	0
10	+1
11	+2
12	+3
13	+4
14	+5
15	+6
16	+7
17	+8
18	0
19	+1
20	+2
21	+3
22	+4
23	+5
24	+6
25	+7
26	+8
27	0
28	+1
29	+2
30	+3
31	+4
32	+5
33	+6
34	+7
35	+8
36	0
37	+1
38	+2
39	+3
40	+4
41	+5
42	+6
43	+7
44	+8
45	+0
46	+1
47	+2
48	+3
49	+4
50	+5
51	+6
52	+7
53	+8
54	0
55	+1
56	+2
57	+3
58	+4
59	+5
60	+6
61	+7
62	+8
63	0
64	+1

Therefore 9 will be written as 0 hereafter.

Fractions of 8 and its powers expressed as decimals and digits integrated.			Indig (integration of digits)
1/8	.125		.8
2/8	.250		.7
3/8	.375		.6
4/8	.500		.5
5/8	.625		.4
6/8	.750		.3
7/8	.875		.2
8/8	1.000		.1
9/8	1.125	9/	0
10/8	1.250		8

Indicate that first octave of single digits are minus or fractional or decimal 8

Fractions of 64 1/8x8 or 1/8² expressed as decimals with digits integrated		Indigs of 64ths	Fractions of 512 or 1/8x8x8 or 1/8³ expressed as decimals. First 20 only.	Indigs of 512ths	Fractions of 4096ths or 1/8x8x8x8 or 1/8⁴ expressed as decimals. First 20 only.	Indigs of 4096ths
1/64	.015625	1	.001953125	8	.000244140625	1
2/64	.031250	2	.003906250	7	.000488281250	2
3/64	.046875	3	.005859375	6	.000732421875	3
4/64	.062500	4	.007812500	5	.000976562500	4
5/64	.078125	5	.009765625	4	.001220703125	5
6/64	.093750	6	.011718750	3	.001464843750	6
7/64	.109375	7	.013671875	2	.001708984375	7
8/64	.125000	8	.015625000	1	.001953125000	8
9/64	.140625	9/0	.017578125	9/0	.002197265625	9/ 0
10/64	.156250	1	.019531250	8	.002441406250	1
11/64	.171875	2	.021484375	7	.002685546875	2
12/64	.187500	3	.023437500	6	.002929687500	3
13/64	.203125	4	.025390625	5	.003173828125	4
14/64	.218750	5	.027343750	4	.003417968750	5
15/64	.234375	6	.029296875	3	.003662109375	6
16/64	.250000	7	.031250000	2	.003906250000	7
17/64	.265625	8	.033203125	1	.004150390625	8
18/64	.281250	0	.035156250	0	.004394531250	0
19/64	.296875	1	.037109375	8	.004638671875	1
20/64	.312500	2	.039062500	7	.004882812500	2

Order of Octaves Inverted with Each Power of Fraction

Numbers 1 to 64 squared with digits integrated		Indigs of N²	Numbers 1 to 64 cubes with digits integrated		Indigs of N³
1	1	1	1	1	1
2	4	4	2	8	8
3	9	0	3	27	0
4	16	7	4	64	1
5	25	7	5	125	8
6	36	0	6	216	0
7	49	4	7	343	1
8	64	1	8	512	8
9	81	0	9	729	0
10	100	1	10	1000	1
11	121	4	11	1331	8
12	144	0	12	1728	0
13	169	7	13	2197	1
14	196	7	14	2744	8
15	225	0	15	3375	0
16	256	4	16	4096	1
17	289	1	17	4913	8
18	324	0	18	5832	0
19	361	1	19	6859	1
20	400	4	20	8000	8
21	441	0	21	9621	0
22	484	7	22	10648	1
23	529	7	23	12167	8
24	576	0	24	13824	0
25	625	4	25	15625	1
26	676	1	26	17576	8
27	729	0	27	19683	0
28	784	1	28	21952	1
29	841	4	29	24389	8
30	900	0	30	27000	0
31	961	7	31	29791	1
32	1024	7	32	32768	8
33	1089	0	33	35937	0
34	1156	4	34	39304	1
35	1225	1	35	42875	8
36	1296	0	36	46656	0
37	1369	1	37	50653	1

Note final digits repeat in phrases of ten - Nine digits and one zero with 5 inside

9 4 9 6 5 6 9 4 1 0 1 4 9 6 5 6 9 4 1

Note final digits repeat in phrases of ten - 9 digits and 1 zero with 5 in middle.

All numbers except 3 and 6 and cubed numbers have balanced 2 and 4 part octaves. In 3 and 6 based octaves the interval 0 occurs at 4-point intervals, as it does in the square as well.

EFFECTIVE FUNCTIONS OF DIGITS AND COMPOUNDS THEREOF

Only first 8 digits operative; 9 is functional zero.

Base 9 - no change

Base 10 - Adds 1 (left); Base 8 - Subtracts 1 (right)

1	2	3	4	5	6	7	8	0
10	11	12	13	14	15	16	17	0
19	20	21	22	23	24	25	26	0
28	29	30	31	32	33	34	35	0
37	38	39	40	41	42	43	44	0
46	47	48	49	50	51	52	53	0
55	56	57	58	59	60	61	62	0
64	65	66	67	68	69	70	71	0
73	74	75	76	77	78	79	80	0
82	83	84	85	86	87	88	89	0
91	92	93	94	95	96	97	98	0
100	101	102	103	104	105	106	107	0
109	110	111	112	113	114	115	116	0

Fig. 1223.12 *Indig Tables.*

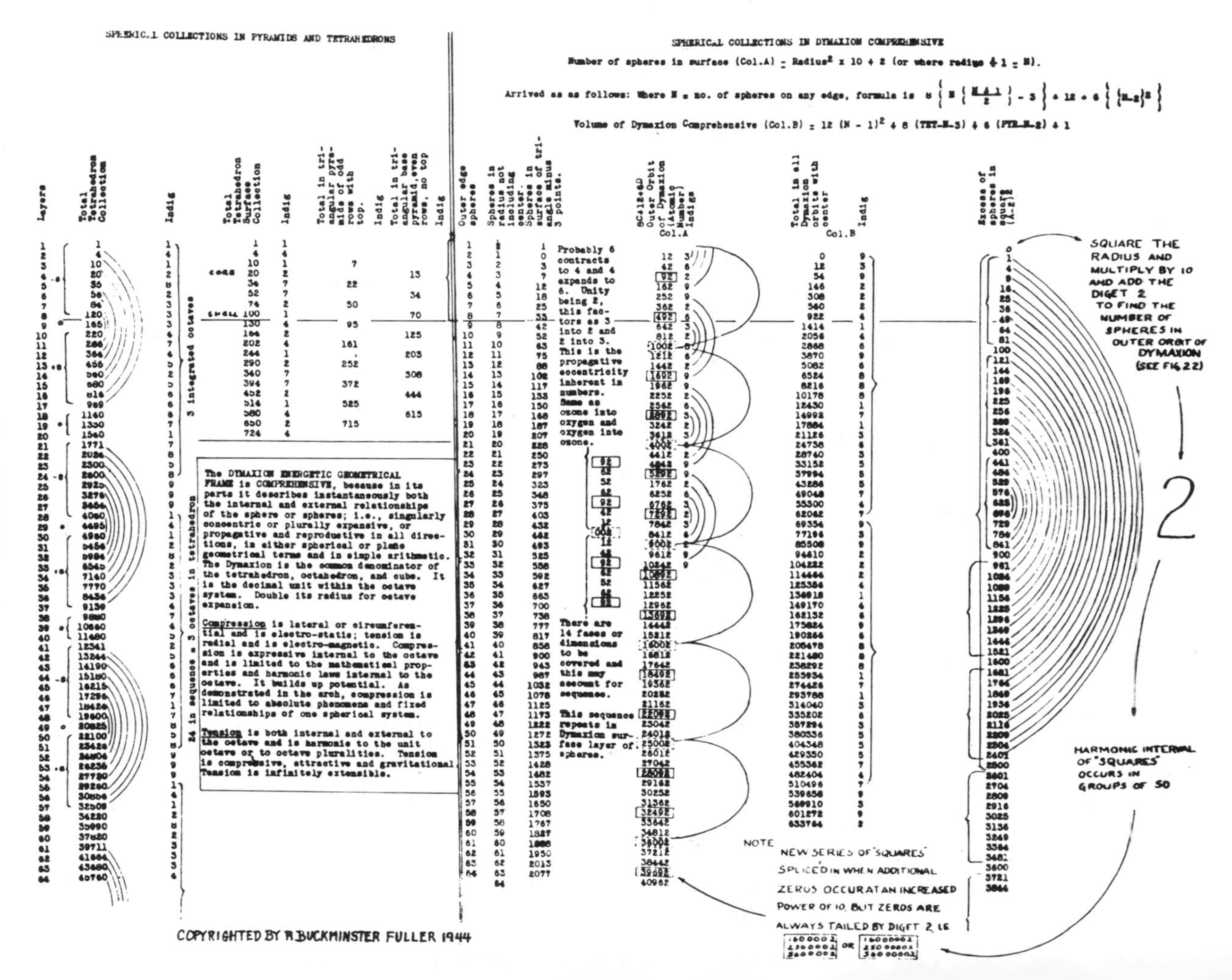

Fig. 1223.12 (cont.).

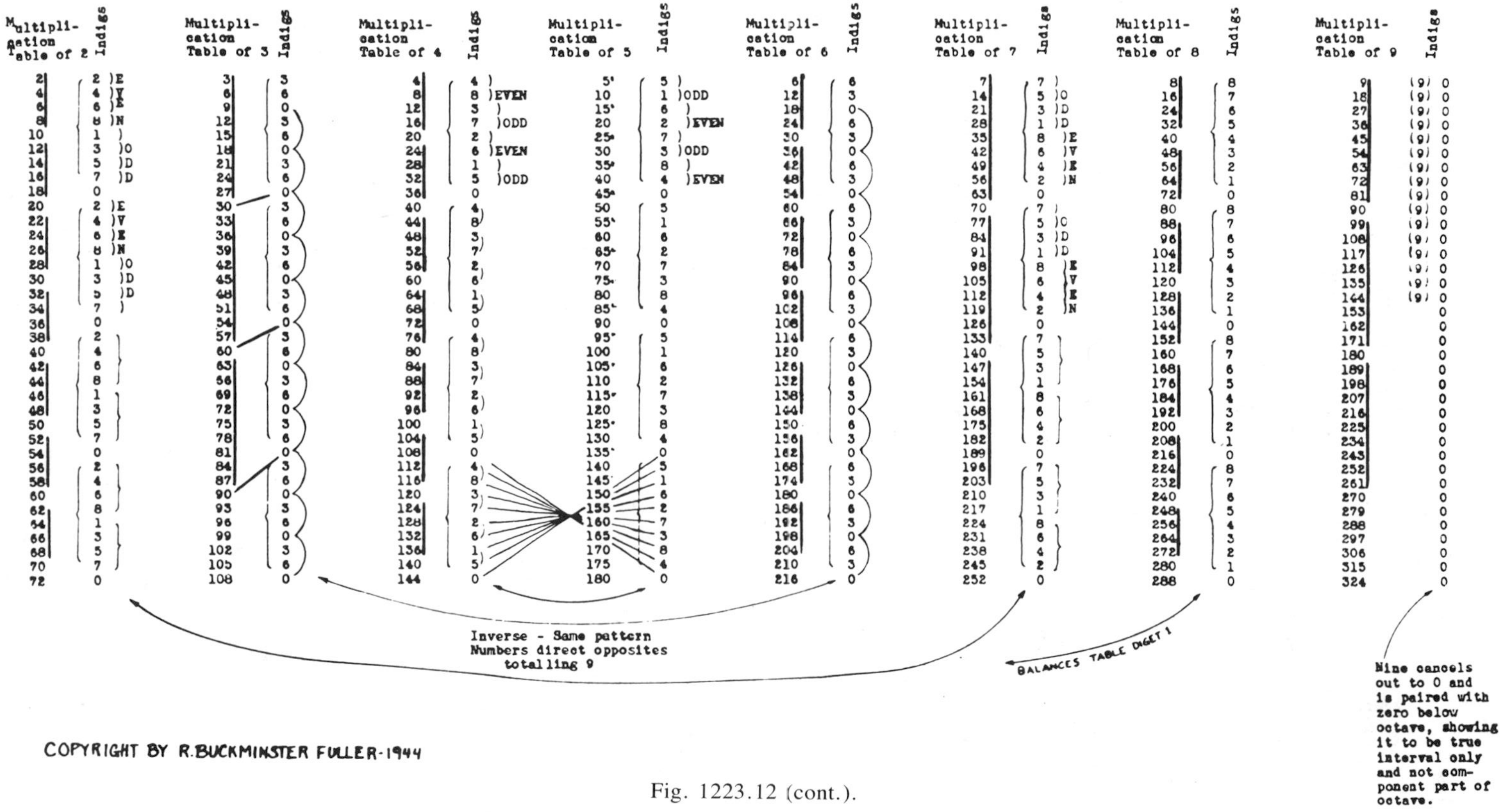

Multiplication Table of 2	Indigs	Multiplication Table of 3	Indigs	Multiplication Table of 4	Indigs	Multiplication Table of 5	Indigs	Multiplication Table of 6	Indigs	Multiplication Table of 7	Indigs	Multiplication Table of 8	Indigs	Multiplication Table of 9	Indigs
2	2	3	3	4	4	5	5	6	6	7	7	8	8	9	(9) 0
4	4	6	6	8	8	10	1	12	3	14	5	16	7	18	(9) 0
6	6	9	0	12	3	15	6	18	0	21	3	24	6	27	(9) 0
8	8	12	3	16	7	20	2	24	6	28	1	32	5	36	(9) 0
10	1	15	6	20	2	25	7	30	3	35	8	40	4	45	(9) 0
12	3	18	0	24	6	30	3	36	0	42	6	48	3	54	(9) 0
14	5	21	3	28	1	35	8	42	6	49	4	56	2	63	(9) 0
16	7	24	6	32	5	40	4	48	3	56	2	64	1	72	(9) 0
18	0	27	0	36	0	45	0	54	0	63	0	72	0	81	(9) 0
20	2	30	3	40	4	50	5	60	6	70	7	80	8	90	(9) 0
22	4	33	6	44	8	55	1	66	3	77	5	88	7	99	(9) 0
24	6	36	0	48	3	60	6	72	0	84	3	96	6	108	(9) 0
26	8	39	3	52	7	65	2	78	6	91	1	104	5	117	(9) 0
28	1	42	6	56	2	70	7	84	3	98	8	112	4	126	(9) 0
30	3	45	0	60	6	75	3	90	0	105	6	120	3	135	(9) 0
32	5	48	3	64	1	80	8	96	6	112	4	128	2	144	(9) 0
34	7	51	6	68	5	85	4	102	3	119	2	136	1	153	0
36	0	54	0	72	0	90	0	108	0	126	0	144	0	162	0
38	2	57	3	76	4	95	5	114	6	133	7	152	8	171	0
40	4	60	6	80	8	100	1	120	3	140	5	160	7	180	0
42	6	63	0	84	3	105	6	126	0	147	3	168	6	189	0
44	8	66	3	88	7	110	2	132	6	154	1	176	5	198	0
46	1	69	6	92	2	115	7	138	3	161	8	184	4	207	0
48	3	72	0	96	6	120	3	144	0	168	6	192	3	216	0
50	5	75	3	100	1	125	8	150	6	175	4	200	2	225	0
52	7	78	6	104	5	130	4	156	3	182	2	208	1	234	0
54	0	81	0	108	0	135	0	162	0	189	0	216	0	243	0
56	2	84	3	112	4	140	5	168	6	196	7	224	8	252	0
58	4	87	6	116	8	145	1	174	3	203	5	232	7	261	0
60	6	90	0	120	3	150	6	180	0	210	3	240	6	270	0
62	8	93	3	124	7	155	2	186	6	217	1	248	5	279	0
64	1	96	6	128	2	160	7	192	3	224	8	256	4	288	0
66	3	99	0	132	6	165	3	198	0	231	6	264	3	297	0
68	5	102	3	136	1	170	8	204	6	238	4	272	2	306	0
70	7	105	6	140	5	175	4	210	3	245	2	280	1	315	0
72	0	108	0	144	0	180	0	216	0	252	0	288	0	324	0

Brackets beside the first rows of the Indigs columns:

- Table of 2: 2, 4, 6, 8 marked EVEN; 1, 3, 5, 7 marked ODD (repeated for 20–34).
- Table of 4: 4, 8 marked EVEN; 3, 7 marked ODD; 2, 6 marked EVEN; 1, 5 marked ODD.
- Table of 5: 5, 1 marked ODD; 6, 2 marked EVEN; 7, 3 marked ODD; 8, 4 marked EVEN.
- Table of 7: 7, 5, 3, 1 marked ODD; 8, 6, 4, 2 marked EVEN (repeated for 70–126).

Inverse - Same pattern
Numbers direct opposites
totalling 9

BALANCES TABLE DIGIT 1

Nine cancels out to 0 and is paired with zero below octave, showing it to be true interval only and not component part of octave.

Fig. 1223.12 (cont.).

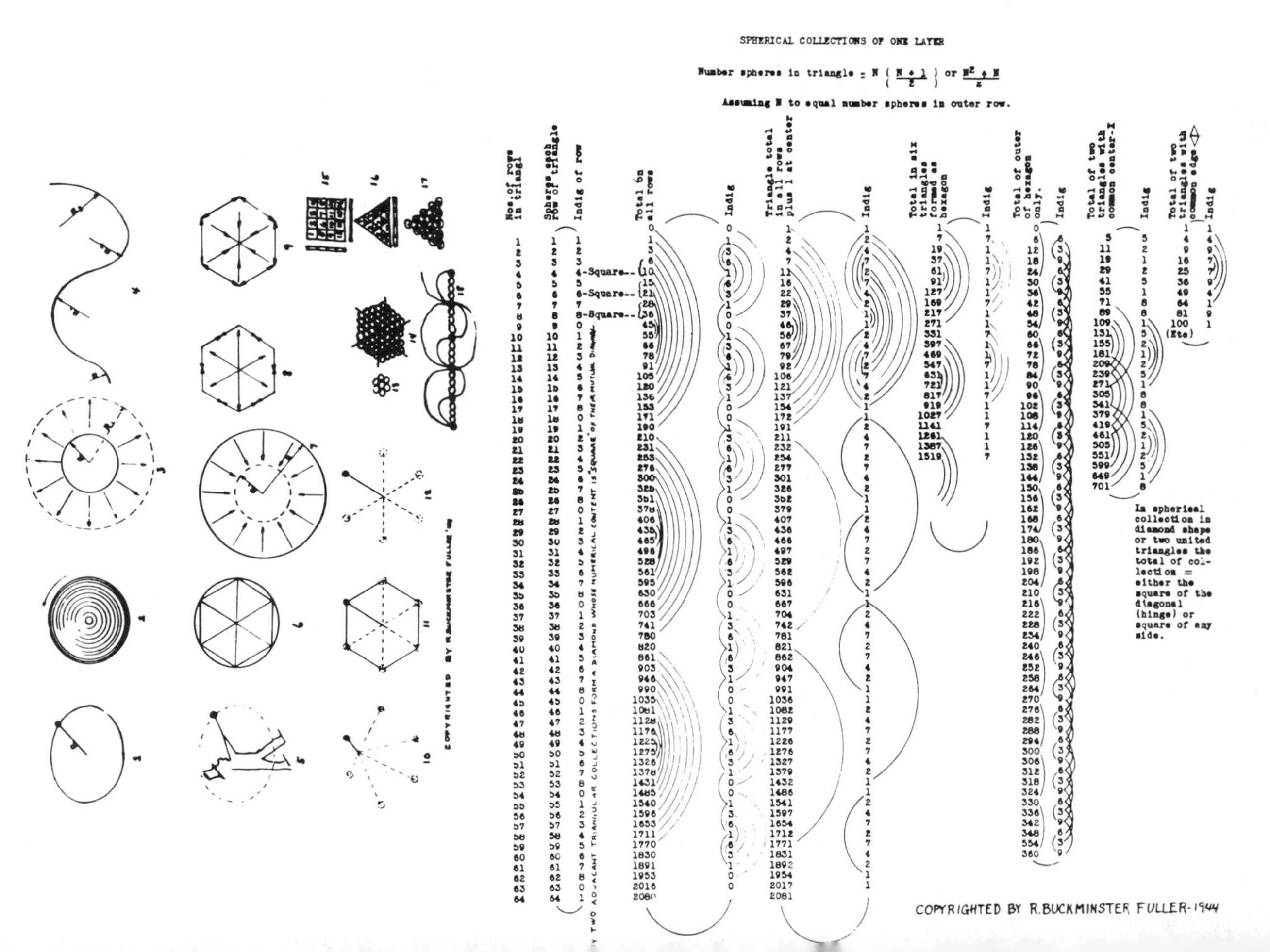

SPHERICAL COLLECTIONS OF ONE LAYER
Number spheres in triangle = N (N + 1 / 2) or N2 + N / 2
Assuming N to equal number spheres in outer row.
Nos. of rows in triangl
Spheres each row of triangle
Indig of row
4-Square--
6-Square--
8-Square--
TWO ADJACENT TRIANGULAR COLLECTIONS FORM A DIAMOND WHOSE NUMERICAL CONTENT IS SQUARE OF THE MUTUAL DIAGN.
Total 6n all rows
Indig
Triangle total in all rows plus 1 at center
Indig
Total in six triangles formed as hexagon
Indig
Total of outer of hexagon only.
Indig
Total of two triangles with common center-X
Indig
Total of two triangles with common edge
Indig
(Etc)
In spherical collection in diamond shape or two united triangles the total of collection = either the square of the diagonal (hinge) or square of any side.
COPYRIGHTED BY R. BUCKMINSTER FULLER-1944
COPYRIGHTED BY R. BUCKMINSTER FULLER '44

CHART A

noninterfering zero points, or indeed by the varifrequencied waves through both one another's internal and external zero intervals. (See Illus. 1012.13 A and B.)

1223.16 If the readers would like to do some of their own indig exploration they may be instructively intrigued by taking a book of mathematical tables and turning to the table of second powers of integers. If they undertake to indig each of those successively listed second-power numbers they will discover that, for the first 100 numbers listed, a unique sequence of 24 integers will appear that peaks at 25, reverses itself, and bottoms at one, only to turn again and peak at 50, bottom at 75, and peak again at the 100th number which, when analyzed, manifests a $2 \times 2 \times 2 = 8 = 2^3 \times 3 = 24$ four-dimensional wave. This four-dimensional wave is only comprehendible when we discover (see Sec. 982.62) the three-frequency reality of $F^3 \times 2^1/_2$, 3, 4, 5, 6, the a priori, initially-volumed, ergo three-dimensional reality multiplied by the third power of omnidirectional growth rate.

1230.00 Scheherazade Numbers

1230.10 Prime-Number Accommodation: Integration of Seven: The Babylonians did not accommodate a prime number like 7 in their mathematics. Plato had apparently been excited by this deficiency, so he multiplied 360 by 7 and obtained 2,520. And then, seeing that there were always positives and negatives, he multiplied 2,520 by 2 and obtained 5,040. Plato apparently intuited the significance of the number 5,040, but he did not say why he did. I am sure he was trying to integrate 7 to evolve a comprehensively rational circular dividend.

1230.11 H_2O is a simple low number. As both chemistry and quantum physics show, nature does all her associating and disassociating in whole rational numbers. Humans accommodated the primes 1, 2, 3, and 5 in the decimal and duodecimal systems. But they left out 7. After 7, the next two primes are 11 and 13. Humans' superstition considers the numbers 7, 11, and 13 to be bad luck. In playing dice, 7 and 11 are "crapping" or drop-out numbers. And 13 is awful. But so long as the comprehensive cyclic dividend fails to contain prime numbers which may occur in the data to be coped with, irrational numbers will build up or erode the processing numbers to produce irrational, ergo unnatural, results. We must therefore realize that the tables of the trigonometric functions include the first 15 primes 1, 2, 3, 5, 7, 11, 13, 17, 19, 23, 29, 31, 41, 43.

1230.12 We know 7×11 is 77. If we multiply 77 by 13, we get 1,001. Were there not 1,001 Tales of the Arabian Nights? We find these numbers always involved with the mystical. The number 1,001 majors in the name of the storytelling done by Scheherazade to postpone her death in the *Thousand and One Nights*. The number 1,001 is a binomial reflection pattern: one, zero, zero, one.

1230.20 **SSRCD Numbers:** If we multiply the first four primes, we get 30. If we multiply 30 times 7, 11, and 13, we have $30 \times 1{,}001$ or 30,030, and we have used the first seven primes.

1230.21 We can be intuitive about the eighth prime since the octave seems to be so important. The eighth prime is 17, and if we multiply 30,030 by 17, we arrive at a fantastically simple number: 510,510. This is what I call an SSRCD Number, which stands for *Scheherazade Sublimely Rememberable Comprehensive Dividend*. As an example we can readily remember the first eight primes factorial—510,510! (Factorial means successively multiplied by themselves, ergo $1 \times 2 \times 3 \times 5 \times 7 \times 11 \times 13 \times 17 = 510{,}510$.)

1230.30 **Origin of Scheherazade Myth:** I think the Arabian priest-mathematicians and their Indian Ocean navigator ancestors knew that the binomial effect of 1,001 upon the first four prime numbers 1, 2, 3, and 5 did indeed provide comprehensive dividend accommodation of all the permutative possibilities of all the ''story-telling-taling-tallying,'' or computational systems of the octave system of integers.

1230.31 The function of the grand vizier to the ruler was that of mathematical wizard, the wiz of wiz-dom; and the wiz-ard kept secret to himself the mathematical navigational ability to go to faraway strange places where he alone knew there existed physical resources different from any of those occurring ''at home,'' then voyaging to places that only the navigator-priest knew how to reach, he was able to bring back guaranteed strange objects that were exhibited by the ruler to his people as miracles obviously producible only by the ruler who secretly and carefully guarded his vizier's miraculous power of wiz-dom.

1230.32 To guarantee their own security and advantage, the Mesopotamian mathematicians, who were the overland-and-overseas navigator-priests, deliberately hid their knowledge, their mathematical tools and operational principles such as the mathematical significance of $7 \times 11 \times 13 = 1{,}001$ from both their rulers and the people. They used psychology as well as outright lies, combining the bad-luck myth of the three prime integers with the mysterious inclusiveness of the *Thousand and One Nights*. The priests warned that bad luck would befall anyone caught using 7s, 11s, or 13s.

1230.33 Some calculation could only be done by the abacus or by positioning numbers. With almost no one other than the high priests able to do any calculation, there was not much chance that anyone would discover that the product of 7, 11, and 13 is 1,001, but "just in case," they developed the diverting myth of Scheherazade and her postponement of execution by her *Thousand and One Nights*.

1231.00 *Cosmic Illions*

1231.01 Western-world humans are no longer spontaneously cognizant of the Greek or Latin number prefixes like *dec-*, or *non-*, or *oct-*, nor are they able spontaneously to formulate in appropriate Latin or Greek terms the larger numbers spoken of by scientists nowadays only as *powers of ten*. On the other hand, we are indeed familiar with the Anglo-American words *one, two,* and *three,* wherefore we may prefix these more familiar designations to the constant *illion,* suffix which we will now always equate with a set of three successive zeros. (See Table 1238.80.)

1231.02 We used to call 1,000 *one thousand*. We will now call it *one-illion*. Each additional set of three zeros is recognized by the prefixed number of such three-zero sets. 1,000,000 = two-illion. 1,000,000,000 is 1 three-illion. (This is always hyphenated to avoid confusion with the set of subillion enumerators, e.g., 206 four-illions.) The English identified illions only with six zero additions, while the Americans used illions for every three zeros, starting, however, only *after 1,000,* overlooking its three zeros as common to all of them. Both the English and American systems thus were forced to use awkward nomenclature by retaining the initial word *thousand* as belonging to a different concept and an historically earlier time. Using our consistent illion nomenclature, we express the largest experientially conceivable measurement, which is the diameter of the thus-far-observed Universe measured in diameters of the nucleus of the atom, which measurement is a neat 312 fourteen-illions. (See Sec. 1238.50.)

1232.00 *Binomial Symmetry of Scheherazade Numbers*

1232.10 **Exponential Powers of 1,001:** As with all binomials, for example $A^2 + 2AB + B^2$, the progressive powers of the 1,001 Scheherazade Number produced by $7 \times 11 \times 13$, the product of which, multiplied by itself in successive stages, provides a series of symmetrical reflection numbers. They are not only sublimely rememberable but they resolve themselves into a symmetrical mirror pyramid array:

$1001^2 = 1{,}00\mathbf{2}{,}001$
$1001^3 = 1{,}00\mathbf{3}{,}00\mathbf{3}{,}001$
$1001^4 = 1{,}00\mathbf{4}{,}006{,}00\mathbf{4}{,}001$
$1001^5 = 1{,}00\mathbf{5}{,}010{,}010{,}00\mathbf{5}{,}001$
$1001^6 = 1{,}00\mathbf{6}{,}015{,}020{,}015{,}00\mathbf{6}{,}001$

$1001^7 = 1{,}007{,}021{,}035{,}035{,}021{,}007{,}001$

$1001^8 = 1{,}00\mathbf{8}{,}028{,}056{,}070{,}056{,}028{,}00\mathbf{8}{,}001$
$1001^9 = 1{,}00\mathbf{9}{,}036{,}084{,}126{,}126{,}084{,}036{,}00\mathbf{9}{,}001$

$1001^{10} = 1{,}0\mathbf{10}{,}045{,}120{,}210{,}252{,}210{,}120{,}045{,}0\mathbf{10}{,}001$

1232.11 The binomial symmetry expands all of its multiples in both left and right directions in reflection balance. Note that the exponential power to which the 1,001 Scheherazade Number is raised becomes the second whole integer from either end. As with $(A+B)^2 = A^2 + 2AB + B^2$, the interior integers consist of expressions and products of the exponent power.

1232.20 **Cancellation of "Leftward Spillover":** In the pyramid array of 1,001 Scheherazade Numbers (see Sec. 1232.10), we observe that *due to the double-symbol notation of the number 10,* the symmetry seems to be altered by the introduction of the leftward accommodation of the two integers of 10 in a single-integer position. For instance,

$1001^5 = 1{,}005{,}010{,}010{,}005{,}001$

10 10
$= 1{,}005{,}000{,}000{,}005{,}001$

1 1
$= 1{,}005{,}000{,}000{,}005{,}001$

1
Ten could be written vertically as 0 instead of 10, provided we always assumed that the vertically superimposed integer was to be spilled into the addition of the next leftward column, for we build leftward positively and rightward negatively from our decimal *zero-zero.*

1232.21 The abacus with its wires and beads taught humans how to fill a column with figures and thereafter to fill additional columns, by convention to the left. The Arabic numerals developed as symbols for the content of the columns. They filled a column and then they emptied it, but the cipher prevented them from using the column for any other notation, and the excess—by convention—was moved over to the left. This "spillover" can begin earlier or later, depending on the modulus employed. The spillover to the next column begins later when we are employing Modulo 12 than when we are employing Modulo 10. To disembarrass the symmetry of the leftward spillover, the spillover number in the table has been written vertically.

1232.22 The table of the ten successive powers of the 1,001 Scheherazade Number accidentally discloses a series of progressions:

(1) in the extreme right-hand column, a progression of zeros;

(2) in the fourth column from the right, an arithmetical progression of $\frac{N^2-N}{2}$, which we will call triangular; and

(3) in the seventh column from the right, a tetrahedral progression.

1232.23 The tetrahedron can be symmetrically or asymmetrically altered to accommodate the four unique planes that produce the fourth-dimensional accommodation of the vector equilibrium. The symmetry disclosed here may very well be four-dimensional symmetry that we have simply expressed in columns in a plane.

1232.24 The number 1,001 looks exciting because we are very close to the binary system of the computers. (We remember that Polynesians only counted to one and two.) The binary yes-no sequence looks so familiar. The Scheherazade Number has all the numbers you have in the binary system. The 1,001-ness keeps persisting throughout the table.

1232.25 The numbers $7 \times 11 \times 13 \times 17$ included in the symmetric dividend 510,510 may have an important function in atomic nucleation, since it accommodates all the prime numbers involved in the successive periods.

1232.26 Many mathematicians assume that the integer 1 is not to be counted as a prime. Thus 2, 3, 5, 7, 11, and 13 make a total of six effective primes that may be identified with the fundamental vector edges of the tetrahedron and the six axes of conglomeration of 12 uniradius spheres closest packed around one nuclear sphere, and the fundamental topological abundance of universal lines that always consist of even sets of six.

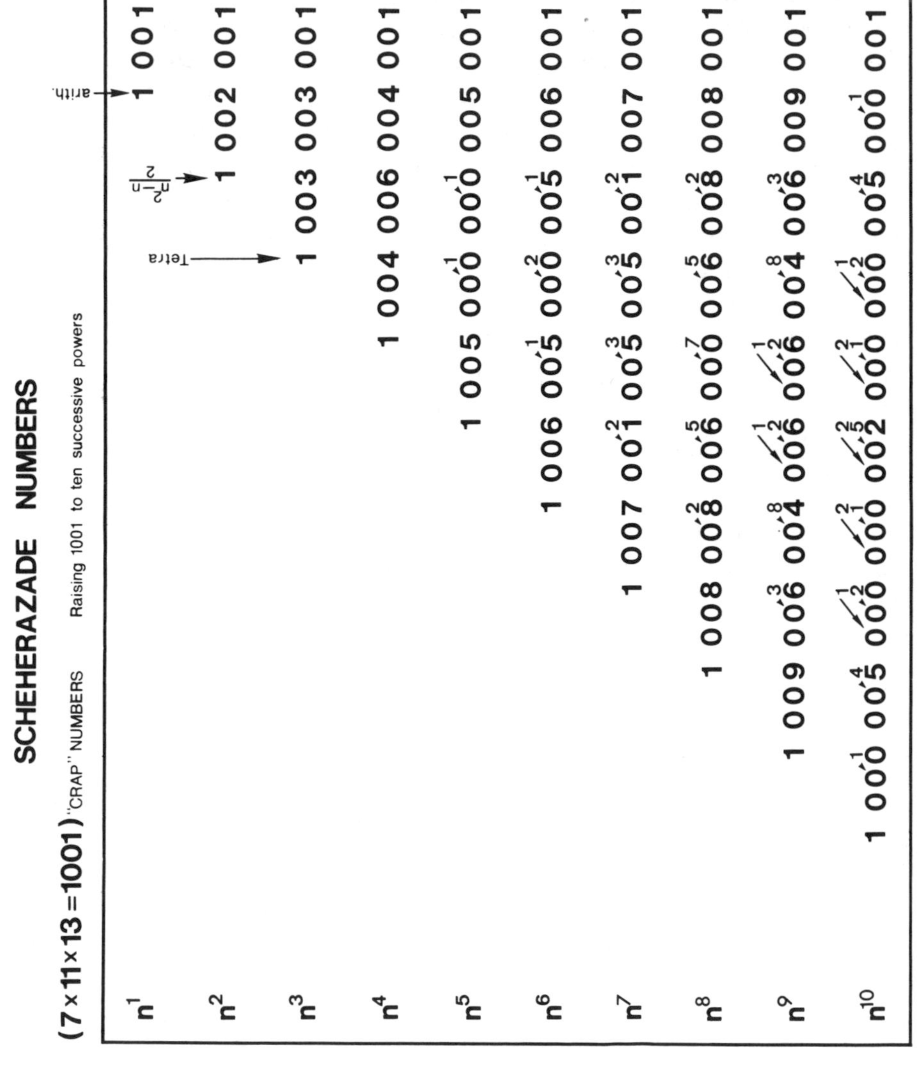

SCHEHERAZADE NUMBERS
(7 × 11 × 13 = 1001) "CRAP" NUMBERS
Raising 1001 to ten successive powers
arith.
Tetra
$\frac{n^2-n}{2}$
n1 1 001
n2 1 002 001
n3 1 003 003 001
n4 1 004 006 004 001
n5 1 005 000 000 005 001
n6 1 006 005 000 005 006 001
n7 1 007 001 005 005 001 007 001
n8 1 008 008 006 000 006 008 008 001
n9 1 009 006 004 006 006 004 006 009 001
n10 1 000 005 000 000 002 000 000 005 000 001
We witness here, basic reflective symmetry,
plus compounding of first five prime numbers.

Table 1232.21 *Cancellation of "Leftward Spillover" to Disclose Basic Reflection Symmetry of Successive Powers of the Scheherazade Numbers:* Raising 1001 to ten successive powers, we recognize basic reflective symmetry plus compounding of five basic primes. $7 \times 11 \times 13 = 1001$.

1232.30 **Scheherazade Reflection Patterns:**

$1 \cdot 2 \cdot 3 \cdot 5$	30
$7 \cdot 11 \cdot 13$	1,001
$1 \cdot 2 \cdot 3 \cdot 5 \cdot 7 \cdot 11 \cdot 13$	30,030
$(1 \cdot 2 \cdot 3 \cdot 5 \cdot 7 \cdot 11 \cdot 13)^2$	901,800,900
$(1 \cdot 2 \cdot 3 \cdot 5 \cdot 7 \cdot 11 \cdot 13)^2 \cdot 5$	4,509,004,500
$(1 \cdot 2 \cdot 3 \cdot 5 \cdot 7 \cdot 11 \cdot 13)^3$	27,081,081,027,000
$(1 \cdot 2 \cdot 3 \cdot 5 \cdot 7 \cdot 11 \cdot 13)^3 \cdot 9$	243,729,729,243,000
$(1 \cdot 2 \cdot 3 \cdot 5 \cdot 7 \cdot 11 \cdot 13)^4$	813,244,863,240,810,000
$(1 \cdot 2 \cdot 3 \cdot 5 \cdot 7 \cdot 11 \cdot 13)^4 \cdot 3$	2,439,734,589,722,430,000
$(1 \cdot 2 \cdot 3 \cdot 5 \cdot 7 \cdot 11 \cdot 13)^5$	24,421,743,243,121,524,300,000
$1 \cdot 2 \cdot 3 \cdot 5 \cdot 7 \cdot 11 \cdot 13 \cdot 17$	510,510
$(1 \cdot 2 \cdot 3 \cdot 5 \cdot 7 \cdot 11 \cdot 13 \cdot 17)^2$	260,620,460,100
$1 \cdot 2 \cdot 3 \cdot 5 \cdot 7 \cdot 11 \cdot 13 \cdot 17 \cdot 19$	9,699,690
$1 \cdot 2 \cdot 3 \cdot 5 \cdot 7 \cdot 11 \cdot 13 \cdot 17 \cdot 19 \cdot 23 \cdot 29$	6,469,693,230
$1 \cdot 2 \cdot 3 \cdot 5 \cdot 7 \cdot 11 \cdot 13 \cdot 17 \cdot 19 \cdot 23 \cdot 29 \cdot 31$	200,560,490,130
$(1 \cdot 2 \cdot 3 \cdot 5 \cdot 7 \cdot 11 \cdot 13) \cdot (1 \cdot 2 \cdot 3 \cdot 5 \cdot 7 \cdot 11 \cdot 13 \cdot 17)$	153,306,153
$(1 \cdot 2 \cdot 3 \cdot 5 \cdot 7 \cdot 11 \cdot 13) \cdot (1 \cdot 2 \cdot 3 \cdot 5 \cdot 7 \cdot 11 \cdot 13 \cdot 17) \cdot 3$	459,918,459

1234.00 Seven-illion Scheherazade Number

1234.01 The Seven-illion Scheherazade Number includes the first seven primes, which are:

$(1 \cdot 2 \cdot 3 \cdot 5 \cdot 7 \cdot 11 \cdot 13)^5$ . . . to the fifth power.

It reads,

24,421,743,243,121,524,300,000

1234.02 In the first days of electromagnetics, scientists discovered fourth-power energy relationships and Einstein began to find fifth-power relationships having to do with gravity accommodating fourth- and fifth-powering. The first seven primes factorial is a sublimely rememberable number. It is a big number, yet rememberable. When nature gives us a number we can remember, she is putting us on notice that the cosmic communications circuits are open: you are connected through to many sublime truths!

1234.03 Though factored by seven prime numbers, it is expressible entirely as various-sized increments of three to the fifth power. There is a four-place overlapping of one. Three to the fifth power means five-dimensionality triangulation, which means that five-dimensional structuring as triangulation *is* structure.

1234.04 When it is substituted as a comprehensive dividend for 360° 00′ 00″ to express cyclic unity in increments equal to one second of arc, while recalculating the tables of trigonometric functions, it is probable that *many,* if not *most,* and possibly *all* the function fractions will be expressible as whole rational numbers. Thc use of 24,421,743,243,121,524,300,000 as cyclic unity will eliminate much cumulative error of the present trigonometric-function tables.

1234.10 Seven-illion Scheherazade Number: Symmetrical Mirror Pyramid Array

```
                     2 4 3 2 4 3
              1 2 1 5            1 2 1 5
          2 4 3                          2 4 3
0 0 0 0 0 0                                   0 0 0 0 0 0
0 0 0 0 0 2 4 4 2 1 7 4 3 2 4 3 1 2 1 5 2 4 3 0 0 0 0 0
-(5·2)^5                                      +(5·2)^5
           3^5                             3^5
               5·3^5               5·3^5
                         3^5   3^5
```

where $3^5 = 243$, $5 \cdot 3^5 = 1215$, $-(5 \cdot 2)^5 =$ five zero prefix, $+(5 \cdot 2)^5 =$ five zero sufix

1236.00 Eight-illion Scheherazade Number

1236.01 The Eight-illion Scheherazade Number is

$1 \cdot 2 \cdot 3 \cdot 5 \cdot 7 \cdot 11 \cdot 13$

$1 \cdot 2 \cdot 3 \cdot 5 \cdot 7 \cdot 11 \cdot 13 \cdot 17$

$1 \cdot 2 \cdot 3 \cdot 5 \cdot 7 \cdot 11 \cdot 13 \cdot 17 \cdot 19 \cdot 23 \cdot 29 \cdot 31$

Which is:

$1^n \cdot 2^3 \cdot 3^8 \cdot 5^5 \cdot 7^4 \cdot 11^3 \cdot 13^3 \cdot 17^2 \cdot 19 \cdot 23 \cdot 29 \cdot 31$

It reads:

1,452,803,177,020,770,377,302,500

1236.02 The Eight-illion Scheherazade Number accommodates all trigonometric functions, spherical and planar, when unity is 60 degrees; its halfway turnabout is 30 degrees. It also accommodates the octave-nine-zero of the icosahedron's corner angles of 72 degrees, one-half of which is 36 degrees (ergo, 31 is the greatest prime involved), which characterizes maximum spherical excess of the vector equilibrium's sixty-degreeness.

1237.00 Nine-illion Scheherazade Number

1237.01 The Nine-illion Scheherazade Number includes the first 12 primes, which are:

$1^n \cdot 2^{10} \cdot 3^8 \cdot 5^8 \cdot 7^4 \cdot 11^3 \cdot 13^3 \cdot 17^2 \cdot 19 \cdot 23 \cdot 29 \cdot 31$

It reads:

185,958,806,658,658,608,294,720,000,000

It is full of mirrors:

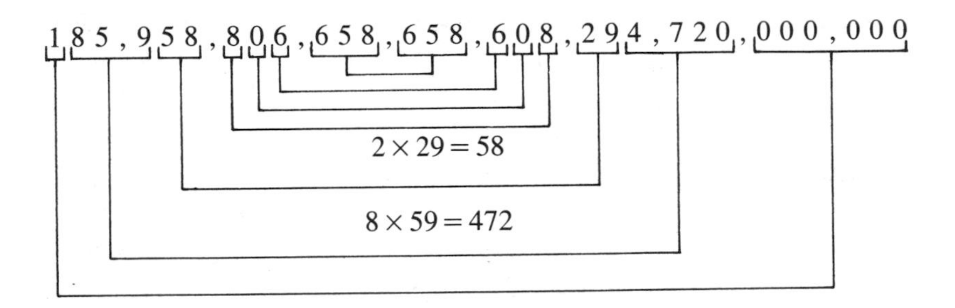

1238.00 *Fourteen-illion Scheherazade Number*

1238.20 **Trigonometric Limit: First 14 Primes:** The Fourteen-illion Scheherazade Number accommodates all the omnirational calculations of the trigonometric function tables whose largest prime number is 43 and whose highest common variable multiple is 45 degrees, which is one-eighth of unity in a Universe whose polyhedral systems consist always of a minimum of four positive and four negative quadranted hemispheres.

1238.21 45 degrees is the zero limit of covarying asymmetry because the right triangle's 90-degree corner is always complemented by two corners always together totalling 90 degrees. The smallest of the covarying, 90-degree complementaries reaches its maximum limit when both complementaries are 45 degrees. Accepting the concept that one is not a prime number, we have 14 primes—2, 3, 5, 7, 11, 13, 17, 19, 23, 29, 31, 37, 41, 43—which primacy will accommodate all the 14 unique structural faceting of all the crystallography, all the biological cell structuring, and all bubble agglomerating: the 14 facets being the polar facets of the seven and only seven axes of symmetry of Universe, which are the 3-, 4-, 6-, 12-great circles of the vector equilibrium and the 6-, 10-, 15-great circles of the icosahedron.

1238.30 **Cosmic Commensurability of Time and Size Magnitudes:** Each degree of 360 degrees of circular arc is subdivided into 60 minutes and the minutes into 60 seconds each. There are 1,296,000 seconds of arc in a circle of 360 degrees.

1238.31 One minute of our 8,000-mile-diameter planet Earth's great circle arc = one nautical mile = 6,076 feet approximately. A one-second arc of a great circle of Earth is 6,076/60 = 101.26 feet, which means one second of great-circle arc around Earth is approximately 100 feet, or the length of one tennis court, or one-third of the distance between the opposing teams' goalposts on a football field. We can say that each second of Earth's great circle of arc equals approximately 1,200 inches (or 1,215.12 ''exact''). There are $2^1/_2$ trillion atomic-nucleus diameters in one inch. A hundredth of an inch is the smallest interval clearly discernible by the human eye. There are 25 billion atomic-nucleus diameters in the smallest humanly visible ''distance'' or linear size increment. A hundredth of an inch equals 1/120,000th of a second of great-circle arc of our spherical planet Earth. This is expressed decimally as .0000083 of a second of great-circle arc = .01 inch; or it is expressed scientifically as .01 inch $= 83 \times 10^{-7}$. A hundredth of an inch equals the smallest humanly visible dust speck; therefore: minimum dust speck $= 83 \times 10^{-7}$ seconds of arc, which equals 25 billion atomic-nucleus diameters—or $2^1/_2$ million angstroms. This is to say that it requires seven places to the right of the decimal to express the fractional second of the great-circle arc of Earth that is minimally discernible by the human eye.

1238.40 **Fourteen-illion Scheherazade Number:** The Fourteen-illion Scheherazade Number includes the first 15 primes, which are:

$$1^n \cdot 2^{12} \cdot 3^8 \cdot 5^6 \cdot 7^6 \cdot 11^6 \cdot 13^6 \cdot 17^2 \cdot 19 \cdot 23 \cdot 29 \cdot 31 \cdot 37 \cdot 41 \cdot 43$$

It reads:

3,128,581,583,194,999,609,732,086,426,156,130,368,000,000

1238.50 **Properties:** The 3 fourteen-illion magnitude Scheherazade Number has 3×10^{43} whole-number places, which is 10^{37} more integer places than has the 1×10^6 number expressing the 1,296,000 seconds in 360 degrees of whole-circle arc, and can therefore accommodate rationally not only calculations to approximately 1/100th of an inch (which is the finest increment resolvable by the human eye), but also the 10^{-7} power of that minimally visible magnitude, for this 3×10^{43} SSRCD has enough decimal places to express rationally the 22-billion-light-years-diameter of the omnidirectional, celestial-sphere limits thus far observed by planet Earth's humans expressed in linear units measuring only 1/1,000ths of the diameter of one atomic nucleus.

1238.60 Size Magnitudes

An Atomic Nucleus Diameter = A.N.D. = 

			Atomic Nucleus Diameters:
10,000	⊖	= 1 Angstrom (One atomic diameter)	= 10 one-illion
$1 \cdot 10^4$	⊖	= 1 Angstrom	= 10 one-illion
$25 \cdot 10^9$	⊖	= 1 Speck of Dust (= One hair's breadth)	= 25 three-illion
$25 \cdot 10^{11}$	⊖	= 1 Inch	= 2½ four-illion
$3 \cdot 10^{13}$	⊖	= 1 Foot	= 30 four-illion
$1 \cdot 10^{14}$	⊖	= 1 Meter	= 100 four-illion
$10 \cdot 10^{16}$	⊖	= 1 Kilometer	= 100 five-illion
$18 \cdot 10^{16}$	⊖	= 1 Mile (Nautical)	= 180 five-illion
$144 \cdot 10^{19}$	⊖	= 1 Diameter of Earth	= 1.44 seven-illion
$144 \cdot 10^{21}$	⊖	= 1 Diameter of Sun	= 144 seven-illion
$144 \cdot 10^{25}$	⊖	= 1 Diameter of Solar System	= 1½ nine-illion
$108 \cdot 10^{28}$	⊖	= 1 Light Year (6 trillion miles)	= 1 ten-illion
$2^1/_3 \cdot 10^{40}$	⊖	= Diameter of astro observed sweepout (22 billion light years)	= 23 thirteen-illion

1238.70 Time Magnitudes: Heartbeats (Seconds of Time)

$1 \cdot 10^6$	= 1 million	= 1 two-illion heartbeats ago	= 2 weeks
$31 \cdot 10^6$	= 31 million	= 31 two-illion heartbeats ago	= 1 year
$5 \cdot 10^8$	= 500 million	= 500 two-illion heartbeats ago	= 16 years (college)
$1 \cdot 10^9$	= 1 billion	= 1 three-illion heartbeats ago	= 32 years (prime life)
$2 \cdot 10^9$	= 2 billion	= 2 three-illion heartbeats ago	= Average lifetime

$42 \cdot 10^9$	= 42 billion	= 42 three-illion heartbeats ago	= Mohammed
$60 \cdot 10^9$	= 60 billion	= 60 three-illion heartbeats ago	= Christ
$78 \cdot 10^9$	= 78 billion	= 78 three-illion heartbeats ago	= Buddha
$200 \cdot 10^9$	= 200 billion	= 200 three-illion heartbeats ago (8,000 years ago)	= Earliest Egypt
$500 \cdot 10^9$	= 500 billion	= 500 three-illion heartbeats ago (15,000 years ago)	= Earliest artistic culture (Thailand)
$1 \cdot 10^{12}$	= 1 trillion	= 1 four-illion heartbeats ago	= 30,000 years ago (Last Ice Age)
$75 \cdot 10^{12}$	= 75 trillion	= 75 four-illion heartbeats ago	= Leakey: Earliest human skull: 2½ million years ago
$75 \cdot 10^{12}$	= 75 trillion	= Capital Wealth of World	
$1 \cdot 10^{17}$	= 100 quadrillion	= 100 five-illion heartbeats ago	= Age of our planet Earth
$3 \cdot 10^{17}$	= 300 quadrillion	= 300 five-illion heartbeats ago	= Known limit age of Universe (10 billion years ago)

1238.80 **Number Table: Significant Numbers** (see Table 1238.80)

1239.00 Limit Number of Maximum Asymmetry

1239.10 **Powers of Primes as Limit Numbers:** Every so often out of an apparently almost continuous absolute chaos of integer patterning in millions and billions and quadrillions of number places, there suddenly appears an SSRCD rememberable number in lucidly beautiful symmetry. The exponential powers of the primes reveal the beautiful balance at work in nature, which does not secrete these symmetrical numbers in irrelevant capriciousness. Nature endows them with functional significance in her symmetrically referenced, mildly asymmetrical, structural formulations. The SSRCD numbers suddenly appear as unmistakably as the full Moon in the sky.

1239.11 There is probably a number limit in nature that is adequate for the rational, whole-number accounting of all the possible general atomic systems' permutations. For instance, in the Periodic Table of the Elements, we find 2, 8, 8, 18. These number sets seem familiar: the 8 and the 18, which is twice 9, and the twoness is perfectly evident. The largest prime number in 18 is 17. It could be that if we used all the primes that occur between 1 and 17, multiplied by themselves five times, we might have all the possible number accommodations necessary for all the atomic permutations.

1239.20 **Pairing of Prime Numbers:** I am fascinated by the fundamental interbehavior of numbers, especially by the behavior of primes. A prime cannot be produced by the interaction of any other numbers. A prime, by definition, is only divisible by itself and by one. As the integers progress, the primes begin to occur again, and they occur in *pairs*. That is, when a prime number appears in a progression, another prime will appear again quite near to it. We can go for thousands and thousands of numbers and then find two primes appearing again fairly close together. There is apparently some kind of companionship among the primes. Euler, among others, has theories about the primes, but no one has satisfactorily accounted for their behavior.

1239.30 **Maximum Asymmetry:** In contrast to all the nonmeaning, the Scheherazade Numbers seem to emerge at remote positions in numerical progressions of the various orders. They emerge as meaning out of nonmeaning. They show that nature does not sustain disorder indefinitely.

1239.31 From time to time, nature pulses inside-outingly through an omnisymmetric zerophase, which is always our friend vector equilibrium, in which condition of sublime symmetrical exactitude nature refuses to be caught by temporal humans; she refuses to pause or be caught in structural stability. She goes into progressive asymmetries. All crystals are built in almost-but-not quite-symmetrical asymmetries, in positive or negative triangulation stabilities, which is the maximum asymmetry stage. Nature pulsates torquingly into maximum degree of asymmetry and then returns to and through symmetry to a balancing degree of opposite asymmetry and turns and repeats and repeats. The maximum asymmetry probably is our minus or plus four, and may be the fourth degree, the fourth power of asymmetry. The octave, again.

1238.80 Number Table: Significant Numbers

Number of Zeros	*Power of Ten*	*American System*	*British Sys*
3	10^3	Thousand	Thousand
6	10^6	Million	Million
9	10^9	Billion	Milliard
12	10^{12}	Trillion	Billion
15	10^{15}	Quadrillion	——
18	10^{18}	Quintillion	Trillion
21	10^{21}	Sextillion	——
24	10^{24}	Septillion	Quadrillion
27	10^{27}	Octillion	——
30	10^{30}	Nonillion	Quintillion
33	10^{33}	Decillion	——
36	10^{36}	Undecillion	Sextillion
39	10^{39}	Duodecillion	——
42	10^{42}	Tredecillion	Septillion
45	10^{45}	Quattuordecillion	——
48	10^{48}	Quindecillion	Octillion
51	10^{51}	Sexdecillion	——
54	10^{54}	Septendecillion	Nonillion

ergetics System	*Significant Numbers* (*One Nuclear Diameter Taken as Unit Measurement*)
ne-illion	1×10^4: One Angstrom in atomic-nuclei diameters
vo-illions	
ree-illions	1×10^{10}: 10 three-illions: One meter in atomic-nuclei diameters
our-illions	
ve-illions	16×10^{16}: 160 five-illions: One mile in atomic-nuclei diameters
x-illions	
even-illions	
ight-illions	
ine-illions	96×10^{28}: 960 nine-illions: One light-year in atomic-nuclei diameters
en-illions	
leven-illions	
welve-illions	2×10^{40}: 23 thirteen-illions: Diameter of explored Universe in atomic-nuclei diameters
hirteen-illions	
ourteen-illions	
ifteen-illions	
ixteen-illions	3×10^{41}: fourteen-illion SSRCD: 3,128, 581,583,194,999,609,732,086,426,156, 130,368,000,000. Contains factorially all of the prime numbers in 45 degrees, which is one "octant" of trigonometry, which covers all general systems interrelationships.
eventeen-illions	
ighteen-illions	

Afterpiece

The phenomenon lag *is simply due to the limited mechanism of the brain.*

We have to wait for the afterimage in order to realize.

The norm of Einstein is absolute speed instead of at rest. ''At rest'' was what we called instantaneous in our innocence of yesterday. We evolute toward ever lesser brain-comprehension lags—ergo, toward ever diminishing error; ergo, ever diminishing misunderstandings; ergo, ever diminishing fear, and its brain-lagging painful errors of objectivity; wherefore we approach eternal instantaneity of absolute and total comprehension.

The eternal instantaneity of no lag at all. However, we have now learned from our generalizations of the great complexity of the interactions of principles—as we are disembarrassed of our local, exclusively physical chemistry of information-sensing devices—that what is approached is eternal and instant awareness of absolute reality of all that ever existed.

All the great metaphysical integrity of all the individuals, which is potential and inherent in the complex interactions of generalized principles, will always and only coexist eternally.

Drawings Section

A Note to the Reader: Each of the drawings in this section is presented in sequence according to the Section number of the text to which the drawing pertains.

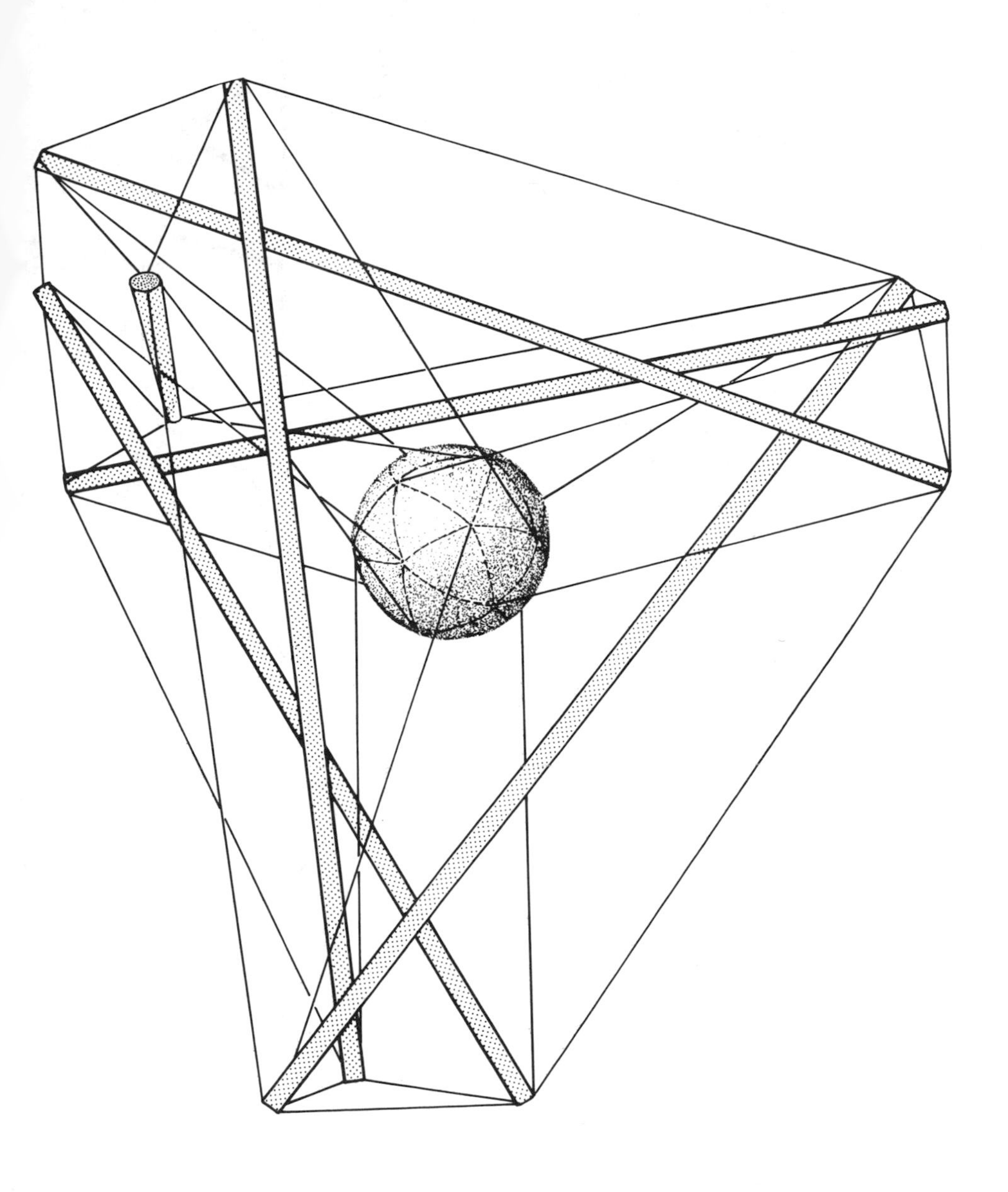

Fig. 401.00 *Tensegrity Tetrahedron with "Me" Ball Suspended at Center of Volume of the Tetrahedron:* Note that the six solid compression members are the acceleration vectors trying to escape from Universe at either end, by action and reaction; whereas the ends of each would-be escapee are restrained by three tensors, one long and two short; while the ball at the center is restrained from local torque and twist by three triangulated tensors tangentially affixed from each of the four corners.

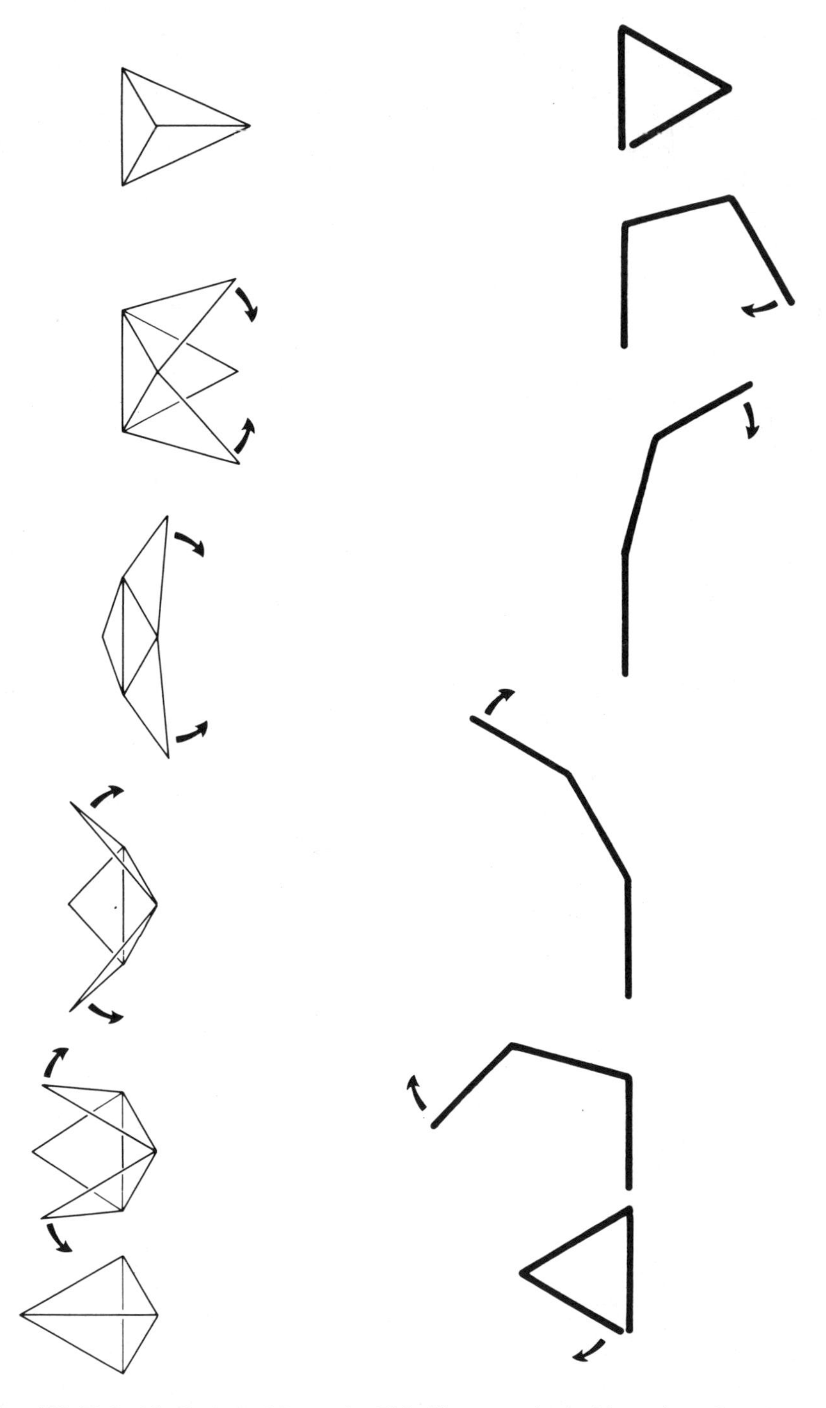

Fig. 453.02 *Inside-Outing of Triangle:* This illustrates the inside-outing of a triangle, which transformation is usually misidentified as "left vs. right" or "positive and negative" or as "existence vs. annihilation" in physics. The inside-outing is four-dimensional and often complex. The inside-outing of the rubber glove explains "annihilation" and demonstrates complex into-extroverting.

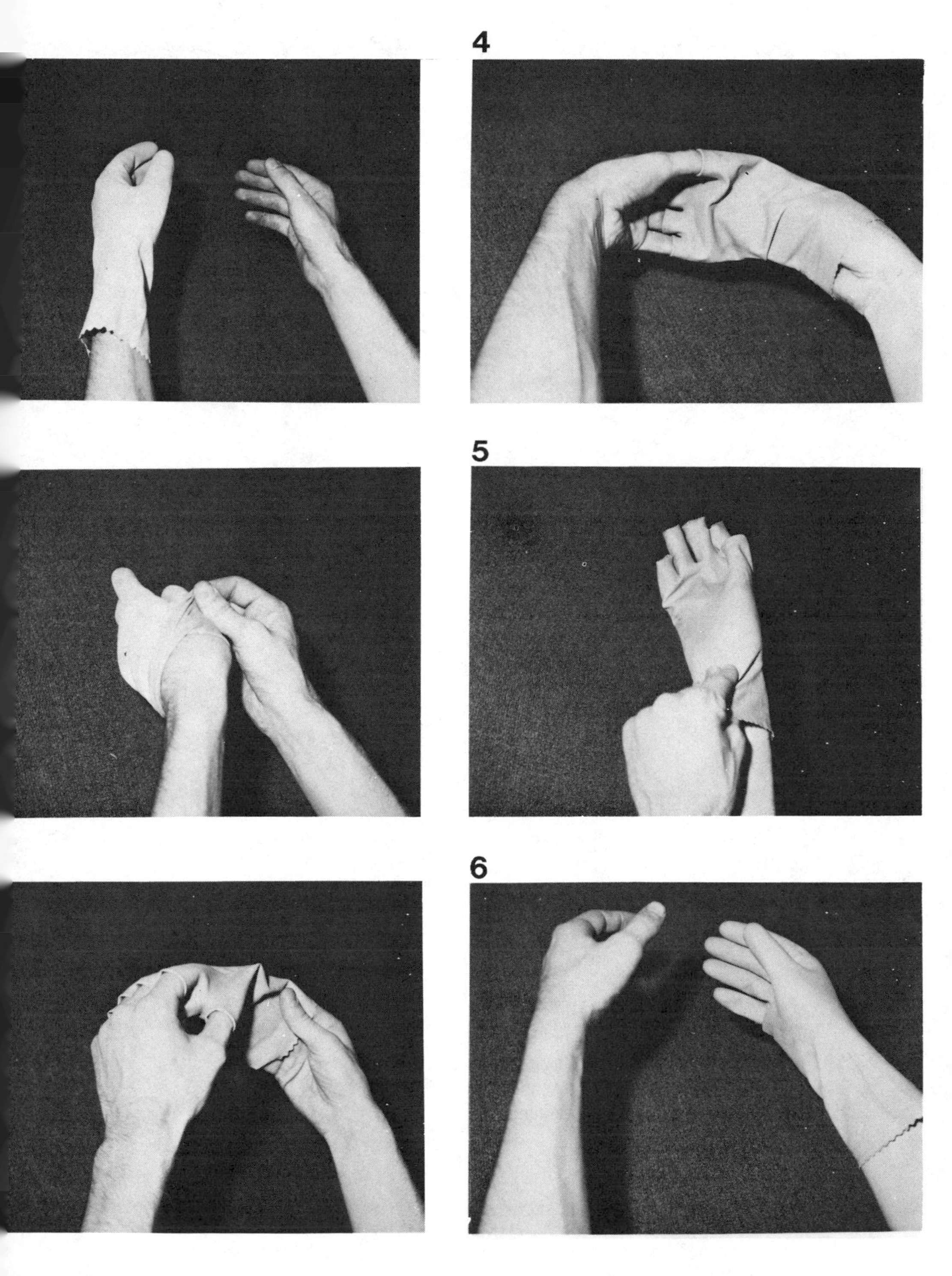
4
5
6

Fig. 454.01A The six great circles of the vector equilibrium disclose the spherical tetrahedra and the spherical cube and their chordal, flat-faceted, polyhedral counterparts.

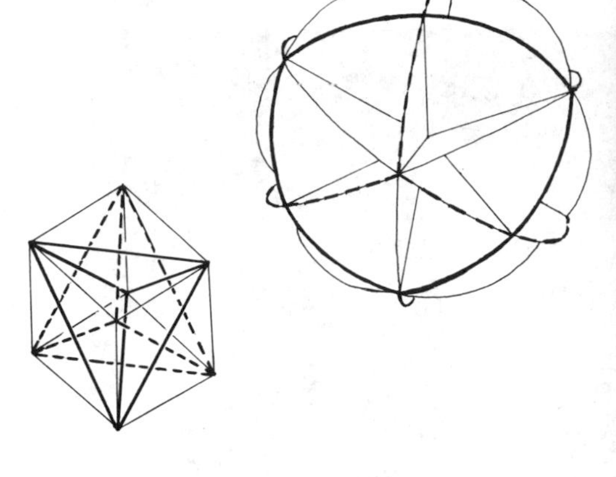

Fig. 454.01B The six great circles of the vector equilibrium disclose the six square faces of the spherical cube facets whose eight vertexes are centered in the areal centers of the vector equilibrium's eight spherical triangles.

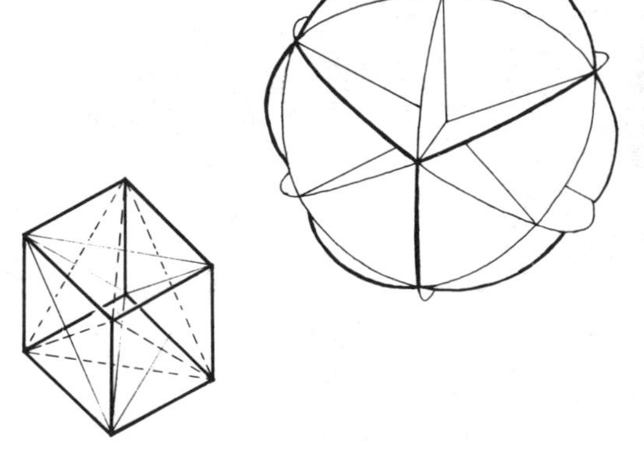

Fig. 454.01C The six great circles of the vector equilibrium disclose the 12 rhombic diamond facets (cross-hatching) of the rhombic dodecahedron, whose centers are coincident with the 12 vertexes (dots) of the vector equilibrium.

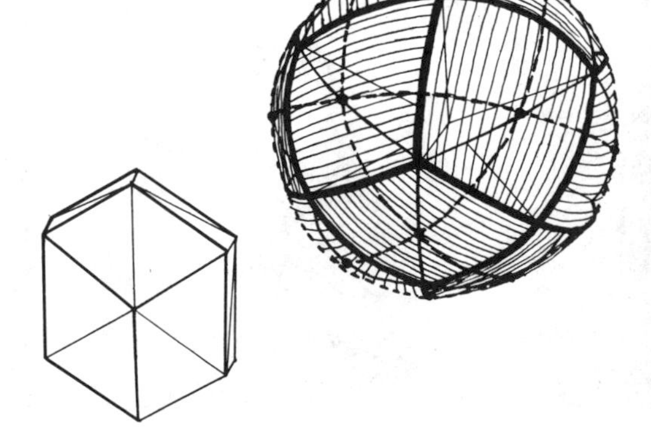

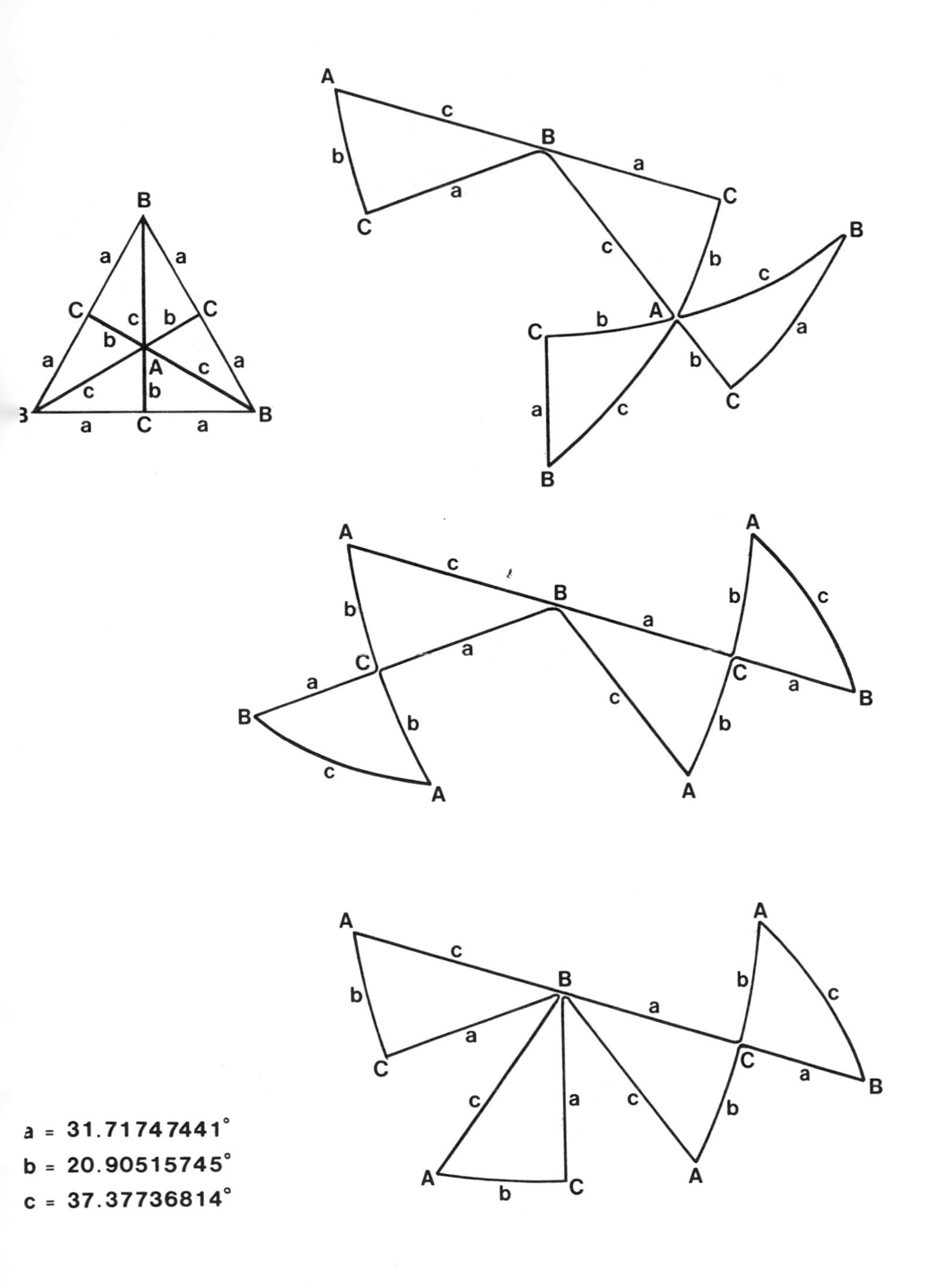

Fig. 459.01 *Great Circle Foldabilities of Icosahedron.*

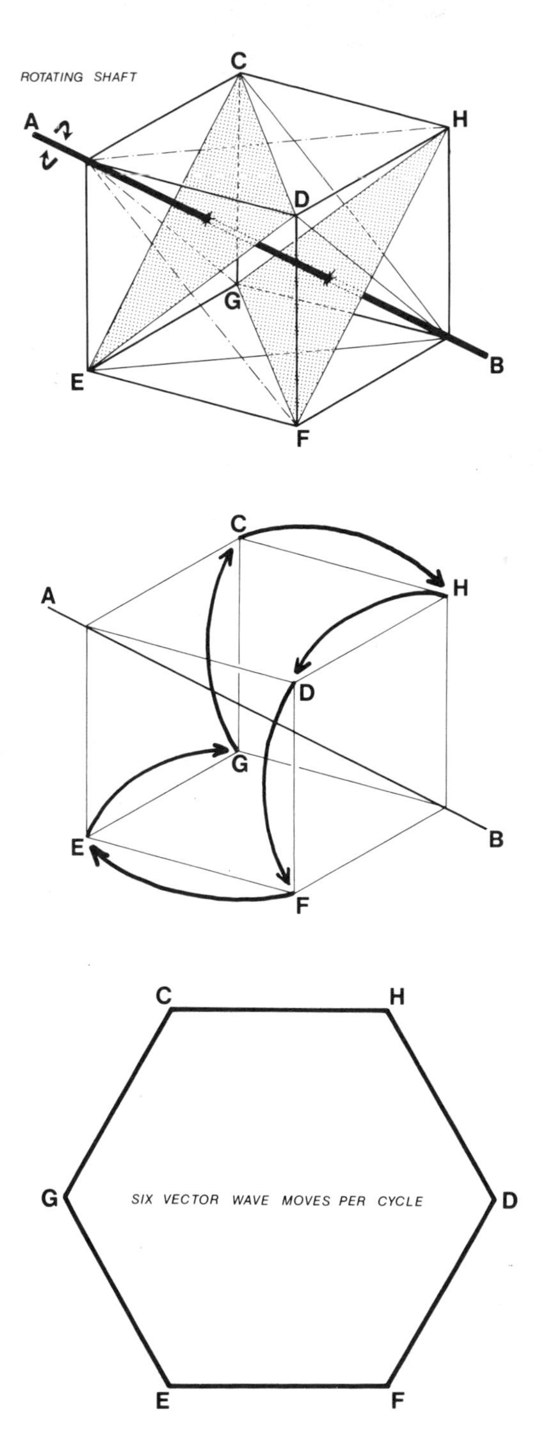

Fig. 464.01 *Triangle in Cube as Energetic Model:* The rotating shaft is labeled *AB*. The model demonstrates that there are six vector wave moves per cycle.

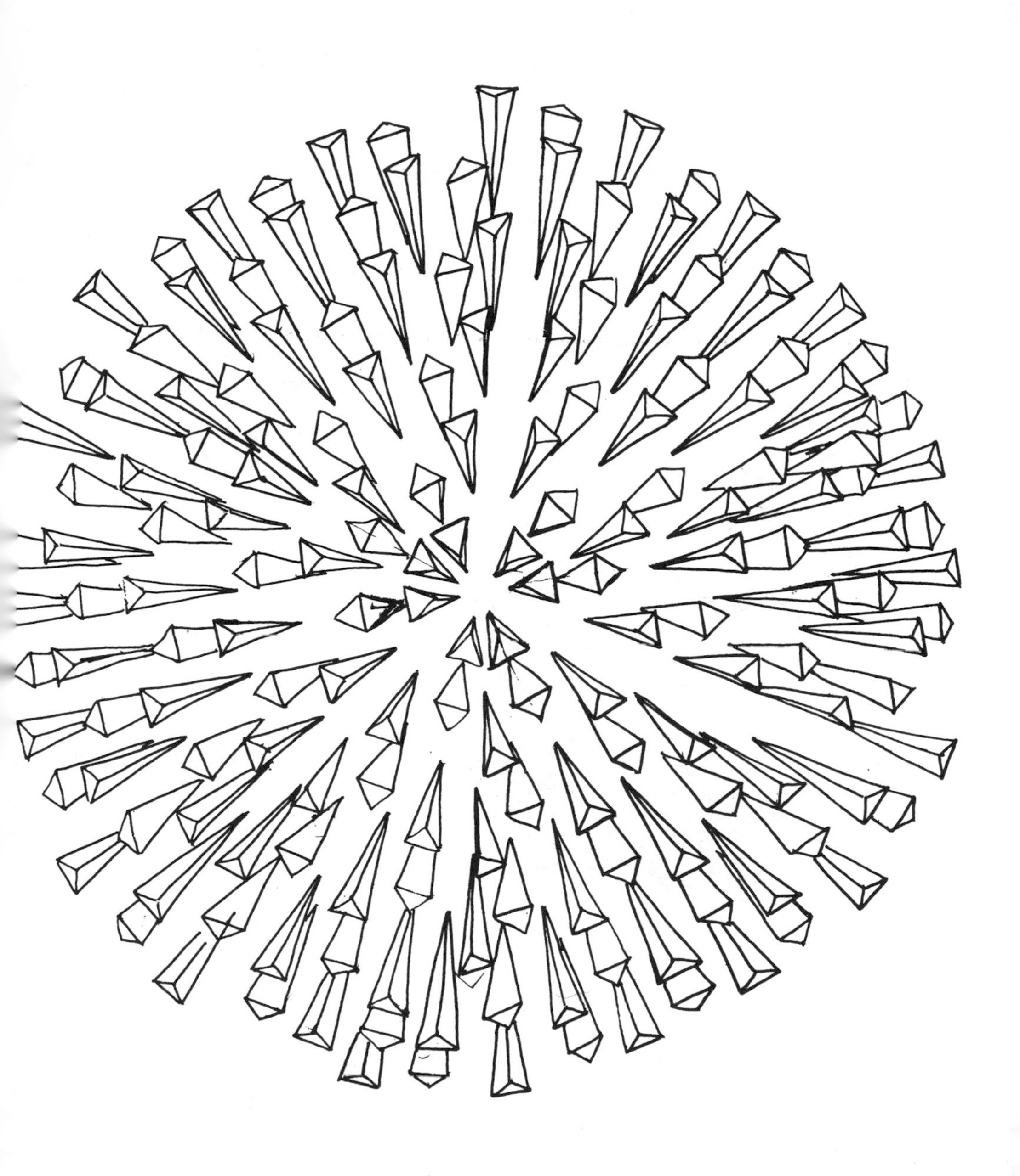

Fig. 541.00 *Energy Separated Out into Tetrahedral Photon Packages.*

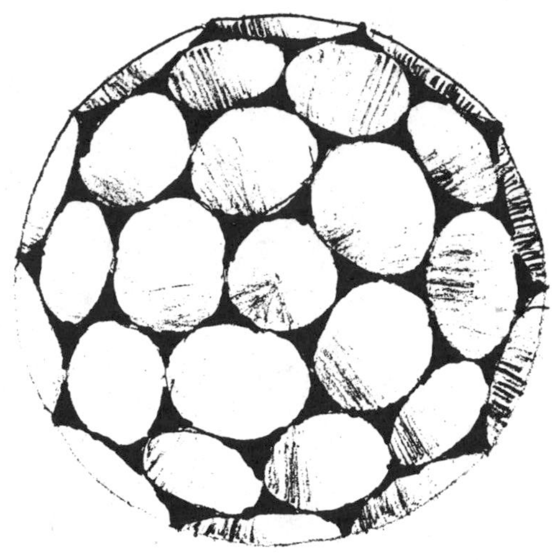

Fig. 541.30H *Circular Cornucopia Assembled Around Interior Points to Form a Spherical Array.* The tangent circle areas as well as concave triangle interstices constitute the total spherical surface.

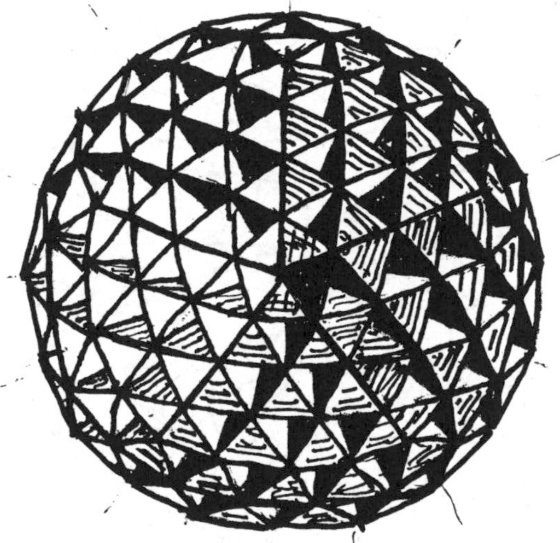

Fig. 541.30I *Three-Sided, Triangular Cornucopia Subdivide the Total Sphere.*

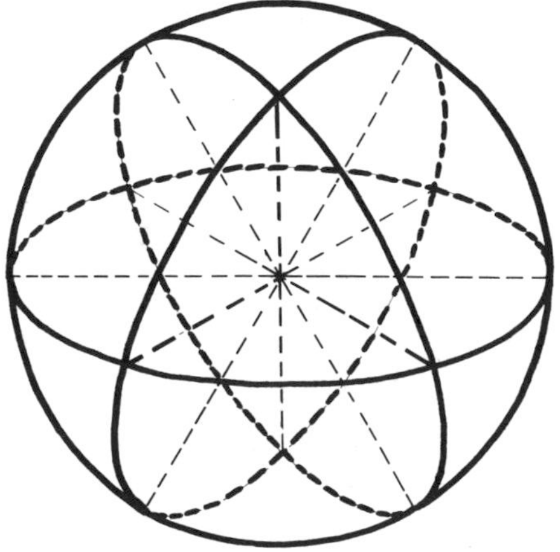

Fig. 541.30M Gradually the four tetrahedron-defining, vertexial components of the photon package's spiralling results in an equilibrium-seeking interaction of their four separate interattractions, which generates the four great circles of the vector equilibrium and establishes its tactical energy center as the four planes of the zerophase tetrahedron.

THE 120 DISEQUILIBRIOUS L.C.D. 

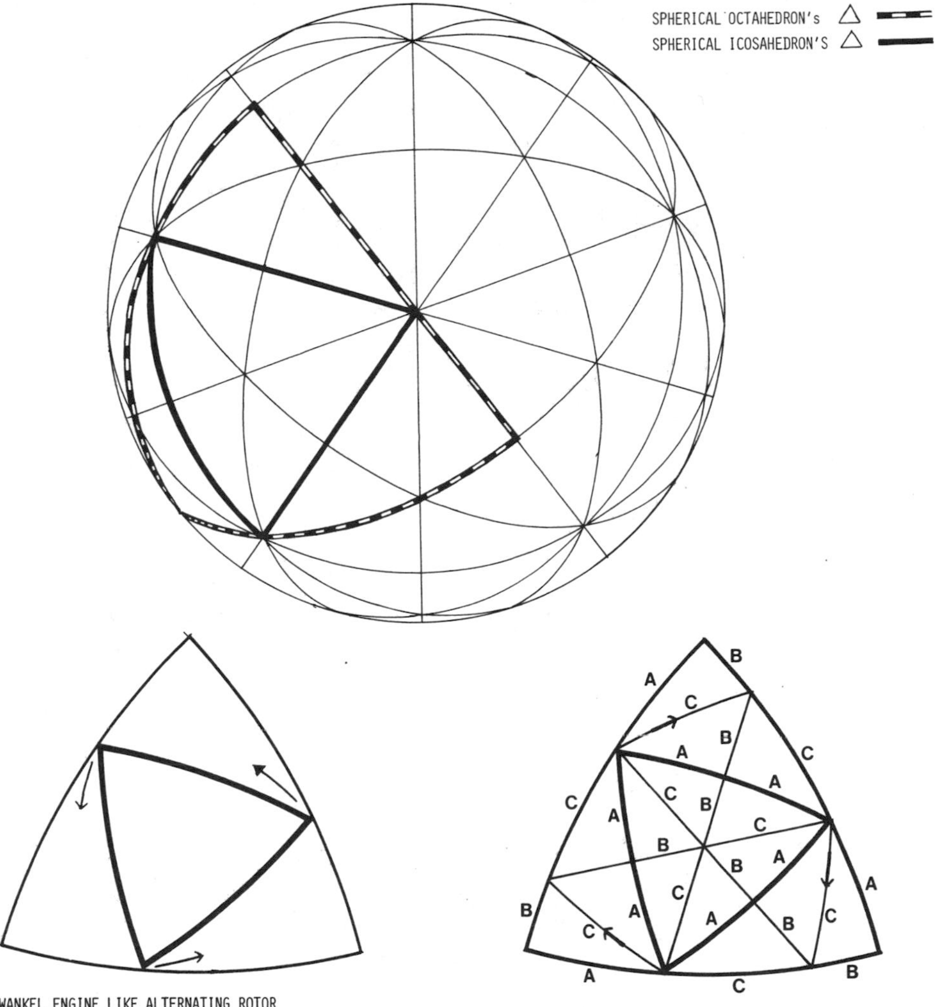

A = 31.71747441°
B = 20.90515745°
C = 37.37736814°

Fig. 901.03 *The Basic Disequilibrium 120 LCD Triangle:*

12 vertexes surrounded by 10 converging angles	$12 \times 10 = 120$
20 vertexes surrounded by 6 converging angles	$20 \times 6 = 120$
30 vertexes surrounded by 4 converging angles	$30 \times 4 = 120$
	360 converging angles

The 360 convergent angles must share the 720° reduction from absolute sphere to chorded sphere: $\frac{720}{360} = 2°$ per each corner; 6° per each triangle.

All of the spherical excess 6° has been massaged by the irreducibility of the 90° and 60° corners into the littlest corner. $\therefore$ 30→36.

In reducing 120 spherical triangles described by the 15 great circles to planar faceted polyhedra, the spherical excess 6° would be shared proportionately by the 90°–60°–30° flat triangle relationship = 3 : 2 : 1.

The above tells us that freezing the 60-degree center of the icosa triangle and sharing the 6-degree spherical excess find the *A* Quanta Module angles exactly congruent with the icosa's 120 interior angles.

Each of these groups is 1/12 th of a Rhombic Dodecahedron.
The on their bases are the external faces of the
144 module Rhombic Dodecahedron.

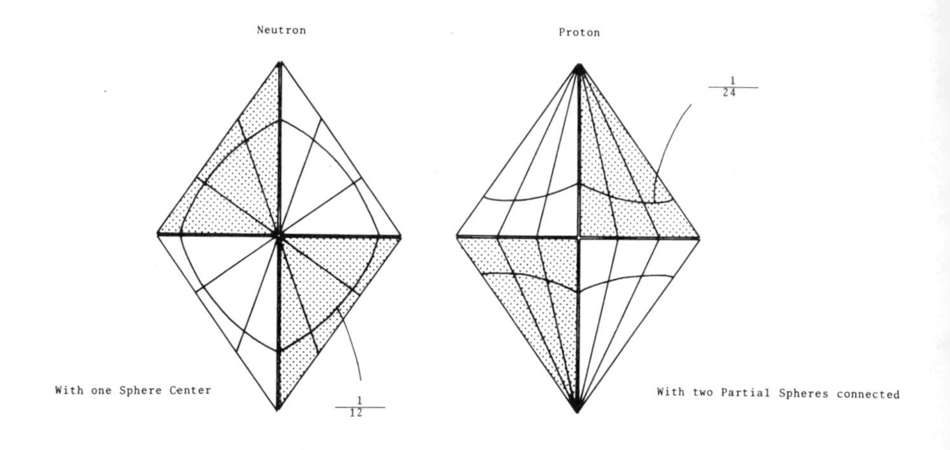

Fig. 943.00A *Quanta Module Orientations as Neutron and Proton 1/24-Sphere Centers: A* and *B* Quanta Modules.

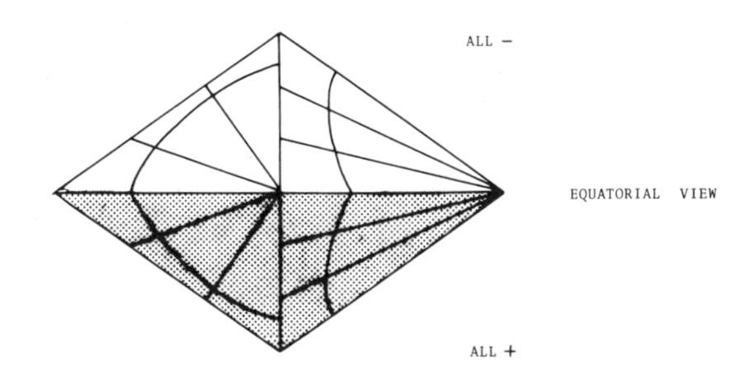

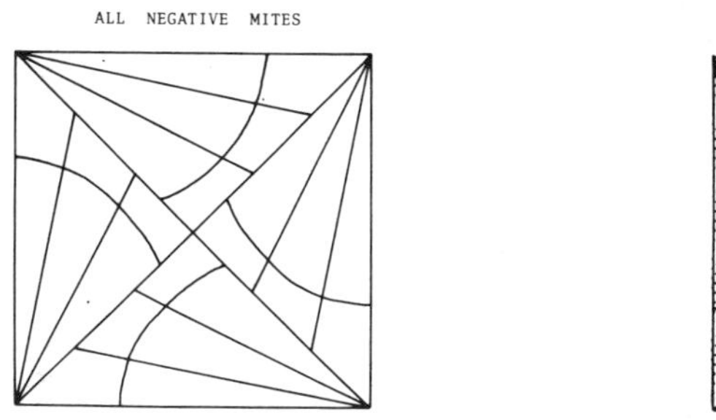

Fig. 943.00B *Hierarchy of Quanta Module Orientations: A* and *B* Quanta Modules.

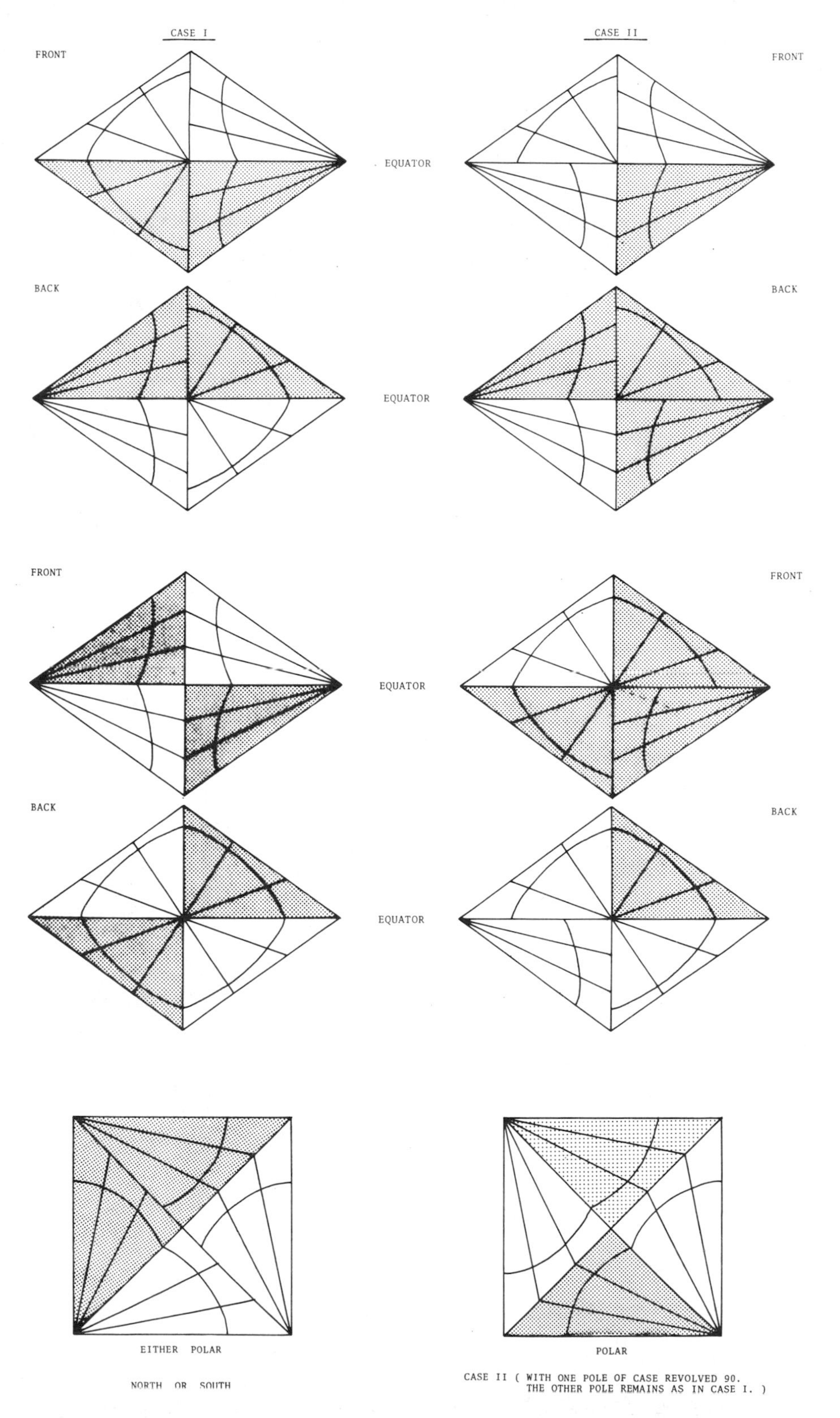

Fig. 943.00B *Hierarchy of Quanta Module Orientations: A* and *B* Quanta Modules.

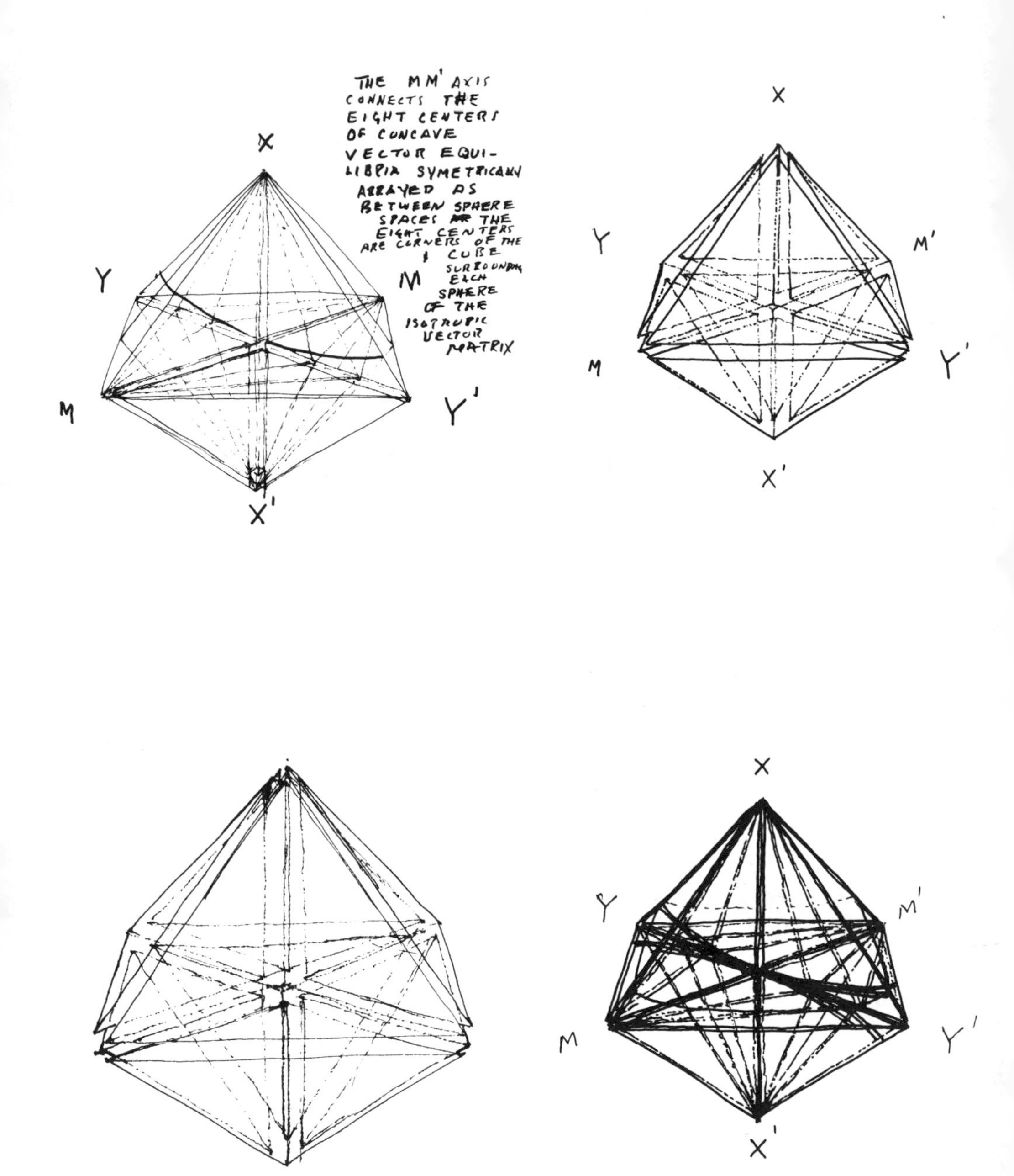
THE MM' AXIS
CONNECTS THE
EIGHT CENTERS
OF CONCAVE
VECTOR EQUI-
LIBRIA SYMETRICALLY
ARRAYED AS
BETWEEN SPHERE
SPACES THE
EIGHT CENTERS
ARE CORNERS OF THE
CUBE
SURROUNDING
EACH
SPHERE
OF THE
ISOTROPIC
VECTOR
MATRIX
X
Y
M
M
Y'
X'
X
Y
M'
M
Y'
X'
X
Y
M'
M
Y'
X'

Fig. 954.00A A *and* B *Quanta Module Orientations*

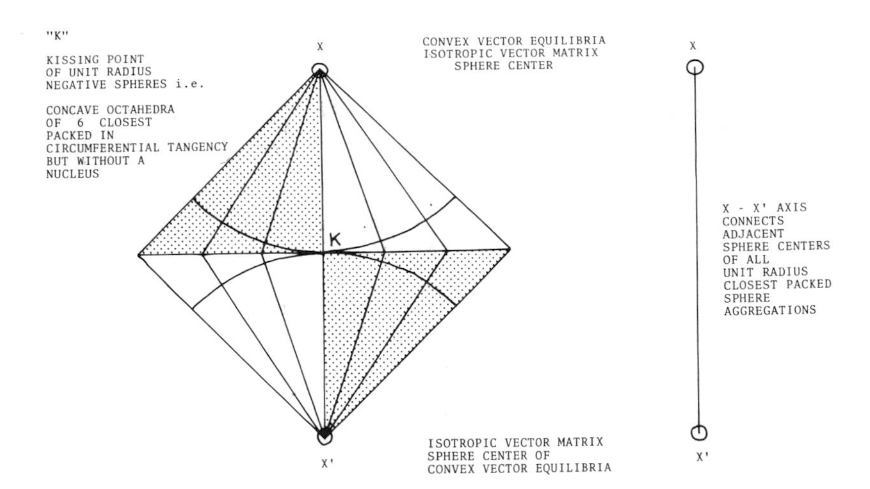

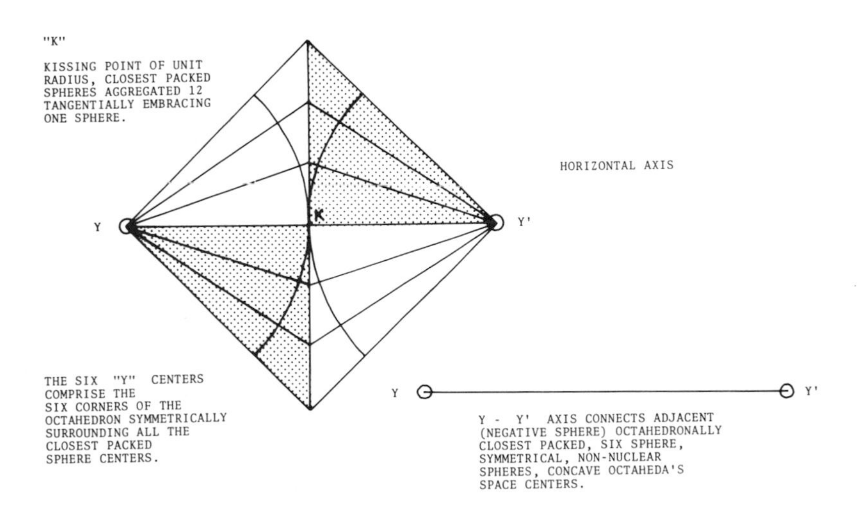

"K"
KISSING POINT OF UNIT
RADIUS, CLOSEST PACKED
SPHERES AGGREGATED 12
TANGENTIALLY EMBRACING
ONE SPHERE.
HORIZONTAL AXIS
Y
K
Y'
THE SIX "Y" CENTERS
COMPRISE THE
SIX CORNERS OF THE
OCTAHEDRON SYMMETRICALLY
SURROUNDING ALL THE
CLOSEST PACKED
SPHERE CENTERS.
Y
Y'
Y - Y' AXIS CONNECTS ADJACENT
(NEGATIVE SPHERE) OCTAHEDRONALLY
CLOSEST PACKED, SIX SPHERE,
SYMMETRICAL, NON-NUCLEAR
SPHERES, CONCAVE OCTAHEDA'S
SPACE CENTERS.

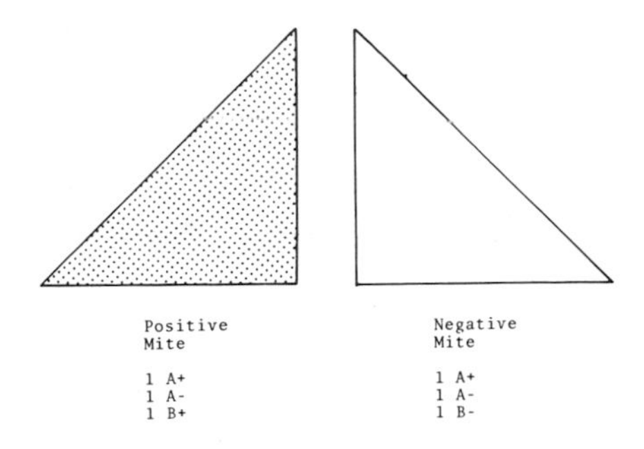

Fig. 954.00A A *and* B *Quanta Module Orientations*

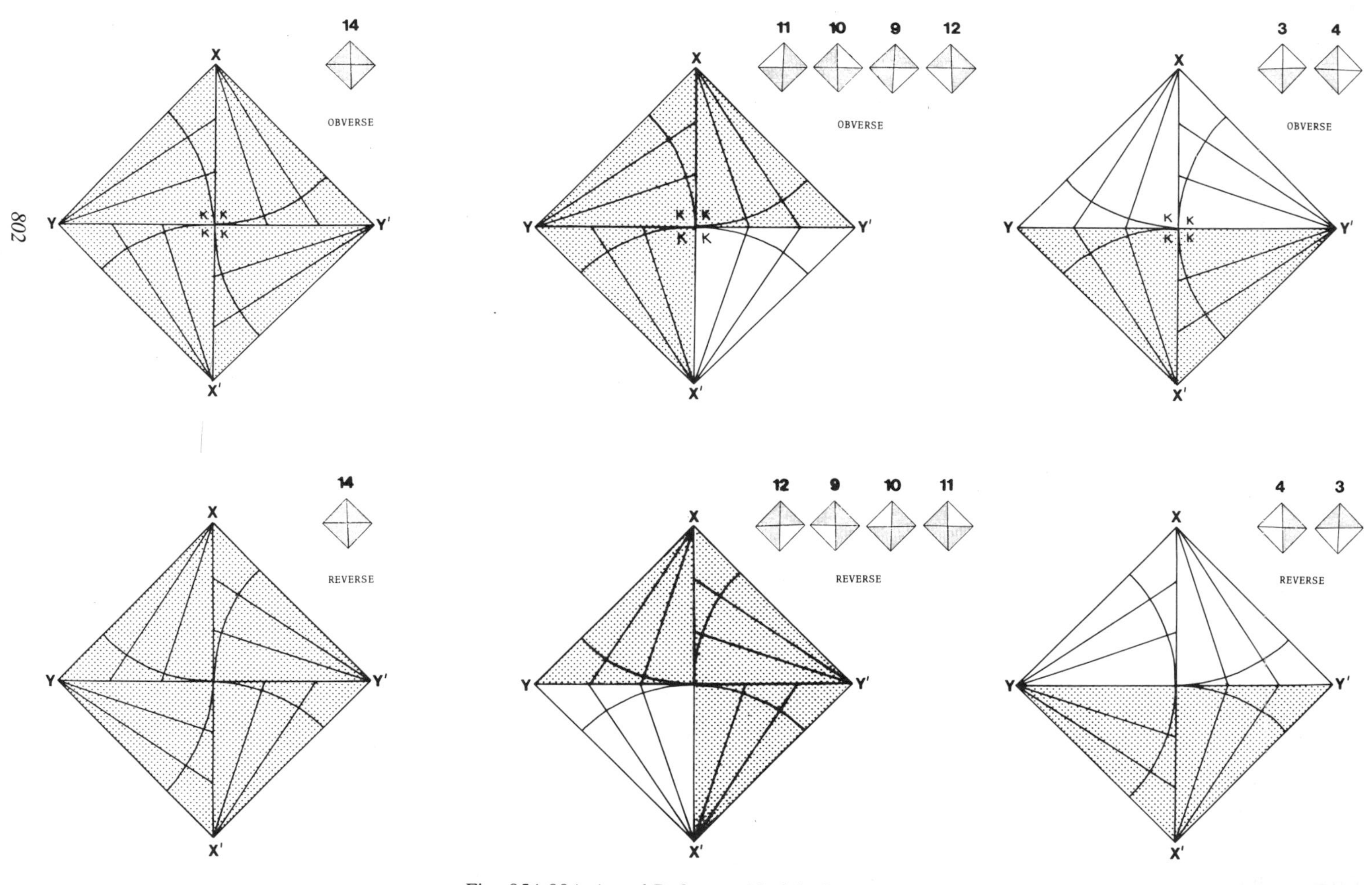

Fig. 954.00A A *and* B *Quanta Module Orientations*

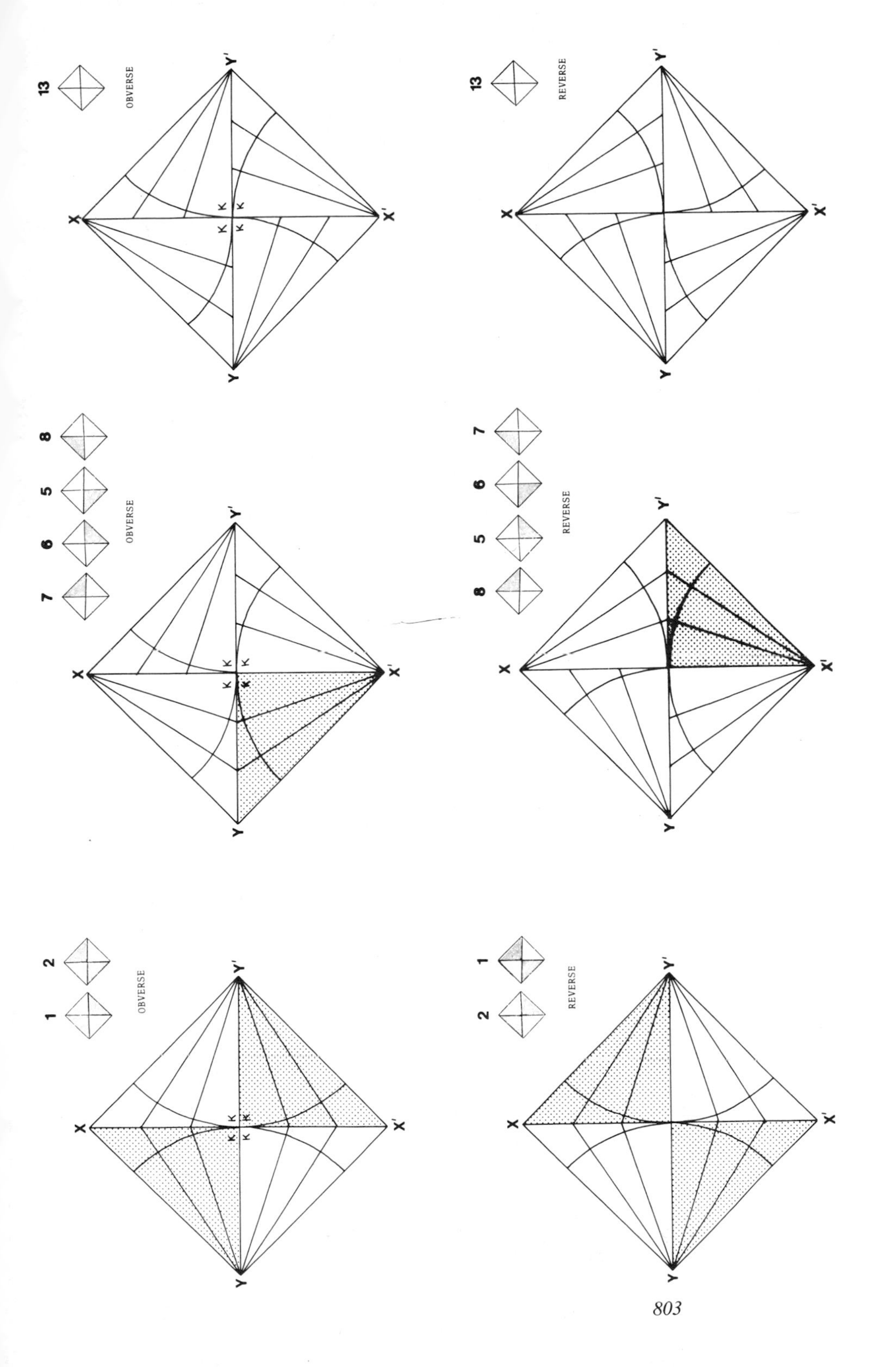
13
OBVERSE
13
REVERSE
7
6
5
8
OBVERSE
8
5
6
7
REVERSE
1
2
OBVERSE
2
1
REVERSE
X
X'
Y
Y'
K

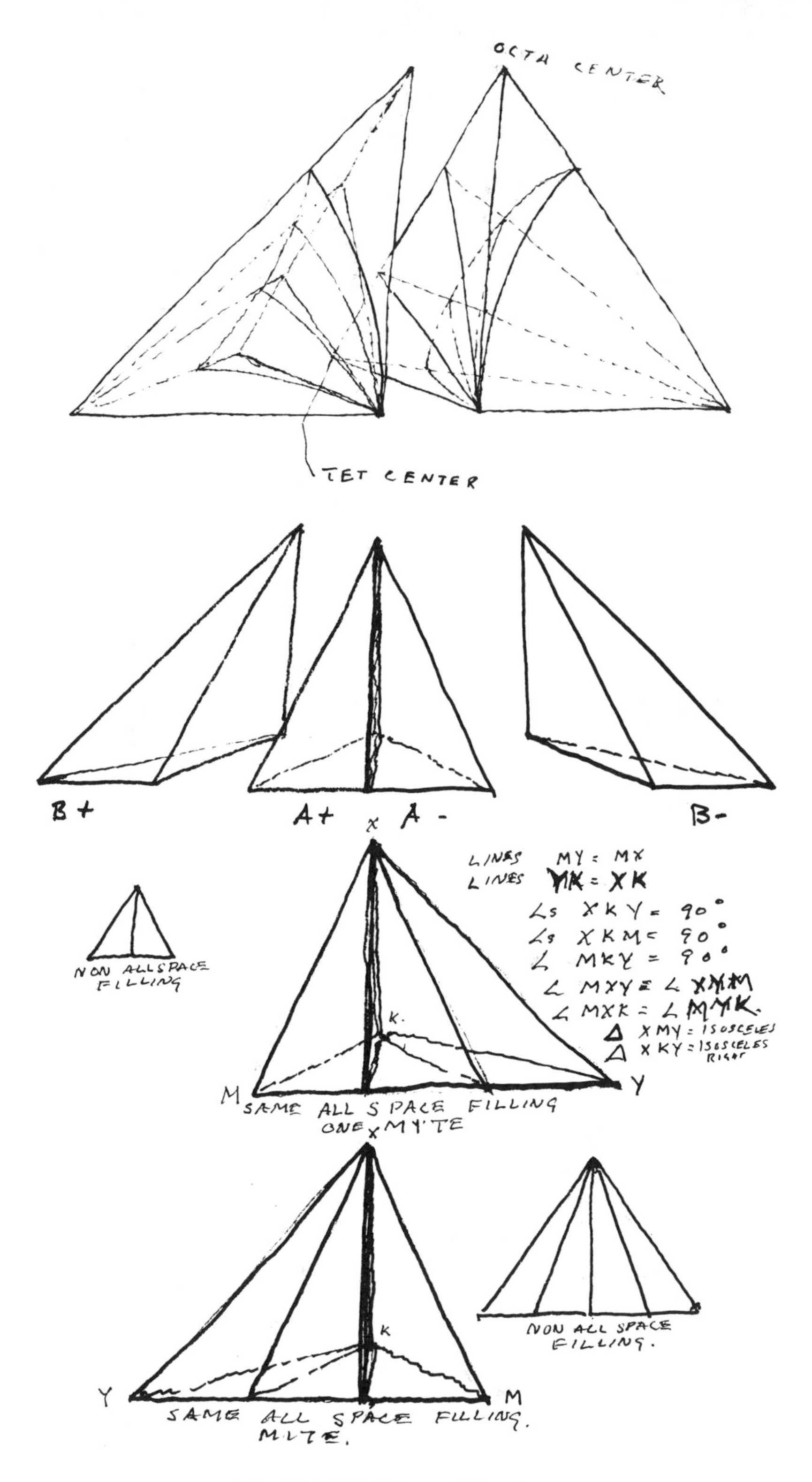

Fig. 954.00B *Mites and Couplers*

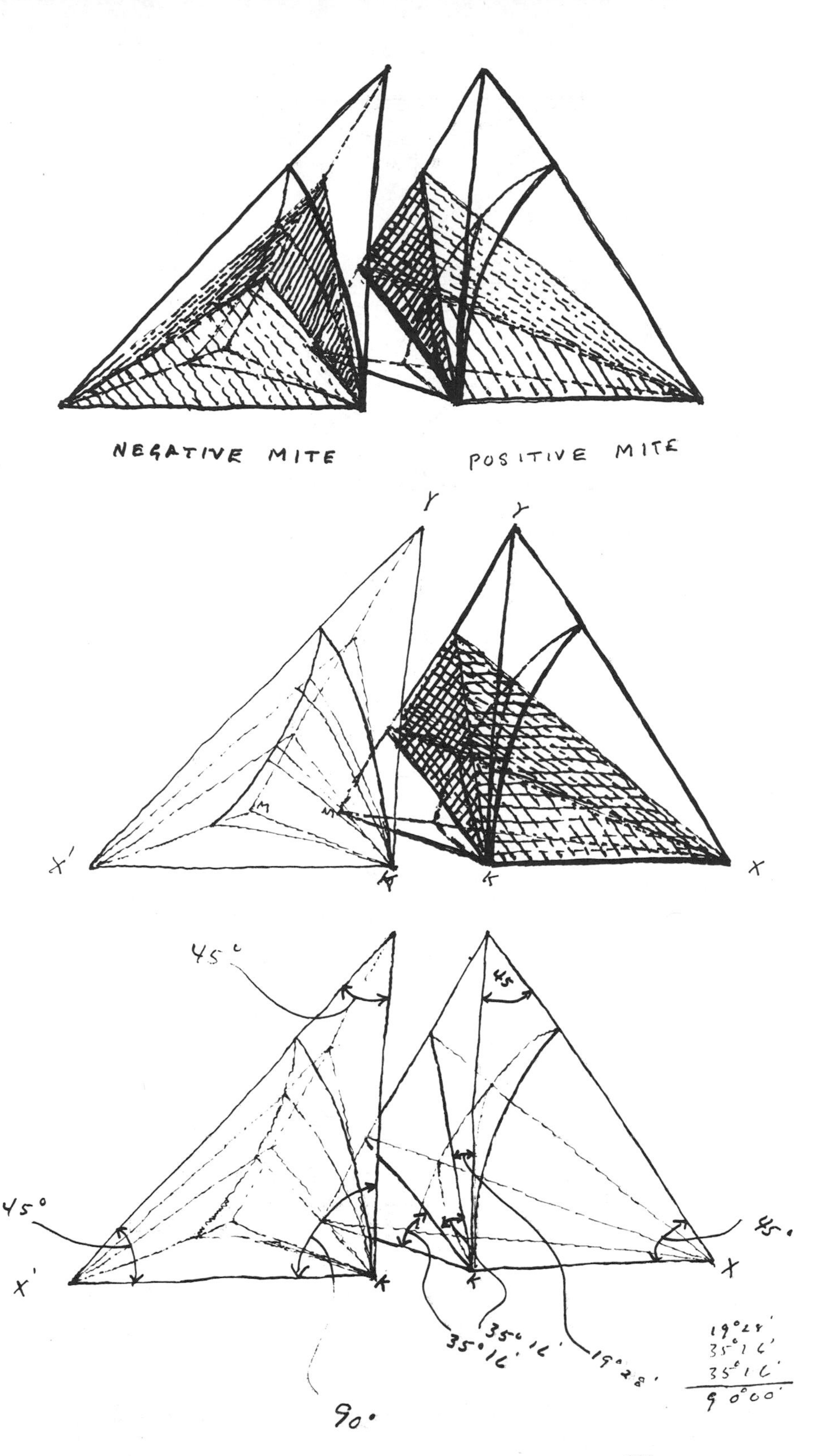
NEGATIVE MITE
POSITIVE MITE
Y
Y
M
M
X'
K
K
X
45°
45°
45°
45°
X'
K
K
X
35°16'
35°16'
19°28'
90°
19°28'
35°16'
35°16'
90°00'

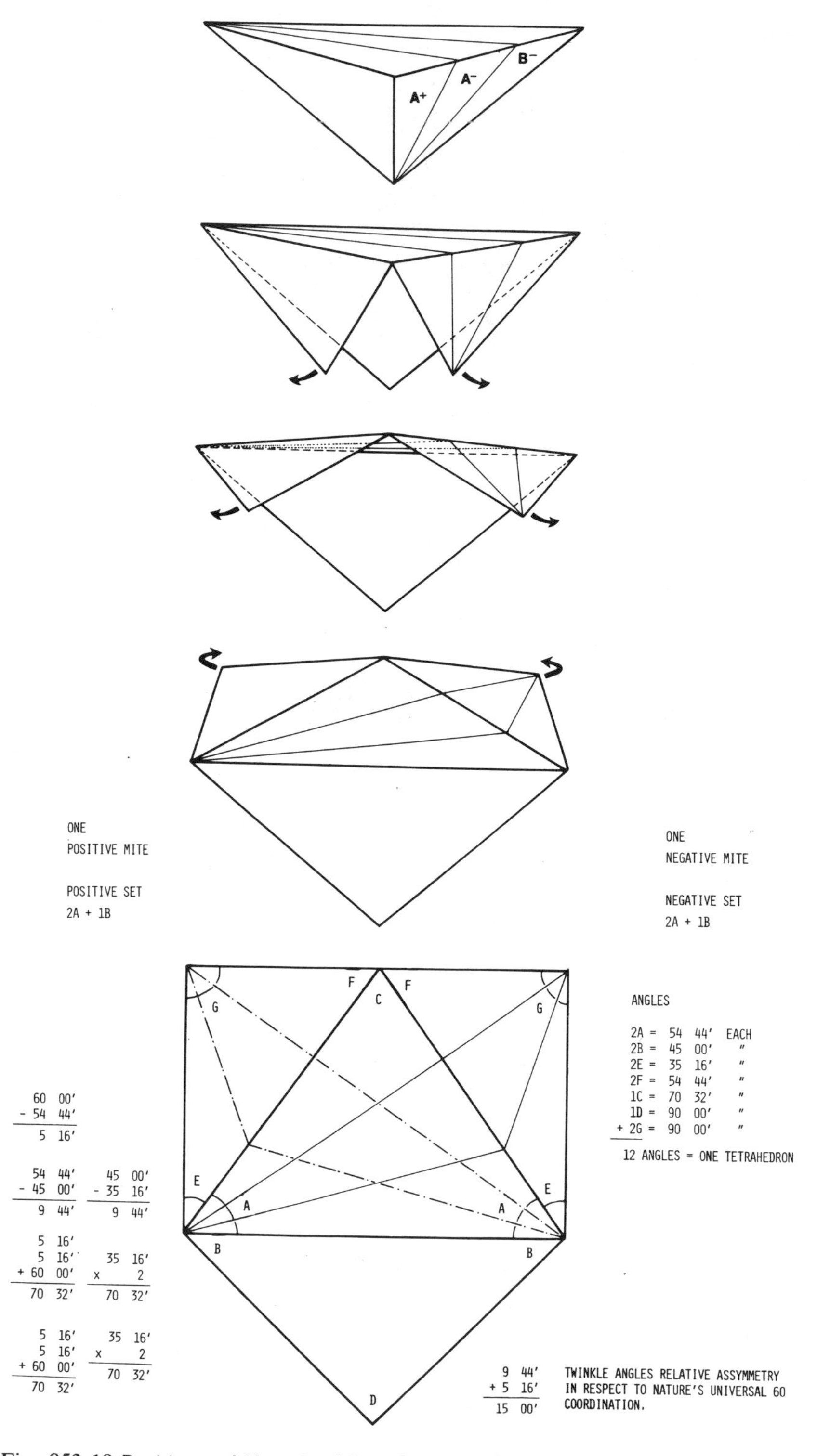

Fig. 953.10 *Positive and Negative Mites Constituted of Two* A *Quanta Modules and One* B *Quanta Module.*

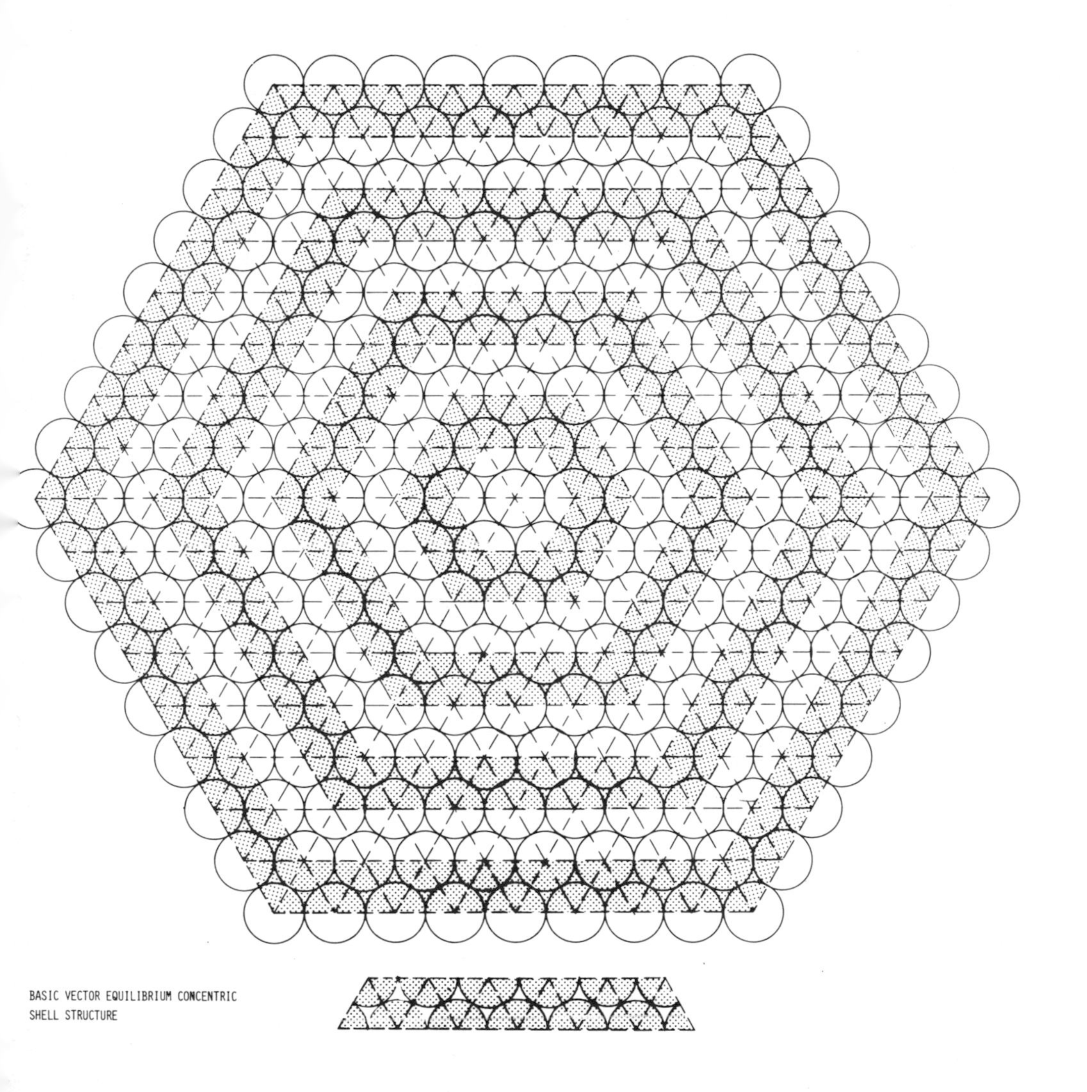

Fig. 970.20 *Basic Vector Equilibrium Concentric Shell Structure:* The legend at the bottom illustrates the interstitial between-sphere spaces.

1	2	3	4	5	6	7	8	9	10	11
	Cumulative number of spheres in triangle of frequency equal to F-1=Q	Cumulative number of spheres in tetrahedron of frequency equal to F-2=P							Cumulative volume (Tetrahedron = 1) of Vector Equilibrium of frequency = F	Shell volume of Vector Equilibrium of frequency equal to F+1 = R
F	$\frac{Q^2-Q}{2}$	$\frac{P^3-P}{6}$	+ 0 =		x 6		F^3	x 20 =	$20F^3$	$\left[\left(\left[\frac{R^2-R}{2}\right] \times 12\right) + 2\right] \times 10$
0	0	0	+ 0 =	0	x 6	+ 0 =	0	x 20 =	0	20 = [(0 x 12) + 2] x 10
1	1	0	+ 0 =	0	x 6	+ 1 =	1	x 20 =	20	140 = [(1 x 12) + 2] x 10
2	3	1	+ 0 =	1	x 6	+ 2 =	8	x 20 =	160	380 = [(3 x 12) + 2] x 10
3	6	4	+ 0 =	4	x 6	+ 3 =	27	x 20 =	540	740 = [(6 x 12) + 2] x 10
4	10	10	+ 0 =	10	x 6	+ 4 =	64	x 20 =	1,280	1,220 = [(10 x 12) + 2] x 10
5	15	20	+ 0 =	20	x 6	+ 5 =	125	x 20 =	2,500	1,820 = [(15 x 12) + 2] x 10
6	21	35	+ 1 =	36	x 6	+ 0 =	216	x 20 =	4,320	2,540 = [(21 x 12) + 2] x 10
7	28	56	+ 1 =	57	x 6	+ 1 =	343	x 20 =	6,860	3,380 = [(28 x 12) + 2] x 10
8	36	84	+ 1 =	85	x 6	+ 2 =	512	x 20 =	10,240	4,340 = [(36 x 12) + 2] x 10
9	45	120	+ 1 =	121	x 6	+ 3 =	729	x 20 =	14,580	5,420 = [(45 x 12) + 2] x 10
10	55	165	+ 1 =	166	x 6	+ 4 =	1000	x 20 =	20,000	6,620 = [(55 x 12) + 2] x 10
11	66	220	+ 1 =	221	x 6	+ 5 =	1330	x 20 =	26,620	7,940 = [(66 x 12) + 2] x 10
12	78	286	+ 2 =	288	x 6	+ 0 =	1728	x 20 =	34,560	9,380 = [(78 x 12) + 2] x 10
13	91	364	+ 2 =	366	x 6	+ 1 =	2197	x 20 =	43,940	10,940 = [(91 x 12) + 2] x 10
14	105	455	+ 2 =	457	x 6	+ 2 =	2744	x 20 =	54,880	12,620 = [(105 x 12) + 2] x 10
15	120	560	+ 2 =	562	x 6	+ 3 =	3375	x 20 =	67,500	14,420 = [(120 x 12) + 2] x 10
16	136	680	+ 2 =	682	x 6	+ 4 =	4096	x 20 =	81,920	16,340 = [(136 x 12) + 2] x 10
17	153	816	+ 2 =	818	x 6	+ 5 =	4913	x 20 =	98,260	18,380 = [(153 x 12) + 2] x 10
18	171	969	+ 3 =	972	x 6	+ 0 =	5832	x 20 =	116,640	20,540 = [(171 x 12) + 2] x 10
19	190	1140	+ 3 =	1143	x 6	+ 1 =	6859	x 20 =	137,180	22,820 = [(190 x 12) + 2] x 10
20	210	1330	+ 3 =	1333	x 6	+ 2 =	8000	x 20 =	160,000	

Fig. 971.01 *Chart: Table of Concentric, Sphere-Shell Growth Rates*

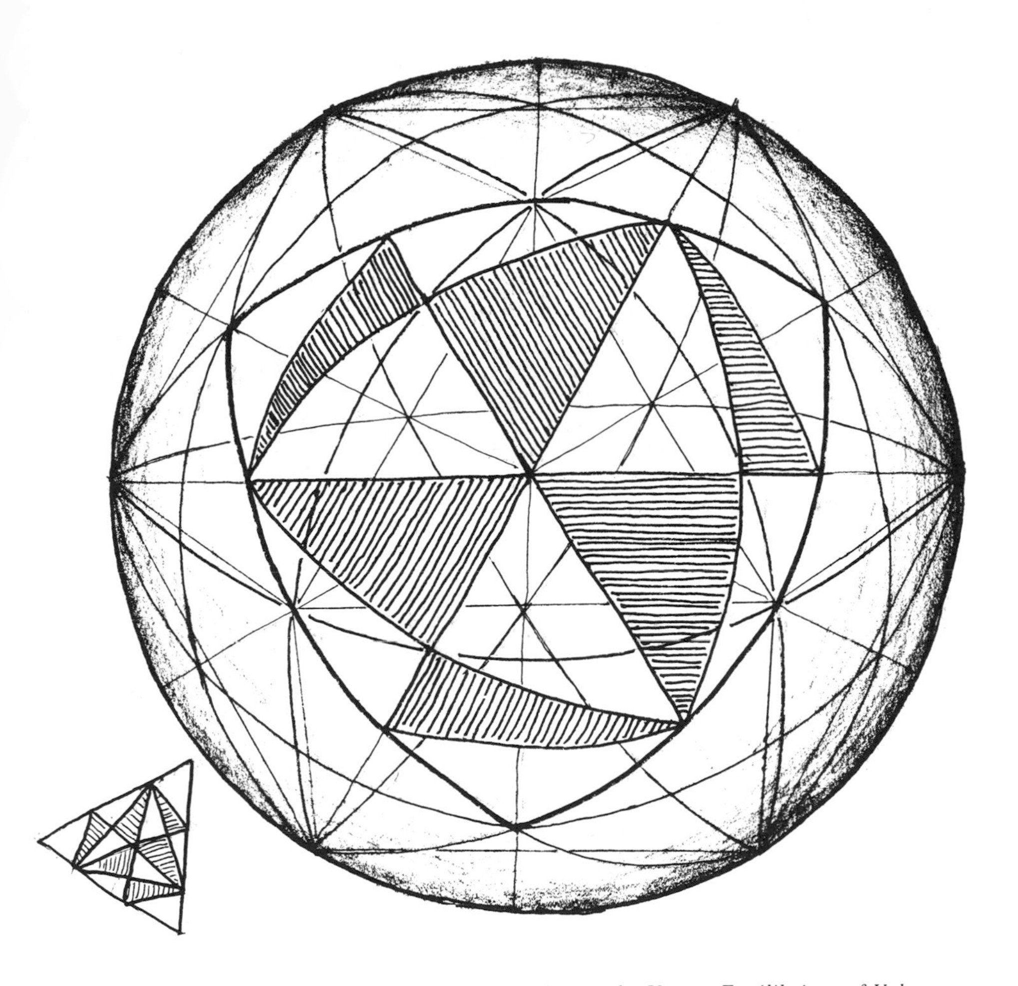

Fig. 982.58 *Nuclear Sphere of Volume 5 Enclosing the Vector Equilibrium of Volume $2^1/_2$ with the Vector Equilibrium's Vertexes Congruent with the Nuclear Sphere:* Shown are 15 of the Basic Disequilibrium 120 LCD triangles per sphere which transform as *A* Quanta Module tetrahedra. In the 25-great-circle subdividing of the vector equilibrium's sphere, the three great-circles produce the spherical octahedron, one of whose eight spherical triangles is shown here. As was shown on the icosahedron, the 120 triangles of the 15 great circles divide the sphere in such a way that the spherical octahedron's triangle can be identified exactly with 15 of the Basic Disequilibrium 120 LCD Triangles. Here we show the 15 disequilibrium triangles on the spherical octahedron of the vector equilibrium: $8 \times 15 = 120$ spherical right triangles which tangentially accommodate—closely but not exactly—the 120 *A* Quanta Modules folded into tetrahedra and inserted, acute corners inward to the sphere's center, which could not be exactly accommodated in the shallower icosahedral phase because of nuclear collapse and radius shortening in the icosahedron.

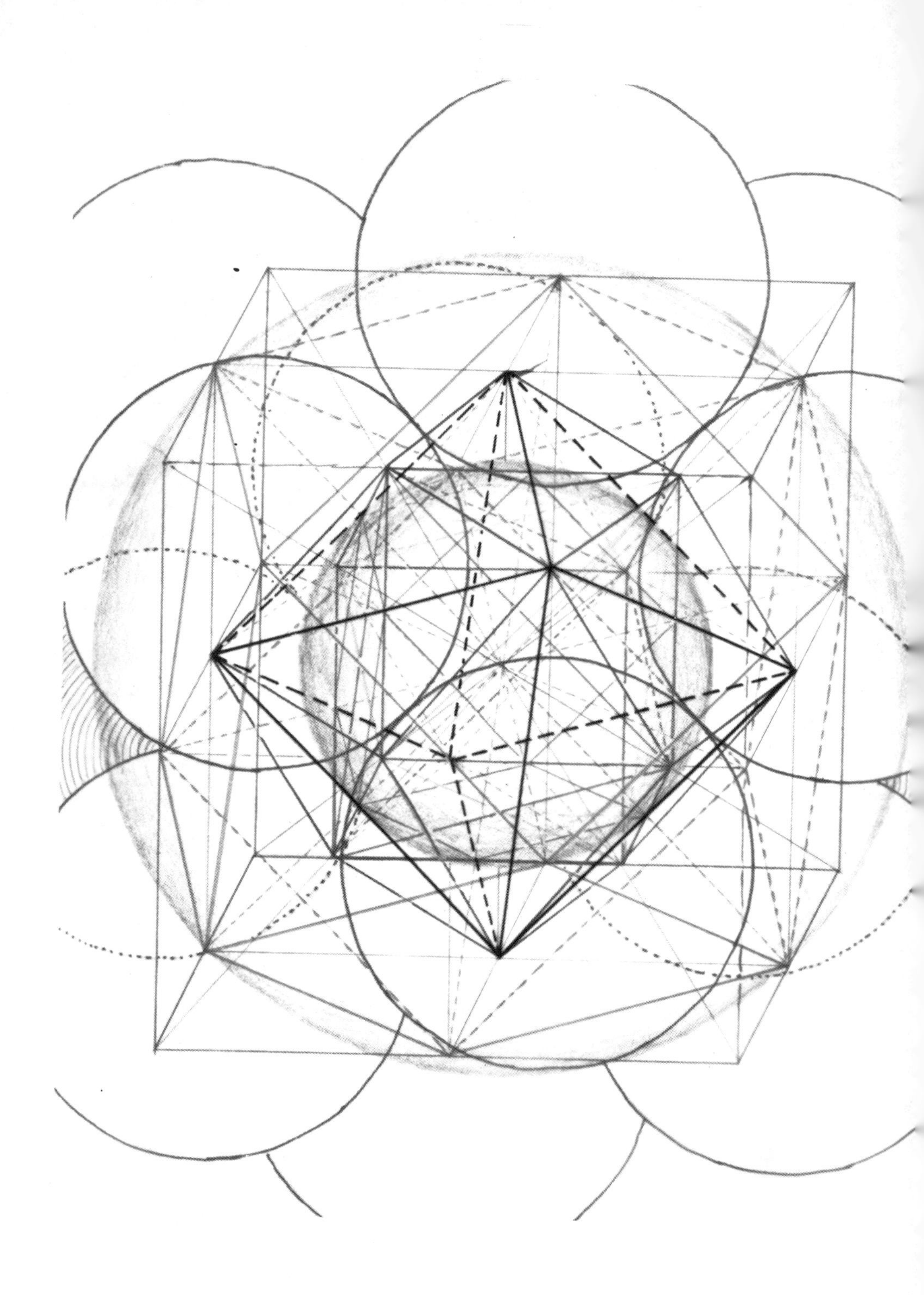

Fig. 982.61 *Synergetics Isometric of the Isotropic Vector Matrix:* See text for full legend. Note the twelve-around-one, closest-packed spheres.

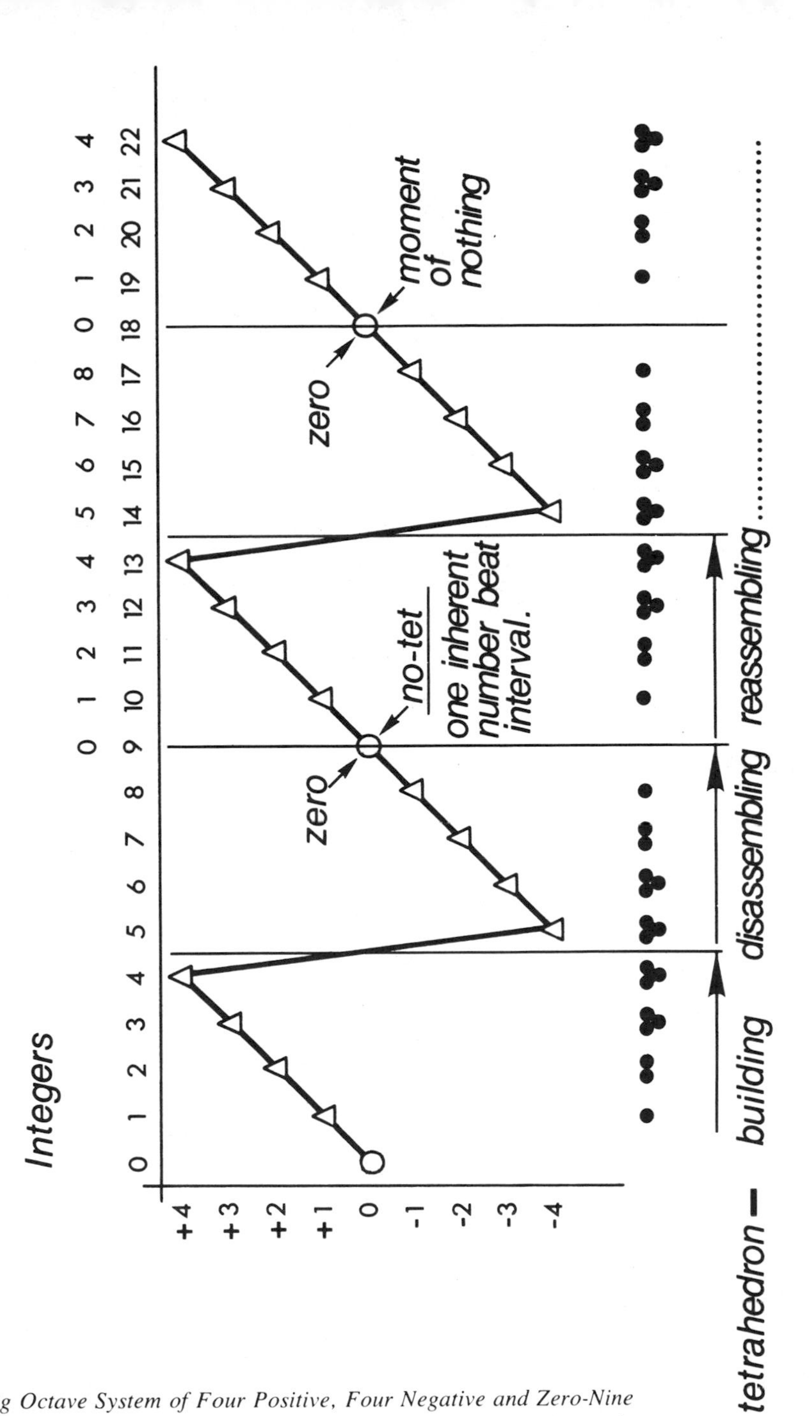

Fig. 1012.14A *Indig Octave System of Four Positive, Four Negative and Zero-Nine Wave Pattern of Experiential Number:* This basic discontinuous wave disclosure is intimately related to inherent octavization through tension-chord halving discovered by the Pythagoreans, and the major-minor-mode "fifthing" obtained by tension-chord thirding of length. These inherent additive-subtractive, alternate pulsing effects of number produce positive waves, but *not continuously* as had been misassumed. Zero—or "No-tet, None, Nine" intrudes. Waves are discontinuous and confirm unit quantation, one tetrahedron inherently constituting the basic structural system of Universe. The star Sun's combining of four hydrogen atoms into helium atoms generates quanta radiation.

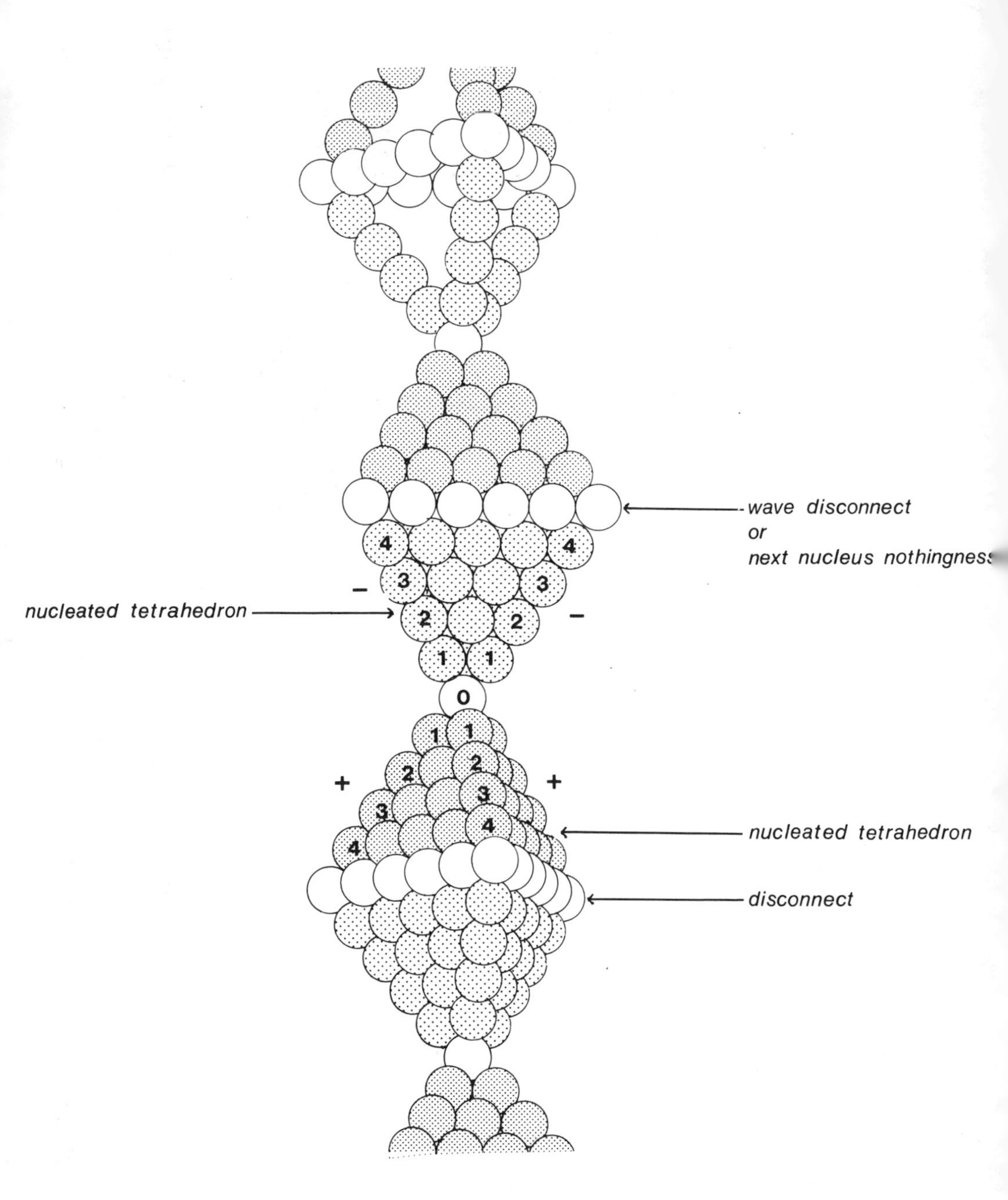

Fig. 1012.14B *Wave, Quanta, Indigs, Unity-Is-Plural Bow Ties:* This works for any pair of the ten pairs of tetrahedra in the vector equilibrium, of which only four pairs are active at any one time. The bow-tie waves illustrate the importance of zero. They come into phase with one another and with physics and chemistry. (Note the "Wave Disconnect" or next-nucleus nothingness.)

Evolution of Synergetics: a portfolio of six drawings made by the author in 1948.

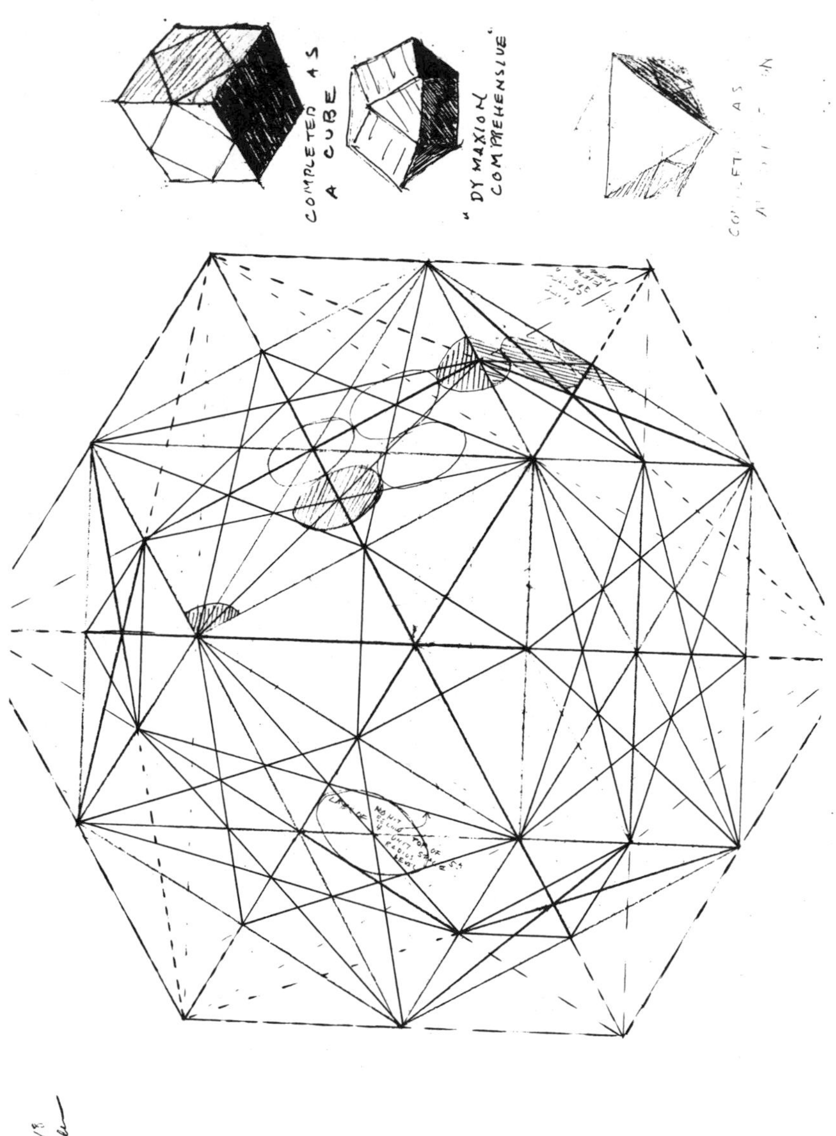

Transformations of Dymaxion Comprehensive (8 February 1948).

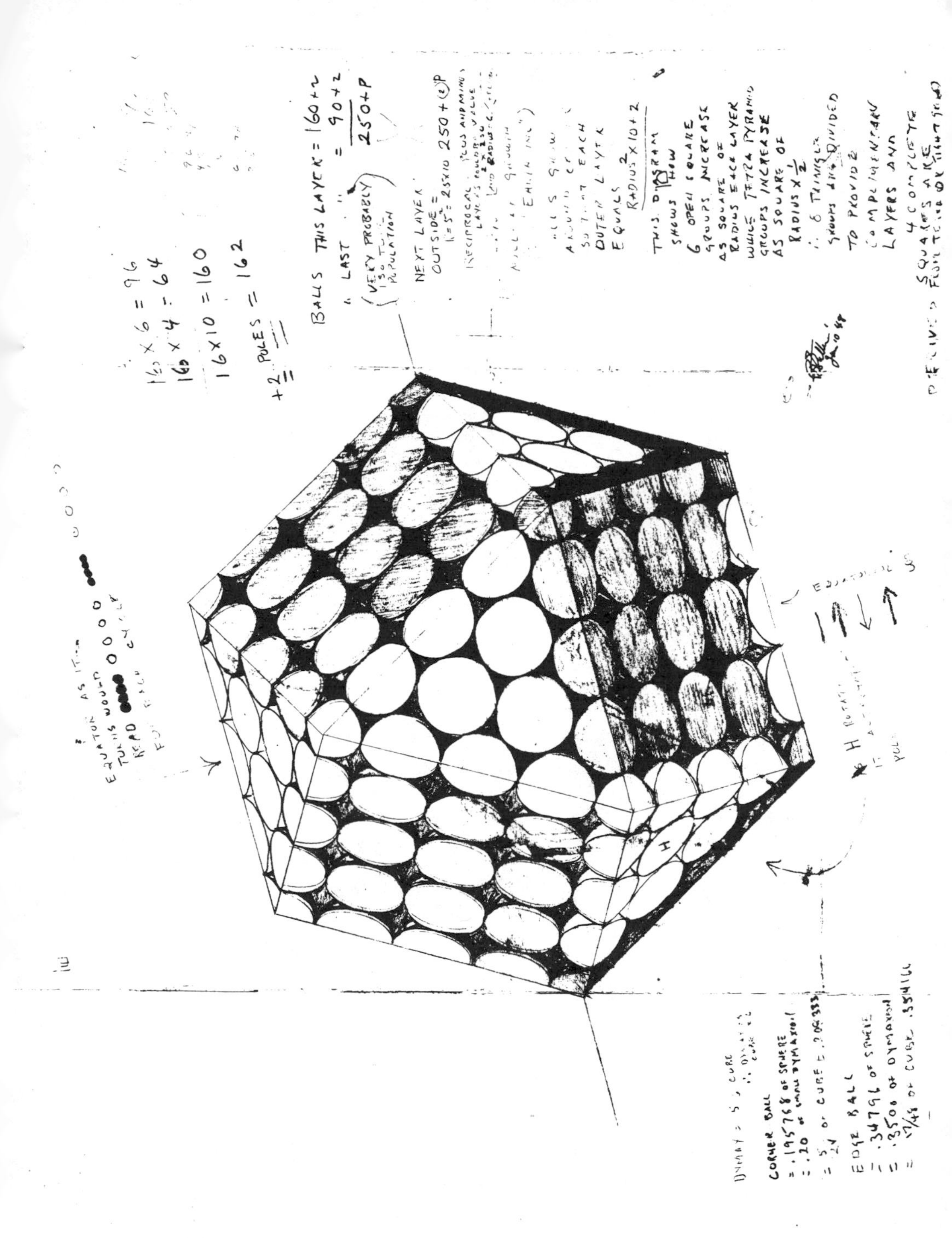

Dymaxion Nuclear Growth (10 June 1948).

#95

R. Buckminster Fuller
5/7/1948

SUM OF BALLS IN ANY ONE LAYER
$10R^2+2$

centerball (+2) is infinite neutral axis

2 congruent spheres = unity
2 spherical 180° triangles = unity

$6R^2+2$ = MASS OF BALL ENCLOSING LAYER WHERE UNITY = 1 AND R = RADIUS OF CENTER OF GRAVITY OF ANY TWO COMBINED LAYERS.

26
98
218
386
602
866
1178

1, 7, 19, 37, 61, 91, 127, 169, 217, 271, 331, 397, 469, 547, 631, 721

56, 152, 296, 488, 728, 1016, 1352

RATE OF GROWTH OF HEXAGON OF BALLS IN ONE LAYER

WHERE CENTER ROW OR BALL = 0

this no = this no when ÷ 20

INDICATES RATIONAL NUMBER RELATIONSHIP OF SPHERE GROWTH AND ABSOLUTE SPACE GROWTH i.e. EACH BALL = $\frac{\sqrt{R}}{10}$

Whole-Number or Rational Basic Relationship of Dymaxion Comprehensive System to Light Quanta Particle Growths (7 May 1948).

...LE NUMBER OR RATIONAL RELATIONSHIP OF DYMAXION ...HENSIVE SYSTEM TO LIGHT QUANTA, (C^2) GROWTHS, GRAVITY QUANTA, SPEED OF LIGHT TO GRAVITATIONAL FORCE (R^2) TO MOLECULAR DIFFUSION TO SPIN, TO EXPANSION, AND ELECTRO MAGNETIC WAVE

$16^3 \times 20$
$15^3 \times 20$
$14^3 \times 20$
$13^3 \times 20$
$12^3 \times 20$
$11^3 \times 20$
$10^3 \times 20$
$9^3 \times 20$
$8^3 \times 20$
$7^3 \times 20$
$6^3 \times 20$
$5^3 \times 20$
$4^3 \times 20$
$3^3 \times 20$
$2^3 \times 20$
$1^3 \times 20$

← CUMULATIVE VOLUME OR MASS OF SUCCESSIVE DYMAXIONS $= R^3 \times 20$ (WHERE CENTRAL MASS = 20)

$[(0 \times 12) + 2] \times 10$
$+1 = [(1 \times 12) + 2] \times 10$
$+2 = [(3 \times 12) + 2] \times 10$
$+3 = [(6 \times 12) + 2] \times 10$
$+4 = [(10 \times 12) + 2] \times 10$
$+5 = [15 \times 12 + 2] \times 10$
$+6 = [21 \times 12 + 2] \times 10$
$+7 = [28 \times 12 + 2] \times 10$
$+8 = [36 \times 12 + 2] \times 10$
$+9 = [45 \times 12 + 2] \times 10$
$+10 = [55 \times 12 + 2] \times 10$
$+11 = [66 \times 12 + 2] \times 10$
$+12 = [78 \times 12 + 2] \times 10$
$+13 = [91 \times 12 + 2] \times 10$
$+14 = [105 \times 12 + 2] \times 10$
$+15 = [120 \times 12 + 2] \times 10$

← = PROGRESSION OF SUM OF INTEGRALS
= " " " OF "TRIANGULAR NUMBERS"
(I.E. SUCCESSIVE LAYERS OF TETRA-PYRAMID OF SPHERES IN WHICH ANY TWO LAYERS COMPOSITE TO EQUAL SQUARE OF EDGE OF THE LARGER OF THE TRIANGLES = 1, 3, 6, 10, 15, 21, 28, ETC)

OUTER (MASS OR VOLUME) LAYER OF DYMAXION IN TERMS OF MASS OF CENTRAL DYMAXION EQUALS 20 UNITS

PROGRESSION OF SUM

$$= \left[\frac{(R-1)^2 + (R-1)}{2} \times 120\right] + 20$$

VALUE OF OUTER LAYER

$$= 60\left[(R-1)^2 + (R-1)\right] + 20$$
$$= 60(R-1)^2 + 60(R-1) + 20$$

...ION $= \frac{\sqrt{R}}{6}$ = AS $\sqrt{R} = 1$ = B $\frac{1}{10}$ = D = $\frac{1}{2}$ OF UNIT OF SPACE
∴ IF UNIT OF SPACE = 120 DYMAXION = 20, SPACE 12

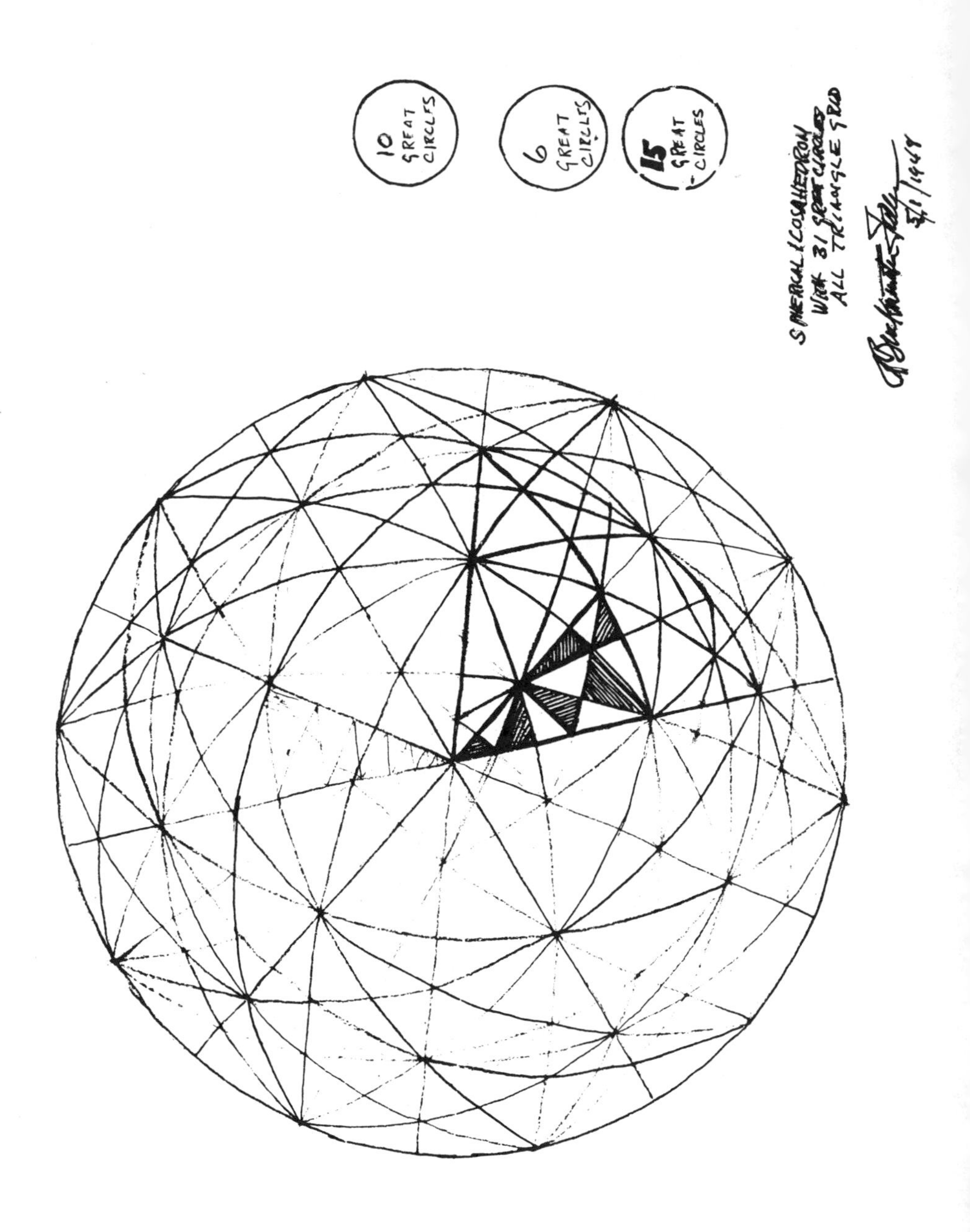

Thirty-one Great-Circle Triangular Grid of Icosahedron (1 May 1948).

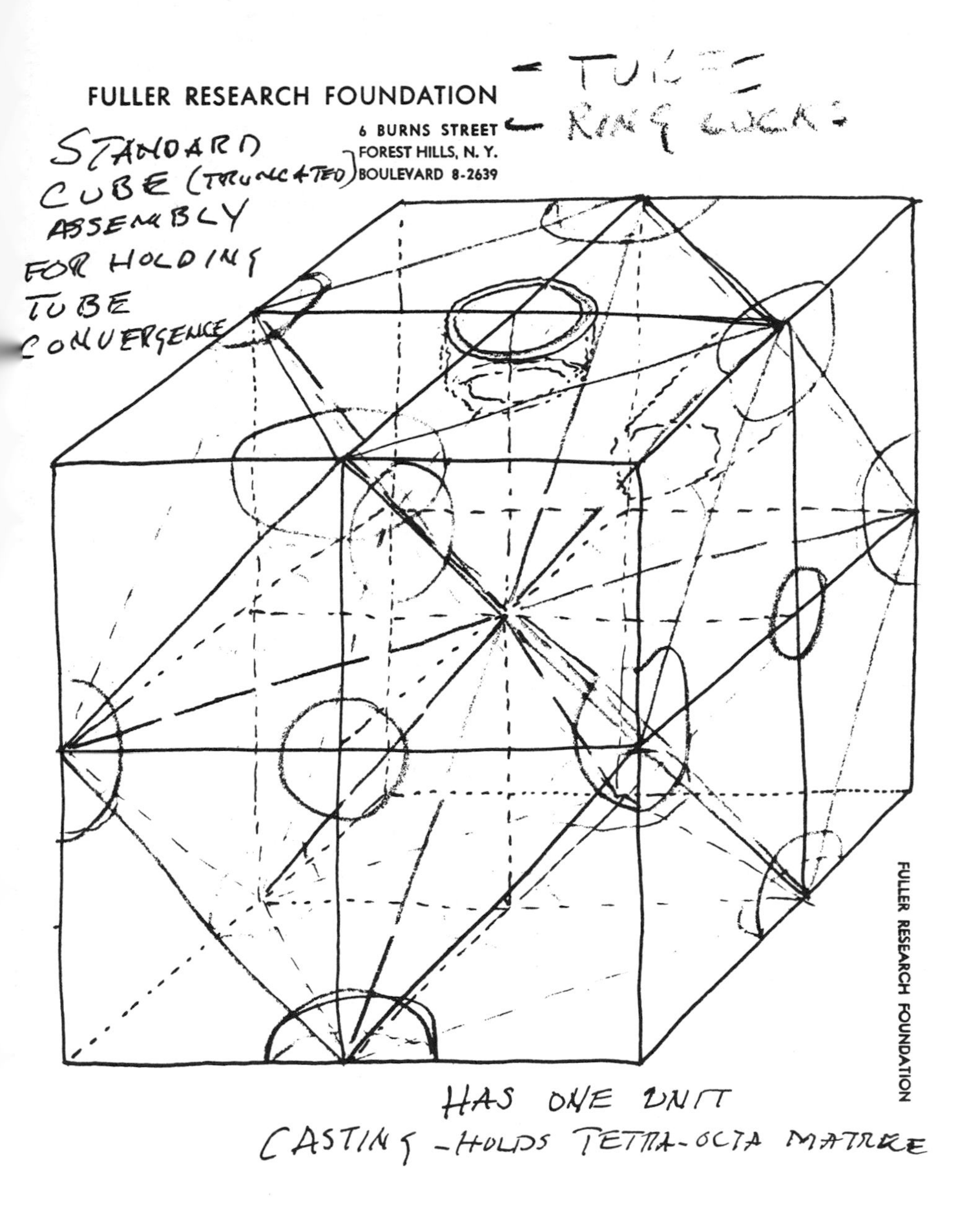

FULLER RESEARCH FOUNDATION
6 BURNS STREET
FOREST HILLS, N. Y.
BOULEVARD 8-2639
STANDARD
CUBE (TRUNCATED)
ASSEMBLY
FOR HOLDING
TUBE
CONVERGENCE
FULLER RESEARCH FOUNDATION
HAS ONE UNIT
CASTING - HOLDS TETRA-OCTA MATRICE

Standard Cube (truncated) Assembly for Holding Tube Convergence (Undated, 1948).

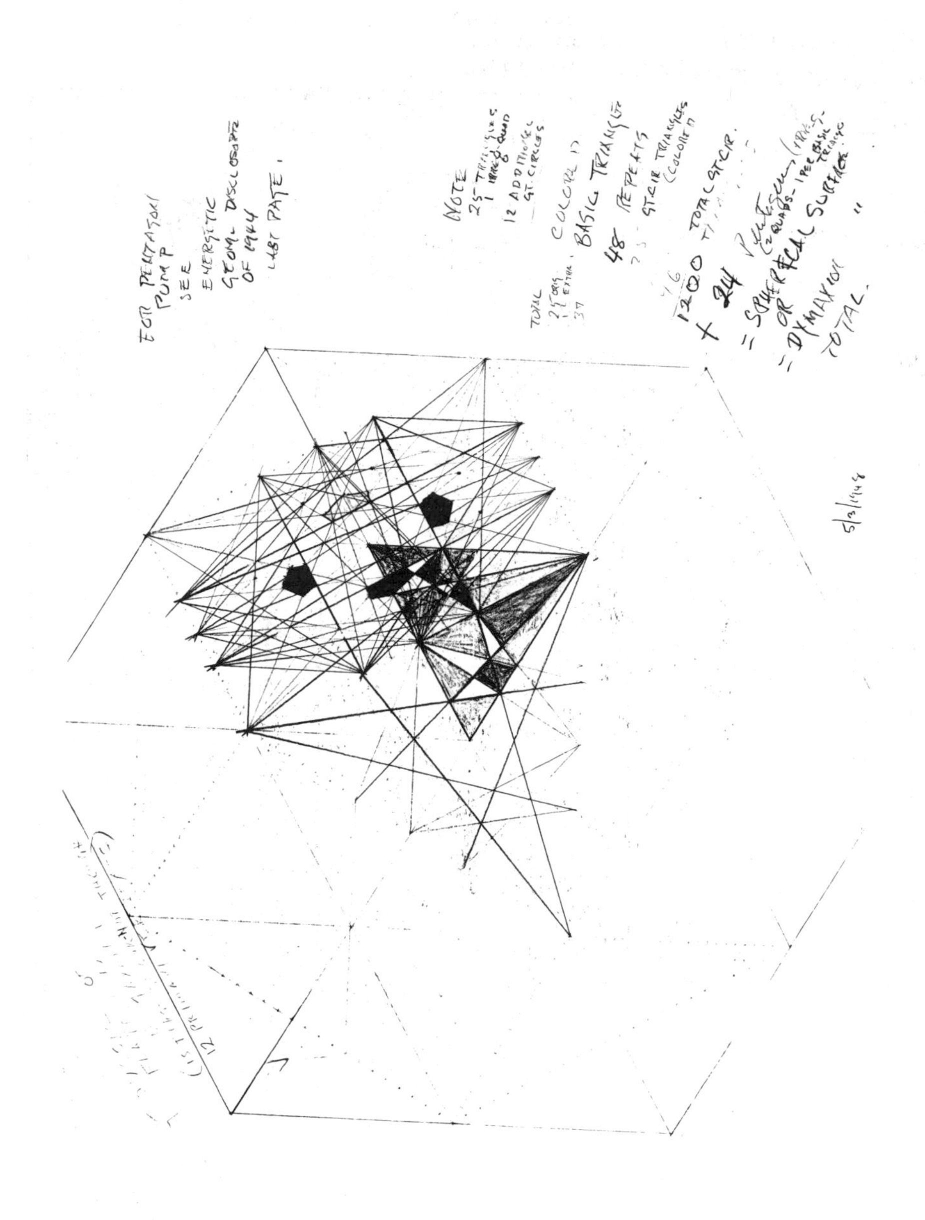

Geodesic Flattened Dymaxion (3 May 1948).

Contribution to *Synergetics*

Arthur L. Loeb

Department of Visual and Environmental Studies,
Carpenter Center

HARVARD UNIVERSITY

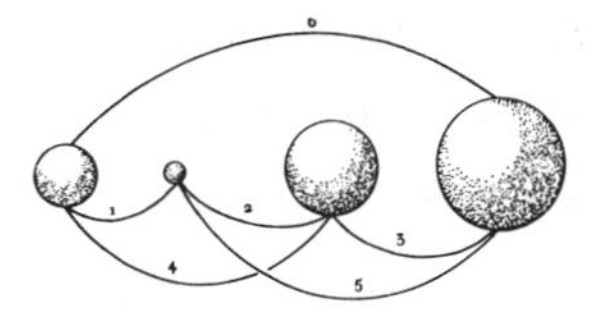

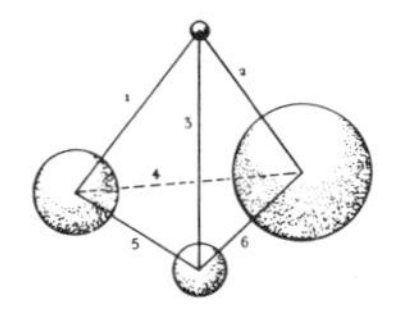

A INTRODUCTION

This portion of *Synergetics* is intended to complement R. Buckminster Fuller's text in several ways. The most extensive section, "Vector Equilibrium in Crystals," summarizes the author's own investigation of spatial order on a microscopic scale, and discloses an architecture of crystal structures that has many features in common with Fuller's macroscopic coordinate systems. The question has been raised whether such structural similarities on such different scales are fortuitous, or whether the organizing forces are comparable over such different distances. Such a question ignores the geometrical properties of space: space is not a passive vacuum, but has properties which impose powerful constraints on any structure that inhabits it. These properties of space, unlike the various specific interactive forces between atoms or dome joints, are the same on every scale, and override any specific interactions.

Fuller's experimental mathematical approach is characterized by an uncanny sensitivity to significant numerical as well as geometrical patterns. Sometimes the underlying origin of these patterns is easily established, whereas in other cases it is still obscure. In a number of cases I have presented a proof or derivation from fundamental principles. While aware of R.B.F.'s disdain of proofs I am of the opinion that these proofs constitute important bridges between Fuller's discoveries and established bodies of information and knowledge. I have attempted in these proofs to adhere to Fuller's experimental intuitive style without sacrificing rigor, stressing the concrete and geometric rather than the abstract and symbolic.

These sections were originally planned as interpolations in Fuller's principal exposition. The difference in style and trains of thought made such interruptions inappropriate. These contributions do not, for this reason, pretend to great continuity, but should be read as marginal notes to Fuller's exposition.

B Proof of Euler's Equation

Euler's equation is $V+F=E+2$ where $V=$ number of vertices, $F=$ number of faces, E = number of edges. For a tetrahedron (not necessarily regular), $V=4$, $F=4$, $E=6$; hence Euler's equation is satisfied. From this tetrahedron, we can "grow" any other triangulated polyhedron by the following procedure. Choose a point outside a given face of the tetrahedron, and join that point to the vertices of that face by three straight lines. The polyhedron thus generated has *one* more vertex than did the tetrahedron and *three* additional edges. The *original* face was replaced by *three* new faces, so that the new polyhedron has *two* faces more than did the tetrahedron. The same procedure can be repeated indefinitely, each time growing a tetrahedron from a triangular face. With each step the number of vertices is increased by *one,* the number of faces by *two,* the number of edges by *three.*

A polyhedron having V vertices can thus be created by adding to the tetrahedron (four vertices) in succession $(V-4)$ vertices. The resulting polyhedron has a number of edges equal to that of the original tetrahedron (6) plus $3(V-4)$; the number of its faces equals that of the original tetrahedron (4) plus $2(V-4)$. Therefore, $E=6+3(V-4)=3V-6$, $F=4+2(V-4)=2V-4$. $\therefore$ $E+2=3V-4$; $V+F=3V-4$; $\therefore$ $V+F=E+2$, q.e.d.

We have thus proven Euler's equation for a polyhedron that has only triangular faces. Suppose that we have a polyhedron having a face that is a polygon with n vertices, called an "n-gon"; all its other faces are triangles. Call the number of vertices of this polyhedron v, its number of faces f, and its number of edges e. Choose a point outside the n-gon, and join it by straight lines to the n vertices of the n-gon. The polyhedron so created has only triangular faces; we have already proven that Euler's equation is valid for it. In creating the new polyhedron from the original one we added a single vertex, so that this new polyhedron has $(v+1)$ vertices. We replaced the original n-gon by n triangular faces; as a result, the total number of faces of the new polyhedron is $(f+n-1)$. The new number of edges is $(e+n)$. Since, for the new polyhedron, Euler's equation holds,

$$(v+1)+(f+n-1)=e+n+2$$
$$\therefore\ v+f=e+2$$

We see, therefore, that Euler's equation also holds for the polyhedron that had the n-gonal face. Since each n-gonal face can be replaced in the same fashion by triangular faces, we have proven Euler's equation for polyhedra having any kind of polygonal faces.

C Doubly Connected Polyhedra

A doughnut is a doubly connected body (Fig. A1). By this we mean that we can choose two points in a doughnut, connect them by two ropes, and pull the ropes taut without having the ropes coincide.

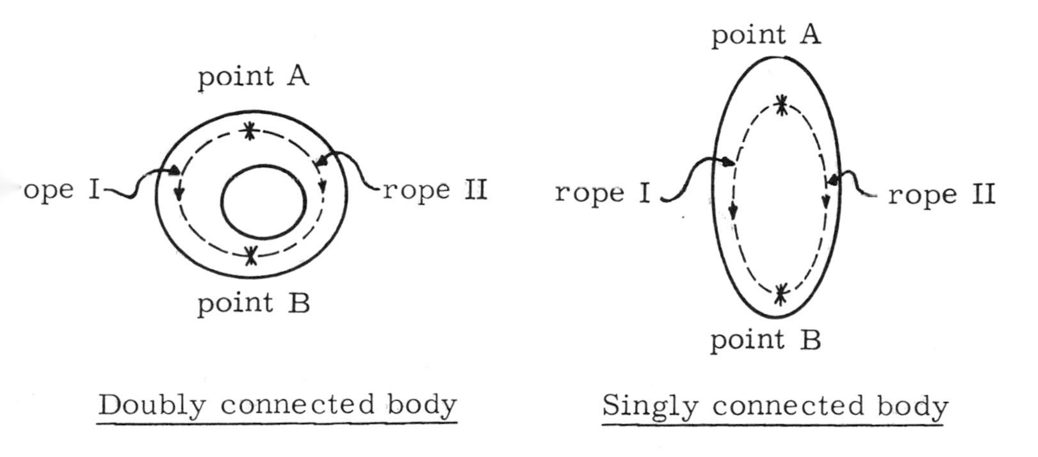

Fig. A1. Doubly and singly connected figures.

The creation of a polyhedron from a tetrahedron, as described above, will generally result in a singly connected polyhedron (Fig. A1). However, in some cases, multiply connected bodies can result, and in this case Euler's equation must be modified. Let us suppose that a polyhedron has a horseshoelike outline. This might be thought of as a horseshoe magnet, with a very small gap between north and south poles and a faceted surface (Fig. A2). Suppose, now, that the faces on both sides of the gap are identical n-gons. For this polyhedron, Euler's equations are valid: it could be grown from a tetrahedron by the procedure just discussed. Now if we let the gap width go to zero, the north and south poles annihilate each other, and a doubly connected body results. When this happens, the two faces on both sides of the gap

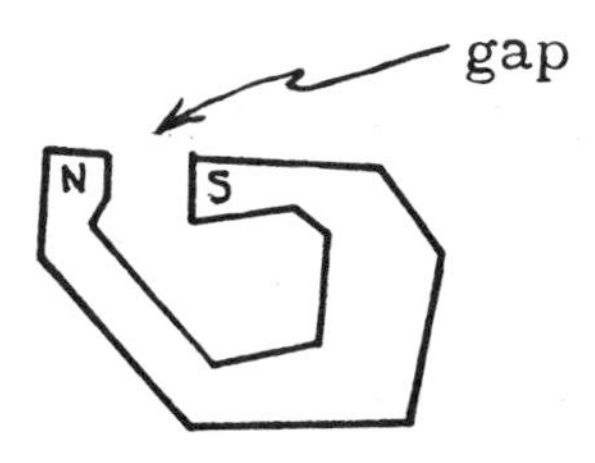

Fig. A2. Polyhedron with horseshoelike outline.

vanish. Moreover, the n vertices of one face merge with the n vertices of the opposite face, so that $2n$ vertices merge into only n vertices, with a resulting loss of n vertices. Similarly, there is a loss of n edges. Thus the balance sheet is as follows:

	Horseshoe-Shaped Polyhedron	Doubly Connected Body
No. of vertices:	v	$V = v - n$
No. of faces:	f	$F = f - 2$
No. of edges:	e	$E = e - n$

From this balance sheet it follows that for the doubly connected body: $V + F - E = v - n + f - 2 - e + n = v + f - e - 2$. Since for the horseshoe-shaped polyhedron $v + f = e + 2$, $V + F - E = 0$. For a doubly connected body, Euler's equation is therefore: $V + F = E$. Fuller would then associate the "2" in the equation for singly connected polyhedra with the two faces on either side of the gap, which are destroyed when the doubly connected body is created. The argument also holds in reverse; any multiply connected body can be turned into a singly connected one by "sawing through" each connection and thus creating a pair of faces.

D The Sum of the Surface Angles of a Singly Connected Polyhedron

Consider a polyhedron whose faces are general n-gons. In each face, choose a point, and connect this point by straight lines to the vertices of the n-gon. The surface of the polyhedron is then triangulated.

Two triangles meet at each edge; the sum of the angles of the triangles in the surface of the entire polyhedron is therefore $2E \times 180°$. However, this sum includes the angles that meet at the points chosen in each face, which add up to 360° in each face. The sum of the surface angles of the polyhedron therefore equals $(2E - 2F) \times 180°$.

For singly connected polyhedra, $E - F = V - 2$, so that the sum of the surface angles is: $V \times 360° - 720°$.

For doubly connected polyhedra, $E - F = V$, so that the sum of the surface angles is: $V \times 360°$.

E Duals

If we choose a single point in each face of a polyhedron, then these points may be considered as vertices of a new polyhedron. Two polyhedra so related are called each other's duals. The number of faces of a polyhedron equals the number of vertices of its dual. The sum of the number of faces and the number of vertices is therefore the same for any polyhedron as for its dual. From Euler's equation, it follows that dual polyhedra have identical numbers of edges.

To find the dual of a regular polyhedron, we choose points at the geometric centers of its faces. Let us consider first the regular, *triangulated* solids, which Fuller considers fundamental. The simplest of these is the tetrahedron: $V = 4, F = 4, E = 6$. Since the number of vertices of the tetrahedron equals its number of faces, its dual is again a tetrahedron. The tetrahedron is its own dual; it is unique in this respect. Note that Fuller's positive and negative tetrahedra are each other's duals.

The octahedron has $V = 6$, $F = 8$, $E = 12$. Its dual must have $V = 8$, $F = 6$, $E = 12$. This is the cube.

The icosahedron has $V = 12$, $F = 20$, $E = 30$. Its dual has $V = 20$, $F = 12$, $E = 30$; it is the pentagonal dodecahedron. We see, therefore, that the two nontriangulated Platonic solids are just the duals of two triangulated solids. The numbers of vertices, edges, and faces for the Platonic solids are thus summarized as follows:

↓	V	E	F	↑
Tetrahedron	4	6	4	Tetrahedron
Octahedron	6	12	8	Cube
Icosahedron	12	30	20	Pentagonal Dodecahedron
	F	E	V	

F Close Packing

Consider for a moment a planar array of closely packed identical circular disks. Imagine a bee inside each circular cell, producing a wax cylinder. If the bees all produce at equal rate but do not stop at the point when the cylinders just touch, then the well-known honeycomb pattern is produced (Fig. A3).

We call the hexagonal cell around each bee his "domain"; the plane is divided equally between all bees. *Every* point within a cell is closer to its central bee than to any other bee.

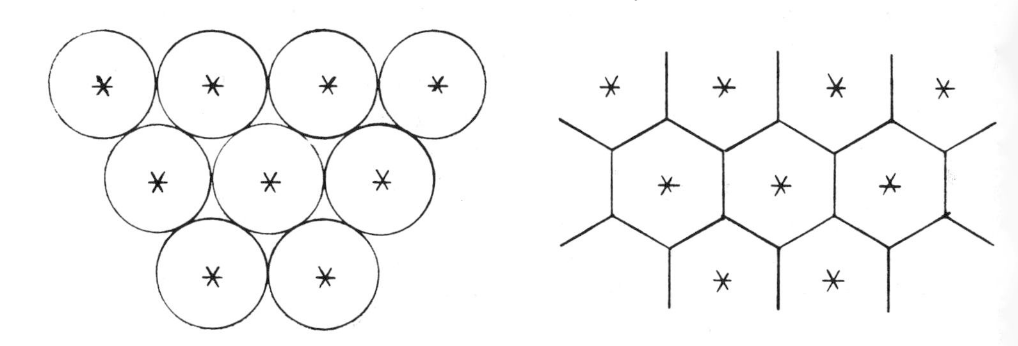

Fig. A3. Honeycomb pattern.

The edges of the hexagonal domains perpendicularly bisect the lines joining neighbor bees; every point of such an edge is equidistant between these bees. Edges meet at vertices that are equidistant from *three* bees located at the corners of an equilateral triangle.

Let us apply the same idea to a three-dimensional, cubically close-packed array of identical spheres. Suppose there to be a creature called a "cee," which spins wax spheres around itself at a steady radial rate. Suppose that these cees locate themselves most efficiently—namely, at the centers of cubically close-packed identical spheres. Like the bees, the cees continue their activity beyond the point where their spheres just make contact. What is the resulting domain for each cee? It is the so-called "rhombohedral dodecahedron" (cf. Fig. A19), which has 12 diamond-shaped faces. Because each face has one acute and one obtuse angle, this is not a regular polyhedron, the latter having all edges and all angles equal. The rhombododecahedron is a space filler (cf. Sec. I). For the rhombododecahedron, $V = 14$, $F = 12$, $E = 24$. Its dual has $V = 12$, $F = 14$, $E = 24$; it is our old friend the cuboctahedron, or in Fuller's terms, the "vector equilibrium." The duality between "vector equilibrium" and rhombododecahedron is easily visualized as follows: Each "cee" in the center of a cubically close-packed sphere has 12 neighbor cees at the corners of a "vector-equilibrium" polyhedron. Each "radial" from a cee to one of its 12 neighbors is perpendicularly bisected by a face of its domain. Therefore, to each *vertex* of the "vector equilibrium," there corresponds a *face* of the rhombododecahedron, and vice versa; this observation establishes the duality relation between the two polyhedra.

Being only a "semiregular" polyhedron, the rhombododecahedron has two different kinds of vertices. There are six vertices at which four acute surface angles meet. In addition, there are eight vertices at which three obtuse surface angles meet. The six vertices form the vertices of an octahedron around each cee in the center of the rhombododecahedron and are themselves in turn octahedrally surrounded by cees. The eight vertices form a cube around the cee at the center of the rhombododecahedron, and are themselves tetrahedrally surrounded by cees. Any creature desirous of maintaining a maximum distance from as many cees as possible would locate itself preferably on one of the octahedrally surrounded vertices of the domain, with second choice going to the tetrahedrally surrounded vertices. Although "cees" are only fictional animals, we find that the configuration of ions in many crystalline solids actually do obey these geometrical principles (cf. Sec. P).

G The Icosahedron and the Golden Section

The icosahedron is the most complex of the three triangulated Platonic solids. Its thirty edges belong to five distinct sets of six edges each, such that all six edges within one set are mutually parallel or perpendicular. We say, therefore, that the 30 edges belong to five orthogonal sets. The centers of the six edges belonging to the same orthogonal set form the vertices of a regular octahedron. Therefore, *five* regular octahedra can be inscribed in an icosahedron, having their vertices on the centers of the edges of the icosahedron.

A straight line joining any vertices that are not adjacent is called a "diagonal." The icosahedron has diagonals of only two different lengths. The *longer diagonals* pass through the center of the icosahedron, joining vertices on opposite sides of the center. The *shorter diagonals* join vertices that are second-nearest neighbors; going along the *surface* of the icosahedron, it would require two edge lengths to go from one end of a "shorter diagonal" to the other. If we make a cross section (Fig. A4) through the icosahedron along a plane perpendicular to a long diagonal and containing five vertices, a pentagon results. The edges of this pentagon are all edges of the icosahedron, its vertices are vertices of the icosahedron, and its diagonals are shorter diagonals of the icosahedron. The edges and shorter diagonals of the icosahedron are related by the golden section, as will be shown below.

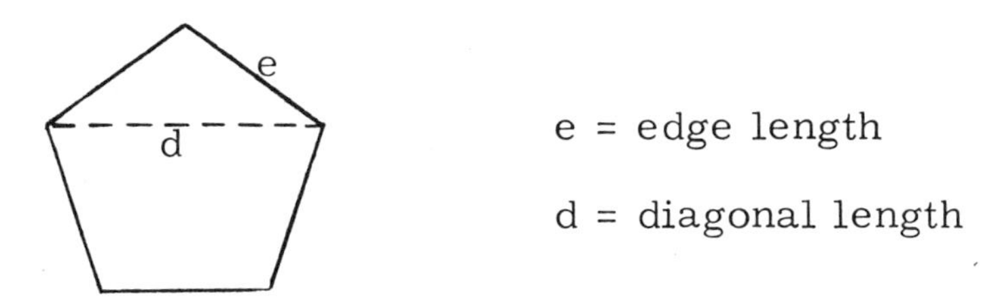

Fig. A4. Pentagonal cross section through an icosahedron.

The golden section is the division of a whole into two unequal parts such that the smaller part is related to the larger as the larger is to the whole. This section has a perfection that has appealed to designers for centuries; its relation to the most perfect of Platonic solids, the icosahedron, is therefore worth investigating. Below (Fig. A5), we have drawn a circle around the pentagonal cross section of the icosahedron, as well as a second diagonal that intersects the first at point X. The five vertices divide the circumference of the circle into five equal arcs, with the result that each angle subtended at the circumfer-

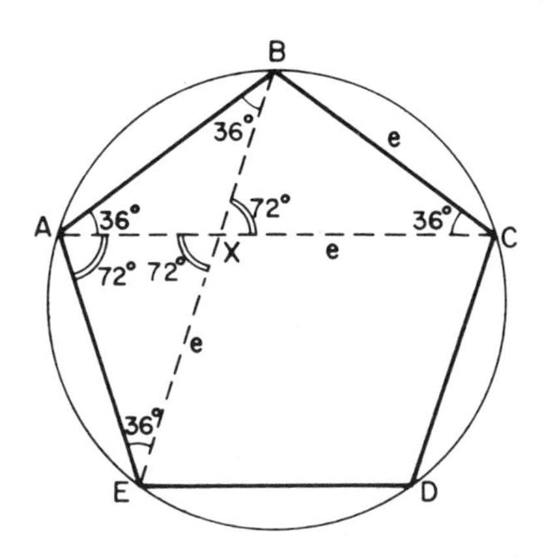

Fig. A5. Regular pentagon and the golden section.

ence of the circle by each arc equals 36 degrees. From the equality of the angles $\angle BXC$ and $\angle XBC$, it follows that $\overline{XC} = \overline{BC}$, so that the point X divides the diagonal AC into two portions, of which the larger equals the edge length. It similarly divides diagonal BE into two unequal portions, of which the larger equals the edge length.

As a result of the similarity of $\triangle AXB$ and $\triangle ABC$:

$$\overline{AX} \div \overline{AB} = \overline{AB} \div \overline{AC}$$

Therefore, $(d-e) \div e = e \div d$. This proves that the shorter diagonals of the icosahedron intersect at points that divide each diagonal according to the golden section, and that the larger portion of each diagonal just equals the edge length of the icosahedron.

If we draw in all the diagonals of the pentagon (Fig. A6), a new pentagon is generated whose vertices are U, V, W, X, Y. This pentagon can be consid-

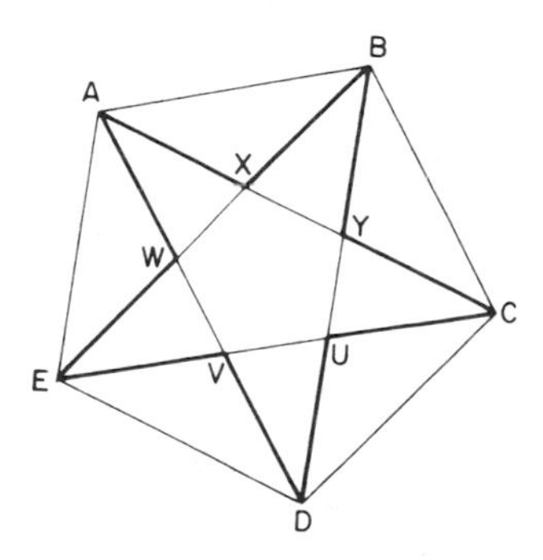

Fig. A6. Generation of smaller pentagon from larger one by the golden section.

ered a cross section through an icosahedron whose edges are portions of the shorter diagonals of the original icosahedron. There is also a "magical" relation between the edge lengths of the larger and smaller icosahedra: because of the similarity of $\triangle BXY$ and $\triangle BED$, $XY \div ED = BX \div BE$. Therefore, $XY \div e = (d-e) \div d$. The ratio of the edge lengths of the smaller and larger icosahedra is therefore the same as that of the *smaller* portion to the whole in a golden section.

Diagonals of the pentagon $UVWXY$ generate a smaller pentagon, just as $UVWXY$ was generated from $ABCDE$, and the process can be repeated indefinitely; each time the edge length is decreased by a factor $(d-e)/d$.

The polygon $AXBYCUDVEW$ is a pentagonal star, whose remarkable geometrical properties derived above were often thought to give it magic power.

The age of reason rejected the magical power of geometrical forms, and rightly so. However, in rejecting this power, it lost interest in the forms per se; we are now discovering that there is distinct physical significance in many geometrical forms.

H Remarks on Some Elementary Volume Relations Between Familiar Solids*

The volume of a regular octahedron equals exactly four times the volume of a regular tetrahedron of the same edge length.† A tetrahedron inscribed in a cube such that the edges of the tetrahedron constitute face diagonals of the cube occupies exactly one third of the volume of the cube. As we are, in the calculation of areas and volumes, concerned basically with problems in affine geometry, these relations are not limited to regular solids, but apply equally well to irregular tetrahedra and octahedra having congruent faces, and to an irregular tetrahedron inscribed in a general parallelepiped. When these relations are derived with the aid of the usual formula for the volume of a pyramid, $V = \frac{1}{3}Ah$, a good many irrational numbers are involved, and the simple integral ratios emerge almost incidentally. Somehow these simple integral values of the volume ratios of common solids are not part of our scientific culture, and a lack of familiarity with them frequently leads to unnecessarily cumbersome computations. It appears that a bias of our culture to orthogonal Cartesian coordinates has obscured these relations.

Instead of accepting and mechanically applying some geometric formulas, let us follow the common line of reasoning of physics, namely, to generalize from some physical observations. We shall then find that the common geometrical formulas are based on an *arbitrary* assignment of units that happen to simplify calculations in orthogonal Cartesian coordinates, but complicate work in other coordinates.

In Fig. A7 we show two geometrically similar pairs of figures; in Fig. A7*a* these figures are squares, while in Fig. A7*b* they are irregular triangles. It is observed that in *both* cases the areas of these pairs of figures are in ratio 4:1, whereas their edge lengths are in ratio 2:1.

* This section has been adapted from an article by the author in *The Mathematics Teacher* (May 1965) with permission of the editors.

† Cf. Fig. 950.31. Cf. also Loeb and Pearsall, *Am. J. Phys. 31*, 191.

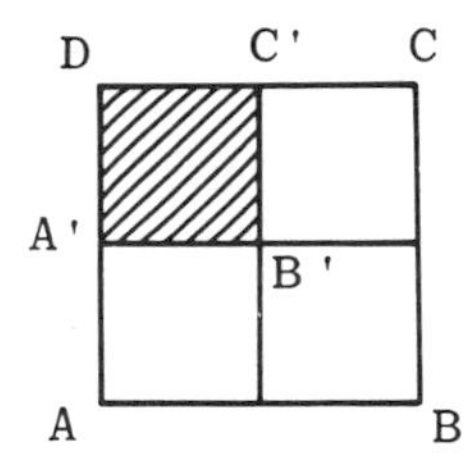

Fig. A7a. Square *ABCD* is similar to square *A'B'C'D*.

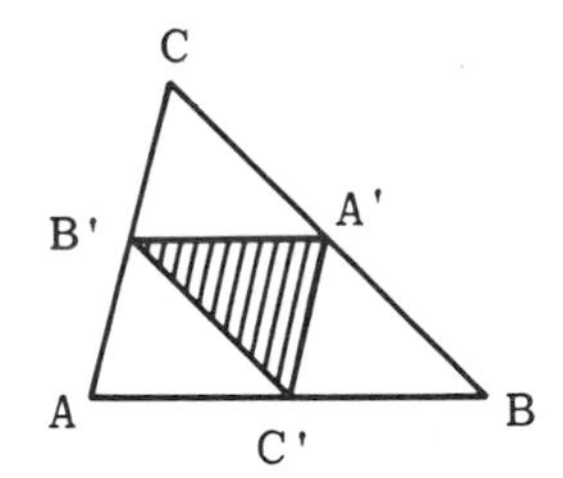

Fig. A7b. Triangle *ABC* is similar to triangle *A'B'C'*.

Lemma I: The areas, A, of geometrically similar figures are proportional to the square of the length of corresponding lines in these figures, d:

$$A = kd^2 \tag{1}$$

where k is a constant depending on the shape of the figure.

Equation (1) has fundamental significance; the numerical value of the proportionality constant k depends on the choice of units for A and d. In orthogonal Cartesian coordinates it is convenient to choose units such that a square of unit length has unit area. There is no fundamental objection, though, to choosing a system of units such that an equilateral triangle of unit edge length has a unit area. In fact, someone working with the geometry of close-packed identical spheres might consider such a system of units "rationalized"!

Similar considerations lead to:

Lemma II: The volumes, V, of geometrically similar solids are proportional to the cube of the length of corresponding lines in these solids, d:

$$V = Kd^3 \tag{2}$$

where K is a constant depending on the shape of the *solid*.

It should be observed that Loeb and Pearsall's derivation of the volume ratios of an octahedron and a tetrahedron was based on *comparing* volumes having the same "shape constant" K, so that assignment of units did not enter the problem at all. Application of the "familiar" formula $V = \frac{1}{3}Ah$ involves the numerical determination of the constant K, which is irrational for a regular tetrahedron, and eventually cancels out again.

The twelve lines joining the centers of the edges of a (generally irregular) tetrahedron constitute the edges of a (generally irregular) octahedron (Fig. A8). The volume of the tetrahedron, V_T, equals that of the octahedron, V_o, plus the combined volume of four tetrahedra similar to the original tetrahedron but having half its edge length. If the volume of each smaller tetrahedron is V_t:

$$V_T = 4V_t + V_o.$$

From Lemma II:

$$V_T = 8V_t$$

$$\therefore\ V_o = 4V_t. \tag{3}$$

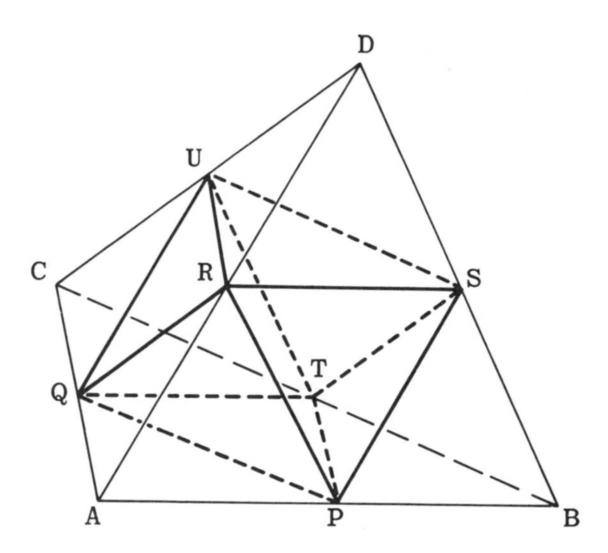

Fig. A8. The tetrahedron *ABCD* (volume V_T) consists of:

In Fig. A9 there is a parallelepiped with an inscribed (generally irregular) tetrahedron. The parallelepiped contains, besides the inscribed tetrahedron, four octants of an octahedron whose faces are congruent with those of the inscribed tetrahedron. If the volume of the parallelepiped is denoted by V_p:

$$V_p = V_t + \frac{4}{8}V_o.$$

Substitute Equation 3:

$$V_p = 3V_t. \tag{4}$$

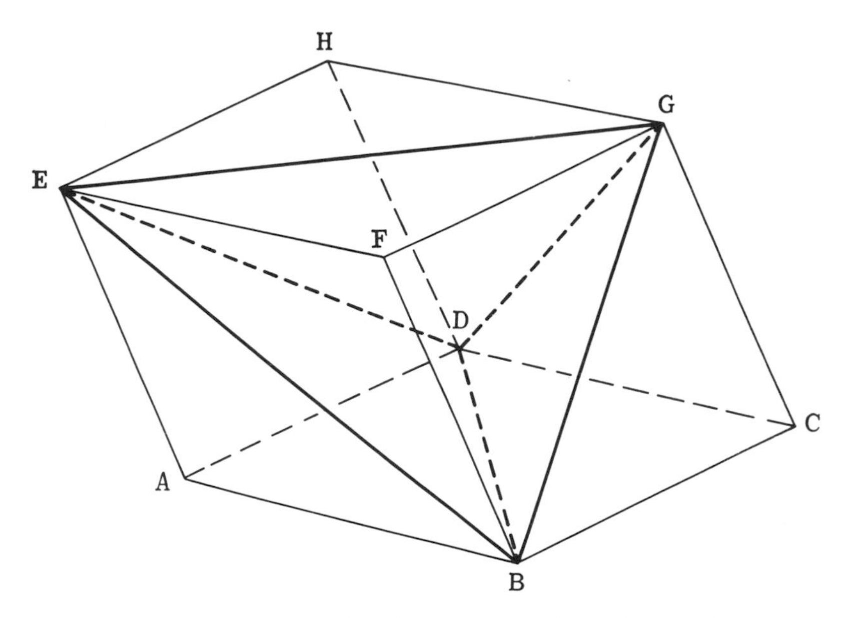

Fig. A9. Tetrahedron *EBDG* inscribed in parallelepiped *ABCDEFGH*. (Another tetrahedron, *ACFH*, enantiomorphic with *EBDG*, can be inscribed in the same parallelepiped.)

Hence, a tetrahedron inscribed in a parallelepiped occupies exactly one-third the volume of that parallelepiped. In general, there are two enantiomorphic tetrahedra of equal volumes that can be inscribed in a parallelepiped.

Professor Philippe LeCorbeiller has pointed out that one of the proofs * of the Law of Pythagoras flows directly from Lemma I. In Figure A10, $BC \perp AC$, $CD \perp AB$.

$$\triangle ABC \sim \triangle ACD \sim \triangle CBD.$$

Considering the hypotenuse as a "corresponding line" in each of these triangles, one concludes from Lemma I:

$$\text{Area } (\triangle ABC) = k\overline{AB}^2$$
$$\text{Area } (\triangle ACD) = k\overline{AC}^2$$
$$\text{Area } (\triangle CBD) = k\overline{CB}^2$$

* Georges Bouligand, *La Causalité des Théories Mathématiques* (Paris: Hermann & Cie, Editeurs, 1934), p. 25; T. L. Heath, *The Thirteen Books of Euclid's Elements* (London: Cambridge University Press, 1908; New York: Dover Publications, Inc., 1956), I, 353, and II, 270.

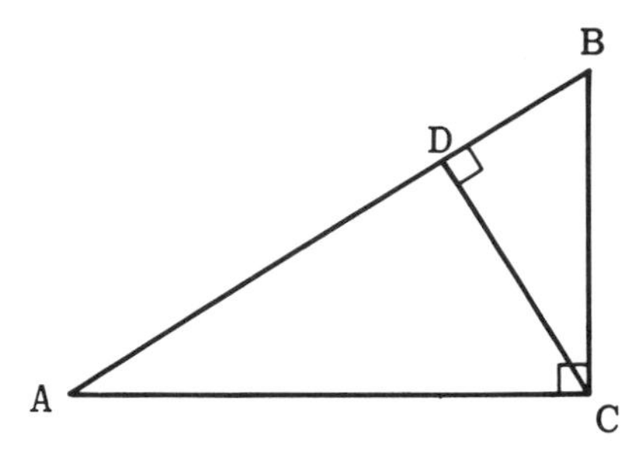

Fig. A10. Proof of Pythagoras' theorem.

Since Area $(\triangle ABC) = \text{Area } (\triangle ACD) + \text{Area } (\triangle CBD)$,

$$\overline{AB}^2 = \overline{AC}^2 + \overline{CB}^2, \text{ q.e.d.}$$

Conclusion

Uncritical acceptance of geometrical formulas as fundamental laws, particularly in systems that do not naturally fit orthogonal Cartesian coordinates, frequently leads to unnecessarily clumsy calculations and tends to obscure fundamental relationships. It is well to avoid instilling too rigid a faith in the orthogonal system into students of tender and impressionable age!

I Space-Filling Polyhedra

Place two identical cubes next to each other, with a pair of faces in contact. Place an identical pair of cubes on top of this original pair. The result is a slab two cubes high, two cubes wide, a single cube deep. When two such identical slabs are placed with their square faces adjoining, a cube results whose volume equals eight times (linear dimension twice) that of the original cube (Fig. A11). The procedure can be repeated indefinitely until all of space is occupied by cubes. We therefore say that a cube is a space-filling polyhedron, or simply a space filler.

It is not possible to fill space with regular tetrahedra * only; four tetrahedra

* A regular polyhedron has all edges of equal length and all surface angles equal to each other.

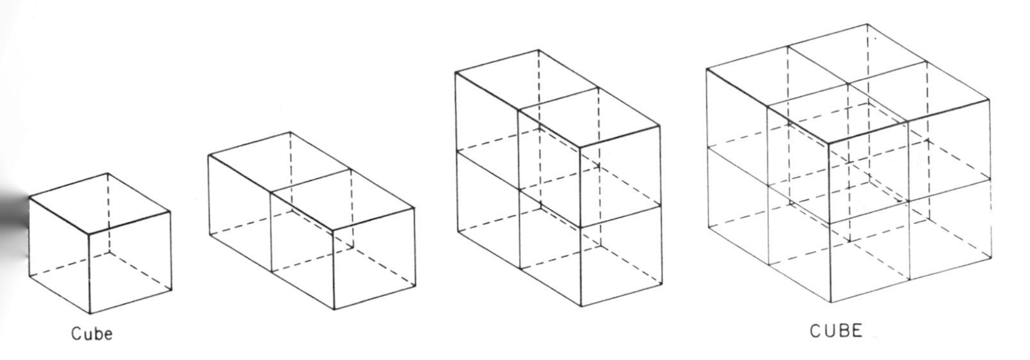

Fig. A11. Filling space with cubes.

and an octahedron are necessary to produce a tetrahedron having eight times the volume (twice the linear dimension) of the original tetrahedron (Fig. A12). Conversely, six octahedra *and eight tetrahedra* are necessary to produce an octahedron having eight times the volume (twice the linear dimension) of the original octahedron (Fig. A13). In turn, the larger octahedra and tetrahedra can be combined in the appropriate proportions to produce even larger tetrahedra and octahedra until all of space can be filled. Therefore, whereas octahedra and tetrahedra cannot by themselves fill all of space, they can be packed together to do so.

In Fig. A14, we place two tetrahedra on opposite faces of an octahedron. The result is a parallelepiped that has six identical faces, each being a rhomb or diamond with angles 60 degrees and 120 degrees. Such a particular parallelepiped is called a rhombohedron. Identical rhombohedra can be placed next to

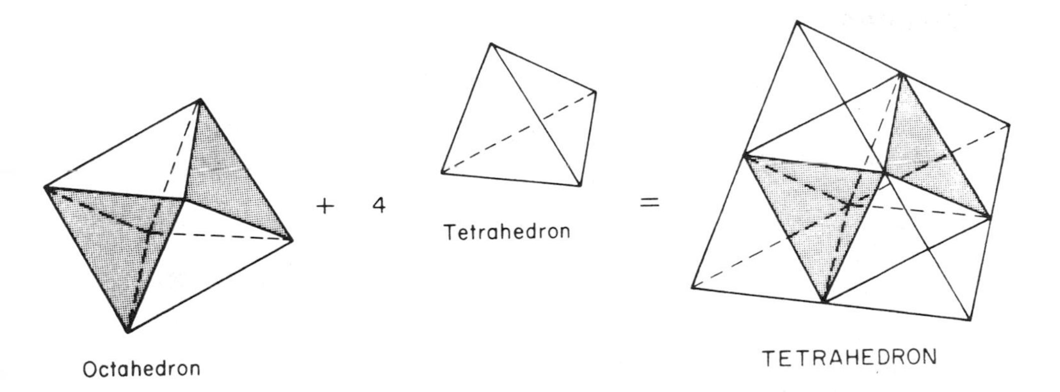

Fig. A12. Scaling the tetrahedron by a linear factor of two.

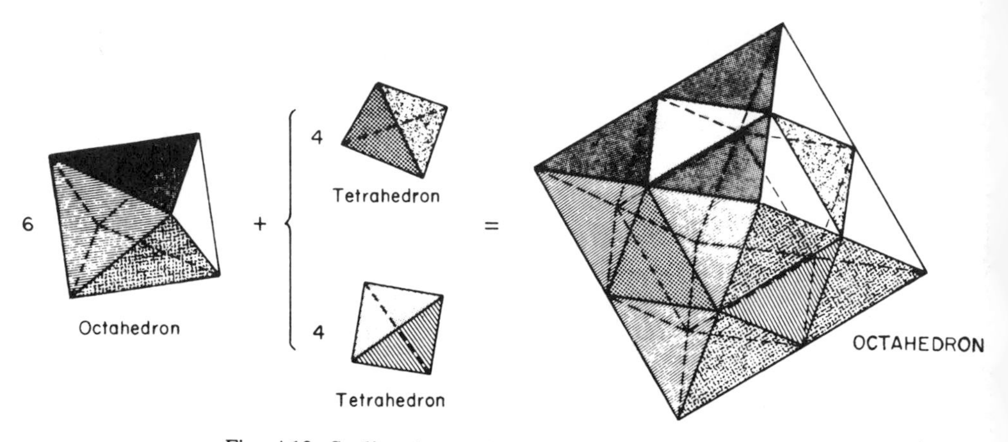

Fig. A13. Scaling the octahedron by a linear factor of two.

each other, in a fashion analogous to that shown in Fig. A11, to fill all of space. Since the rhombohedron is a space filler, and can be built from an octahedron and two tetrahedra (in opposite orientations), we may conclude that in order to fill space, each octahedron must be accompanied by a pair of tetrahedra in opposite orientation. We shall later test several polyhedra and combinations of polyhedra for space filling. Since these polyhedra can

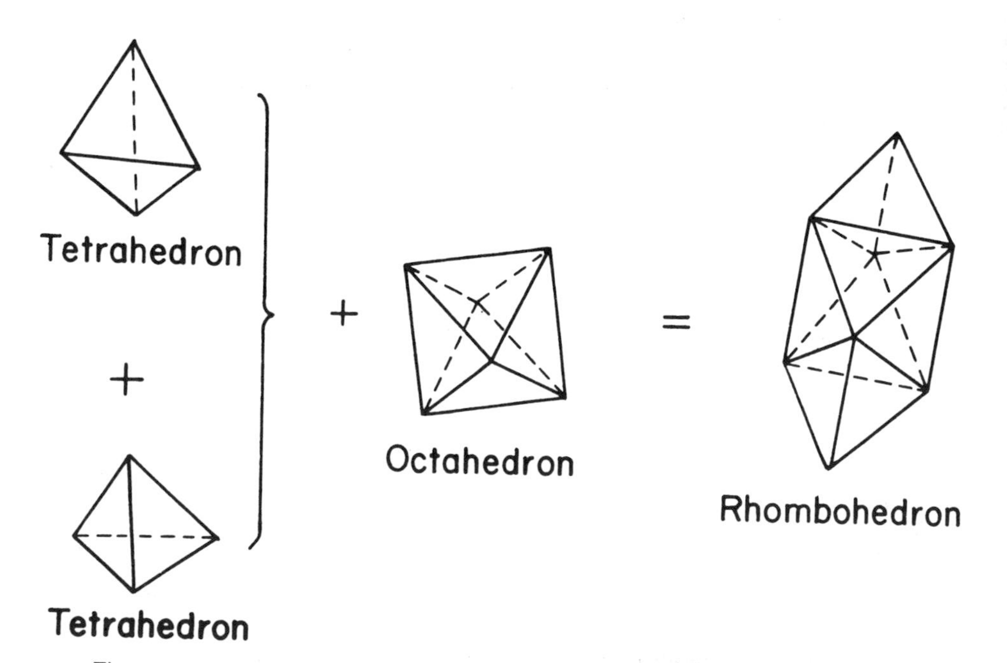

Fig. A14. Generation of rhombohedron from an octahedron and a pair of tetrahedra.

frequently be expressed as combinations of tetrahedra and octahedra, we shall require that each time, as a necessary condition for space filling, each octahedron occur in conjunction with a pair of oppositely oriented tetrahedra.

As an example of this necessary condition, let us check whether the large tetrahedron formed in Fig. A12 and the large octahedron formed in Fig. A13 can indeed fill space. Suppose that we did not know the ratio in which they must be mixed. Let us try a tetrahedra and b octahedra, and determine the ratio of a to b. We then have the following balance sheet:

	No. of Small Tetrahedra	No. of Small Octahedra
a Tetrahedra:	$4a$	a
b Octahedra:	$8b$	$6b$
a Tetrahedra + b Octahedra:	$4a+8b$	$a+6b$

The necessary condition for space filling is then: $4a + 8b = 2(a + 6b)$. Hence $a = 2b,$ so that the large tetrahedron and octahedron also must mix in the $2 \div 1$ ratio, as would be expected.

Since the cube is a space filler, it too should be expressible in terms of two tetrahedra per octahedron. This is done in Fig. A15, where a tetrahedron has one *octant* of an octahedron attached to each of its four faces to form a cube. Hence:

$$\text{Cube} = \text{Tetrahedron} + {}^4/_8 \text{ Octahedron} = \text{Tetrahedron} + {}^1/_2 \text{ Octahedron}$$

Therefore, the cube contains a tetrahedron and a half-octahedron, hence satisfies the necessary condition for space filling.

We have now surveyed all of the regular solids, except the icosahedron and its dual, the pentagonal dodecahedron. These polyhedra are not space fillers

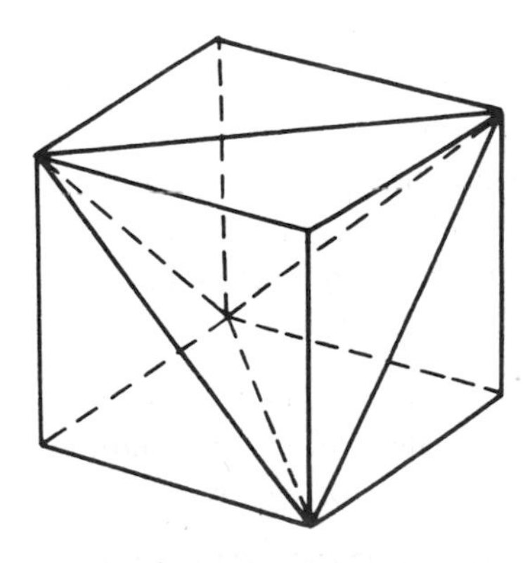

Fig. A15. The cube expressed in terms of octahedra and tetrahedra.

because they have fivefold rotational symmetry: a fundamental theorem in symmetry theory precludes the existence of parallel fivefold axes of rotational symmetry.*

We next turn our attention to semiregular solids, i.e., solids having all edges of the same length but having various surface angles. The first is a truncated octahedron. Eight of its faces are regular hexagons, six are squares. It is obtained from the cube by bisecting each of its octants, as shown in Fig. A16*a*. Since this solid is obtained by a simple bisection of the cube, it is itself a space filler.

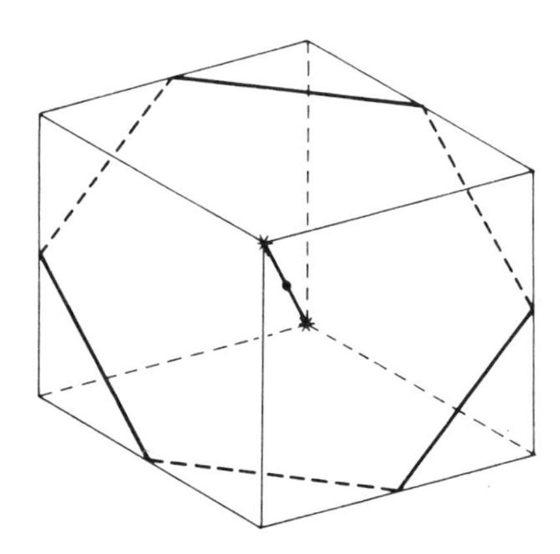

Fig. A16a. One octant of the cube: the plane of the hexagon perpendicularly bisects the body diagonal.

Figure A16*c* shows how the truncated octahedron can be related to octahedra and tetrahedra. A "supercube" is bisected as before: *eight* small cubes (such as shown in Fig. A16*b*) here produce *one octant* of a truncated superoctahedron. The octahedra and tetrahedra are differentiated by shading, as indicated in the drawing. Six of the eight small cubes are cut by the bisecting plane; one cube lies entirely on the reader's side of the plane; the remaining one lies beyond that plane. The bisecting plane cuts a single octant of an octahedron off each of six cubes: in three of these cubes, this octant lies on the reader's side of the plane; in the remaining three, it lies beyond this plane. Therefore, there are *beyond* the plane:

1 tetrahedron and $^4/_8$ octahedron in a *single* cube.
1 tetrahedron and $^3/_8$ octahedron in *three* cubes.
$^1/_8$ octahedron in *three* cubes.

* A. L. Loeb, *Color and Symmetry,* p. 33 (John Wiley and Sons, 1971).

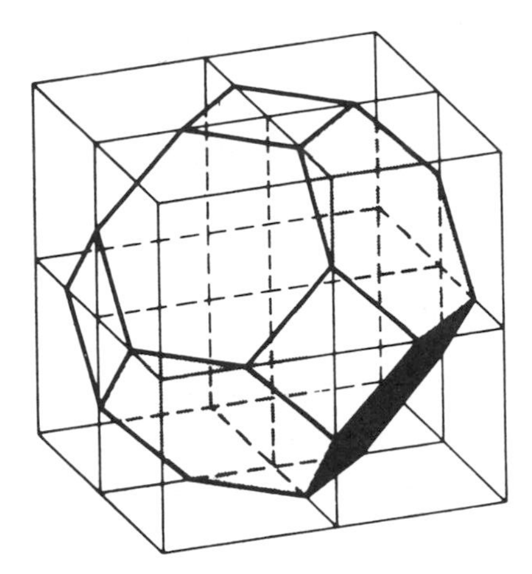

Fig. A16b. The entire cube with inscribed truncated octahedron.

Therefore, the portion of the supercube *beyond* the plane contains in total (1 + 3) tetrahedra and ($^4/_8 + ^9/_8 + ^3/_8$) octahedra, i.e., 4 tetrahedra and 2 octahedra. Of course, the portion of the cube on the reader's side of the plane also contains 4 tetrahedra and 2 octahedra. From this construction we conclude:

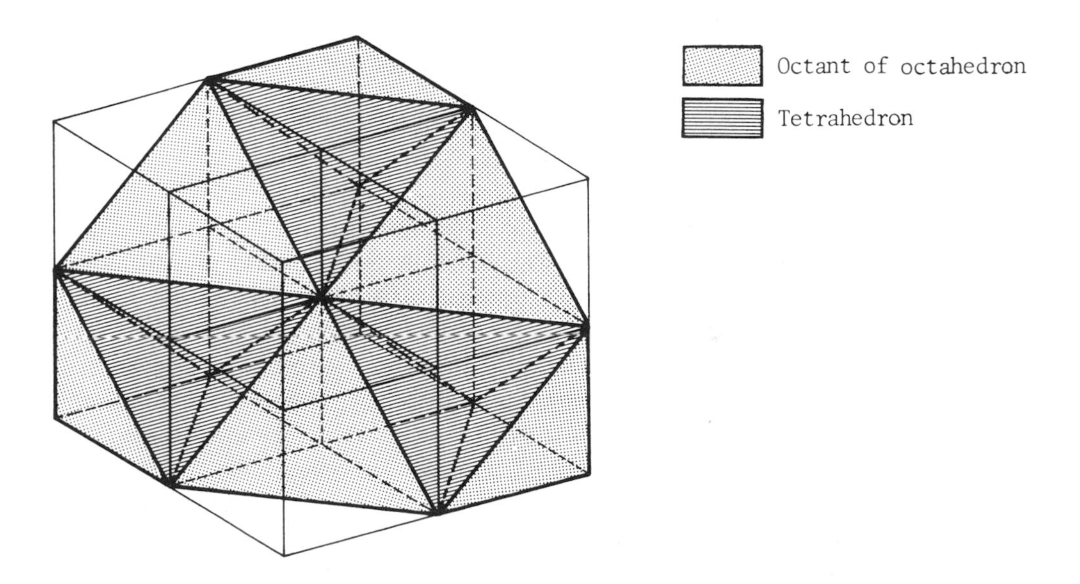

Fig. A16c. Relation of truncated octahedron to octahedra and tetrahedra.

a) The supercube contains 8 tetrahedra and 4 octahedra.
b) The truncated superoctahedron contains 32 tetrahedra and 16 octahedra.

It should be noted that in each of these figures the condition for space filling is obeyed.

Next we consider the cuboctahedron, Buckminster Fuller's "vector equilibrium." This solid has six square faces and eight triangular ones; its twelve vertices are the midpoints of the twelve edges of a cube (Fig. A17). In terms of octahedra and tetrahedra, each square face is the equatorial square of an octahedron, and each triangular face is the base of a tetrahedron. There are, therefore, eight tetrahedra and six half-octahedra, or a ratio of 8 tetrahedra to 3 octahedra. The cuboctahedron is therefore one octahedron short of meeting

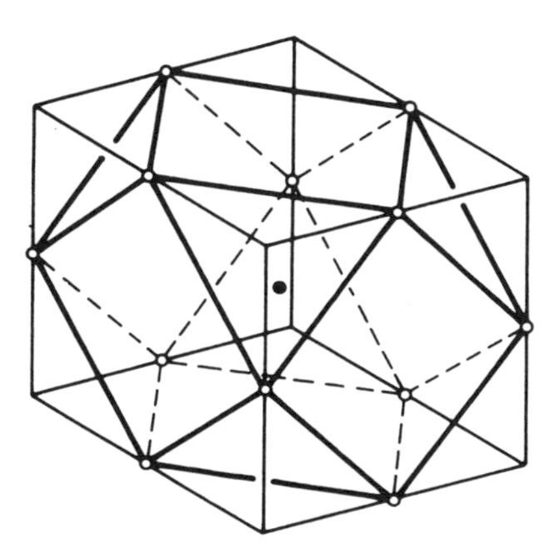

Fig. A17. Cuboctahedron.

the necessary space-filling condition. When an octant of an octahedron is added to each of its eight triangular faces, the cuboctahedron is transformed into a cube (Fig. A17), hence a space filler. Cuboctahedra and octahedra thus together fill space in ratio 1 ÷ 1.

Next is the truncated tetrahedron (Fig. A18*a*). It has four triangular and four hexagonal faces. It can be constructed by placing an octahedron on each of the four faces of a tetrahedron; the triangular faces of the truncated tetrahedron are faces of these octahedra (Fig. A18*b*). The dotted edges in Fig. A18*b* are supplied by tetrahedra: one at the center and six at the dotted edges. The ratio is therefore 7 tetrahedra ÷ 4 octahedra, 1 tetrahedron short of satisfying the space-filling condition. Therefore, truncated tetrahedra together with tetrahedra fill space in a ratio 1 ÷ 1.

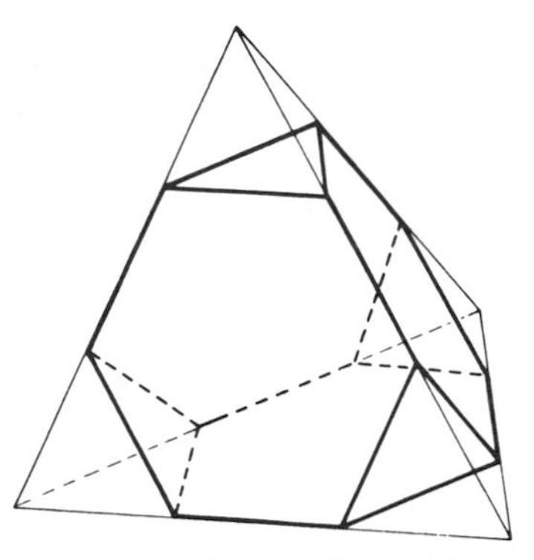

Fig. A18a. The truncated tetrahedron with equal edges.

Since truncated tetrahedra are lacking in tetrahedra, and the cuboctahedron has an excess of tetrahedra for meeting the space-filling condition, it might be interesting to see whether they could be admixed appropriately to fill space. The following would be the balance sheet:

	No. of Tetrahedra	No. of Octahedra
a cuboctahedra	$8a$	$3a$
b truncated tetrahedra	$7b$	$4b$
a cuboctahedra $+ b$ truncated tetrahedra	$8a + 7b$	$3a + 4b$

The space-filling condition therefore requires: $8a + 7b = 2(3a + 4b) \therefore b = 2a$. Thus the necessary condition is fulfilled if there are two truncated tetrahedra per cuboctahedron. However, this condition is not sufficient, for the two solids do not have compatible faces; the square faces of the cuboctahedron cannot be adjacent to the hexagonal faces of the truncated tetrahedron. What

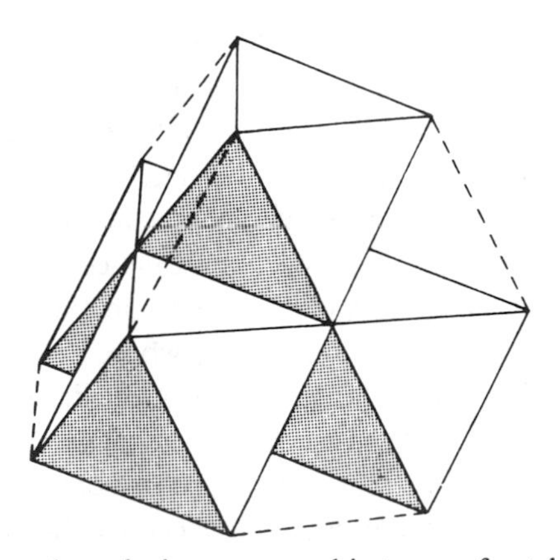

Fig. A18b. The truncated tetrahedron expressed in terms of octahedra and tetrahedra.

we need is yet another solid, which provides hexagonal and square faces, as a bridge between the other two solids. Such a solid is the truncated octahedron considered above (Fig. A16). We make up *two* balance sheets: one for the ratio of tetrahedra to octahedra, the other for matching faces:

	No. of Tetrahedra	No. of Octahedra
a cuboctahedra	$8a$	$3a$
b truncated tetrahedra	$7b$	$4b$
c truncated octahedra	$\frac{1}{2}c$	$\frac{1}{4}c$
	$8a + 7b + \frac{1}{2}c$	$3a + 4b + \frac{1}{4}c$

$\therefore 8a + 7b + \frac{1}{2}c = 2(3a + 4b + \frac{1}{4}c)$, $\therefore b = 2a$. The balance for the truncated octahedra follows from the fact that the truncated octahedron is obtained by bisecting the cube. It is observed that the number c cancels out in the balancing equation; this happens because the truncated octahedron is itself a space filler, and may be admixed without disturbing the balance. The balance sheet for matching faces is as follows:

	No. of Square Faces	No. of Triangular Faces	No. of Hexagonal Faces
a cuboctahedra	$6a$	$8a$	—
b truncated tetrahedra	—	$4b$	$4b$
c Truncated octahedra	$6c$	—	$8c$

Since the number of square faces of the cuboctahedra must equal the number of square faces of the truncated octahedra (assuming that no two cuboctahedra have square faces in contact), $a = c$. Similarly, the balance of triangular faces requires $8a = 4b$, $\therefore b = 2a = 2c$. We have already seen that the condition $b = 2a$ satisfies the tetrahedra $\div$ octahedra balance. The equation $b = 2c$ gives the appropriate balance of hexagonal faces. Therefore, both conditions are satisfied if cuboctahedra, truncated tetrahedra, and truncated octahedra are combined in the ratio $1 \div 2 \div 1$. We have here an example of *three* semiregular solids combining to fill all of space.

A regular solid can be subdivided into congruent pyramids, the base of each being a polygonal face of the solid, each apex the center of the solid. For example, the regular tetrahedron can be divided into four congruent pyramids, each having as its base an equilateral triangle whose corners subtend angles of approximately 109 degrees at the apex. An octahedron can be divided into eight pyramids, each also having an equilateral triangle as base, but the angles around the apex are here 90 degrees. For a cube there are six pyramids, each having a square base. The pyramids so produced can be combined in interesting ways to produce semiregular solids.

One particularly important solid, already encountered in connection with close packing, is the rhombohedral dodecahedron, or briefly rhombododecahedron (Fig. A19). This solid is the *dual* of the cuboctahedron. It can be generated in two different ways. First take two identical cubes. Divide one of these cubes into its six component pyramids. Place each of these pyramids with its square face against a face of the second cube. The result is the rhombododecahedron, which has a volume twice that of each cube from which it is generated. Secondly, take an octahedron and two tetrahedra. Divide both tetrahedra into their four component pyramids. Place each of the resulting eight pyramids against one of the faces of the octahedron. The result is again the rhombododecahedron. This solid has eight corners where three angles of about 109 degrees meet, and six corners at each of which four angles of about 71 degrees meet.

From both constructions, it is obvious that the rhombododecahedron is a space filler; the component tetrahedra and octahedra occur in the desired 2 ÷ 1 ratio, and all faces are identical lozenges that adjoin in space filling. It should be noted that whereas a cube is a space filler; its dual the octahedron is *not*.

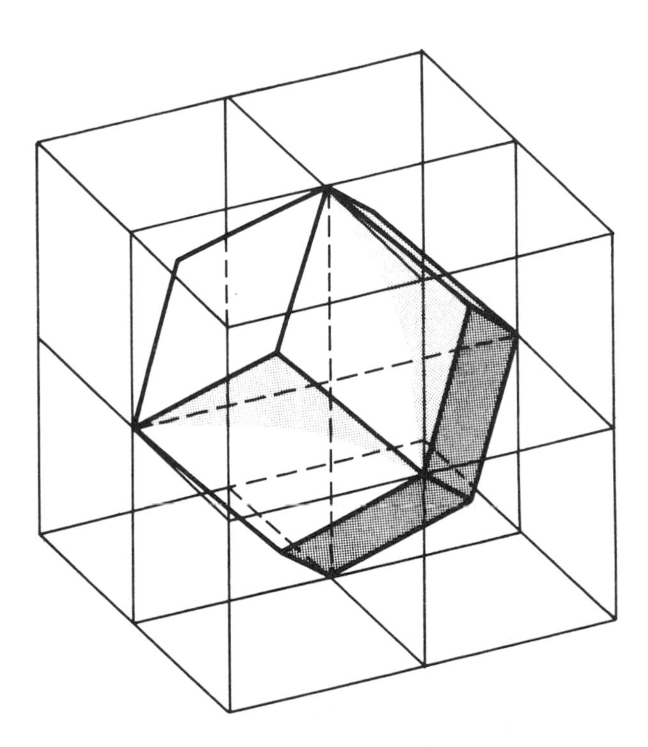

Fig. A19. Rhombododecahedron.

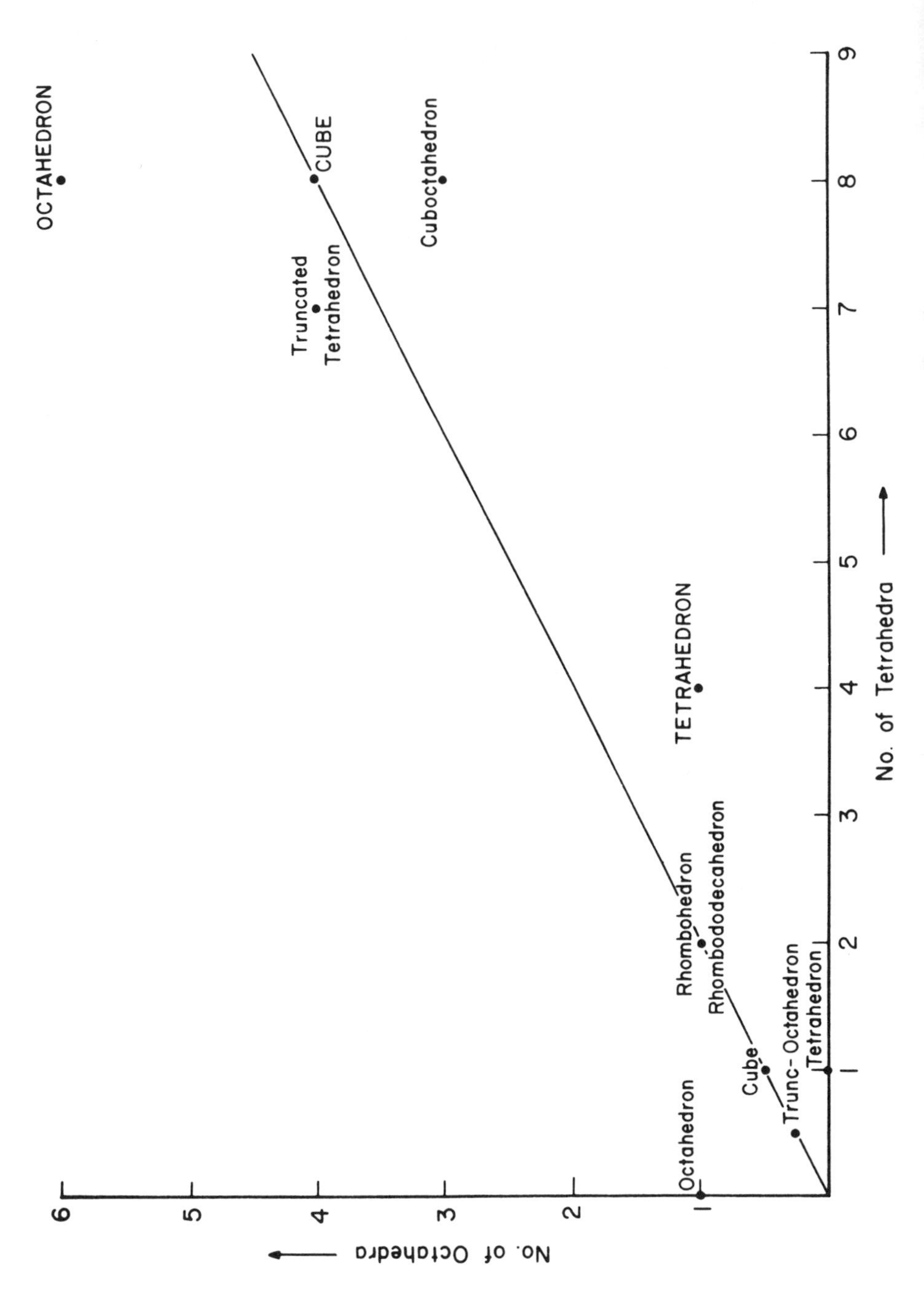

Fig. A20. Regular and semiregular solids expressed in terms of tetrahedra and octahedra.

The rhombododecahedron fills space; its dual the cuboctahedron does not. Generally, then, the duals of space fillers do not fill space. It should also be noted that *only one* of the regular (Platonic) solids is also a space filler, and that is the cube.

Fig. A20 summarizes the proportions of octahedra and tetrahedra in each of the solids discussed here. The straight line represents the condition of space filling. Such a diagram suggests that each polyhedron can be represented by a two-dimensional vector whose components express the ''tetrahedral and octahedral natures'' admixed in each polyhedron.

When several polyhedra are packed together, a necessary condition for space filling is that the sum of their vectors lies along the solid line. A polyhedron possessing too much ''octahedrality'' for space filling lies above the solid line; one possessing too much ''tetrahedrality'' (too little octahedrality) lies below the solid line.

J The A *and* B *Modules and the Trirectangular Tetrahedron*

The feeling for balance of space between octahedrality and tetrahedrality led Fuller to devise his so-called *A* and *B* modules. These particles (cf. Fig. 951.10) are derived from the regular octahedron and regular tetrahedron as follows: Take an octant of the octahedron, and subdivide the face of the octahedron symmetrically into six mutually equivalent triangles (Fig. A21).

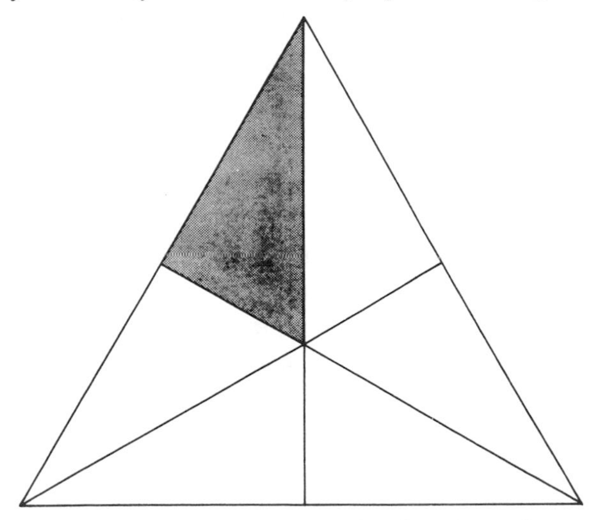

Fig. A21. Subdivision of triangular face into six equal parts.

The tetrahedron whose base is one sixth of the octahedral face and whose apex is the center of the octahedron (corner of an octant) is called the *asymmetric unit* of the octahedron; its volume equals $^1/_6 \times {}^1/_8 = {}^1/_{48}$ of that of the octahedron.

The asymmetric unit of the tetrahedron is obtained similarly: the tetrahedron is first subdivided into four equivalent parts, each having as base a face of the tetrahedron and as apex its center of gravity. Next the base is subdivided into six equivalent parts according to Fig. A21: each asymmetric unit of the tetrahedron has one of the right triangles as base, and the center of the original tetrahedron as apex. Its volume is $^1/_6 \times {}^1/_4$ of that of the octahedron.

The volume ratio of the octahedral and tetrahedral asymmetric units equals:

$$({}^1/_{48} \text{ volume of octahedron}) \div ({}^1/_{24} \text{ volume of tetrahedron}) =$$

$$^{24}/_{48} \times \frac{\text{volume of octahedron}}{\text{vol. of tetrahedron}} = {}^{24}/_{48} \times 4 = 2$$

The *A* module is defined as the asymmetric unit of the regular tetrahedron; the *B* module is the portion of the asymmetric unit of the octahedron that is left over after the *A* module has been scooped out of it. These, then, are the units in terms of which Fuller expresses his space. Since they cannot be symmetrically subdivided, they are true modular quanta. The *A* and *B* modules have equal volumes, but are not commensurable: they are not expressible in terms of a common unit.

It should be noted that there are actually *four* fundamental units, two being the mirror images of the other two. We shall denote the mirror images by asterisks: *A** and *B**. Since the regular and semiregular polyhedra all have mirror symmetry, we can expect their constituent *A* and *B* modules always to be paired to their respective mirror images *A** and *B**.

In Fig. A22 we have redrawn Fig. A20 in terms of the basic units *A* and *B*: this plot is somewhat comparable to a plot of imaginary numbers, in which case the units are also non-commensurable quantities of equal magnitude, namely 1 and $\sqrt{-1}$. The volume of each polyhedron plotted equals (in terms of the volume of a single *A* or *B* module) the sum of its absissa and its ordinate in Fig. A22.

Buckminster Fuller would not fail to remark on the fact that all polyhedra represented in Fig. A22 are composed of exact *dozens* of *A*s and *B*s, the number *12* having special significance in the cubic system. (All polyhedra discussed have so-called cubic symmetry, i.e., the same symmetry as the cube.) In this system every asymmetric unit is repeated three times around each axis perpendicular to triangular faces or cross-sections, and four times around each axis perpendicular to a square face or cross-section, hence at least twelve times.

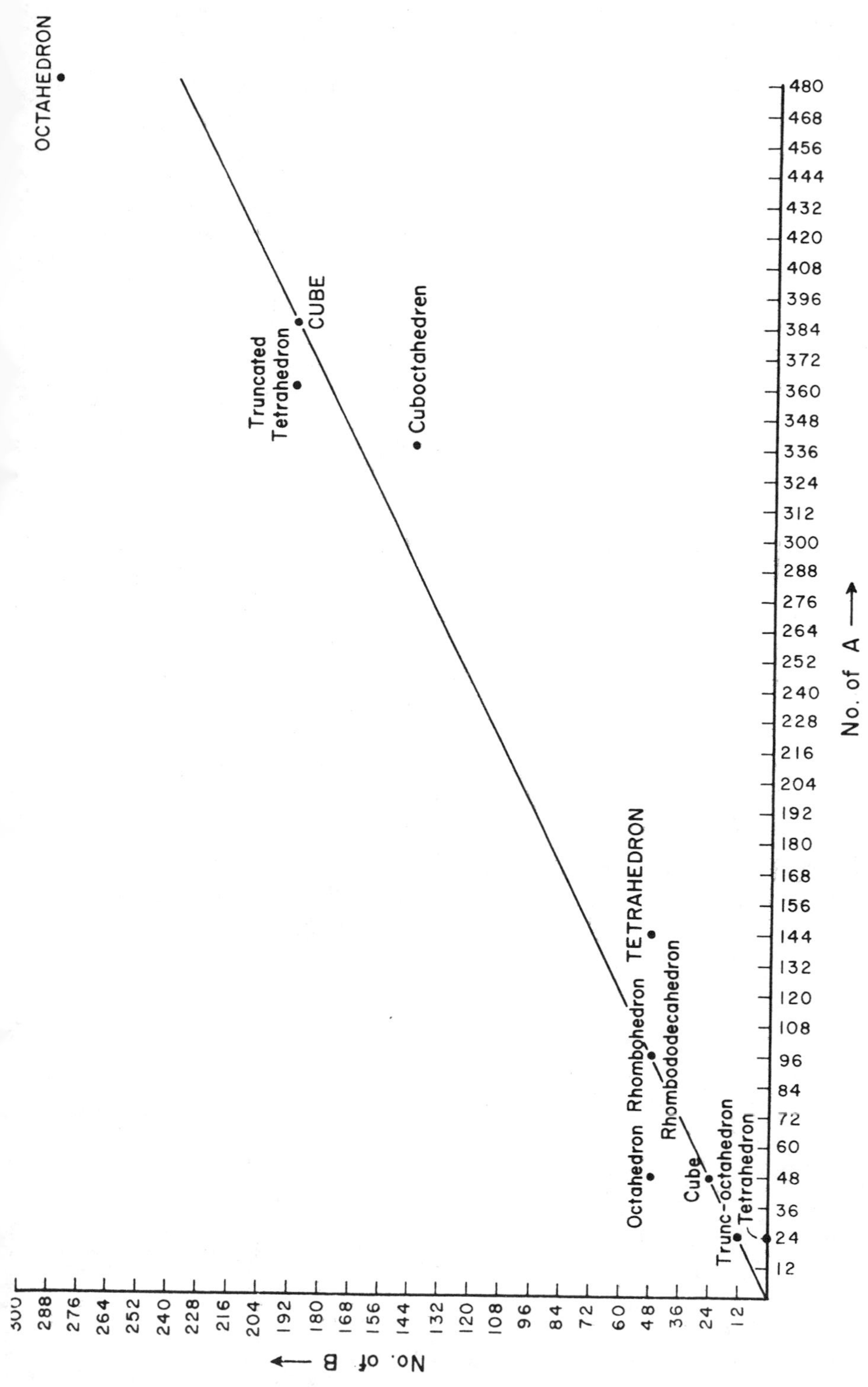

Fig. A22. Polyhedra expressed in terms of *A* and *B* units.

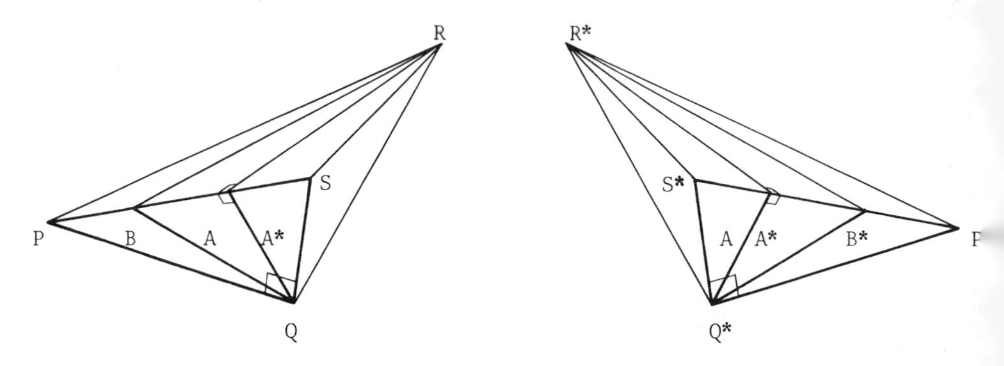

Fig. A23. Space filling with AA^*B and A^*AB^* modules.

The trirectangular tetrahedron modules shown in Fig. 951.10 both consist of an A module, its mirror image, and a single B module. If an asterisk is used to denote a mirror image, then one module can be denoted AA^*B, its mirror image A^*AB^*. These two modules are shown in Fig. A23 and again in Fig. A24; these two figures demonstrate the two ways in which these modules may be stacked to fill space. In Fig. A23, triangles QRS and Q^*R^*S are brought in contact, whereas in Fig. A24, triangles PQS and $P^*Q^*S^*$ are brought in contact for space filling. Since B has no mirror symmetry, B and B^* are oppositely congruent, so that AA^*B would be oppositely congruent with A^*AB^*.

In Fuller's demonstration that AA^*B and A^*AB^* together can fill space in two distinct ways, something quite remarkable happened, namely that the two modules exchanged places (cf. Figs. A23 and A24). Since one module can exactly occupy the space previously occupied by the other, albeit in a rotated position, the following logical conclusions may be drawn:

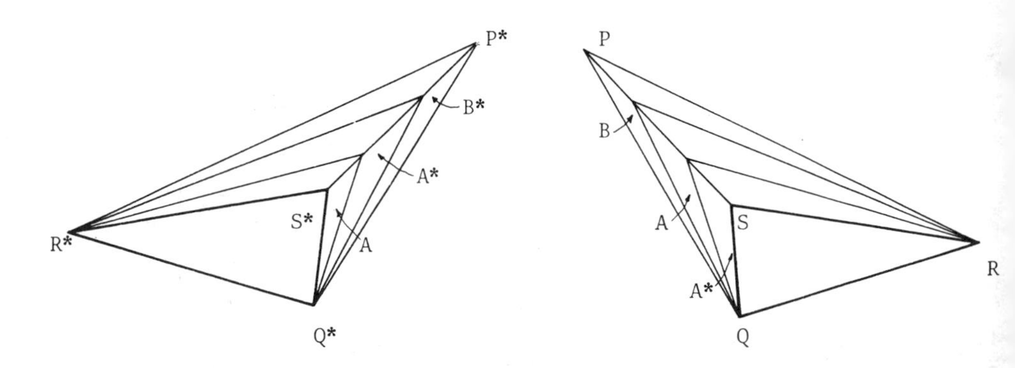

Fig. A24. The alternate arrangement of space filling with AA^*B and A^*AB^* modules.

1. *AA*B* is directly congruent with *A*AB**.
2. Since the two modules are *directly* congruent, only *one* of them is required to fill all of space.
3. The mirror-symmetrical module *AA** can combine with the *asymmetrical* module *B* in 1 : 1 ratio to fill all of space. The module *AA** could equally well combine with *B** to fill all of space.

These conclusions are quite remarkable because so far we have only encountered space fillers that have mirror symmetry, such as octahedra plus tetrahedra, cubes, or rhombododecahedra. Fuller himself uses *AA*B in conjunction with* its enantiomorph (mirror image) to fill space. Nevertheless, we shall presently see that the conclusions listed above are entirely correct, and hinge on the *direct* congruency of *AA*B* and *A*AB**.

We already know that *AA*B* and *A*AB** are *oppositely* congruent (Figs. A23 and A24). Two objects that are simultaneously *directly* and *oppositely* congruent are necessarily *each* mirror symmetrical. The three above-mentioned conclusions can therefore only be valid if we locate a plane of mirror sym-

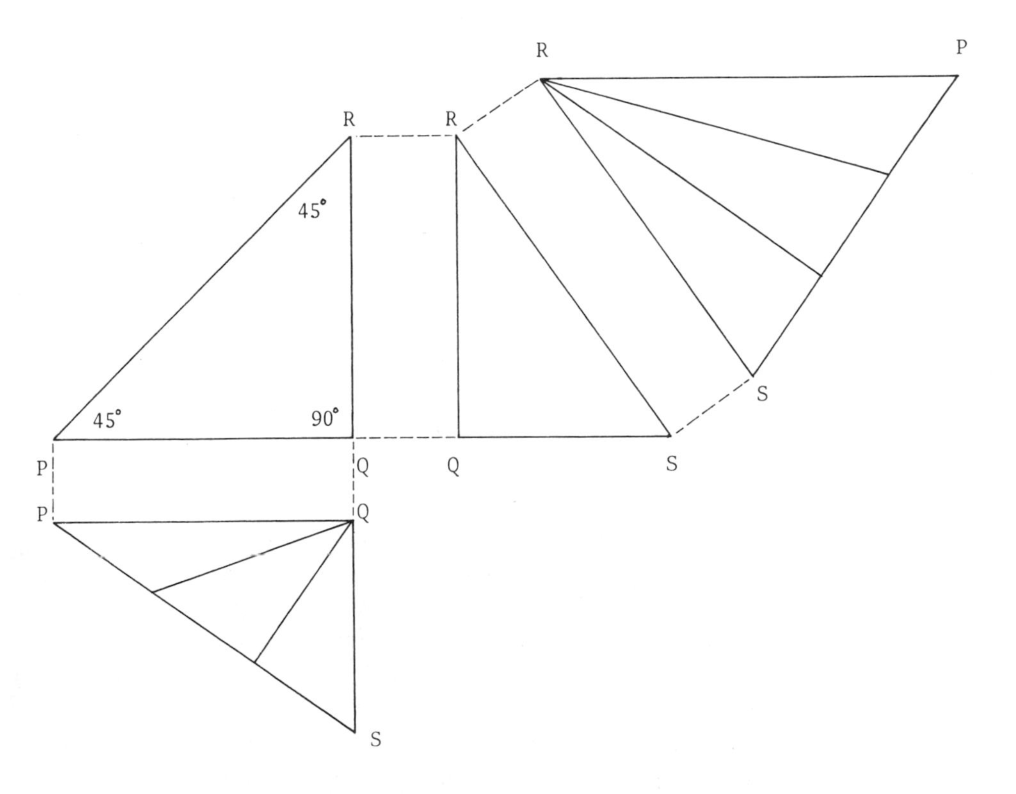

Fig. A25. The *AA*B* or *A*AB** module unfolded.

metry in each of the modules AA^*B and A^*AB^*. To this purpose we have, in Fig. A25, folded the faces of AA^*B out, and laid them flat beside each other. Module AA^*B^*, being oppositely congruent to AA^*B, would yield the same layout. The triangle PQR is a face of the B module. Since AB was defined as the asymmetrical unit of a regular octahedron, $\triangle PQR$ represents just one quadrant of an equilateral plane of such an octahedron. Therefore, $\overline{PQ} = \overline{RQ}$, so that $\triangle PQS$ is congruent with $\triangle RQS$. From this congruency, we conclude (cf. Fig. A23 or A24) that module AA^*B has indeed a plane of mirror symmetry; this plane contains the edge QS, and bisects $\angle PQR$. Module A^*AB^* is similarly bisected by a mirror plane through edge Q^*S^*.

Space filling with *only* the module AA^*B is illustrated in Fig. A26: here, $\triangle QRS$ of one module is brought into coincidence with $\triangle QPS$ of an adjacent

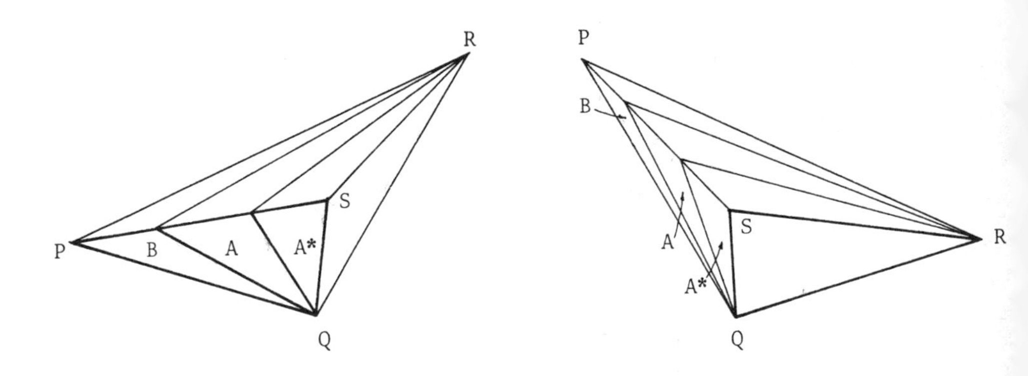

Fig. A26. The module AA^*B in two orientations, ready for space filling.

module. This arrangement is, in fact, a *third* alternative to the two already proposed by Fuller (Figs. A23 and A24); it is, of course, complemented by an entirely equivalent packing of modules A^*AB^*.

There is a direct relation between the AA^*B module and the rhombododecahedron. It was shown (see p. 813) that the rhombododecahedron can be thought of as a regular octahedron with a quarter-tetrahedron attached to each of its eight faces. The rhombododecahedron can be divided into eight octants, each of which consists of an octant of an octahedron plus a quarter-tetrahedron. One such octant is shown in Fig. A27. In turn, the octant can be subdivided into six equivalent modules, of which three are enantiomorphic to the other three. As a matter of fact, three are AA^*B modules, the other three are A^*AB^* modules. Of the AA^*B modules, the AB portion constitutes the asymmetric unit of the octahedron, while the A^* portion constitutes the asym-

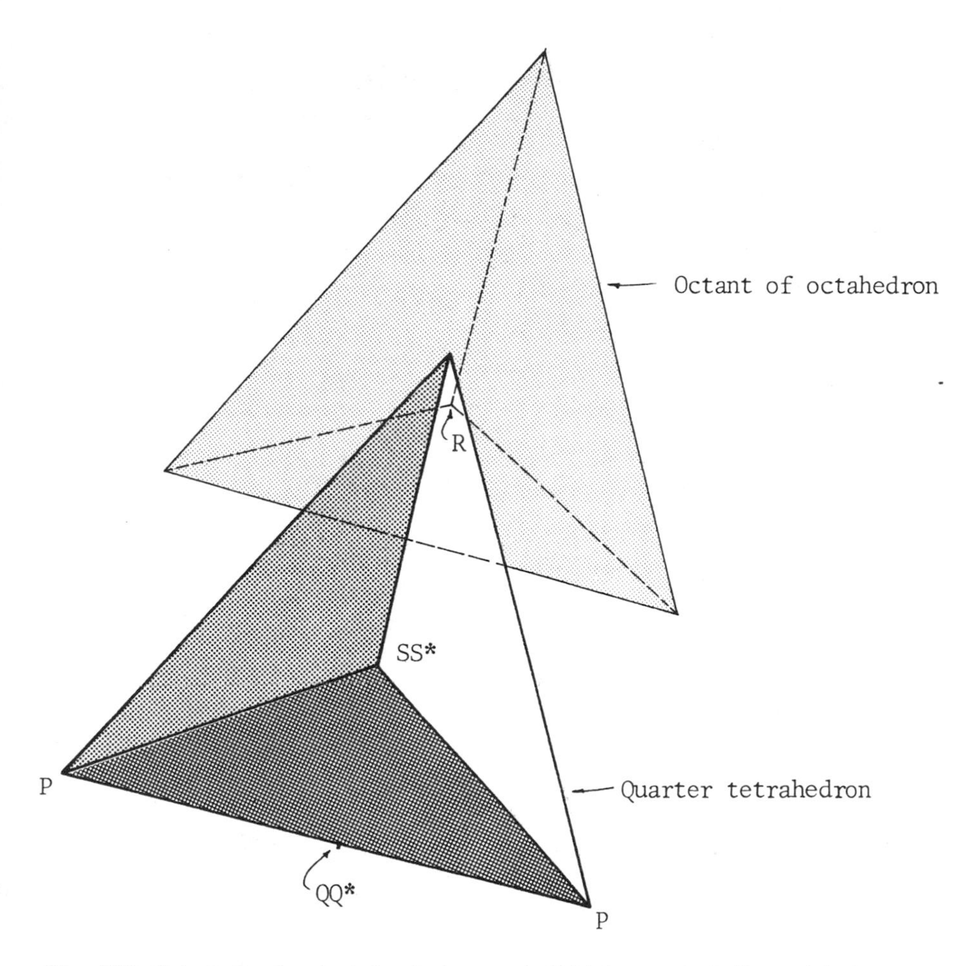

Fig. A27. Octant of a rhombododecahedron exploded into an octant of an octahedron plus a quarter-tetrahedron.

metric unit of the quarter-tetrahedron. Thus the rhombododecahedron can be constructed from 24 *AA**B* modules and 24 *A*AB** modules. Since, however, these modules were shown to be directly congruent with each other, 48 of either kind of module will also form a rhombododecahedron.

The rhombododecahedron can be constructed in two ways from 24 *AA*B* and 24 *A*AB** modules, which correspond to Fuller's two space-filling modes. In Fig. A28, we labeled some vertices to correspond to the orientation of the modules shown in Fig. A23. The orientation of Fig. A24 might equally well have been used, however. In the first case, the *R*-vertices of the modules all coincide at the center of the rhombododecahedron, whereas in the second, the *R*-vertices constitute the acute vertices of the rhombododecahedron. The

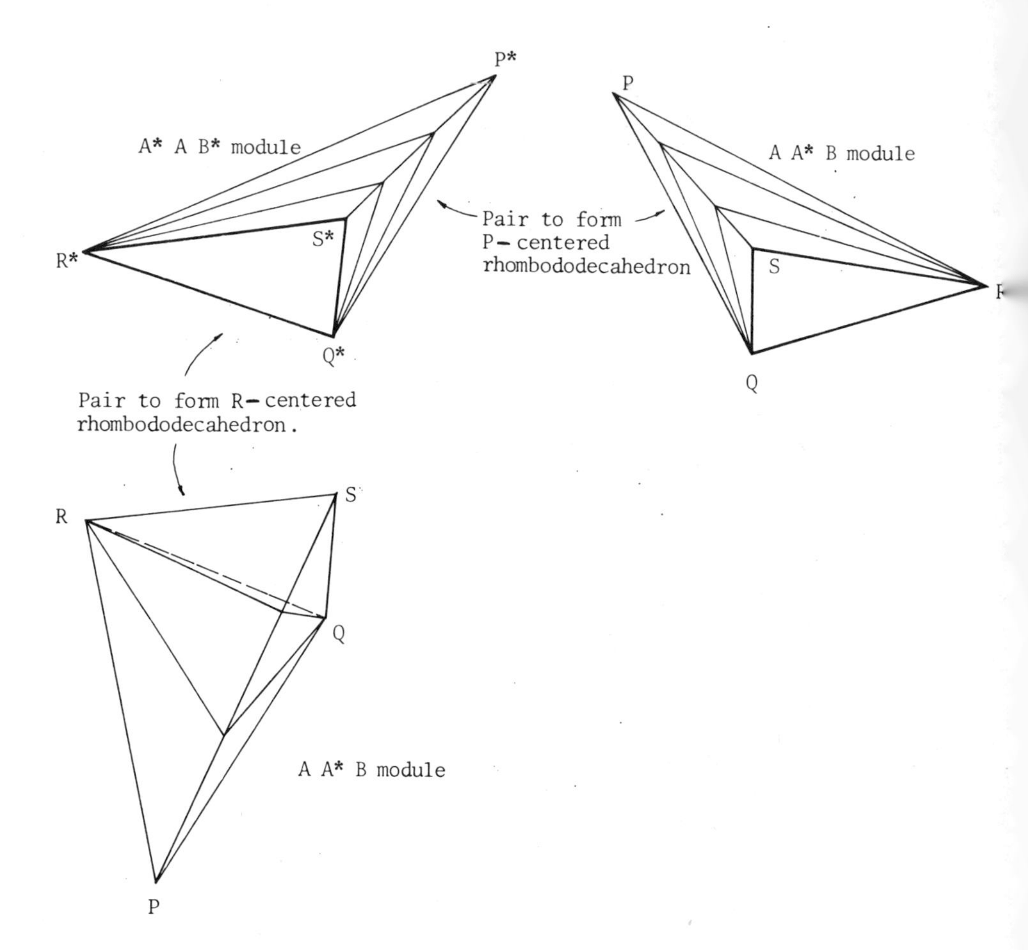

Fig. A28. Ambivalence of the modes of Figs. A23 and A24 in space filling.

orientation of Fig. A26 leads to the construction of a rhombododecahedron from 48 AA^*B modules; we shall return to this construction later on.

I want to raise first the question whether or not the two modes represented by Figs. A23 and A24 respectively are indeed distinct modes of space filling. There is no doubt about their representing two different modes of constructing a rhombododecahedron. As the rhombododecahedron is a space filler, the problem would appear to be solved right here. However, when rhombododecahedra are stacked together to fill all of space, the centers of the rhombododecahedra form a close-packed cubic lattice, as do their acute vertices (cf. section F). One might therefore fill space with rhombododecahedra having R-vertices at their centers, but slice space up again into equivalent rhombododecahedra whose centers are P-vertices. This ambivalence is illustrated by Fig. A28, in which an A^*AB^* module is juxtaposed with two adjacent AA^*B modules in a space-filling array. One might equally well pair A^*AB^* with one AA^*B neighbor to form an R-centered rhombododecahedron as with the other to form a P-centered rhombododecahedron: in actual space filling, it does not matter at all which vantage point one takes. We conclude, therefore, that the orientations of Figs. A23 and A24 lead to different finite polyhedra, but are equivalent in filling *all* of space. However, the orientation of Fig. A26, which uses B without B^*, is fundamentally different, so that there are two fundamentally different arrangements for space filling, namely the one in which AA^*B and A^*AB^* are used together, and the one in which only AA^*B or A^*AB^* is used.

K The Rectangular Cross Section of the Regular Tetrahedron

A cross section through a regular tetrahedron made parallel to a pair of nonintersecting edges is a rectangle. Every rectangle so generated from a given tetrahedron has the same circumference. This latter observation is proven conveniently by unfolding the tetrahedron, keeping the trace of the rectangular cross section on the faces of the tetrahedron (Fig. A29). The faces of the tetrahedron can be unfolded into a parallelogram having one pair of edges as long as the edge length of the tetrahedron, the second pair twice as long. The trace of the rectangular cross section on the surface of the tetrahedron appears in the unfolded figure as a line parallel to the longer edge of the parallelogram. The length of this line equals twice the edge length of the tetrahedron regardless of where the cross section is made. Therefore, the circumference of any rectangular cross section of a regular tetrahedron equals twice the edge length of the tetrahedron.

L Symmetry and Great Circles of Solids

A solid has rotational symmetry if it does not appear changed after rotation around a straight line through an angle smaller than 360 degrees. The straight line is called an *axis of rotational symmetry,* or simply a *symmetry axis.*

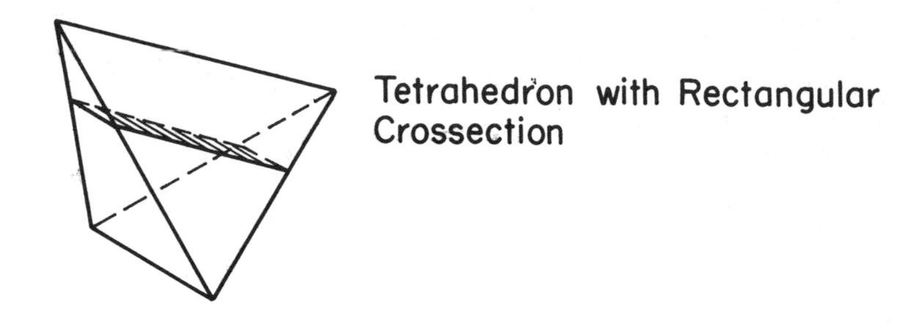

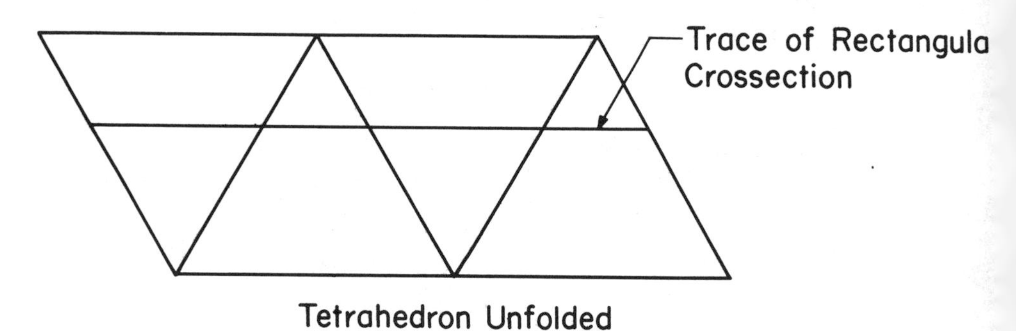

Fig. A29. Cross sections of tetrahedra.

When a solid is rotated 360 degrees around a symmetry axis, the same pattern makes its appearance several times: the number of times it appears through a 360 degrees rotation is called the *symmetry value* of the axis. A body diagonal of a cube is a three-fold symmetry axis; lines joining centers of opposite faces of a cube are four-fold symmetry axes, and lines joining centers of opposite edges of a cube are two-fold axes.

If the geometrical center of a polyhedron is collinear with and equidistant from a pair of vertices, this polyhedron has inversion symmetry. A cube and an octahedron have inversion symmetry, but a tetrahedron does not.

If a solid can be divided by a plane into two halves that are each other's mirror image, then that solid has *reflection symmetry;* the plane is called a *mirror plane*. When a solid possesses several mirror planes, any line of intersection of these planes is an axis of rotational symmetry.

The cuboctahedron (''vector equilibrium'') has the following axes of rotational symmetry:

Six two-fold axes joining the six pairs of opposite vertices.

Four three-fold axes joining the centers of opposite triangular faces.

Three four-fold axes joining the centers of opposite square faces.

In addition, there are lines joining the centers of opposite edges; these have a novel symmetry property. When the cuboctahedron is rotated through 180 degrees around such a line, the original pattern reappears, but upside down rather than in its original orientation. These lines are called roto-reflection axes: rotation through such an axis followed by reflection in a plane perpendicular to the axis reproduces the original polyhedron. There are twelve such roto-reflection axes.

Each of the 25 great circles of the cuboctahedron is defined by a plane perpendicular to one of the symmetry axes. The six circles perpendicular to the two-fold axes lie in reflection planes of the cuboctahedron. The four great circles perpendicular to the three-fold axes are *not* mirror planes: one half of the cuboctahedron, upon being reflected into the plane of such a great circle, must also be rotated 60 degrees to reproduce the other half. Rotation of 60 degrees of one half of the cuboctahedron with respect to the other half produces a polyhedron associated with hexagonal close packing; such a polyhedron has not four, but only a single three-fold axis, and a mirror plane perpendicular to it.

The three great circles perpendicular to the four-fold axes lie in mirror planes. The 12 great circles perpendicular to the roto-reflection axes do *not* lie in mirror planes.

The icosahedron has 30 edges; the centers of opposite edges are pairwise joined by *15* two-fold axes. There are 15 great circles perpendicular to these axes, each lying in a mirror plane of the icosahedron. There are 20 (triangular) faces, whose centers are pairwise joined by *ten* three-fold axes. The 10 great circles perpendicular to these axes do *not* lie in mirror planes. There are 12 vertices, pairwise joined by *six* five-fold axes; the six great circles perpendicular to the five-fold axes do not lie in mirror planes. A cross section made through the icosahedron perpendicular to a five-fold axis equidistant from opposite vertices makes a trace on the surface of the icosahedron that has the shape of a regular decagon. This cross section divides the icosahedron into two equivalent halves, which are oriented 36 degrees from mirror symmetry with respect to each other.

M Proof of the Equation N = $10\nu^2 + 2$

The cuboctahedron ("vector equilibrium," cf. Fig. 222.01) has six square faces and eight triangular ones, totaling 14 faces. There are 24 edges and 12 vertices; Euler's equation is satisfied. To find the total number of spheres in each layer, we must add up the total number of spheres in all faces, but allow for the fact that each sphere in an edge is counted twice (two faces meet at an edge), and each sphere at a vertex is counted four times (four faces meet at each vertex). Thus the total number of spheres in each layer, N, is given by:

$N =$ (no. of square faces) × (number of spheres in a square face) +
+ (no. of triangular faces) × (number of spheres in a triangular face) +
− (no. of edges) × (number of spheres on an edge but not on a vertex) +
− 3 (no. of vertices) =

= 6 (number of spheres in a square face) +
+ 8 (number of spheres in a triangular face) +
− 24 (number of spheres on an edge but not on a vertex) +
− 36.

Here it is assumed that the frequency is at least unity. If the frequency is denoted by ν, the number of spheres in a square face is $(\nu+1)^2$: a unit-frequency square has four spheres, a square of frequency two has nine spheres, etc. The number of spheres in a triangular face is certainly a quadratic function of frequency, expressible in the general form $(a\nu^2 + b\nu + c)$. We can easily determine experimentally that the unit-frequency triangle contains three spheres, that the triangle of frequency two contains six spheres, and that the triangle of frequency three has 10 spheres. From these data it follows that $a = {}^1/_2$, $b = {}^3/_2$, and $c = 1$. Therefore:

$$N = 6(\nu+1)^2 + 8({}^1/_2\nu^2 + {}^3/_2\nu + 1) - 24(\nu - 1) - 36 = 10\nu^2 + 2, \text{ q.e.d.}$$

N Parallel and Series Connections

The statement "A chain is as strong as its weakest link" represents the so-called series connection of a set of links (Fig. A30*a*). When a single link gives way, the loading capacity of the entire chain is destroyed. Each link here carries the full load of the chain.

Conversely, the use of links in parallel connection (Fig. A30*b*) could be expressed as "This system is as strong as its strongest link," for here the load is shared by the links, and the strongest link alone might be able to bear the entire load. A series connection is very economical, but not very safe, a parallel connection is safe, but not very economical. In general, an optimum combination can be found (Fig. A30*c*).

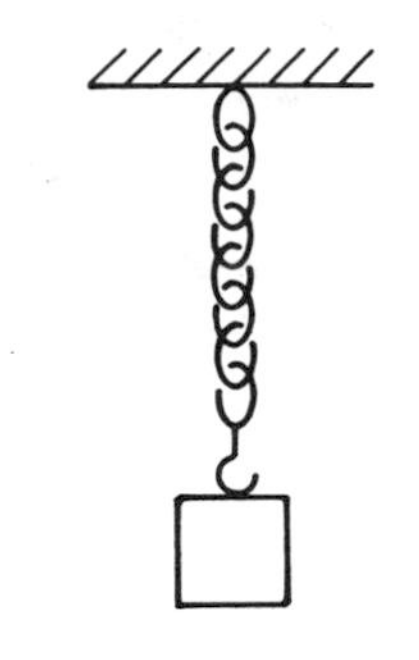

Series Connection

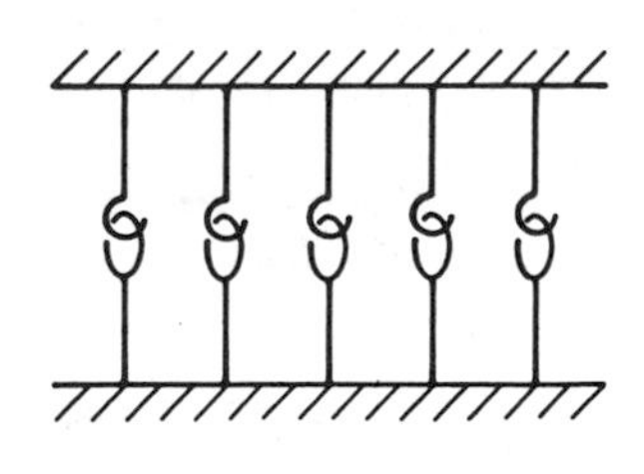

Parallel Connection

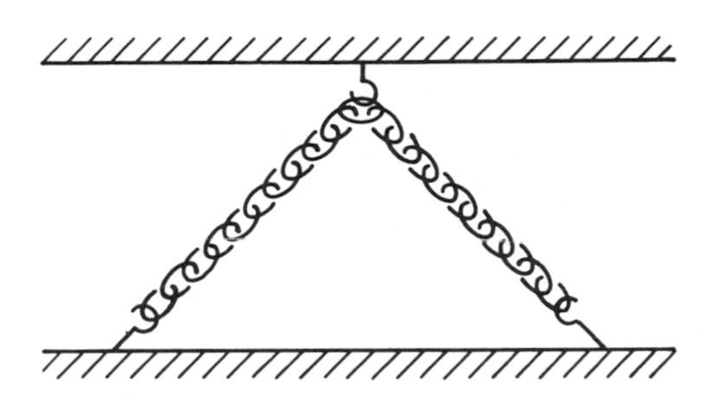

Series-Parallel Connection

Fig. A30. Connections.

O The Number of Connections Between Events

Fuller uses the expression $^1/_2(N^2 - N)$ for the number of connections between N "events." This formula can be proven by the so-called method of mathematical induction. Accordingly, we must prove: (a) the formula is valid for at least one value of N; (b) if the formula is valid for a single integer value of N, say N_0, then it is also valid for $N_1 \equiv N_0 + 1$. Once (a) and (b) are both proven, then the expression is necessarily proven for every integer value of N.

Part (a) has already been proven by Fuller, e.g., for $N = 3$ (three connections). Part (b) is proven as follows: The addition of a single event to N_0 existing events necessitates the addition of N_0 connections, for each of the N_0 original events must be connected to the newcomer. If the number of connections between N_0 events is $^1/_2(N_0^2 - N_0)$, then the number of connections between N_1 events, where $N_1 \equiv N_0 + 1$, must be:

$$^1/_2(N_0^2 - N_0) + N_0 = {}^1/_2(N_0^2 + N_0) = {}^1/_2(N_0 + 1)N_0 = {}^1/_2 N_1(N_1 - 1) = {}^1/_2(N_1^2 - N).$$

Therefore we have shown that the expression $^1/_2(N^2 - N)$ holds for $N_1 \equiv N_0 + 1$ if it is valid for N_0, q.e.d.

P Vector Equilibrium in Crystals

It is interesting to note the similarity between the arrays formed by rods in tension-compression equilibrium and those formed by identical efficiently packed spheres. The former form the basis for the Fuller domes, the latter provide a model for crystal structures. Because rods and spheres have opposite geometrical properties, it is difficult to imagine a fundamental connection between Fuller's macrostructures and nature's microstructures. Yet Fuller's sensitivity to patterns gave him an interest in sphere packing in connection with his tensegrities. Conversely, this author has discovered that a tension-compression model for crystals has broader applicability than does the usual sphere-packing model. We shall see that the geometrical shapes encountered in both the macrostructures and the microstructures are less the result of properties of the individual components than of the geometrical constraints on the system of components as a whole.

A sphere has infinite symmetry: *any* direction is equivalent to every other direction. A rod, on the other hand, has two unique (oppositely directed) directions. The most stable configuration of three identical spheres is one in which the spheres are in contact with each other, their centers the vertices of an equilateral triangle. Three identical rods are in a stable configuration when they form the sides of an equilaterial triangle. Accordingly, both the spheres and the rods stabilize in a triangular array, even though the individual components have totally different geometries. The reason is that the geometrical constraints are the same: the centers of the sphere have a minimum distance of approach equal to the sphere diameter, and the ends of the rods are rigidly separated by the length of the rods. We see that, although the sphere has infinite symmetry, the symmetry of the aggregate of spheres is not higher than that of the aggregate of rods.

The highest number of identical spheres that can be in contact with each other and with an identical central sphere is 12. The centers of these 12 spheres constitute the vertices of a cuboctahedron (cf. section I), i.e., of Fuller's "vector equilibrium." The compression strength of such a "close-packed" array of spheres is due to the forces acting between the spheres at the points of contact. These forces are directed toward the centers of the spheres, i.e., along the directions of the "vector equilibrium." The problem with a model of packed spheres is that it has compression strength but not tensional strength; we shall see presently that this lack of tensional strength is a weakness of the sphere-packing model of crystal structures.

In many metals, e.g., copper, the atoms arrange themselves as if they were closely packed spheres. They form layers with a triangular configuration, and these layers stack in such a way that any atom is cuboctahedrally surrounded by its nearest neighbors (Fig. A31). Because the cuboctahedron has the same

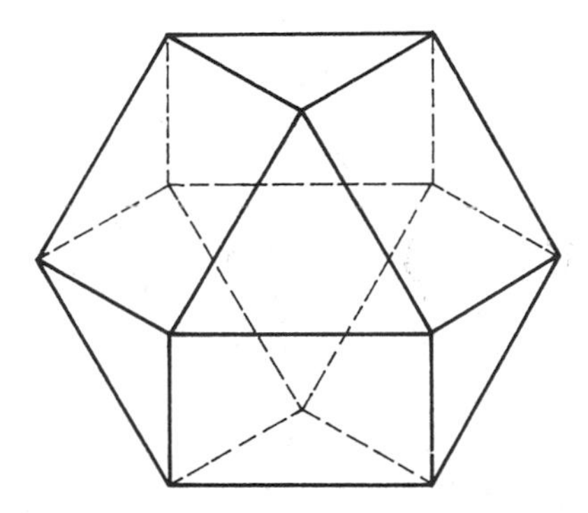

Fig. A31. Cuboctahedron configuration viewed perpendicularly to triangular layers.

symmetry as the cube, such an array is called "cubically close-packed" (ccp). Equally efficient packing of the spheres is obtained when the lower half of the cuboctahedron, as seen in Fig. A31, is twisted 60 degrees with respect to the upper half; the resulting array is called hexagonally close-packed (hcp). Here each sphere is also surrounded by 12 identical spheres, but less symmetrically than in the ccp case. Many metals, e.g., cobalt, are hexagonally close-packed. For all of these metals, the sphere-packing model is very satisfactory. A fly in the ointment, though, is the fact that more metals assume a so-called body-centered cubic configuration than the ccp one. This body-centered array will be discussed presently; we shall first continue to follow the successes of the sphere-packing model.

Many crystals are represented by a model in which large, closely packed spheres correspond to larger ions, and small spheres in the voids between the closely packed ones correspond to smaller ions. There are two kinds of voids between closely packed spheres: those bounded by eight spherical surfaces (octahedral voids) and those bounded by four spherical surfaces (tetrahedral voids). For each close-packed sphere, there are two tetrahedral voids and a single octahedral one. In Fig. A32, there is a schematic representation of the most common minerals whose structure can be represented as a distribution of small ions over the voids between closely packed large ions.

This author has invented four models called Moduledra Crystal Building Blocks * to represent structures such as those of Fig. A32. Two of these blocks are tetrahedra, two are octahedra (Fig. A33); they represent the voids between spheres. Their vertices represent the centers of closely packed ions,

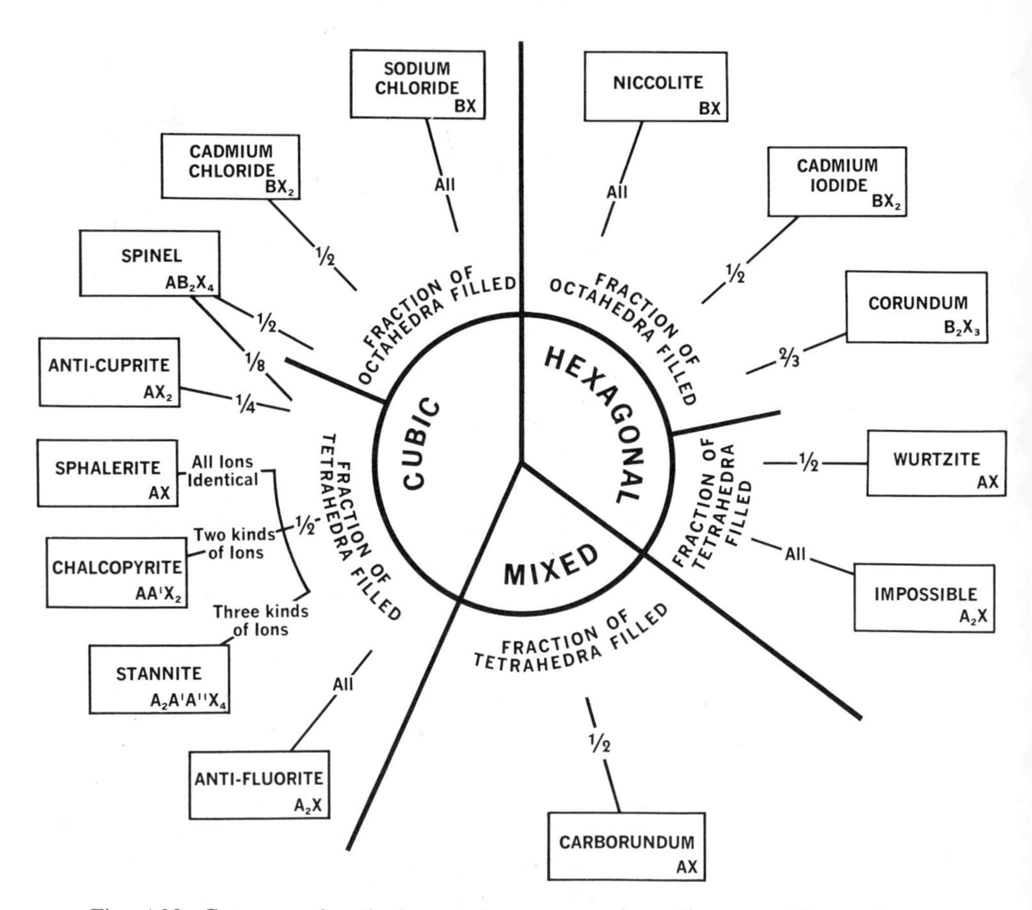

Fig. A32. Common minerals that can be represented as close-packed large spheres, with smaller spheres distributed over the voids between the large spheres.

* Trade Mark Registered

and a sphere in the center represents occupancy of a void. The closely packed spheres are not shown explicitly.

We have already seen that octahedra and tetrahedra together can fill all of space in a ratio $1 \div 2$. Different permutations of the four blocks produce models of about 80 percent of all inorganic crystals (Fig. A34).

The success of the sphere-packing model is thus surprisingly great when one considers how different an atom is from a rigid sphere. We have already noted, however, that the sphere model has compression strength but no ten-

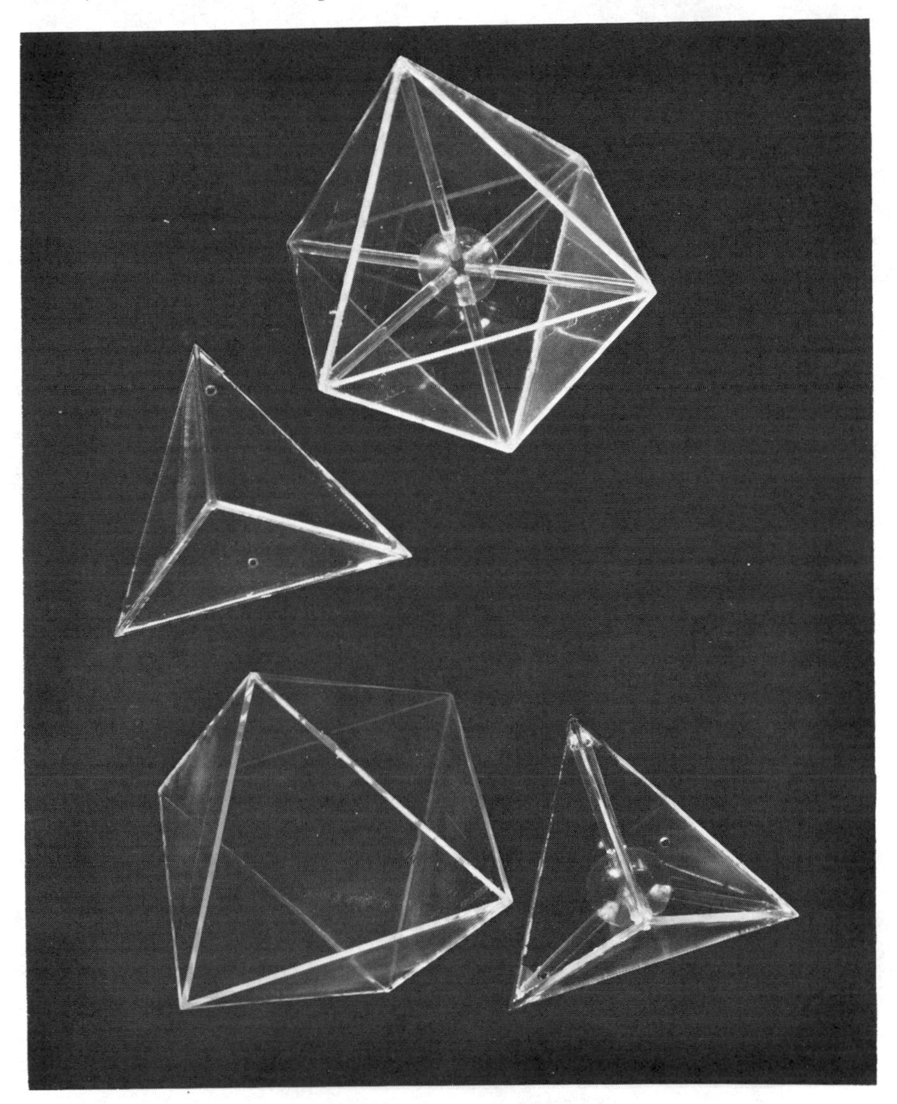

Fig. A33. Moduledra Crystal Building Blocks.

Fig. A34. Some structures modeled with Modulera Crystal Building Blocks.

sion strength. For this reason, it works well for the most compact structures, but it totally fails to represent less compact but just the same very prevalent structures. One of such is the body-centered cubic (bcc) structure, which is almost twice as prevalent as the ccp structure among metals. In this structure, each atom is surrounded by eight equivalent atoms at the corners of a cube. Since the number of near neighbors is here smaller than that in the ccp and hcp structures, it is less compact, and not accounted for by the sphere-packing model. Some examples where the sphere-model is almost absurdly successful are presented by several alloys where metal ions are tetrahedrally or octahedrally surrounded without being small enough to fit in voids between the other metal ions. Although such alloys can be conveniently represented by the Moduledra Blocks, it makes no sense to think here of close-packed spheres: the framework is the same, but the explanation must be different.

Curious also is the fact that the array of metal ions in the copper-magnesium alloy known as a "Laves phase" (after the German-Swiss crystallographer F. Laves) is exactly the same as that of the metal ions in the mineral spinel. In the latter structure, the metal ions do fit in the voids between closely packed oxygen ions, but in the Laves phase, there is no closely packed framework guiding the location of these ions.

We are therefore faced with the problem of devising a model that is consistent with the results of the sphere-packing model where the latter is applicable, but that will also apply to less compact structures such as the bcc arrays. Let us first imagine two atoms or ions that exert a force on each other, a force that depends only on the distance between their centers and is independent of the particular shape or orientation of the atoms. In the case of rigid spheres, this force would be infinite for distances less than the sphere diameter, zero for distances greater than the sphere diameter (Fig. A35*a*). In the case of a tension-compression equilibrium, the force is positive (repulsive) for small distances, negative (attractive) for large distances. At a certain distance, the forces just balance (Fig. A35*b*).

The two atoms will form a stable duo if there is a distance at which the repulsive and attractive forces just balance, just as in the case of tension-compression equilibrium. If the forces between the atoms are repulsive for all distances, the atoms will fly apart, never to meet again. However, if they are attractive for all distances, the atoms will move toward each other and become one forevermore. In the case of a stable equilibrium, the two atoms will remain at (or oscillate around) an equilibrium distance. Conversely, stable systems must have forces similar to those of Fig. A35*b*: the exact nature of the repulsive and attractive forces is immaterial, as long as they balance at a certain distance.

When a third identical atom is added to the first two, we can assume that

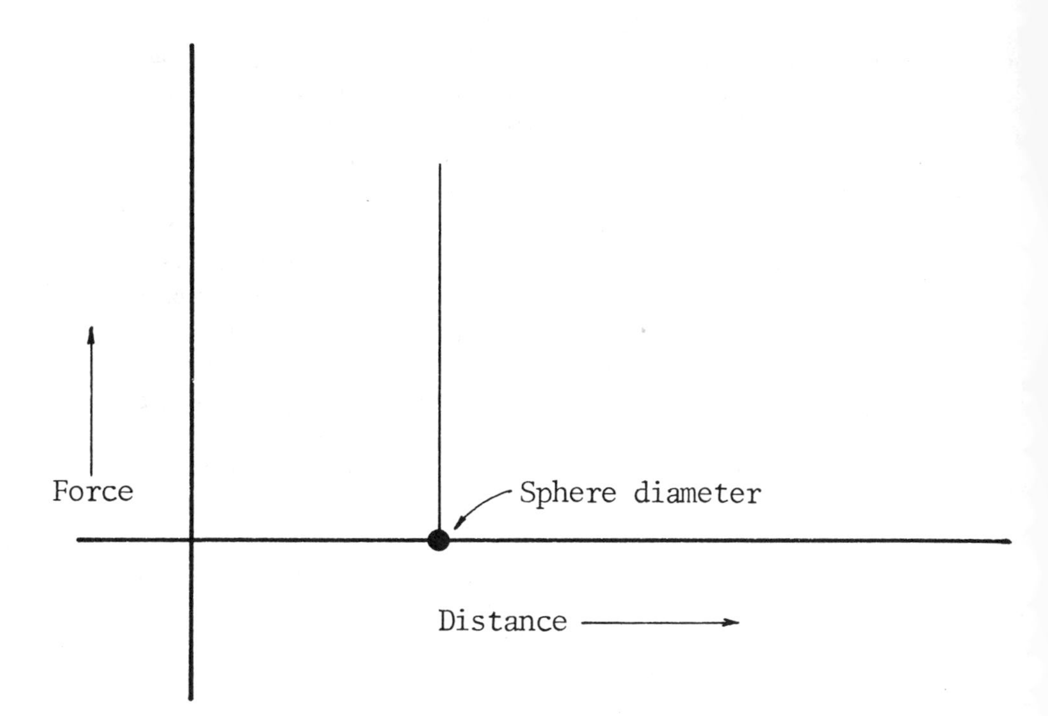

Fig. A35a. Distance-dependence of force between two rigid spheres.

the force between individual pairs is unaffected by the presence of the third one. In this case, the three identical atoms will arrange themselves at the corners of an equilateral triangle, so that the environment of each is identical with that of the other two. Any attempt to place one of the atoms in a unique position, e.g., on a line with the other two, disturbs the equilibrium, and the atoms will rearrange themselves to form the equilateral triangle once again. Four atoms will prefer a tetrahedral arrangement to a square one, because in the tetrahedral array, each atom has the same distance from each of the other three, whereas in the square array, one atom (the one diagonally across) is farther away than the other two. If the nearer two in the square are at the equilibrium distance, the two pairs of atoms diagonally across from each other will experience an attraction, and the square will pucker to form a tetrahedron. Conversely, if the diagonally opposite atoms are at the equilibrium distance, the adjacent ones are closer together, hence in the repulsive-distance region. As a result, the square will buckle to form a tetrahedron. Basic to the new model is this fact that *identical atoms* or ions want as much as possible to establish *identical environments* so that a maximum number of them can be at equlibrium distance from each other. This requirement is put in the form of a postulate:

The Vector Equilibrium Postulate
Crystal structures tend to assume configurations in which a maximum number of identical atoms or ions are equidistant from each other. If more than a single type of atom or ion is present, then each atom or ion tends to be equidistant from as many as possible of each type of atoms or ions.

This postulate (to be referred to as the VEP) is purely phenomenological, but therein lies its power. It establishes a condition for equilibrium configurations, assuming only that a distance of equilibrium exists for each pair of atoms.

Having substituted this postulate for the rigid-sphere model, we must investigate: (a) whether it produces results consistent with this sphere model where the latter is applicable, and (b) whether it agrees with experimental observations where the sphere model does not apply, in particular whether it is applicable to the bcc array.

The cubically and hexagonally close-packed structures conform ideally to the VEP: every atom is identically surrounded by the maximum of 12 identical atoms. Every atom also has six next-nearest neighbors, at a distance 41 percent greater than that of its nearest neighbors. In ccp structures, these six

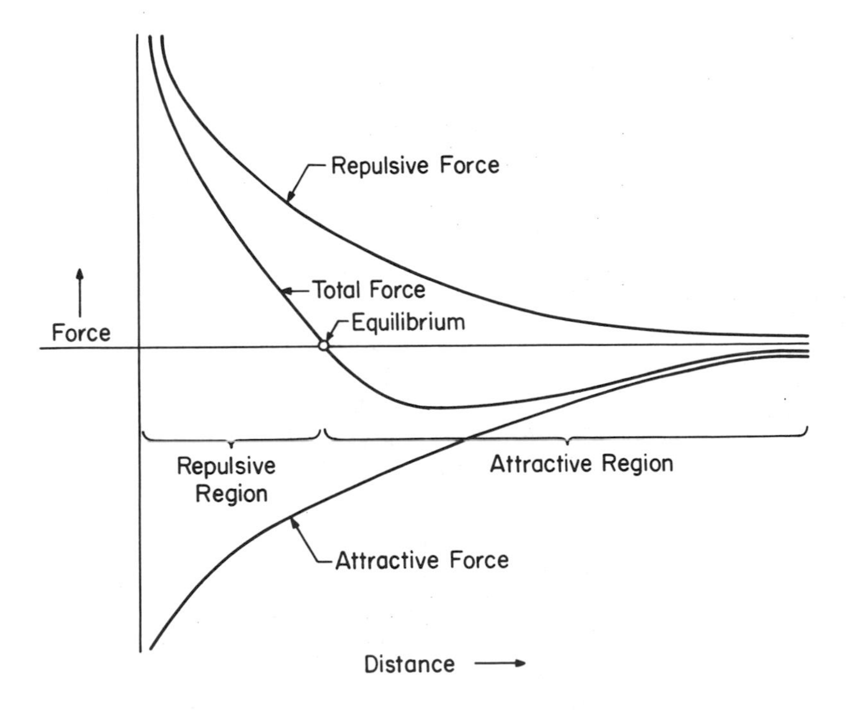

Fig. A35b. Tension-compression equilibrium.

next-nearest neighbors form the vertices of a regular octahedron; in hcp ones, they are located at the vertices of a triangular prism. The relative distances of nearest and next-nearest neighbors in ccp structures are shown in Fig. A36*a*.

By comparison, the bcc structure, which does not fit the sphere-packing model, fares quite well by the VEP. As shown in Fig. A36*b*, an atom in a bcc array has eight nearest neighbors, compared to 12 in ccp and hcp arrays, but its six next nearest neighbors are only 15 percent farther away, instead of 41 percent in ccp. Thus a bcc atom has 14 rather near neighbors. When the forces between atoms have appropriate dependence on distance, the relative scarcity of nearest neighbors in the bcc structure may be offset by the relative proximity of next-nearest neighbors, resulting in a stable bcc structure.

Because we have abandoned the spheres, we have also lost the voids between them. The VEP must therefore provide a substitute, which must be based on a sufficiently general concept that we can find something equivalent to voids in the bcc arrays as well.

Starting with a ccp, hcp, bcc, or any other regular array of points, we must locate positions of vector equilibrium between them. Instead of concentrating separately on ccp, hcp, and bcc arrays, we shall first approach the problem generally, and apply the result to each of these arrays later on. In general, we

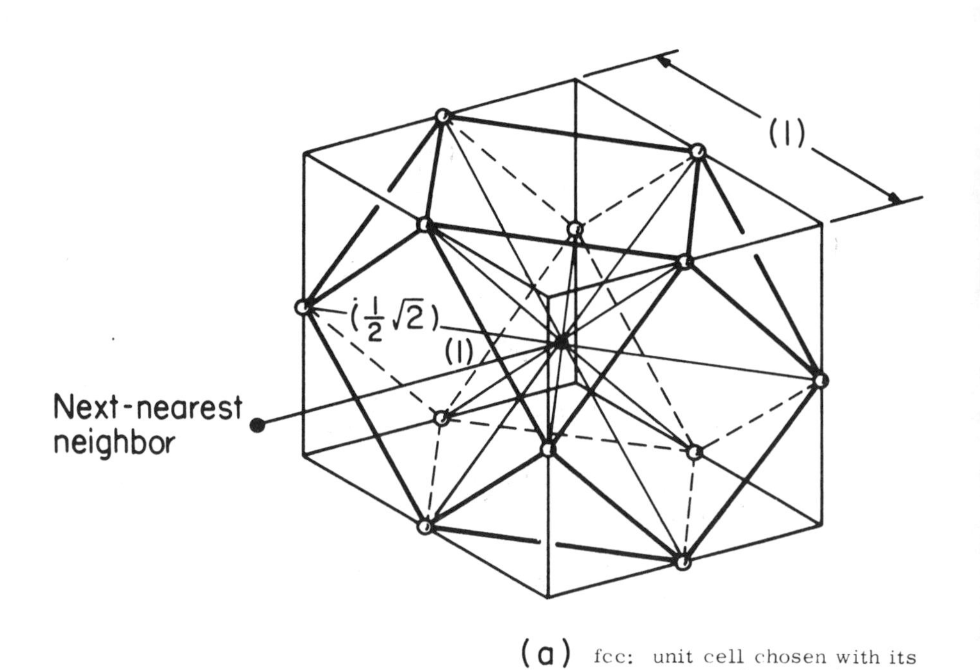

Fig. A36a. Neighbors of a cubically close-packed atom.

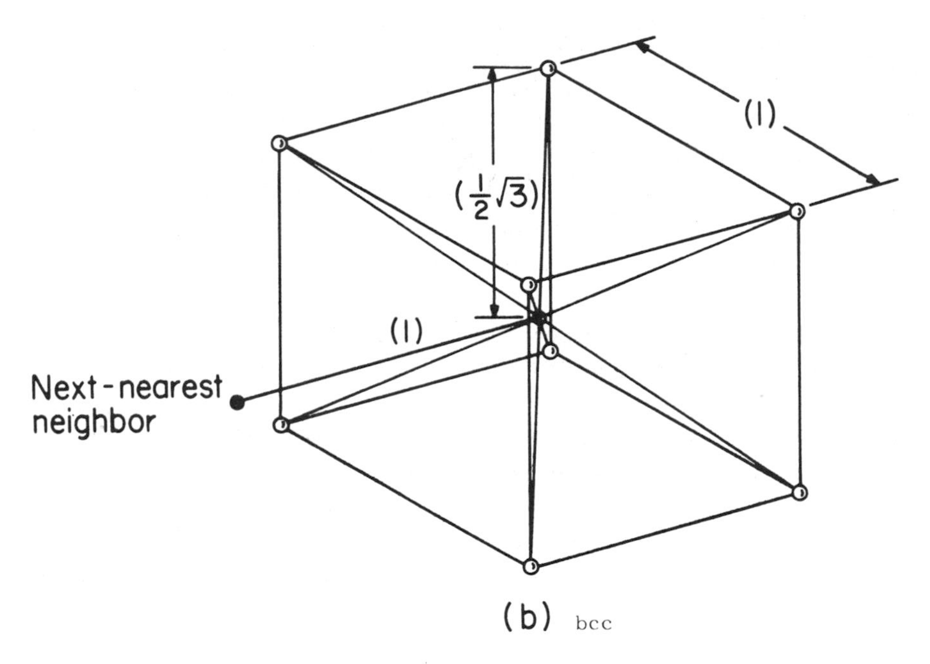

Fig. A36b. Neighbors of an atom in a body-centered cubic array.

shall call the regular array, as exemplified by the ccp, hcp, and bcc structure, a Point Complex. The points of vector equilibrium are points equidistant from a maximum of constituents of the Point Complex.

The locus of all points in space equidistant from *two* given points is the *plane* that perpendicularly bisects the line segment joining these points. The locus of all points equidistant from three given points is a *line* perpendicular to the plane of these points, passing through that plane at the intersection of the perpendicular bisectors of the sides of the triangle formed by the three points (cf. Fig. 401.01). The locus of the points equidistant from any *four* given points in space is the *center* of a sphere passing through these four given points. There is thus an infinity of points equidistant from two or three points, but in general only a single point equidistant from any four points not in the same plane. To locate vector-equilibrium points in a Point Complex, one needs, therefore, to specify at least four points of the Complex.

To arrive at a general construction of vector-equilibrium locations, the concept of a *Domain* is helpful. A Domain of a Point Complex is a region within which all points are closer to *a given* constituent of the Point Complex than to any other constituent of that Complex. The given constituent is at the center of the Domain. Consider, for example, the bees constructing their honeycomb (cf. section F). These bees were located at the centers of closely packed identical cylinders or circular disks, which expanded beyond the point of tangency to form hexagonal cells. These cells are two-dimensional Domains in the plane of the honeycomb: each point inside the cell is closer to the bee at its center than to any other bee. The boundaries of this Domain are equidistant from two bees; the vertices of the hexagonal cells would be vector-equilibrium positions (in the plane of the honeycomb) in the Point Complex formed by the bees. We shall find out presently which would be the vector-equilibrium points in the Point Complex formed by the hypothetical creatures we called "cees."

For the general construction of a Domain, perpendicularly bisect all line segments joining the center of the Domain to all other points in the Complex. The region enclosed by the *innermost* polyhedron formed by the bisector planes is the Domain. Ordinarily, only the nearest neighbors and at most next-nearest neighbors actually contribute to the Domain, because more distant points give rise to planes outside the innermost polyhedron. The vertices of the Domain, being the intersection of three perpendicular bisector-planes, are equidistant from four constituents of the Point Complex, hence vector-equilibrium points in the Point Complex!

It is concluded, therefore, that the vertices of the Domain of a Point Complex constitute vector-equilibrium positions in that Complex. Since each face of a Domain perpendicularly bisects a line segment joining the center of that Domain to another constituent of the Point Complex, each face of the Domain corresponds to a neighboring point in the Point Complex. These neighbors in the Point Complex form what is called a Coordination Polyhedron around the central point.

In the ccp array, each constituent is surrounded by *twelve* points at the corners of a cuboctahedron: this cuboctahedron is the Coordination Polyhedron for ccp arrays. The Domain for the ccp array has accordingly 12 faces: it is the rhombododecahedron (see sec. I) shown in Fig. A37. The rhombododecahedron has two kinds of vertices: the "acute" ones, where four acute angles meet, and the "obtuse" ones, where three obtuse angles meet. We have already seen that this polyhedron is a space filler. When packed together to fit all of space, four of those dodecahedra meet at the obtuse vertices and six meet at the acute vertices. Accordingly, each obtuse vertex is symmetrically surrounded by *four,* each acute vertex by *six* constituents of the ccp Point Complex. Each obtuse vertex is therefore exactly equivalent to a tetrahedral interstice, each acute vertex to an octahedral interstice in the sphere-packing model. We have thus proven that the VEP leads to results entirely equivalent to those of the sphere-packing model in the ccp case.

In the hcp array, the Domain is one that can be visualized as a rhombododecahedron that is first bisected by a plane perpendicular to a diagonal

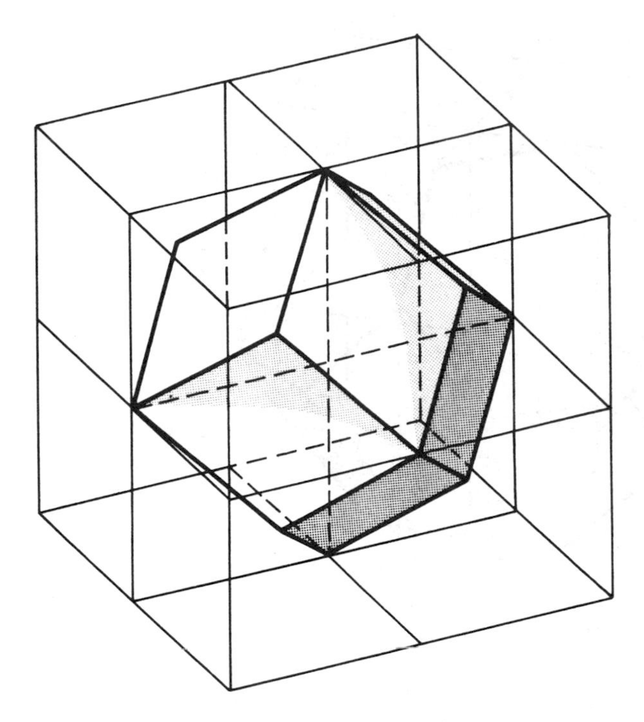

Fig. A37. Rhombododecahedron.

joining opposite obtuse vertices, after which the two halves are twisted 120 degrees with regard to each other. This procedure does not alter either the space-filling property or the vertices, with the result that again the VEP and sphere-packing model are in agreement.

With confidence gained by this agreement between VEP and sphere-packing model, we can now proceed to the bcc array, where the sphere-packing model utterly failed. Here the Domain involves nearest and next-nearest neighbors: the eight nearest neighbors give rise to eight hexagonal faces, and the six next-nearest neighbors to six square faces. The Domain of the bcc array is accordingly the truncated octahedron (see sec. I), as shown in Fig. A38. The vertices of this polyhedron would be points of vector equilibrium in the bcc array. Fig. A39 shows that, indeed, a very important structure, "β-tungsten" is made up of a bcc array of atoms accompanied by an array of atoms that occupy one half of the vector-equilibrium positions! These vector-equilibrium sites are occupied symmetrically, in accordance with the VEP: any other distribution of atoms over the available sites would bring pairs of atoms too close together.

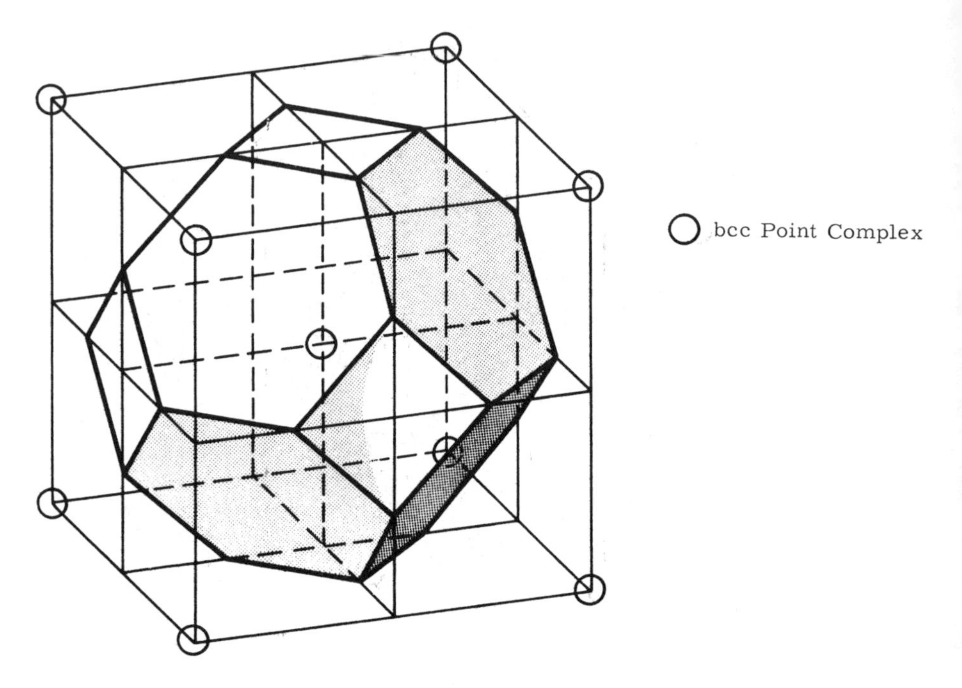

Fig. A38. Domain of a bcc Point Complex.

The β-tungsten structure is usually described as a bcc array with linear chains, perpendicular to each other but never intersecting, weaving in between the bcc points. That these chains do in fact exist can be seen by stacking cubes identical to Fig. A38 together in three dimensions, all in identical orientations. The existence of these chains was not understood, and the connection of the chains with the bcc array (they always seem to occur together) was even more of a mystery. The VEP accounts for both phenomena and establishes a fundamental relation between the ccp and hcp systems on the one hand, and the bcc one on the other hand. It is interesting to compare the β-tungsten structure (Fig. A39) with the six-strut tensegrity icosahedron (Figs. 717.01 and 740.21): both structures can now be understood on the basis of vector equilibrium.

In β-uranium hydride and in a gold-zinc alloy, one type of atom (uranium, resp. gold) forms a bcc array and also occupies half of the vector-equilibrium positions, in effect forming a β-tungsten structure. The other type of ions occupy sites that on closer investigation turn out to be exactly equidistant from one bcc point and three of its vector-equilibrium positions. This means that the β-tungsten structure, which is itself made up of a bcc Point Complex plus half its vector-equilibrium positions, can be in turn considered as a Point Complex in which some of its own vector-equilibrium positions are occupied!

Interesting also is the above-mentioned Laves phase. Here copper ions form a diamond Point Complex, so called because, just as in diamond, each copper atom is surrounded by four equivalent copper atoms at the corners of a regular tetrahedron. Such "low coordination" (i.e., low number of neighbors) is not easy to understand in the light of the VEP. In the diamond structure, the VEP does not apply because the assumption that the force between atoms depends on distance and not on orientation is not applicable here. On the contrary, each carbon atom in diamond has four chemical bonds directed toward the corners of a regular tetrahedron: in diamond, these bonds lock in place, resulting in the tetrahedral coordination. The Domain in the diamond structure is a truncated tetrahedron plus four quarter-tetrahedra. It should be recalled that the truncated tetrahedron has four hexagonal faces and four triangular ones. It is not a space filler, but it can combine with tetrahedra in 1 ÷ 1 ratio to fill all of space (cf. sec. I and Figs. A14*a* and A14*b*). By its very definition, a Domain must be a space filler: every point in space must be closer to *some*

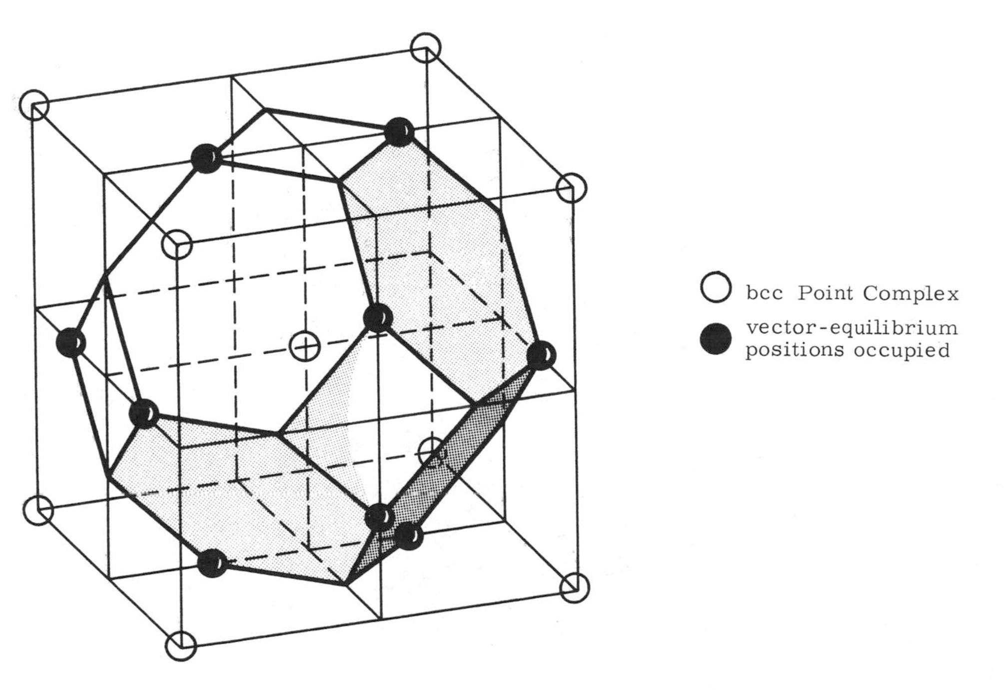

Fig. A39. Relation of the β-tungsten structure to the Domain of the bcc Point Complex.

constituent than to any other constituent of the Point Complex. The truncated tetrahedron and tetrahedron can fill space by stacking in such a way that the tetrahedron shares each of its faces with a triangular face of a truncated tetrahedron (Fig. A40). The Domain of the diamond Point Complex consists of the truncated tetrahedron, with a quarter-tetrahedron attached to each triangular face. This Domain has two kinds of vertices: 12 vertices of the truncated tetrahedron, and four vertices that were the centers of the tetrahedra between the truncated tetrahedra.

We can now understand how a diamond Point Complex can occur in metallic alloys, where there are no chemical directed bonds as in diamond. In the Laves phase, the copper atoms are not each other's *nearest* neighbors, but rather *next* nearest neighbors. The nearest neighbors of copper atoms are 12 magnesium atoms at the vertices of a truncated tetrahedron! The magnesium atoms occupy vector-equilibrium positions in the Point Complex occupied by copper atoms, thus stabilizing the system. The two types of vertices of the Domain of the diamond Point Complex are too close together to be occupied

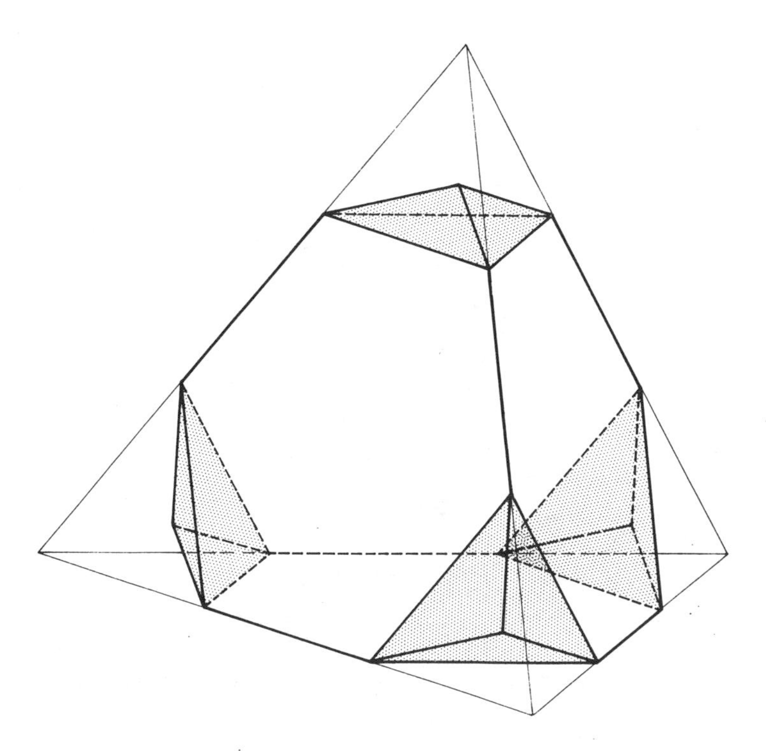

Fig. A40. Domain of the diamond complex; shaded faces represent quarters of a tetrahedron, which are added on to the triangular faces of a truncated tetrahedron.

simultaneously. The vertices at the center of the tetrahedra together form a diamond Point Complex congruent to the one formed by the Domain centers. The two diamond complexes together form a bcc array!

In conclusion, then, we find that there are geometrical relations between the various polyhedra that are more profound than the simple, though very serviceable, sphere-packing model. The definition of concepts such as vector equilibrium, domains, and coordination polyhedra are useful in establishing these relationships. They help us recognize patterns (such as the one formed by hydrogens in β-uranium hydrides), and explain the natural co-occurrence of arrays that can be interpreted as Point Complexes together with their vector-equilibrium positions. The relationship between Point-Complex and vector-equilibrium positions is actually a mutual one, but the simpler one of the two is usually taken as the Point Complex starting point.

Q Bibliography

B. G. Bagley, ''A dense packing of hard spheres with five-fold symmetry,'' Nature *208,* 674–675, Nov. 13, 1965

P. A. Beck, ''Survey of possible layer stacking structures,'' Z. f. Krist. *124,* 101–114 (1967)

G. B. Bokii, B. K. Vul'f, and N. L. Smirnova, ''Crystal structures of ternary metallic compounds,'' Z. Strukt. Khimii *2,* No. 1, 74–111, Jan–Feb 1961

M. J. Buerger, ''Derivative crystal structures,'' J. Chem. Phys. *15,* 1–16 (1947)

Keith Critchlow, ''Order in space,'' The Viking Press, New York (1965)

J. D. H. Donnay, E. Hellner, and A. Niggli, ''Coordination polyhedra,'' Z. f. Krist. *120,* 364–374 (1964)

W. Fischer, H. Burzlaff, E. Hellner, J. D. H. Donnay, ''Space groups and lattice complexes,'' U.S. Department of Commerce, National Bureau of Standards Monograph 134 (1973)

F. C. Frank and J. S. Kasper, ''Complex alloy structures regarded as sphere packings,'' Acta Cryst. *11,* 184–190 (1958), and *12,* 483–499 (1959)

W. G. Gehman, ''Standard ionic crystal structures,'' J. Chem. Ed. *40,* 54–60 (1963)

W. G. Gehman, ''Translation-permutation operation algebra for the description of crystal structures,'' I, Acta Cryst. *17,* 1561–1567 (1964)

W. G. Gehman and S. B. Austerman, ''Translation-permutation operation algebra for the description of crystal structures,'' II, Acta Cryst. *18,* 375–380 (1965)

E. W. Gorter, ''Some structural relationship of ternary transition metal oxides,'' XVIIth International Congress of Pure and Appl. Chem., 303–328 (Butterworths, London, 1960)

Shuichi Iida, ''Layer structures of magnetic oxides,'' J. Phys. Soc. Japan *12,* 222–233 (1957)

F. Jellinek, "Der NiAs Strukturtyp und seine Abarten," Osterr. Chem-Zeit. *60,* 311–321 (1959)

J. S. Kasper, "Atomic and magnetic ordering in transition metal structures," Trans. Am. Soc. Metals *48A,* 264–278 (1956)

P. J. Kripyakevich, "The systematics of structure types of intermetallic compounds," Z. Strukt. Khimii *4,* 117–136 (1963)

F. Laves, "Crystal structure and atomic size," Trans. Am. Soc. Metals *48A,* 124–198 (1956)

J. Lima-de-Faria, "A condensed way of representing inorganic close-packed structures," Z. f. Krist. *122,* 345–358 (1965)

J. Lima-de-Faria, "Systematic derivation of inorganic close-packed structures: AX and AX_2 compounds, sequence of equal layers," Z. f. Krist. *122,* 359–374 (1965)

A. L. Loeb, "A binary algebra describing crystal structures with closely packed anions," Acta Cryst. *11,* 469–476 (1958)

I. L. Morris and A. L. Loeb, "A binary algebra describing crystal structures with closely packed anions, Part II: A common system of reference for cubic and hexagonal structures," Acta Cryst. *13,* 434–443 (1960)

A. L. Loeb, "A modular algebra for the description of crystal structures," Acta Cryst. *15,* 219–226 (1962)

A. L. Loeb and G. W. Pearsall, "Moduledra crystal models: A teaching and research aid in solid state physics," Am. J. Phys. *31,* 190–196 (1963)

A. L. Loeb, "The subdivision of the hexagonal net and the systematic generation of crystal structures," Acta Cryst. *17,* 179–182 (1964)

A. L. Loeb, "A systematic survey of cubic crystal structures," J. Solid State Chemistry, *1,* 237–267 (1970)

A. L. Loeb, "The architecture of crystals," contribution to Module, Proportion, Symmetry, Rhythm, of the Vision and Value series. Gyorgy Kepes, editor (Braziller, New York, 1966)

A. L. Loeb, "Color and symmetry," John Wiley and Sons, New York (1971)

A. L. Mackay, "A dense non-crystallographic packing of equal spheres," Acta Cryst. *15,* 916–918 (1962)

E. Mooser and W. B. Pearson, "On the crystal chemistry of normal valence compounds," Acta Cryst. *12,* 1015–1022 (1959)

W. B. Pearson, "The crystal structures of semiconductors and a general valence rule," Acta Cryst. *17,* 1–15 (1964)

W. B. Pearson, "Handbook of lattice spacings and structures of metals," Vols. 1 and 2, Pergamon Press (1967)

N. L. Smirnova, "An analytical method of representation of inorganic crystal structures," Acta Cryst. 21, *A34* (1966)

A. F. Wells, "The geometrical basis of crystal chemistry,"
Parts I and II, Acta Cryst. *7,* 535–554 (1954);
Parts III and IV, *7,* 842–853 (1954);
Part V, *8,* 32–36 (1955);
Part VI, *9,* 23–28 (1956)

A. F. Wells, "The third dimension in chemistry," Clarendon Press, Oxford (1956)

A. F. Wells, "Structural inorganic chemistry," Oxford University Press, 3rd ed. (1962)

R. E. Williams, "Natural structure," Eudaemon Press (Calif.) (1971)

The attention of the reader is directed to the detailed Table of Contents for each chapter which has been supplied by the author in lieu of an index.